Oberwolfach Seminars

Volume 37

Hemodynamical Flows

Modeling, Analysis and Simulation

Giovanni P. Galdi
Rolf Rannacher
Anne M. Robertson
Stefan Turek

Birkhäuser
Basel · Boston · Berlin

Authors:

Giovanni P. Galdi
Anne M. Robertson
Department of Mechanical Engineering
University of Pittsburgh
641 Benedum Engineering Hall
Pittsburgh, PA 15261
USA
e-mail: galdi@engr.pitt.edu
         annerob@engr.pitt.edu

Rolf Rannacher
Institute of Applied Mathematics
University of Heidelberg
Im Neuenheimer Feld 293/294
D-69120 Heidelberg
Germany
e-mail: rannacher@iwr.uni-heidelberg.de

Stefan Turek
Department of Mathematics
University of Dortmund
Vogelpothsweg 87
D-44227 Dortmund
Germany
e-mail: Stefan.Turek@math.uni-dortmund.de

2000 Mathematics Subject Classification: 76Zxx, 74F10, 74L15, 76D05, 76M10, 92C35

Library of Congress Control Number: 2007938510

Bibliographic information published by Die Deutsche Bibliothek
Die Deutsche Bibliothek lists this publication in the Deutsche Nationalbibliografie;
detailed bibliographic data is available in the Internet at <http://dnb.ddb.de>.

ISBN 978-3-7643-7805-9 Birkhäuser Verlag, Basel – Boston – Berlin

© 2008 Birkhäuser Verlag AG
Basel · Boston · Berlin
P.O. Box 133, CH-4010 Basel, Switzerland
Part of Springer Science+Business Media
Printed on acid-free paper produced from chlorine-free pulp. TCF ∞

ISBN 978-3-7643-7805-9

ISBN 978-3-7643-7806-6 (eBook)

9 8 7 6 5 4 3 2 1

www.birkhauser.ch

Contents

Preface

Hemodynamics is the study of the forces and physical mechanisms associated with blood flow in the cardiovascular system. Due to the fact that blood is a suspension of flexible particles in plasma and to the coupling between motion of blood and the vessel wall, necessarily this subject includes both fluid and solid mechanical processes. Hemodynamic features such as flow separation, flow recirculation, and low and oscillatory wall shear stress are believed to play important roles in the localization and development of vascular diseases such as atherosclerosis, cerebral aneurysms, post-stenotic dilations and arteriovenous malformations. Therefore, modeling, mathematical analysis and numerical simulation of these processes can ultimately contribute to improved clinical diagnosis and therapeutic planning.

However, the circulatory system is extremely complex and so researchers are faced with the need to formulate the numerical or mathematical problem in a form which is sufficiently simple to be tractable, yet maintains enough complexity to be relevant. For example, rather than modeling the entire circulatory system, isolated segments of the circulation are studied, introducing the need to choose appropriate inflow and outflow boundary conditions and possibly take a multi-scale approach. The blood vessel wall is an inhomogeneous, nonlinear, material capable of growth and remodeling, and blood is a concentrated suspension of deformable cellular elements in plasma. The modeler needs to choose suitable constitutive models for the wall and blood. The diameter of vessels in the circulatory system ranges from the order of centimeters in the larger arteries to microns in the capillaries. It is therefore appropriate to model blood as a single phase continuum in some parts of the circulatory system, while in others, it is necessary to model blood as a suspension. The chapters in this book address these and other topics from different perspectives.

The present volume is a collection of six chapters which are based on a series of lectures delivered by Anne M. Robertson (University of Pittsburgh), Giovanni P. Galdi (University of Pittsburgh), Rolf Rannacher (University of Heidelberg), and Stefan Turek (University of Dortmund) at the Oberwolfach Seminar "Hemodynamical Flows: Aspects of Modeling, Analysis and Simulation", during the period November 20–26, 2005.

These lectures focused on various aspects of hemodynamics from different angles, including physical modeling, mathematical analysis and numerical simulation. Accordingly, this volume addresses the following main topics:

- General background in continuum mechanics;
- Multiphase nature of blood;
- Rheological data for blood;
- Newtonian and non-Newtonian constitutive models for blood;
- Mechanical models for blood vessel walls;
- Numerical methods for flow simulation;
- Aspects of mesh and model adaptivity;
- Particle transport in viscous flows;

- Flows through systems of pipes;
- Fluid-structure interaction in blood vessels.

The above topics are organized as follows:

In the first chapter, *Review of Relevant Continuum Mechanics*, by A.M. Robertson, the basic kinematical and dynamical issues that are at the foundation of continuum mechanics used in the book are surveyed. The constitutive theory for Newtonian fluids, general nonlinear viscous fluids, yield stress "fluids", viscoelastic fluids and thixotropic fluids are covered. In preparation for a discussion of experimental data on blood in the second chapter, viscometric flows and commonly used rheometers are discussed. Finally, the fundamentals of nonlinear elastic solids are introduced.

The second chapter, *Hemorheology*, by A.M. Robertson, A. Sequeira, and M.V. Kameneva, is dedicated to constitutive models for blood, based on phenomenological considerations. Experimental data on the multiphase properties of blood are considered as well as the relationship between these properties and the mechanical behavior of blood. These mechanical properties include shear thinning viscosity, yield stress behavior and viscoelasticity. The significance of these non-Newtonian behaviors in the circulation are addressed. The subject of blood coagulation is considered, motivated by its importance in cardiovascular device design. The chapter concludes with sections on the effect of gender and certain diseases states on the mechanical response of blood.

The third chapter, *Mathematical Problems in Fluid Mechanics*, by Giovanni P. Galdi, discusses some, of the many, topics which are at the foundation of the analysis of models for blood flow, and points out directions for future research. Specifically, it focuses on the following three different problems: pipe flow of a Navier-Stokes liquid, flow of non-Newtonian and, in particular, viscoelastic liquids, and liquid-particle interaction. This analysis has two main objectives. The first is the study of the well-posedness of the relevant problems, whereas the second is to provide a rigorous explanation of some fundamental experiments. In particular, special attention is given to the investigation of the dependence of even qualitative features of the on the non-Newtonian properties of the liquid.

The fourth chapter, *Methods for Numerical Flow Simulation*, by Rolf Rannacher, introduces the computational methods for the simulation of PDE based models of laminar hemodynamical flows. Space and time discretization is discussed with emphasis on operator-splitting and finite-element Galerkin methods because of their flexibility and rigorous mathematical basis. Special attention is paid to the simulation of pipe flow and the related question of artificial outflow boundary conditions. Further topics include efficient methods for the solution of the resulting algebraic problems, techniques of sensitivity-based error control and mesh adaptation, as well as flow control and model calibration. The analysis is restricted to laminar flows, where all relevant spatial and temporal scales can be resolved, and no additional modeling of turbulence effects is required. This covers most of the relevant situations of hemodynamical flows.

In the fifth chapter, *Numerics of Fluid-Structure Interaction*, by Sebastian Bönisch, Thomas Dunne, and Rolf Rannacher, numerical methods for simulating the interaction of viscous liquids with rigid or elastic bodies are described. General examples of fluid-solid/structure interaction (FSI) problems are flow transporting rigid or elastic particles (particulate flow), flow around elastic structures (airplanes, submarines) and flow in elastic structures (hemodynamics, transport of fluids in closed containers). A common variational description of FSI is developed as the basis of a consistent Galerkin discretization with a-posteriori error control and mesh adaptation, as well as the solution of optimal control problems based on the Euler-Lagrange approach.

The sixth chapter, *Numerical Techniques for Multiphase Flow with Liquid-Solid Interaction*, by Jaroslav Hron and Stefan Turek, discusses numerical methods for simulating multiphase flows with liquid-solid interaction based on the incompressible NavierStokes equations combined with constitutive models for nonlinear solids. More precisely, it addresses the following three topics. The first concerns finite-element discretization and corresponding solver techniques for the resulting algebraic systems. The second regards a fully monolithic finite-element approach for fluid-structure interactions with elastic materials which is applied to several benchmark configurations. Finally, the third section introduces the concept of FEM fictitious boundary techniques, together with operator-splitting approaches for particulate flow. This latter is especially designed for the efficient simulation of systems with many solid particles of different shapes and sizes.

This work is aimed at a diverse readership. For this reason, an effort was made to keep every topic as self-contained as possible. In fact, whenever details are not explicitly given, the reader is referred to the appropriate literature.

Last, but not least, the authors would like to convey their sincere thanks to all participants, who, with their questions and insights, helped to maintain a lively and stimulating scientific atmosphere, typical of all Oberwolfach meetings.

Hemodynamical Flows. Modeling, Analysis and Simulation
Oberwolfach Seminars, Vol. 37, 1–62
© 2008 Birkhäuser Verlag Basel/Switzerland

Review of Relevant Continuum Mechanics

Anne M. Robertson

Introduction

In this chapter we review the basic continuum mechanics at the foundation of
the technical material in this book. Readers interested in further information are
referred to monographs on this subject including Truesdell and Noll [65], Chadwick [10], Gurtin [27], Galdi [21, 22], Holzapfel [32], Temam and Miranville [61],
and Spencer [59]. The material is organized as follows:

Contents

Acknowledgements

The authors would like to thank Dr. Tomas Bodnar of the Czech Technical University (CTU) for his assistance on the graphics for some of the figures in this chapter.

Notation

Before proceeding further, we discuss some of the notation used in this chapter. We make use of Cartesian coordinates and the standard Einstein summation convention. For clarity, relations are often given in both coordinate-free notation as well as component form.

In general, lower-case letters (Greek and Latin) are used for scalar quantities, boldface lower-case letters are used for first-order tensors (vectors) and boldface upper-case letters are used for second-order tensors. The components of tensors relative to a fixed orthonormal basis (rectilinear) (e_1, e_2, e_3) are denoted using Latin subscripts. Standard summation convention is used whereby repeated Latin indices imply summation from one to three. The inner product between two arbitrary vectors u and v is denoted using the "$\cdot$" notation and defined in this chapter as

$$\boldsymbol{u} \cdot \boldsymbol{v} \;=\; u_i \, v_i. \tag{0.1}$$

The linear transformation formed by the operation of a second-order tensor $\boldsymbol{A}$ on a vector $\boldsymbol{u}$ to generate a vector $\boldsymbol{v}$ is written as $\boldsymbol{v} = \boldsymbol{A} \cdot \boldsymbol{u}$. The dot product of two second-order tensors $\boldsymbol{A} \cdot \boldsymbol{B}$ generates another second-order tensor $\boldsymbol{C}$,

$$\boldsymbol{C} \;=\; \boldsymbol{A} \cdot \boldsymbol{B} \qquad \text{or} \qquad C_{ij} \;=\; A_{ik} \, B_{kj}. \tag{0.2}$$

The inner product of two second-order tensors $\boldsymbol{A}$ and $\boldsymbol{B}$ is denoted by $\boldsymbol{A} : \boldsymbol{B}$ and defined in this chapter as

$$\boldsymbol{A} : \boldsymbol{B} \;=\; \text{trace}(\boldsymbol{A}^T \cdot \boldsymbol{B}) \qquad \text{or} \qquad \boldsymbol{A} : \boldsymbol{B} \;=\; A_{ij} \, B_{ij}. \tag{0.3}$$

1. Kinematics

In this section, we introduce the kinematics necessary for describing the motion of the solids and liquids discussed in this book.

1.1. Description of motion of material points in a body

First consider a physical body, which we identify by the symbol $\mathfrak{B}$. In this work, we take a continuum approach and assume $\mathfrak{B}$ is composed of a continuous set of material particles. We identify an arbitrary material point in $\mathfrak{B}$ by its position $\boldsymbol{X}$ in a chosen reference configuration, which we denote as κ_0 [1]. Namely, we assume the body can be embedded in a three-dimensional Euclidean space, Figure 1. For example, κ_0 could be the configuration of the body at time zero, though it is not necessary that the configuration of the body ever coincided with reference configuration κ_0. The mapping of each material particle of the body from κ_0 to Euclidean 3-space at an arbitrary time t will be called the configuration of $\mathfrak{B}$ at time t, denoted by $\kappa(t)$. This mapping is assumed to be one-to-one, invertible and differentiable as many times as necessary for all time.

The region occupied by the entire body in κ_0 is denoted by $\mathcal{R}_0$ with closed boundary $\partial \mathcal{R}_0$, Figure 1. The corresponding regions and boundaries for the body in $\kappa(t)$ are $\mathcal{R}$ and $\partial \mathcal{R}$, respectively. An arbitrary material region within $\mathcal{B}$ in κ_0 will be denoted as $\mathcal{V}_0$ ($\mathcal{V}_0 \subseteq \mathcal{R}_0$) with boundary $\partial \mathcal{V}_0$. The corresponding subregion in the current configuration $\kappa(t)$ is denoted as $\mathcal{V}$ ($\mathcal{V} \subseteq \mathcal{R}$) with boundary $\partial \mathcal{V}$.

During the deformation (or motion) of $\mathfrak{B}$, an arbitrary material particle located at position $\boldsymbol{X}$ in reference configuration κ_0 will move to position $\boldsymbol{x}$ in configuration $\kappa(t)$. It is assumed that the deformation of all points in the body can be described through a relationship of the form

$$\boldsymbol{x} \;=\; \boldsymbol{\chi}_{\kappa_0}(\boldsymbol{X}, t). \tag{1.1}$$

---

[1] The use of the upper case symbol for the position vector in κ_0 is an exception to the notation for vectors introduced above.

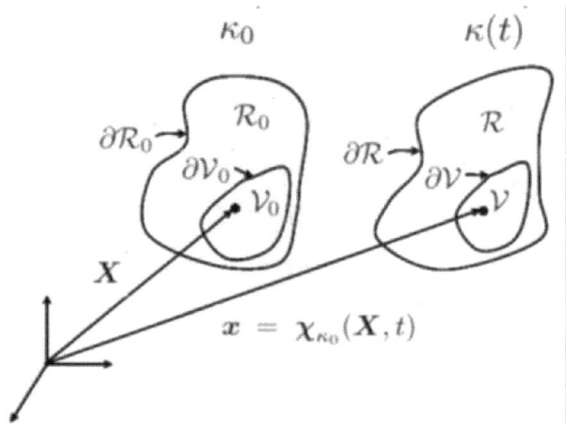

FIGURE 1. Schematic of notation used to identify material points and regions in an arbitrary body $\mathfrak{B}$ in the reference configuration κ_0 and current configuration $\kappa(t)$.

We further assume $\chi_{\kappa_0}(\boldsymbol{X}, t)$ is differentiable as many times as necessary in space and time and possesses an inverse,

$$\boldsymbol{X} = \chi_{\kappa_0}^{-1}(\boldsymbol{x}, t). \tag{1.2}$$

For our current purposes, it suffices to define one reference configuration for the body. Hence, in further discussions, we drop the subscript κ_0 and it will be understood that the functions $\chi(\boldsymbol{X}, t)$ and $\chi^{-1}(\boldsymbol{x}, t)$ depend on the choice of reference configuration. In the discussion of viscoelastic fluids with fading memory in Section 5, we will use the current configuration as the reference configuration. The relevant kinematics for viscoelastic fluids will be discussed in Section 5.1.

1.2. Referential and spatial descriptions

Field variables such as density, ρ, will either be written as a function of $\boldsymbol{X}$ and t (the *referential* or *Lagrangian* description) or as a function of $\boldsymbol{x}$ and t (the *spatial* or *Eulerian* description),

$$\rho = \bar{\rho}(\boldsymbol{X}, t) = \hat{\rho}(\boldsymbol{x}, t). \tag{1.3}$$

Note that the Eulerian description $\hat{\rho}(\boldsymbol{x}, t)$ is independent of information about the position of individual material particles and is typically used for the motion of viscous fluids. Clearly these two descriptions can be related using (1.1) and (1.2).

The velocity $\boldsymbol{v}$ and acceleration $\boldsymbol{a}$ of a material particle can then be defined with respect to the motion (1.1) through

$$\boldsymbol{v} = \frac{\partial \chi(\boldsymbol{X}, t)}{\partial t}, \qquad \boldsymbol{a} = \frac{D\boldsymbol{v}}{Dt} = \frac{\partial^2 \chi(\boldsymbol{X}, t)}{\partial t^2}, \tag{1.4}$$

where the notation $D(\cdot)/Dt$ is used to denote the material derivative of a quantity. It is literally the time derivative of the Lagrangian description of a quantity, holding the material particle fixed. Namely, it is understood in evaluating the material derivative in (1.4), that $\boldsymbol{X}$ is held fixed. Sometimes the material derivative is called the *total derivative* or *substantial derivative*.

Frequently in fluid mechanics, the Eulerian description of the field variables is of interest rather than the material description. Using the chain rule, the material derivative can be written for the spatial formulation of a field variable. For example,

$$\frac{D\rho}{Dt} = \frac{\partial \bar{\rho}(\boldsymbol{X},t)}{\partial t} = \frac{\partial \hat{\rho}(\boldsymbol{x},t)}{\partial t} + v_i \frac{\partial \hat{\rho}(\boldsymbol{x},t)}{\partial x_i}. \tag{1.5}$$

Using (1.5), it follows that the acceleration, defined in (1.4), can be written with respect to the spatial representation of the velocity field,

$$a_i = \frac{\partial \hat{v}_i(\boldsymbol{x},t)}{\partial t} + v_j \frac{\partial \hat{v}_i(\boldsymbol{x},t)}{\partial x_j}. \tag{1.6}$$

1.3. Deformation gradient and measures of stretch and strain

1.3.1. Deformation gradient. The *deformation gradient* or *displacement gradient* of the motion relative to the reference configuration κ_0 is a second-order tensor defined by

$$\boldsymbol{F} = \frac{\partial \boldsymbol{\chi}(\boldsymbol{X},t)}{\partial \boldsymbol{X}} \quad \text{or} \quad F_{iA} = \frac{\partial \chi_i(\boldsymbol{X},t)}{\partial X_A}. \tag{1.7}$$

It is the fundamental kinematic variable describing the change in length and orientation of an infinitesimal material element $d\boldsymbol{X}$ in κ_0 to $d\boldsymbol{x}$ in $\kappa(t)$ during the deformation. It can be shown that the relationship between $d\boldsymbol{x}$ and $d\boldsymbol{X}$ is

$$d\boldsymbol{x} = \boldsymbol{F} \cdot d\boldsymbol{X}, \quad \text{or} \quad dx_i = F_{iA} dX_A. \tag{1.8}$$

In future discussions, we denote the Jacobian of the transformation (1.1) by J. Namely, the Jacobian is equivalent to the determinant of $\boldsymbol{F}$,

$$J \equiv \det \frac{\partial \chi_i(\boldsymbol{X},t)}{\partial X_A} = \det(\boldsymbol{F}). \tag{1.9}$$

We restrict attention to deformations for which $\boldsymbol{\chi}(\boldsymbol{X},t)$ has an inverse, so necessarily $J \neq 0$. From (1.7), $\boldsymbol{F} = \boldsymbol{I}$ in κ_0. Since the deformation is taken to be continuous and $J \neq 0$, it follows that $0 < J < \infty$. The relationship between the infinitesimal volume occupied by material points in the current configuration dv is related to the infinitesimal volume dV occupied by the same material points in the reference configuration,

$$dv = J \, dV. \tag{1.10}$$

It will be useful in the discussion of the governing equations to note the following result for the material derivative of the Jacobian of the transformation (1.1),

$$\frac{DJ}{Dt} = J \operatorname{div} \boldsymbol{v}, \quad \text{or} \quad \frac{DJ}{Dt} = J \frac{\partial v_i}{\partial x_i}. \tag{1.11}$$

1.3.2. Measures of deformation and strain. While the deformation gradient is a fundamental kinematic variable, it will be shown in the later sections of this chapter, that the dependence of the stress tensor on $\boldsymbol{F}$ is restricted if invariance requirements are to be satisfied. As a result, the left Cauchy–Green tensor,

$$\boldsymbol{B} = \boldsymbol{F} \cdot \boldsymbol{F}^T, \quad \text{or} \quad B_{ij} = F_{iA}F_{jA}, \tag{1.12}$$

and the right Cauchy–Green tensor,

$$\boldsymbol{C} = \boldsymbol{F}^T \cdot \boldsymbol{F}, \quad \text{or} \quad C_{AB} = F_{iA}F_{iB}, \tag{1.13}$$

will be useful in discussions of constitutive models for solids. It follows from (1.12), (1.13) and the properties of J that $\boldsymbol{B}$ and $\boldsymbol{C}$ are positive definite, symmetric tensors.

For rigid motions, $\boldsymbol{B}$ and $\boldsymbol{C}$ are equal to the identity tensor. It is a simple matter to used these variables to define measures of strain which vanish for rigid motions. For example, the *Lagrangian* and *Eulerian* strains are respectively defined as

$$\boldsymbol{E} \equiv \frac{1}{2}(\boldsymbol{C} - \boldsymbol{I}), \quad \text{and} \quad e \equiv \frac{1}{2}(\boldsymbol{I} - \boldsymbol{B}^{-1}). \tag{1.14}$$

1.3.3. Physical significance of stretch and strain measures. In order to better understand the physical significance of the four tensors defined in (1.12)–(1.14), we now consider the change in orientation and magnitude of an infinitesimal material element $\boldsymbol{dX}$ in reference configuration κ_0 that transforms to $\boldsymbol{dx}$ during the deformation, (1.8). We will denote the magnitude of $\boldsymbol{dx}$ and $\boldsymbol{dX}$ as ds and dS, respectively and let $\boldsymbol{m}$ and $\boldsymbol{M}$ be the unit vectors tangent to $\boldsymbol{dx}$ and $\boldsymbol{dX}$, respectively. It then follows from (1.8) that

$$\frac{ds^2}{dS^2} = C_{AB} M_A M_B. \tag{1.15}$$

If we now consider the special case where $\boldsymbol{dX}$ is parallel to one of the coordinate axis, for example, $\boldsymbol{dX} = dS\boldsymbol{e}_1$, it follows from (1.15) that

$$C_{11} = \frac{ds^2}{dS^2}, \quad \text{and} \quad E_{11} = \frac{1}{2}\frac{ds^2 - dS^2}{dS^2}. \tag{1.16}$$

Namely, C_{11} is equal to the square of the stretch and E_{11} is a measure of the Lagrangian strain of an infinitesimal material element which was aligned with $\boldsymbol{e}_1$ in the reference configuration. Similar results hold for the other diagonal components of $\boldsymbol{C}$ and $\boldsymbol{E}$.

In parallel with (1.15), it can be shown that

$$dS^2 = B_{ij}^{-1} m_i m_j ds^2. \tag{1.17}$$

Therefore, if an infinitesimal element $\boldsymbol{dx}$ is aligned with $\boldsymbol{e}_1$ in the *current* configuration,

$$B_{11}^{-1} = \frac{dS^2}{ds^2} \quad \text{and} \quad e_{11} = \frac{1}{2}\frac{ds^2 - dS^2}{ds^2}. \tag{1.18}$$

Thus, B_{11}^{-1} is equal to the square of the inverse of the stretch and e_{11} is a measure of the Eulerian strain of an infinitesimal material element which is aligned with e_1 in the current configuration.

1.4. Velocity gradient and the rate of deformation tensor

While the fundamental kinematic variable for elastic solids is the deformation gradient tensor, for viscous fluids, this quantity is the velocity gradient, L. The components of L with respect to rectangular coordinates are [2]

$$L = \operatorname{grad} v, \quad \text{or} \quad L_{ij} = \frac{\partial v_i(x,t)}{\partial x_j}. \tag{1.19}$$

As will be discussed later in this chapter, the dependence of the stress tensor on L must be restricted if invariance requirements are to be satisfied. As a result, we are generally interested in the symmetric part of the velocity gradient, D, often referred to as the rate of deformation tensor,

$$D = \frac{1}{2}\left(L + L^T\right), \quad \text{or} \quad D_{ij} = \frac{1}{2}\left(\frac{\partial v_i}{\partial x_j} + \frac{\partial v_j}{\partial x_i}\right). \tag{1.20}$$

1.4.1. Physical significance of the components of D.

The physical significance of D can be studied by considering the rate of change in magnitude of an infinitesimal material element dx of length ds. Using (1.4), (1.7), (1.8), and (1.19), it can be shown that

$$\frac{D(dx)}{Dt} = L \cdot dx \quad \text{or} \quad \frac{Ddx_i}{Dt} = L_{ij}dx_j. \tag{1.21}$$

Therefore,

$$\frac{D(ds^2)}{Dt} = 2\,D_{ij}dx_idx_j. \tag{1.22}$$

For example, we see from (1.22) that the rate of change of magnitude of an infinitesimal element dx, which at time t is parallel to the e_1 axis is

$$D_{11} = \frac{1}{ds}\frac{Dds}{Dt}. \tag{1.23}$$

The other diagonal elements of D can be interpreted in a similar way.

Now consider two infinitesimal material elements dx and dy at time t. It follows from (1.20) and (1.21), that

$$\frac{D(dx \cdot dy)}{Dt} = 2\,dx \cdot D \cdot dy \quad \text{or} \quad \frac{D(dx \cdot dy)}{Dt} = 2\,D_{ij}dx_idy_j. \tag{1.24}$$

If at time t, dx and dy are parallel to the e_1 and e_2 axis, respectively, then, from (1.24),

$$D_{12} = -\frac{1}{2}\frac{D\beta}{Dt}, \tag{1.25}$$

where β is the angle formed by dx and dy. The other off-diagonal elements can be interpreted similarly. Note that this physical meaning of the components of

---

[2]The definition of the gradient of a vector varies in the literature. In some works, it is the transpose of that used here.

D does not require knowledge of the history of deformation of specific material elements. Rather the components of $D(x,t)$ are related to the rate of change of material elements only at time t.

1.5. Special motions

1.5.1. Rigid motions. A rigid motion is one in which the distance between arbitrary material points remains constant. Therefore, for rigid motions, the material derivative of ds is zero for all points in the body, for all time during the motion. We see from (1.22) that a necessary and sufficient condition for a motion to be rigid is that D be identically zero at all points in the body for all time.

One can show that the most general rigid motion can be written as

$$x = x_o(t) + Q(t) \cdot X, \tag{1.26}$$

where Q is a proper orthogonal, second-order tensor.

1.5.2. Isochoric motions. Isochoric motions are those in which the volume occupied by fixed material particles is unchanged during the motion. A material does not have to be incompressible to undergo isochoric motions. We see from the relation (1.10), that if a motion is isochoric, then the value of J is one at all points in the body, throughout the motion. In this case, it follows from (1.11) that the divergence of v is equal to zero and hence the trace of D is zero. In summary, the following are each necessary and sufficient conditions for a motion to be isochoric,

$$
\begin{aligned}
J = 1, \qquad dv = dV, \qquad \frac{DJ}{Dt} = 0, \\
\frac{\partial v_i}{\partial x_i} = 0, \qquad \text{trace}(D) = 0,
\end{aligned}
\tag{1.27}
$$

at all points in the body and throughout the motion.

1.5.3. Simple shear. A flow field that is of great significance in fluid mechanics is simple shear flow, sometimes called Couette flow. In this flow, the velocity field is steady, fully developed and uni-directional. The magnitude of the velocity depends linearly on the spatial component with axis perpendicular to the direction of flow. For example, if rectangular coordinates x_i are chosen such that the flow direction is parallel to the x_1 axis, then the orientation and origin of the coordinate system can be chosen such that the velocity field can be written as

$$\textbf{Simple Shear} \qquad v = \dot{\gamma}_0\, x_2\, e_1, \tag{1.28}$$

where $\dot{\gamma}_0$ is a constant. Clearly simple shear is an example of an isochoric motion. Using (1.4) and the initial condition $x = X$ at $t = 0$, we obtain,

$$\textbf{Simple Shear} \qquad x = X + \dot{\gamma}_0\, X_2\, t e_1. \tag{1.29}$$

It is possible to show that simple shear flow can be generated between parallel plates for a wide class of fluids called simple fluids (see Section 5.1). For this reason, some rheometers are designed to generate approximations to this flow, Section 7. With this in mind, we consider simple shear flow between two parallel plates separated by a distance h, when the upper plate is moving with speed U in the

e_1 direction and the origin of the coordinate system coincides with a point on the bottom plate, Figure 2. Using the no-slip boundary conditions at the top surface with (1.28) it follows that $\dot\gamma_0 = U/h$.

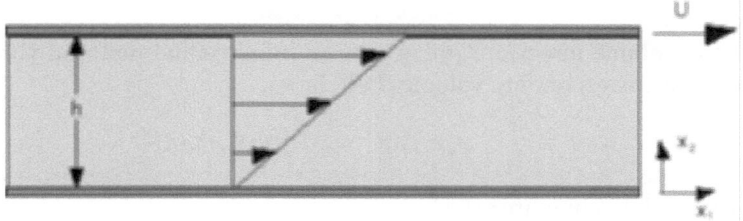

FIGURE 2. Schematic of the velocity field in simple shear flow between two parallel plates, driven by the motion of the upper plate.

From (1.28) and (1.29), as well as the definitions of the kinematic tensors introduced earlier in this section, it is straightforward to show that for simple shear,

$$
[\boldsymbol{F}] = \begin{bmatrix} 1 & \dot\gamma_0 t & 0 \\ 0 & 1 & 0 \\ 0 & 0 & 1 \end{bmatrix}, \qquad
[\boldsymbol{F}^{-1}] = \begin{bmatrix} 1 & -\dot\gamma_0 t & 0 \\ 0 & 1 & 0 \\ 0 & 0 & 1 \end{bmatrix},
$$

$$
[\boldsymbol{B}] = \begin{bmatrix} 1+\dot\gamma_0^2 t^2 & \dot\gamma_0 t & 0 \\ \dot\gamma_0 t & 1 & 0 \\ 0 & 0 & 1 \end{bmatrix}, \qquad
[\boldsymbol{C}] = \begin{bmatrix} 1 & \dot\gamma_0 t & 0 \\ \dot\gamma_0 t & 1+\dot\gamma_0^2 t^2 & 0 \\ 0 & 0 & 1 \end{bmatrix},
$$

$$
[\boldsymbol{E}] = \frac{1}{2}\begin{bmatrix} 0 & \dot\gamma_0 t & 0 \\ \dot\gamma_0 t & \dot\gamma_0^2 t^2 & 0 \\ 0 & 0 & 0 \end{bmatrix}, \qquad
[e] = \frac{1}{2}\begin{bmatrix} 0 & \dot\gamma_0 t & 0 \\ \dot\gamma_0 t & -\dot\gamma_0^2 t^2 & 0 \\ 0 & 0 & 0 \end{bmatrix}, \tag{1.30}
$$

$$
[\boldsymbol{D}] = \frac{\dot\gamma_0}{2}\begin{bmatrix} 0 & 1 & 0 \\ 1 & 0 & 0 \\ 0 & 0 & 0 \end{bmatrix}.
$$

2. Governing equations

In this section, we discuss governing equations applicable to both fluids and solids. Specific constitutive equations are discussed in the following sections. It is useful to first recall the transport theorem which is used below to obtain local forms of the governing equations from integral forms.

2.1. The transport theorem

Consider an arbitrary subset of the body $\mathcal{B}$ that occupies the region $\mathcal{V}_0$ with boundary $\partial\mathcal{V}_0$ in κ_0 and region $\mathcal{V}$ with boundary $\partial\mathcal{V}(t)$ in $\kappa(t)$, Figure 1. Recall that $\mathcal{V}(t)$ is a material region of the fluid: a region occupied by a fixed set of material particles in the body, that may change in time. The region $\mathcal{V}_0$ is in general different than $\mathcal{V}(t)$. Let ϕ be any scalar or tensor-valued field with the following representations,

$$\phi = \bar{\phi}(\boldsymbol{X}, t) = \hat{\phi}(\boldsymbol{x}, t). \qquad (2.1)$$

Consider the volume integral, $\mathcal{I}$, of ϕ over an infinitesimal material volume dv in κ and over the corresponding volume $J\,dV$ in κ_0,

$$\mathcal{I} \equiv \int_{\mathcal{V}(t)} \hat{\phi}(\boldsymbol{x}, t)\,dv = \int_{\mathcal{V}_0} \bar{\phi}(\boldsymbol{X}, t) J\,dV. \qquad (2.2)$$

The transport theorem states that

$$\frac{d\mathcal{I}}{dt} = \int_{\mathcal{V}(t)} \left[\frac{D\phi}{Dt} + \hat{\phi}(\boldsymbol{x}, t)\,\mathrm{div}\hat{\boldsymbol{v}}(\boldsymbol{x}, t) \right] dv. \qquad (2.3)$$

Alternatively, using the divergence theorem, it is sometimes convenient to write the transport theorem as,

$$\frac{d\mathcal{I}}{dt} = \int_{\mathcal{V}(t)} \frac{\partial \hat{\phi}(\boldsymbol{x}, t)}{\partial t}\,dv + \int_{\partial V(t)} \hat{\phi}(\boldsymbol{x}, t)\,\hat{\boldsymbol{v}}(\boldsymbol{x}, t) \cdot \boldsymbol{n}\,da \qquad (2.4)$$

where $\boldsymbol{n}$ is the outward unit normal to $\partial \mathcal{V}$.

2.2. Conservation of mass

The mass $\mathcal{M}$ of a fixed subset of material particles of the body occupying a sub-region $\mathcal{V}(t)$ of region $\mathcal{R}(t)$ at time t, Figure 1, is

$$\mathcal{M} = \int_{\mathcal{V}(t)} \rho(\boldsymbol{x}, t)\,dv, \qquad (2.5)$$

where $\rho(\boldsymbol{x}, t)$ is the mass density of the fluid in $\kappa(t)$. The mass $\mathcal{M}$ of the same material region in the reference configuration κ_0 is

$$\mathcal{M} = \int_{\mathcal{V}_0} \rho_0(\boldsymbol{X})\,dV, \qquad (2.6)$$

where $\rho_0(\boldsymbol{X})$ is the mass density of the same material region in κ_0. The principle of conservation of mass is the postulate that the mass of this fixed set of material particles does not change in time,

$$\frac{d\mathcal{M}}{dt} = 0. \qquad (2.7)$$

In stating (2.7), we have assumed there are no mass sinks or sources. Making use of the transport theorem (2.3) as well as (2.5) and (2.7), we can write the principle of conservation of mass with respect to the spatial (Eulerian) representation of the field variables,

$$0 = \int_{\mathcal{V}(t)} \left(\frac{D\rho}{Dt} + \rho\,\mathrm{div}\,\boldsymbol{v} \right) dv. \qquad (2.8)$$

Making suitable assumptions about continuity of the field variables and making use of the requirement that (2.8) holds for all subregions $\mathcal{V}(t)$ of $\mathcal{R}(t)$, we obtain the **local form of conservation of mass**:

$$\frac{D\rho}{Dt} + \rho\operatorname{div}\boldsymbol{v} = 0 \quad\text{or}\quad \frac{D\rho}{Dt} + \rho\frac{\partial\hat{v}_i(\boldsymbol{x},t)}{\partial x_i} = 0. \tag{2.9}$$

It follows from (1.10) with (2.5)–(2.7) that

$$\rho J = \rho_0. \tag{2.10}$$

2.2.1. Implications of conservation of mass for incompressible fluids. The volume of the region occupied by material elements of an incompressible fluid cannot change in time (the fluid cannot be compressed). Therefore, incompressible fluids can only undergo isochoric motions and the conditions in (1.27) must be satisfied for all motions of an incompressible fluid. It therefore follows from (2.9) that the density of a material element of an incompressible fluid cannot change in time,

$$\frac{D\rho}{Dt} = 0 \quad\text{and therefore,}\quad \rho = \rho_0. \tag{2.11}$$

Alternatively, this last result follows directly from (2.10) with $J = 1$.

2.3. Balance of linear momentum

The postulate of balance of linear momentum is the statement that the rate of change of linear momentum of a fixed mass of the body is equal to the sum of the forces acting on the body,

$$\frac{d}{dt}\int_{\mathcal{V}(t)} \rho\boldsymbol{v}\,dv = \int_{\mathcal{V}(t)} \rho\boldsymbol{b}\,dv + \int_{\partial\mathcal{V}(t)} \boldsymbol{t}\,da \tag{2.12}$$

where $\boldsymbol{b}$ is the body force per unit mass, $\boldsymbol{t} = \hat{\boldsymbol{t}}(\boldsymbol{x},t,\boldsymbol{n})$ is the force per unit surface area in $\kappa(t)$ and $\boldsymbol{n}$ is the unit normal to this same surface at time t. The vector $\boldsymbol{t}$ is often called the Cauchy stress vector. The first and second integrals on the right-hand side of (2.12) represent the net force arising from body forces and surface forces, respectively. Note, the stress vector depends on position, time and the unit normal to the surface at $\boldsymbol{x}$. Recall Cauchy's lemma, which follows from suitable smoothness assumptions on the function $\boldsymbol{t}(\boldsymbol{x},t,\boldsymbol{n})$ and the balance of linear momentum (2.12)

$$\hat{\boldsymbol{t}}(\boldsymbol{x},t,\boldsymbol{n}) = -\hat{\boldsymbol{t}}(\boldsymbol{x},t,-\boldsymbol{n}). \tag{2.13}$$

Building on this last result, the existence of a second-order tensor, $\boldsymbol{T}$, called the Cauchy stress tensor, can be shown (e.g., [59]), where

$$\boldsymbol{t} = \hat{\boldsymbol{t}}(\boldsymbol{x},t,\boldsymbol{n}) = \hat{\boldsymbol{T}}(\boldsymbol{x},t)\cdot\boldsymbol{n}. \tag{2.14}$$

Significantly, $\boldsymbol{T}$ is independent of $\boldsymbol{n}$.

The local form of the equation of linear momentum can be obtained by using the transport theorem, the divergence theorem, and (2.14) to write (2.12) as a single volume integral. Then, making suitable assumptions about the continuity of the field variables, using the transport theorem and (2.8), we obtain the **local Eulerian form of the balance of linear momentum,**

$$\rho\,\boldsymbol{a} \;=\; \operatorname{div}\boldsymbol{T} \,+\, \rho\boldsymbol{b}, \qquad \text{or} \qquad \rho\,a_i \;=\; \frac{\partial T_{ij}}{\partial x_j} \,+\, \rho b_i. \tag{2.15}$$

It is sometimes convenient to represent $\boldsymbol{T}$ as the sum of a deviatoric and spherical part,

$$\boldsymbol{T} \;=\; \boldsymbol{\tau} + \bar{t}\,\boldsymbol{I}, \tag{2.16}$$

where

$$\operatorname{tr}\boldsymbol{\tau} \;=\; 0 \qquad \text{and therefore} \qquad \bar{t} \;=\; \frac{1}{3}\operatorname{tr}(\boldsymbol{T})$$

$$\text{or} \tag{2.17}$$

$$\tau_{ii} \;=\; 0, \qquad \text{and} \qquad \bar{t} \;=\; \frac{1}{3}T_{kk}.$$

The tensor $\boldsymbol{\tau}$ is often referred to as the deviatoric part of $\boldsymbol{T}$ and $\bar{t}\boldsymbol{I}$ as the spherical part. When the Cauchy stress tensor is decomposed in this way, $-\bar{t}$ is often called the *pressure* and is denoted by p. Using the decomposition, (2.16), the balance of linear momentum can be written as

$$\rho\,\boldsymbol{a} \;=\; -\operatorname{grad}p + \operatorname{div}\boldsymbol{\tau} + \rho\boldsymbol{b} \qquad \text{or} \qquad \rho\,a_i \;=\; -\frac{\partial p}{\partial x_i} + \frac{\partial \tau_{ij}}{\partial x_j} + \rho b_i. \tag{2.18}$$

For compressible fluids, p is a thermodynamic pressure. An equation of state relating pressure to other thermodynamic variables such as mass density and temperature will be necessary. For incompressible fluids, p is a mechanical pressure arising from the constraint of incompressibility. No equation of state is necessary, rather, p will be determined as part of the solution to the governing equations and boundary conditions.

2.3.1. Lagrangian form of balance of linear momentum. Recall that $\boldsymbol{t}(\boldsymbol{x},t,\boldsymbol{n})$ is the force per unit area in the *current configuration*. In some cases, it is more useful to consider an alternate stress vector defined as the force in the current configuration per unit area in the *reference configuration*, which we will denote by $\boldsymbol{p}$ (first Piola–Kirchhoff stress vector). The stress vectors $\boldsymbol{p}$ and $\boldsymbol{t}$ are then related through

$$\int_{\partial\mathcal{V}_0} \boldsymbol{p}(\boldsymbol{X},t,\boldsymbol{N})\,dA \;=\; \int_{\partial\mathcal{V}} \boldsymbol{t}(\boldsymbol{x},t,\boldsymbol{n})\,da, \tag{2.19}$$

where $\boldsymbol{N}$ and $\boldsymbol{n}$ are the unit normals to surfaces $\partial\mathcal{V}_0$ and $\partial\mathcal{V}$, respectively, at the same material point. In addition, dA and da are the corresponding infinitesimal material areas on $\partial\mathcal{V}_0$ and $\partial\mathcal{V}$, respectively. Nanson's formula relates these geometric quantities,

$$\boldsymbol{n}\,da = J\boldsymbol{F}^{-T}\boldsymbol{N}\,dA \qquad \text{or} \qquad n_i\,da \;=\; J\,F_{Ai}^{-1}\,N_A\,dA. \tag{2.20}$$

Following arguments similar to those used to obtain (2.14), the existence of a second-order tensor $\boldsymbol{P}$ (first Piola–Kirchhoff tensor) can be shown,

$$p(\boldsymbol{X}, t, \boldsymbol{N}) = \boldsymbol{P}(\boldsymbol{X}, t) \cdot \boldsymbol{N}. \tag{2.21}$$

It follows from (2.19)–(2.21),

$$\boldsymbol{P} = J\boldsymbol{T} \cdot \boldsymbol{F}^{-T} \quad \text{or} \quad P_{iA} = JT_{ij}F_{Aj}^{-1}. \tag{2.22}$$

Using these results for $\boldsymbol{p}$ and $\boldsymbol{P}$, we can rewrite (2.12) over surface and volume integrals in κ_0 to obtain,

$$\int_{V_0} \rho_0 \, \boldsymbol{a} \, dV = \int_{V_0} \rho_0 \, \boldsymbol{b} \, dV + \int_{\partial V_0} \boldsymbol{p} \, dA. \tag{2.23}$$

The corresponding **local Lagrangian form of balance of linear momentum** is

$$\rho_0 \boldsymbol{a} = \operatorname{grad}_0 \cdot \boldsymbol{P} + \rho_0 \, \boldsymbol{b} \quad \text{or} \quad \rho_0 a_i = \frac{\partial P_{iA}}{\partial X_A} + \rho_0 b_i. \tag{2.24}$$

2.4. Balance of angular momentum

The balance of angular momentum is the statement that the rate of change of angular momentum of a fixed material region arises from the combined torques on the body. In the absence of body couples, the integral form of the balance of angular momentum can be written as

$$\frac{d}{dt} \int_{V(t)} \rho \boldsymbol{x} \times \boldsymbol{v} \, dv = \int_{V(t)} \rho \boldsymbol{x} \times \boldsymbol{b} \, dv + \int_{\partial V(t)} \boldsymbol{x} \times \boldsymbol{t} \, da. \tag{2.25}$$

The integral on the left-hand side of (2.25) is the angular momentum of the material body at time t. The first and second integrals on the right-hand side of (2.25) are the resultant torques due to body and surface forces, respectively. We can also write a corresponding integral form over surfaces and volumes in the reference configuration. In the interest of space we do not do so here. Making use of (2.9) and (2.18) as well as suitable continuity assumptions, it can be shown that (2.25) reduces to the requirement that the Cauchy stress tensor must be symmetric,

$$\boldsymbol{T} = \boldsymbol{T}^T \quad \text{or} \quad T_{ij} = T_{ji}. \tag{2.26}$$

It follows from (2.26) and (2.22) that

$$\boldsymbol{P} \cdot \boldsymbol{F}^T = \boldsymbol{F} \cdot \boldsymbol{P}^T. \tag{2.27}$$

Clearly, $\boldsymbol{P}$ is not in general symmetric. Sometimes a third stress tensor, the second Piola–Kirchhoff stress tensor, $\boldsymbol{S}$, is defined through

$$\boldsymbol{S} = \boldsymbol{F}^{-1} \cdot \boldsymbol{P} \quad \text{or} \quad S_{AB} = F_{Ai}^{-1} P_{iB}, \tag{2.28}$$

and therefore,

$$\boldsymbol{S} = J\boldsymbol{F}^{-1} \cdot \boldsymbol{T} \cdot \boldsymbol{F}^{-T} \quad \text{or} \quad S_{AB} = JF_{Ai}^{-1}T_{ij}F_{Bj}^{-1}. \tag{2.29}$$

It follows from (2.27) and (2.28) that $\boldsymbol{S}$ is symmetric. This characteristic makes it useful in some ways, though it does not have as clear a physical meaning as $\boldsymbol{P}$.

2.5. Mechanical energy equation

It is sometimes useful to consider the *Mechanical Energy Equation* which can be obtained by taking the inner product of the velocity vector and the equation of linear momentum (2.15) with (1.6),

$$\frac{1}{2}\rho\frac{D}{Dt}(v_i v_i) = \frac{\partial T_{ij}}{\partial x_j}v_i + \rho b_i v_i. \tag{2.30}$$

It should be emphasized that (2.30) is *not* derived from the equation of balance of energy. An integral form of the mechanical energy equation can be obtained by integrating (2.30) over the material region $\mathcal{V}(t)$ with surface $\partial\mathcal{V}(t)$, to obtain

$$\frac{d}{dt}\int_{\mathcal{V}(t)}\frac{1}{2}\rho\boldsymbol{v}\cdot\boldsymbol{v}\,dv = \int_{\partial\mathcal{V}(t)}\boldsymbol{t}\cdot\boldsymbol{v}\,da + \int_{\mathcal{V}(t)}\rho\boldsymbol{b}\cdot\boldsymbol{v}\,dv - \int_{\mathcal{V}(t)}\boldsymbol{T}:\boldsymbol{D}\,dv, \tag{2.31}$$

where we have made use of the divergence theorem, the transport theorem and the conservation of mass. The first term in (2.31) is the rate of change of kinetic energy in the material region $\mathcal{V}(t)$, the second term is the rate of work done by surface forces on the surface $\partial\mathcal{V}(t)$, and the third integral is rate of work done on the material region $\mathcal{V}(t)$ by body forces. The scalar, $\boldsymbol{T}:\boldsymbol{D} = \text{tr}(\boldsymbol{T}^T\cdot\boldsymbol{D}) = T_{ij}D_{ij}$ is the rate of work by stresses per unit volume of the body. The last integral in (2.31) is called the *stress power* or *rate of internal mechanical work*.

2.6. Balance of energy

We have already discussed the ability of the body to store energy in the form of mechanical energy and for energy to enter the body through work done by surface and body forces. Here, we extend this discussion and consider thermal energy.

If we consider a part of the body $\mathcal{V}(t)$ in the current configuration, we can hypothesize the existence of a scalar functional called the specific internal energy, $u = u(\boldsymbol{x},t)$ (internal energy per unit mass). The internal energy for the part $\mathcal{V}(t)$ of the body will then be

$$\int_{\mathcal{V}(t)}\rho\,u\,dv. \tag{2.32}$$

We further hypothesize that thermal energy may enter the body through the surface $d\mathcal{V}(t)$ of the body with outward unit normal $\boldsymbol{n}$. It can be shown that this thermal energy ("heat") flux per unit surface area can be represented as the scalar product of a vector $\boldsymbol{q}$ and the normal to the surface $\boldsymbol{n}$, where $\boldsymbol{q}\cdot\boldsymbol{n}$ positive is associated with heat leaving the surface and $\boldsymbol{q}\cdot\boldsymbol{n}$ negative is associated with heat entering the surface. In addition, thermal energy may enter the body as a specific heat supply per unit time, $r = r(\boldsymbol{x},t)$ (the heat entering the body per unit mass per unit time). Therefore the rate at which thermal energy enters the region $\mathcal{V}(t)$ of the body is

$$-\int_{d\mathcal{V}(t)}\boldsymbol{q}\cdot\boldsymbol{n}da + \int_{\mathcal{V}(t)}\rho r dv. \tag{2.33}$$

The balance of energy is a statement that the rate of increase in energy (both internal and kinetic) in region $\mathcal{V}(t)$ of the body is equal to the rate of work by body forces and contact forces plus thermal energy entering the body per unit time. We can write this statement as

$$\frac{d}{dt}\int_{\mathcal{V}(t)}\rho(u+\frac{1}{2}\boldsymbol{v}\cdot\boldsymbol{v})dv = \int_{d\mathcal{V}(t)}\boldsymbol{t}\cdot\boldsymbol{v}da + \int_{\mathcal{V}(t)}\rho\,\boldsymbol{b}\cdot\boldsymbol{v}dv$$
$$- \int_{d\mathcal{V}(t)}\boldsymbol{q}\cdot\boldsymbol{n}da + \int_{\mathcal{V}(t)}\rho\,r\,dv. \tag{2.34}$$

Eq. (2.34) is sometimes called the **first law of thermodynamics**. Making suitable assumptions about continuity of the field variables, we can obtain the local form of (2.34),

$$\rho\left(\frac{Du}{Dt}+\boldsymbol{v}\cdot\frac{D\boldsymbol{v}}{Dt}\right) = \boldsymbol{T}:\boldsymbol{D} + \boldsymbol{v}\cdot(\mathrm{div}\,\boldsymbol{T}) + \rho\boldsymbol{v}\cdot\boldsymbol{b} - \mathrm{div}\boldsymbol{q} + \rho r. \tag{2.35}$$

Using results from the mechanical energy equation, we can rewrite (2.35) as

$$\rho\frac{Du}{Dt} = \boldsymbol{T}:\boldsymbol{D} - \mathrm{div}\boldsymbol{q} + \rho r, \tag{2.36}$$

or in indicial form,

$$\rho\frac{Du}{Dt} = T_{ij}D_{ij} - \frac{\partial q_i}{\partial x_i} + \rho r. \tag{2.37}$$

2.7. Restrictions on constitutive equations

Thus far, we have discussed governing equations fundamental to any continuum material. To close this system of governing equations, we need to select constitutive models for the material of interest. We choose to take a classical approach to this subject, whereby we start with general forms of the constitutive equation and then use fundamental principles to reduce the class of acceptable constitutive models. Prior to turning attention to the constitutive theories, we briefly summarize requirements imposed on constitutive models in order that they be deemed "physically reasonable". Here, and in the remainder of this chapter, we focus attention on purely mechanical theories, where, for example, the effect of temperature variations are negligible. We will disregard any non-mechanical influences and assume the state of the body is determined solely by the kinematical history, (e.g., [65], pages 56–68). Motivated by applications to blood flow, in later sections, attention will be concentrated on incompressible materials.

2.7.1. Principle of coordinate invariance. The constitutive equations must be independent of the coordinate system used to describe the motion of the body.

2.7.2. Principle of determinism for the stress. The stress in a body at the current time is determined by the history of motion of that body and independent of any aspect of its future behavior, [45].

2.7.3. Principle of local action. The determination of the stress for a given particle in the body is independent of the motion outside an arbitrary neighborhood of that particle (see [65], page 57 for a mathematical description of this principle). This principle was originally combined with the previous principle [45].

2.7.4. Principle of equipresence. Under this principle, a quantity which appears as an independent variable in one constitutive equation should be present in all others for that material unless it violates some law of physics or rule of invariance (e.g., [65], pages 359–360, for an example in the context of thermoelasticity and a historical discussion of this principle).

2.7.5. Principle of material frame indifference. There are two separate principles which embody the concept that the response of the material should be unaffected by its location and orientation. In the first, the mechanical response of a body is required to be unchanged under a superposed rigid body motion of the body if the change in orientation and position of the body is accounted for (e.g., [26] and pages 484–486 of [44]). The second principle is the requirement that the material response should be invariant under an arbitrary change of observer. For historical reasons, Truesdell and Noll refer to the first of these principles as the *Hooke–Poisson–Cauchy form* and the second as the *Zaremba–Jaumann form*. Strictly speaking, these two principles are different, the second being more restrictive since it includes improper orthogonal transformations, such as reflections. Truesdell and Noll provide an interesting discussion of the history of these two principles in [65], pages 45-47.

As will be discussed in the remainder of this chapter, invariance requirements play an important role in continuum mechanics in restricting the form of constitutive equations.

2.7.6. Thermodynamic restrictions. The second law of thermodynamics is the restriction that the total entropy production for all thermodynamic processes is never negative. In the remainder of this chapter, we restrict attention to purely mechanical theories for which thermal effects are negligible. In this case, this restriction can be reduced to the statement that the stress power be non-negative [39, 63, 64],

$$\boldsymbol{T} : \boldsymbol{D} \geq 0. \tag{2.38}$$

2.7.7. Well-posedness. The initial value problem associated with the governing equations for the purely mechanical theory arising from the conservation of mass, balance of linear momentum and the constitutive equation for the stress tensor should be well-posed. By this we mean existence, uniqueness and continuous dependence of the solution on the data can be shown (see, e.g., [28]).

2.7.8. Stability of the rest state. One of the methods used to evaluate the range of physically reasonable parameters for a material is to evaluate the conditions under which the rest state is stable. It seems physically reasonable to exclude ranges of material parameters for which the rest state is unstable to infinitesimal disturbances. This criterion has been used, for example, for viscoelastic fluids (see, e.g., [15, 36, 24]) as well as fiber fluid mixtures [25].

2.7.9. Attainability. An additional test, which is relatively straightforward and does not require formulation within the context of thermodynamics, is to evaluate the *attainability* of solutions for chosen benchmark flows for fluids or equilibrium deformations for solids. A constitutive equation would seem to be physically unreasonable if a chosen steady or time-periodic motion (e.g., steady Couette flow) is unattainable, no matter how gradually the driving mechanism is ramped to a constant value and no matter how small this constant value is. Attainability of "physically reasonable" steady flows has been studied for Newtonian fluids (e.g., [17, 31, 23] and the literature there cited) and well as for some viscoelastic fluids [56]. We emphasize that attainability of a given solution should not be confused with an examination of the stability of this solution (e.g., [17]).

2.7.10. Mechanical response of real materials. Experiments on real materials also provide restrictions on the range of material parameters that are physically reasonable. Since we cannot test every material in existence, strictly speaking, we cannot actually "prove" an experimentally-based restriction on a constitutive equation be necessary. Rather, experimental results for certain categories of real materials (e.g., polymeric fluids) provide guidelines for defining a "reasonable" range of parameters for a given material.

By way of example, in this subsection, we turn attention to some restrictions we might impose on a class of fluids called incompressible simple fluids, which will be discussed in more detail in Section 7.1. Briefly, an incompressible simple fluid is a material for which the stress at a point and the current time is determined up to a pressure once the strain of each past configuration relative to the present configuration is known (e.g., [12]). Namely, unlike solid materials, we do not need to know the strain relative to some inherent "natural" configuration. As discussed earlier in this chapter, the density of incompressible materials is unchanged during the deformation.

It can be shown that the mechanical behavior of a chosen (but arbitrary) incompressible simple fluid is completely determined in some flows, for example, simple shear flow, once three material functions are known for that fluid. This result is somewhat unexpected since an incompressible simple fluid may have many more than three material functions or constants. By material functions, we mean functions that depend only on the nature of the material, not, for example, on the experimental conditions. For reasons to be described below, we will refer to these three functions as *viscometric functions*.

The three viscometric functions can be defined relative to the rectangular components of the Cauchy stress tensor T_{ij} given in (1.28) for simple shear,

$$\tau(\dot{\gamma}_0) = T_{12} \qquad \mathcal{N}_1(\dot{\gamma}_0) = T_{11} - T_{22} \qquad \mathcal{N}_2(\dot{\gamma}_0) = T_{22} - T_{33}. \qquad (2.39)$$

We will refer to the functions $\tau(\dot{\gamma}_0), \mathcal{N}_1(\dot{\gamma}_0), \mathcal{N}_2(\dot{\gamma}_0)$ as the *shear stress function, first normal stress difference* and the *second normal stress difference*, respectively [3].

---

[3]There is some variation in the literature for the definitions of normal stress and even the sign of the stress tensor. See page 71 of [58] for a nice discussion of this issue.

It can be shown that $\tau(\kappa)$ is an odd function while $\mathcal{N}_1(\kappa)$ and $\mathcal{N}_2(\kappa)$ are even functions (e.g., pages 70–71 of [58]).

Alternatively, we can consider the viscometric functions η, ψ_1, ψ_2,

$$\eta(\dot{\gamma}_0) = \frac{T_{12}}{\dot{\gamma}_0} \qquad \psi_1(\dot{\gamma}_0) = \frac{T_{11} - T_{22}}{\dot{\gamma}_0^2} \qquad \psi_2(\dot{\gamma}_0) = \frac{T_{22} - T_{33}}{\dot{\gamma}_0^2}, \qquad (2.40)$$

referred to as the *viscosity, first normal stress coefficient* and *second normal stress coefficient*, where $\dot{\gamma}_0 \neq 0$. For Newtonian fluids, ψ_1 and ψ_2 are identically zero.

It turns out that simple shear is not the only motion for which the behavior of an incompressible simple fluid is completely determined once η, ψ_1 and ψ_2 are known. While most flows do not meet this requirement, there are a number of other flows, called *viscometric flows*, which do (see, e.g., [12]). Viscometric flows include steady, fully developed flow in a straight pipe of constant circular cross section (sometimes called Poiseuille flow) and steady, unidirectional flow between two concentric circular cylinders driven by the rotation of one or both of the cylinders about their common axis (sometimes called Couette flow). Most rheometers are designed to generate viscometric flows and can be used to measure one or more of these material functions.

Based on experimental data for real polymeric fluids, the sign of ψ_1 is expected to be non-negative and the sign of ψ_2 to be non-positive [7]. In addition, the ratio of the magnitude of the second normal stress coefficient to the first normal stress coefficient is commonly believed to be less than one half (e.g., [38, 53]).

2.7.11. Further Comments. Roughly speaking, the five principles given in this subsection restrict the general functional form of the constitutive equation (e.g., [65] for further details and historical information). The latter requirements impose restrictions on the range of parameters for specific constitutive models. In the following sections, applications of these restrictions to particular constitutive equations are considered.

3. Nonlinear viscous fluids

Due to its relevance to blood, we now determine the most general physically reasonable form of constitutive equations for *incompressible viscous fluids*,

$$\boldsymbol{T} = -p\boldsymbol{I} + \boldsymbol{\tau}(\boldsymbol{L}), \qquad (3.1)$$

where p is the Lagrange multiplier arising from the incompressibility constraint and $\boldsymbol{\tau}$ is not necessarily the deviatoric stress defined in (2.16) and (2.17). Inherent in the form of the stress tensor given in (3.1) is the assumption that the current state of stress depends only on the velocity gradient at the current time and not on any previous deformation the fluid might have undergone. Clearly (3.1) satisfies the first four principles given above. We now turn attention to the fifth principle.

3.1. Restrictions due to invariance requirements

Using invariance of the stress tensor under a superposed rigid body motion (see, Section 2.7.5), and a representation theorem for symmetric isotropic tensor functions, it can be shown that the most general form of (3.1) which satisfies invariance requirements is (e.g., [2])

$$\boldsymbol{\tau} = \phi_0(\mathrm{II_D}, \mathrm{III_D})\boldsymbol{I} + \phi_1(\mathrm{II_D}, \mathrm{III_D})\boldsymbol{D} + \phi_2(\mathrm{II_D}, \mathrm{III_D})\boldsymbol{D}^2, \qquad (3.2)$$

where $\mathrm{II_D}, \mathrm{III_D}$ are the second and third principal invariants of $\boldsymbol{D}$ and can be written in the following form for isochoric motions,

$$\mathrm{II_D} = -1/2\,\mathrm{tr}\,(\boldsymbol{D}^2), \quad \mathrm{III_D} = \det \boldsymbol{D}. \qquad (3.3)$$

The first invariant, $\mathrm{I_D} = \mathrm{tr}\boldsymbol{D}$, is identically zero for isochoric motions. The function, ϕ_0 can be absorbed into the Lagrange multiplier arising from the incompressibility constraint to obtain

$$\boldsymbol{T} = -p\boldsymbol{I} + \phi_1(\mathrm{II_D}, \mathrm{III_D})\boldsymbol{D} + \phi_2(\mathrm{II_D}, \mathrm{III_D})\boldsymbol{D}^2. \qquad (3.4)$$

Incompressible fluids with stress tensor of the form (3.4) are called *Reiner–Rivlin fluids*. The Navier–Stokes fluid is a special Reiner–Rivlin fluid with ϕ_2 equal to zero and ϕ_1 constant.

3.2. Restrictions on the Reiner–Rivlin equation due to behavior of real fluids

In this section, we calculate the viscometric functions from solutions for the Reiner–Rivlin fluid in simple shear in order to determine restrictions on the material functions ϕ_1 and ϕ_2 arising from knowledge of the behavior of real fluids. It follows from (1.30), (3.2) and (3.3), that the only non-zero components of the extra stress tensor $\boldsymbol{\tau}$ in simple shear are

$$\tau_{12} = \tau_{21} = \phi_1\frac{\dot{\gamma}_0}{2}, \qquad \tau_{11} = \tau_{22} = \phi_0 + \phi_2\frac{\dot{\gamma}_0^2}{4}, \qquad \tau_{33} = \phi_0. \qquad (3.5)$$

Therefore, from (3.5) and (2.40), the viscometric functions for a Reiner–Rivlin fluid are

$$\eta = \frac{\phi_1}{2}, \qquad \psi_1 = 0, \qquad \psi_2 = \frac{1}{4}\phi_2. \qquad (3.6)$$

Hence, the physical meanings of ϕ_1 and ϕ_2 are

$$\phi_1 = 2\,\eta, \qquad \phi_2 = 4\,\psi_2. \qquad (3.7)$$

However, there is no evidence of real fluids exhibiting a zero value for ψ_1 and a non-zero value for ψ_2 [2]. For this reason, attention is typically confined to a reduced form of the Reiner–Rivlin equation with ϕ_2 equal to zero. In short, based on invariance requirements and the behavior of real fluids, the most general "reasonable" incompressible constitutive equation of the form $\boldsymbol{T} = \hat{\boldsymbol{T}}(\boldsymbol{L})$ is

$$\boldsymbol{T} = -p\boldsymbol{I} + 2\,\eta(\mathrm{II_D}, \mathrm{III_D})\,\boldsymbol{D}. \qquad (3.8)$$

The functional dependence of η on $\mathrm{III_D}$ is often neglected. As discussed in [2], page 54, there is some evidence that this may be reasonable for real fluids. In any case, the quantity $\mathrm{III_D}$ is identically zero in simple shear (and other viscometric flows) and therefore the functional dependence of η on III_D cannot be determined in most rheometers. In particular, the nonlinear viscosity functions used for blood are assumed to depend only on II_D. Since II_D is not a positive quantity for isochoric motions, it is useful to introduce a positive metric of the rate of deformation, denoted by $\dot{\gamma}$,

$$\dot{\gamma} \equiv \sqrt{2\,tr\,(\boldsymbol{D}^2)} = \sqrt{-4\text{II}_D}. \tag{3.9}$$

For example, in simple shear flow, this metric coincides with the shear rate $\dot{\gamma}_0$ introduced in (1.28). In summary,

Incompressible, generalized Newtonian fluid

$$\boldsymbol{T} = -p\boldsymbol{I} + 2\eta(\dot{\gamma})\,\boldsymbol{D} \quad \text{where} \quad \dot{\gamma} \equiv \sqrt{2\,tr\,(\boldsymbol{D}^2)}, \tag{3.10}$$

and p is the mechanical pressure. We emphasize that the use of the representation (3.10) does not restrict attention to simple shear or other viscometric flows.

From (3.6) it is clear that, unlike the viscoelastic constitutive equations which will be discussed in Section 5, the generalized Newtonian constitutive model (3.10), is not capable of modeling non-zero normal stress effects in real fluids.

3.3. Restrictions on generalized Newtonian fluids due to thermodynamic considerations

In this section, we use the second law of thermodynamics to deduce restrictions on the choices of the material functions $\eta(\text{II}_D, \text{III}_D)$ in (3.8) (see, e.g., [63, 64, 11, 39]). For the generalized Newtonian fluid (3.8),

$$\boldsymbol{T} : \boldsymbol{D} \equiv \text{tr}\,(\boldsymbol{T}^\text{T} \cdot \boldsymbol{D}) = 2\,\eta\,\text{tr}\,(\boldsymbol{D}^2), \tag{3.11}$$

where we have made use of the condition of incompressibility. Therefore, if the stress power is assumed to be non-negative as in (2.38), we obtain the familiar requirement that the viscosity be non-negative

$$\eta \geq 0. \tag{3.12}$$

3.4. Examples of generalized Newtonian fluids

As can be seen from (3.10), the generalized Newtonian fluids differ through their choice of the viscosity function. One of the simplest viscosity functions is the power-law model,

Power-Law Model $\quad \eta(\dot{\gamma}) = K\,\dot{\gamma}^{(n-1)}, \tag{3.13}$

where n and K are termed the power-law index and consistency, respectively. The power-law model includes the **Newtonian fluid** as a special case (n=1), where the viscosity η is then a constant, often denoted by μ. For n < 1, the power-law is shear thinning (decreasing viscosity with shear rate), while for n > 1 it is shear thickening, Figure 3. The shear thinning power-law model is often used for blood, even though it predicts an unbounded viscosity at zero shear rate and zero viscosity as the shear rate tends to infinity.

In order to address the limitations of the power-law model, a number of viscosity functions have been defined with bounded and non-zero limiting values

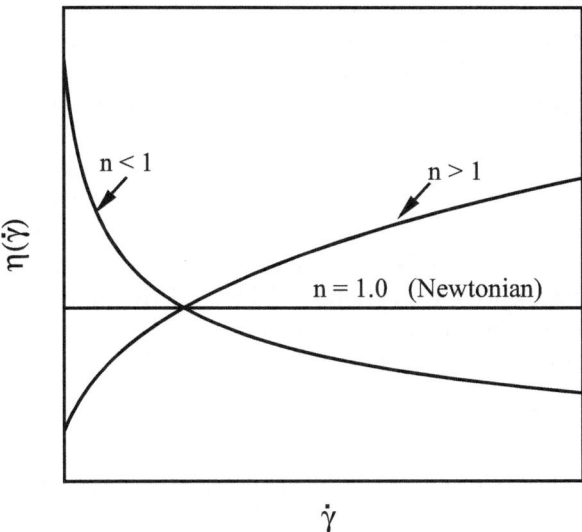

FIGURE 3. Viscosity as a function of shear rate for power-law fluids which are (a) shear thinning (n < 1), (b) constant viscosity (n = 1) and (c) shear thickening (n > 1). Here, a finite range of shear rates $\dot{\gamma}$ is considered with $\dot{\gamma} > 0$.

of viscosity. Specific forms for the viscosity function for blood are discussed in Section 7 of [57].

4. Yield stress "fluids"

When some fluid suspensions are placed between parallel plates and gradually loaded by an applied force on the plates, Figure 2, the material appears to resist flow until a finite level of stress is reached. Above this level, the suspensions appear to flow like a fluid. This behavior led researchers to hypothesize the existence of a material property called the yield value or yield stress, τ_Y, of the material. Namely, a critical stress level is required for the material to flow. As discussed in Section 7.1 of [57], the existence of a true yield stress is somewhat of a controversial issue (see, for example, [5, 42]).

In formulating a mathematical description of yield stress materials, and in particular a yield criterion, it is necessary to first select a metric of the stress tensor. In simple shear flow (1.28), this is trivial because there are only two non-zero components of the extra stress tensor and they are identical, $\tau_{12} = \tau_{21}$. In this case, the metric of the stress magnitude is just $|\tau_{12}|$ and the yield criterion can be simply written as $|\tau_{12}| = \tau_Y$. For $|\tau_{12}| < \tau_Y$, the material can either behave rigidly (no deformation) or display a non-rigid behavior (for example, deform elastically). For $|\tau_{12}| \geq \tau_Y$, the material will flow and can display shear thinning or even viscoelastic properties. Strictly speaking, yield stress materials are not fluids since they require a finite level of applied stress to flow.

While it is typically not discussed in the blood literature (see pages 74–75 in [19] for an exception), it should be understood that when the flow is more complex than simple shear, a more general metric of shear stress is needed. For example, the yield criterion $|\tau_{12}| = \tau_Y$ does not satisfy the principle of coordinate invariance discussed in Section 2.7. Instead, we will use a more general function of the extra stress tensor $\mathfrak{f}(\boldsymbol{\tau})$ as a metric of stress magnitude and write the yield criterion as $\mathfrak{f}(\boldsymbol{\tau}) = \tau_Y$. In doing so, we have assumed yielding of the material is independent of the mechanical pressure. It is straightforward to show the most general yield criterion of this form which satisfies the principal of material invariance can be represented as $\mathfrak{f}(\mathrm{I}_\tau, \mathrm{II}_\tau, \mathrm{III}_\tau) = \tau_Y$, where

$$\mathrm{I}_\tau = \operatorname{tr} \boldsymbol{\tau}, \qquad \mathrm{II}_\tau = 1/2\left((\operatorname{tr} \boldsymbol{\tau})^2 - \operatorname{tr} \boldsymbol{\tau}^2\right) \qquad \mathrm{III}_\tau = \det \boldsymbol{\tau}. \qquad (4.1)$$

A simple form for the yield function, which is consistent with our physical expectations from simple shear, is

$$\mathfrak{f}(\mathrm{I}_\tau, \mathrm{II}_\tau, \mathrm{III}_\tau) = \sqrt{|\mathrm{II}_\tau|}. \qquad (4.2)$$

The corresponding yield criterion is

$$\sqrt{|\mathrm{II}_\tau|} = \tau_Y \qquad \text{Yield Criterion.} \qquad (4.3)$$

In simple shear, the shear stress and therefore the stress metric are constant throughout the flow field, so that (4.3) is satisfied everywhere or nowhere. However, in more complex flows, such as fully developed flow in a straight pipe, the shear stress varies throughout the flow field and the yield criterion will not be met simultaneously throughout the fluid domain. Rather, there will be regions where the yield criterion is reached (and the fluid is *flowing*) while in other regions the value of $\sqrt{|\mathrm{II}_\tau|}$ is below τ_Y and the fluid does not flow ($\boldsymbol{D} = 0$ in that region). One such example is the well known plug flow discussed below for steady, fully developed flow of a yield stress fluid in a straight pipe, Section 7.1.1.

4.1. Bingham model

The simplest yield stress model is the Bingham material, which behaves rigidly until the yield criterion (4.3) is reached, after which it behaves like an incompressible Newtonian fluid (constant viscosity), Figure 4. For a Bingham fluid in simple shear (Section 1.5.3),

$$\begin{aligned}
|\tau_{12}| < \tau_Y &\implies \dot{\gamma}_0 = 0 \\
|\tau_{12}| \geq \tau_Y &\implies \tau_{12} = \tau_Y + \mu \dot{\gamma}_0,
\end{aligned} \qquad (4.4)$$

where μ is the constant viscosity in the region of flow. For simplicity, the prescribed shear rate is non-negative ($\dot{\gamma}_0 \geq 0$) here and in the remainder of this section.

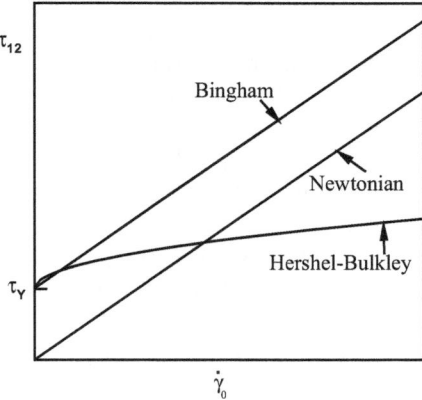

FIGURE 4. Behavior of representative Newtonian and yield stress fluids in simple shear. Shear stress τ_{12} versus shear rate $\dot{\gamma}_0$ curves for three models: (i) Newtonian, (ii) Bingham and (iii) Shear thinning Herschel–Bulkley.

It is useful to have a Bingham model which is more generally valid than (4.4), yet reduces to (4.4) in simple shear. A mathematical statement of a three-dimensional Bingham model is (e.g., [16, 51])

$$\sqrt{|\mathrm{II}_\tau|} < \tau_Y \quad \Longrightarrow \quad D_{ij} = 0$$

$$\sqrt{|\mathrm{II}_\tau|} \geq \tau_Y \quad \Longrightarrow \quad \begin{cases} D_{ij} = \dfrac{1}{2\mu}\left(1 - \dfrac{\tau_Y}{\sqrt{|\mathrm{II}_\tau|}}\right)\tau_{ij} \\[3mm] \tau_{ij} = 2\left(\mu + \dfrac{\tau_Y}{2\sqrt{|\mathrm{II_D}|}}\right)D_{ij}, \end{cases} \qquad (4.5)$$

where μ is the constant viscosity attained once the material flows.

Materials which have been modeled by the Bingham constitutive model include pastes, margarine, mayonnaise, ketchup and blood. The Bingham fluid equation is important because it appears to model some real fluids well over a range of applied stress and yet it is possible, in some cases, to obtain analytical solutions for the flow field. The review article by Bird, Dai and Yarusso [8] contains a useful collection of analytical solutions and an extensive reference list for the Bingham model. Early studies of this constitutive equation can be found in [47, 48, 54, 51].

4.2. Modified Bingham model

While Bingham fluids are relatively simple in form, they can be challenging to model numerically due to the difficulty in tracking the yield surfaces in the flow field. Papanastasiou [50] introduced a generalized Newtonian constitutive model which eliminates the discontinuity in viscosity at zero shear rate, Figure 5,

$$\eta(\dot{\gamma}) = \mu + \frac{\tau_Y \left(1 - e^{-c\,\dot{\gamma}^2/2}\right)}{\dot{\gamma}^2}, \qquad (4.6)$$

where c, μ, and τ_Y are material constants. We will refer to this model as the modified Bingham fluid. The value of c determines the rate of exponential increase to the constant viscosity μ, Figure 5.

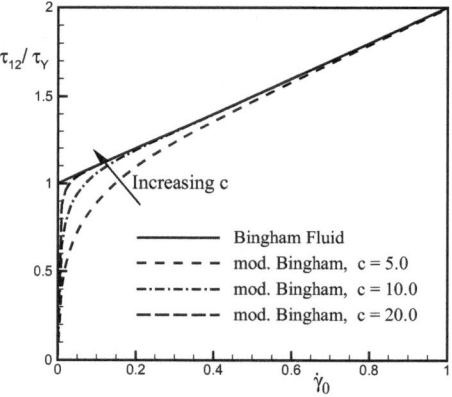

FIGURE 5. Shear stress versus shear rate in simple shear for the Bingham fluid and modified Bingham fluid for $c = 5, 10, 20$.

4.3. Herschel–Bulkley model

Like the Bingham model, the Herschel–Bulkley model [29, 30] behaves rigidly until a critical level of the stress metric is reached. However, the Herschel–Bulkley model, as well as the Casson model discussed next, can capture a nonlinear dependence of viscosity on shear rate. The Herschel–Bulkley model displays a power-law viscosity once it begins to flow, Figure 4. In simple shear,

$$\begin{aligned}
|\tau_{12}| < \tau_Y &\implies \dot{\gamma}_0 = 0 \\
|\tau_{12}| \geq \tau_Y &\implies \tau_{12} = \tau_Y + K\,\dot{\gamma}_0^{\mathrm{n}}.
\end{aligned} \qquad (4.7)$$

The three-dimensional Herschel–Bulkley model can be written in a form similar to that of the Bingham model (4.5) with μ replaced by $\eta(\dot\gamma) = K\dot\gamma^{n-1}$,

$$\sqrt{|\text{II}_\tau|} < \tau_Y \implies D_{ij} = 0$$

$$\sqrt{|\text{II}_\tau|} \geq \tau_Y \implies
\begin{cases}
D_{ij} = \dfrac{1}{2K\dot\gamma^{n-1}} \left(1 - \dfrac{\tau_Y}{\sqrt{|\text{II}_\tau|}}\right) \tau_{ij} \\[4mm]
\tau_{ij} = 2 \left(K\dot\gamma^{n-1} + \dfrac{\tau_Y}{2\sqrt{|\text{II}_D|}}\right) D_{ij}.
\end{cases}
\tag{4.8}$$

4.4. Casson model

Casson [9] developed a theoretical expression for capturing the yield stress and shear thinning behavior of fluid suspensions formed by dispersing various types of pigments in lithographic varnish. In this theoretical model, he used high aspect ratio rods to model the chain-like groups of pigments arising due to attractive forces between the pigments. With increasing shear rate, the disruptive hemodynamic forces increase, causing a decrease in rod size, which gives rise to the shear thinning behavior. Within this formulation, the suspension was treated as dilute. The resulting Casson equation, like the last two yield stress models, behaves rigidly until the yield criterion (4.3) is satisfied, after which it displays a shear thinning behavior. The Casson constitutive model is typically presented for simple shear flow,

$$|\tau_{12}| < \tau_Y \implies \dot\gamma_0 = 0$$
$$|\tau_{12}| \geq \tau_Y \implies \tau_{12}^{1/2} = \tau_Y^{1/2} + (\mu_N \dot\gamma_0)^{1/2}.
\tag{4.9}$$

This equation reduces to the Newtonian model when τ_Y is set to zero, in which case, μ_N is the Newtonian viscosity. While Eq. (4.9) is only applicable to simple shear, it is straightforward to write a three-dimensional constitutive equation that reduces to (4.9) in simple shear,

$$\sqrt{|\text{II}_\tau|} < \tau_Y \implies D_{ij} = 0$$

$$\sqrt{|\text{II}_\tau|} \geq \tau_Y \implies
\begin{cases}
D_{ij} = \dfrac{1}{2\mu_N} \left(1 - \dfrac{\sqrt{\tau_Y}}{\sqrt[4]{|\text{II}_\tau|}}\right)^2 \tau_{ij} \\[4mm]
\tau_{ij} = 2 \left(\sqrt{\mu_N} + \dfrac{\sqrt{\tau_Y}}{\sqrt[4]{4|\text{II}_D|}}\right)^2 D_{ij}.
\end{cases}
\tag{4.10}$$

5. Viscoelastic fluids

Viscoelastic liquids have the ability to store as well as dissipate energy. This property can lead to fascinating behaviors not seen in inelastic fluids, including rod-climbing, tubeless siphons and elastic recoil (see, e.g., [7]). As will be discussed in the next two chapters, the viscoelastic nature of human blood is believed to arise from the ability of three-dimensional aggregates of RBC (and to a lesser degree individual RBCs) to store and release energy during flow.

The theoretical, experimental and numerical studies of polymeric fluids during the last several decades have significantly increased our understanding of viscoelastic fluids and provide a strong foundation on which to discuss this subject. There is not sufficient space in this section to provide a detailed coverage of the theory of viscoelastic fluids. Rather, the reader is referred to Truesdell and Noll's classical reference *Non-Linear Field Theories of Mechanics* [65] as well as some of the original references by Coleman and Noll, cited in this chapter. The reader is also referred to the texts by Astarita and Marucci [2], Bird, Armstrong and Hassager [7], Joseph [37], Owens and Phillips [49], and Schowalter [58].

In this section, we provide a very brief overview of some theoretical aspects of viscoelastic fluids, beginning with a discussion of simple materials and simple fluids. We then turn attention to simple fluids with fading memory, including two approximate theories for simple fluids: ordered fluids arising from the retarded motion expansion and linear viscoelastic fluids.

5.1. Simple fluids

Many of the continuum models for blood are specific cases of incompressible simple fluids which is a subcategory of a class of substances called simple materials. A simple material is a body in which the stresses for a material element are determined by the cumulative history of the deformation gradient for that element. The concept of simple materials including simple fluids was originally developed by W. Noll [45].

5.1.1. Kinematics. Prior to defining simple materials in mathematical terms, it is necessary to introduce some relevant kinematics. Recall that material points in a body can be identified by specifying their location in some reference configuration. In elastic materials, it is often convenient to choose this configuration as the zero stress state. For viscoelastic fluids with fading memory, it is convenient to choose the reference configuration to be the configuration of the body at current time t, denoted by κ_t. The current time is sometimes called the *time of observation*. We then identify an arbitrary material point in the body by its position, $\boldsymbol{x}$, in κ_t. The position of these material points during the period prior to time t, at some time $\tau \leq t$ will be denoted as $\boldsymbol{\xi}$ and can be expressed as,

$$\boldsymbol{\xi} = \boldsymbol{\chi}_t(\boldsymbol{x}, \tau) \qquad \tau \in (-\infty, 0]. \tag{5.1}$$

We denote the configuration at arbitrary prior time τ by κ_τ. The function $\boldsymbol{\chi}_t(\boldsymbol{x}, \tau)$ is called the *relative deformation function* and is defined so that at time t ,

$$\boldsymbol{\chi}_t(\boldsymbol{x}, t) = \boldsymbol{x}. \tag{5.2}$$

We define the rectangular components of the relative deformation gradient $\boldsymbol{F}_t$ as

$$F_{t_{ij}} = \frac{\partial \chi_{ti}(\boldsymbol{x}, \tau)}{\partial x_j}, \tag{5.3}$$

where F_t can be shown to be a second-order tensor. An infinitesimal material element dx in κ_t will be mapped to $d\xi$ in κ_τ through

$$d\xi \ = \ F_t \cdot dx. \tag{5.4}$$

We will assume F_t is suitably continuous such that it is invertible, and hence,

$$\det F_t \neq 0. \tag{5.5}$$

It follows from (5.3) that $F_t(x, t) \ = \ I$, so that at time t the $\det F_t$ is equal to 1. Since the motion is assumed to be continuous and $\det F_t$ is never equal to zero, we have that $0 < \det F_t < \infty$. For isochoric motions, $\det F_t$ is equal to 1 for all τ.

It is sometimes convenient to use an inverted time scale $s = t - \tau$. Hence, s is equal to zero at the current time and increases in magnitude as we go back in time.

We now turn attention to the example of simple shear flow, (1.29). For these flows, the reduced history can be written as

$$\xi \ = \ x \ - \ s \, \dot{\gamma}_0 \, x_2 \, e_1. \tag{5.6}$$

Then from (5.3), the relative deformation gradient in simple shear is

$$\left[F_{tij}\right] \ = \ \begin{bmatrix} 1 & -s \, \dot{\gamma}_0 & 0 \\ 0 & 1 & 0 \\ 0 & 0 & 1 \end{bmatrix}. \tag{5.7}$$

For use in future discussions, we define the right relative Cauchy–Green tensor C_t, the Cauchy strain tensor G_t and the Finger strain tensor H_t,

$$C_t \ = \ F_t^T \cdot F_t, \qquad G_t \ = \ C_t - I, \qquad H_t \ = \ C_t^{-1} - I. \tag{5.8}$$

The strain measures G_t and H_t are defined such that they are equal to 0 at the current time t. Another kinematic quantity of interest is the kth Rivlin–Ericksen tensor A_k, which can be defined for any G_t which is n times differentiable with respect to s at time $s = 0$ [4],

Rivlin–Ericksen Tensors
$$A_k \ = \ (-1)^k \frac{d^k G_t(s)}{d s^k}\Big|_{s=0}, \qquad k = 1, 2, 3, \ldots \tag{5.9}$$

For example,

$$A_1 \ = \ \mathrm{grad} v \ + \ (\mathrm{grad} v)^T. \tag{5.10}$$

Note that A_k are functions of the current time t as well as position x. A recurrence formula can be used to calculate the other Rivlin–Ericksen tensors,

$$A_{N+1} \ = \ \frac{D A_N}{Dt} \ + \ L^T \cdot A_N \ + \ A_N \cdot L. \tag{5.11}$$

---

[4]For simplicity, in what follows, we will only explicitly write the dependence of a function on s. The dependence on the current time t and reference position x will be understood.

These Rivlin–Ericksen tensors are of particular interest because they arise when, under suitable continuity assumptions, we consider a Taylor series expansion of G_t about the current time ($s = 0$),

$$G_t(s) = \sum_{N=1}^{\infty} (-1)^N \frac{1}{N} A_N s^N, \qquad (5.12)$$

or

$$G_t(s) = -sA_1 + \frac{s^2}{2} A_2 - \frac{s^3}{3} A_3 + \cdots . \qquad (5.13)$$

The result (5.12) will be quite useful in the discussion below on the retarded motion expansion.

Simple materials. Employing the kinematics given above, we can then define the constitutive equation for a **simple material** as

$$\textbf{Simple Material } \quad T = \underset{s=0}{\overset{\infty}{\mathfrak{H}}} \, [F_t(s); \rho], \qquad (5.14)$$

where the symbol $\underset{s=0}{\overset{\infty}{\mathfrak{H}}} [\cdot]$ is used to denote a second-order tensor-valued functional of the history of the relative deformation gradient.

Simple fluids are simple materials with the maximal possible isotropy group and are in fact isotropic. It can be shown [45], that without loss of generality, the Cauchy stress tensor for an incompressible simple fluid that satisfies invariance requirements can be written as

$$\textbf{Incompressible Simple Fluid } \quad T = -pI + [\underset{s=0}{\overset{\infty}{\mathfrak{S}}} G_t(s)],$$

$$\text{subject to } \quad Q(t) \cdot \left(\underset{s=0}{\overset{\infty}{\mathfrak{S}}} G_t(s) \right) \cdot Q^T(t) = \underset{s=0}{\overset{\infty}{\mathfrak{S}}} [Q(t) \cdot G_t(s) \cdot Q^T(t)], \qquad (5.15)$$

for all proper orthogonal tensors, $Q(t)$, where p is the Lagrange multiplier associated with the incompressibility constraint. This constraint can be written with respect to G_t as

$$\det(G_t + I) = 1. \qquad (5.16)$$

For incompressible materials, the extra stress tensor is determined to within a scalar multiple of the identity tensor. In order to remove this indeterminacy, we specify that

$$\text{tr} \, \underset{s=0}{\overset{\infty}{\mathfrak{S}}} \, [(G_t)] = 0. \qquad (5.17)$$

Under this normalization, p can be seen to be the mechanical pressure $p = -1/3 \, \text{tr} \, T$, defined previously.

An important flow history, which we will make use of below, is the rest history $G_t(s) = 0$ for all $s \in [0, \infty)$. Using $(5.15)_2$, it can be shown if a simple fluid has only experienced the rest history, the extra stress is proportional to the identity tensor, namely, the stress is isotropic. It therefore follows that if a simple fluid does experience an anisotropic stress, it must flow. For this reason, materials displaying a yield stress are not simple fluids.

5.1.2. Simple fluids with fading memory. The idea of simple fluids with fading memory was introduced by Coleman and Noll in 1960 based on the expectation that deformations in the recent past should have more influence on the current state of stress than deformations in the distant past, [13]. Namely, the functional in (5.15) should be more sensitive to values of G_t for small s, than large s. Physically, this means the "memory" of the system will "fade away" in time. This sense of fading memory then introduces the idea of a time scale inherent in the material that sets the memory of the material, often referred to as the "natural time". In fact, what is really important is the ratio of this natural time, say λ, to a characteristic time of observation associated with the flow, for example, T. This important ratio is the Deborah number,

$$De = \frac{\lambda}{T},\qquad(5.18)$$

which was first introduced by Reiner in 1964 [55]. Reiner named this ratio after the prophetess Deborah who stated, "The mountains flowed before the Lord". As noted by Reiner, Deborah understood that even mountains are "fluid-like" if viewed on a sufficiently long time scale.

The role of the Deborah number is clearly illustrated in well-known experiments with "silly putty", or unvulcanized rubber (e.g., [7], page 95). The time constant of the silly putty is on the order of seconds. If silly putty is rolled into a sphere and dropped to the floor, it will bounce. There is insufficient time during this motion for the fluid to "forget" its original shape ($T << \lambda$). If the sphere of silly putty is set on a table, it will flow into a puddle over a time period on the order of tens of minutes. In this case, $T >> \lambda$. The value of λ is the same in both cases. It is the time of observation which leads to a large De and elastic behavior, in the first case, and low De and inelastic behavior, in the latter.

While simple fluids are more general than simple fluids with fading memory, in practice we would not be able to sensibly test materials that depend significantly on their *entire history* of deformation ($s \to \infty$). For materials with fading memory, we can in principal design rheometric protocols such that mechanical measurement of the material properties are independent of the history of the material prior to initiating the experiment.

Coleman and Noll defined fading memory more precisely within the context of a Banach space $\mathcal{L}$ with a norm that can be used to measure the distance of a given reduced history $G_t(s)$ from the rest history ($G_t = 0$). This norm should place more weight on values of $G_t(s)$ for which s is small than for large s.

Coleman and Noll [13] introduced the $\mathcal{L}_{h,p}$-*norm*, which is characterized by a real constant p with $1 \leq p \leq \infty$ and an *influence function*, $h(s)$, which is a real-valued function that approaches zero 'rapidly' as $s \to \infty$,

$$\begin{aligned}
||G_t(s)||_{h,p} &= \sqrt[p]{\int_0^\infty h(s)\,|G_t(s)|^p\,ds} \quad &&\text{if } 1 \leq p < \infty \\
||G_t(s)||_{h,\infty} &= \sup_{s \geq 0} |G_t(s)|\,h(s) \quad &&\text{if } p = \infty,
\end{aligned}\qquad(5.19)$$

and $|\boldsymbol{G}_t(s)|$ is the norm of the instantaneous value of $\boldsymbol{G}_t(s)$,

$$|\boldsymbol{G}_t(s)| \;=\; (G_{tik}(s)G_{tik}(s))^{1/2}. \tag{5.20}$$

The influence function, $h(s)$ is chosen to satisfy the following conditions,

1. $h(s)$ is a positive, real-valued function, continuous over $s \in [0, \infty)$.
2. $h(s)$ is normalized such that $h(0) = 1$.
3. $h(s)$ tends to zero in such a way that $\lim_{s \to \infty} s^r h(s) = 0$, monotonically for large s.

The value of r is often called the order of the influence function and determines the strength of the fading memory. The set of all histories $\boldsymbol{G}_t(s)$ with finite $\mathcal{L}_{h,p}$-norm form a Banach space, $\mathcal{L}_{h,p}$. Under this norm, two strain histories $\boldsymbol{G}_t(s)$, which are the same for small s, but differ greatly for large s, will have nearly the same norm.

For example, the function

$$h(s) = \exp(-\beta s), \tag{5.21}$$

with β a positive constant, is an influence function of any order. Namely, the first two conditions are satisfied for (5.21) and the third condition is satisfied no matter how large a value is chosen for r. It follows from (5.19) that the value of $\|\boldsymbol{G}_t\|$ is more greatly influenced by events in the recent past than those in the distant past.

We have yet to mathematically define simple fluids with fading memory. Depending on the choice for r we can define fluids with weak or strong fading memory. In this work, we confine attention to simple fluids with strong fading memory so that we can expand the stress functional about the **rest history** or **zero history**. As noted earlier, the rest history is a special flow history where the fluid has been at rest for all time, so that $\boldsymbol{F}_t = \boldsymbol{I}$ or, $\boldsymbol{G}_t = \boldsymbol{0}$ for all $s \in [0, \infty)$.

The **Stronger Principles of Fading Memory** can be stated as (see, e.g., [65], page 104),

There exists an influence function of order greater than $n + 1/2$ such that the response functional $\overset{\infty}{\underset{s=0}{\mathfrak{S}}} \,(\boldsymbol{G}_t(s))$ is defined and n times Fréchet-differentiable in a neighborhood of the zero history of the function space $\mathcal{L}_{hp}$.

It should be emphasized that this definition does not require that $\boldsymbol{G}_t$ be continuous to be admissible. For example, we can still consider the step functions of strain used in stress relaxation experiments as admissible for simple fluids with strong fading memory.

An example of an incompressible simple fluid with strong fading memory is the *lower convected Maxwell fluid*,

$$\boldsymbol{T} \;=\; -p\boldsymbol{I} \;+\; \int_o^\infty a\,e^{-s/\lambda}\,\boldsymbol{G}_t(s)\,ds. \tag{5.22}$$

In (5.22), a and λ are material constants and λ is a relaxation time, or a measure of how quickly the "material forgets". The "memory" of the material increases with increasing λ.

5.2. Approximations for simple fluids with fading memory

For viscometric flows, it is possible to obtain valuable results for the behavior of general simple fluids without a-priori selection of a particular history functional, (e.g., [12])[5]. However, viscometric flows are a very restricted class of flows. With the goal of obtaining general results for simple fluids, without restriction to viscometric flows, we turn attention to two important approximate theories for simple fluids. The theory of "ordered fluids" was obtained by Coleman and Noll as a hierarchy of approximations to simple fluids with fading memory for "slow flows". The theory of infinitesimal linear viscoelasticity is based on the approximation that the strain relative to the rest history is small. In this section, we provide a brief introduction to these two approximate theories.

Coleman and Noll [14] proved that the general constitutive theory of incompressible simple fluids with fading memory can be written as

$$\textbf{Finite Linear Viscoelasticity}\quad \boldsymbol{T} \;=\; -p\boldsymbol{I} \;-\; \int_0^\infty m(s)\,\boldsymbol{G}_t(s)\,ds, \qquad (5.23)$$

in the limit of

$$\|\boldsymbol{G}_t(s)\| \to 0 \qquad (5.24)$$

with an error approaching zero faster than $\|\boldsymbol{G}_t(s)\|$. Namely, (5.23) is a first-order approximation for all incompressible, simple fluids with fading memory. In order for this material to display fading memory, m(s) must tend to zero as $s \to \infty$, at a rate that is consistent with the strength of the fading memory,

$$\int_0^\infty |m(s)|^2\,\frac{1}{h^2}\,ds < \infty. \qquad (5.25)$$

The function $m(s)$ is called the memory function and can be related to the commonly measured **relaxation function** $\mathcal{G}(s)$ through

$$\mathcal{G}(s) \;=\; \int_s^\infty m(u)\,du \qquad \text{so} \qquad m(s) \;=\; -\frac{d\mathcal{G}(s)}{ds}. \qquad (5.26)$$

We can therefore write the Cauchy stress tensor given in (5.23) as

$$\boldsymbol{T} \;=\; -p\boldsymbol{I} \;-\; \int_0^\infty \mathcal{G}(s)\,\frac{\partial \boldsymbol{G}_t(s)}{\partial s}\,ds, \qquad (5.27)$$

and therefore knowledge of $\mathcal{G}(s)$ is sufficient for determining the mechanical behavior of finite linear viscoelastic materials.

---

[5]Viscometric flows were introduced in Section 2.7.10 and will be further discussed in more detail in Section 7.1

Some examples of relaxation moduli are:

Maxwell Model $\mathcal{G}(s) = \mathcal{G}_o e^{-s/\lambda}$

General Linear
Viscoelastic Model $\mathcal{G}(s) = \sum_{k=1}^{N} \mathcal{G}_k e^{-s/\lambda_k}$ (5.28)

Relaxation Spectrum $\mathcal{G}(s) = \int_0^{\infty} \frac{H(\lambda)}{\lambda} e^{-s/\lambda} d\lambda.$

The variable λ with or without a subscript denotes the relaxation times for the material. The general linear viscoelastic model is just a superposition of simple Maxwell type models to allow for multiple relaxation times in the material. In principle, these relaxation times can be determined experimentally from oscillatory shear experiments which are discussed in more detail in Section 7.2.1.

The finite linear viscoelastic equation was obtained as an approximation for small $|\boldsymbol{G}_t(s)|$. We could equally well have started with a different relative strain measure. For example, $\boldsymbol{H}_t = \boldsymbol{C}_t^{-1} - \boldsymbol{I}$. In this case, we would have obtained,

$$\boldsymbol{T} = -p\boldsymbol{I} + \int_0^{\infty} m(s)\boldsymbol{H}_t(\boldsymbol{x}, s)\, ds. \qquad (5.29)$$

It is interesting to note that while the integral model (5.23) was derived from the simple fluid model under the condition (5.24), it satisfies invariance requirements and can be considered a constitutive model for finite strain problems. This is also true for (5.29). Namely, while they will not represent the behavior of general simple fluids at large strain, they can be treated as specific simple fluid models. Their usefulness for large strain then depends on the existence of real materials that display behaviors, at least in some categories of flows, that are consistent with the predictions of (5.23) or (5.29). In fact, one such example of (5.29) is the Upper Convected Maxwell (or Maxwell B) fluid. We will come back to this point in the next section.

5.2.1. Infinitesimal linear viscoelasticity. One way the condition (5.24) will be satisfied is if the kinematics of the simple fluid are restricted to motions in which the norm of the strain relative to the rest history is small for all past times (see, e.g., [12]). We define

$$\epsilon = \sup_{s \in [0,\infty)} (|\boldsymbol{F}_t(s) - \boldsymbol{I}|). \qquad (5.30)$$

Then a deformation history $\boldsymbol{F}_t(s)$ will be considered infinitesimal for all past times if $\epsilon << 1$. In infinitesimal linear viscoelasticity, only terms of order ϵ or larger are considered and so all nonlinear effects such as shear thinning are ignored. An example of a infinitesimal deformation history is small amplitude oscillatory shear which will be discussed in Section 7.2.1. We can define the infinitesimal strain tensor $\boldsymbol{E}_t(s)$ as

$$\boldsymbol{E}_t(s) = \frac{1}{2}\left(\boldsymbol{F}_t + \boldsymbol{F}_t^T - 2\boldsymbol{I}\right) \qquad (5.31)$$

so that,

$$G_t = 2\,E_t + \mathcal{O}(\epsilon^2). \qquad (5.32)$$

Therefore from (5.23), (5.27) and (5.32) (see [12] for details),

Infinitesimal Linear Viscoelasticity

$$T = -pI - 2\int_0^\infty m(s)\,E_t(s)\,ds. \qquad (5.33)$$

To the order of the approximation of the infinitesimal theory, the simple Maxwell model can be written as

$$\tau + \lambda\frac{\partial\tau}{\partial t} = 2\,\eta\,D. \qquad (5.34)$$

Note that neither (5.33) nor (5.34) satisfy general invariance requirements.

5.2.2. Retarded motion expansion. Due to the fading memory of the material, the relative strain does not necessarily have to be small for all past times in order that (5.24) be met. Instead, it needs only to be small in the recent past. Coleman and Noll introduced the idea of retarded motions which are, in a sense to be made precise below, flows which are "slow enough" that condition (5.24) is met and $G_t(s)$ can be expanded about the zero history, (when some additional smoothness requirements are made (e.g., [65], page 108).

Coleman and Noll showed that for these retarded motions, the functionals for all simple fluids with strong fading memory can be approximated by a multi-linear function of the present values of the time derivatives of $G_t(s)$. The resulting hierarchy of approximations to the general history functional are the so-called "ordered fluids". Solutions obtained, for example, for second-order fluids are representative of the behavior we would find for any simple fluid if the flow is "sufficiently slow".

To quantify this notion of slow flows, Coleman and Noll defined a retardation functional $\Gamma_\alpha[\cdot]$ with a retardation factor α, as a linear transformation of G_t,

$$(\Gamma_\alpha G_t)(s) = G_t(\alpha s), \qquad 0 < \alpha \le 1. \qquad (5.35)$$

They showed that Γ_α maps the space $\mathcal{L}_{h,p}$ into itself. So defined, the retardation replaces a history $G_t(s)$ with a history which is similar, but slower.

In order to expand the history functional about the zero history, Coleman and Noll proved an approximation theorem that permits the asymptotic approximation of a memory functional for "slow histories", using a polynomial function of the Rivlin–Ericksen tensors (5.9).

Using the approximation theorem, Nth-order approximations to slow flows of simple fluids with fading memory can be obtained. By Nth order it is understood that the norm of the remainder is of order α^{N+1}. The first-order expansion for slow flows is just a linear viscous fluid,

First-order fluid $\quad T = -p\,I + \eta A_1. \qquad (5.36)$

The second-order expansion (called a second-order fluid) is

Second-order fluid $\quad T = -\tilde{p}\,I + \eta A_1 + \alpha_1 A_2, + \alpha_2\,A_1^2, \qquad (5.37)$

where η, α_1 and α_2 are constants. The tilde notation is used to emphasize that $\tilde{p}$ is in general different from the mechanical pressure $p = -1/3\text{tr } \boldsymbol{T}$. Higher-order fluids can also be obtained, but we do not discuss these here.

The ordered fluids can then be thought of as an expansion about an incompressible Newtonian fluid. As the lowest ordered fluid displaying viscoelastic behavior, the second-order fluid has historically received a great deal of attention. Since all simple fluids behave like the ordered fluids for "slow enough" flows, these models have been used with great success to develop physical insight about viscoelastic fluids. However, it should be cautioned that these are approximate models. As will be discussed in Galdi [20], some of the ordered fluids have been shown to display unsettling behavior such as instability of the rest state for ranges of material parameters which are realistic for real fluids. See [7] for examples of exact solutions for the ordered fluids and a discussion of the limitations and appropriate practical applications of these fluids. As noted on page 104 of [65], although the ordered fluids are obtained as approximations to simple fluids with strong fading memory, they do not themselves satisfy the conditions for strong fading memory.

5.3. Finite viscoelastic models

Truesdell and Noll [65] subdivided viscoelastic constitutive equations into three categories: (i) differential, (ii) integral, and (iii) rate type. In differential constitutive models, the extra stress tensor can be written as an explicit function of a finite number of time derivatives of the temporal derivatives of $\boldsymbol{G}_t$ or other suitable strain measures, at the *current time t*. Viscoelastic fluids of the integral type are simple fluids for which the history functional can be written as one or more integrals. An example was given in Equation (5.22). Rate-type constitutive models include one or more time derivatives of the extra stress tensor. They do not appear as an explicit expression for the stress tensor. These categories are not necessarily distinct. For example, as will be discussed below, the Maxwell B model can be written as both an integral and rate type model.

Many of the commonly used nonlinear viscoelastic models are special cases of **quasi-linear rate-type viscoelastic models** called Maxwell models (e.g., [37]),

$$\boldsymbol{\tau} + \lambda\frac{\mathcal{D}\boldsymbol{\tau}}{\mathcal{D}t} = 2\eta\boldsymbol{D}, \qquad (5.38)$$

where the definition of the operator $\mathcal{D}(\ )/\mathcal{D}t$ is not unique and will be discussed below. It is chosen so that the operator on $\boldsymbol{\tau}$ is objective under a superposed rigid body motion and the resulting second-order tensor is symmetric (we consider cases where there is no body couple). The following operators can all be shown to be objective choices for $\mathcal{D}\boldsymbol{\tau}/\mathcal{D}t$,

Upper convected derivative $\quad \overset{\triangledown}{\boldsymbol{\tau}} = \dfrac{D\boldsymbol{\tau}}{Dt} - \boldsymbol{L}\cdot\boldsymbol{\tau} - \boldsymbol{\tau}\cdot\boldsymbol{L}^T,$

Lower convected derivative $\quad \overset{\triangle}{\boldsymbol{\tau}} = \dfrac{D\boldsymbol{\tau}}{Dt} + \boldsymbol{\tau}\cdot\boldsymbol{L} + \boldsymbol{L}^T\cdot\boldsymbol{\tau},$ $\qquad(5.39)$

Co-rotational Derivative $\quad \overset{\circ}{\boldsymbol{\tau}} = \dfrac{1}{2}\left(\overset{\triangledown}{\boldsymbol{\tau}} + \overset{\triangle}{\boldsymbol{\tau}} \right),$

where $\boldsymbol{L}$ is the velocity gradient defined in (1.19). To avoid confusion, we also write the upper and lower convected derivatives in indicial form,

$$
\begin{aligned}
\overset{\triangledown}{\tau}_{ij} &= \frac{\partial \tau_{ij}}{\partial t} + v_k \frac{\partial \tau_{ij}}{\partial x_k} - \frac{\partial v_i}{\partial x_k}\tau_{kj} - \tau_{ik}\frac{\partial v_j}{\partial x_k} \\
\overset{\triangle}{\tau}_{ij} &= \frac{\partial \tau_{ij}}{\partial t} + v_k \frac{\partial \tau_{ij}}{\partial x_k} + \tau_{ik}\frac{\partial v_k}{\partial x_j} + \frac{\partial v_k}{\partial x_i}\tau_{kj}.
\end{aligned}
\qquad(5.40)
$$

5.3.1. Johnson–Segalman model. Any superposition of objective operators is also an objective operator. One such commonly used operator is

$$
\overset{\square}{\boldsymbol{\tau}} = \left(1 - \frac{\zeta}{2}\right) \overset{\triangledown}{\boldsymbol{\tau}} + \frac{\zeta}{2} \overset{\triangle}{\boldsymbol{\tau}}.
\qquad(5.41)
$$

When the operator defined in (5.41) is used in (5.38) and a constant viscosity inelastic contribution is added (often called the solvent contribution), we obtain the four-constant Johnson–Segalman (J–S) model [35],

$$
\boldsymbol{\tau} = \boldsymbol{\tau}^{(1)} + \boldsymbol{\tau}^{(2)}
\qquad(5.42)
$$

with

$$
\boldsymbol{\tau}^{(1)} + \lambda \overset{\square}{\boldsymbol{\tau}}^{(1)} = 2\,\eta_1\,\boldsymbol{D}, \qquad \boldsymbol{\tau}^{(2)} = 2\,\eta_2\,\boldsymbol{D}.
\qquad(5.43)
$$

This model can equivalently be written as

$$
\boldsymbol{\tau} + \lambda \overset{\square}{\boldsymbol{\tau}} = 2\,(\eta_1 + \eta_2)\left(\boldsymbol{D} + \lambda\,\frac{\eta_2}{\eta_1 + \eta_2}\overset{\square}{\boldsymbol{D}} \right).
\qquad(5.44)
$$

There is a wealth of literature evaluating the J–S model and we do not discuss this here. Instead, we focus on special cases of the J–S model.

5.3.2. Convected Maxwell models. The lower convected Maxwell (LCM) and upper convected Maxwell models (UCM) are special cases of the J–S model with ζ equal to 2 and zero, respectively as well as $\eta_2 = 0$. Both models also have an integral representation (see, e.g., pages 14–17 of [37]),

$$
\textbf{LCM:}\quad \boldsymbol{\tau} + \lambda\overset{\triangle}{\boldsymbol{\tau}} = 2\,\eta\,\boldsymbol{D} \quad \text{or} \quad \boldsymbol{\tau} = \int_0^\infty \frac{\eta}{\lambda^2} e^{-s/\lambda}\,(\boldsymbol{I} - \boldsymbol{C}_t)\,ds
$$

$$
\qquad(5.45)
$$

$$
\textbf{UCM:}\quad \boldsymbol{\tau} + \lambda\overset{\triangledown}{\boldsymbol{\tau}} = 2\,\eta\,\boldsymbol{D} \quad \text{or} \quad \boldsymbol{\tau} = \int_0^\infty \frac{\eta}{\lambda^2} e^{-s/\lambda}\,(\boldsymbol{C}_t^{-1} - \boldsymbol{I})\,ds.
$$

5.3.3. Oldroyd-A and Oldroyd-B models. The Oldroyd-A and B models are special cases of the J-S model with ζ equal to 2 and zero, respectively. As will be discussed below, the viscometric functions of the Oldroyd-A model have not been found to match the behavior of real fluids. For this reason we turn attention to the Oldroyd-B model.

The differential form of the three-constant **constant Oldroyd-B** model is the sum of contributions from a Maxwell B model $\boldsymbol{\tau}^{(1)}$ and a linear viscous fluid $\boldsymbol{\tau}^{(2)}$,

$$\boldsymbol{\tau} = \boldsymbol{\tau}^{(1)} + \boldsymbol{\tau}^{(2)} \tag{5.46}$$

with

$$\boldsymbol{\tau}^{(1)} + \lambda \overset{\nabla}{\boldsymbol{\tau}}{}^{(1)} = 2\,\eta_1\,\boldsymbol{D}, \qquad \boldsymbol{\tau}^{(2)} = 2\,\eta_2\,\boldsymbol{D}, \tag{5.47}$$

or, equivalently,

$$\boldsymbol{\tau} + \lambda \overset{\nabla}{\boldsymbol{\tau}} = 2\,(\eta_1 + \eta_2)\left(\boldsymbol{D} + \lambda\,\frac{\eta_2}{\eta_1 + \eta_2}\overset{\nabla}{\boldsymbol{D}}\right). \tag{5.48}$$

When the Oldroyd-B equation is used to model the behavior of polymeric solutions, the contributions (1) and (2) are often referred to as the polymeric and solvent contributions. For small amplitude oscillatory flow, $\overset{\nabla}{\boldsymbol{\tau}}_{(1)}$ reduces to $\partial\boldsymbol{\tau}^{(1)}/\partial t$ and, as a result, the Maxwell B models $(5.47)_1$ reduces to the linear Maxwell model (5.34). To the order of the second-order retarded motion expansion, the Oldroyd-B model is equivalent to a second-order fluid with $\psi_2 = 0$, or $\alpha_2 = -2\alpha_1$.

6. Thixotropic fluids

In his illuminating review article on thixotropy, Barnes [3] provides a historical back drop for the field of thixotropy. We summarize some of the relevant points here. Peterfi introduced the term thixotropy to describe the intriguing experimental results of Schalek and Szegvari, in which iron oxide gels appeared to completely liquefy under gentle shaking and then resolidify when left at rest for sufficient time [3]. The accepted explanation for their experimental results is that iron oxide particles form an association (microstructure) which is weak enough to be destroyed by shaking but can re-establish itself when the fluid is left to stand. The word thixotropy is appropriately a combination of the Greek words thixis (stirring or shaking) and trepo (turning or changing).

Barnes emphasizes the tremendous variation in published definitions of thixotropy. While those defined in industrial applications are typically directed at this apparent gel-sol transition under shaking, those in many other communities focus on the related issue of the time dependence of rheological properties (e.g., viscosity, normal stress effects) arising from the finite time required for the breakdown and build-up in microstructure. For example, Bauer and Collins provide the following definition.

When a reduction in magnitude of rheological properties of a system, such as elastic modulus, yield stress, and viscosity, for example, occurs reversibly and isothermally with a distinct time dependence on application of shear strain, the system is described as thixotropic [6].

All liquids with microstructure have the potential to demonstrate thixotropy. The microstructure may correspond, for example, to flocs, junctions in polymer solutions, alignment of particles or arrangement of suspension microstructure. In the case of blood at low shear rates, the microstructure is the three-dimensional structure formed by red blood cells. This structure requires a finite time to form and break leading to a dependence of viscosity and viscoelasticity on the duration of applied shear. The shape of the particles will also affect the thixotropy. An early rheologist Pryce-Jones noted that thixotropy is more pronounced in systems composed of non-spherical particles [3]. Blood clearly falls in this category.

The theoretical formulation for thixotropic materials is far less developed than that for simple fluids. Barnes [3] described three categories of thixotropic models. In the first category, *indirect microstructural theories*, a scalar parameter is used to describe the level of microstructure. An evolution equation is then introduced to describe the rate of change of this parameter as a result of competing effects driving build-up or breakdown in microstructure. In the second category, *direct structural theories*, an approximate physical model is used to develop an evolution equation describing temporal changes in actual microstructure. The third category is more phenomenological in nature. Data for changes in material properties such as viscosity as a function of time are directly incorporated into the model without introducing a structure parameter.

There is not sufficient space to discuss all three theories. Rather, we refer the reader to [3] and the references cited there. Here, we will briefly cover one representative indirect structural models that captures some of the major physical features of thixotropic materials. This model will provide a framework for discussing the experimental data for blood thixotropy. In the indirect microstructural theories, a metric of the level of microstructure, λ, is used [6]. Typically, λ is normalized such that $\lambda \in [0, 1]$ and $\lambda = 0$ refers to the absence of structure (e.g., only individual red blood cells remain) and $\lambda = 1$ refers to the largest level of structure possible under the given physical conditions. A representative evolution equation is

$$\frac{d\lambda}{dt} = g(\dot{\gamma}, \lambda) \quad \text{where} \quad g(\dot{\gamma}, \lambda) = \frac{1}{\tau}(1 - \lambda) - \alpha\,\lambda\,\dot{\gamma}, \quad\quad (6.1)$$

and τ is a characteristic time of build-up of the structure and α is a second positive material constant reflecting the rate of breakdown. The variable $\dot{\gamma}$ is a non-negative metric of shear such as the second invariant of the symmetric part of the velocity gradient. The first term of $g(\dot{\gamma}, \lambda)$ is non-negative and leads to microstructure build-up, while the second results in breakdown due to fluid shear. The net outcome of these competing effects is then reflected in the sign of $g(\dot{\gamma}, \lambda)$. When $g(\dot{\gamma}, \lambda)$ is

---

[6]The use of the notation λ as a metric of microstructure is restricted to this section and Section 7.3. In the remainder of this chapter it is used as a time constant for viscoelastic fluids.

positive, the net effect is build-up, when it is negative, the microstructure tends to breakdown. When these two effects balance, $(d\lambda/dt = 0)$, and an equilibrium value for λ is obtained,

$$\lambda_{eq} = \frac{1}{1 + \alpha\dot{\gamma}\tau}. \tag{6.2}$$

Note that the equilibrium value depends on the applied shear rate. As the applied shear rate tends to zero, the value of λ_{eq} tends to 1 (maximum microstructure), while λ_{eq} tends to zero (complete breakdown) in the limit of $\dot{\gamma}$ tending to infinity.

In real fluids, build-up of the microstructure requires the particles to travel sufficiently close for their bonds to reform. This motion can arise due to Brownian motion or, more effectively, via convection from fluid motion, though at the expense of imposing forces that will tend to breakdown the structure. The model above includes the effect of fluid motion on breakdown through $\dot{\gamma}$, but not build-up. The time scales for changes in microstructure can range from seconds to hours. Typically, the time scale for restructuring is longer than that for breakdown.

The second part of the indirect microstructural theory is a constitutive model describing the dependence of the mechanical properties (e.g., viscosity, viscoelasticity) on the structural parameter. By way of illustration consider a relatively simple form for the viscosity function that demonstrates a decrease in viscosity with decreasing level of microstructure (one type of thixotropy),

$$\eta = \frac{\eta_\infty}{(1 - K\lambda)^2} \quad \text{and} \quad K = 1 - (\frac{\eta_\infty}{\eta_o})^{1/2}. \tag{6.3}$$

The model (6.3) was introduced by Baravian, Quemada and Parker (see [3]). The material constants η_o and η_∞ can be seen to correspond to zero and infinite microstructure, respectively,

$$\eta_o = \eta(\lambda = 1), \qquad \eta_\infty = \eta(\lambda = 0). \tag{6.4}$$

When (6.3) is used in conjunction with an evolution equation of the form (6.1), η_o and η_∞ can be seen to correspond to the limiting equilibrium viscosities at low and high shear rates, respectively.

7. Rheometrical flows

In this section, we discuss several categories of flows which are generated in commonly used rheometers.

7.1. Viscometric flows

While the rheological behavior of incompressible Newtonian fluids is completely determined by the constant viscosity η, the situation for more general fluids is much more complicated. Fortunately, it has been shown that the behavior of any simple fluid in a broad class of flows called *viscometric flows* can be completely determined once three material functions are determined. Appropriately, these three functions are called *viscometric material functions* and are intrinsic properties of the fluid. Though these viscometric functions were already introduced in Section 2.7.10, we return to the subject to provide a more precise definition of viscometric flows and discuss some specific examples (see [12, 2] for a more detailed discussion).

Recall that for simple shear flow, (5.7), the relative deformation gradient $\boldsymbol{F}_t$ is of the form

$$\boldsymbol{F}_t(s) \; = \; (\boldsymbol{I} - s\,\boldsymbol{M}), \qquad (7.1)$$

where the components of $\boldsymbol{M}$ in rectangular coordinates are

$$[\boldsymbol{M}] \; = \; \begin{bmatrix} 0 & \kappa & 0 \\ 0 & 0 & 0 \\ 0 & 0 & 0 \end{bmatrix}, \qquad (7.2)$$

where κ is a constant equal to $\dot{\gamma}_0$ for simple shear. From the point of view of the fluid element, any viscometric flow is indistinguishable from steady simple shear flow when it is described in terms of suitably chosen local Cartesian coordinates. In fact, the class of viscometric flows can be considered as a generalization of these simple shear flows (7.1), for which $\boldsymbol{F}_t$ is of the following form,

$$\boldsymbol{F}_t(s) \; = \; \boldsymbol{R}(s) \cdot (\boldsymbol{I} - s\,\boldsymbol{M}), \qquad (7.3)$$

where $\boldsymbol{R}(s)$ is orthogonal for all s and normalized so that $\boldsymbol{R}(0) = \boldsymbol{I}$. Furthermore, the components of $\boldsymbol{M}$ take the form (7.2) for any orthonormal basis $\boldsymbol{b}^i$ not just a rectangular basis. In simple shear flow $\boldsymbol{M}$ is independent of the material point and time. In contrast, in viscometric flows, $\boldsymbol{M}$ may depend on both these quantities. In particular, $\boldsymbol{R}(s), \kappa$ and $\boldsymbol{b}^i$ may be functions of $\boldsymbol{x}$ and t.

Examples of viscometric flows include simple shear flow as well as steady, fully developed flows in channels and pipes of constant cross section. For the reader's convenience, we briefly recall the viscometric functions given in (2.40), which were defined relative to simple shear flow (expressed in a Cartesian frame) with shear rate $\dot{\gamma}_0$,

Viscosity or Shear Viscosity	$\eta(\kappa) \equiv T_{12}/\kappa,$
First normal stress coefficient	$\psi_1(\kappa) \equiv (T_{11} - T_{22})/\kappa^2,$
Second normal stress coefficient	$\psi_2(\kappa) \equiv (T_{22} - T_{33})/\kappa^2,$

(7.4)

where $\kappa = \dot{\gamma}_0 \neq 0$ and T_{ij} are the rectangular components of the Cauchy stress tensor [7].

The coefficients ψ_1 and ψ_2 are zero for both Newtonian and generalized Newtonian fluids. If we consider the Johnson–Segalman model in simple shear, it follows

---

[7]As noted earlier, alternative definitions of the first and second normal stress definitions are sometimes used, though they can be formed from linear combinations of these given here.

directly that

$$\eta(\kappa) = \left(\frac{\eta_1}{1 + 2\,\zeta\lambda^2\,(1 - \zeta/2)\kappa^2} + \eta_2 \right),$$

$$\Psi_1(\kappa) = \left(\frac{2\,\eta_1\,\lambda}{1 + 2\,\zeta\lambda^2\,(1 - \zeta/2)\kappa^2} \right), \tag{7.5}$$

$$\Psi_2(\kappa) = -\left(\frac{\zeta\,\eta_1\,\lambda}{1 + 2\,\zeta\lambda^2\,(1 - \zeta/2)\kappa^2} \right).$$

It is clear from (7.5), that in general the viscometric functions η, Ψ_1 and Ψ_2 are not constants. For $\zeta \in (0, 2)$ the viscometric functions in the Johnson–Segalman model are shear thinning, while for $\zeta = 0$ or $\zeta = 2$ they are constant. The viscometric functions for special cases of the Johnson–Segalman as well as the second-order fluid are given in Table 1.

Model	η	ψ_1	ψ_2
Newtonian Fluid, Eq. (5.36)	η	0	0
Second-Order Fluid, Eq. (5.37)	η	$-2\,\alpha_1$	$(2\,\alpha_1 + \alpha_2)$
Maxwell A (LCM), Eq. (5.43) $\zeta = 2, \eta_2 = 0$	η_1	$2\eta_1\lambda$	$-2\eta_1\lambda$
Maxwell B (UCM), Eq. (5.43) $\zeta = 0, \eta_2 = 0$	η_1	$2\eta_1\lambda$	0
Oldroyd-A, Eq. (5.43) $\zeta = 2$	$\eta_1 + \eta_2$	$2\eta_1\lambda$	$-2\eta_1\lambda$
Oldroyd-B, Eq. (5.43) $\zeta = 0$	$\eta_1 + \eta_2$	$2\eta_1\lambda$	0

TABLE 1. Viscometric functions for some non-Newtonian fluids

Based on measurements of polymer solutions and melts, it is commonly assumed that the ratio ψ_2/ψ_1 is negative with a magnitude less than $1/2$. As a consequence, the Maxwell A (or Lower Convected Maxwell, LCM) and Oldroyd-A models are rarely used. It follows from the results in Table 1 that we would therefore expect $\alpha_1 \leq 0$ and $\alpha_2 \leq -2\,\alpha_1$ for the second-order fluid. Unfortunately, when physically reasonable values of these material constants are used, the second-order fluid is shown to illustrate anomalous behavior in a number of unsteady flows, see, e.g.,

Galdi [20]. For this reason, applications of second-order fluids are generally confined to steady flows.

7.1.1. Fully developed steady flow in a pipe.

Due to its importance in capillary rheometers, we turn attention to a specific viscometric flow: fully developed steady flow in pipes of circular cross section of radius R under an imposed steady axial pressure gradient. It is convenient to use a cylindrical coordinate system (r, θ, z), with z axis coincident with the pipe centerline. We look for velocity fields of the form $\boldsymbol{v}(r) = v_z(r)\boldsymbol{e}_z$ which satisfy the governing equations and boundary condition $\boldsymbol{v} = 0$ at $r = R$. For such flows, the only non-zero components of $\boldsymbol{D}$ are $D_{rz} = D_{zr} = (dv_z/dr)/2$ and therefore, $\dot{\gamma} = |dv_z/dr|$, (3.11).

The incompressibility condition is identically satisfied by this velocity field and the balance of linear momentum for a simple fluid reduces to (e.g., [12])

$$
\begin{aligned}
0 &= -\frac{\partial p}{\partial r} + \frac{\partial \tau_{rr}}{\partial r} + \frac{\tau_{rr} - \tau_{\theta\theta}}{r}, \\
0 &= -\frac{1}{r}\frac{\partial p}{\partial \theta}, \\
0 &= -\frac{\partial p}{\partial z} + \frac{1}{r}\frac{d}{dr}\left(r\tau_{rz}\right),
\end{aligned}
\tag{7.6}
$$

where we have used the fact that the extra stress tensor for a simple fluid can only be a function of r for the given form of the velocity field. It follows from (7.6), that for these flows, $\partial p/\partial z$ is a constant. Therefore, after simple integration $\partial p/\partial z = \Delta P/L$, where L is the length of the tube and ΔP is the difference between the cross-sectional average pressures between the pipe inlet and outlet. It then follows from $(7.6)_3$ that

$$
\tau_{rz} = -\frac{1}{2}\frac{\Delta P}{L}r,
\tag{7.7}
$$

where we have required the shear stress to be bounded at $r = 0$.

It is useful in capillary rheometers to be able to determine the viscosity function from measured values of ΔP and volumetric flow rate Q in pipes of known R and L without selecting a constitutive model a priori. Prior to discussing such a relationship, we rephrase the problem as one of obtaining η as a function of a characteristic shear rate $\dot{\gamma}_a$, wall shear rate $\dot{\gamma}_w$, and wall shear stress τ_w,

$$
\dot{\gamma}_a = \frac{4Q}{\pi R^3}, \qquad \tau_w = -\frac{1}{2}\frac{\Delta P}{L}R, \qquad \dot{\gamma}_w = \dot{\gamma}(\tau_w).
\tag{7.8}
$$

In writing $(7.8)_3$, we have assumed the function $\tau_{rz} = \tau_{rz}(\dot{\gamma})$ is invertible. Beginning with the definition of Q for this flow, integrating by parts, using the condition $v_z = 0$ at $r = R$ gives,

$$
Q = 2\pi \int_0^R v_z\, r\, dr = 2\pi \int_0^R r^2\, \dot{\gamma}\, dr.
\tag{7.9}
$$

After making a change of variables from r to τ, and using the invertibility of $\tau_{rz} = \tau_{rz}(\dot\gamma)$, we find,

$$\frac{Q\tau_w^3}{\pi R^3} = \int_0^{\tau_w} \tau_{rz}^2\,\dot\gamma(\tau_{rz})\,d\tau. \tag{7.10}$$

After differentiating with respect to τ_w and rearranging, we obtain the Mooney–Rabinowitsch equation [43, 52],

$$\frac{\dot\gamma_w}{\dot\gamma_a} = \frac{1}{4}\left[3 + \frac{\partial \ln \dot\gamma_a}{\partial \ln \tau_w}\right]. \tag{7.11}$$

It follows from the definition of viscosity, that $\eta(\dot\gamma_w) = \tau_w/\dot\gamma_w$, so from (7.11),

$$\eta(\dot\gamma_w) = \frac{\tau_w}{\dot\gamma_a}\frac{4n'}{3n'+1} = \frac{\pi R^4}{2Q}\frac{\Delta P}{L}\frac{n'}{3n'+1}, \tag{7.12}$$

where

$$n' = \frac{\partial \ln \tau_w}{\partial \ln \dot\gamma_a}. \tag{7.13}$$

Experiments can be run at different pressure drops to obtain a curve of $\dot\gamma_a$ as a function of τ_w (recall (7.8)). If these results are plotted on a log scale, the value n' can be obtained from the slope of the curve, which will in general be a function of τ_w.

While the relationship (7.12) is quite useful for viscometry, it is of interest to include some of the exact solutions for steady, fully developed flows of simple fluids in straight pipes of constant cross section. For simple fluids for which η is a constant (e.g., Newtonian, Oldroyd-B, second-order fluid), it is a simple matter to obtain the velocity field from (7.7). The well-known solution for the velocity field and axial pressure drop for a constant viscosity fluid can be written as

Constant Viscosity
$$\begin{cases} \boldsymbol{v} = \dfrac{2Q}{\pi R^2}\left(1 - \dfrac{r^2}{R^2}\right)\boldsymbol{e}_z \\[2mm] \dfrac{dp}{dz} = -\dfrac{8Q\,\mu}{\pi R^4}, \end{cases} \tag{7.14}$$

where μ is the constant fluid viscosity. For the fluids with power-law-type viscosity functions (3.13), we can similarly obtain an explicit expression for the velocity field,

Fluid with power-law viscosity
$$\begin{cases} \boldsymbol{v} = \dfrac{Q}{\pi R^2}\dfrac{3+1/n}{1+1/n}\left[1 - (\dfrac{r}{R})^{1+1/n}\right]\boldsymbol{e}_z \\[2mm] \dfrac{dp}{dz} = -\dfrac{2K}{R}(\dfrac{3+1/n}{\pi R^3})^n Q^n. \end{cases} \tag{7.15}$$

For the same flow rate, the velocity field in the shear thinning power-law model is flatter than for the constant viscosity case, Figure 6.

There are also closed form solutions for fully developed, steady flow for some of the yield stress fluids. Recall that for a yield stress fluids such as Bingham or

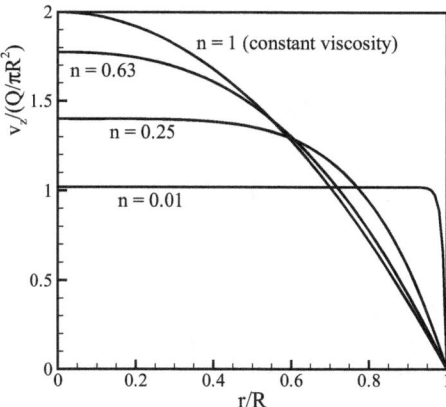

FIGURE 6. Nondimensional axial velocity profile, $v_z/(Q/\pi R^2)$, as a function of nondimensional radius r/R for steady, fully developed flow of fluids with a power-law viscosity in a straight pipe, Eq. (7.15). Here, the profiles are for the same flow rate but with different values of n.

Casson, the value of $\boldsymbol{D}$ is zero at all points where $\sqrt{|\mathrm{II}_\tau|}$ is less than the yield stress. For the flow under consideration,

$$\sqrt{|\mathrm{II}_\tau|} = |\tau_{rz}| = \frac{r\Delta P}{2L}. \qquad (7.16)$$

Clearly, $\sqrt{|\mathrm{II}_\tau|}$ will be largest at the largest radial position: $r = R$ and so when $\Delta PR/2L < \tau_Y$, the yield criterion will not be met anywhere in the fluid. In this case, $\boldsymbol{D}$ will be zero throughout the fluid. From the boundary conditions, it then follows that the velocity is zero everywhere. If the driving pressure gradient is increased so that $\Delta P/L > 2\tau_Y/R$, there will be a region $r \in [0, 2L\tau_Y/\Delta P]$, where the yield criterion is not met and $\boldsymbol{D} = 0$ (fluid moves with constant velocity). The radius bounding this region is denoted as r_Y,

$$r_Y = \frac{2L\tau_Y}{\Delta P}. \qquad (7.17)$$

For $r \in (r_Y, R]$ the material will flow.

The exact solution for steady, fully developed flow of a Bingham fluid (4.5) in a pipe of constant radius R is:

Bingham Fluid

$$v = \begin{cases} -\dfrac{dp}{dz}\dfrac{R^2}{4\mu}\left(1-\dfrac{r_Y}{R}\right)^2 e_z & \text{for} \quad r \leq r_Y \\[2ex] -\dfrac{dp}{dz}\dfrac{R^2}{4\mu}\left((1-\dfrac{r^2}{R^2}) - \dfrac{2r_Y}{R}(1-\dfrac{r}{R})\right) e_z & \text{for} \quad r \geq r_Y. \end{cases} \tag{7.18}$$

This velocity field corresponding to (7.18) is shown in Figure 7. The plug region with $D = 0$ is clearly seen for $r \in [0, r_Y]$. For comparison, the corresponding velocity field for a Newtonian fluid flowing under the same axial pressure gradient and with the same viscosity is shown as well. The resulting flow rates will be different for these two fluids.

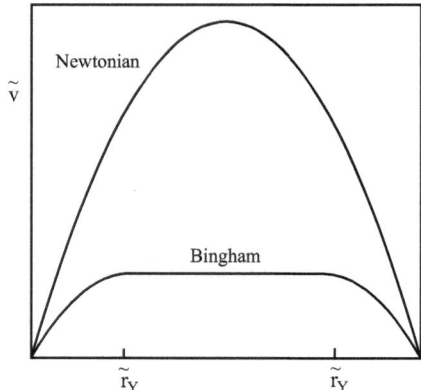

FIGURE 7. Nondimensional axial velocity profile, $\tilde{v} = -v_z(4\mu)/(R^2 dp/dz)$, as a function of nondimensional radius $\tilde{r} = r/R$ for steady, fully developed flow of Newtonian and Bingham fluids in a straight pipe, Eq. (7.18). The profiles shown are for the same axial pressure drop, viscosity and pipe radius.

The corresponding velocity field for flow of a Casson fluid (4.10) is (e.g., [46], page 44):

Casson Fluid

$$
v = \begin{cases}
-\dfrac{dp}{dz}\dfrac{R^2}{4\mu_N}\left(1 - \dfrac{r_Y^{1/2}}{R^{1/2}}\right)^3\left(1 + \dfrac{1}{3}\dfrac{r_Y^{1/2}}{R^{1/2}}\right) e_z & \text{for } r \leq r_Y \\[4ex]
-\dfrac{dp}{dz}\dfrac{R^2}{4\mu_N}\left((1 - \dfrac{r^2}{R^2}) - \dfrac{8}{3}\dfrac{r_Y^{1/2}}{R^{1/2}}(1 - \dfrac{r^{3/2}}{R^{3/2}})\right. \\[2ex]
\left. \qquad\qquad + 2\dfrac{r_Y}{R}(1 - \dfrac{r}{R})\right) e_z & \text{for } r \geq r_Y.
\end{cases}
$$

$$\tag{7.19}$$

7.2. Periodic flows

7.2.1. Small amplitude oscillatory shear flows. Most measurements of linear viscoelastic properties of blood are based on small amplitude oscillatory shear experiments. Consider, for example, oscillatory flow between parallel plates separated by a gap h, driven by the oscillatory motion of the upper plate. This is a generalization of the steady shear flow, Eq. (1.28). For that flow it is convenient to used rectangular coordinates such that the plates are parallel to the x_1 axis and the bottom plate is located at $x_2 = 0$, Figure 2. Using this coordinate system, we write the velocity of the top plate as $v = U\cos(\omega t)e_1$. The bottom plate is held fixed. We represent the position of material points in the fluid at arbitrary time t by

$$x_1(t) = X_1 + \gamma_0 X_2 \sin(\omega t), \qquad x_2(t) = X_2, \qquad x_3(t) = X_3, \tag{7.20}$$

where X is the position of the particle at time zero and $x_1, x_3 \in (-\infty, \infty)$, $x_2 \in [0, h]$. The shear displacement relative to the initial position is then

$$x(t) - X = \gamma_0 X_2 \sin(\omega t)e_1, \tag{7.21}$$

and the corresponding velocity field is

$$v = \gamma_0 \omega X_2 \cos(\omega t)e_1. \tag{7.22}$$

It follows from (7.21) that particles at the top plate have the maximum amplitude of motion equal to $\gamma_0 h$. The shear displacement and velocity are 90° out of phase. The only non-zero components of D are then $D_{12} = D_{21} = \gamma_0 \omega \cos(\omega t)/2$. In this case, it follows from (3.9) that $\dot\gamma = \gamma_0 \omega |\cos\omega t|$. If we use $\dot\gamma_M$ to denote the maximum of $\dot\gamma$ during the cycle (its amplitude), then $\dot\gamma_M = \gamma_0 \omega$. It follows from (7.22) and the no-slip boundary condition at the upper plate that $\gamma_0 = U/(\omega h)$ and therefore, $\dot\gamma_M = U/h$.

As discussed in Section 5.1, when analyzing the response of simple fluids with fading memory, it is convenient to write the position of a material particle at arbitrary past time τ with respect to the current position. Using results from

Section 5.1.1 with (7.21), we find that for oscillatory shear flow the relationship between the current position of a particle x and its past location ξ at time τ is

$$\xi(x,\tau) = x + \gamma_0 x_2 (\sin(\omega\tau) - \sin(\omega t))e_1. \tag{7.23}$$

Furthermore, for simplicity of the following calculations and ensuing discussion, we represent (7.23) as

$$\xi = x + \mathcal{R}e[x_2 i\gamma_0 e^{i\omega t}(1 - e^{-i\omega s})]e_1, \tag{7.24}$$

where $\mathcal{R}e$ denotes the real part of a complex entity and it should be recalled that $s = t - \tau$. The only non-zero components of G_t for oscillatory shear (7.24) are

$$G_{t12} = G_{t21} = \gamma_o \mathcal{R}e\left[i\,e^{i\omega t}\left[1 - e^{-i\omega s}\right]\right], \qquad G_{t22} = G_{t12}^2. \tag{7.25}$$

To ensure the validity of the infinitesimal linear viscoelasticity approximations for all times, we require $\gamma_0 << 1$ or $\dot{\gamma}_M/\omega << 1$. Then, from (7.25), to the order of this linearized theory G_{t22} can be neglected.

It follows from (5.27), to the first order of approximation the only non-zero components of the extra stress tensor are

$$\tau_{12} = \tau_{21} = \mathcal{R}e[\tau^* e^{i\omega t}], \tag{7.26}$$

where the complex shear stress τ^* is

$$\tau^* = \gamma_0 \omega \int_0^\infty \mathcal{G}(s)\, e^{-i\omega s} ds. \tag{7.27}$$

In writing (7.27) we assume, to the order of the approximation, the motion has been oscillatory for all previous times. Normal stress effects are only of second order and therefore cannot be captured in this first-order theory. By convention, we define the complex modulus G^* as

$$G^* = i\omega \int_0^\infty \mathcal{G}(s)\, e^{-i\omega s} ds \tag{7.28}$$

which can be split into its real and imaginary parts,

$$G^* = G'(\omega) + i\,G''(\omega) \tag{7.29}$$

where G' and G'' are both real numbers are referred to as the *storage modulus* and *loss modulus*. It is sometimes more convenient to write G^* with respect to the loss tangent δ,

$$G^* = |G^*|e^{i\delta}. \tag{7.30}$$

Clearly we can relate δ to G' and G'' through

$$\tan\delta = G''/G'. \tag{7.31}$$

Using the notation, (7.28) we can rewrite (7.27) as

$$\tau* = -\gamma_0\, i\, G^*. \tag{7.32}$$

Then, using (7.26) and (7.29), we have,

$$\tau_{12} = \gamma_0\left[G'\sin(\omega t) + G''\cos(\omega t)\right]. \tag{7.33}$$

For purely elastic materials, the shear stress and shear strain are in phase, in which case $G'' = 0$ and G' plays the role of the elastic shear modulus. For purely viscous materials, the velocity gradient and shear stress are in phase and $G' = 0$. The storage modulus G' is thus named because of its association with the ability of an elastic material to store elastic energy while the term loss modulus is due to the energy dissipation associated with viscous fluids.

The complex viscosity η^* is often reported in the literature on blood viscoelasticity,

$$\eta^* = \eta' - i\eta'' \tag{7.34}$$

where the viscous component η' and the elastic component η'' are real. The relationship between the complex viscosity and the complex modulus is

$$\eta^* = G^*/i\omega, \qquad \eta' = G''/\omega, \qquad \eta'' = G'/\omega. \tag{7.35}$$

From (7.32) and (7.35), we have that

$$\tau_{12} = \gamma_0 \omega \left[\eta' \cos(\omega t) + \eta'' \sin(\omega t) \right]. \tag{7.36}$$

For a purely viscous fluid, η'' is zero while for a purely elastic solid η' is zero. For this reason, η' and η'' are sometimes referred to as the viscosity and elasticity.

7.2.2. Maxwell fluids in infinitesimal oscillatory shear. Using the definition of the relaxation modulus for a Maxwell fluid given in $(5.28)_1$, we can calculate the storage and loss moduli for a Maxwell fluid using (7.28) and (7.29),

$$G'(\omega) = \frac{\eta \omega^2 \lambda}{1 + \omega^2 \lambda^2}, \qquad G''(\omega) = \frac{\eta \omega}{1 + \omega^2 \lambda^2}, \tag{7.37}$$

where we have used the result $\eta = \mathcal{G}_0 \lambda$ and η is the shear viscosity defined in (7.4). Therefore,

$$\delta = \tan^{-1}(\frac{1}{\lambda\omega}). \tag{7.38}$$

Results for G' and G'' as a function of $De = \omega \lambda$ are shown in Figure 8. It follows from (7.37), that these curves cross at the point

$$\omega \lambda = 1. \tag{7.39}$$

The value of $1/\omega$ at this intersection point is sometimes used as an approximate relaxation time. However, typically more than one relaxation time is needed to fit the data well and a more complex model such as the general linear viscoelastic model is used.

The real and imaginary parts of the complex viscosity for a simple Maxwell fluid follow from (7.35) and (7.37),

$$\eta' = \frac{\eta}{1 + \omega^2 \lambda^2}, \qquad \eta'' = \frac{\eta \omega \lambda}{1 + \omega^2 \lambda^2}. \tag{7.40}$$

The dimensionless shear moduli η' and η'' are shown in Figure 9. Recall, that as the value of De tends to zero, unsteady viscoelastic effects become less and less important. As expected, in this limit, the value of η' tends to the steady state viscosity η.

A.M. Robertson

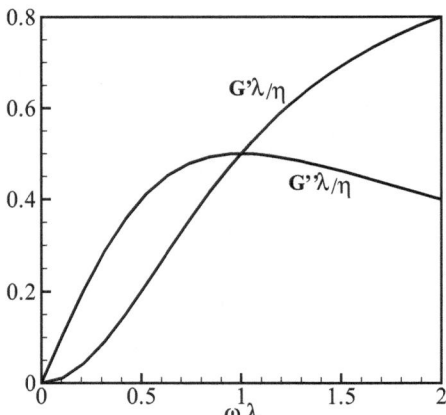

FIGURE 8. Dimensionless small amplitude oscillatory shear moduli $G'\lambda/\eta$ and $G''\lambda/\eta$ as a function of the Deborah number $De = \omega\lambda$ for Maxwell fluids, Eq. (7.37). Here, η' and η'' are the real and imaginary parts of the complex viscosity and η is the shear viscosity, Eq. (7.4).

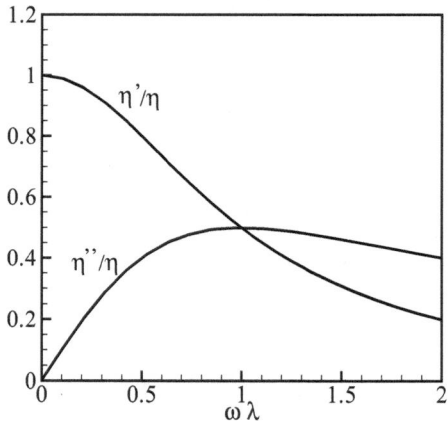

FIGURE 9. Dimensionless small amplitude oscillatory shear moduli η'/η and η''/η as a function of the Deborah number $De = \omega\lambda$ for Maxwell fluids, Eq. (7.37).

7.2.3. Fully developed oscillatory flow in a pipe. Much of the data on blood viscoelasticity is obtained from oscillatory flow experiments in capillary rheometers. Motivated by this application, we now consider flow through a straight pipe of circular cross section of radius R, driven by a pressure gradient which is periodic in time. In particular, we look for solutions for the velocity $\boldsymbol{v}$ and pressure p which satisfy the governing equations for all $r \in [0, R]$ with $\boldsymbol{v} = 0$ at $r = R$. Attention is restricted to solutions $\boldsymbol{v}$ which are unidirectional, fully developed and axisymmetric,

$$\boldsymbol{v} = v_z(r, t)\boldsymbol{e}_z. \tag{7.41}$$

For velocity fields of the form (7.41), the incompressibility condition is identically satisfied, the components of the extra stress tensor depend at most on r and t, so the balance of linear momentum, reduces to the following form,

$$
\begin{aligned}
0 &= -\frac{\partial p}{\partial r} + \frac{\partial \tau_{rr}}{\partial r} + \frac{\tau_{rr} - \tau_{\theta\theta}}{r}, \\
0 &= -\frac{1}{r}\frac{\partial p}{\partial \theta}, \\
\rho\frac{\partial v_z}{\partial t} &= -\frac{\partial p}{\partial z} + \frac{1}{r}\frac{\partial}{\partial r}\left(r\tau_{rz}\right),
\end{aligned}
\tag{7.42}
$$

with boundary conditions $v_z = 0$ on $r = R$. It follows (7.42) that the axial component of the pressure gradient $\partial p/\partial z$ is at most dependent on time. We now consider the specific case of an applied oscillatory axial pressure gradient of the form, $\partial p/\partial z = -K_1 \cos \omega t$ where K_1 is a real constant.

For some constitutive models, the equation for τ_{rz} is linear and decoupled from the other components of stress when the velocity field is of the form (7.41). Some examples are,

$$
\begin{aligned}
\textbf{Newtonian} &\qquad \tau_{rz} = \eta\,\frac{\partial v_z}{\partial r}, \\[2mm]
\textbf{UCM, Eq. (5.45)} &\qquad \tau_{rz} + \lambda\frac{\partial \tau_{rz}}{\partial t} = \eta\,\frac{\partial v_z}{\partial r}, \\[2mm]
\textbf{Oldroyd-B, Eq. (5.48)} &\qquad \tau_{rz} + \lambda\frac{\partial \tau_{rz}}{\partial t} = \eta\,\frac{\partial v_z}{\partial r} + \eta_2\lambda\frac{\partial^2 v_z}{\partial r\partial t}, \\[2mm]
\textbf{Maxwell, Eq. (5.34)} &\qquad \tau_{rz} + \lambda\frac{\partial \tau_{rz}}{\partial t} = \eta\,\frac{\partial v_z}{\partial r}.
\end{aligned}
\tag{7.43}
$$

It should be recalled that the Maxwell model is an approximate model for small strain histories. In all these cases, the material parameters η, η_2, λ are constant. As a result of the linearity, it is also possible to obtain the velocity field for pulsatile flow as a superposition of the velocity fields generated from steady and oscillatory axial pressure gradients.

We will consider the dependent variables as composed of a real and imaginary part, for example,

$$\frac{\partial p}{\partial z}(t) = \mathcal{R}e[-K_1\,e^{i\omega t}]. \tag{7.44}$$

We then look for solutions of the form

$$
\begin{aligned}
v_z(r,t) &= \mathcal{R}e[v_z^*(r)e^{i\omega t}], & Q(t) &= \mathcal{R}e[Q^*e^{i\omega t}], \\
\tau_{rz}(r,t) &= \mathcal{R}e[\tau_{rz}^*(r)e^{i\omega t}], & \tau_{rr}(r,t) &= \mathcal{R}e[\tau_{rr}^*(r)e^{i\omega t}],
\end{aligned}
\tag{7.45}
$$

where v_z^*, Q^* and τ_{rz}^* will, in general, be complex. After substituting the representations (7.45) in (7.43), we see that can obtain an explicit form for τ_{rz}^*,

Newtonian
$$
\tau_{rz}^* = \eta \, \frac{\partial v_z^*}{\partial r},
$$

UCM and **Maxwell**
$$
\tau_{rz}^* = \frac{\eta}{1+i\lambda\omega} \, \frac{\partial v_z^*}{\partial r},
\tag{7.46}
$$

Oldroyd-B
$$
\tau_{rz}^* = \frac{(\eta + i\,\eta_2\,\lambda\,\omega)}{1+i\lambda\omega} \, \frac{\partial v_z^*}{\partial r}.
$$

We see, from (7.46), that in all cases, we obtain an explicit expression of the form

$$
\tau_{rz}^*(r) = \eta^* \, \frac{dv_z^*}{dr}
\tag{7.47}
$$

where η^* is a complex constant. Making use of this last result we see that $(7.42)_3$ decouples from the other equations,

$$
i\,\omega\,\rho v_z^* = -K_1 + \frac{\eta^*}{r} \frac{\partial}{\partial r}\left(r\frac{\partial v_z^*}{\partial r}\right).
\tag{7.48}
$$

Equation (7.48) is of the same form as that for a Newtonian fluid with η replaced by η^*. Making use of the well-known solution for a Newtonian fluid, we can write the solution to (7.48) as

$$
v_{rz}^* = -\frac{K_1 i}{\rho\omega}\left[1 - \frac{J_o(\beta r/R)}{J_o(\beta)}\right], \qquad \beta = R\sqrt{\frac{-i\rho\omega}{\eta^*}}.
\tag{7.49}
$$

The constant β can be related to the variable Y used in the work by Thurston on this subject (e.g., [62]),

$$
\beta = Y e^{i\sigma}, \quad \text{with} \quad
\begin{cases}
Y = R\sqrt{\dfrac{\rho\omega}{|\eta^*|}} \\[2mm]
\eta^* = |\eta^*|e^{-i\phi}, \quad \sigma = \dfrac{\phi}{2} - \dfrac{\pi}{4}.
\end{cases}
\tag{7.50}
$$

We see that Y is similar to the Womersely number with η replaced by η^*. However, Y is no longer a measure of the ratio of unsteady inertial to viscous effects since it also includes viscoelastic effects. The corresponding complex flow rate is

$$
Q^*(t) = \frac{i\,K_1\pi\,R^2}{\rho\,\omega}\left[1 - \frac{2J_1(\beta)}{\beta\,J_0(\beta)}\right].
\tag{7.51}
$$

We can expand this result for small β (equivalently, small Y), to obtain

$$Q^*(t) = \frac{i\,K_1\,\pi\,R^2}{\rho\omega}\,\frac{\beta^2}{8}\left[1 + \frac{1}{6}\beta^2 + \mathcal{O}(\beta^4)\right].\tag{7.52}$$

Splitting (7.52) into its real and imaginary parts and using (7.46), we obtain

$$Q(t) = \frac{K_1\pi R^4}{8\,|\eta^*|}\left[\cos(\phi + \omega t) + \frac{1}{6}Y^2\sin(2\phi + \omega t) + \mathcal{O}(Y^4)\right].\tag{7.53}$$

7.3. Non-periodic unsteady flows

For thixotropic experiments, one of the following types of non-periodic unsteady functions for either shear rate (or shear stress) are often used:

- A step function in either shear rate or shear stress.
- A linear increase in shear rate followed directly by a linear decrease.
- Oscillatory shear flow.

We have already discussed oscillatory shear, so, in this section we turn attention to the first two experiments. In both cases, we consider idealized flows which are time-dependent homogeneous shear fields with velocity fields of the form $\mathbf{v} = v(x_2, t)\mathbf{e}_1$. For example, flow generated between two plates parallel to the x_1 axis, driven by the motion of upper plate in the x_1 direction. The applied shear stress necessary to generate this flow will be denoted by τ_{12}. The symmetric part of the velocity gradient then has two identical non-zero components, which we denote as $\dot{\gamma}(t)$ and refer to as the shear rate.

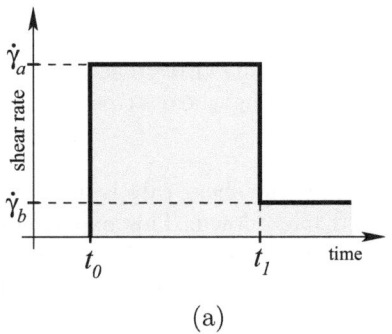
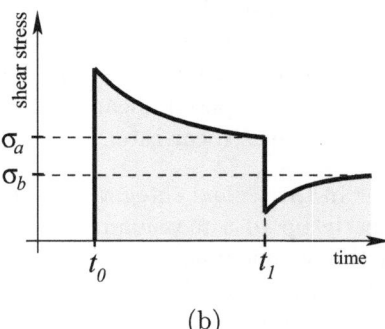

(a) (b)

FIGURE 10. Qualitative features of the dependence of shear stress on time for a thixotropic fluid experiencing a step increase, then decrease in shear rate. (a) Imposed shear rate versus time. (b) Resulting temporal dependence of shear stress.

In the first type of experiment, at $t = t_0$ a step increase in shear rate from zero to $\dot{\gamma} = \dot{\gamma}_a$ is applied. The shear rate is then held constant until time $t = t_1$ when the shear rate is abruptly dropped to a constant value $\dot{\gamma}_b < \dot{\gamma}_a$, Figure 10. In the case that our test fluid has maximum structure, $\lambda = 1$, at the onset of the experiment, the microstructure will break down over time, tending towards an equilibrium value denoted by $\lambda_{eq}(\dot{\gamma}_a)$. As a result of the diminishing microstructure, the viscosity will also decrease in time, tending toward $\eta_{eq}(\dot{\gamma}_a)$. In turn, the applied stress will initially be large, but as time passes and the viscosity drops, the shear stress necessary to maintain this fixed shear rate will drop as well until the equilibrium value is reached $\sigma_a = \eta_{eq}(\dot{\gamma}_a)\,\dot{\gamma}_a$. In the second part of the experiment, $\dot{\gamma}$ is dropped at time t_1, from $\dot{\gamma}_a$ to $\dot{\gamma}_b$. Over time, the microstructure will begin to reform, with λ tending to the value of $\lambda_{eq}(\dot{\gamma}_b)$. There will be a corresponding increase in viscosity towards $\eta_b = \eta_{eq}(\lambda_b)$ and applied shear stress towards $\sigma_b = \eta_{eq}(\dot{\gamma}_b)\dot{\gamma}_b$ (see, [3]). The rate at which these changes occur will depend on the time constants for build-up and breakdown, which are material properties of the fluid.

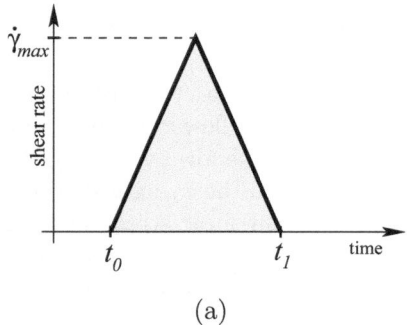
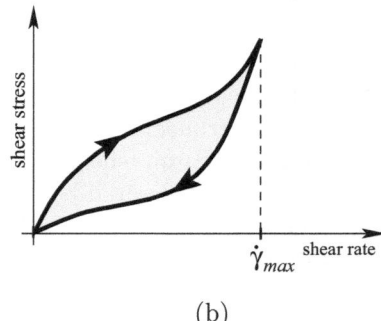

(a)   (b)

FIGURE 11. Qualitative features of the dependence of shear stress on shear rate for a representative thixotropic fluid experiencing a linear increase in shear rate followed directly by a linear decrease. (a) Imposed shear rate versus time. (b) Resulting shear stress versus shear rate.

In the second category of experiments, Figure 11, the shear rate is increased linearly up to a maximum value and then ramped back down. This experiment for thixotropic fluids is more complicated because $\dot{\gamma}$ and hence the microstructure will continuously change during the experiment. If the stress tensor depends only on the current shear rate (the fluid does not display viscoelasticity or thixotropy), the curved of shear stress as a function of shear rate will be identical for increasing and decreasing shear. However, the curves for viscoelastic and/or thixotropic materials is much more complex. In the following discussion, we consider the effects of thixotropy and viscoelasticity separately. As before, assume the material has the maximum microstructure at time t_0. As the shear rate is increased, the microstructure breaks down, diminishing the viscosity of the material. As a result, during the phase of diminishing shear rate which takes place at a later time, the viscosity will be lower than that at the same shear rate during the phase of increasing shear rate. The resulting loop traced out on the stress-shear rate curve is called a hysteresis loop. The size of this loop is often used as a measure of thixotropy.

The previous categories of flows can be used to obtain information about specific material functions without selecting a particular constitutive model (e.g., viscometric functions, storage and loss moduli). Unfortunately, there are no universal "thixotropic material functions". Rather, these non-periodic time-dependent flows are used to provide information about material functions for a specific constitutive model or used as a probe of qualitative features of the mechanical behavior. Even this qualitative interpretation can be confounded by combined viscoelastic and thixotropic effects.

8. Rheometers

The three most commonly used rheometers for measuring the bulk properties of blood and other complex fluids are the concentric cylinder rheometer (Couette rheometer), the cone and plate rheometer and the capillary rheometer, Figure 12. These rheometers can be used to generate various special cases of the viscometric, periodic and unsteady non-periodic flows described in the last section. The viscosity and linear viscoelastic material functions can be measured in all three rheometers. While the cone and plate rheometer is used to measure the first normal stress coefficient for some fluids (e.g., [4]), to date this has not been successfully performed for blood. These three rheometers are briefly considered below. In the interest of space, we confine our discussion to steady flows in the rheometers. The reader is referred to texts such as [12] and [40] for further discussion of these devices. It the following discussion, the fluid (e.g., blood) is treated as incompressible and the material properties are assumed to be independent of pressure variations.

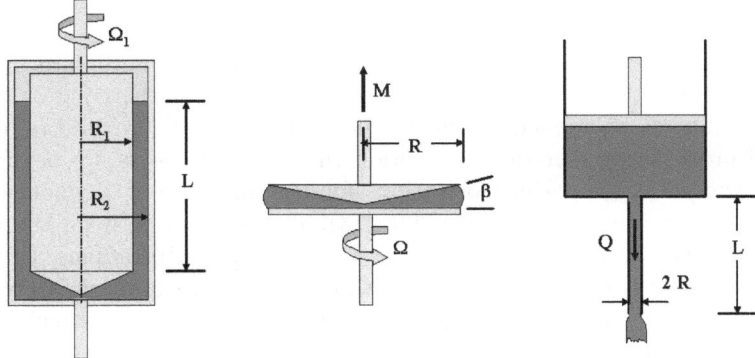

FIGURE 12. Schematic of three rheometers commonly used for blood: (a) Couette rheometer, (b) cone and plate rheometer, (c) capillary rheometer.

8.1. Couette rheometer

In a Couette rheometer, the test fluid is placed between two concentric cylinders with inner and outer radius R_i and R_o, respectively, and length L, Figure 12. A torque M is applied to rotate the cylinders generating a relative rotation rate between the outer and inner cylinder $\Delta\Omega = |\Omega_o - \Omega_i|$. In obtaining the material properties from measured quantities, flow is assumed to be axisymmetric and purely circumferential with negligible end effects (e.g., [12]). For a finite-gap cylinder, the shear rate varies across the gap and so a constitutive model must be chosen a priori to obtain an expression for the viscosity. For example, for an incompressible Newtonian fluid under steady rotation there is an explicit solution for the viscosity as a function of the experimental parameters,

$$\eta = \frac{M(R_o^2 - R_i^2)}{4\,\pi\,L\,R_o^2 R_i^2 \Delta\Omega}. \tag{8.1}$$

A narrow-gap approximation is typically invoked to avoid a-priori selection of a constitutive model. For example, if the inner cylinder is rotated and the outer cylinder is held fixed, the following relationship is valid irrespective of the specific simple fluid,

$$\eta = \frac{M}{4\pi R_i^2 L\,\Omega_i}\left(1 - \frac{R_i^2}{R_o^2}\right). \tag{8.2}$$

It is recommended that the narrow-gap approximation only be used for very small gaps ($R_i/R_o \geq 0.99$), [40]. Higher-order approximations for the viscosity relation based on a McLaurin series can also be used for gaps with $0.5 < R_i/R_o < 0.99$ (e.g., [40]).

8.2. Cone and plate rheometer

The cone and plate rheometer consists of a cone atop a flat plate, Figure 12. The test fluid between is driven by the motion of either the cone or the plate. Measurements of the applied torque M and rotation rate Ω are used to obtain the material properties under the assumption that inertial effects are negligible and the free surface is spherical. Corrections for the influence of secondary motions are available (see, e.g., [40], pages 209–213). The angle β between the cone and plate is assumed to be small. Typically, β is less than 0.10 radians. Under these assumptions, the stress and shear rate within the fluid are approximately constant and the viscosity and first normal stress coefficient ψ_1 take a particularly simple form. For example, in steady flow experiments,

$$\eta = \frac{3M}{2\pi R^3}\frac{\beta}{\Omega}, \qquad \Psi_1 = \frac{2F}{\pi R^2}\frac{\beta^2}{\Omega^2}, \tag{8.3}$$

where R is the cone radius and F is the resultant upward force on the cone due to normal stress effects.

8.3. Capillary rheometer

The capillary rheometer is a popular instrument for measuring blood viscosity and viscoelastic properties, due to its relative simplicity and ease of use. Some researchers also found the similarity between the rheometer geometry and that of some blood vessels appealing. In a capillary rheometer, the test fluid is driven by gravity, compressed gas, or a piston from a reservoir through a cylindrical rigid tube of radius R and length L, Figure 12. The underlying assumptions are that the flow is fully developed and unidirectional. One source of error in the capillary rheometer, even for single phase fluids, are end effects arising from increased pressure losses at the entrance of tube from the reservoir and at the exit into a second reservoir or air. Correction factors have been introduced for these artifacts (e.g., [40]).

The capillary rheometry is frequently used to measure viscosity and linear viscoelastic properties of liquids. By way of illustration, here we focus on applications to measurements of viscosity in steady flow. A discussion of the use of capillary rheometers for measurements of viscoelastic properties in oscillatory flows is covered in Section 8.1 in [57] of this volume. The shear rate varies with the radial position in a capillary rheometer, even for steady flow of a constant viscosity fluid. Therefore, the viscosity of shear thinning fluids will vary over the cross section. An explicit expression for the viscosity as a function of the volumetric flow rate Q and pressure drop ΔP only exists for a small number of cases, constant viscosity and power-law viscosity fluids being the most obvious examples (Section 7.1). In the constant viscosity case, we see from Eq. (7.14),

$$\textbf{Constant Viscosity Fluids} \qquad \eta = (\pi R^4 \Delta P)/(8QL). \qquad (8.4)$$

The Weissenberg–Rabinowitsch relationship provides a means of obtaining the viscosity function from measurements of Q and ΔP without choosing a constitutive equation a priori ([52, 43], see, also for example, pages 238-242 of Macosko [40]),

$$\eta = \frac{\pi R^4 \Delta P}{8QL} \left(\frac{4n'}{3n' + 1} \right) \qquad \text{where} \qquad \frac{1}{n'} = \frac{d\ln Q}{d\ln \Delta P}. \qquad (8.5)$$

However, (8.5) is typically not used in blood rheology. Only an approximation to the viscosity is calculated using the expression for Newtonian fluids, (8.4), rather than the more precise results (8.5), even in regimes where shear thinning effects are important. Alternatively, a measure of viscosity is calculated assuming a form of the viscosity function a priori [41, 1]. In these cases it is important to identify the range of shear rates over which the constitutive model provides a good match for the data.

9. Nonlinear elastic solids

In order to provide some background for the equations used in the study of fluid-solid coupling discussed in the contribution of Turek and Hron [66] of this volume, we now turn attention to some fundamental issues in nonlinear elasticity. There are several recent texts and review articles on the subject of the biomechanics of the arterial wall (see, e.g., [32, 33, 34, 60]). In the interest of space, we therefore only very briefly introduce nonlinear elasticity and by way of example, discuss a few nonlinear isotropic models which are used later in this volume. The reader is referred to [65, 32, 10, 27, 59] for additional background material on nonlinear elasticity.

9.1. Introduction to hyperelastic materials

In a purely mechanical theory, an elastic body is one where the response depends on the deformation gradient or another appropriate measure of strain. For hyperelastic materials, we will assume the existence of a **strain energy density** function, $\Sigma(\boldsymbol{X}, t)$, sometimes called the stored energy per unit mass, such that

$$\rho \frac{D\Sigma}{Dt} = \boldsymbol{T} : \boldsymbol{D} \quad \text{or} \quad \rho \frac{D\Sigma}{Dt} = T_{ij} D_{ij}. \tag{9.1}$$

In other words, the change in strain energy per unit mass of the body arises from work done on the body by internal stresses. The total strain energy per unit mass of the body will be denoted by $\mathcal{U}$ and is therefore,

$$\mathcal{U} = \int_{\mathcal{V}(t)} \rho \, \Sigma \, dv. \tag{9.2}$$

After integrating (9.1) over an arbitrary volume $\mathcal{V}(t)$ and using (2.30) with (9.2), it follows that

$$\frac{d}{dt} (\mathcal{K} + \mathcal{U}) = \int_{\partial \mathcal{V}(t)} \boldsymbol{t} \cdot \boldsymbol{v} da + \int_{\mathcal{V}(t)} \rho \boldsymbol{b} \cdot \boldsymbol{v} dv, \tag{9.3}$$

where $\mathcal{K}$ is the kinetic energy in the body,

$$\mathcal{K} = \frac{1}{2} \int_{\mathcal{V}(t)} \rho \boldsymbol{v} \cdot \boldsymbol{v} dv. \tag{9.4}$$

We see that for elastic bodies, work on the body is directly converted to kinetic or stored energy. Equation (9.3) is the statement that the rate of change of kinetic energy plus the rate of change of strain energy equals the rate of work by surface and body forces.

In classical hyperelasticity, it is assumed that the strain energy at each material point and for all time depends on the deformation gradient at the same point and time,

$$\Sigma = \tilde{\Sigma}(\boldsymbol{F}). \tag{9.5}$$

For homogeneous materials, the form of the function given in (9.5) will be the same at all points. A normalization condition is typically applied to the strain energy function, so that the strain energy vanishes in the reference configuration where $\boldsymbol{F} = \boldsymbol{I}$,

$$\tilde{\Sigma}(\boldsymbol{I}) = 0. \tag{9.6}$$

It then follows from (9.1) and (9.5) that

$$T = \rho \frac{\partial \Sigma}{\partial F} \cdot F^T \quad \text{or} \quad T_{ij} = \rho \frac{\partial \Sigma}{\partial F_{iA}} F_{jA}. \tag{9.7}$$

Using the definitions (2.22) and (2.28) with (9.7),

$$P = \rho_0 \frac{\partial \Sigma}{\partial F} \quad \text{or} \quad P_{iA} = \rho_0 \frac{\partial \Sigma}{\partial F_{iA}} \tag{9.8}$$

and

$$S = \rho_0 F^{-1} \cdot \frac{\partial \Sigma}{\partial F} \quad \text{or} \quad S_{AB} = \rho_0 F_{Ai}^{-1} \frac{\partial \Sigma}{\partial F_{iB}}. \tag{9.9}$$

Sometimes a strain energy per unit volume in the *reference* configuration W is introduced,

$$\mathcal{U} = \int_{\mathcal{V}_o} W \, dV, \tag{9.10}$$

and therefore, from (9.2), it follows that, $W = \rho J \Sigma = \rho_o \Sigma$. For example, we can write (9.7) as

$$T = \frac{1}{J} \frac{\partial W}{\partial F} \cdot F^T \quad \text{or} \quad T_{ij} = \frac{1}{J} \frac{\partial W}{\partial F_{iA}} F_{jA}. \tag{9.11}$$

Using (2.22) and (2.28) with (9.11),

$$P = \frac{\partial W}{\partial F} \quad \text{or} \quad P_{iA} = \frac{\partial W}{\partial F_{iA}} \tag{9.12}$$

and

$$S = F^{-1} \cdot \frac{\partial W}{\partial F} \quad \text{or} \quad S_{AB} = F_{Ai}^{-1} \frac{\partial W}{\partial F_{iB}}. \tag{9.13}$$

9.2. Invariance restrictions

Invariance requirements restrict the form of the functional dependence of the strain energy function on F. Without loss in generality, the most general form of the strain energy function that satisfies invariance can be written as $\Sigma = \hat{\Sigma}(C)$. It then follows from (9.7) and the definitions (1.7) and (1.13) that

$$T = \rho F \cdot \left(\frac{\partial \Sigma}{\partial C} + \frac{\partial \Sigma}{\partial C^T} \right) \cdot F^T \quad \text{or} \quad T_{ij} = \rho F_{iA} F_{jB} \left(\frac{\partial \Sigma}{\partial C_{AB}} + \frac{\partial \Sigma}{\partial C_{BA}} \right), \tag{9.14}$$

where $\hat{\Sigma}(C)$ is symmetric in C_{AB}. In some works, rather than (9.14) the following result is used,

$$T = 2 \rho F \cdot \frac{\partial \Sigma}{\partial C} \cdot F^T \quad \text{or} \quad T_{ij} = 2\rho F_{iA} F_{jB} \frac{\partial \Sigma}{\partial C_{AB}}. \tag{9.15}$$

Care must be taken in using (9.15). See pages 212–213 of [60] for a straightforward comparison of these two representations.

9.3. Example: incompressible, isotropic hyperelastic materials

When we say a hyperelastic material is isotropic, we require that

$$\bar{\Sigma}(QCQ^T) \; = \; \bar{\Sigma}(C) \tag{9.16}$$

where Q is an arbitrary proper orthogonal second-order tensor. Namely, the mechanical response is the same even if the material undergoes an arbitrary rotation in the reference configuration. It then follows from (9.16) and a representation theorem for invariant scalar functions of symmetric tensors, that Σ can be written as a function of the invariants of C,

$$\mathrm{I_C} \; = \; \mathrm{tr}\, C, \qquad \mathrm{II_C} \; = \; \frac{1}{2}((\mathrm{tr}\, C)^2 - \mathrm{tr}\,(C^2)), \qquad \mathrm{III_C} \; = \; \det C. \tag{9.17}$$

For an incompressible material $\mathrm{III_C}$ is constant and we simply use $\Sigma = \Sigma(\mathrm{I_C}, \mathrm{II_C})$. The components of the Cauchy stress tensor for an isotropic, incompressible hyperelastic material can then be reduced from (9.15) to

$$T = -pI + 2\,\rho_0\,\frac{\partial \Sigma}{\partial I_B}\,B - 2\,\rho_0\,\frac{\partial \Sigma}{\partial II_B}\,B^{-1}, \tag{9.18}$$

where p is the Lagrange multiplier arising from incompressibility and we have used the fact that $\mathrm{I_B} = \mathrm{I_C}, \mathrm{II_B} = \mathrm{II_C}$. Alternatively, using the relation $W = \rho_0\Sigma$,

$$T = -pI + 2\,\frac{\partial W}{\partial I_B}\,B - 2\,\frac{\partial W}{\partial II_B}\,B^{-1}. \tag{9.19}$$

Common examples of nonlinear, isotropic, incompressible, hyperelastic materials are the (i) Mooney–Rivlin material, (ii) a special type of Mooney–Rivlin material called a Neo-Hookean material and (iii) an exponential model which has been used for biological materials. For a **Mooney–Rivlin material**, the strain energy function can be written as

$$W \; = \; \frac{\alpha}{2}(I_B - 3) \; + \; \frac{\beta}{2}(II_B - 3) \tag{9.20}$$

where α and β are constants. It follows from (9.19) that the corresponding Cauchy stress tensor for the Mooney–Rivlin material is

$$T \; = \; -pI \; + \; \alpha B \; - \; \beta B^{-1}. \tag{9.21}$$

The **neo-Hookean material** is a special Mooney–Rivlin material for which $\beta = 0$,

$$W \; = \; \frac{\alpha}{2}(I_B - 3), \qquad T \; = \; -pI \; + \; \alpha B. \tag{9.22}$$

A special type of hyperelastic material with an **exponential** dependence on the first invariant of B was introduced by Fung to model the nonlinearly elastic response of biological tissue, [18],

$$W \; = \; \frac{\alpha}{2\gamma}\left(e^{\gamma(I_B-3)} - 1\right), \qquad T \; = \; -pI \; + \; \alpha B\,e^{\gamma(I_B-3)} \tag{9.23}$$

where α and γ are material constants. A more recent application of this model can be found in [67]. Notice that for the exponential model, the coefficient for B

is not a constant as it was for the Mooney–Rivlin and Neo-Hookean models. In addition, both the Neo-Hookean and exponential models are independent of II_B.

References

[1] T. Alexy, R.B. Wenby, E. Pais, L.J. Goldstein, W. Hogenauer, and H.J. Meiselman. An automated tube-type blood viscometer: Validation studies. *Biorheology*, 42:237-247, 2005.

[2] G. Astarita and G. Marrucci. *Principles of Non-Newtonian Fluid Mechanics*. McGraw Hill, 1974.

[3] H.A. Barnes. Thixotropy – a review. *J. Non-Newtonian Fluid Mech.*, 70:1–33, 1997.

[4] H.A. Barnes, J.F. Hutton, and K. Walters. *An Introduction to Rheology*. Elsevier, 1989.

[5] H.A. Barnes. The yield stress – a review or '$\pi\alpha\nu\tau\alpha\ \rho\epsilon\iota$' – everything flows? *J. Non-Newtonian Fluid Mech.*, 81:133–178, 1999.

[6] W.H. Bauer and E.A. Collins. Thixotropy and dilatancy. In F.R. Eirich, editor, *Rheology: Theory and Applications*, volume 4. Academic Press, 1967.

[7] R.B. Bird, R.C. Armstrong, and O. Hassager. *Dynamics of Polymeric Liquids, Volume I: Fluid Mechanics*. John Wiley & Sons, second edition, 1987.

[8] R.B. Bird, G.C. Dai, and B.J. Yarusso. The rheology and flow of viscoplastic materials. *Reviews in Chemical Engineering*, 1:1–70, 1983.

[9] N. Casson. A flow equation for pigment-oil suspensions of the printing ink type. In *Rheology of Disperse Systems*, pages 84–102. Pergamon, 1959.

[10] P. Chadwick. *Continuum Mechanics*. John Wiley & Sons, 1976.

[11] B.D. Coleman. Kinematical concepts with applications in the mechanics and thermodynamics of incompressible viscoelastic fluids. *Arch. Rational Mech. Anal.*, 9:273–300, 1962.

[12] B.D. Coleman, H. Markovitz, and W. Noll. *Viscometric Flows of Non-Newtonian Fluids*. Springer-Verlag, 1966.

[13] B.D. Coleman and W. Noll. An approximation theorem for functionals, with approximations in continuum mechanics. *Arch. Rational Mech. Anal.*, 6:355–370, 1960.

[14] B.D. Coleman and W. Noll. Foundations of linear viscoelasticity. *Rev. Mod. Phys.*, 33(2):239–249, Apr 1961.

[15] J.E. Dunn and R.L. Fosdick. Thermodynamics, stability and boundedness of fluids of complexity 2 and fluids of second grade. *Arch. Rational Mech. Anal.*, 56:191–252, 1974.

[16] G. Duvant and J.L. Lions. *Inequalities in Mechanics and Physics*. Springer-Verlag, 1982.

[17] R. Finn. Stationary solutions of the Navier-Stokes equations. *Symp. Appl. Math.*, 17:121–153, 1965.

[18] Y.C. Fung. Elasticity of soft tissue in simple elongation. *Am. J. Physiology*, 213:1532–1544, 1967.

[19] Y.C. Fung. *Biomechanics: Mechanical Properties of Living Tissues*. Springer-Verlag, 1993.

[20] G.P. Galdi. Mathematical problems in classical and non-Newtonian fluid mechanics. In *Hemodynamical Flows. Modeling, Analysis and Simulation*. Oberwolfach-Seminars, Vol. 37, p. 121–273. Birkhäuser Verlag, 2008.

[21] G.P. Galdi. An introduction to the mathematical theory of the Navier-Stokes equations, Vol. I: Linearized steady problems. In *Springer Tracts in Natural Philosophy*. Springer-Verlag, (revised edition), 1998.

[22] G.P. Galdi. An introduction to the mathematical theory of the Navier-Stokes equations, Vol. II: Non-linear steady problems. In *Springer Tracts in Natural Philosophy*. Springer-Verlag (revised edition), 1998.

[23] G.P. Galdi, J.G. Heywood, and Y. Shibata. On the global existence and convergence to steady state of Navier–Stokes flow past an obstacle that is started from rest. *Arch. Rational Mech. Anal.*, 138:307–318, 1997.

[24] G.P. Galdi, M. Padula, and K.R. Rajagopal. On the conditional stability of the rest state of a fluid of second grade in unbounded domains. *Arch. Rational Mech. Anal.*, 109(2):173–182, 1990.

[25] G.P. Galdi and B.D. Reddy. Well-posedness of the problem of fiber suspension flows. *J. Non-Newtonian Fluid Mech.*, 83(3):205–230, 1999.

[26] A.E. Green and P.M. Naghdi. A note on invariance under superposed rigid body motions. *J. Elasticity*, 9:1–8, 1979.

[27] M.E. Gurtin. *Introduction to Continuum Mechanics*. Academic Press, 1981.

[28] J. Hadamard. *Lectures on Cauchy's Problem in Linear Partial Differential Equations*. Yale University Press, 1923.

[29] W.H. Herschel and R. Bulkley. Konsistenzmessungen von Gummi-Benzol-Lösungen. *Kolloid Z.*, 39:291–300, 1926.

[30] W.H. Herschel and R. Bulkley. Measurement of consistency as applied to rubber-benzene solutions. *Proc. ASTM, Part II*, 26:621–629, 1926.

[31] J.G. Heywood. *On non-stationary problems for the Navier-Stokes equations and the stability of stationary flows*. PhD thesis, Stanford University, 1969.

[32] G.A. Holzapfel. *Nonlinear Solid Mechanics: A Continuum Approach for Engineering*. Wiley, 2000.

[33] J.D. Humphrey. *Cardiovascular Solid Mechanics: Cells, Tissues, and Organs*. Springer-Verlag, 2002.

[34] J.D. Humphrey. Review paper: Continuum biomechanics of soft biological tissues. *Proceedings of the Royal Society A: Mathematical, Physical and Engineering Sciences*, 459:3–46, 2003.

[35] M.W. Johnson Jr. and D. Segalman. A model for viscoelastic fluid behavior which allows non-affine deformation. *J. Non-Newtonian Fluid Mech.*, 2:255–269, 1977.

[36] D.D. Joseph. Instability of the rest state of fluids of arbitrary grade greater than one. *Arch. Rational Mech. Anal.*, 75:251–256, 1981.

[37] D.D. Joseph. *Fluid Dynamics of Viscoelastic Liquids*. Springer-Verlag, 1990.

[38] M.M. Keentok, A.G. Georgescu, A.A. Sherwood, and R.I. Tanner. The measurement of the second normal stress difference for some polymer solutions. *J. Non-Newtonian Fluid Mech.*, **6**:303–324, 1980.

[39] D.C. Leigh. Non-Newtonian fluids and the second law of thermodynamics. *Physics of Fluids*, 5(4):501–502, 1962.

[40] C.W. Macosko. *Rheology: Principles, Measurements and Applications*. VCH Publishers, Inc., 1994.

[41] G.N. Marinakis, J.C. Barbenel, A.C. Fisher, and S.G. Tsangaris. A new capillary viscometer for whole blood viscometry. *Biorheology*, 36:311–318, 1999.

[42] P.C.F. Moller, J. Mewis, and D. Bonn. Yield stress and thixotropy: on the difficulty of measuring yield stress in practice. *Soft Matter*, 2:274–288, 2006.

[43] M. Mooney. Explicit formulas for slip and fluidity. *J. Rheol.*, 2:210–221, 1931.

[44] P.M. Naghdi. Mechanics of solids. In C. Truesdell, editor, *Handbuch der Physik*, volume VIa/2, chapter The Theory of Shells and Plates, pages 425–640. Springer-Verlag, 1972.

[45] W. Noll. A mathematical theory of the mechanical behavior of continuous media. *Arch. Rational Mech. Anal.*, 2(3):197–226, 1958.

[46] S. Oka. *Cardiovascular Hemorheology*. Cambridge University Press, 1981.

[47] J.G. Oldroyd. A rational formulation of the equations of plastic flow for a Bingham solid. *Proc. Camb. Philos. Soc.*, 43:100–105, 1947.

[48] J.G. Oldroyd. Two-dimensional plastic flow of a Bingham solid. *Proc. Camb. Philos. Soc.*, 43:383–395, 1947.

[49] R.G. Owens and T.N. Phillips. *Computational Rheology*. Imperial College Press, 2005.

[50] T.C. Papanastasiou. Flows of materials with yield. *J. of Rheology*, 31:384–404, 1987.

[51] W. Prager. *Introduction to Mechanics of Continua*. Dover Phoenix Edition, 1961.

[52] B. Rabinowitsch. Über die viskosität und elastizität von solen. *Z. Physik Chem.*, A145:1–26, 1929.

[53] S. Ramachandran, H.W. Gao, and E.B. Christiansen. Dependence of viscoelastic flow functions on molecular structure for linear and branched polymers. *Macromolecules*, 18:695–699, 1985.

[54] M. Reiner. Elasticity and Plasticity. In S. Flügge, editor, *Handbuch der Physik*, Volume VI. Chapter: Rheology, p. 434-550. Springer-Verlag, 1958.

[55] M. Reiner. The Deborah number. *Physics Today*, 17:62, 1964.

[56] A.M. Robertson. On the attainability of steady viscometric flows. *Quaderni di Matematica*, 10:219–245, 2003.

[57] A.M. Robertson, A. Sequeira, and M.V. Kameneva. Hemorheology. In *Hemodynamical Flows. Modeling, Analysis and Simulation*. Oberwolfach-Seminars, Vol. 37, p. 63–120. Birkhäuser Verlag, 2008.

[58] W.R. Schowalter. *Mechanics of Non-Newtonian Fluids*. Pergamon Press, 1978.

[59] A.J.M. Spencer. *Continuum Mechanics*. Dover Publications, Inc, 2004.

[60] L.A. Taber. *Nonlinear Theory of Elasticity: Applications in Biomechanics*. World Scientific Publishing Co., 2004.

[61] R. Temam and A. Miranville. *Mathematical Modeling in Continuum Mechanics*. Cambridge University Press, 2001.

[62] G.B. Thurston. Viscoelastic properties of blood and blood analogs. *Advances in Hemodynamics and Hemorheology*, 1:1–30, 1996.

[63] C. Truesdell. A new definition of a fluid, II. The Maxwellian fluid. *J. Math Pures Appl.*, 30:111–155, 1951.

[64] C. Truesdell. The mechanical foundations of elasticity and fluid dynamics. *J. Rational Mech. Anal.*, 1:125–300, 1952.

[65] C. Truesdell and W. Noll. Non-linear field theories of mechanics. In S.Flugge, editor, *Handbuch der Physik*, volume III/3. Springer-Verlag, 1965.

[66] S. Turek and J. Hron. Numerical techniques for multiphase flow with liquid-solid interaction. In *Hemodynamical Flows. Modeling, Analysis and Simulation.* Oberwolfach-Seminars, Vol. 37, p. 379–501. Birkhäuser Verlag, 2008.

[67] R. Wulandana and A.M. Robertson. An Inelastic Multi-Mechanism Constitutive Equation for Cerebral Arterial Tissue. *Biomechanics and Modeling in Mechanobiology*, 4(4):235–248, 2005.

Anne M. Robertson
Department of Mechanical Engineering and Materials Science
McGowan Institute for Regenerative Medicine
Center for Vascular Remodeling and Regeneration (CVRR)
University of Pittsburgh
641 Benedum Engineering Hall
Pittsburgh, PA 15261
USA
e-mail: annerob@engr.pitt.edu

Hemodynamical Flows. Modeling, Analysis and Simulation
Oberwolfach Seminars, Vol. 37, 63–120
© 2008 Birkhäuser Verlag Basel/Switzerland

Hemorheology

Anne M. Robertson, Adélia Sequeira and Marina V. Kameneva

Introduction

Hemorheology is the science of deformation and flow of blood and its formed elements. This field includes investigations of both macroscopic blood properties using rheometric experiments as well as microscopic properties *in vitro* and *in vivo*. Hemorheology also encompasses the study of the interactions among blood components and between these components and the endothelial cells that line blood vessels.

Blood performs the essential function of delivering oxygen and nutrients to all tissues, removing waste products and defending the body against infection through the action of antibodies. The blood circulation in the cardiovascular system depends not only on the driving force of the heart and the architecture and mechanical properties of the vascular system, but also on the mechanical properties of blood itself. Whole blood is a concentrated suspension of formed cellular elements including red blood cells (erythrocytes), white blood cells (leukocytes) and platelets. The non-Newtonian behavior of blood is largely due to three aspects of erythrocyte behavior: their ability to aggregate and form a branched three-dimensional (3D) microstructure at low shear rates, their deformability, and their tendency to align with the flow field at high shear rates [29, 118]. An understanding of the coupling between the blood composition and its physical properties is essential for developing suitable constitutive models to describe blood behavior.

Hemodynamic factors such as flow separation, flow recirculation, and low and oscillatory wall shear stress are recognized as playing important roles in the localization and development of vascular diseases. Therefore, mathematical and numerical simulations of blood flow in the vascular system can ultimately contribute to improved clinical diagnosis and therapeutic planning (see, e.g., [43, 80]). However, meaningful hemodynamic simulations require constitutive models that can accurately model the rheological response of blood over a range of physiological flow conditions. Experimental studies of the instability of the 3D microstructure of erythrocytes (RBCs) suggests that it is probably reasonable to treat the blood viscosity as constant in most parts of the arterial system of healthy individuals, due to the high shear rates found in these vessels and the length of time necessary for the blood microstructure to form. However, in disease states in which the stability of the aggregates is enhanced or for diseases in which the arterial geometry has been altered to include regions of recirculation (e.g., saccular aneurysms), this simplifying assumption may need to be relaxed and a more complex blood constitutive model should be used. In addition, even in healthy patients, the non-Newtonian characteristics of blood can play an important role in parts of the venous system.

We begin this chapter with a brief summary of blood components followed by a discussion of relevant parameters in the cardiovascular system. In Section 3, we turn attention to experimental data on the multiphase properties of blood and the relationship between these properties and the mechanical response of blood. A short introduction to the important subject of blood coagulation is then provided in Section 4, motivated by its importance in medical devices such as stents and artificial hearts as well as the pressing need for theoretical, experimental and numerical efforts to further develop and validate these models. In Section 5, we then turn to the subject of mechanical measurements of blood properties and challenges in blood rheometry. In Section 6, the viscosity of whole blood is considered, including experimental data and generalized Newtonian models for blood viscosity. The following section covers the interesting, though controversial subject of yield stress behavior of blood. In Section 8, the viscoelastic behavior of blood is briefly discussed. The chapter ends in Section 9 with a discussion of the relationship between some disease states and mechanical properties of blood.

The constitutive models discussed in this chapter are phenomenological in nature. Looking toward the future, we expect that more and more quantitative experiments will be performed at the scale of the red blood cell, providing a rational basis to extend these models to include microstructural aspects of blood. For example, data on the time constants associated with the formation and breakup of RBC microstructure in representative flow regimes and shear fields are extremely important. Ultimately, it is expected that these microstructural models will make it possible to predict hemodynamic features such as the inhomogeneous spatial distribution of the red cells in vessels and the thixotropic behavior of blood.

The material of this chapter is organized as follows.

Contents

Acknowledgements

The authors would like to thank doctoral student Michael Hill of the University of Pittsburgh for performing the nonlinear regression analysis presented in Section 6.4.2 and doctoral student Dalong Li of the University of Pittsburgh for preparing some of the figures.

1. Blood components

Whole blood is a concentrated suspension of formed cellular elements that includes red blood cells (RBCs) or erythrocytes, white blood cells (WBCs) or leukocytes, and platelets or thrombocytes, Table 1. The average person has between 4.5 and 6 L of blood, accounting for approximately 6 – 8% of the body weight in healthy individuals. During a singleton pregnancy, the blood volume increases by 50%.

The formed elements represent approximately 45% by volume of the normal human blood. The process by which all formed elements of the blood are produced (hematopoiesis), occurs mostly in the bone marrow, where cells mature from primitive stem cells. Important factors in regulating blood cell production include: the environment of the bone marrow, interactions among the cells, and secreted chemicals called growth factors.

Cell	Number per mm^3	Unstressed shape and dimensions (μm)	Volume concentration (%) in blood
Erythrocytes	$4 - 6 \times 10^6$	Biconcave disc $8 \times 1 - 3$	45
Leukocytes			
Total	$4 - 11 \times 10^3$		
Granulocytes			
Neutrophils	$1.5 - 7.5 \times 10^3$	Roughly	
Eosinophils	$0 - 4 \times 10^2$	spherical	
Basophils	$0 - 2 \times 10^2$	$7 - 22$	1
Lymphocytes	$1 - 4.5 \times 10^3$		
Monocytes	$0 - 8 \times 10^2$		
Platelets	$250 - 500 \times 10^3$	Rounded or oval $2 - 4$	

TABLE 1. Quantity, shape, size and concentration of cellular components in normal human blood, from page 159 of [17].

1.1. Plasma

Plasma consists primarily of water (approximately 90 – 92% by weight) in which inorganic and organic substances (approximately 1 – 2%), various proteins (mostly albumin, globulins, and fibrinogen) as well as many other components are dissolved. Its central physiological role is to transport these dissolved substances, nutrients, wastes and the formed cellular elements throughout the circulatory system. Plasma also plays an important role in homeostasis (constant colloid-osmotic pressure), buffer function and coagulation.

1.2. Red blood cells (Erythrocytes)

Erythrocytes, or red blood cells, are highly flexible cells filled with an almost saturated solution (approximately 32% by weight) of hemoglobin in water (65%) as well as inorganic elements (K, Na, Mg and Ca). Hemoglobin is the protein inside RBC that gives blood its red color and is primarily involved in oxygen and carbon dioxide transport between the lungs and tissues of the body. As will be discussed throughout this chapter, erythrocytes, which are the most numerous of the formed elements (about 98%), have the largest influence on the mechanical properties of blood. The properties of an individual RBC changes as it ages and the normal life-span of a RBC in human blood is 100 – 120 days.

The shape of a normal unsheared erythrocyte is a biconcave discoid with a diameter of 6 – 8 μm, surface area of approximately 130 μm^2 and volume of approximately 98 μm^3, Figure 1. This shape can be changed as a result of mechanical, chemical or thermal effects. As elaborated on in Section 3, under normal hemodynamic loading, the shape of the RBC changes dramatically from a biconcave disk to an ellipsoid. Significantly, a biconcave disk has a greatly increased surface area to volume ratio compared to spheres. For example, the surface area of a sphere of volume of 98 μm^3 is 103 μm^2 compared with a surface area of 130 μm^2 for a biconcave RBC of an equivalent volume, Figure 1. This makes it possible for the red blood cell to greatly deform without significant strain compared to a spherical shaped cell, (e.g., Chapter 4 of [53]). In fact, RBC membranes cannot withstand more than 5 – 10% increase in area without hemolysis [16].

1.2.1. Hematocrit (Ht). The volume concentration of RBCs in whole blood is called the hematocrit, Ht. It is measured by centrifuging blood in a hematocrit tube and measuring the volume occupied by the packed cells in the bottom of the tube. Since some plasma remains trapped in the packed cells, the true hematocrit is about 96% of this measured value [60]. Burton calculated the highest hematocrit possible for undeformed RBC based on geometric arguments for the closest packing of biconcave human RBCs of thickness 2.7 μm and radius 8.1 μm. He concluded RBCs must deform when the hematocrit is higher than 63%, (page 41 of [16]). Normal hematocrit levels for women and men are below this level, with an average about 40% and 45%, respectively. In addition to gender, the hematocrit level can vary due to disease state, activity level and the altitude in which the individual lives. For example, severe blood loss and diseases such as haemolytic anemia, sickle cell anemia and thalassemia can cause Ht to drop as low as 10%. Elevated hematocrit levels can arise due to high altitude, dehydration or diseases characterized by excessive RBC production. In this latter case, hematocrit levels can rise as high as 60 – 70% [60]. The mechanical properties of blood are strongly dependent on the hematocrit level.

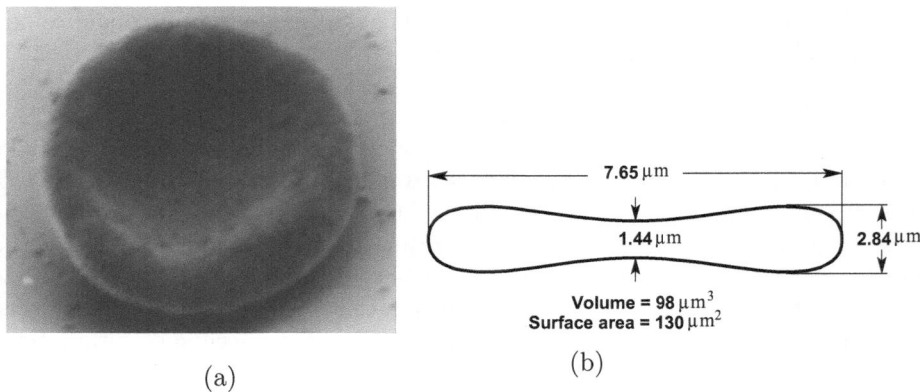

(a)   (b)

FIGURE 1. (a) Scanning electron microscope image of a red blood cell (with permission from Dr. K.B. Chandran); (b) Schematic of a RBC profile with average geometric parameters from 14 healthy subjects (values from [54]).

1.3. White blood cells (Leukocytes)

WBCs are much less numerous than erythrocytes (less than 1% of the volume of blood). They are normally roughly spherical in shape with diameters ranging from about 7 – 22 μm, (page 167 of [17]). Leukocytes play a vital role in fighting infection in the body both through (i) the destruction of bacteria and viruses (via phagocytosis) and (ii) the formation of antibodies and sensitized lymphocytes. There are five morphologically different types of WBC: basophils, eosinophils and neutrophils (collectively called granulocytes) and also monocytes and lymphocytes, Table 1. The term granular arises from the granular appearance of these cells. WBCs originate partially in the bone marrow (granulocytes, monocytes and some lymphocytes) and partially in the lymph tissue such as the lymph glands and tonsils (most lymphocytes). While there is a constant supply of WBCs circulating in the blood stream, about three times this number are stored in the marrow. These cells can be rapidly transported to areas of infection and inflammation. We refer the reader to classical physiology texts and modern references for more information on these processes (e.g., [60]). Leukocytes are believed to have little influence on the rheology of blood, except in extremely small vessels like capillaries or in disease conditions, (e.g., [112, 78]). Their diameter is larger than the average diameter of a capillary and, as a result, they must undergo large deformations in order to pass through the systemic or pulmonary microcirculation.

In some cases, the response of the leukocytes can lead to a cascade of activities which are deleterious rather than beneficial to the body. It is now known, that the adherence of leukocytes to the endothelial monolayer is one of the first steps in the formation of atherosclerotic lesions and can be induced by an atherogenic diet, (e.g., [77]). The leukocytes maintain adhesion to the endothelium through rolling. There is a growing body of work directed at modeling and better understanding this process from a biomechanical point of view as well as the general behavior of the WBCs (e.g., [5, 23, 47, 71, 112]).

1.4. Platelets (Thrombocytes)

Platelets are small discoid non-nucleated cell fragments that are much smaller than erythrocytes (approximately 6 μm^3 in volume as compared to 98 μm^3). They form a small fraction of the particulate matter in human blood, Table 1. Platelets are vital for the prevention of blood loss. When they come in contact with a damaged vascular surface or foreign substance they dramatically change their physical form and release chemicals that activate nearby platelets and cause them to adhere to each other. This can lead to a platelet plug which is sufficient to stop bleeding in small injuries. For larger injuries, the formation of a blood clot (coagulation) is essential. Platelets play an important role in coagulation, which is discussed in Section 4.

2. Relevant parameters for flow in the human cardiovascular system

Prior to discussing the physical behavior of RBC in flowing blood, we provide typical values of physical parameters in the vessels of the human body, Tables 2 and 3 and corresponding non-dimensional parameters. For pulsatile flow of a generalized Newtonian fluid in a pipe, the important flow parameters are a characteristic diameter (D), the fluid density (ρ), a characteristic fluid viscosity (η_c), a characteristic velocity (V), and a characteristic angular frequency (ω). For example, the characteristic velocity could be the value of velocity averaged over both the cross section and one period of the pulsatile waveform ($\bar{V}$), and the characteristic angular frequency could be the fundamental angular frequency of the flow. Recall, the angular frequency is related to the frequency (f) through $\omega = 2\pi f$. There may be other important geometric length scales besides the diameter arising, for example, in the axial direction due to variations in cross section. When the vessel diameter is on the order of the diameter of RBC or characteristic length scale of the 3D microstructure formed by RBCs it is no longer appropriate to model blood as a single phase continuum. In this case, additional parameters will become important.

Vessel	Diameter (mm)	Volumetric Flow Rate (mL/s)			Spatial Mean Velocity (mm/s)		
		min.	max	mean	min.	max	mean
Ascending Aorta	23.0 – 43.5	—	—	364	—	—	245 – 876
Femoral Artery	5.0	-6.9	23.1	3.7	-350	1175	188
Common Carotid	5.9	2.7	10.6	5.1	99	388	187
Carotid Sinus	5.2	1.8	6.9	3.3	85	325	156
External Carotid	3.8	0.9	3.7	1.8	83	327	157
Thoracic inferior	20.0	—	—	34 – 50	—	—	107 – 160

TABLE 2. Physiological flow parameters for the human circulation from [57].

Vessel	Diameter (mm)	Time average of spatial mean velocity (mm/s)	Peak of spatial mean velocity (mm/s)
Ascending Aorta	20	—	630
Descending Aorta	16	—	270
Large Arteries	2 – 6	—	200 – 500
Capillaries	0.005 – 0.01	0.5 – 1	—
Veins	5 – 10	150 – 200	—
Vena Cava	20	110 – 160	—

TABLE 3. Physiological flow parameters compiled for the human circulation, from page 93 of [141]. Mean velocity is the spatial average of velocity over the cross section.

The *Reynolds number* (*Re*) and the *Womersley number* (α) can be formed from these five parameters,

$$Re = \frac{\rho \bar{V} D}{\eta_c}, \qquad \alpha = \frac{D}{2}\sqrt{\frac{\rho \omega}{\eta_c}}. \tag{2.1}$$

The Reynolds and Womersley numbers are measures of the significance of steady and unsteady inertial effects to viscous effects, respectively. These parameters are important for several reasons. First, they play a role in determining when it is physically reasonable to make simplifications to the governing equations. In the limit of small *Re*, it may be reasonable to neglect convective acceleration terms while in the limit of small α, it may be suitable to treat even unsteady flows as quasi-static, (e.g., pages 57–60 of [17]). For example, in the capillary bed the Reynolds number is much less than one and the diameters of the capillaries are on the order of the diameter of the RBC. Therefore, when modeling flow in the capillary bed, we can ignore inertial effects in the balance of linear momentum, but it will be important to model the motion of individual RBC, (e.g., [99]).

The choice of characteristic viscosity requires some thought for non-Newtonian fluids such as blood because blood viscosity varies with shear rate (see Section 6). Frequently, the asymptotic high shear viscosity is used, though this is not possible for power-law fluids. Alternatively, the characteristic viscosity can be chosen as the viscosity at a representative shear rate, such as the average velocity $\bar{V}$ divided by the pipe diameter D. Using this definition for a power-law fluid (see formula (3.13) of the article [107] in this volume),

$$Re = \frac{\rho \bar{V} D}{\eta_c} = \rho \bar{V}^{2-n} D^n / K. \tag{2.2}$$

We now turn attention to estimating the range of Re and α in the human circulatory system. As will be discussed in Section 6, blood viscosity strongly depends on shear rate, temperature and hematocrit. Typical values for normal human blood density and viscosity at 37°C for shear rates greater than approximately 400 s^{-1} are

$$\begin{array}{lll}
\text{Blood density} & \rho & = 1.06 \times 10^3 \text{ kg} \cdot \text{m}^{-3}, \\
\text{High shear-rate blood viscosity [75]} & \eta & = 3 - 5.5 \text{ mPa} \cdot \text{s}.
\end{array} \tag{2.3}$$

We now calculate representative Reynolds and Womersley numbers using the diameters and mean velocities in Tables 2 and 3 with $\rho = 1.06 \times 10^3$ kg $\cdot$ m^{-3} and $\eta_c = 3.5$ mPa $\cdot$ s, assuming a fundamental frequency of 1 Hz. For later discussion of the non-Newtonian behavior of blood, we also calculate a representative value for the mean wall shear rate $(\dot{\gamma}_w)$ using the mean velocity and diameter. In these calculations, the velocity field is approximated as parabolic. Therefore, the reasonableness of the wall shear rate estimate depends on how close the profile is to parabolic.

The estimated range of Reynolds numbers in the ascending aorta are well above the magnitude typically associated with transition to turbulence of a Newtonian fluid in a straight pipe. Dintenfass (pages 129–132 of [44]) includes a thought provoking discussion of various factors which would tend to suppress the development of turbulent flow in the circulatory system including the thixotropy of the fluid, existence of polymeric materials in the blood, and pulsatility of the flow. To this list, vessel curvature can be added. As noted by Dintenfass, certain disease states will diminish the effectiveness of some of these factors.

When the viscoelasticity of blood is considered, other material parameters enter the problem such as λ, a characteristic time associated with the memory of the fluid. A non-dimensional time scale called the Deborah number (see Section 5.1.2 of the article [107] in this volume) can then be used,

$$De = \lambda \omega. \tag{2.4}$$

If finite viscoelasticity is considered, non-dimensional first and second normal stress coefficients can also be considered (see Section 7.1 of the article [107] in this volume).

Vessel	Reynolds Number Re	Womersley Number α	Mean Wall Shear Rate (1/s)
Ascending Aorta[a]	$3200 - 6100$	$16 - 30$	$45 - 300$
Femoral Artery[a]	280	3.4	300
Common Carotid[a]	330	4.1	250
Carotid Sinus[a]	245	3.6	240
External Carotid[a]	180	2.6	330
Capillaries[b]	0.0015	$0.003 - 0.007$	$400 - 1600$
Large Veins[b]	$300 - 450$	3.4-6.9	$120 - 320$
Vena Cava[b]	$670 - 970$	13.8	$44 - 64$
Thoracic inferior[a]	$650 - 970$	13.8	$43 - 64$

TABLE 4. Estimates of parameters Re, α, and wall shear rate $\dot{\gamma}_w$ based on data in Tables 2 and 3. The superscripts "a" and "b" next to the vessel names are used to denote data from Tables 2 and 3, respectively. Definitions of Re and α are given in (2.1). The wall shear rate is estimated using a parabolic velocity profile so that $\dot{\gamma}_w = 8\bar{V}/D$. Values of density, viscosity, and characteristic frequency are $\rho = 1.06 \times 10^3$ kg $\cdot$ m^{-3}, $\eta = 3.5$ mPa $\cdot$ s, and 1 Hz, respectively.

3. Multiphase behavior of blood in shear flows

The non-Newtonian behavior of blood is largely due to three characteristics of the erythrocytes: their tendency to form aggregates when at rest or at low shear rates, their deformability, and their tendency to align in the flow direction at high shear rates (e.g., [29, 118, 134]). The shape of the RBC can change (deform) both due to in plane stretching of the RBC membrane as well as from bending. The high deformability of RBC is due to the absence of a nucleus, to the elastic and viscous properties of its membrane and also to geometric factors such as the shape, volume and membrane surface area [24].

As will be elaborated on in the following sections, blood displays a shear thinning viscosity, viscoelasticity, thixotropy and possibly a yield stress. In trying to understand the mechanisms behind these bulk mechanical properties, we first discuss experiments made over the years to categorize the response of RBC to shear flows in three flow regimes. As discussed below, at low shear rates, the RBC form a complex three-dimensional microstructure, while at high shear rates, this microstructure is lost and flow-induced radial migration may lead to a non-homogeneous distribution of RBC. A transition in microstructure is found between these two regimes.

To better interpret and analyze the experimental data on blood it is helpful to turn to the literature on the rheology of particle suspensions. For rigid particles, a vast amount of published literature exists (see, e.g., [108]). Aspects of this subject are discussed in detail in Galdi [55]. However, the study of suspensions of multiple, interacting and highly deformable particles such as blood, has received

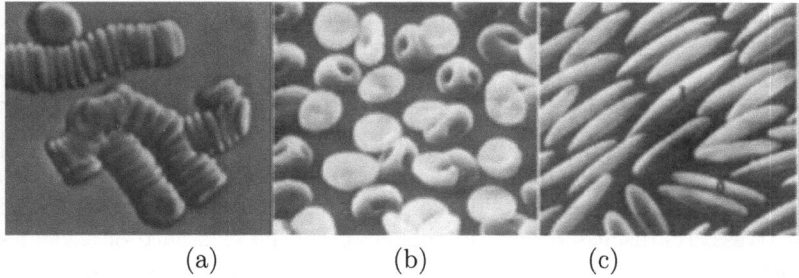

(a)   (b)   (c)

FIGURE 2. View of erythrocytes from normal human blood. In
(a), they are seen forming rouleaux (interference microscopy),
from p. 141 of [10] (with permission from Springer Verlag). In (b)
and (c), they have been fixed with glutaraldehyde while sheared
in a viscometer under a shear stress of (b) 10 Pa and (c) 300 Pa,
from [17] (with permission of the author and Oxford University
Press).

less attention and presents a challenge for both theoretical and computational fluid
dynamacists, (e.g., [41]).

3.1. Low shear behavior: aggregation and disaggregation of erythrocytes

In the presence of fibrinogen and large globulins (proteins found in plasma) erythrocytes have the ability to form a primary aggregate structure of rod-shaped
stacks of individual cells called *rouleaux*. At very low shear rates, the rouleaux align
themselves in an end-to-side and side-to-side fashion and form a secondary structure consisting of branched three-dimensional aggregates of the rouleaux [116],
Figure 2(a). Erythrocyte aggregation is a reversible dynamic phenomenon, considered as the main factor responsible for the shear thinning behavior at low shear
rates, [28]. It is observed both *in vitro* and *in vivo*. These stacks will not form if
the erythrocytes have been hardened or in the absence of fibrinogen and globulins
[85, 139]. In fact, suspensions of erythrocytes in plasma strongly demonstrate a
shear thinning behavior, while in albumin Ringer or isotonic salt solution (with
no fibrinogen or globulins) this behavior is greatly diminished (see Section 6.2).

Fåraheus demonstrated rouleaux formation at low shear rates in both physiological and pathological conditions [48]. It is believed that in the majority of
the circulation, under healthy conditions, the shear rates are too high to allow
for the appearance of rouleaux. Flow in some veins and venules are exceptions
[11, 59, 99, 124]. As will be elaborated on in Section 9, in some disease states the
tendency for blood to form a three-dimensional microstructure of RBCs and the
strength of this microstructure are significantly increased.

Fundamental advances in our understanding of how the microstructure of
blood influences its mechanical properties were made possible with the invention
of the rheoscope [117]. In this rheometer, a transparent cone and plate are placed

upon an inverted microscope, enabling visualization of the changing microstructure of blood under shear. While prior experimental work provided indirect evidence of the principal role of erythrocyte aggregation and disaggregation in determining the rheological properties of whole blood at low shear (less than about 50 s^{-1}), experiments using the rheoscope provided direct evidence of this relation [59].

For blood at rest, or at very low shear rates, the three-dimensional structure formed by the RBC appears solid-like, resisting flow, suggesting blood may have a yield stress. As discussed later in this chapter, the existence of a yield stress is somewhat of a controversial issue (see, for example, [7, 89]). Under increasing loads, the blood begins to flow and the solid-like structure breaks up into 3D networks of various sizes which appear to move as individual units and reach an equilibrium size. This size is dependent on the shear rate. Increases in the shear rate lead to a breakdown of the aggregates and a consequent reduction in equilibrium size. The smaller aggregates result in a lower effective viscosity, leading to the shear thinning behavior of blood. In the studies by Schmid-Schönbein [116] at shear rates between 5.8 and 46 s^{-1}, each doubling of the shear rate resulted in a decrease in aggregate size of approximately 50%. He defined a critical shear rate $\dot{\gamma}_{max}$, as the shear rate at which there are no more aggregates larger than 15 μm in constant shear experiments. In whole blood from healthy human subjects, differing values are reported for this critical shear rate, largely in the range of 5 – 100 s^{-1} (e.g., [59, 43]). In diseased states, $\dot{\gamma}_{max}$ can increase substantially and will in turn have a large impact on the mechanical properties of the blood. In blood samples from patients with acute myocardial infarctions, the critical shear rate for dispersion was found to be greater than approximately 250 s^{-1} and the average aggregate size was larger than in controls for all shear rates [59, 116]. Rouleaux in blood samples from a cardiac patient are shown in Figure 3 with normal blood shown for comparison. The orderly three-dimensional network of RBC seen in normal blood is replaced by three-dimensional clumps of heavily aggregated RBCs. The relationship between blood properties and disease are further discussed in Section 9.

The process of disaggregation under increasing shear is reversible. When the shear rate is quasi-statically stepped down to lower and lower values, the individual cells form shorter chains, then longer rouleaux and eventually a 3D microstructure [59]. The finite time necessary for equilibrium of the structure to be reached (both during aggregation and disaggregation) is the central reason for the thixotropic behavior of blood, (see Section 6 in [107] of this volume for a general discussion of thixotropy). The time constants for aggregation and disaggregation are functions of the shear rate. The equilibria are found to be reached more rapidly at higher shear rates and more gradually at lower shear rates (e.g., [43]). A time interval of 20 to 200 seconds was required for rouleaux to reform in a cone and plate viscometer for shear rates between 0.01 and 1.0 s^{-1}.

Of the blood proteins, fibrinogen is believed to have the most important effect on blood microstructure, [83, 121, 56]. As fibrinogen levels are increased from normal (2.5 – 3.9 g/l) to pathological levels (5 – 10 g/l) the aggregates increase in size and strength, and larger shear rates are needed to break them down into

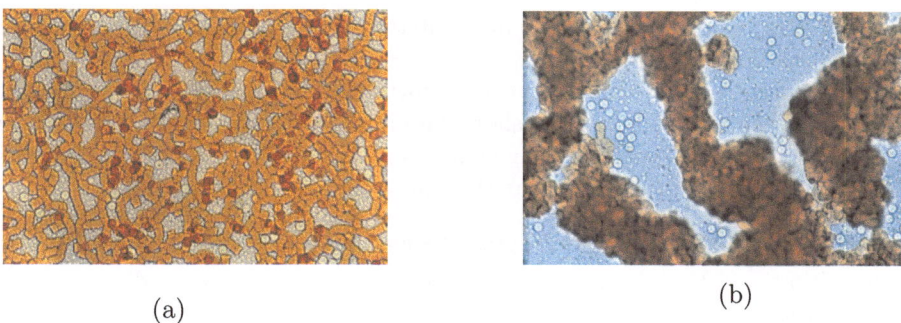

(a)   (b)

FIGURE 3. Three-dimensional microstructure of RBC aggregates
in human blood from (a) a healthy donor and (b) a cardiac patient. Rouleaux formed by rod-shaped stacks of RBCs can clearly
be seen in (a). The isolated darker circles on top of the rouleaux
arise from rouleaux branching off these stacks, forming the three-
dimensional microstructure of RBC aggregates. These branches
are less transparent and therefore darker. In (b), the heavily aggregated RBCs form large three-dimensional clumps. In both figures,
the large light circles are white blood cells while the much smaller
light circles are platelets. Magnification 100X, (images from M.
Kameneva, with permission).

small rouleaux or a single cell suspension [30, 56, 96, 98]. Microrheological studies
presented by Merrill et al. in [86, 85, 84] indicate the process of red blood cell
aggregation and the resulting "apparent" yield shear stress and low shear viscosity
are influenced by additional factors such as temperature and hematocrit level,
red cell shape and deformability and most of all by the presence of other large
molecular weight plasma proteins such as immunoglobulins.

3.2. High shear rate behavior: deformation, tumbling and realignment of erythrocytes

Schmid-Schönbein and Wells used their rheoscope to study the shape and orientation of RBC dispersed (no microstructure) in a viscous liquid as a function of
the shear rate [118, 115]. They found the behavior of a RBC to be similar to that
of a liquid droplet. At very low shear rates of 1 s^{-1}, the RBC were biconcave in
shape and could be seen to tumble. As the shear rate was increased to 2.3 s^{-1}, the
cells began to align with the flow direction and tumble less. At 11.5 s^{-1}, the cells
began to take on the shape of prolate ellipsoids with long axis parallel to the flow
direction and the red cell membrane could be seen to rotate about the internal
liquid, similar to the motion of the tread around the wheel of a tank. Theoretical
studies for this tank-treading motion can be found in [104, 50, 70]. The cells continued to elongate with increasing shear rate. When the shear rate was increased

above 400 s^{-1}, the RBCs deformed into outstretched ellipsoids with major axes parallel to the flow direction.

This process was reversible. At each stepwise reduction of the shear rate, the major axes of the ellipsoids become shorter and progressively the red cells recover their biconcave shape. This increasing elongation, orientation and membrane tank-treading were associated with a shear thinning behavior [28, 113].

3.3. Spatial distribution of erythrocytes in shear flows

In some situations, the distribution of red blood cells across the radius of blood vessels or other conduits is far from uniform due to geometric, gravitational and fluid dynamic effects. This inhomogeneity can lead to some important physiological phenomena and unusual features such as the dynamic hematocrit of blood, plasma skimming and the Fåhraeus–Lindqvist effect, examined later in Section 3.4. Avoiding inhomogeneities in the spatial distribution of RBC in rheometers is one of the biggest challenges in obtaining accurate measurements of the mechanical properties of blood.

3.3.1. Geometric packing effects. From purely a geometric packing point of view, the density of RBC in a tube will be diminished at the wall compared with the bulk of the fluid. This arises because the center of the RBC must be at least half the RBC thickness away from the wall of a tube. This is sometimes called the "excluded volume effect" or "Vand effect". The viscosity near the tube wall is thus diminished due to this decrease in effective RBC density. A discussion of equations used to predict the drop in viscosity due to the excluded volume effect can be found in [20]. For studies under conditions for which a signficant blood microstructure composed of RBC aggregates exists, the characteristic length is determined by the size of the blood microstructure rather than the RBC diameter, increasing these geometric effects.

3.3.2. Particle sedimentation. A source of inhomogeneity which occurs even in the absence of flow is RBC sedimentation. The specific gravity of RBCs is 1.10 and that of plasma is 1.03, resulting in a very slow sedimentation rate for individual RBC on the order of a few mm/hr (page 15 of [16]). However, the sedimentation rate will increase with RBC aggregation and rouleaux formation. This sedimentation rate is affected by the age, gender, pregnancy state, and disease state of the donor [48, 59]. The increased aggregate size associated with some diseases can substantially enhance gravitational effects [59]. In fact, the sedimentation rate is used as a means of monitoring disease states, Section 9.

3.3.3. Radial migration: dilute suspensions. When particles are placed in shear flows, they experience flow-derived lift and drag forces in addition to gravitational forces. These forces will in general not be balanced, leading to migration of the particles. For dilute suspensions of rigid, neutrally buoyant, spherical particles, the equilibrium position was experimentally shown to be 0.6 R from the axis (where R is the tube radius) . This effect is called *Segré–Silberberg effect*, and is covered in

[55] in this volume. Interestingly, the equilibrium radial position was found to be independent of the particle radius, fluid density and fluid viscosity. The nature of the particle distribution will in general depend on the shape and flexibility of the particle as well as the shear rate of the flow. Recent analytical and experimental results for radial migration of rigid particles are discussed in detail in [55]. Palmer and Betts compared the radial migration of fresh and hardened RBC and found the fresh RBC shifted radially much more than the hardened cells [94].

3.4. Outcomes of non-homogeneous distribution of erythrocytes

3.4.1. Fåhraeus effect. Fåhraeus studied the flow of blood from large feeding tubes into long, narrow glass tubes (capillary tubes) of diameters between 0.05 and 1.5 mm. He found that in capillary tubes with diameters below 0.3 mm, the ratio of the tube hematocrit to that in the feeding tube decreased with decreasing diameter, [48, 49]. This effect, known as the Fåhraeus effect or dynamic hematocrit, is attributed to the pronounced axial migration of the erythrocytes. Fåhraeus conjectured that as the RBC migrate toward the center of the capillary tube, their average velocity increases. For the mass flow rate of RBC in the larger feeding tube to be equal to that in the capillary tube, the density of RBC must be lower in the capillary tube, (see [58] for a detailed discussion of this effect and Fåhraeus's contributions in general). This is a separate effect from any change in the tube hematocrit due to entrance or "screening" effects as blood flows from the feeding tube into the capillary tube.

3.4.2. Fåhraeus–Lindqvist effect. Fåhraeus and Lindqvist [49] measured the relationship between flow rate and pressure drop in long tubes with diameters between 0.04 and 0.5 mm (a type of capillary rheometer). The shear rates they considered were significantly larger than $\dot{\gamma}_{max}$, so the effect of the RBC microstructure would be negligible. They found that for tubes of diameters less than 0.3 mm, the relative viscosity appeared to strongly decrease with decreasing radius. For example, the relative viscosity of normal blood was 30% lower in a 0.05 mm diameter capillary tube than in a 1 mm diameter tube, [48]. This effect is commonly referred to as the Fåhraeus–Lindqvist effect, [58, 100]. The blood viscosity is a material property and clearly is not dependent on the size of the device used to test it. Rather, some of the assumptions used in calculating the viscosity from the measured pressure and flow rate render the use of formula (8.4) of the article [107] in this volume invalid. For example, the assumption that the RBCs are homogeneously distributed in the capillary tube is violated. In addition, as noted by Fåhraeus and Lindqvist, the decreased dynamic hematocrit, or the Fåhraeus effect, results in lower values of Ht in the smaller capillary rheometers, which leads to lower viscosities (see also [36]). Fåhraeus and Lindqvist [49] noted this seeming decrease in viscosity would result in an advantageous decrease in flow resistance in arterioles and small veins.

3.4.3. Plasma skimming. As discussed above, geometric packing effects and radial migration of RBC can act to lower the hematocrit adjacent to the vessel wall. In some circumstances, when blood flows from this vessel into small lateral vessels, the fluid entering the smaller vessels will be preferentially drawn from this region of lowered hematocrit. This effect, often referred to as *plasma skimming*, results in a diminished hematocrit in small side branches compared with the parent vessel. See pages 119–122 of [43] for a summary of early work on this subject exploring the impact of various factors on plasma skimming such as branch angle, degree of RBC aggregation, shear rate, and ratio of vessel diameters.

4. Platelet activation and blood coagulation

When a blood vessel is injured, a complex physiological process called *hemostasis* is set into action. Hemostasis literally means the stopping of blood and includes the following steps which are elaborated on below:

- *Vasoconstriction*, whereby the blood vessel diameter is diminished, slowing bleeding.
- *Primary hemostasis*, during which platelets bind to the exposed collagen in the wall, resulting in the formation of a hemostatic plug seconds after the injury occurs.
- *Secondary hemostasis or coagulation*, a complex cascade of activities leading to the strengthening of the platelet plug with fibrin strands and the formation of a clot.
- *Wall repair and clot dissolution*.

4.1. Vasoconstriction

Following endothelial disruption, there is an immediate reaction that causes the smooth muscle in the wall to contract (vasoconstriction), decreasing the vessel diameter and diminishing blood loss. Vasoconstriction slows blood flow, enhancing platelet adhesion and activation.

4.2. Primary hemostasis

Platelets (Section 1.4) play a central role in primary hemostasis. The surface of platelets are coated with glycoproteins, which prevent adherence of the platelets to a healthy endothelium, yet lead to adherence to injured endothelial cells and especially to exposed collagen.

Primary hemostasis begins when organelles contained in the platelet cytoplasm come in contact with collagen exposed by arterial damage. The platelets are then activated and release multiple factors from their cytoplasmic granules. This in turn leads to activation of additional platelets which adhere to the originally activated platelets. This process continues, resulting in a platelet plug and concluding the primary step in hemostasis.

When the concentration of activators exceeds a threshold value, platelet aggregates that are formed by this process can break up, damaging the platelets and causing aggregation at locations other than at the site of damage. Blood platelets can also be activated by prolonged exposure to high or rapid increases in shear stress [37, 73].

4.3. Secondary hemostasis or clot formation

The final hemostatic mechanism is coagulation (clot formation) which involves a complex cascade of enzymatic reactions. Thrombin is the bottom enzyme of the coagulation cascade. Prothrombin activator converts prothrombin to thrombin. The primary role of thrombin is to convert fibrinogen, a blood protein, into polymerized fibrin, stabilizing the adhered platelets and forming a blood clot (or thrombus) [111, 105], Figure 4.

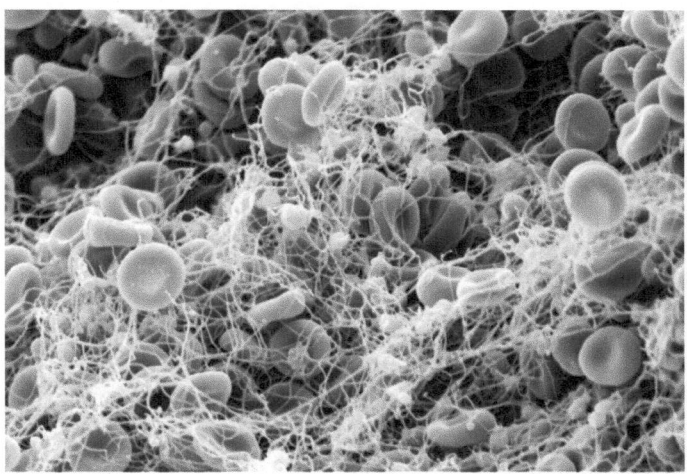

FIGURE 4. SEM image of fibrin binding to RBCs, (Obtained from [92], with permission from the Electron Microscope Unit, University of Cape Town).

4.4. Wall repair and clot dissolution

The clot attracts and stimulates the growth of fibroblasts and smooth muscle cells within the vessel wall, and begins the repair process that ultimately results in fibrinolysis and the dissolution of the clot (clot lysis). Clot dissolution can also occur due to mechanical factors such as high shear stress [106]. In practice a blood clot can be continuously formed and dissolved. Generally, many factors affect its structure, including the concentration of fibrinogen, thrombin, albumin, platelets and red blood cells. At the end of the hemostatic process, normal blood flow conditions are restored. However, some abnormal hemodynamic and biochemical conditions can lead to pathologies such as thromboembolic or bleeding disorders, which are of great clinical importance.

4.5. Models of activation and blood coagulation

The process of platelet activation and blood coagulation is quite complex and not yet well understood. Recent reviews detailing the structure of the blood coagulation system are available, for example in [2, 111]. Numerous experimental studies recognize that thrombus formation rarely occurs in regions of parallel flow, but primarily in regions of stagnation point flows within blood vessel bifurcations, branching and regions of strong curvature. Moreover, internal cardiovascular devices such as prosthetic heart valves, ventricular assist devices and stents, generally harbor high hemodynamic shear stresses that can cause platelet activation and result in coagulation. Thrombotic deposition encountered in these devices is a major cause of their failure.

Reliable phenomenological models that can predict regions of platelet activation and deposition have the potential to help optimize design of internal cardiovascular devices and can also be used to identify regions of the arterial tree susceptible to the formation of blood clots (e.g., [123, 143]). A number of researchers have attempted to tackle the challenging problem of developing models of this kind (e.g., [51, 74, 138, 6]). Anand et al. recently introduced a model for clot formation and lysis in flowing blood that attempts to extend these existing models to integrate more of the biochemical, physiologic and rheological factors [2, 3]. Preliminary 3D numerical results for this model can be found in [12]. There remains a pressing need for further experimental data to validate and develop these models.

5. Special considerations in rheometry of blood

In the remainder of this chapter, we turn attention to the measurement of the mechanical properties of blood and constitutive models to describe this behavior. The three most commonly used rheometers for blood are the concentric cylinder rheometer (Couette rheometer), the cone and plate rheometer, and the capillary rheometer, Figure 12 of the article [107] in this volume. The theory behind these rheometers was previously discussed in Section 8 of [107]. These devices can be used in steady and oscillatory modes to measure the viscosity and linear viscoelastic properties of blood, respectively. In the cone and plate rheometer, it is also possible to measure the first normal stress coefficient, though to date this has not been successfully performed for blood.

We should expect these rheometers to provide mechanical properties of whole blood that are in agreement with each other over a wide range of shear rates. In practice, the range of each rheometer is limited, so that more than one rheometer is used when blood properties are needed over a wide range of shear rates (e.g., $0.01 - 500$ s^{-1}). To aid the reader in a rational interpretation of experimental data in the literature, we first discuss some of the central challenges in blood rheometry.

5.1. Inhomogeneous distribution of particles

The application of all three rheometers to blood is nearly always based on the assumption that the blood is a single-phase, homogeneous material. A necessary condition for modeling blood as a homogeneous continuum is that the smallest length scale of the device is large compared to the largest length scale of the RBC (or the characteristic length of the RBC aggregate). However, like other suspensions, this condition is not a guarantee of homogeneity. The cells may be non-uniformly distributed due to the geometric, gravitational and fluid-dynamical mechanisms discussed in Section 3.3. These factors are in turn dependent on the aggregate size which is strongly influenced by the concentration of fibrinogen and high molecular weight globulins, hematocrit level and shear rate [97]. As a result, measurement of the mechanical properties of blood is in general more difficult at low shear rates, where aggregates are found (in blood from healthy donors, less than about 50 s^{-1}). These challenges are heightened by physiological conditions that lead to increased size of RBC aggregates, for example, pathological levels of fibrinogen.

RBC migration limits the smallest tube diameter that can be used in capillary rheometers. When the tube diameter is below approximately 300 μm, the RBC migration discussed in Section 3.3 leads to a layer near the tube wall that is nearly deplete of RBCs (e.g., [49]). This inhomogeneous distribution of RBC is outside the scope of the equations used to obtain viscosity from capillary rheometers, Section 8 of [107]. As discussed in Section 3.4, an apparent reduction in viscosity will result as well as a dependence of viscosity on capillary diameter.

Sedimentation can also generate an inhomogeneous distribution of RBC in the reservoirs of capillary rheometers, which will then alter the concentration of RBC in the capillary tube [49, 84]. Sedimentation is more of a problem in horizontal tube viscometers and can be diminished by premixing the blood and using wider capillaries to reduce measurement time.

In applications of cone and plate rheometers to suspensions such as blood, a truncated cone is required, chosen such that the gap between the cone tip and plate exceeds the characteristic length of particles in the suspension by a factor of ten [91]. In addition, sedimentation can generate a RBC free layer adjacent to the upper plate [84], which will lead to an under prediction of the viscosity. The lift on particles due to fluid-dynamic mechanisms can diminish the importance of gravitational effects at higher flow rates. Therefore, one method used to mitigate sedimentation is to mix blood prior to testing. Sedimentation is more pronounced at shear rates where RBC aggregation is stronger and hemodynamic effects are lessened.

Couette viscometers are not as sensitive to RBC sedimentation as cone and plate rheometers [82, 97]. However, in constant shear rate experiments below 1 – 10 s^{-1}, a non-monotonic torque transient is observed [34]. In particular, after initiation of the constant shear, the torque is seen to reach a maximum value and then decays to a steady-state value. A controversy remains over which torque to use for blood viscosity calculations. Some authors use the torque obtained by back extrapolation to zero time, while others use the final steady-state value [34, 1]. The torque decay has been lessened by roughening the walls of the cylinders [34, 84, 97]. The source of this torque transient is believed to arise from an observed non-homogeneous distribution of RBC in the radial direction [34, 84]. This distribution has been attributed to geometric hindrance [34] and/or radial migration [97].

5.2. Thixotropy

In blood, thixotropic effects are due to the finite time required for the formation and breakdown of the 3D microstructure formed by the RBC aggregates [1]. This lag time can play an important role in blood rheometry. For example, if blood leaves the reservoir of a capillary rheometer in a state of lower/higher microstructure than the expected equilibrium value, the fluid will have to flow a finite distance down the tube before the microstructure associated with the imposed shear rate is formed. If this distance is large enough, the entrance effects in the tube will be significant and the viscosity will be substantially different from the equilibrium value. While entrance effects are not an issue in rotational devices such as the cone and plate and Couette viscometers, thixotropy can affect rheometry in other ways [22]. For example, it seems likely that thixotropic effects play an important role in the torque decay discussed previously.

5.3. Biochemical effects

Outside the body, blood has a tendency to coagulate after a few minutes. Various additives are effective at preventing coagulation, though possibly at the expense of altering the mechanical properties of blood from its native state. Merrill and co-workers found the rheological properties of blood were similar between native blood and blood with heparin added, as long as the tests were performed within eight hours of withdrawal [86]. Similarly, Copley et al. [39] found viscosity was insensitive to the addition of heparin and EDTA over a wide range of shear rates $(0.0009 - 1000 \text{ s}^{-1})$.

5.4. Other considerations

Another consideration in the selection of rheometers is the need to assume a particular constitutive model for blood to interpret the data. A major advantage of both the Couette rheometer and the cone and plate rheometer is that under certain assumptions (respectively, those of small gap and small angle) the shear rate is constant throughout the fluid domain. The viscosity can then be determined without an a-priori selection of a blood constitutive model, (see Section 8 in [107] of this volume). For example, for a narrow gap Couette rheometer in which the inner cylinder is rotated and the outer cylinder is held fixed, it is reasonable to use the following relationship to calculate the fluid viscosity irrespective of the specific fluid,

$$\eta = \frac{M}{4\pi R_i^2 L \, \Omega_i} \left(1 - \frac{R_i^2}{R_0^2} \right) \quad \text{for narrow gaps } R_i/R_0 \geq 0.99, \qquad (5.1)$$

where R_i and R_o are the radii of the inner and outer cylinder, respectively, L is the cylinder length, M is the magnitude of the torque applied to the inner cylinder and Ω_i is the rotation rate of the inner cylinder. For a small angle cone and plate

---

[1] A general discussion of thixotropy is covered in Section 6 in [107] of this volume and the references cited therein.

viscometer, the viscosity can be calculated from

$$\eta = \frac{3M}{2\pi R^3} \frac{\beta}{\Omega} \quad \text{for small angles } \beta < 0.10 \text{ rad}, \tag{5.2}$$

where R is the cone radius, β is the angle between the cone and plate and M is the magnitude of the applied torque. Equations (5.1) and (5.2) are discussed in more detail in [107] of this volume.

When the RBC aggregate size is elevated, it may not be possible to use narrow gap rheometers. In practice, the relation (5.1) is sometimes used to calculate an approximate viscosity in rheometers with wider gaps. In other cases, an approximate viscosity is calculated using the finite-gap formula for Newtonian fluids given in (8.1) of [107] even though the blood displays a shear thinning viscosity.

When selecting a rheometer, an additional consideration is the minimum sample size, particularly for human newborns or small animal studies. For example, the entire blood volume of a rat can be less than 20 – 25 ml. An advantage of the cone and plate rheometer is the relatively small sample size (about 0.5 – 1.0 ml).

6. Viscosity of whole blood

The most well studied non-Newtonian characteristic of blood is its diminishing viscosity with increasing shear rate. This shear thinning behavior is the subject of this section. For clarity, we begin with nomenclature for blood viscosity. We then turn attention to a discussion of experimental data for the equilibrium measurements of blood viscosity as a function of shear rate. Next, the relationship between blood microstructure and the qualitative form of this viscosity dependence on shear rate is discussed. Following this, the significance of shear thinning in the circulatory system is considered. As a result of the time necessary for the 3D microstructure of blood to form, the strong shear thinning behavior measured for normal blood *in-vitro* does not play a role in much of the circulatory system. We then turn attention to specific functional forms used to describe blood shear thinning. Finally, the dependence of blood viscosity on other factors such as hematocrit level and temperature is considered.

6.1. Nomenclature for blood viscosity

It should be emphasized that the nomenclature for viscosity is not used uniformly within the literature on blood rheology. For clarity, we begin this section with comments on this nomenclature and a statement of the definitions that will be used in this chapter.

Viscosity. When we refer to blood viscosity, we mean the viscosity function defined in (7.4) of the article [107] in this volume. We will sometimes call this the *material viscosity* to emphasize that it is a material property. Different rheometers should predict the same value for this quantity if they are used under the same conditions (e.g., temperature and shear rate).

Apparent Viscosity. There is some variability in the use of the term apparent viscosity. We define apparent viscosity as the approximate viscosity obtained using relations between viscosity and measured properties that are strictly only valid for constant viscosity fluids, (e.g., Section 5.2 of Fung [53] and page 13 of [20]). For example, for capillary rheometers, the apparent viscosity is calculated from the data using (8.4) of [107], whereas the material viscosity is calculated using (8.5) of [107]. As will be discussed shortly, the viscosity of blood from healthy human donors at shear rates greater than 200 s^{-1} is approximately constant. At these higher shear rates, the blood microstructure formed by aggregates of RBCs in the test region of the capillary rheometer is likely insignificant, even though the shear rate varies from zero to a maximum across the capillary tube. If this is the case, the apparent viscosity will be close to the material viscosity at these higher shear rates. In flow regimes where the non-constant viscosity is important, the apparent viscosity is not an intrinsic material property of the fluid. In other works, the term apparent viscosity is used for the material viscosity defined above and what we are referring to as the apparent viscosity is called the *effective viscosity* (e.g., [36]).

Relative viscosity is the ratio of the suspension viscosity to the viscosity of the suspending fluid alone or some other reference viscosity. In the hemorheology literature, it is defined as the ratio of either the material or apparent viscosity of whole blood to that of either plasma or water at the same temperature. Sometimes the terminology *relative apparent viscosity* is used to emphasize the relative viscosity is the ratio of apparent viscosities.

6.1.1. Units of viscosity.

While much of the scientific and engineering literature makes use of SI units, important works on blood rheology use the CGS unit of centipoise (cP). The poise was named after Jean Louis Marie Poiseuille and is equivalent to 1 g/(cm · s). A centipoise is just 1/100 of poise. A common SI unit of viscosity is the Pascal second: 1 Pa · s = 1 kg/(m · s), where

$$1 \text{ cP} = 10^{-3} \text{ Pa} \cdot \text{s} = 1 \text{ mPa} \cdot \text{s}. \tag{6.1}$$

6.2. Experimental data for whole blood viscosity

Shown in Figure 5 are representative data for the apparent viscosity as a function of the shear rate for whole human blood with a hematocrit of 40% at 23° C. The data were obtained using a Couette viscometer at shear rates from 0.06 to 128 s^{-1} and with a capillary viscometer at higher shear rates.

As discussed above, experiments on blood at low shear rates are exceedingly difficult to perform. As a result, there remains a controversy over the behavior of blood in the limit of shear rate tending to zero. In the absence of a yield stress, the viscosity would tend to a finite value, denoted here by η_o. However, the concept of a yield stress remains controversial, even for simpler suspensions [7]. We will delay further discussion of this issue until Section 7.

The behavior of the viscosity for shear rates on the order of 1 s^{-1} and larger are less controversial. As the shear rate is increased above this range, there is a steep decrease in viscosity until a plateau in viscosity is apparently reached,

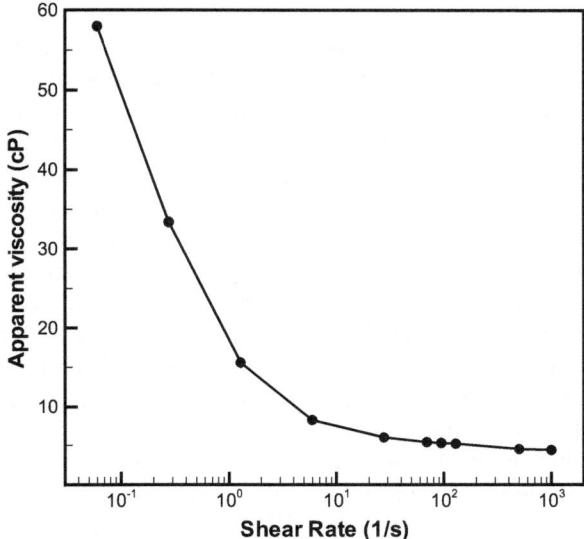

FIGURE 5. Apparent viscosity as a function of the shear rate for whole blood obtained from a 25 year old female donor with Ht = 40%, $T = 23°C$. Obtained using a Contraves LS30 (Couette) viscometer at shear rates of $\dot{\gamma} \in [0.06, 128]$ s^{-1}) and a Cannon–Manning Semi-Micro (capillary) viscometer, (Cannon Instrument Co.) at shear rates of $\dot{\gamma} \in [300, 1000]$ s^{-1} (unpublished data from M. Kameneva, with permission).

Figure 5. This plateau value is often referred to as the asymptotic blood viscosity and denoted by η_∞.

While the definitions for η_0 and η_∞ are clear from a mathematical perspective,

$$\eta_0 = \lim_{\dot{\gamma} \to 0} \eta(\dot{\gamma}), \quad \eta_\infty = \lim_{\dot{\gamma} \to \infty} \eta(\dot{\gamma}), \tag{6.2}$$

in practice, η_o can only be approximated from experimental data and the definition of η_∞ is only a mathematical construct. The maximum value of viscosity that can be measured is limited by the challenges of measuring blood viscosity at low shear rates. The definition for η_∞ given in (6.2) is a mathematical convenience since at shear rates on the order of $10^4 - 10^5$ s^{-1} the RBC are lysed, [76]. In practice, this is not important since a representative plateau value is easily measured if the blood sample is properly anticoagulated, tested as soon as possible after withdrawal, well mixed and well defined (hematocrit and temperature). As will be described below, most generalized Newtonian constitutive viscosity models for blood include both these constants. They can be determined as part of the nonlinear regression analysis used to determine the material constants from the experimental data. Alternatively, η_∞ can be obtained from measurements of the plateau viscosity at a sufficiently high shear rate.

Classical references on blood viscosity include [34, 30, 84, 39, 63]. Cho and Kensey [33] provide a compilation of viscosity data for blood at 37°C and fit nine generalized Newtonian models to the data. However, these data include both human and canine blood and the generalized Newtonian models are fit to a combined data set with a relatively wide distribution of hematocrits (33 – 45%). Unfortunately, some references for the source of the data are incomplete. More recently, Picart et al. provided shear stress versus shear rate data for shear rates as low as 10^{-3} s^{-1} and up to 10 s^{-1} using a Couette-type rheometer with roughened walls. Separate data sets were provided for hematocrits of 54, 66 and 74% [97, 98].

Chien demonstrated that RBC aggregation is necessary to attain the strong shear thinning effect shown in Figure 5, [28]. He compared the relative viscosities of human RBCs in three solutions: normal RBCs in heparinized plasma (NP), normal RBCs in 11% albumin Ringer solution (NA) and hardened RBCs in 11% albumin Ringer solution (HA). All solutions were adjusted to the same hematocrit (45%). The 11% albumin solution was adjusted to have the same viscosity (1.2 mPa · s) as the plasma but did not cause RBC aggregation.

In the NP suspension, the RBCs could aggregate, deform and align. The associated viscosity dropped by a factor of approximately 45 as the shear decreased from 0.01 to 500 s^{-1}, Figure 6. In the NA suspension, the RBCs could deform and align but could not form aggregates. In this case, the drop in viscosity was only a factor of 3.5 over the same range of shear rates. Consistent with these latter results, Chmiel and Walitza (page 97 of [31]) reported a three-fold shear thinning effect over the same range of shear rates for normal RBC under conditions of non-aggregation (RBC in an isotonic salt solution). Chmiel and Walitza used a 50% by volume suspension.

It is also interesting to note that the NP and NA curves are indistinguishable for shear rates higher than approximately 6 s^{-1}. This value is in the low end of the range commonly reported for $\dot{\gamma}_{max}$. Presumably at higher shear rates, further drop in viscosity is due largely to deformation and alignment of individual RBCs. For normal RBCs in plasma, the majority of the drop in viscosity (approximately 95%) occurs in this low shear range ($\dot{\gamma} \leq [0.01, 6]$ s^{-1}) where aggregation is playing a role. Comparing the curves with and without aggregation, it can be seen that this drop is diminished by 92% in the absence of aggregation.

6.3. Significance of shear thinning in the circulatory system

Often, viscosity functions are fit to data such as that in Figure 5 and used in computational fluid-dynamic simulations of flow in the vasculature. However, it should be emphasized that each of the data points in the viscosity curve represents an *equilibrium viscosity*, obtained after the viscometer was run at a fixed shear rate for several minutes at low shear rates and 30 – 40 seconds at higher shear rates. Since the time for a drop of dye in the circulation to make its way through the

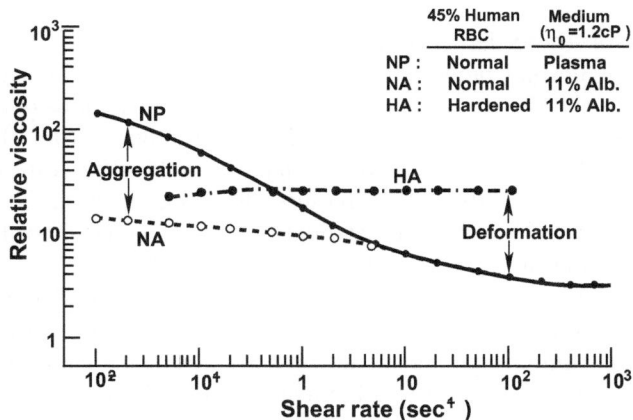

FIGURE 6. Variation in relative viscosity as a function of the shear rate for three types of RBC suspensions at 37°C: normal RBCs in heparinized plasma (NP), normal RBC in 11% albumin Ringer solution (NA) and hardened RBCs in 11% albumin Ringer solution (HA). The solutions were all adjusted to a RBC volume of 45%. The plasma and albumin Ringer solution both had a viscosity of 1.2 mPa · s. From [28], (reprinted with permission from AAAS).

entire circulatory system has been estimated to be on the order of a minute, it is therefore important to consider under what circumstances shear thinning will play a role in the circulation.

Cokelet (see [35], pages 144–148), summarized results for several important time constants for human blood including: aggregate formation, aggregate dis-aggregation and the recovery time for RBC deformation. Schmid-Schönbein and co-workers studied the kinetics of RBC aggregation for normal and pathological blood samples and found the half-time for aggregate formation in blood to be $3 - 5$ seconds for normal blood and $0.5 - 1.5$ seconds for pathological blood samples, [114]. The aggregation time was measured for samples of blood in which the shear rate was dropped abruptly from 460 s^{-1} to approximately zero. Experiments were run for blood with Ht = 45% at a temperature of 37°C. On the other hand, disaggregation is expected to be much more rapid. Based on results from micropipette-derived deformations of RBC, the half-time for a mechanically deformed RBC to relax to half its initial stretch is estimated to be on the order of 0.06 seconds (see, e.g., [35], page 146). More recently, Thurston has estimated an aggregation time on the order of a minute after the sudden cessation of oscillatory shear at a shear rate of 500 s^{-1} and shear strain of 1.77 (RMS). The disaggregation and alignment characteristic of the high shear state was found to take only a few seconds [133].

If it takes on the order of seconds to minutes for the 3D structure of RBC to form, then for normal blood this structure will only exist in segments of the circulation where there is a stable recirculation regime or regions of stagnant flow with shear rates significantly lower than 1 s^{-1}. It follows from the discussion above, that 95% of the drop in viscosity occurs over the shear rate range of 0.01 to 6 s^{-1} and of this drop about 92% requires RBC aggregation. Therefore, with few exceptions, it is incorrect to use strongly shear thinning viscosity models for studies of the circulatory system in healthy patients. Either a constant viscosity model should be employed or a much weaker shear thinning model appropriate for individual RBCs in plasma should be used (e.g., the NA curve in Figure 6). Even this weaker shear thinning model should be used with some caution since part of this shear thinning effect is due to alignment of the RBC in the flow which may happen over longer time frames than experienced *in-vivo* and have varying effects in the various vessel geometries.

With some exceptions, the reader should therefore be discouraged from evaluating whether non-Newtonian features of normal blood are important by comparing hemodynamic results using Newtonian models with those obtained using shear thinning models fit to data such as that in Figure 5. A more relevant question is to ask whether this microstructure will exist in the region of interest. Namely, has the flow experienced low shear for a sufficient time for the 3D aggregate structure to form?

For normal blood, this aggregate structure may have time to form in regions of stable recirculation. For example, shear thinning may be important in a stable vortex downstream of a stenosis, in a stable vortex inside a saccular aneurysm, or in some anastomoses of the cerebral vasculature. In some regions of the arterial system, the flow is nearly stagnant for extended periods of time, for example in parts of the venous system, and so it is likely the aggregate structure can form in these regions as well [11, 59, 99, 124]. In patients with pathological conditions that increase the strength of the RBC aggregates, shear thinning may be significant in extensive regions of the circulation. Some examples of these conditions are given in Section 9 of this chapter.

In addition to modeling flow in the circulatory system, there are other valuable reasons to quantify the dependence of blood viscosity on shear rate. The viscosity function can be used to define metrics such as the low shear viscosity, the asymptotic or high shear viscosity, as well as the the steepness and location of the transition between these two regions. As elaborated on in Section 9, these viscosity metrics can then be used to quantify blood properties such as aggregation strength and RBC deformability. Abnormal values of these metrics are in turn associated with certain disease states.

6.4. Viscosity models for blood

In this section, we discuss shear thinning models for blood viscosity. Yield stress models are discussed in Section 7.

6.4.1. Constant viscosity models.
As just discussed, the shear thinning nature of normal blood is expected to play a minor role in the majority of the arterial circulation. For this reason, the blood viscosity can often be approximated by the *constant* infinite shear viscosity, η_∞ or possibly a *constant* intermediate shear rate for which the aggregates of RBC are completely dispersed.

6.4.2. Generalized Newtonian models.
Using a nonlinear regression analysis, viscosity functions of the form $\eta(\dot\gamma)$ are often fit to whole blood viscosity data such as that shown in Figure 5.

The power-law model has frequently been used for blood viscosity,

$$\textbf{Power-Law Model} \qquad \eta(\dot\gamma) = K\,\dot\gamma^{(n-1)}, \tag{6.3}$$

where n and K are termed the power-law index and consistency, respectively, and n is chosen as less than one to reflect the shear thinning properties of blood [2]. The shear thinning power-law model predicts an unbounded value for η_0 and zero viscosity as the shear rate tends to infinity ($\eta_\infty = 0$). While the behavior of blood in the limit of zero shear rate is still a subject of debate, the high shear asymptotic behavior is unphysical and limits the range of shear rates over which the power-law model is reasonable for blood. Despite this limitation, the power-law model is frequently used due to the number of analytical solutions which can be obtained (e.g., Eq. (7.15) of [107] in this volume).

Most other viscosity functions for blood have a finite value for both η_o and η_∞ and can be written in the form

$$\eta = \eta_\infty + (\eta_o - \eta_\infty)f(\dot\gamma), \tag{6.4}$$

or, in non-dimensional form as

$$\frac{\eta - \eta_\infty}{\eta_o - \eta_\infty} = f(\dot\gamma). \tag{6.5}$$

The Carreau–Yasuda model contains several other models as special cases, so we briefly comment on this model here. The viscosity function for this model is

$$\frac{\eta - \eta_\infty}{\eta_o - \eta_\infty} = \frac{1}{[1 + (\lambda\dot\gamma)^a]^{(1-n)/a}}, \tag{6.6}$$

where $n < 1$ for a shear thinning model. For small values of $(\lambda\dot\gamma)^a$ (the zero-shear-rate region), the viscosity tends to a plateau of constant η_o. In the limit of large $(\lambda\dot\gamma)^a$ (the power-law region), this model tends to a power-law-type model with non-zero η_∞. For this reason, n is referred to as the "power-law exponent". The

---

[2]In some publications $\dot\gamma$ is used to represent the shear rate in simple viscometric flows such as simple shear. It should be recalled from formula (3.9) of [107] that $\dot\gamma^2 = 2\text{tr}\,(\mathbf{D}^2) = -4II_D$ where $\mathbf{D}$ is the symmetric part of the velocity gradient. Namely, in this chapter and in [107], $\dot\gamma$ is a scalar invariant that has meaning in any flow field.

value of a therefore determines the size of the transition region between the zero-shear-rate and power-law regions. In the special case of a Carreau model, $a = 2$, while in the Cross-model a $= 1-$n. In the Simplified Cross, a is set equal to 1 and n is set to zero.

Examples of $f(\dot{\gamma})$ used to model blood viscosity are given in Table 5 along with material constants obtained from a nonlinear least squares analysis of the data presented in Figure 5. In each case, the values of η_o and η_∞ were obtained along with the other material constants as part of the regression analysis. The corresponding viscosity functions are shown with the original data in Figure 7. The power-law constants were obtained in a similar manner for these data, and are

$$\textbf{Power-Law model} \qquad \text{n}= 0.628, \; K= 20.2 \, \text{mPa} \cdot \text{s}^{\text{n}}. \qquad (6.7)$$

The R^2 values for the analysis were calculated from

$$R^2 \;=\; \sum_{i=1}^{n} \left[(\eta^i - \eta^i_{fit})^2 \right] / \sum_{i=1}^{n} \left[(\eta^i - \eta^i_{mean})^2 \right], \qquad (6.8)$$

where n is the total number of data points, η is the measured viscosity, η_{fit} is the viscosity obtained from the functional approximation, and η_{mean} is the average of the measured viscosity values. All models except the power-law and Powell–Eyring had excellent fits, with R^2 values greater than 0.998. The power-law model shows a large deviation with the data at larger shear rates, Figure 7. This is expected since $\eta_\infty = 0$ for this model. The R^2 values, for the power-law and Powell–Eyring are 0.987 and 0.753, respectively.

6.5. Dependence of blood viscosity on factors other than shear rate

Blood viscosity is quite sensitive to a number of factors besides shear rate. These include physical factors such as (i) hematocrit and concentrations of other blood components, (ii) temperature, and (iii) plasma viscosity and composition, as well as physiological factors such as (i) gender, (ii) disease state, (iii) natural age of RBCs, and (iv) exercise level. Some of these factors will be discussed below. Due to the number of factors which influence viscosity, care must be taken in interpreting rheological data and selecting blood parameters.

6.5.1. Effect of hematocrit and blood components on viscosity.
As can be seen in Figure 8, blood viscosity increases dramatically as the hematocrit increases (see also [27, 30, 122, 135]). These data were obtained for human blood at a fixed shear rate of $128 \, \text{s}^{-1}$ and the hematocrit level was controlled by dilution of blood with autologous plasma.

Brooks et al. found a strong dependence of the shear thinning properties of blood on hematocrit [14] in their study of human RBC suspensions in ACD-plasma, Figure 9. Interestingly, the shear thinning behavior is no longer discernible below a critical hematocrit (Ht between 13% and 30%).

Model	$\dfrac{\eta - \eta_\infty}{\eta_0 - \eta_\infty}$	Material Constants
Powell–Eyring	$\sinh^{-1}(\lambda\dot\gamma)$	$\eta_0 = 60.2\,\text{mPa·s}$, $\eta_\infty = 64.9\,\text{mPa·s}$, $\lambda = 1206.5$ s,
Modified Powell–Eyring	$\dfrac{\ln(1 + \lambda\dot\gamma)}{(\lambda\dot\gamma)^m}$	$\eta_0 = 57.46\,\text{mPa·s}$, $\eta_\infty = 4.93\,\text{mPa·s}$, $\lambda = 5.97$ s, $m = 1.16$
Simplified Cross	$\dfrac{1}{1 + \lambda\dot\gamma}$	$\eta_0 = 73.0\,\text{mPa·s}$, $\eta_\infty = 5.18\,\text{mPa·s}$, $\lambda = 4.84$ s
Cross Model	$\dfrac{1}{1 + (\lambda\dot\gamma)^m}$	$\eta_0 = 87.5\,\text{mPa·s}$, $\eta_\infty = 4.70\,\text{mPa·s}$, $\lambda = 8.00$ s, $m = 0.801$
Carreau	$\dfrac{1}{[1 + (\lambda\dot\gamma)^2]^{(1-n)/2}}$	$\eta_0 = 63.9\,\text{mPa·s}$, $\eta_\infty = 4.45\,\text{mPa·s}$, $\lambda = 10.3$ s, $n = 0.350$
Carreau–Yasuda	$\dfrac{1}{[1 + (\lambda\dot\gamma)^a]^{(1-n)/a}}$	$\eta_0 = 65.7\,\text{mPa·s}$, $\eta_\infty = 4.47\,\text{mPa·s}$, $\lambda = 10.4$ s, $n = 0.34$, $a = 1.76$

TABLE 5. Representative generalized Newtonian models for blood viscosity with corresponding material constants. Constants were obtained using a nonlinear regression analysis of experimental data shown in Figure 5. Data are from blood of a 25 year old female donor with Ht $= 40\%$, $T = 23°\text{C}$. Obtained using a Contraves LS30 (Couette) viscometer ($\dot\gamma \in [0.06, 128]\ \text{s}^{-1}$) and a Cannon–Manning Semi-Micro (capillary) viscometer, (Cannon Instrument Co.) at $\dot\gamma \in [300, 1000]\ \text{s}^{-1}$ (unpublished data from M. Kameneva, with permission). The power-law constants for the same data are, $n = 0.628$, $K = 20.2\ \text{mPa} \cdot \text{s}^n$. Note that 1 cP $= 1$ mPa·s

The strong nonlinear increase in viscosity with hematocrit at low shear rates is thought to be due to an increase in rouleaux density, length, and cell-cell interaction with increasing RBC concentration [17].

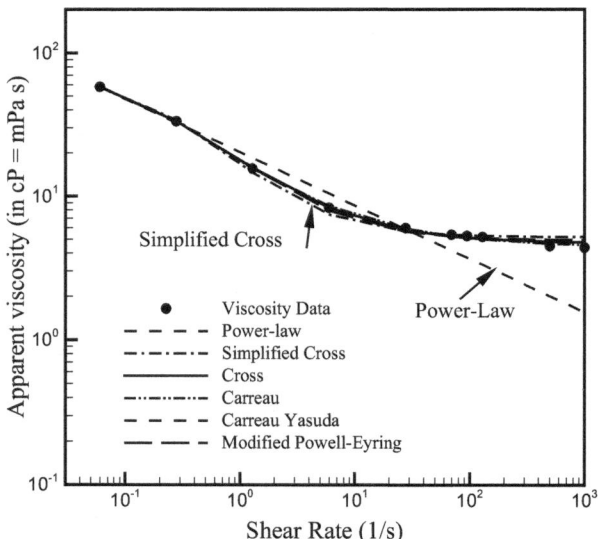

FIGURE 7. Apparent blood viscosity as a function of shear rate. The viscosity data is that shown in Figure 5. Also shown are curves for five generalized Newtonian constitutive models fit to this same data set. The definitions of viscosity functions and associated material constants are given in Table 5.

Walburn and Schneck [137] extended the power-law model, Eq. (3.13) of [107]. They evaluated the relative importance of shear rate, hematocrit, albumin, total lipid, and TPMA (total protein minus albumin) on blood viscosity using a cone and plate rheometer. The material constants in the original power-law model were generalized to include a dependence on various combinations of these five parameters. For a one-variable model, the shear rate is the most statistically significant variable while for a two-variable model, shear rate and hematocrit were statistically the most important variables. The two-variable power-law model they proposed is

$$\eta(\dot\gamma) = K\,\dot\gamma^{(n-1)} \qquad \begin{matrix} K = C_1 \exp(C_2\,Ht) \\ n = C_4 - C_3\,Ht \end{matrix} \right\} \quad \begin{matrix} \text{2-parameter} \\ \text{W-S Model,} \end{matrix} \qquad (6.9)$$

where $\dot\gamma \in [23.28, 232.80]\mathrm{s}^{-1}$, $Ht \in [35, 50]\%$, $T = 37°C$. Here, Ht is given as a percentage, (e.g., Ht=45, when the hematocrit is 45%). Walburn and Schneck [137] reported values of the constants as

$$[C_1, C_2, C_3, C_4] = [1.48\ \mathrm{mPa \cdot s}^n,\ 0.0512,\ 0.00499,\ 1.00], \qquad (6.10)$$

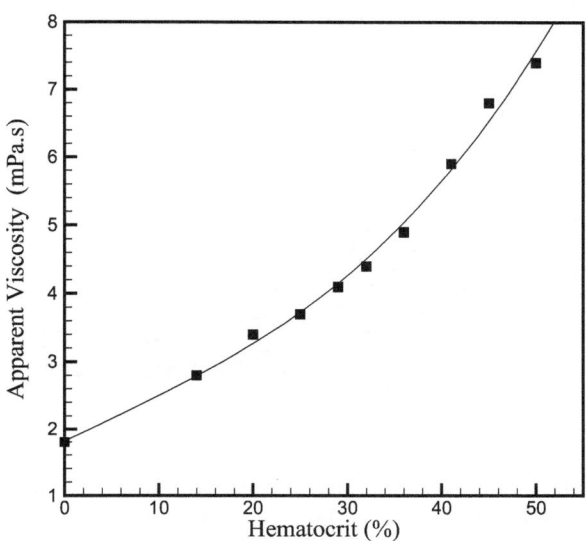

FIGURE 8. Relationship between blood viscosity and hematocrit for human blood diluted with autologous plasma at 21°C and a shear rate $\dot{\gamma} = 128$ s^{-1}. Data obtained using a Contraves LS30 (Couette) viscometer, (unpublished data from M.V. Kameneva, with permission).

(see [109] for other constants for this model). Walburn and Schneck noted this model was previously introduced by Sacks [110].

Walburn and Schneck [137] found the three most important parameters for a 3-constant model are shear rate, Ht and TPMA levels in decreasing order of significance and developed the following model,

$$\eta(\dot{\gamma}) = K\,\dot{\gamma}^{(n-1)} \quad \begin{array}{l} K = D_1 \exp((D_2 Ht)) \exp((D_4 TPMA/Ht^2)) \\ n = 1 - D_3\,Ht \end{array} \left.\begin{array}{l} \\ \\ \end{array}\right\} \begin{array}{l} \text{3-parameter} \\ \text{W-S Model,} \end{array}$$

(6.11)

where $[D_1, D_2, D_3, D_4] = [7.97$ mPa $\cdot$ s^n, 0.0608, 0.00499, 145.85 dl/g].

Interestingly, the normal hematocrit varies substantially across species. For example, it is 27% in camel, 32% in a sheep, 33% in a goat, and 46% in dog [125]. In turn, the dependence of viscosity on hematocrit varies across these species, likely due to differences in factors such as shape, size and flexibility of the RBCs [125]. For example, the RBCs of camels and llamas are elliptical in contrast to the biconcave disk shape found in many other mammals. Stone et al. [125] made the interesting observation that the hematocrit of animals appears to be in some sense

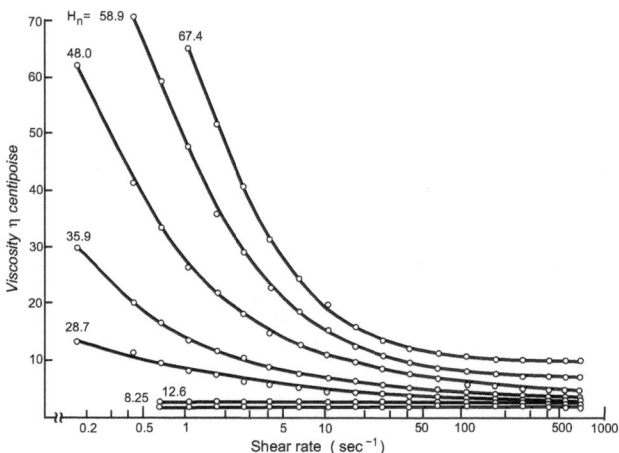

FIGURE 9. Relationship between human blood viscosity and shear rate for RBCs suspended in ACD-plasma (acid-citrate-dextrose anticoagulant) at 25°C for various volume concentration of RBC defined through H_n. Here, H_n is defined as the hematocrit times 0.96. Viscosity obtained using a Couette rheometer (reproduced from [14], with permission from the American Physiological Society).

optimal for transporting oxygen for a given dependence of viscosity on shear rate in each animal. Namely, for a fixed pressure drop, increasing the RBC concentration will increase the oxygen transport rate up to a critical level, beyond which further increases in hematocrit actually diminish oxygen transport due to the cost of transport at increased viscosity. They evaluated the hematocrit in a number of species and found that it falls near the optimal level. Similar results were found for blood from human females (but not males) in a study of blood from 47 pre-menopausal women and 51 age-matched men [66].

6.5.2. Effect of temperature on blood viscosity. Like many other liquids, the viscosities of both plasma and whole blood are strongly dependent on temperature. For example, when the temperature of blood with Ht = 40% is decreased from body temperature (37°C) to room temperature (22°C), the viscosity at a shear rate of 212 s^{-1} increases from 3.8 mPa·s to 6.3 mPa·s, an increase of 66% (n $\geq$ 10) [102]. Under the same temperature drop, plasma viscosity of males age 44 – 45 increased more than 45%, from 1.2 to 1.76 mPa·s, (n=125), Table 6.

Merrill et al. found the dependence of the ratio of whole blood viscosity to water viscosity to be relatively insensitive to temperature for T $\in [10, 40]°$ C and $\dot{\gamma} \in [1, 100]$s^{-1}, [86]. As a result, blood viscosity is often reported relative to the

	n	Plasma Viscosity (in cP = mPa · s)	Temperature
Males (age 44 – 45), [44]	125	1.760 ± 0.134	20°C
	125	1.229 ± 0.086	37°C
	125	1.150 ± 0.076	40°C
Male (age 22 – 37), [44] athletes	9	1.630 ± 0.045	20°C
	9	1.183 ± 0.021	37°C
	9	1.132 ± 0.019	40°C

TABLE 6. Plasma viscosity for men at different temperatures and ages. Ten males in the 44 – 45 age group showed cardiovascular disorders within one year of the test [44]. For comparison, the viscosity of water is 1.01 mPa · s and 0.69 mPa · s at 20°C and 37°C, respectively.

viscosity of water at the same temperature. Merrill at al. proposed an Arrhenius type relationship for blood with the same activation energy E_a (cal/mol) as used for water [86], which can be written in the following form,

$$\eta(\dot{\gamma}, T) = \eta(\dot{\gamma}, T_0) \exp(-\frac{E_a}{R}(\frac{1}{T} - \frac{1}{T_0})) \quad \begin{cases} T \in [10, 40]°\text{C} \\ \dot{\gamma} \in [1, 100]\text{s}^{-1}, \end{cases} \quad (6.12)$$

where R is the gas constant, T is the temperature (°K), and T_0 is a chosen reference temperature (°K) at which the dependence of blood viscosity on shear rate, $\eta(\dot{\gamma}, T_0)$, is known. Stoltz et al. use $E_a/R = 2.01 \pm 0.03 \times 10^3$ ° K (page 20 of [124]).

The relationship (6.12) can be used to normalize blood viscosity data obtained at different temperatures. If the temperature dependence of blood is modeled through (6.12), then if follows from (6.4), with the exception of η_o and η_∞, that the material constants for a generalized Newtonian fluid will be independent of temperature. If, as suggested by Merrill, we use the value of E_a (cal/mol) for water, we can easily determine the viscosity at different temperatures through

$$\eta(\dot{\gamma}, T) = \eta(\dot{\gamma}, T_0) \frac{\eta_{H_2O}(T)}{\eta_{H_2O}(T_0)}. \quad (6.13)$$

For example, using (6.13), the high and low shear viscosities for the Carreau–Yasuda model at 37° C can easily be obtained from those given in Table 5 at 20° C and are shown in Table 7.

Though the temperature of the human body is typically carefully maintained at approximately 37°C, the extremities of the body can drop substantially below this. Some medical conditions such as Raynaud's syndrome and medical treatments such as induced hypothermia during cardio-pulmonary bypass [69] involve lowered blood temperature in-vivo. The relation (6.13) is useful for determining appropriate blood viscosities for studies under these low temperature conditions.

	T = 20°C	T = 37°C
η_o (mPa · s = 1 cP)	65.7	45.1
η_∞ (mPa · s = 1 cP)	4.47	3.07

(6.14)

TABLE 7. Asymptotic (low and high shear) viscosities for the Carreau–Yasuda model at 20°C and 37°C using results from Table 5 and Eq. (6.13). In evaluating Eq. (6.13), the viscosity of water was taken as $\eta_{H_2 0} = (1.01, 0.694)$ mPa · s at $(20,37)$°C.

6.5.3. Effect of plasma viscosity on blood viscosity. Although the molecular weight of some plasma proteins exceeds 10^6 Da, the plasma viscosity does not show a measurable dependence on shear rate. Early reports of a non-Newtonian plasma viscosity were later attributed to the formation of a surface film at the plasma/air interface in the rheometer (see, e.g., [43], p. 41).

Even in the healthy population, there is a variation in plasma viscosity which is enhanced by some diseases. It has been found to depend on various factors including temperature, age, gender and level of physical fitness, Table 6, [44]. These variations are important because plasma viscosity influences whole blood viscosity. In diseases in which the concentrations of fibrinogen and/or immunoglobulins are increased, the plasma viscosity can increase significantly.

7. Yield stress behavior of blood

We recall from Section 4 of [107] that yield stress materials require a finite shear stress to commence flowing. In formulating a constitutive model for such materials, it is necessary to define a yield criterion. This criterion must be met for the material to flow. As discussed in Section 4 of [107], a relatively simple, physically relevant yield criterion is

$$\sqrt{|\mathrm{II}_\tau|} = \tau_Y, \tag{7.1}$$

where τ_Y is a material property of the fluid called the yield stress and II_τ is a scalar invariant of the extra stress tensor, $\boldsymbol{\tau}$,

$$\mathrm{II}_\tau = 1/2\left((\mathrm{tr}\,\boldsymbol{\tau})^2 - \mathrm{tr}\,\boldsymbol{\tau}^2\right). \tag{7.2}$$

Therefore, for $\sqrt{|\mathrm{II}_\tau|} < \tau_Y$, the fluid will not flow.

Measurements of the yield stress τ_Y are typically performed using rheometers which control either kinematic variables (e.g., cylinder or cone rotation rate) or stress-related variables (torque, pressure drop). In shear-rate-controlled experiments, the yield stress cannot be measured directly, rather it is obtained by back extrapolation to zero shear rate, possibly using a specific constitutive model. An extensive description of methods for measuring yield stress is given in [91]. Copley et al. [38] summarize early experimental evidence for blood yield stress in both *in-vitro* and in-vivo experiments.

7.1. Yield stress data for blood

As discussed previously, measurements on blood at low shear rates are notoriously difficult, contributing to the large variation in yield stress values for blood reported in the literature. These values have ranged from 0.20 to 40 mPa (see, e.g., [43]). This spread is also seen in Figure 10, obtained from [98], and has been attributed to artifacts arising from interactions between the RBCs and surfaces of the rheometer [20] as well as the experimental method used to measure (approximate) the yield stress. Picart et al. roughened the walls of a Couette rheometer in an effort to

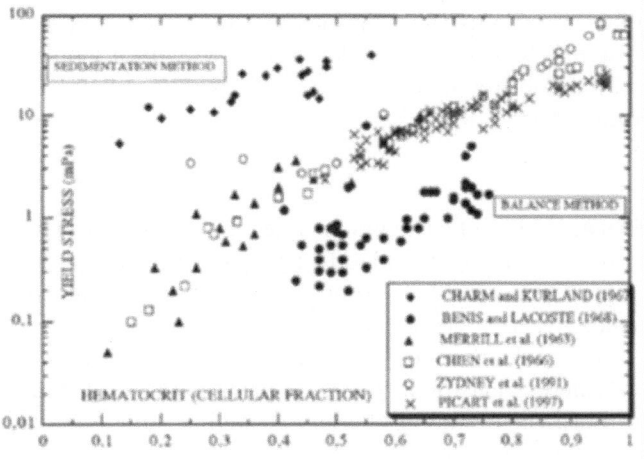

FIGURE 10. Yield stress as a function of hematocrit reported in the literature using different experimental methods: sedimentation method [19]; viscometric balance method [9], back extrapolation from shear-rate-controlled experiments [86, 30, 145, 98]. A large variability in yield stress values can be seen. Reproduced from [98] with permission from The Society of Rheology.

diminish slip at the inner cylinder during experiments at shear rates down to 1×10^{-3} s^{-1}. When they compared their results with data obtained in smooth surface rheometers they found nearly an order of magnitude difference, Figure 11, [98]. They concluded the slip was an important artifact at these lower shear rates. In that work, they defined the yield stress as the value of stress measured at a shear rate of 1×10^{-3} s^{-1}. Namely, they did not use back extrapolation even though the shear stress was still decreasing with decreasing shear rate, Figure 11.

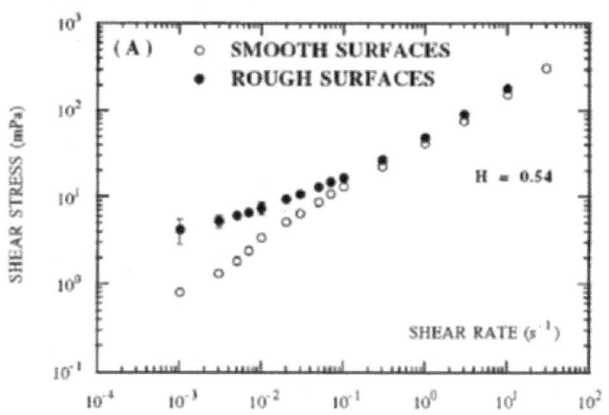

FIGURE 11. Comparison of shear stress as a function of shear rate for RBCs in plasma at low shear rates in a Couette cylinder with smooth and roughened walls, T = 25°C and Ht = 54%. Reproduced from [98] with permission from The Society of Rheology.

Dintenfass appears to be the first to question the appropriateness of the yield stress as a material constant for blood (e.g., page 82 of [43]). He suggested that rather than treating the yield stress as a constant, it should be considered as a function of time. A time dependence was also noted in [21]. In fact, due to the thixotropic nature of blood, we expect measurements of the yield stress to be quite sensitive to the microstructure of the blood prior to yielding, which is expected to be sensitive to both the shear rate history as well as time (e.g., [89]).

The large range in yield stress seen for blood is consistent with results for other fluids where a large spread in yield values is attributed to the experimental methodology, the criterion used to define the yield stress, and the length of time over which the experiment is run [91]. A true material constant should be independent of these factors and these results have called into serious question the treatment of the yield stress as a material parameter [91, 7, 89].

Barnes and Walters [8] point out the dilemma in trying to verify the existence of a yield stress if it is defined as a stress below which no unrecoverable flow occurs.

"Such a definition effectively rules out experimental proof of the existence of a yield stress, should such exist, since it would require an infinite time to show that the shear rate, at any given stress is actually zero".

Furthermore, in a compelling review article on the subject of yield stress, Barnes notes several examples where the data for the viscosity appeared to grow unbounded as the shear rate was reduced to zero, suggesting the existence of a yield stress [7]. However, when experiments were run at smaller values of shear rate, the viscosity plateaued to a finite value.

Having said this, we turn to Barnes comments on the usefulness of the yield stress [7],

> "Although we have shown that, as a physical property describing a critical yield stress below which no flow takes place, yield stresses do not exist, we can, without any hesitation, say that the concept of a yield stress has proved – and, used correctly, is still proving – very useful in a whole range of applications, once the yield stress has been properly defined. This proper definition is as a mathematical curve-fitting constant, used along with the other parameters to produce an equation to describe the flow curve of a material over a limited range of shear rates".

In this light, we turn attention to yield stress models that can be useful in the low shear rate region for blood and briefly discuss a few of these models.

7.2. Yield stress constitutive models for blood

One of the most commonly used yield stress models for blood is Casson's model (see also Section 4 in [107] of this volume). Casson materials behave rigidly until (7.1) is satisfied, after which they display a shear thinning behavior. The typically cited Casson equation, is only applicable for simple shear flow. This is discussed in greater detail in Section 4 in [107] of this volume. It is straightforward to write a three-dimensional constitutive equation that reduces to Casson's equation in simple shear,

$$\sqrt{|\mathrm{II}_\tau|} < \tau_Y \implies D_{ij} = 0$$

$$\sqrt{|\mathrm{II}_\tau|} \geq \tau_Y \implies \begin{cases} D_{ij} & = \dfrac{1}{2\mu_N}\left(1 - \dfrac{\sqrt{\tau_Y}}{\sqrt[4]{|\mathrm{II}_\tau|}}\right)^2 \tau_{ij} \\[3mm] \tau_{ij} & = 2\left(\sqrt{\mu_N} + \dfrac{\sqrt{\tau_Y}}{\sqrt[4]{4|\mathrm{II}_\mathrm{D}|}}\right)^2 D_{ij}. \end{cases} \tag{7.3}$$

The Newtonian constitutive equation is a special case of (7.3) with τ_Y set to zero, in which case μ_N is the Newtonian viscosity.

Scott-Blair was the first to apply the Casson model to blood [120, 103]. He compared the predictions of Casson's equation with capillary rheometer data for whole blood, plasma and serum from humans, cows and pigs and found an excellent fit for capillary diameters greater than 0.15 mm. The Casson model is still frequently used for whole blood as well as the Bingham model (4.4) of [107] and the Herschel–Bulkley model (4.7) of [107]. Oka [93] generalized these equations to allow gradual disruption of bonds with shear rate. Charm found that Casson's model gives the best fit to blood data [20], though there is some argument that these data do not match the Casson model well below 1 s^{-1}, [84]. Charm [20] suggested this might be due to problems in viscometry at this low shear rate.

Merrill et al. found a relatively simple relationship between hematocrit and yield stress to be valid for Ht up to about 50% [86],

$$\tau_y^{1/3} = A\,(\mathrm{Ht} - \mathrm{Ht}_c)/100 \qquad A \approx 0.8 \pm 0.2\,\mathrm{mPa}, \tag{7.4}$$

where Ht_c is a critical value of hematocrit below which there is no yield behavior. This material parameter varies between individuals. In the five donors considered in [86], Ht_c ranged from 1.3% to 6.5%.

8. Viscoelasticity of blood

Viscoelastic fluids are viscous fluids which have the ability to store and release energy (see Section 5 of [107] for background material on viscoelastic fluids). The viscoelasticity of blood at normal hematocrits is largely due to its ability to store and release energy from its branched 3D microstructure [25, 32]. As will be discussed shortly, the ability of the microstructure to store energy is affected by the properties of the RBC membrane and the bridging mechanisms within the 3D structure. Elastic energy can also be stored in the deformation of individual RBC, though this is not believed to play an appreciable role unless the RBC concentration is significantly elevated above normal physiological levels [25].

Using a rheoscope, Schmid-Schönbein obtained fascinating results for the elastic deformations of the three-dimensional aggregate structure, [116]. Under increasing applied shear rate, the aggregate length could be seen to increase up to three fold [116]. Elongation of the rouleaux aggregates under applied shear was found to arise from several mechanisms: (1) realignment of the individual cells (sliding of the cells from a parallel stack to a sheared stack), (2) trapezoidal deformation of cells located at the branch point of two rouleaux, and (3) deformation of individual cells within an individual rouleaux (ellipsoidal and eventually prolate deformation), Figure 12(a)–(c). Under increasing shear, contact between some of cells was eventually lost and the resulting segments were seen to recoil, Figure 12(f)–(g). If the shear rate was decreased prior to reaching this rupture level, recoil of the structure could also be observed, Figure 12(d)–(e). The extension, rupture and recoil were only observed in rouleaux that bridged larger secondary structures such as shown in Figure 12.

In [40] and [136], deformations of the rouleaux structure were evaluated under low frequency shear flow (0.01 Hz). Cyclic elastic deformations of the microstructure were observed to occur at the same frequency as the imposed shear rate (Ht = 45%). This was consistent with the larger magnitudes in blood viscoelasticity measured under those same conditions. As the frequency was increased to 10 Hz, the RBC formed irregular clumps and the aggregated network was no longer continuous.

8.1. Measurements of blood viscoelasticity

The magnitude of viscoelastic effects in blood are relatively small and, as a result, successful measurements of the first and second normal stress differences for blood have not been reported, (e.g., [40]). To date, the viscoelastic properties of blood have been measured by generating oscillatory (or sometimes pulsatile) flows and step transient flows in the rheometers discussed in Section 8 of [107]. Thurston was the first to measure the viscoelastic properties of blood and has contributed

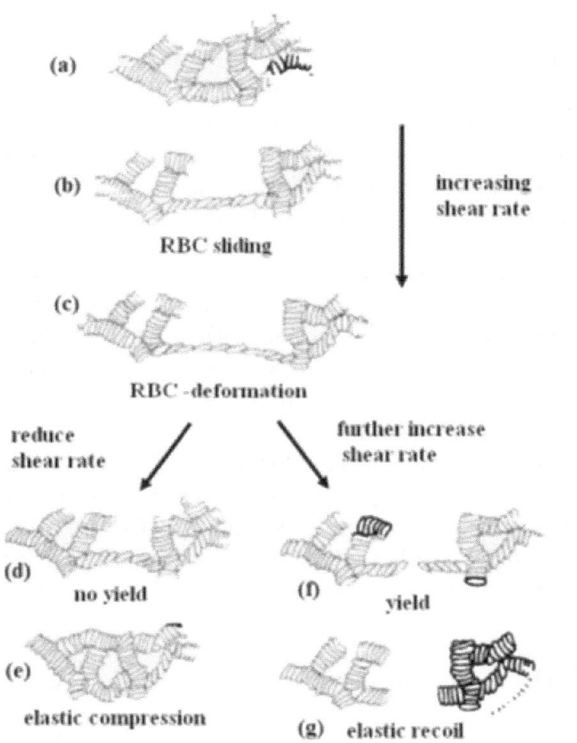

(a)

(b)

RBC sliding

increasing
shear rate

(c)

RBC -deformation

reduce
shear rate

further increase
shear rate

(d)

no yield

(f)

yield

(e)

elastic compression

(g)   elastic recoil

FIGURE 12. Schematic of the deformation of three-dimensional blood aggregates with increasing shear rate, as observed by Schmid-Schönbein using his rheoscope. Reproduced from [116] (with permission from Springer Science and Business Media).

the largest body of experimental work in this area of blood rheology (e.g., [134] and references cited therein).

The viscoelastic data for blood is often reported in terms of the complex viscosity, η^* which can be decomposed into its real and imaginary parts,

$$\eta^* = \eta' - i\,\eta'' = |\eta^*|e^{-i\phi}. \tag{8.1}$$

For a perfectly viscous fluid, $\eta'' = 0$ and $\eta = \eta'$. For fluids with a mechanism to store elastic energy, η'' is non-zero. For this reason, the material constants η' and η'' are often referred to as the viscosity and elasticity, respectively.

As detailed in Section 7.2.1 of [107], if the periodic deformations are of small amplitude, the linear viscoelastic rheological properties η' and η'' can be obtained independent of the selection of a constitutive model. However, these linear viscoelastic functions take on a different meaning and relevance in nonlinear regimes of deformation such as finite strain oscillatory flow and the pulsatile flow found in the circulatory system. For both finite-amplitude oscillations and pulsatile flows, an exact solution for fully developed periodic flow in a straight pipe of constant circular cross section exists for some quasi-linear viscoelastic models such as the Oldroyd-B model. These models can be used to interpret data obtained for large-amplitude experiments in capillary rheometers, including pulsatile flow. However, these models have constant shear viscosity and exact solutions are not available for periodic flows of viscoelastic materials with shear thinning viscosity.

8.1.1. Oscillatory flow in capillary rheometers. Most measurements of blood viscoelasticity are made for oscillatory or pulsatile flows generated in capillary rheometers. We therefore begin this section with a brief summary of the theoretical foundation behind the measurement of viscoelastic parameters in capillary rheometers. Further details can be found in Section 7.2.1 of [107], Chapter 5 of [31], Appendix B of [134], [126], and [127].

Consider flow of an incompressible fluid in a pipe of constant radius R which is driven by an oscillatory axial pressure gradient of the form $\partial p/\partial z = -K_1 \cos(\omega t)$, where p is the mechanical pressure, t is time, ω is the angular frequency of oscillation, and the z axis is parallel to the pipe centerline. As in Section 7.2.1 of [107], for ease of analysis, we rewrite the dependent variables as the real part of complex counterparts,

$$
\begin{aligned}
\frac{\partial p}{\partial z}(t) &= \mathcal{Re}[-K_1\, e^{i\omega t}], & v_z(r,t) &= \mathcal{Re}[v_z^*(r)e^{i\omega t}], \\
Q(t) &= \mathcal{Re}[Q^* e^{i\omega t}], & \tau_{rz}(r,t) &= \mathcal{Re}[\tau_{rz}^*(r)e^{i\omega t}],
\end{aligned}
\tag{8.2}
$$

where v_z^*, Q^* and τ_{rz}^* are in general complex and K_1 is real. Following Thurston (e.g., Appendix B of [134], [126], and [127]), we restrict attention to constitutive equations for which τ_{rz}^* is linear in dv_{rz}^*/dr,

$$
\tau_{rz}^* = \eta^*\, dv_{rz}^*/dr
\tag{8.3}
$$

where η^* is assumed to be independent of the shear rate. In this case, it is possible to obtain an explicit solution relating the flow rate to the applied pressure gradient and ultimately determine expressions for η' and η'' in terms of these measured quantities. However, this assumption forces us to restrict attention to either infinitesimal deformations or consider large amplitude oscillations but for specific quasi-linear constitutive models which do not display shear thinning behavior. Some examples are Newtonian fluids and the constant viscosity Oldroyd-B fluid, (see, e.g., Section 7.2.3 of [107]). For example, for the Upper Convected Maxwell fluid and Oldroyd-B fluid (see, e.g., Section 7.2 of [107]):

UCM $\qquad \eta^* = \dfrac{\eta}{1+i\,\lambda\,\omega}, \qquad \eta' = \dfrac{\eta}{1+\lambda^2\,\omega^2}, \qquad \eta'' = \dfrac{\eta\lambda\omega}{1+\lambda^2\,\omega^2},$

Oldroyd-B $\quad \eta^* = \dfrac{(\eta + i\,\eta_2\,\lambda\,\omega)}{1+i\,\lambda\omega}, \quad \eta' = \dfrac{\eta + \eta_2\,(\lambda\omega)^2}{1+\lambda^2\,\omega^2}, \quad \eta'' = \dfrac{(\eta - \eta_2)\,(\lambda\omega)}{1+\lambda^2\,\omega^2}.$

$$
\tag{8.4}
$$

The explicit solution for the complex flow rate for fluids which satisfy (8.3) is (see Section 7.2.3 of [107] for details),

$$Q^*(t) = \frac{i K_1 \pi R^2}{\rho \omega} \left[1 - \frac{2 J_1(\beta)}{\beta J_0(\beta)} \right],$$ (8.5)

where

$$\beta = R \sqrt{\frac{-i \rho \omega}{\eta^*}}.$$ (8.6)

For small values of β, (8.5) can be approximated as,

$$Q^*(t) = \frac{i K_1 \pi R^2}{\rho \omega} \frac{\beta^2}{8} \left[1 + \frac{1}{6} \beta^2 + \mathcal{O}(\beta^4) \right].$$ (8.7)

Using an electrical analogy, a complex impedance per unit length (Z^*) can be defined as,

$$Z^* = \frac{K_1}{Q^*}.$$ (8.8)

It follows from (8.7) and (8.8) that,

$$Z^* = \frac{8\eta''}{\pi R^4} + i \left(\frac{4}{3} \frac{\rho \omega}{\pi R^2} - \frac{8\eta''}{\pi R^4} \right) + \mathcal{O}(\beta^2).$$ (8.9)

It is useful to consider the following representations for Z,

$$Z^* = \mathcal{R}e[Z^*] + i\mathcal{I}m[Z^*] = |Z^*| e^{-i\theta}.$$ (8.10)

It is then clear from (8.8), that θ is the phase shift between the pressure gradient and volumetric flow rate. We see from (8.7) and (8.8) that through $\mathcal{O}(\beta^2)$,

$$\eta' = \frac{\pi R^4}{8} \mathcal{R}e[Z^*], \qquad \eta'' = \frac{1}{6} \rho \omega R^2 - \frac{\pi R^4}{8} \mathcal{I}m[Z^*].$$ (8.11)

Therefore, using (8.2), (8.8), (8.10), and (8.11), the values of η' and η'' can be determined from measurements of K_1, $Q(t)$ and the phase difference between them in a capillary rheometer of known geometry.

For linear viscoelastic materials, η' and η'' depend only on the frequency ω. However, data for η' and η'' are often reported outside this linear range where they also depend on shear rate and shear strain. Since the shear rate and strain vary with the radius in a capillary rheometer, representative values of these quantities must be chosen. A characteristic shear rate is often defined to be the amplitude of the shear rate at the wall ($\dot{\gamma}_w$). If follows from the explicit solution for the velocity field (see Section 7.2.3 of [107]),

$$\dot{\gamma}_w = |\mathcal{R}e \left[\frac{\tau_{rz}^*(R)}{\eta^*} \right]|,$$ (8.12)

where, τ_{rz}^* can be obtained from a control volume analysis of the balance of linear momentum for the tube,

$$\tau_{rz}^*(R) = \frac{K_1 R}{2} - \frac{i \rho \omega Q^*}{2 \pi R}.$$ (8.13)

A measure of the strain (γ_w) is then defined relative to $\dot{\gamma}_w$ through

$$\dot{\gamma}_w = \gamma_w\,\omega. \tag{8.14}$$

The subscript "w" is often dropped in the literature.

8.1.2. Data on blood viscoelasticity. Representative data for η' and η'' for whole human blood are shown in Figs. 13 and 14. It can be seen that η' and η'' both

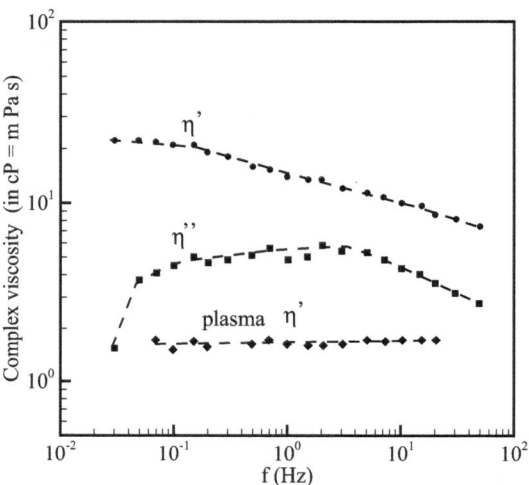

FIGURE 13. The dependence of the viscous (η') and elastic (η'') components of the complex viscosity on the frequency of oscillation compared with the viscous component of complex viscosity for plasma. Reprinted from [128] (with permission from IOS Press).

depend on frequency, though η'' is relatively independent of ω over the frequency range associated with a normal human pulse (0.5 to 20 Hz) with values of $\eta'' \in [3.6, 5.8]$ mPa·s for frequencies in that range [132]. The qualitative shape of the dependence on frequency [128], is similar to that shown in Figure 8 of [107] for a simple Maxwell fluid, suggesting a multiple relaxation Maxwell model could be used to fit the blood data.

Shown in Figure 14 are the complex components of blood viscosity from healthy donors as a function of the RMS value of oscillatory shear rate at the wall, [131] (from oscillatory flow in a capillary rheometer). For comparison, the viscosity obtained in steady-flow experiments in a Contraves LS30 (Couette) viscometer are shown (labeled η_s for emphasis). Both components of the complex viscosity η'' can be seen to have relatively constant values for shear rates below 1.5 s^{-1}. In this range, the elastic and viscous component are approximately 3.9 mPa·s and 11.5 mPa·s, respectively. As the shear rate is increased beyond this level, blood displays a nonlinear viscoelastic behavior (η' and η'' are dependent on shear rate). This range of shear rates is outside the linear viscoelastic range. The value of η'' starts dropping rapidly as the shear rate is increased beyond this level, diminishing to 0.1 mPa·s by 16 s^{-1}. This sharp decrease is tied to the breakdown of the blood microstructure formed by RBC aggregates. As expected after this region, there is a merging of the η' and η_s curves.

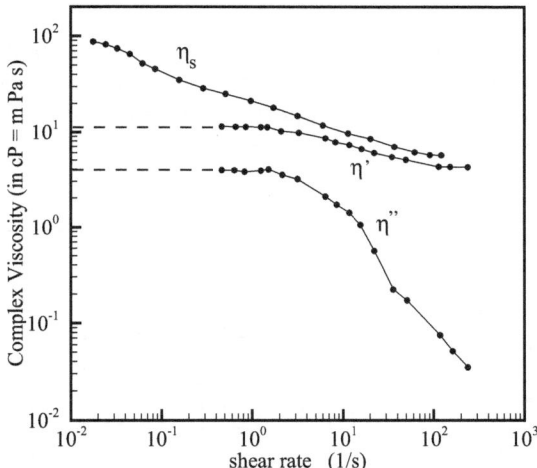

FIGURE 14. Dependence of the steady flow viscosity (η_s) on shear rate (Contraves LS30 viscometer) and the dependence of the viscous (η') and elastic (η'') components of the complex viscosity on the RMS value of wall shear rate (capillary rheometer). Human blood from healthy donors for Ht = 43% and T = 22°C. Oscillatory data measured at a frequency of 2 Hz. Reproduced from [131] with permission from The Society of Rheology.

8.2. Dependence of viscoelasticity on other factors

The complex viscosity is strongly dependent on hematocrit and temperature. It is of interest to determine a functional dependence of complex viscosity on these variables, so that blood data obtained at one hematocrit and temperature can be used for applications at other values. Shown in Figure 15 are η' and η'' for human blood in which the hematocrit was adjusted by removing cells or plasma. Both components increase steeply with increasing hematocrit, though the elastic component was found to be more sensitive. The general character of these curves

	$\dot{\gamma}_w = 1s^{-1}$	$\dot{\gamma}_w = 10s^{-1}$
$\eta'(1 \text{ cP} = 1 \text{ mPa} \cdot \text{s})$	$10 - 12.9$	$7.7 - 9.4$
$\eta''(1 \text{ cP} = 1 \text{ mPa} \cdot \text{s})$	$3.9 - 4.5$	$1.5 - 2.2$

TABLE 8. Average values of the complex viscosity components η' and η'' for blood from 25 young healthy male donors at T = 22°C and a frequency of 2 Hz. Normalized to a hematocrit of 45%. Data from [130].

is unchanged by hematocrit level, except at the highest hematocrit, where the curves can be seen to turn up at large shear rates. Thurston determined that for Ht $\in [35, 50]\%$, η' is approximately proportional to Ht^2 and η'' to Ht^3, [130].

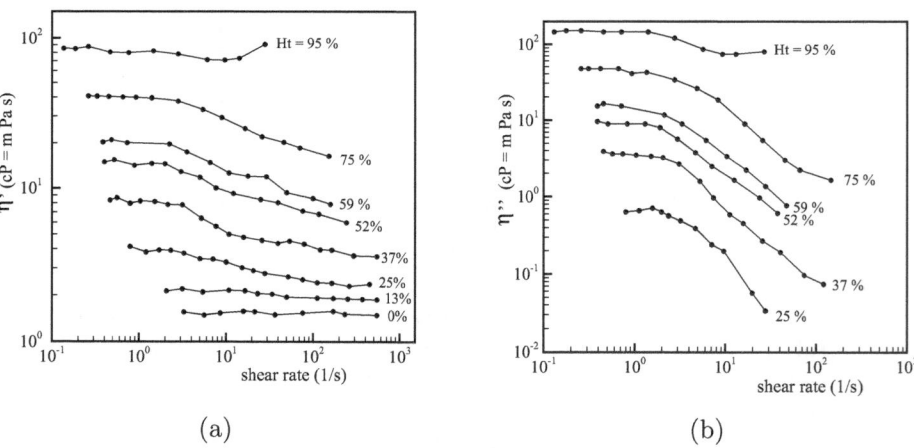

(a)   (b)

FIGURE 15. The dependence of the (a) viscous (η') and (b) elastic (η'') components of the complex viscosity on the hematocrit and RMS value of wall shear rate (capillary rheometer). Human blood from healthy donors with T = 22°C, f = 2 Hz. Reprinted from [130] with permission from IOS Press.

Using these results, Thurston normalized complex viscosity data for blood from healthy young males (n=25) to a hematocrit of 45% at 22°C, and obtained the average values shown in Table 8. Thurston also explored the temperature dependence and found that, like viscosity (Eq. (6.12)), both η' and η'' displayed an Arrhenius-type relationship (see, e.g., [134]).

Additional factors which affect the viscoelasticity are plasma components, cell rigidity and tendency for RBC to aggregate. We refer the reader to [134] for a review of the dependence of blood viscoelasticity on these factors.

8.3. Significance of viscoelasticity in the circulatory system

As discussed at the beginning of this section, the viscoelastic behavior of blood is largest in magnitude under conditions where the 3D microstructure of blood exists. Once this structure is broken down to the level of individual cells or groups of cells, the capacity of normal blood to store elastic energy is greatly diminished as reflected in the greatly diminished magnitude of η'' and the approach of η' to η. Therefore, many of the issues which were considered in the discussions of the shear thinning viscosity of blood in Section 6.3 are also important here. Briefly, for normal human blood there are few regions of the circulation where the 3D microstructure is stable. Some exceptions are parts of the venous system and regions of stable recirculation downstream of a stenosis or inside some aneurysm sacs. Pathological blood conditions exist for which the microstructure is greatly strengthened. In these cases, blood viscoelasticity may be significant in extensive regions of the circulation and could be an important factor in arterial flow for patients with these conditions.

As discussed in detail in [134], an assessment of blood viscoelasticity has merit as a sensitive indicator of the state of the blood and as a result, patient health. For example, the value of η'' has also been found to be quite sensitive to pathologies such as myocardial infarction, peripheral vascular disease, and diabetes (e.g., [134]), and hence the viscoelastic properties of blood have been used as a method of probing the pathological state of blood.

8.4. Viscoelastic constitutive models

While the linear viscoelastic functions are relatively straightforward to obtain and useful as a metric of the state of the blood components, it should be emphasized that most real motions of blood in the circulatory system are not in the small strain regime. Therefore, while the linear elastic material constants are indicative of aspects of the microstructure, nonlinear viscoelastic models are needed if the finite viscoelastic behavior of blood is to be modeled.

A significant challenge in developing nonlinear viscoelastic constitutive models for blood is selecting the simplest model that captures the frequency and shear rate dependence found in physiologically relevant flows. A number of models have already been considered that include both shear thinning and viscoelastic effects. Phillips and Deutsch [95] employed the following rate-type shear thinning model,

$$\boldsymbol{\tau} + \lambda_1 \overset{\triangledown}{\boldsymbol{\tau}} + \mu_o \left(\mathrm{tr}\boldsymbol{\tau}\right) \boldsymbol{D} = 2\eta_0 \boldsymbol{D} + 2\eta_0 \lambda_2 \overset{\triangledown}{\boldsymbol{D}}, \qquad (8.15)$$

in steady flows. They noted this model could predict the torque overshoot observed experimentally for blood. See Chapter 4 of [31] for additional discussion of this model. They further tested this model in a Couette device, comparing predicted and measured values of the critical Taylor number for instability, [42]. Based on their results, they extended the model (8.15) to include higher order terms, [42].

Yeleswarapu et al. [144] introduced a five-constant generalization of the Oldroyd-B model given in Eq. (5.48) of [107]. In this model, they replaced the constant viscosity $\eta = \eta_1 + \eta_2$ with a generalized Newtonian viscosity of the form

$$\eta(\dot\gamma) = \eta_\infty + (\eta_0 - \eta_\infty) \left[\frac{1 + \ln(1 + \Lambda\dot\gamma)}{(1 + \Lambda\dot\gamma)} \right]. \tag{8.16}$$

They obtained data for steady, capillary flow and used it to determine the constants in their model and found, $\eta_o = 200$ mPa·s, $\eta_\infty = 6.5$ mPa·s and $\Lambda = 11.14$ s. A steady flow of this kind, can only be used to evaluate the viscosity function and can not differentiate between this model and a shear thinning inelastic model with the same viscosity function. It is not clear why a logarithmic function rather than a simpler generalized Newtonian model was used. In this work, comparisons with other models were confined to the constant viscosity case.

More recently, Anand and Rajagopal considered a finite viscoelastic model [4] developed in the context of a more general thermodynamic framework, [101]. This model, along with the Yeleswarapu model, and a different generalization of the Oldroyd-B model were evaluated against oscillatory shear data from a capillary rheometer study [129]. However, the Anand and Rajagopal model has not yet been evaluated for transient viscoelastic experiments (e.g., [15]).

9. Disease states and mechanical properties of blood

A variety of diseases are associated with pathological changes in the mechanical properties of blood such as blood and plasma hyperviscosity, diminished RBC deformability, and increased RBC aggregation. For example, diabetes, myocardial infarction, malignant and rheumatic diseases are accompanied by an increase in blood viscosity [81]. Acute psychological stress was also shown to increase blood viscosity [90]. As discussed earlier, blood viscoelasticity is altered in myocardial infarction, peripheral vascular disease, and diabetes [134]. Below, we elaborate on a few of these pathological changes to blood properties.

9.1. Increased hematocrit

One of the most common hemorheological disorders is significantly elevated hematocrit (Ht > 50%) or polycythemia, which increases both low shear and asymptotic blood viscosity and causes harmful circulatory effects. For example, high hematocrit and blood viscosity levels associated with hypoxic pulmonary vasoconstriction cause a strong elevation in pulmonary vascular resistance and may lead to right heart failure. Primary and secondary polycythemias increase mortality of patients with heart and/or lung diseases [80].

9.2. Plasma hyperviscosity

As discussed in Section 6.5.3, blood viscosity is strongly affected by plasma viscosity. Therefore, an elevation in plasma viscosity may markedly raise vascular resistance. Heightened plasma viscosity, which is usually accompanied by anomalous RBC aggregation, may be caused by increased concentrations of proteins, particularly those with high molecular weight and/or nonspheroid shape (fibrinogen and/or certain serum globulins).

9.3. Hyperaggregation of red blood cells

Increased RBC aggregation is observed in a variety of clinical states such as infections, trauma, burns, diabetes mellitus, postinfarction, malignant and rheumatic diseases, AIDS, macroglobulinemia, myeloma, etc. [80]. This is clinically manifested by the presence in blood of large, three-dimensional aggregates of RBCs and abnormally high erythrocyte sedimentation rates. Figure 3(a) shows the RBC aggregate structure characteristic of normal human blood under no flow conditions. Figure 3(b) shows an example of pathological RBC aggregation in blood of a cardiac patient under similar conditions. The 3D aggregates of RBC from the cardiac patient are significantly more resistant to breakup under shear stress loading than those from normal human blood.

Hyperaggregation of RBCs causes a pathological elevation of blood viscosity under low flow (low shear rate). This process is particularly important for pathological conditions such as circulatory collapse and shock. Since the magnitude of blood flow in postcapillary venules is less than in precapillary arterioles, RBC aggregation, blood viscosity and vascular resistance can be preferentially elevated in the postcapillary vessels. As a result, intracapillary pressure is increased and the liquid phase is filtrated from capillaries into the interstitium. This, in turn, causes a rise in local hematocrit and hence an additional increase in RBC aggregation and blood viscosity. This process can also be of considerable significance in low blood flow states, such as found in patients during cardiac failure.

9.4. Decreased RBC deformability

A decrease in RBC deformability is a major hemorheological disorder which leads to a decrease in the number of functioning capillaries, impaired tissue perfusion and oxygenation and deteriorated removal of metabolites from tissue and organs. A significant decrease in RBC deformability can be caused by hematological diseases such as malaria or sickle cell disease.

Malaria is a group of diseases caused by plasmodia – a parasitic protozoan, which are transferred to the human blood by mosquitoes and which occupy and destroy red blood cells. Blood viscosity and RBC deformability are altered, causing severe anemia in people with this disease [45, 46].

Sickle cell disease is a hereditary disease caused by a defective form of hemoglobin. Upon deoxygenation, polymerization of abnormal sickle hemoglobin can occur causing the RBC to become sickle-shaped. Sickled cells demonstrate reduced deformability [13] and a lower tendency to aggregate [26]. The viscosity of blood containing irreversibly sickled cells exceeds that of normal blood under all experimental conditions and it can be of the order of fifty times greater [26]. The sickled RBCs obstruct blood flow through the microvessels resulting in lowered tissue perfusion and oxygenation which in turn causes severe abdominal and muscle pain, lung tissue damage (pneumonia, acute chest syndrome), strokes, hematuria, infection and many other complications. They also cause damage to most organs including the spleen, kidneys and liver. Sickle cells are rapidly destroyed in the body causing severe anemia.

A decrease in RBC deformability can also result from exposure of blood to abnormal hydrodynamic (e.g., elevated shear stresses) and environmental (e.g., hemodilution, temperature stress) conditions. Such conditions can arise, for example, during extracorporeal blood circulation in cardio-pulmonary bypass surgery and in patients implanted with heart valves and cardiac-assist devices or treated with other blood-contacting artificial devices [68, 69, 88]. Deformability is the major parameter which defines the ability of RBCs to enter and pass through the smallest capillaries and provide adequate gas transport. A decrease in RBC deformability may lead to diminished tissue oxygen supply and waste removal.

10. Gender and the mechanical properties of blood

Kameneva et al. [64, 65] studied the effect of gender on hematocrit, plasma and blood viscosity using a Couette viscometer for low shear values and a capillary viscometer for the high shear values. The gender-related variation in blood-rheological parameters was studied in connection with the remarkable difference in mortality from ischemic heart disease and, especially, from myocardial infarction between men and pre-menopausal women. The major hypothesis was that the reduced morbidity and mortality from cardiovascular diseases in women of reproductive age in comparison to age-matched men might be associated with the difference in age distribution of RBCs in blood due to monthly bleeding and the subsequent difference in mechanical properties of blood of pre-menopausal women and men.

Approximately 0.8% of the total number of RBCs is renewed every 24 hours [119]. Due to monthly blood loss, female blood is expected to contain more young and fewer old RBCs. In fact, it was shown that age distribution of RBCs is significantly different in males and females [87]. The authors demonstrated that premenopausal female blood has 80% more young RBCs and 85% fewer old RBCs than male blood. Elevated levels of old RBCs have a marked effect on blood viscosity, increasing both low and high shear viscosity values.

Results for the differences in mean values of hemorheological parameters for male and female blood is shown in Table 9. The data were obtained for 98 healthy volunteers: 47 premenopausal women (age 25.7 ± 4.8) and 51 men (age 26.2 ± 5.1) [66]. The ability of RBCs to deform was evaluated using a RBC rigidity index which is the inverse value of the RBC deformability. This index is defined as the ratio of the high shear rate asymptotic viscosity to the plasma viscosity η_∞/η_p.

The high shear asymptotic blood viscosity was measured at a standard hematocrit of Ht=40% and taken to be the blood viscosity in the high shear plateau region [79]. Also shown in this table, is the erythrocyte sedimentation rate (ESR) which is a measure of the level of RBC aggregation. The ESR was obtained by measuring the degree of RBC sedimentation after one hour in Wintrobe tubes, for blood with a standard hematocrit of 40% [142]. The level of RBC aggregation is also reflected in the blood viscosity at low shear rates [81].

	Women (n=47)	Men (n=50)	Level of Statistical Significance
Hematocrit,%	40.0 ± 2.4	45.8 ± 2.7	$p < 0.001$
Plasma viscosity, mPa·s	1.73 ± 0.09	1.74 ± 0.08	n.s.
Low Shear Blood Viscosity, mPa·s $\dot{\gamma} = 0.277\,s^{-1}$, Original Ht	36.1 ± 4.9	58.3 ± 12.1	$p < 0.001$
Asymptotic Blood Viscosity, mPa·s $\dot{\gamma} > 400\,s^{-1}$, Original Ht	4.8 ± 0.4	6.0 ± 0.8	$p < 0.001$
Low Shear Blood Viscosity, mPa·s $\dot{\gamma} = 0.277\,s^{-1}$, Ht $= 40\%$	35.59 ± 4.2	41.6 ± 4.4	$p < 0.001$
Asymptotic Blood Viscosity, mPa·s $\dot{\gamma} > 400\,s^{-1}$, Ht $= 40\%$	4.84 ± 0.4	5.3 ± 0.5	$p < 0.001$
Erythrocyte Sedimentation Rate mm/hr, Ht $= 40\%$	8.4 ± 3.1	10.8 ± 4.1	$p < 0.005$
RBC Rigidity Index, Ht $= 40\%$	2.8 ± 0.1	3.1 ± 0.1	$p < 0.001$

TABLE 9. Rheological parameters of male and female blood (data are presented as mean value $\pm$ a standard deviation for each parameter, unpaired student's t-test). All tests were performed at 23°C, from [66].

This study demonstrated a statistically significant difference in most hemorheological parameters of male versus female blood. Hematocrit level, blood viscosity at both high and low shear rates and for both original and standard hematocrit, RBC sedimentation rate (ESR) and RBC rigidity index were found to be significantly higher in male than in female blood, whereas plasma viscosity did not differ significantly. This latter finding is consistent with a much earlier thorough study of many aspects of human plasma viscosity by Harkness [61]. The higher values of both low shear and high shear viscosity in male blood is consistent with the larger values of hematocrit and RBC aggregation as well as the lower RBC deformability (inversely proportional to the Rigidity Index).

Significantly, increased hematocrit, increased blood viscosity, increased RBC aggregability and decreased RBC deformability have been shown to be risk factors for development of cardiovascular diseases [67, 81, 72]. Therefore, the measured differences in the mechanical properties of male and female blood place men at a much higher risk of cardiovascular diseases than pre-menopausal women.

References

[1] T. Alexy, R.B. Wenby, E. Pais, L.J. Goldstein, W. Hogenauer, and H.J. Meiselman. An automated tube-type blood viscometer: Validation studies. *Biorheology*, 42:237-247, 2005.

[2] M. Anand, K. Rajagopal, and K.R. Rajagopal. A model incorporating some of the mechanical and biochemical factors underlying clot formation and dissolution in flowing blood. *J. of Theoretical Medicine*, 5:183–218, 2003.

[3] M. Anand, K. Rajagopal, and K.R. Rajagopal. A model for the formation and lysis of blood clots. *Pathophysiol. Haemost. Thromb.*, 34:109–120, 2005.

[4] M. Anand and K.R. Rajagopal. A shear thinning viscoelastic fluid model for describing the flow of blood. *Int. J. Cardiovascular Medicine and Science*, 4(2):59–68, 2004.

[5] A.M. Artoli, A. Sequeira, A.S. Silva-Herdade and C. Saldanha. Leukocytes rolling and recruitment by endothelial cells: Hemorheological experiments and numerical simulations. *Journal of Biomechanics*, 40:3493–3502, 2007.

[6] F.I. Ataullakhanov and M.A. Panteleev. Mathematical modeling and computer simulation in blood coagulation. *Pathophysiol. Haemost. Thromb*, 34:60–70, 2005.

[7] H.A. Barnes. The yield stress – a review or παντα ρει - everything flows? *J. Non-Newtonian Fluid Mech.*, 81:133–178, 1999.

[8] H.A. Barnes and K. Walters. The yield stress myth? *Rheol. Acta*, 24(4):323–326, 1985.

[9] A.M. Benis and J. Lacoste. Study of erythrocyte aggregation by blood viscosity at low shear rates using a balance method. *Circ. Res.*, 22:29–42, 1968.

[10] M. Bessis. *Living Blood Cells and Their Ultrastructure*. Springer-Verlag, 1973.

[11] J.J. Bishop, A.S. Popel, M. Intaglietta, and P.C. Johnson. Rheological effects of red blood cell aggregation in the venous network: a review of recent studies. *Biorheology*, 38:263–274, 2001.

[12] T. Bodnár and A. Sequeira. Numerical simulation of the coagulation dynamics of blood. *Computational and Mathematical Methods in Medicine*, to appear.

[13] M.M. Brandão, A. Fontes, M.L. Barjas-Castro, L.C. Barbosa, F.F. Costa, C.L. Cesar, and S.T. Saad. Optical tweezers for measuring red blood cell elasticity: application to the study of drug response in sickle cell disease. *Eur J Haematol*, 70(4):207–211, 2003.

[14] D.E. Brooks, J.W. Goodwin, and G.V.F. Seaman. Interactions among erythrocytes under shear. *J. Appl. Physiol.*, 28(2):172–177, 1970.

[15] M. Bureau, J.C. Healy, D. Bourgoin, and M. Joly. Rheological hysteresis of blood at low shear rate. *Biorheology*, 17:191–203, 1980.

[16] A.C. Burton. *Physiology and the Biophysics of the Circulation.* Year Book Medical Publishers Inc., 1972.

[17] C.G. Caro, T.J. Pedley, R.C. Schroter, and W.A. Seed. *The Mechanics of the Circulation.* Oxford University Press, 1978.

[18] K.B. Chandran, A.P. Yoganathan, and S.E. Rittgers. *Biofluid Mechanics: The Human Circulation.* CRC Press, 2006.

[19] S.E. Charm and G.S. Kurland. Static method for determining blood yield stress. *Nature,* 216:1121–1123, 1967.

[20] S.E. Charm and G.S. Kurland. *Blood Flow and Microcirculation.* John Wiley & Sons, 1974.

[21] D.C.-H. Cheng. Yield stress: a time dependent property and how to measure it. *Rheol. Acta,* 25(5):542–554, 1986.

[22] D.C.-H. Cheng. Characterisation of thixotropy revisited. *Rheol Acta,* 42(4):372–382, 2003.

[23] S. Chien. White blood cell rheology. In:*Clinical Blood Rheology,* Vol. I, G.D.O. Lowe, Ed., CRC Press, Boca Raton, Florida, 1998, pp. 87-109.

[24] S. Chien. Red cell deformability and its relevance to blood flow. *Ann. Rev. Physiology,* 49:177–192, 1987.

[25] S. Chien, R.G. King, R. Skalak, S. Usami, and A.L. Copley. Viscoelastic properties of human blood and red cell suspensions. *Biorheology,* 12:341–346, 1975.

[26] S. Chien, S. Usami, and J.F. Bertler. Abnormal rheology of oxygenated blood in sickle cell anemia. *J. Clin. Invest.,* 49(4):623–634, 1970.

[27] S. Chien, S.Usami, R.J. Dellenback, and C.A. Bryant. Comparative hemorheology–hematological implications of species differences in blood viscosity. *Biorheology,* 8(1):35–57, 1971.

[28] S. Chien. Shear dependence of effective cell volume as a determinant of blood viscosity. *Science,* 168:977–979, 1970.

[29] S. Chien, S. Usami, R.J. Dellenback, and M.I. Gregersen. Shear-dependent deformation of erythrocytes in rheology of human blood. *Am. J. Physiol.,* 219(1):136–142, 1970.

[30] S. Chien, S. Usami, H.M. Taylor, J.L. Lundberg, and M.I. Gregersen. Effect of hematocrit and plasma proteins on human blood rheology at low shear rates. *J. Appl. Physiol.,* 21(1):81–87, 1966.

[31] H. Chmiel and E.Walitza. On the Rheology of Blood and Synovial Fluid. Research Studies Press, 1980.

[32] H. Chmiel and I. Anadere and E. Walitza. The determination of blood viscoelasticity in clinical hemorheology. *Biorheology,* 27:883-894, 1990.

[33] Y.I. Cho and K.R. Kensey. Effects of the non-Newtonian viscosity of blood on flows in a diseased arterial vessel. Part I: Steady flows. *Biorheology,* 28:241–262, 1991.

[34] G.R. Cokelet, E.W. Merrill, E.R. Gilliland, H.Shin, A.Britten, and R.E. Wells,Jr. The rheology of human blood - Measurement near and at zero shear rate. *Trans. Soc. Rheol.,* 7:303–317, 1963.

[35] G.R. Cokelet. Dynamics of Red Blood Cell Deformation and Aggregation, and In Vivo Flow. In: *Erytrocyte Mechanics and Blood Flow*, Kroc Foundation Series Vol. 13, Eds. G.R. Cokelet, H.J. Meiselman, D.E. Brooks, Alan R. Liss, Inc., New York, 1980.

[36] G.R. Cokelet. Viscometric, *in vitro* and *in vivo* blood viscosity relationships: how are they related? *Biorheology*, 36:343–358, 1999.

[37] G. Colantuoni, J.D. Hellums, J.L. Moake, and C.P. Alfrey. The response of human platelets to shear stress at short exposure times. *Trans. Amer. Soc. Artificial Int. Organs*, 23:626–631, 1977.

[38] A.L. Copley, L.C. Krchma, and M.E. Whitney. I. Viscosity studies and anamolous flow properties of human blood systems with heparin and other anticoagulents. *J. Gen. Physiol.*, 26:49–64, 1942.

[39] A.L. Copley, C.R. Huang, and R.G. King. Rheogoniometric studies of whole human blood at shear rates from 1000 to 0.0009 sec^{-1} Part I - Experimental findings. *Biorheology*, 10(1):17–22, 1973.

[40] A.L. Copley, R.G. King, S. Chien, S. Usami, R. Skalak, and C.R. Huang. Microscopic observations of viscoelasticity of human blood in steady and oscillatory shear. *Biorheology*, 12:257–263, 1975.

[41] V. Cristini and G.S. Kassab. Computer modeling of red blood cell rheology in the microcirculation: A brief overview. *Ann Biomed Eng*, 33(12):1724–1727, 2005.

[42] S. Deutsch and W.M. Phillips. The use of a Taylor-Couette stability problem to validate a constitutive equation for blood. *Biorheology*, 14:253–266, 1977.

[43] L. Dintenfass. *Blood Microrheology - Viscosity Factors in Blood Flow, Ischaemia and Thrombosis*. Butterworth, 1971.

[44] L. Dintenfass. *Blood Viscosity, Hyperviscosity and Hyperviscosaemia*. MTP Press Ltd, 1985.

[45] A.M. Dondorp, B.J. Angus, K. Chotivanich, K. Silamut, R. Ruangveerayuth, M.R. Hardeman, P.A. Kager, J. Vreeken, and N.J. White. Red blood cell deformability as a predictor of anemia in severe falciparum malaria. *Am. J. of Trop. Med. & Hyg.*, 60(5):733–737, 1999.

[46] A.M. Dondorp, P.A. Kager, J. Vreeken, and N.J. White. Abnormal blood flow and red blood cell deformability in severe malaria. *Parasitol Today*, 16(6):228–232, 2000.

[47] C. Dong and X.X. Lei. Biomechanics of cell rolling: shear flow, cell-surface adhesion and cell deformability. *J. Biomech*, 33(1):35–43, 2000.

[48] R. Fåhraeus. The suspension stability of the blood. *Physiol. Rev.*, 9(2):241–274, 1929.

[49] R. Fåhraeus and T. Lindqvist. The viscosity of the blood in narrow capillary tubes. *Am. J. Physiol.*, 96:562–568, 1931.

[50] T.M. Fischer, M. Stöhr-Lissen, and H. Schmid-Schönbein. The red cell as a fluid droplet: Tank tread-like motion of the human erythrocyte membrane in shear flow. *Science*, 202:894–896, 1978.

[51] A.L. Fogelson. Continuum models of platelet aggregation: formulation and mechanical properties. *SIAM J. Appl. Math.*, 52(4):1089–1110, 1992.

[52] M. Fröhlich, H. Schunkert, H.W. Hense, A. Tropitzsch, P. Hendricks, A. Döring, G.A. Riegger, and W. Koenig. Effects of hormone replacement therapies on fibrinogen and plasma viscosity in postmenopausal women. *British Journal of Haematology*, 100(3):577–581, 1998.

[53] Y.C. Fung. *Biomechanics: Mechanical Properties of Living Tissues*. Springer-Verlag, 1993.

[54] Y.C. Fung, W.C. Tsang, and P. Patitucci. High-resolution data on the geometry of red blood cells. *Biorheology*, 18:369–385, 1981.

[55] G.P. Galdi. Mathematical problems in classical and non-Newtonian fluid mechanics. In *Hemodynamical Flows. Modeling, Analysis and Simulation*. Oberwolfach-Seminars, Vol. 37, p. 121–273. Birkhäuser Verlag, 2008.

[56] L. Game, J.C. Voegel, P. Schaaf, and J.F. Stoltz. Do physiological concentrations of IgG induce a direct aggregation of red blood cells: comparison with fibrinogen. *Biochim Biophys Acta*, 1291(2):138–142, 1996.

[57] H.L. Goldsmith. Poiseuille medal award lecture: From papermaking fibers to human blood cells. *Biorheology*, 30:165–190, 1993.

[58] H.L. Goldsmith, G.R. Cokelet, and P. Gaehtgens. Robin Fåhraeus: evolution of his concepts in cardiovascular physiology. *Am. J. Physiol.*, 257:H1005–H1015, 1989.

[59] J. Goldstone, H. Schmid-Schönbein, and R.E. Wells. The rheology of red blood cell aggregates. *Microvas. Res.*, 2:273–286, 1970.

[60] A.C. Guyton and J.E. Hall. *Textbook of Medical Physiology*. W.B. Saunders Company, (Tenth Edition), 2000.

[61] J. Harkness. The viscosity of human blood plasma; its measurement in health and disease. *Biorheology*, 8(3):171–193, 1971.

[62] R.M. Hochmuth, C.A. Evans, H.C. Wiles, and J.T. McCown. Mechanical measurement of red cell membrane thickness. *Science*, 220:101–102, 1983.

[63] C.R. Huang, R.G. King, and A.L. Copley. Rheogoniometric studies of whole human blood at shear rates down to 0.0009 sec^{-1} II - Mathematical interpretation. *Biorheology*, 10(1):23–28, 1973.

[64] M.V. Kameneva, K.O. Garrett, M.J. Watach, and H.S. Borovetz. Red blood cell aging and risk of cardiovascular diseases. *Clinical Hemorheology and Microcirculation*, 18(1):67–74, 1998.

[65] M.V. Kameneva, M.J. Watach, and H.S. Borovetz. Gender difference in rheologic properties of blood and risk of cardiovascular diseases. *Clinical Hemorheology and Microcirculation*, 21:357–363, 1999.

[66] M.V. Kameneva, M.J. Watach, and H.S. Borovetz. In: *Advances in Experimental Medicine & Biology. Oxygen Transport to Tissue XXIV*, volume 530. Kluwer Academic/Plenum Publishers, 2003.

[67] M.V. Kameneva. Effect of hematocrit on the development and consequences of some hemodynamic disorders. In *Contemporary Problems of Biomechanics*, pages 111–126. CRC Press, 1990.

[68] M.V. Kameneva, J.F. Antaki, H.S. Borovetz, B.P. Griffith, K.C. Butler, K.K. Yeleswarapu, M.J. Watach, and R.L. Kormos. Mechanisms of red blood cell trauma in assisted circulation. Rheologic similarities of red blood cell transformations due to natural aging and mechanical stress. *ASAIO Journal*, 41(3):M457–M460, 1995.

[69] M.V. Kameneva, A. Ündar, J.F. Antaki, M.J. Watach, J.H. Calhoon, and H.S. Borovetz. Decrease in red blood cell deformability caused by hypothermia, hemodilution, and mechanical stress: factors related to cardiopulmonary bypass. *ASAIO Journal*, 45(4):307–310, 1999.

[70] S.R. Keller and R. Skalak. Motion of a tank-treading ellipsoidal particle in a shear flow. *J. Fluid Mech.*, 120:27–47, 1982.

[71] M.R. King and D.A. Hammer. Multiparticle adhesive dynamics: Hydrodynamic recruitment of rolling leukocytes. *PNAS*, 98(26):14919–14924, 2001.

[72] W. Koenig and E. Ernst. The possible role of hemorheology in atherothrombogenesis. *Atherosclerosis*, 94:93–107, 1992.

[73] M.H. Kroll, J.D. Hellums, L.V. McIntire, A.I.Schafer, and J.L. Moake. Platelets and shear stress. *Blood*, 88(5):1525–1541, 1996.

[74] A.L. Kuharsky and A.L. Fogelson. Surface-mediated control of blood coagulation: the role of binding site densities and platelet deposition. *Biophys. J.*, 80(3):1050–1074, 2001.

[75] C. Lentner, editor. *Geigy Scientific Tables, 8th edition*, volume 3. Medical Educational Division, CIBA-GEIGY Corporation, 1984.

[76] L.B. Leverett, J.D. Hellums, C.P. Alfrey, and E.C. Lynch. Red blood cell damage by shear stress. *Biophysical Journal*, 12(3):257–273, 1972.

[77] P. Libby, P.M. Ridker, and A. Maseri. Inflammation and Atherosclerosis. *Circulation*, 105:1135–1143, 2002.

[78] M.A. Lichtman. Rheology of leukocytes, leukocyte suspensions, and blood in leukemia. Possible relationship to clinical manifestations. *J. Clin. Invest.*, 52(2):350–358, 1973.

[79] G.D.O. Lowe. Blood viscosity and cardiovascular disease. *Thromb. Haemost.*, 67 (5):494–498, 1992.

[80] G.D.O. Lowe, Ed., *Clinical Blood Rheology, Vol. I and II*. CRC Press, Boca Raton, Florida, 1998.

[81] G.D.O. Lowe and J.C. Barbenel. Plasma and blood viscosity. In: *Clinical Blood Rheology*, Vol. I, G.D.O. Lowe, Ed., CRC Press, Boca Raton, Florida, 1998, pp. 11–44.

[82] C.W. Macosko. *Rheology: Principles, Measurements and Applications*. VCH Publishers, Inc., 1994.

[83] N. Maeda, M. Seike, Kume, T. Takaku, and T. Shiga. Fibrinogen-induced erythrocyte aggregation: erythrocyte-binding site in the fibrinogen molecule. *Biochimica Biophysica Acta*, 904(1):81–91, 1987.

[84] E.W. Merrill. Rheology of blood. *Physiol. Rev.*, 49(4):863–888, 1969.

[85] E.W. Merrill, G.C. Cokelet, A. Britten, and R.E. Wells. Non-Newtonian rheology of human blood.– Effect of fibrinogen deduced by "subtraction". *Circulat. Res.*, 13:48–55, 1963.

[86] E.W. Merrill, E.R. Gilliland, G. Cokelet, H. Shin, A. Britten, and R.E. Wells, Jr.. Rheology of human blood, near and at zero flow. Effects of temperature and hematocrit level. *Biophys. J.*, 3:199–213, 1963.

[87] V. Micheli, A. Taddeo, A.L. Vanni, L. Pecciarini, M. Massone, and M.G. Ricci. Distribuzione in gradiente di densita' degli eritrociti umani: differenze lagate al sesso. *Boll. Soc. Italiana Biol. Speriment.*, LX(3):665–671, 1984.

[88] T. Mizuno, T. Tsukiya, Y. Taenaka, E. Tatsumi, T. Nishinaka, H. Ohnishi, M. Oshikawa, K. Sato, K. Shioya, Y. Takewa, and H. Takano. Ultrastructural alterations in red blood cell membranes exposed to shear stress. *ASAIO Journal*, 48(6):668–670, 2002.

[89] P.C.F. Møller, J. Mewis, and D. Bonn. Yield stress and thixotropy: on the difficulty of measuring yield stress in practice. *Soft Matter*, 2:274–288, 2006.

[90] M.F. Muldoon, T.B. Herbert, S.M. Patterson, M.V. Kameneva, R. Raible, and S.B. Manuck. Effects of acute psychological stress on serum lipids, hemoconcentration and blood viscosity. *Arch. Intern. Med.*, 155(6):615–620, 1995.

[91] Q.D. Nguyen and D.V. Boger. Measuring the flow properties of yield stress fluids. *Annu. Rev. Fluid Mech.*, 24:47–88, 1992.

[92] University of Cape Town Electron Microscope Unit. Web site: Retrieved December 9, 2006, from http://sbio.uct.ac.za/Webemu/gallery/descriptions.php, 2006.

[93] S. Oka. An approach to a unified theory of flow behavior of time independent non-Newtonian suspensions. *Jap. J. Appl. Physics*, 10(3):287–291, 1971.

[94] A.A. Palmer and H.J. Jedrzejczyk. The influence of rouleaux on the resistance to flow through capillary channels at various shear rates. *Biorheology*, 12:265–270, 1975.

[95] W.M. Phillips and S. Deutsch. Toward a constitutive equation for blood. *Biorheology*, 12:383 –389, 1975.

[96] C. Picart. *Rhéometrie de cisaillement et microstructure du sang: applications cliniques. Thèse de Génie Biologique et Medical.* PhD thesis, Université Joseph Fourier, Grenoble, France, 1997.

[97] C. Picart, J.-M. Piau, H. Galliard, and P. Carpentier. Blood low shear rate rheometry: influence of fibrinogen level and hematocrit on slip and migrational effects. *Biorheology*, 35:335–353, 1998.

[98] C. Picart, J.-M. Piau, H. Galliard, and P. Carpentier. Human blood shear yield stress and its hematocrit dependence. *J. Rheol.*, 42(1):1–12, 1998.

[99] A.S. Popel and P.C. Johnson. Microcirculation and hemorheology. *Annu. Rev. Fluid Mech.*, 37:43–69, 2005.

[100] A.R. Pries, D. Neuhaus, and P. Gaehtgens. Blood viscosity in tube flow: dependence on diameter and hematocrit. *Am. J. Physiol.*, 263:H1770–H1778, 1992.

[101] K.R. Rajagopal and A.R. Srinivasa. A thermodynamic framework for rate type fluid models. *J. Non-Newtonian Fluid Mech.*, 88(3):207–227, 2000.

[102] P.W. Rand, E. Lacombe, H.E. Hunt, and W.H. Austin. Viscosity of normal human blood under normothermic and hypothermic conditions. *J. Appl. Physiol.*, 19(1):117–122, 1964.

[103] M. Reiner and G.W. Scott Blair. The flow of blood through narrow tubes. *Nature*, 1843:354–?, 1959.

[104] E. Richardson. Deformation and haemolisys of red cells in shear flow. *Proc. R. Soc. Lond.*, A.338:129–153, 1974.

[105] P. Riha, F. Liao, and J.F. Stoltz. Effect of fibrin polymerzation on flow properties of coagulating blood. *J. Biol. Phys.*, 23:121–128, 1997.

[106] P. Riha, X. Wang, F. Liao, and J.F. Stoltz. Elasticity and fracture strain of whole blood clots. *Clin Hemorheol and Microcirc*, 21(1):45–49, 1999.

[107] A. Robertson. Review of relevant continuum mechanics. In *Hemodynamical Flows. Modeling, Analysis and Simulation.* Oberwolfach-Seminars, Vol. 37, p. 1–62. Birkhäuser Verlag, 2008.

[108] M.C. Roco, editor. *Particulate Two-Phase Flow.* Series in Chemical Engineering. Butterworth-Heinemann Publ., 1993.

[109] C.M. Rodkiewicz, P. Sinha, and J.S. Kennedy. On the application of a constitutive equation for whole human blood. *J. Biomech. Eng.*, 112:198–206, 1990.

[110] A.H. Sacks, K.R. Raman, and J.A. Burnell, and E.G. Tickner. Report no. 119. Technical report, VIDYA, 1963.

[111] M. Schenone, B.C. Furie, and B. Furie. The blood coagulation cascade. *Curr. Opin. Hematol.*, 11(4):272–277, 2004.

[112] H. Schmid-Schönbein, J. Barroso-Aranda, and R. Chavez-Chavez. Microvascular leukocyte kinetics in the flow state. In H. Boccalon, editor, *Vascular Medicine*, pages 349–352. Elsevier, 1993.

[113] H. Schmid-Schönbein. Microrheology of erythrocytes, blood viscosity and the distribution of blood flow in the microcirculation. *Internat. Rev. Physiol. Cardiovasc. Physiol. II*, 9:1–62, 1976.

[114] H. Schmid-Schönbein, E. Volger and H.J. Klose. Microrheology and light transmission of blood. II The photometric quantification of red cell aggregation formation and dispersion in flow. *Pflüggers Arch*, 333:140–155, 1972.

[115] H. Schmid-Schönbein, R. Wells, and J. Goldstone. Model experiments in red cell rheology: the mammalian red cell as a fluid drop. In H.H. Hartert and A.L. Copley, editors, *Theoretical and Clinical Hemorheology.* Springer-Verlag, 1971.

[116] H. Schmid-Schönbein and R.E. Wells. Rheological properties of human erythrocytes and their influence upon the "anomalous" viscosity of blood. *Ergeb. Physiol.*, 63:146–219, 1971.

[117] H. Schmid-Schönbein, R. Wells, and E.R. Schildkraut. Microscopy and viscometry of blood flowing under uniform shear (rheoscopy). *J. Appl. Physiol.*, 26:674–678, 1969.

[118] H. Schmid-Schönbein and R. Wells. Fluid drop-like transition of erythrocytes under shear. *Science*, 165:288–291, 1969.

[119] R.F. Schmidt and G. Thews, editors. *Human Physiology.* Springer-Verlag, 1983.

[120] G.W. Scott Blair. An equation for the flow of blood, plasma and serum through glass capillaries. *Nature*, 183:613–614, 1959.

[121] T. Shiga, K. Imaizumi, N. Harada, and M. Sekiya. Kinetics of rouleaux formation using TV image analyzer. I. Human erythrocytes. *Am. J. Physiol.*, 245:H252–H258, 1983.

[122] R. Skalak, P.H. Chen, and S. Chien. Effect of hematocrit and rouleaux on apparent viscosity in capillaries. *Biorheology*, 9(2):67–82, 1972.

[123] E.N. Sorensen, G.W. Burgreen, W.R. Wagner, and J.F. Antaki. Computational simulation of platelet deposition and activation: I. Model development and properties. *Ann Biomed Eng*, 27(4):436–448, 1999.

[124] J.-F. Stoltz, M. Singh, and P. Riha. *Hemorheology in Practice*. IOS Press, 1999.

[125] H.O. Stone, H.K. Thompson, Jr., and K. Schmidt-Nielsen. Influence of erythrocytes on blood viscosity. *Amer. J. Physiol.*, 214(4):913–918, 1968.

[126] G.B. Thurston. Period flow through circular tubes. *The Journal of the Acoustical Society of America*, 24(6):653–656, 1952.

[127] G.B. Thurston. Measurement of the acoustic impedence of a viscoelastic fluid in a circular tube. *The Journal of the Acoustical Society of America*, 33(8):1091–1095, 1961.

[128] G.B. Thurston. Frequency and shear rate dependence of viscoelasticity of human blood. *Biorheology*, 10:375–381, 1973.

[129] G.B. Thurston. Elastic effects in pulsatile blood flow. *Microvasc. Res.*, 9:145–157, 1975.

[130] G.B. Thurston. Effects of hematocrit on blood viscoelasticity and in establishing normal values. *Biorheology*, 15:239–249, 1978.

[131] G.B. Thurston. Erythrocyte rigidity as a factor in blood rheology: Viscoelastic dilatancy. *J. Rheology*, 23(6):703–719, 1979.

[132] G.B. Thurston. Significance and methods of measurement of viscoselastic behavior of blood. In D.R. Gross and E.H.C. Hwang, editors, *The Rheology of Blood, Blood Vessels and Associated Tissue*, pages 236–256. Sijthoff and Noordhoff, 1981.

[133] A. Gaspar-Rosas and G.B. Thurston. Erythrocyte aggregate rheology by transmitted and reflected light. *Biorheology*, 25:471–487, 1988.

[134] G.B. Thurston. Viscoelastic properties of blood and blood analogs. *Advances in Hemodynamics and Hemorheology*, 1:1–30, 1996.

[135] S. Usami, S. Chien, and M.I. Gregersen. Viscometric characteristics of blood of the elephant, man, dog, sheep, and goat. *Am. J. Physiol.*, 217(3):884–90, 1969.

[136] S. Usami, R.G. King, S. Chien, R. Skalak, C.R. Huang, and A.L. Copley. Microcinephotographic studies of red cell aggregation in steady and oscillatory shear – A note. *Biorheology*, 12(5):323–325, 1975.

[137] F.J. Walburn and D.J. Schneck. A constitutive equation for whole human blood. *Biorheology*, 13:201–210, 1976.

[138] N.-T. Wang and A.L. Fogelson. Computational methods for continuum models of platelet aggregation. *J. Comput. Phys.*, 151(2):649–675, 1999.

[139] R.E. Wells, T.H. Gawronski, P.J. Cox, and R.D. Perera. Influence of fibrinogen on flow properties of erythrocyte suspensions. *Amer. J. Physiol.*, 207:1035–1040, 1964.

[140] S.R.F. Whittaker and F.R. Winton. The apparent viscosity of blood flowing in the isolated hindlimb of the dog, and its variation with corpuscular concentration. *J. Physiol.*, 78(4):339–369, 1933.

[141] R.L. Whitmore. *Rheology of the Circulation*. Pergamon Press, 1968.

[142] M.M. Wintrobe. *Clinical Hematology*. Lea & Febiger, 6th edition, 1967.

[143] D.M. Wootton, C.P. Markou, S.R. Hanson, and D.N. Ku. A mechanistic model of acute platelet accumulation in thrombogenic stenoses. *Ann Biomed Eng*, 29(4):321–329, 2001.

[144] K.K. Yeleswarapu, M.V. Kameneva, K.R. Rajagopal, and J.F. Antaki. The flow of blood in tubes: Theory and experiment. *Mechanics Research Communications*, 25(3):257–262, 1998.

[145] A.L. Zydney, J.D. Oliver, and C.K. Colton. A constitutive equation for the viscosity of stored red cell suspensions: Effect of hematocrit, shear rate, and suspending phase. *J. Rheol.*, 35:1639–1680, 1991.

Anne M. Robertson
Department of Mechanical Engineering and Materials Science
McGowan Institute for Regenerative Medicine
Center for Vascular Remodeling and Regeneration (CVRR)
University of Pittsburgh
641 Benedum Engineering Hall
Pittsburgh, PA 15261
USA
e-mail: annerob@engr.pitt.edu

Adélia Sequeira
Departamento de Matemática
Centro de Matemática e Aplicações
Instituto Superior Técnico
Av. Rovisco Pais, 1
1049-001 Lisboa
Portugal
e-mail: adelia.sequeira@math.ist.utl.pt

Marina V. Kameneva
McGowan Institute for Regenerative Medicine
University of Pittsburgh
100 Technology Drive, Suite 200
Pittsburgh, PA 15219-3138
USA
e-mail: kamenevamv@upmc.edu

Hemodynamical Flows. Modeling, Analysis and Simulation
Oberwolfach Seminars, Vol. 37, 121–273
© 2008 Birkhäuser Verlag Basel/Switzerland

Mathematical Problems in Classical and Non-Newtonian Fluid Mechanics

Giovanni P. Galdi

This work was partially supported by the National Science Foundation, Grants DMS–0404834 and DMS–0707281.

Introduction

Blood flow *per se* is a very complicated subject. Thus, it is not surprising that the mathematics involved in the study of its properties can be, often, extremely complex and challenging.

The role of mathematics in the investigation of blood flow properties – as in the most part of applied sciences – is twofold and is directed toward the accomplishment of the following objectives. The first one, of a more theoretical nature, is the validation of the models proposed by engineers, and consists in securing conditions under which the governing equations possess the fundamental requirements of well-posedness, such as existence and uniqueness of corresponding solutions and their continuous dependence upon the data. The second one, of a more applied character, is to prove that these models give a satisfactory interpretation of the observed phenomena. In general, both tasks present serious difficulties in that they require the study of several different, and frequently combined, topics that include, among others, Navier–Stokes equations, non-Newtonian fluid models, non-linear elasticity, fluid-structure interaction and multi-phase flow. It must be added that some of these topics are still at the beginning of a systematic mathematical research, whereas some others are in a continuous growth.

As a matter of fact, the initiation or the methodical investigation of several of the above research areas was just motivated by problems arising in blood flow. Moreover, blood flow can also pose challenging questions in "classical" topics, questions that, in the past, happened to receive little or no attention at all. A typical example is provided by the problem of the flow of a Navier–Stokes liquid in an unbounded piping system, under a given time-periodic flow-rate, that has been "discovered" only in 2005, thanks to the work of H. Beirão da Veiga; see [7].

It seemed to me hopeless to present and to describe in this article all relevant aspects of the mathematical analysis related to blood flow, even at an introductory level. Therefore, I preferred to concentrate on some, of the many, topics which are at the foundation of this analysis, and to point out directions for future research. More specifically, I focused on three different subjects which are the content of as many separate chapters.

The first chapter deals with the study of some fundamental properties of the flow of a Navier–Stokes liquid in a piping system, which can be either unbounded or bounded. Here, I have concentrated the analysis mostly on steady-state and time-periodic motions, and on their attainability. There are several reasons for this choice. On the one hand, because these types of motions are the most "elementary" to occur in the arterial and venuous system, and, on the other hand, because the initial-boundary value problem in an unbounded piping system has been investigated in full detail in the most recent article [86], to which I refer the interested reader.

The second chapter is dedicated to the mathematical analysis of certain non-Newtonian fluid models, including generalized Newtonian, second-order and Oldroyd-B. Basically, I have focused my attention on the well-posedness of the problems related to these models.

In the third chapter, I analyze two problems of fluid-particle interaction. The first is related to the orientation of symmetric rigid bodies sedimenting in viscoelastic liquids, while the second deals with the lateral migration of a rigid sphere in the shear flow of a viscoelastic liquid in a horizontal channel. These two problems provide examples of how a mathematical analysis is able to furnish a rigorous explanation of the observed phenomena. In this respect, it should be observed that the method employed in the treatment of both problems requires one particle at a time interacting with the liquid. There are different approaches to the study of a system of particles moving into a liquid. However, the results obtained with these latter are only of basic nature and regard, mainly, well-posedness of the problems associated with the relevant equations. For this different, but very important, aspect of particulate flow we refer the reader to [31] and to literature cited therein.

A significant topic that I have left out is the problem of interaction between a fluid and an elastic structure. The reason for this choice is because, I believe, this area is still at the beginning, and, moreover, in order to describe the major contributions, a fair amount of technical prerequisites is needed. The most relevant mathematical works in fluid-structure imteraction – all developed in the last 3 years – can be grouped into two categories. In the first one, where a Navier–Stokes liquid interacts with its bounding walls described either by an elastic string (in the two-dimensional case) or elastic membrane (in the three-dimensional case); see [6, 20, 23]. In the other category, a "bulk" elastic solid is embedded in a Navier–Stokes liquid bounded by a rigid wall [29, 28, 47]. More significant topics in this area remain to be investigated, that are related to blood flow, such as liquid moving in a piping system bounded by elastic walls under prescribed flow-rate, deformable body moving into a liquid in a channel or in a liquid that fills the whole space, to mention a few.

This article is aimed at a diverse readership. For this reason, I tried to make every subject as self-contained as possible. Whenever details are not explicitly given, I refer the reader to the appropriate literature. The contents of this article are organized as follows:

Contents

Notation

As a rule, we shall use the notation of [36]. However, for the reader's convenience, we collect here the most frequently used symbols.

By $\mathbb{N}$ we denote the set of positive integers, while $\mathbb{R}^n$ is the Euclidean n-dimensional space, and $\{e_1, e_2, e_3, \ldots, e_n\} \equiv \{e_i\}$ the associated canonical basis.

Given a second-order tensor $\boldsymbol{A}$ and a vector $\boldsymbol{a}$, of components $\{A_{ij}\}$ and $\{a_i\}$, respectively, in the basis $\{e_i\}$, by $\boldsymbol{a} \cdot \boldsymbol{A}$ [respectively, $\boldsymbol{A} \cdot \boldsymbol{a}$] we mean the vector whose components are given by $A_{ij}a_i$ [respectively, $A_{ij}a_j$]. Moreover, if $\boldsymbol{B} = \{B_{ij}\}$ is another second-order tensor, by the symbol $\boldsymbol{A} \cdot \boldsymbol{B}$ we mean the second-order tensor whose components are given by $A_{il}B_{lj}$. We also set $\boldsymbol{A} : \boldsymbol{B} = \text{trace}(\boldsymbol{A} \cdot \boldsymbol{B}^\mathsf{T})$, where the superscript "T" denotes transpose, and $|\boldsymbol{A}| = \sqrt{\boldsymbol{A} : \boldsymbol{A}}$.

Given a vector field $\boldsymbol{h}(x) \equiv \{h_i(x)\}$, $x \in \mathbb{R}^n$, by $\nabla \boldsymbol{h}$ we denote the second-order tensor field whose components $\{\nabla \boldsymbol{h}\}_{ij}$ in the given basis are given by $\{\partial h_j / \partial z_i\}$.

For any domain $\mathcal{A}$, $C^k(\mathcal{A})$, $k \geq 0$, $L^q(\mathcal{A})$, $W^{m,q}(\mathcal{A})$, $m \geq 0$, $1 < q < \infty$, denote the usual space of functions of class C^k on $\mathcal{A}$, and Lebesgue and Sobolev spaces, respectively. Norms in $L^q(\mathcal{A})$ and $W^{m,q}(\mathcal{A})$ are denoted by $\| \cdot \|_{q,\mathcal{A}}$, $\| \cdot \|_{m,q,\mathcal{A}}$. The duality pairing in $L^q(\mathcal{A})$ is indicated by $(\cdot, \cdot)_{\mathcal{A}}$. The completion of the space $C_0^\infty(\mathcal{A})$, constituted by the infinitely differentiable functions with compact support in $\mathcal{A}$, in the $W^{m,q}(\mathcal{A})$-norm is denoted by $W_0^{m,q}(\mathcal{A})$. The dual space of $W_0^{m,q}(\mathcal{A})$ is indicated by $W_0^{-m,q'}(\mathcal{A})$, $q' := q/(q-1)$. Unless confusion arises, we shall usually drop the subscript "$\mathcal{A}$" in these norms. The trace space on $\partial \mathcal{A}$ for functions from $W^{m,q}(\mathcal{A})$ will be denoted by $W^{m-1/q,q}(\partial \mathcal{A})$ and its norm by $\| \cdot \|_{m-1/q,q,\partial \mathcal{A}}$.

By $D^{k,q}(\mathcal{A})$, $k \geq 1$, $1 < q < \infty$, we indicate the homogeneous Sobolev space of order (m,q) on $\mathcal{A}$, [103] [36], that is, the class of functions u that are (Lebesgue) locally integrable in Ω and with $D^\beta u \in L^q(\mathcal{A})$, $|\beta| = k$, where

$$D^\beta = \frac{\partial^{|\beta|}}{\partial x_1^{\beta_1} \partial x_2^{\beta_2} \partial x_3^{\beta_3}}, \quad |\beta| = \beta_1 + \beta_2 + \beta_3.$$

The natural (semi-norm) in $D^{k,q}(\mathcal{A})$, is given by [1]

$$|u|_{D^{k,q}(\mathcal{A})} := \left(\sum_{|\beta|=k} \int_{\mathcal{A}} |D^\beta u|^q \right)^{1/q}.$$

The completion of $C_0^\infty(\mathcal{A})$ in the norm $|\cdot|_{D^{k,q}(\mathcal{A})}$ is denoted by $D_0^{k,q}(\mathcal{A})$, whereas, setting [2]

$$\mathcal{D}(\mathcal{A}) = \{\varphi \in C_0^\infty(\mathcal{A}) : \nabla \cdot \varphi = 0\},$$

the completion of $\mathcal{D}(\mathcal{A})$ in the norm $|\cdot|_{D^{k,q}(\mathcal{A})}$ is denoted by $\mathcal{D}_0^{k,q}(\mathcal{A})$. The dual spaces of $D_0^{k,q}(\mathcal{A})$ and $\mathcal{D}_0^{k,q}(\mathcal{A})$ are indicated by $D_0^{-k,q'}(\mathcal{A})$ and $\mathcal{D}_0^{-k,q'}(\mathcal{A})$, respectively.

Occasionally, if X is a Banach space, we shall denote its norm by $\| \cdot \|_X$.

Given a Banach space X, and an open real interval (a, b), we denote by $L^q(a, b; X)$ the linear space of (equivalence classes of) functions $f : (a, b) \to X$ whose X-norm is in $L^q(a, b)$. Likewise, for r a non-negative integer and I a real interval, we denote by $C^r(I; X)$ the class of continuous functions from I to X, which are differentiable in I up to the order r included. If $X = \mathbb{R}^n$, we shall simply write $L^q(a, b)$, $C^r(I)$, etc.

---

[1]Typically, we shall omit in the integrals the infinitesimal volume or surface of integration.
[2]Let X be any space of real functions. As a rule, we shall use the same symbol X to denote the corresponding space of vector and tensor-valued functions.

1. Problems in the pipe flow of a Navier–Stokes liquid

In this chapter we shall furnish a mathematical analysis of the motion of a Navier–Stokes liquid in a system of pipes, $\mathcal{P}$, whose bounding walls are rigid and impermeable.

The preliminary question that we would like to investigate is how to model $\mathcal{P}$. We shall be mainly concerned with the following two complementary ways of approaching the problem:

(A1) We assume that the "exiting" pipes (outlets) extend to infinity.

(A2) We truncate the outlets at a finite distance.

Of course, both approaches are – each one in its own way – an idealization of real situations and, as we shall see, both approaches present advantages and disadvantages. At this time we wish to observe only that, in case (A1), the difficulty relies, mostly, in determining the asymptotic behavior at large distances in the outlets, while, in case (A2), the challenging task is that of prescribing the appropriate boundary conditions at the inflow-outflow (open) parts of the boundary which are "physically reasonable" and make the corresponding initial-boundary value and boundary-value problems well set.

Once we have chosen a certain mathematical model and formulated the associated governing equations, the basic questions we shall address include the following ones:

(a) Existence and uniqueness of steady-state motions, when the the driving force is time- independent.

(b) Existence and uniqueness of time-periodic motions when the driving force is time-periodic.

(c) Attainability of the above motions, as time goes to infinity, when the fluid is started from rest.

The investigation of the above properties requires a preliminary study of so called *fully developed flows* in an infinite straight pipe of constant cross-section, which will be accordingly analyzed in full detail.

As it will be shown later on, to date, despite the efforts conveyed by many mathematicians, the answers to the questions (a)–(c) are only partially known, and several basic problems remain still unsettled. Specifically, one is able to produce answers only on condition that the magnitude of the data is sufficiently restricted. However, it is not clear if this is due to a lack of a sufficient mathematical knowledge or, rather, to some hidden physical phenomenon. Unfortunately, this is the typical situation that the mathematician experiences with the Navier–Stokes equations and, probably, it is just for this reason that these equations are so particularly fascinating.

1.1. Fully developed flows

Let us consider a Navier–Stokes fluid in an infinite, straight pipe, Ω, of constant cross-section S (bounded domain of $\mathbb{R}^2$). Thus, assuming that x_1 is parallel to the axis of the pipe, we may write

$$\Omega = \{x \in \mathbb{R}^3 : \ x' := (x_2, x_3) \in S\}\,,$$

and the equations governing the motion of the fluid are given by

$$\left.\begin{array}{c} \dfrac{\partial \boldsymbol{v}}{\partial t} + \boldsymbol{v} \cdot \nabla \boldsymbol{v} = \nu \Delta \boldsymbol{v} - \nabla p \\[2mm] \nabla \cdot \boldsymbol{v} = 0 \end{array}\right\} \quad \text{in } \Omega. \tag{1.1}$$

In these equations ν is the coefficient of kinematical viscosity and $p = P/\rho$, where P is the pressure field of the fluid and ρ is its density.

If we introduce the dimensionless quantities

$$t^* = (d^2/\nu)t\,, \quad x_i^* = x_i/d\,, i = 1, 2, 3\,, \quad \boldsymbol{v}^* = U\boldsymbol{v} \tag{1.2}$$

with d diameter of S and U a reference velocity, it is immediately checked that the system (1.1) admits a (dimensionless) solution of the form (stars omitted)

$$\boldsymbol{v}_P(\boldsymbol{x}, t) = V(x_2, x_3, t)\boldsymbol{e}_1\,, \quad p_P = -G(t)x_1\,, \tag{1.3}$$

with

$$\frac{\partial V}{\partial t} = \left(\frac{\partial^2 V}{\partial x_2^2} + \frac{\partial^2 V}{\partial x_3^2}\right) + G(t)\,, \quad V(x_2, x_3, t)|_S = 0\,. \tag{1.4}$$

Flow described by velocity and pressure fields given in (1.3) are called *fully developed*, in that all kinematical quantities do not depend on the axial coordinate. Of particular significance in many applications are two particular classes of fully developed flow, namely, *steady* flow (*Hagen–Poiseuille flow* [54, 88]) where

$$V = V(x_2, x_3)\,, \quad G = \text{const.}$$

and *time-periodic* flow (*Womersley flow* [116]) where

$$V(x_2, x_3, t + 2\pi) = V(x_2, x_3, t)\,, \quad G(t + 2\pi) = G(t)\,,$$

for all $t \in \mathbb{R}$ and where, without loss of generality, we assume that the period $T = 2\pi$.

In regards to (1.4) one is typically interested to solve two specific classes of problems, that we shall denote by Problem 1 and Problem 2, respectively, and which we formulate next.

Problem 1. Given the axial pressure gradient, G, constant or time-periodic, find the corresponding velocity field V.

Problem 2. Given the flow-rate, Φ, through the cross-section S

$$\Phi := \int_S V\, dS\,, \tag{1.5}$$

constant or time-periodic, find corresponding velocity field V *and* axial pressure gradient G.

It is clear that the choice of solving either Problem 1 or Problem 2 is dictated by the nature of the specific situation we want to address. For example, in the case of blood flow in large arteries, it seems much more appropriate to study Problem 2 [116].

With a look at (1.4), we recognize that the mathematical formulation of the above problems is completely different, in that one is *direct* (Problem 1) and the other is *inverse* (Problem 2). More specifically, let's think of (1.4) as a heat equation problem, where V is the "temperature" and G is the "heat source". Then Problem 1 reduces to the standard (and elementary) situation of determining the "temperature" when the "heat source" is given. In contrast, Problem 2 requires the resolution of the non-standard inverse problem [89] where we prescribe the "average temperature" (the flow-rate) and we have to find pointwise "temperature" *and* "heat source".

In what follows, we show that the solvability of Problem 2 can be reduced to that of Problem 1. This follows from the general result that we prove, namely, that Φ and G are related by an invertible relationship expressed through quantities depending *only* on the cross-section and, therefore, independent of V.

Even though the steady-state case is a particular case of the time-periodic one (see also Remark 1.4), we would like to provide the resolution of the above problems separately for the two situations. This because, in the steady-state case, the resolution is quite immediate. We shall also consider Problems 1 and 2 for the initial-boundary value problem associated to (1.9) in regards to the question of attainability of steady-state and time-periodic solutions.

1.1.1. Steady-state case. In this case, (1.3) reduces to

$$\frac{\partial^2 V}{\partial x_2^2} + \frac{\partial^2 V}{\partial x_3^2} = -G, \quad V(x_2, x_3)|_S = 0. \tag{1.6}$$

Consider the following Dirichlet problem:

$$\frac{\partial^2 \varphi}{\partial x_2^2} + \frac{\partial^2 \varphi}{\partial x_3^2} = -1, \quad \varphi(x_2, x_3)|_{\partial S} = 0. \tag{1.7}$$

It is well known that, if S is of class C^2 or convex, (1.6) has one and only one solution $\varphi \in W^{2,2}(S) \cap W_0^{1,2}(S)$. Multiplying both sides of (1.7) by φ and integrating by parts we find

$$\int_S \varphi = \int_S |\nabla \varphi|^2 > 0.$$

Thus, if we set

$$V(x_2, x_3) = \frac{\Phi}{\displaystyle\int_S \varphi \, dS} \varphi(x_2, x_3), \quad G = \frac{\Phi}{\displaystyle\int_S \varphi \, dS}, \tag{1.8}$$

we obtain at once that these V and G solve (1.6) for any choice of Φ, and that $\Phi = \int_S V \, dS$. Thus, in Problem 1, the solution to (1.6) is given by $V = G\varphi$, while in Problem 2 it is just furnished by (1.8).

Remark 1.1. The function φ depends *only* on S. Therefore, the relation between G and Φ depends only on S and is *independent of the particular velocity field V*.

1.1.2. Time-periodic case. In this case, the problem reduces to the following one

$$\frac{\partial V}{\partial t} = \left(\frac{\partial^2 V}{\partial x_2^2} + \frac{\partial^2 V}{\partial x_3^2}\right) + G(t), \quad V(x_2, x_3, t)|_S = 0,$$

$$V(x_2, x_3, t) = V(x_2, x_3, t + 2\pi), \quad \text{for all } (x_2, x_3) \in S \text{ and } t \in \mathbb{R},$$

(1.9)

where, in Problem 1, the function G is prescribed with $G(t) = G(t + 2\pi)$, whereas in Problem 2 the flow-rate Φ is prescribed with $\Phi(t) = \Phi(t + 2\pi)$ and we have to find time-periodic V and G.

From (1.9) we immediately deduce that the resolution of Problem 1 presents no difficulty, since it is equivalent to the elementary problem of finding time-periodic solutions of the heat equation when the right-hand side G is a *given* time-periodic function. In such a case we have the following (see, e.g., [113])

Theorem 1.1. *Let S be a bounded domain in the plane and let G be a 2π-periodic function with $G \in L^2(-\pi, \pi)$. Then (1.9) has a unique 2π-periodic solution u, such that u is continuous from $[-\pi, \pi]$ in $H_0^1(S)$ and*

$$\frac{\partial V}{\partial t}, \Delta V \in L^2(S \times (-\pi, \pi)).$$

(1.10)

This solution satisfies, in addition, the inequality

$$\max_{t \in [-\pi, \pi]} \|V(t)\|_{1,2}^2 + \int_{-\pi}^{\pi} \left(\left\|\frac{\partial V}{\partial t}\right\|_2^2 + \|\Delta V(t)\|_2^2\right) dt \leq c \int_{-\pi}^{\pi} |G(t)|^2 dt$$

(1.11)

where $c = c(\Omega) > 0$.

We now turn to Problem 2. Its solvability, under very special assumptions on the time-dependence of the flow rate and for S a circle, was first given by Womersley [116]. Much more recently, Beirão da Veiga [7] has solved the problem in its full generality. However, from his approach it is not clear what is the relation between Φ and G and, in fact, it seems to depend on the velocity field V. Successively, Galdi and Robertson [43] provided a different and much more elementary approach that proves, among other things, that Φ and G are related by one-to-one correspondence with coefficients depending *only* on the cross-section and, therefore, independent of V. As a byproduct of this result, they show that the resolution of Problem 2 is *equivalent* to that of Problem 1. In what follows we will present the main ideas of Galdi and Robertson approach, referring to [43] for details.

To this end, consider the following sequence of problems, $n \in \mathbb{N}$, [3]

$$\frac{\partial u_{cn}}{\partial t} = \Delta u_{cn} + \cos(nt) \text{ in } S, \quad u_{cn}|_{\partial S} = 0, \quad u_{cn}(x, 0) = u_{cn}(x, 2\pi), \quad x \in S,$$

$$\frac{\partial u_{sn}}{\partial t} = \Delta u_{sn} + \sin(nt) \text{ in } S, \quad u_{sn}|_{\partial S} = 0, \quad u_{sn}(x, 0) = u_{sn}(x, 2\pi), \quad x \in S.$$

(1.12)

---

[3] For simplicity, the point (x_2, x_3) will be denoted by x, rather than by x'.

The functions u_{cn} and u_{sn} can be viewed as real and imaginary part, respectively, of the function v_n where

$$\frac{\partial v_n}{\partial t} = \Delta v_n + e^{int} \ \text{ in } S, \quad v_n|_{\partial S} = 0, \quad v_n(x,0) = v_n(x, 2\pi), \quad x \in S. \tag{1.13}$$

A solution to (1.13) is given by $v_n(x,t) = V_n(x)e^{int}$, with $V_n = \varphi_n + i\psi_n$, and

$$\left.\begin{array}{r} -n\,\psi_n = \Delta\varphi_n + 1, \\[2mm] n\,\varphi_n = \Delta\psi_n \end{array}\right\} \ \text{ in } S, \quad \varphi_n|_{\partial S} = \psi_n|_{\partial S} = 0. \tag{1.14}$$

By standard methods, we prove that, for each non-negative integer n, the system (1.14) possesses a unique solution $\varphi_n, \psi_n \in W_0^{1,2}(S)$ with $\Delta\varphi_n, \Delta\psi_n \in L^2(S)$. For these properties to hold, S can be an arbitrary bounded domain. We also observe that, obviously, φ_n and ψ_n depend only on n and S. From (1.13) and (1.14) we then conclude that the solutions to (1.12) are given by

$$u_{cn} = \varphi_n \cos(nt) - \psi_n \sin(nt), \quad u_{sn} = \psi_n \cos(nt) + \varphi_n \sin(nt). \tag{1.15}$$

We shall now establish some simple but important properties of solutions to (1.14). Set

$$a_n = \int_S \varphi_n \, dx, \quad b_n = -\int_S \psi_n \, dx \quad n = 0, 1, 2, \dots. \tag{1.16}$$

We emphasize that, for each fixed n, the numbers a_n and b_n depend *only* on S.

Lemma 1.1. *Let $\varphi_n, \psi_n \in W_0^{1,2}(S)$ be a solution to (1.14). Then, the following inequality holds,*

$$\|\Delta\varphi_n\|_2 + \|\Delta\psi_n\|_2 \le |S|^{\frac{1}{2}}, \quad \text{for all } n = 0, 1, 2, \dots, \tag{1.17}$$

where $|S|$ is the area of S. Moreover, the real numbers a_n and b_n defined in (1.16) satisfy the following properties,

(a) $a_n > 0$, *for all* $n = 0, 1, 2, \dots$; $b_0 = 0$, $b_n > 0$ *for all* $n \in \mathbb{N}$;

(b) $a_0 \le \dfrac{|S|^2}{2}$; $a_n \le \dfrac{|S|}{n}$, $b_n \le \dfrac{|S|}{n}$, *for all* $n \in \mathbb{N}$;

(c) $\lim_{n \to \infty} (nb_n) = |S|$.

Proof. See [43, Lemma 2.1]. $\qquad\qquad\square$

We shall next establish the relation between G and Φ. To this end, we write the Fourier series of both quantities:

$$G(t) = \frac{G_{c0}}{2} + \sum_{n=1}^{\infty} \{G_{cn} \cos(n\pi t) + G_{sn} \sin(n\pi t)\},$$

$$\Phi(t) = \frac{\Phi_{c0}}{2} + \sum_{n=1}^{\infty} \{\Phi_{cn} \cos(n\pi t) + \Phi_{sn} \sin(n\pi t)\}. \tag{1.18}$$

The following result holds.

Proposition 1.1. *The Fourier coefficients* (G_{cn}, G_{sn}) *of* G *and those* (Φ_{cn}, Φ_{sn}) *of* Φ *are related to each other according to the following formulas,* $n = 0, 1, 2, \ldots,$

$$\Phi_{cn} = a_n G_{cn} - b_n G_{sn}, \quad \Phi_{sn} = b_n G_{cn} + a_n G_{sn}. \tag{1.19}$$

or, equivalently, by their inverse

$$G_{cn} = \frac{a_n \Phi_{cn} + b_n \Phi_{sn}}{a_n^2 + b_n^2}, \quad G_{sn} = \frac{a_n \Phi_{sn} - b_n \Phi_{cn}}{a_n^2 + b_n^2}. \tag{1.20}$$

Proof. We begin to observe that, obviously, it is enough to show the validity of (1.19), since (1.20) follows directly from this latter. (Notice that, by Lemma 1.1, we have $a_n^2 + b_n^2 > 0$ for all $n = 0, 1, 2, \ldots$). Set

$$\overline{u}_{cn} = \overline{u}_{cn}(x, t) \equiv u_{cn}(2\pi - t), \quad \overline{u}_{sn} = \overline{u}_{sn}(x, t) \equiv u_{sn}(2\pi - t), \quad t \in [-\pi, \pi]. \tag{1.21}$$

From (1.12), it follows that $\overline{u}_{cn}$ satisfies the following problems,

$$\frac{\partial \overline{u}_{cn}}{\partial t} + \Delta \overline{u}_{cn} = -\cos(n(2\pi - t)) \text{ in } S, \ \overline{u}_{cn}|_{\partial S} = 0, \ \overline{u}_{cn}(0) = \overline{u}_{cn}(2\pi),$$

$$\frac{\partial \overline{u}_{sn}}{\partial t} + \Delta \overline{u}_{sn} = -\sin(n(2\pi - t)) \text{ in } S, \ \overline{u}_{sn}|_{\partial S} = 0, \ \overline{u}_{sn}(0) = u_{sn}(2\pi). \tag{1.22}$$

We now multiply both sides of (1.9) by $\overline{u}_{cn}$ and integrate by parts over S. We thus obtain

$$\frac{d}{dt}(V, \overline{u}_{cn}) - (V, \frac{\partial \overline{u}_{cn}}{\partial t} + \Delta \overline{u}_{cn}) = (G, \overline{u}_{cn}).$$

Furthermore, we integrate this relation over $t \in [-\pi, \pi]$. By taking into account (1.5), (1.15)$_1$, (1.16), (1.21)$_1$, (1.22)$_1$ and the fact that $u(x, -\pi) = u(x, \pi)$, we get

$$\int_{-\pi}^{\pi} \Phi(t) \cos[(n(2\pi - t)] \, dt = \int_{-\pi}^{\pi} G(t) \{a_n \cos[n(2\pi - t)] + b_n \sin[n(2\pi - t)]\} \, dt.$$

From this relation we deduce at once

$$\int_{-\pi}^{\pi} \Phi(t) \cos(nt) \, dt = \int_{-\pi}^{\pi} G(t) \{a_n \cos(nt) - b_n \sin(nt)\} \, dt$$

which coincides with the first relation in (1.19). The second relation in (1.19) is obtained exactly by the same procedure, provided we replace in the above argument $\overline{u}_{cn}$ with $\overline{u}_{sn}$. The proposition is, therefore, completely proved. $\square$

The convergence of the series (1.18) and the relation between the norms of G and Φ in $L^2(-\pi, \pi)$ is proved in the following.

Proposition 1.2. *Assume that the numbers* (Φ_{cn}, Φ_{sn}) *and* (G_{cn}, G_{sn}), $n = 0, 1, 2,$ $\ldots$, *satisfy* (1.19) *(or, equivalently,* (1.20))*. Then, if the Fourier series* (1.18)$_1$ *converges to some* $G \in L^2(-\pi, \pi)$, *also the Fourier series* (1.18)$_2$ *converges to some* $\Phi \in L^2(-\pi, \pi)$ *and we have*

$$\|\Phi\|_{L^2(-\pi,\pi)} \leq c_0 |S| \|G\|_{L^2(-\pi,\pi)} \tag{1.23}$$

where $c_0 = \sqrt{\max\{2, |S|^2/4\}}$. *Conversely, if the Fourier series* $(1.18)_2$ *converges to some* $\Phi \in L^2(-\pi, \pi)$ *with* $d\Phi/dt \in L^2(-\pi, \pi)$, *then also the Fourier series* $(1.18)_1$ *converges to some* $G \in L^2(-\pi, \pi)$ *and we have*

$$\|G\|_{L^2(-\pi,\pi)} \le c_1 \|\Phi\|_{L^2(-\pi,\pi)} + \frac{2}{|S|} \left\| \frac{d\Phi}{dt} \right\|_{L^2(-\pi,\pi)}, \qquad (1.24)$$

where c_1 *is a positive constant depending only on* S.

Proof. From (1.19) and Lemma 1.1(a) it follows that

$$\Phi_{c0} = a_0 G_{c0}, \quad |\Phi_{cn}|^2 + |\Phi_{sn}|^2 = (a_n^2 + b_n^2)(|G_{cn}|^2 + |G_{sn}|^2), \quad n \in \mathbb{N}. \quad (1.25)$$

Therefore, from Lemma 1.1(b) we find

$$\frac{|\Phi_{c0}|^2}{2} + \sum_{n=1}^{\infty} \left\{ |\Phi_{cn}|^2 + |\Phi_{sn}|^2 \right\} \le \frac{|S|^4}{4} \frac{|G_{c0}|^2}{2} + 2|S|^2 \sum_{n=1}^{\infty} \left\{ |G_{cn}|^2 + |G_{sn}|^2 \right\}$$

which proves that the series $(1.18)_2$ is converging in $L^2(-\pi, \pi)$, if $G \in L^2(-\pi, \pi)$. Moreover, (1.23) follows from Parseval's equality. Conversely, we notice that (1.25) implies

$$|G_{cn}|^2 + |G_{sn}|^2 \le \frac{1}{b_n^2} \left(|\Phi_{cn}|^2 + |\Phi_{sn}|^2 \right), \quad \text{for all } n \in \mathbb{N}. \quad (1.26)$$

From Lemma 1.1(c) we have that there is a positive integer $\bar{n}$ such that

$$b_n \ge \frac{|S|}{2n}, \quad \text{for all } n \ge \bar{n}.$$

Setting $\bar{b} = \min\{b_1, \dots, b_{\bar{n}}\}$, in view of Lemma 1.1(a), it follows that $\bar{b} > 0$. Thus, from this latter displayed equation, from Lemma 1.1(b) and from (1.26) we find

$$\frac{|G_{c0}|^2}{2} + \sum_{n=1}^{\infty} \left\{ |G_{cn}|^2 + |G_{sn}|^2 \right\} \le \frac{4}{|S|^4} \frac{|\Phi_{c0}|^2}{2} + \frac{1}{\bar{b}^2} \sum_{n=1}^{\infty} \left\{ |\Phi_{cn}|^2 + |\Phi_{sn}|^2 \right\}$$

$$+ \frac{4}{|S|^2} \sum_{n=1}^{\infty} n^2 \left\{ |\Phi_{cn}|^2 + |\Phi_{sn}|^2 \right\}$$

which, by the assumptions on Φ, shows that the series (1.18) converges in $L^2(-\pi, \pi)$. Finally, (1.24) is a consequence of Parseval's equality. $\qquad \square$

An immediate, important consequence of the previous results is the following one which proves the resolution of Problem 2.

Theorem 1.2. *Let* S *be a bounded domain of the plane, and let* Φ *be a* 2π-*periodic function with* $\Phi, d\Phi/dt \in L^2(-\pi, \pi)$. *Then, the problem*

$$\frac{\partial V}{\partial t} = \Delta V + G \quad \text{in } G, \quad V|_{\partial\Omega} = 0, \quad \int_S V(x, t)\, dx = \Phi(t) \qquad (1.27)$$

admits one and only one 2π-periodic solution (V, G), where $V \in C([-\pi, \pi]; W_0^{1,2}(S)$ and satisfies (1.10), while $G \in L^2(-\pi, \pi)$. Furthermore, the solution satisfies the estimate

$$\max_{t \in [-\pi, \pi]} \|V(t)\|_{1,2}^2 + \int_{-\pi}^{\pi} \left(\left\| \frac{\partial V}{\partial t} \right\|_2^2 + \|\Delta V(t)\|_2^2 + |G(t)|^2 \right) dt$$

$$\leq c \int_{-\pi}^{\pi} \left(|\Phi(t)|^2 + |d\Phi/dt|^2 \right) dt, \tag{1.28}$$

where $c = c(S) > 0$.

Proof. Let Φ_{cn}, Φ_{sn} be the Fourier coefficients of Φ, and consider the series on the right-hand side of $(1.18)_1$, with coefficients G_{cn}, G_{sn} given in (1.20). From Proposition 2.2 we know that the series $(1.18)_1$ is convergent to some $G \in L^2(-\pi, \pi)$ and that inequality (1.24) holds. We then solve problem (1.9) with this *given* G. The existence part along with the validity of (1.28) is then a consequence of Theorem 2.1, Proposition 2.1 and of inequality (1.24). As for uniqueness, it suffices to show that if (V, G) solves (1.27) with $\Phi \equiv 0$, then $V \equiv G \equiv 0$. Multiplying both sides of (1.27) by u, integrating over $S \times (-\pi, \pi)$ and using $\Phi \equiv 0$ then furnishes $\int_{-\pi}^{\pi} \|\nabla V\|_2^2 = 0$, that is $V \equiv 0$. By going back to the first equation in (1.27), we then infer $G \equiv 0$, and uniqueness follows. $\qquad\square$

Remark 1.2. Solutions of Theorem 1.2 have the following simple representation,

$$V(x, t) = \frac{G_{c0}}{2} \varphi_0 + \sum_{n=1}^{\infty} \{ [G_{cn} \varphi_n + G_{sn} \psi_n] \cos(nt) + [G_{sn} \varphi_n - G_{cn} \psi_n] \sin(nt) \}, \tag{1.29}$$

where, for $n = 0, 1, 2, \ldots$, the functions φ_n, ψ_n satisfy (1.14) while the numbers G_{cn}, G_{sn} are given in (1.20). This easily follows from (1.12)–(1.15) along with Lemma 1.1.

Remark 1.3. If S is of class C^2 or if it is convex, we can replace the term $\|\Delta V\|_2^2$ in (1.28) with $\|V\|_{2,2}^2$.

Remark 1.4. In the special case when Φ is a constant, then from (1.29), (1.18) and (1.20), we find $V = \Phi \varphi_0 / a_0 = G \varphi_0$ which, by (1.16), coincides with (1.8).

1.1.3. Attainability of steady-state and time-periodic flow. In several applied problems it is of some interest to investigate the rate at which a given unsteady motion, started from rest, approaches, as time goes to infinity, a fully developed steady-state flow or, more generally, time-periodic flow, corresponding to prescribed flow-rate. This is the problem of *attainability* of the given flow. This problem, which in its full generality will be treated in the following Section 1.3.3, requires a preliminary study in the subclass of fully developed flow. This will be the object of the present section. More specifically, assume that the flow-rate of a fully developed flow is smoothly increased from zero to a certain constant or time-periodic function $\overline{\Phi} = \overline{\Phi}(t)$ in the time-interval $[0, 1]$. [4] We shall show that the corresponding velocity field and axial pressure gradient will tend, as $t \to \infty$, exponentially fast to the analogous quantities corresponding to Λ, the constant of decay being proportional to the first eigenvalue of the Laplace operator in S with Dirichlet boundary conditions.

Mathematically, the problem is formulated as follows. Let $\psi = \psi(t)$ be a smooth, non-decreasing "ramping" function defined in $\mathbb{R}$ that is 0 for $t \leq 0$ and is 1 for $t \geq 1$, and set $\Phi(t) = \psi(t)\,\overline{\Phi}(t)$. Moreover, denote by $(\overline{V}, \overline{G})$ the solution corresponding to the flow-rate $\overline{\Phi}$. The attainability problem consists then in finding a solution (V, G) of the following initial-boundary value problem [5]

$$\frac{\partial V}{\partial t} = \left(\frac{\partial^2 V}{\partial x_2^2} + \frac{\partial^2 V}{\partial x_3^2} \right) + G(t)\,, \quad V(x,t)|_S = 0\,,$$

$$V(x,0) = 0\,, \quad x \in S\,, \quad \int_S V(x,t)dx = \Phi(t)\,, \quad t \geq 0\,, \tag{1.30}$$

such that

$$\lim_{t \to \infty} (V(x,t) - \overline{V}(x,t)) = 0\,, \quad \lim_{t \to \infty} (G(t) - \overline{G}(t)) = 0\,. \tag{1.31}$$

We now set

$$u := V - \overline{V}\,, \quad q := G - \overline{G}\,, \quad F := (\psi - 1)\,\overline{\Phi}$$

so that (1.30) and (1.31) reduce to find $u(x,t)$ and $q(t)$ solving the following equations,

$$\frac{\partial u}{\partial t} = \left(\frac{\partial^2 u}{\partial x_2^2} + \frac{\partial^2 u}{\partial x_3^2} \right) + q(t)\,, \quad u(x,t)|_S = 0\,,$$

$$u(x,0) = u_0(x)\,, \quad x \in S\,, \quad \int_S u(x,t)\,dx = F(t)\,, \quad t \geq 0\,, \tag{1.32}$$

$$\lim_{t \to \infty} u(x,t) = 0\,, \quad \lim_{t \to \infty} q(t) = 0\,,$$

where

$$u_0(x) := -\overline{V}(x,0)\,, \text{ and } F(t) = 0 \text{ for all } t \geq 1\,.$$

Our strategy for the resolution of (1.32) is based on the recent work of Galdi, Pileckas and Silvestre [41] and goes as follows. We show that, in a sufficiently smooth class of solutions and for a given u_0, the functions q and F are related by an invertible Volterra linear integral equation of the second kind, with kernel depending only on S; see Proposition 1.3. Thus, given F (sufficiently smooth) and u_0 we can determine the corresponding q and then, by elementary results on the heat equation, we obtain u from (1.32)$_{1,\dots,4}$ in the appropriate function class (Proposition 1.4). Finally, by a general result on solutions to linear integral

---

[4] The number 1 in this interval can be replaced by any number $a > 0$. In such a case, the constants involved in Theorem 1.3 will depend on a as well.

[5] Again for the sake of notational simplicity, we shall denote the point (x_2, x_3) by x, instead of x'.

Volterra equations of the second kind proved in Lemma 1.3, we show that u and q satisfy also $(1.32)_{5,6}$ and find the corresponding rate of decay (Proposition 1.5).

In order to accomplish our goal, let $\theta(x,t)$ be the solution to the following initial boundary value problem in $S^T := S \times (0,T)$, $T > 0$, for the heat equation

$$\begin{cases} \dfrac{\partial}{\partial t}\theta(x,t) = \Delta\theta(x,t)\,, \\[2mm] \theta(x,t)\,|_{\partial S} = 0\,, \quad \theta(x,0) = \varphi(x)\,, \end{cases} \tag{1.33}$$

where φ is the solution to (1.7). We recall that [70], if S is of class C^2 (or convex), then $\varphi \in W^{2,2}(S) \cap \mathring{W}^{1,2}(S)$ and that

$$\|\varphi\|_{W^{2,2}(S)} \leq c_0\,, \tag{1.34}$$

where c_0 depends only on S. Consequently, we conclude that (1.33) has a unique solution θ such that

$$\left\|\frac{\partial}{\partial t}\theta(\cdot,t)\right\|^2_{L^2(S^T)} + \int_0^T \|\theta(\cdot,t)\|^2_{W^{2,2}(S)}\,dt < \infty\,;$$

and that this solution satisfies

$$\left\|\frac{\partial}{\partial t}\theta(\cdot,t)\right\|^2_{L^2(S^T)} + \int_0^T \|\theta(\cdot,t)\|^2_{W^{2,2}(S)}\,dt \leq c\,\|\varphi\|^2_{W^{2,1}(S)}\,, \tag{1.35}$$

where $c_1 = c_1(S) > 0$; see, e.g., [70]. Since $\varphi \in W^{2,2}(S)$, we also deduce that

$$\nabla\frac{\partial\theta}{\partial t} \in L_2(S^T)\,, \quad \frac{\partial\theta}{\partial t} \in C([0,T]; L^2(S))\,,$$

$$\max_{t \in [0,T]}\left(\left\|\frac{\partial}{\partial t}\theta(\cdot,t)\right\|_{L^2(S)} + \|\theta(\cdot,t)\|_{W^{2,2}(S)}\right) + \left\|\nabla\frac{\partial}{\partial t}\theta(\cdot,t)\right\|_{L^2(S^T)} \leq c\|\varphi\|_{W^{2,2}(S)}\,, \tag{1.36}$$

where $c_2 = c_2(S) > 0$; see [70]. Finally, for arbitrary $\delta > 0$, we further find $\frac{\partial^2}{\partial t^2}\theta \in L^2(S \times (\delta,T))$ and $\Delta\frac{\partial}{\partial t}\theta \in L^2(S \times (\delta,T))$. [6]

Now, let $u(x,t)$ be a "sufficiently smooth" solution to (1.32). Multiplying both sides of equation $(1.32)_1$ by $\theta(x,\tau - t)$ and integrating by parts over S, we find

$$\int_S \frac{\partial}{\partial t}u(x,t)\theta(x,\tau-t)\,dx = \int_S \Delta u(x,t)\theta(x,\tau-t)\,dx + q(t)\int_S \theta(x,\tau-t)\,dx,$$

that is,

$$\frac{\partial}{\partial t}\int_S u(x,t)\theta(x,\tau-t)\,dx - \int_S u(x,t)\frac{\partial}{\partial t}\theta(x,\tau-t)\,dx$$

$$= \int_S u(x,t)\Delta\theta(x,\tau-t)\,dx + q(t)\int_S \theta(x,\tau-t)\,dx.$$

---

[6] Note that it is not possible to take $\delta = 0$, because the initial datum $\varphi(x)$ does not satisfy the compatibility condition, i.e., $\Delta\varphi(x)\,|_{\partial S} \neq 0$.

Integrating this last relation with respect to t from 0 to τ and taking into account that

$$\nu\Delta\theta(x,\tau-t) = \frac{\partial}{\partial\tau}\theta(x,\tau-t) = -\frac{\partial}{\partial t}\theta(x,\tau-t),$$

$$u(x,0) = u_0, \quad \theta(x,\tau-t)\big|_{t=\tau} = \theta(x,0) = \varphi(x),$$

we derive

$$\int_S u(x,\tau)\varphi(x)\,dx - \int_S u_0(x)\theta(x,\tau)\,dx = \int_0^\tau H(\tau-t)q(t)\,dt, \tag{1.37}$$

with

$$H(s) := \int_S \theta(x,s)\,dx. \tag{1.38}$$

We next differentiate both sides of (1.37) with respect to τ and employ (1.33) to get

$$\int_S \frac{\partial}{\partial\tau}u(x,\tau)\varphi(x)\,dx - \int_S \Delta u_0(x)\theta(x,\tau)\,dx = H(0)q(\tau) + \int_0^\tau H'(\tau-t)q(t)\,dt.$$

Using equation $(1.32)_1$ and integrating twice by parts we deduce

$$\int_S \frac{\partial}{\partial\tau}u(x,\tau)\varphi(x)\,dx = \int_S \Delta u(x,\tau)\varphi(x)\,dx + q(\tau)\int_S \varphi(x)\,dx$$

$$= \int_S u(x,\tau)\Delta\varphi(x)\,dx + q(\tau)\int_S \varphi(x)\,dx = -\int_S u(x,\tau)\,dx$$

$$+q(\tau)\int_S \varphi(x)\,dx = -F(\tau) + q(\tau)\int_S \varphi(x)\,dx,$$

where $F(\tau)$ is given in $(1.32)_4$. Since $H(0) = \int_S \theta(x,0)\,dx = \int_S \varphi(x)\,dx$, from the two latter displayed equalities, we find that $F(t)$ and $q(t)$ are related by the following Volterra equation of the first kind,

$$-F(\tau) - \int_S \Delta u_0(x)\theta(x,\tau)\,dx = \int_0^\tau H'(\tau-t)q(t)\,dt. \tag{1.39}$$

If we differentiate (1.39) with respect to τ and take into account (1.33), we obtain

$$-F'(\tau) - \int_S \Delta u_0(x)\Delta\theta(x,\tau)\,dx = H'(0)q(\tau) + \int_0^\tau H''(\tau-t)q(t)\,dt, \tag{1.40}$$

and since, by (1.7) and (1.33), we have

$$H'(0) = \int_S \frac{\partial}{\partial\tau}\theta(x,\tau)\Big|_{\tau=0}\,dx = \int_S \Delta\varphi(x)\,dx = -|S|$$

we conclude that (1.40) takes the form

$$q(\tau) = \int_0^\tau K(\tau-t)q(t)\,dt + \Psi(\tau), \tag{1.41}$$

where

$$\Psi(\tau) := \frac{1}{|S|}\left(F'(\tau) + \int_S \Delta u_0(x)\Delta\theta(x,\tau)\,dx\right) \tag{1.42}$$

and

$$K(s) := \frac{1}{|S|} \frac{d^2}{ds^2} \int_S \theta(x, s)\, dx\,. \tag{1.43}$$

Notice that if $u_0 \in W^{1,2}(S)$ and $F \in W^{1,2}(0, T)$, then $\Psi \in L^2(0, T)$ and

$$\|\Psi\|_{L^2(0,T)} \le c_1 \left(\|F\|_{W^{1,2}(0,T)} + \|u_0\|_2 \right)\,, \tag{1.44}$$

with $c_1 = c_1(S) > 0$. This follows by integrating by parts the second term on the right-hand side of (1.42) and by taking into account (1.36). Alternatively, if $F \in C^1([0, T])$ and $u_0 \in W^{2,2}(S)$, then $\Psi \in C([0, T])$ and

$$\max_{t \in [0,T]} |\Psi(t)| \le c_2 \left(\max_{t \in [0,T]} |F'| + \|u_0\|_{2,2} \right)\,, \tag{1.45}$$

with $c_2 = c_2(S) > 0$, as a consequence, again, of (1.42) and (1.36).

Equation (1.41) can be viewed as a Volterra integral equation of the second kind in the unknown function q, for a given Ψ. Notice that the kernel K depends only on S and it is independent of the particular solution u. Furthermore, if $q \in C[0, T]$, then (1.41) and (1.39) are equivalent if and only if $\int_S u_0(x)dx = F(0)$. In fact, from (1.41) we get

$$\frac{d}{d\tau} \left(\int_0^\tau H'(\tau - t)q(t)dt + F(\tau) + (\Delta u_0, \theta(\tau)) \right) = 0\,,$$

which, by (1.38) and (1.36)$_3$, coincides with (1.39) if and only if $F(0) = -(\Delta u_0, \varphi)$ where $\varphi\ (= \theta(x, 0))$ satisfies (1.7). We thus deduce

$$F(0) = -(\Delta u_0, \varphi) = -(u_0, \Delta\varphi) = \int_S u_0(x)dx\,.$$

The existence and uniqueness of a solution q to (1.41) in the class $C[0, T]$ follows from known results, provided $\Psi \in C[0, T]$ and $K \in L^1(0, T)$; see [81]. Thus, if we assume $F' \in C[0, T]$ – namely, $\Lambda' \in C[0, T]$ – from (1.36) we deduce $\Psi \in C[0, T]$. (Recall that, by the results of Sections 1.1.1 and 1.1.2 we may take $u_0 := \overline{V} \in W^{2,2}(\Omega)$.)

Concerning the summability property of K, we have the following result for whose proof we refer to [41, Lemma 2.2].

Lemma 1.2. *Let $\psi_k \in \mathring{W}^{1,2}(S) \cap W^{2,2}(S)$ and $\lambda_k > 0$ be eigenfunctions and eigenvalues of the Laplace operator in S with homogeneous Dirichlet boundary conditions. Then, the kernel $K(t)$ admits the representation*

$$K(t) = \frac{1}{|S|} \sum_{k=1}^\infty \beta_k^2 \lambda_k \exp(-\lambda_k t)\,, \tag{1.46}$$

where $\beta_k = \int_S \psi_k(x)dx$, $k = 1, 2, \ldots$ Thus, $K(t) > 0$ for all $t > 0$ and, moreover,

$$\int_0^\infty K(t)\, dt = 1\,. \tag{1.47}$$

Finally, for all $t \in (0,1]$, the following inequality holds,

$$K(t) \leq C\,t^{-1/2}\,, \tag{1.48}$$

where $C = C(S) > 0$.

We next observe that, as shown in [108], the unique solution q to (1.41) is represented by the formula

$$q(t) = \int_0^t R(t - s)\Psi(s)\,ds + \Psi(t), \quad t \in [0, T], \tag{1.49}$$

where $R(t)$ satisfies the estimate [81]

$$\|R\|_{L_1(0,T)} \leq c(T)\|K\|_{L_1(0,T)}\,.$$

If $F' \in C([0,T])$ and $u_0 \in W^{2,2}(S)$, this inequality combined with (1.36), (1.42), (1.49) and (1.45) readily furnishes that $q \in C([0,T])$ and that

$$\max_{t \in [0,T]} |q(t)| \leq c_3 \left(\max_{t \in [0,T]} |F'(t)| + \|u_0\|_{2,2} \right), \tag{1.50}$$

where $c_3 = c_3(T, S) > 0$. Alternatively, if we only have $F \in W^{1,2}(0,T)$ and $u_0 \in W^{1,2}(S)$, then, from (1.49), from Young's inequality for convolutions and from (1.44) it follows that

$$\|q\|_{L^2(0,T)} \leq c_4 \left(\|F\|_{W^{1,2}(0,T)} + \|u_0\|_{1,2} \right), \tag{1.51}$$

where $c_4 = c_4(T, S) > 0$. From the above considerations we then obtain the following result.

Proposition 1.3. *Let S be of class C^2, $u_0 \in W_0^{1,2}(S)$, and let $F \in W^{1,2}(0,1)$.* [7] *Then, there exists one and only one solution $q \in L^2(0,T)$ to (1.41), for all $T > 0$. Moreover, q satisfies (1.51). If $F \in C^1([0,1])$ and, in addition, $u \in W^{2,2}(S)$, then $q \in C([0,T])$ for all $T > 0$ and estimate (1.50) holds.*

With this result in hand it is easy to prove existence for problem $(1.32)_{1,...,4}$ in a suitable class. Specifically, we prove the following.

Proposition 1.4. *Let S be of class C^2, $F \in W^{1,2}(0,1)$ and let $u_0 \in W_0^{1,2}(S)$ satisfy the condition*

$$\int_S u_0(x)dx = F(0)\,. \tag{1.52}$$

Then, problem $(1.32)_{1,...,4}$ has one and only one solution (u, q) such that, for all $T > 0$,

$$u \in C([0,T]; W_0^{1,2}(S))\,, \quad \frac{\partial u}{\partial t} \in L^2(S^T) \quad u \in L^2((0,T); W^{2,2}(S))\,, \quad q \in L^2(0,T)\,. \tag{1.53}$$

---

[7] Recall that $F(t) = 0$ for all $t \geq 1$.

This solution satisfies, in addition, the inequality

$$\max_{t \in [0,T]} \|u(t)\|_{1,2}^2 + \int_0^T \left(\left\| \frac{\partial u}{\partial t} \right\|_2^2 + \|u(t)\|_{2,2}^2 + |q(t)|^2 \right) dt$$

$$\leq C_1 \left(\|F\|_{W^{1,2}(0,1)}^2 + \|u_0\|_{1,2}^2 \right), \tag{1.54}$$

where $C_1 = C_1(S,T) > 0$. Moreover, if $F' \in C([0,1])$ and, in addition, $u_0 \in W^{2,2}(S)$, then, for all $T > 0$,

$$\frac{\partial u}{\partial t} \in C([0,T]; L^2(S)) \quad u \in C([0,T]; W^{2,2}(S)), \quad q \in C[0,T], \tag{1.55}$$

and u satisfies also the inequality

$$\max_{t \in [0,T]} \left(\left\| \frac{\partial u}{\partial t}(t) \right\|_2 + \|u(t)\|_{2,2} + |q(t)| \right) \leq C_2 \left(\max_{t \in [0,1]} |F'(t)| + \|u_0\|_{2,2} \right), \tag{1.56}$$

where $C_2 = C_2(S,T) > 0$.

Proof. For the given F we pick q as the solution to (1.41), according to Proposition 1.3. We then consider problem $(1.32)_{1,\dots,4}$ with this specific q. The existence part of the field u then follows from classical results on the heat equation, while estimates (1.54) and (1.56) follow from (1.51) and (1.50), respectively. As for uniqueness, assume that u lies in the above class and satisfies the following initial-boundary value problem

$$\frac{\partial u}{\partial t} = \left(\frac{\partial^2 u}{\partial x_1^2} + \frac{\partial^2 u}{\partial x_2^2} \right) + q(t), \quad u(x,t)|_S = 0,$$

$$u(x,0) = 0, \quad x \in S, \quad \int_S u(x,t)dx = 0, \quad t \geq 0. \tag{1.57}$$

Multiplying both sides of $(1.57)_1$ by u, integrating by parts over S and using $(1.57)_4$, we deduce

$$\frac{1}{2} \frac{d}{dt} \|u\|_2^2 + \|\nabla u\|_2^2 = 0,$$

which, in turn, by $(1.57)_3$ implies $u(x,t) \equiv 0$. This, by $(1.57)_1$, furnishes $q(t) \equiv 0$, and uniqueness follows. $\qquad\square$

We shall next show the asymptotic properties of u and q.

Proposition 1.5. *Let S be of class C^2, $F \in W^{1,2}(0,1)$ and let $u_0 \in W_0^{1,2}(S)$ satisfy condition (1.52) Then, the solution (u,q) determined in Proposition 1.4 satisfies the following properties for all $t \geq 1$,*

$$\|u(t)\|_{1,2} \leq \|u(1)\|_{1,2}\, e^{-\lambda_1 t},$$

$$\int_1^\infty \left(\left\| \frac{\partial u}{\partial t} \right\|_2^2 + \|u(t)\|_{2,2}^2 + |q(t)|^2 \right) e^{\lambda_1 t}\, dt \leq C \|\nabla u(1)\|_2^2,$$

where $\lambda_1 = \lambda_1(S)$ is the smallest eigenvalue of the Laplace operator introduced in Lemma 1.2 *and* $C = C(S) > 0$.

Proof. We begin to obtain some estimates on u. Multiplying both sides of $(1.32)_1$ by u and then by $\partial u / \partial t$, and integrating by parts over S we obtain

$$\frac{1}{2}\frac{d}{dt}\|u\|_2^2 + \|\nabla u\|_2^2 = (q, u) = q\,F\,, \tag{1.58}$$

and

$$\frac{1}{2}\frac{d}{dt}\|\nabla u\|_2^2 + \left\|\frac{\partial u}{\partial t}\right\|_2^2 = (q, \frac{\partial u}{\partial t}) = q\,F'\,. \tag{1.59}$$

Since $F(t) = 0$ for all $t \geq 1$, from (1.58), from the Poincaré inequality and from Gronwall's lemma we find that

$$\|u(t)\|_2 \leq \|u(1)\|_2\,e^{-\lambda_1 t}\,, \quad \text{for all } t \geq 1\,, \tag{1.60}$$

where λ_1 is defined in Lemma 1.2. Furthermore, by the Schwarz inequality, from (1.58) we also find that

$$\|\nabla u(t)\|_2^2 \leq \left\|\frac{\partial u}{\partial t}\right\|_2 \|u(t)\|_2\,, \quad \text{for all } t \geq 1\,.$$

If we use Poincaré inequality in this latter relation we deduce

$$\|\nabla u(t)\|_2 \leq \frac{1}{(\lambda_1)^{1/2}}\left\|\frac{\partial u}{\partial t}\right\|_2\,, \quad \text{for all } t \geq 1\,. \tag{1.61}$$

Therefore, replacing (1.61) into (1.59) we obtain

$$\frac{d}{dt}\|\nabla u\|_2^2 + 2\lambda_1\|\nabla u\|_2^2 \leq 0\,, \quad \text{for all } t \geq 1\,.$$

By Gronwall's lemma, this inequality, in turn, furnishes

$$\|\nabla u(t)\|_2 \leq \|\nabla u(1)\|_2\,e^{-\lambda_1 t}\,, \quad \text{for all } t \geq 1\,. \tag{1.62}$$

We next multiply both sides of (1.59) by $e^{\lambda_1 t}$. Integrating the resulting equation from $t = 1$ to $t = \infty$, by a simple calculation we get

$$\int_1^\infty e^{\lambda_1 t}\left\|\frac{\partial u}{\partial t}\right\|_2^2 dt = \frac{e^{\lambda_1}}{2}\|\nabla u(1)\|_2^2 + \frac{\lambda_1}{2}\int_1^\infty e^{\lambda_1 t}\|\nabla u(t)\|_2^2\,dt\,.$$

If we substitute (1.62) into this equation, it follows that

$$\int_1^\infty e^{\lambda_1 t}\left\|\frac{\partial u}{\partial t}\right\|_2^2 dt \leq \frac{e^{\lambda_1} + e^{-\lambda_1}}{2}\|\nabla u(1)\|_2^2\,. \tag{1.63}$$

We now go back to $(1.32)_1$ and multiply both sides by the solution, φ, to (1.7). After an integration by parts, we get

$$A\,q(t) = (\frac{\partial u}{\partial t}, \varphi) + (\nabla u, \nabla\varphi)\,, \tag{1.64}$$

where $A := \int_S \varphi(x) dx > 0$. Consequently, by (1.62) and (1.63), from (1.64) we find

$$\int_1^\infty |q(t)|^2 e^{\lambda_1 t}\, dt \le C \, \|\nabla u(1)\|_2^2, \qquad (1.65)$$

where $C = C(S) > 0$. The proposition then follows from (1.60), (1.62), (1.63), and (1.65). □

If F and u_0 satisfy more regular assumptions than those stated in Proposition 1.5, then we are able to obtain a pointwise decay for q also. This will be a consequence of Proposition 1.5 and of the following general lemma.

Lemma 1.3. *Let τ be an arbitrary positive number, and let $r \in L^q(0,\infty) \cap C[0,\tau]$, $1 \le q < \infty$, be a solution to the integral equation*

$$r(t) = \int_0^t J(t - s)\, r(s)\, ds + G(t) \quad t \in (0,\infty) \qquad (1.66)$$

where $J \in L^{q'}(1,\infty) \cap L^1(0,1)$ and $G \in L^\infty(0,\infty)$. Then $r \in L^\infty(0,\infty)$ and there exists a positive constant κ depending only on ε, q and J such that

$$\|r\|_{L^\infty(1,\infty)} \le \kappa \left(\|r\|_{L^q(0,\infty)} + \|G\|_{L^\infty(0,\infty)} \right). \qquad (1.67)$$

Proof. Let T be an arbitrarily fixed, finite number strictly greater than 1 and suppose that r attains its maximum in $[1,T]$ at some point $\bar{t} \in [1,T]$. For any $\varepsilon \in (0,1)$ we perform the following splitting,

$$\int_0^{\bar{t}} J(\bar{t} - s) r(s)\, ds = \int_0^{\bar{t}-\varepsilon} J(\bar{t} - s) r(s)\, ds + \int_0^\varepsilon J(s) r(\bar{t} - s)\, ds := I_1(\bar{t}) + I_2(\bar{t}). \qquad (1.68)$$

Since $J \in L^1(0,1)$, we have

$$|I_2(\bar{t})| \le |r(\bar{t})| \int_0^\varepsilon |J(s)|\, ds = \gamma(\varepsilon)\, |r(\bar{t})|, \qquad (1.69)$$

where, by the absolute continuity of Lebesgue integral, $\gamma(\varepsilon) \to 0$ as $\varepsilon \to 0$. Moreover,

$$|I_1(\bar{t})| \le \|J\|_{L^{q'}(\varepsilon,\infty)} \|r\|_{L^q(0,\infty)} := M(\varepsilon)\|r\|_{L^q(0,\infty)}. \qquad (1.70)$$

From (1.66)–(1.70) we obtain

$$(1 - \gamma(\varepsilon))\, r(\bar{t}) \le M(\varepsilon)\|r\|_{L^q(0,\infty)} + \|G\|_{L^\infty(0,\infty)}.$$

We now choose ε so small that $\gamma(\varepsilon) < 1/2$, so that this latter inequality furnishes

$$|r(t)| \le 2 \left(M \|r\|_{L^q(0,\infty)} + \|G\|_{L^\infty(0,\infty)} \right) \quad \text{for all } t \in [1,T]. \qquad (1.71)$$

Since $T > 1$ is arbitrary and M is independent of T, (1.71) shows that r is uniformly bounded in $[1,\infty)$ with a bound given by the right-hand side of (1.71). This proves (1.67). Moreover, $r \in C([0,1])$ and, therefore, $r \in L^\infty(0,1)$, which completes the proof of the lemma. □

Combining Proposition 1.5 and Lemma 1.3, we obtain the following.

Proposition 1.6. *In addition to the assumptions of* Proposition 1.5, *suppose that* $F' \in C([0,1])$ *and* $u_0 \in W^{2,2}(S)$. *Then,* $q \in C([0,T]) \cap L^\infty(0,\infty)$, *for all* $T > 0$, *and the following inequality holds for all* $t \geq 1$,

$$|q(t)| \leq C \left(\|\nabla u(1)\|_2 + \|u_0\|_{2,2} \right) e^{-\lambda_1 t/2}. \tag{1.72}$$

Proof. We multiply both sides of (1.41) by $e^{\lambda_1 t/2}$ to find

$$r(t) = \int_0^t J(t-s)\, r(s)\, ds + G(t), \quad t \geq 0, \tag{1.73}$$

where

$$r(t) := q(t)\, e^{\lambda_1 t/2}, \quad J(t) := K(t)\, e^{\lambda_1 t/2}, \quad G(t) := \Psi(t)\, e^{\lambda_1 t/2}.$$

We notice that, from (1.65), we have that $r \in L^2(0,\infty)$. Furthermore, from Lemma 1.2 we easily deduce that $J \in L^2(1,\infty) \cap L^1(0,1)$. Finally, from (1.42) we find

$$|\Psi(t)| \leq C_1 \|u_0\|_{2,2} \|\theta(t)\|_{2,2}, \quad \text{for all } t \geq 1,$$

with $C_2 = C_2(S) > 0$. Therefore, since by classical estimates for the solutions to problem (1.33) we have

$$\|\theta(t)\|_{2,2} \leq C_2 \|\varphi\|_{2,2}\, e^{-\lambda_1 t}, \quad t \geq 0,$$

with $C_1 = C_2(S) > 0$, it follows that

$$\|G\|_{L^\infty(0,\infty)} \leq C_3 \|u_0\|_{2,2}, \tag{1.74}$$

for a suitable $C_3 = C_3(S) > 0$. In conclusion, r, J and G satisfy all the assumptions of Lemma 1.3, and the proposition follows. $\qquad\square$

We now rephrase the results obtained in Proposition 1.4–Proposition 1.6 in terms of the original attainability problem (1.30)–(1.31).

Theorem 1.3. *Let S be of class C^2. The following properties hold.*
Attainability of Steady-State Flow. *Let $(\overline{V}, \overline{G})$ be the solution to the steady-state problem given in (1.8) corresponding to a given constant flow-rate $\overline{\Phi}$. Then, problem (1.30)–(1.31) has one and only one solution (V, G) such that, for all $T > 0$,*

$$\frac{\partial V}{\partial t} \in C([0,T]; L^2(S)) \quad V \in C([0,T]; W^{2,2}(S)), \quad G \in C[0,T]. \tag{1.75}$$

Moreover, there is $C_1 = C_1(S) > 0$ and $C_2 = C_2(S,T) > 0$ such that, for all $t \geq 1$,

$$\|V(t) - \overline{V}\|_{1,2} \leq \|V(1) - \overline{V}(1)\|_{1,2}\, e^{-\lambda_1 t},$$

$$\int_1^\infty \left(\left\| \frac{\partial (V - \overline{V})}{\partial t} \right\|_2^2 + \|(V - \overline{V})(t)\|_{2,2}^2 + |(G - \overline{G})(t)|^2 \right) e^{\lambda_1 t}\, dt$$

$$\leq C_1 \|\nabla(V - \overline{V})(1)\|_2^2, \tag{1.76}$$

$$|G(t) - \overline{G}| \leq C_1 \left(\|\nabla(V - \overline{V})(1)\|_2 + \|\overline{V}\|_{2,2} \right) e^{-\lambda_1 t/2},$$

$$\max_{t \in [0,T]} \left(\left\| \frac{\partial V}{\partial t}(t) \right\|_2 + \|(V - \overline{V})(t)\|_{2,2} + |q(t)| \right) \leq C_2 |\overline{\Phi}|,$$

where $\lambda_1 = \lambda_1(S)$ is the smallest eigenvalue of the Laplace operator introduced in Lemma 1.2.

Attainability of Time-Periodic Flow. *Let $(\overline{V}, \overline{G})$ be the solution to the time-periodic problem given in* Theorem 1.2 *corresponding to a given flow-rate $\overline{\Phi} \in W^{1,2}(-\pi, \pi)$. Then, problem* (1.30)–(1.31) *has one and only one solution (V, G) such that, for all $T > 0$,*

$$V \in C([0,T]; W_0^{1,2}(S)), \quad \frac{\partial V}{\partial t} \in L^2(S^T) \quad V \in L^2((0,T); W^{2,2}(S)), \quad G \in L^2(0,T).$$

Moreover, $V - \overline{V}$ and $G - \overline{G}$ satisfy the estimates (1.76)$_{1,2}$ *and the following one, for all $T > 0$,*

$$\max_{t \in [0,T]} \|(V - \overline{V})(t)\|_{1,2}^2 + \int_0^T \left(\left\| \frac{\partial(V - \overline{V})}{\partial t} \right\|_2^2 + \|(V - \overline{V})(t)\|_{2,2}^2 + |(G - \overline{G})(t)|^2 \right) dt$$
$$\leq C_3 \|\overline{\Phi}\|_{W^{1,2}(-\pi,\pi)}^2,$$

where $C_3 = C_3(S, T) > 0$.

Proof. The proof is an immediate consequence of Proposition 1.4–Proposition 1.6, once we take into account that (i) the compatibility condition (1.52) is satisfied, because $\int_S \overline{V} \, dS = \overline{\Phi}$, and (ii) in the case of steady-state flow, $\overline{V} \in W^{2,2}(S)$. $\quad \square$

1.2. The entry flow problem

One of the most important problems in the theory and application of pipe flow is the so-called *entry flow problem*. A viscous fluid is continuously injected in a straight pipe, at a constant flow-rate Φ. The motion of the fluid is assumed to be steady.

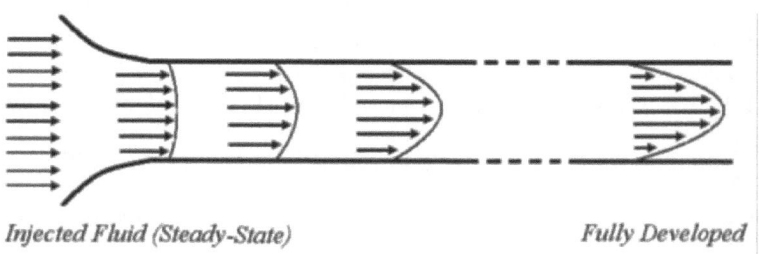

Injected Fluid (Steady-State)   *Fully Developed*

FIGURE 1. Entry Flow Problem.

What is experimentally observed is that there is a critical length, ℓ, depending on the magnitude of Φ (or, more properly, on the Reynolds number), such that after a distance ℓ from the inlet, the flow becomes essentially fully developed, that is, the velocity profile coincides – within a given margin of error – with that of the Poiseuille flow corresponding to the flow-rate Φ. The evaluation of ℓ is, of course, of great relevance because it gives us a measure of where the motion becomes essentially laminar. A sketch of the entry flow problem is given in Figure 1, for a pipe of circular cross-section.

A rigorous mathematical analysis of this problem started in 1978, with the paper of Horgan and Wheeler [59] and continued through the work of many other mathematicians, including C.J. Amick, O.A. Ladyzhenskaya and V.A. Solonnikov and L.E. Payne; see [36, Chapter VI] and [37, Chapter XI].

The objective of this section is to provide a mathematical formulation of the entry-flow problem and to present the basic ideas and the main results. We shall also point out some basic open questions. [8]

Let Ω be a semi-infinite straight pipe of constant cross-section S, defined as follows

$$\Omega = \{x \in \mathbb{R}^3 : x_1 > 0 \text{ and } (x_2, x_3) \in S\}.$$

Furthermore, let (v_P, p_P) be the (time-independent) Poiseuille flow (1.3) corresponding to the flow-rate Φ and let (v, p) be any other steady-state flow in Ω corresponding to the same flow-rate Φ. If we choose in (1.2) $U = \nu/d$, from (1.1) we then obtain that the difference $(u := v - v_P, \phi := p - p_P)$ satisfies the following non-dimensional equations,

$$\left.\begin{aligned} \Delta u - \nabla\phi &= u \cdot \nabla u + v_P \cdot \nabla u + u \cdot \nabla v_P \\ \nabla \cdot u &= 0 \end{aligned}\right\} \text{ in } \Omega, \tag{1.77}$$

$$u|_\Gamma = 0, \quad \int_S u_1 \, dS = 0,$$

where $\Gamma := \partial S \times (0, \infty)$ is the lateral surface of Ω.

Our goal is to investigate the rate of decay to $(0, 0)$ of the "perturbed" flow (u, ϕ).

Before presenting the main results related to this problem, we would like to sketch the basic ideas used for its resolution. In order to not obscure the substance with a number of technical details, due to the presence of the nonlinear term and of the pressure field in $(1.77)_1$, we prefer to introduce these ideas in the simplest "model problem" of Laplace's (scalar) equation:

$$\Delta u = 0 \text{ in } \Omega, \quad u|_\Gamma = 0. \tag{1.78}$$

Thus, suppose that (1.78) has a solution with a finite Dirichlet integral, $\|\nabla u\|_2 < \infty$ (we shall come back later on this assumption). Since $u|_\Gamma = 0$, we may use the following (Poincaré's) inequality,

$$\int_S |u|^2 \, dS \le \mu^2 \int_S |\nabla u|^2 \, dS \tag{1.79}$$

with $\mu := \sqrt{1/\lambda_1}$ (λ_1 defined in Lemma 1.2), to show that, in fact,

$$u \in W^{1,2}(\Omega). \tag{1.80}$$

---

[8] Clearly, the entry-flow problem can be formulated in the more general case of when the injected fluid is driven by a time-periodic flow-rate and the fully developed flow is time-periodic as well; see Remark 1.5

Multiplying both sides of $(1.78)_1$ by u and integrating by parts over $\Omega^{R,\rho} := \{x \in \Omega : 0 \le R < x_1 < \rho\}$, we find

$$\int_{\Omega^{R,\rho}} |\nabla u|^2 = -\int_{S_R} u \frac{\partial u}{\partial x_1} \, dS + \int_{S_\rho} u \frac{\partial u}{\partial x_1} \, dS \,, \qquad (1.81)$$

where $S_a := \{x \in \Omega : x_1 = a, (x_2, x_3) \in S\}$. In view of (1.80), we deduce, along a sequence at least,

$$\lim_{k \to \infty} \int_{S_{\rho_k}} u \frac{\partial u}{\partial x_1} \, dS = 0 \,,$$

so that (1.81) implies

$$\int_{\Omega^R} |\nabla u|^2 = -\int_{S_R} u \frac{\partial u}{\partial x_1} \, dS \,, \qquad (1.82)$$

where $\Omega^a := \Omega^{a,\infty}$, $a > 0$. Using the Schwarz inequality along with (1.79) on the right-hand side of (1.82) we find

$$\int_{\Omega^R} |\nabla u|^2 \le \mu \int_{S_R} |\nabla u|^2 \, dS \,. \qquad (1.83)$$

Therefore, setting $G(R) := \int_{\Omega^R} |\nabla u|^2$, and recalling the definition of μ, this latter inequality furnishes

$$G'(R) \le -\lambda_1^{1/2} G(R) \,, \qquad (1.84)$$

which, in turn, after a simple integration, implies

$$\int_{\Omega^R} |\nabla u|^2 \le e^{-\lambda_1^{1/2} R} \int_\Omega |\nabla u|^2 \,. \qquad (1.85)$$

Moreover, provided Γ is sufficiently smooth (of class C^2, for example) from well known elliptic estimates we have

$$\max_{x \in \Omega^{2R}} (|u(x)| + |\nabla u(x)|) \le M \int_{\Omega^R} |\nabla u|^2 \,, \quad \text{for all } R \ge 1 \,,$$

where the positive constant M is independent of R. This inequality, combined with (1.85), at once furnishes the following decay estimate,

$$|u(x)| + |\nabla u(x)| \le M \, e^{-\lambda_1^{1/2} x_1} \int_\Omega |\nabla u|^2 \,, \quad \text{for all } x_1 \ge 2 \,. \qquad (1.86)$$

It is interesting to observe that the decay constant depends only on the cross-section S and it is easily computed for cross-sections of specific simple shapes, like circles or squares. The quantity $\lambda_1^{-1/2}$, which has the dimension of a length, gives a measure of the "entry length" for problem (1.78). We now turn to the assumption that u has a finite Dirichlet integral and show that it can be fairly weakened. Actually, let us multiply both sides of (1.78) by u and let us integrate by parts over $\Omega_R := \{x \in \Omega : x_1 < R\}$. We thus get

$$g(R) := \int_{\Omega_R} |\nabla u|^2 = \int_{S_0} u \frac{\partial u}{\partial x_1} \, dS - \int_{S_R} u \frac{\partial u}{\partial x_1} \, dS := b - \int_{S_R} u \frac{\partial u}{\partial x_1} \,.$$

Using the Schwarz inequality and (1.79) on the right-hand side of this latter inequality we find

$$g(r) \le b + \lambda_1^{-1/2} \int_{S_R} |\nabla u|^2 \, dS = b + \lambda_1^{-1/2} g'(R) \,. \qquad (1.87)$$

Now, we have the following two possibilities: *either*

$$g(R) \quad \text{uniformly bounded for all } R \ge 0 \,, \qquad (1.88)$$

or

$$\lim_{R \to \infty} g(R) = \infty. \qquad (1.89)$$

In case (1.88), we have $\|\nabla u\|_2 < \infty$ and, consequently, the decay estimate (1.85) holds. In case (1.89), we claim that, necessarily,

$$L := \liminf_{R \to \infty} e^{-\lambda_1^{1/2} R} g(R) > 0 \,. \qquad (1.90)$$

In fact, assume $L = 0$. Then, from (1.87) it follows that

$$\frac{d}{dr} \left(e^{-\lambda_1^{1/2} r} g(r) \right) = e^{-\lambda_1^{1/2} r} (g'(r) - \lambda_1^{1/2} g(r)) \ge -\lambda_1^{1/2} b \, e^{-\lambda_1^{1/2} r} \,.$$

We now integrate both sides of this latter relation from R to $R_1 > R$ to obtain

$$e^{-\lambda_1^{1/2} R_1} g(R_1) - e^{-\lambda_1^{1/2} R} g(R) \ge b \, e^{-\lambda_1^{1/2} R_1} - b \, e^{-\lambda_1^{1/2} R} \,,$$

and so, applying $\liminf_{R_1 \to \infty}$ to both sides of this relation, we get

$$e^{-\lambda_1^{1/2} R} g(R) \le b \, e^{-\lambda_1^{1/2} R} + L = b \, e^{-\lambda_1^{1/2} R} \,,$$

since $L = 0$. Therefore, $g(R)$ would be uniformly bounded, which contradicts (1.89).

The result just showed are summarized in the following.

Lemma 1.4. *Let u be a solution to* (1.78) *such that*

$$\liminf_{R \to \infty} e^{-\lambda_1^{1/2} R} \int_{\Omega_R} |\nabla u|^2 = 0 \,. \qquad (1.91)$$

Then, necessarily, $\|\nabla u\|_2 < \infty$ and the estimate (1.85) *holds. Moreover, if S is of class C^2, the pointwise estimate* (1.86) *is valid.*

The method used in the proof of Lemma 1.4 for the model problem (1.78) can be transposed to the more complicated situation of the Navier–Stokes problem (1.77). However, the results that one is able to obtain in this case are not as complete as in the case of (1.78). This is due, on the one hand, to the presence of the pressure term and, also, to the nonlinear term. On the other hand, and this is the main drawback of the method, one is able to find a decay estimate on the Dirichlet integral of u, namely, an analogous of estimate (1.85), *if the magnitude of the flow-rate Φ is "small enough"*. Another striking difference with the simple case (1.78) is that the decay constant depends not only on the cross-section (like in (1.86)) but also on the flow-rate and on the Dirichlet integral of u. In order to show where all the above restrictions come from, we shall briefly sketch the proof, referring the reader to [36, Chapter VI] and [37, Chapter XI] for full details. Thus, if we assume that $\|\nabla \boldsymbol{u}\|_2 < \infty$, dot-multiplying both sides of $(1.77)_1$ by $\boldsymbol{u}$ and integrating by parts over Ω^R we can show the validity of the following relation (the analog of (1.82)),

$$\int_{\Omega^R} |\nabla \boldsymbol{u}|^2 = -\int_{S_R} \left(\boldsymbol{u} \cdot \frac{\partial \boldsymbol{u}}{\partial x_1} - \phi \, u_1 - \frac{1}{2}|\boldsymbol{u}|^2(u_1 + V) \right) dS - \int_{\Omega^R} u_2 \, V' u_1 \,, \quad (1.92)$$

where the prime denotes differentiation. With the help of (1.79) and of the properties of V we find that

$$\int_{\Omega^R} u_2 \, V' u_1 \leq C(S)|\Phi| \int_{\Omega^R} |\nabla \boldsymbol{u}|^2 \,,$$

where $C(S)$ is a positive constant depending only on the cross-section. Therefore, if $\gamma := 1 - C(S)|\Phi| > 0$, from this latter inequality and from (1.92) we obtain

$$\gamma \int_{\Omega^R} |\nabla \boldsymbol{u}|^2 \leq -\int_{S_R} \left(\boldsymbol{u} \cdot \frac{\partial \boldsymbol{u}}{\partial x_1} - \phi \, u_1 - \frac{1}{2}|\boldsymbol{u}|^2(u_1 + V) \right) dS \,. \quad (1.93)$$

The first term on the right-hand side of (1.93) can be increased exactly as in the model problem to get

$$-\int_{S_R} \boldsymbol{u} \cdot \frac{\partial \boldsymbol{u}}{\partial x_1} \leq \mu \int_{S_R} |\nabla \boldsymbol{u}|^2 \, dS \,, \quad (1.94)$$

while the last term, with the help of (1.79), can be increased as follows,

$$\frac{1}{2} \int_{S_R} V \, |\boldsymbol{u}|^2 \leq C_1(S) \, |\Phi| \int_{S_R} |\nabla \boldsymbol{u}|^2 \, dS \,, \quad (1.95)$$

with $C_1(S) > 0$. Moreover, if Γ is sufficiently smooth, we have (see [37, Lemma XI.4.1])

$$\sup_{x \in \Omega^R} \left(|\boldsymbol{u}(x)| + |\nabla \boldsymbol{u}(x)| \right) \leq C_2 \|\nabla \boldsymbol{u}\|_{2,\Omega^{R-1}} \,, \quad \text{all } R > 1 \,, \quad (1.96)$$

with $C_2 = C_2(S, \Phi, \|\nabla \boldsymbol{u}\|_2) > 0$. Thus, using (1.99) along with (1.79) we obtain

$$\frac{1}{2} \int_{S_R} |\boldsymbol{u}|^2 \, u_1 \, dS \leq C_3 \int_{S_R} |\boldsymbol{u}|^2 \leq C_4 \int_{S_R} |\nabla \boldsymbol{u}|^2 \,, \quad (1.97)$$

where $C_4 = C_4(\Phi, S, \|\nabla \boldsymbol{u}\|_2) > 0$. From (1.93)–(1.95) and (1.97) we deduce

$$\gamma \int_{\Omega^R} |\nabla \boldsymbol{u}|^2 \leq C_5 \int_{S_R} |\nabla \boldsymbol{u}|^2 + \int_{S_R} \phi \, u_1 \, dS \,, \quad (1.98)$$

where $C_5 = C_5(\Phi, S, \|\nabla \boldsymbol{u}\|_2) > 0$. At this point, it is not known (and, probably, not true) whether there are constants $K = K(S, \Phi, \|\nabla \boldsymbol{u}\|_2) > 0$ and $0 \leq \delta < 1$ such that

$$\int_{S_R} \phi \, u_1 \, dS \leq K \int_{S_R} |\nabla \boldsymbol{u}|^2 \, dS + \delta \gamma \int_{\Omega^R} |\nabla \boldsymbol{u}|^2 \,,$$

and so we can not reduce (1.98) to the form of the differential inequality (1.84). However, by using more elaborated tools, one can show (see, e.g., [37, Section XI.4]) that (1.98) implies the following one,

$$\gamma \int_\rho^\infty G(R)\,dR \le -C_5 G'(\rho) + C_6\,G(\rho)\,, \quad \text{all } \rho \ge 1\,, \tag{1.99}$$

where

$$G(R) := \int_{\Omega^R} |\nabla u|^2\,,$$

and $C_6 = C_6(S, \Phi, \|\nabla u\|_2) > 0$. The integro-differential inequality (1.99) can be integrated (see [36, Lemma VI.2.2]) to give

$$G(R) \le K_1 G(0)\,e^{-K_2 R} \quad \text{all } R \ge 1$$

with $K_i = K_i(\Phi, S, \|\nabla u\|_2) > 0$, $i = 1, 2$, that is,

$$\int_{\Omega^R} |\nabla u|^2 \le K_1 \|\nabla u\|_2^2\,e^{-K_2 R} \quad \text{all } R \ge 1\,. \tag{1.100}$$

The estimate (1.100) along with (1.96) allows us to obtain an exponential, pointwise decay for u and ∇u:

$$\sup_{x \in \Omega^R} (|u(x)| + |\nabla u(x)|) \le K_3\,e^{-K_2 R} \quad \text{all } R \ge 1\,, \tag{1.101}$$

with $K_3 = K_3(\Phi, S, \|\nabla u\|_2) > 0$. If Γ is of class C^∞, this decay property can be extended to derivatives of arbitrary order of u and $\nabla\phi$.

Coming back to the assumed condition $\|\nabla u\|_2 < \infty$, we can not prove its validity under a hypothesis as weak as (1.91), and a stronger requirement is needed. Specifically, one can show that if

$$\liminf_{R \to \infty} R^{-3} \int_{\Omega_R} |\nabla u|^2 = 0, \tag{1.102}$$

and if $|\Phi|$ is sufficiently small, then $\|\nabla u\|_2 < \infty$; see [37, Lemma XI.4.3].

The results presented so far can be summarized in the following.

Theorem 1.4. *There exists a constant $c = c(S) > 0$ such that, if $|\Phi| < c$, then all solutions to (1.77) in the class (1.102) must have a finite Dirichlet integral, and decay to zero according to (1.100). Moreover, if Γ is of class C^∞, then u, $\nabla\phi$ and all their derivatives of arbitrary order decay expontially fast to zero with the same decay constant K_2 of (1.100).*

Remark 1.5. The results of Theorem 1.4 suggest the following directions of further research.

1. From a theoretical point of view, the most important question left open by Theorem 1.4 is whether or not the results there established continue to hold without the restriction on the magnitude of the flow-rate.

2. Another important question is to give a *quantitative* lower bound for the decay constant K_2 in (1.100) in terms of S, Φ and of some physical parameters at the inlet of the pipe where the liquid is injected (at $x_1 = 0$, that is), such its maximum velocity, or its total kinetic energy.

3. By a simple argument, based on the spatial analitycity property of solutions to $(1.77)_{1,2}$, one show that the "perturbation" field $\boldsymbol{u}(x)$ can not decay to zero (namely, the flow can not become fully developed) at a *finite* distance from the outlet. Therefore, another not less important problem to investigate is the determination of decay estimates from *below* for the perturbation velocity and pressure fields.

4. Finally, it would be interesting to study the entry flow problem in the more general situation of a time-periodic flow-rate. In this case, one should consider the problem of attainability of the fully developed, time-periodic Poiseuille flow both in space and time; see also Section 1.3.2.

1.3. Mathematical modeling of a piping system. Unbounded domain approach

Assume we have a liquid moving into a piping system, subject to prescribed flow rates $\Phi_i = \Phi_i(t)$ in each outlet Ω_i, $i = 1, 2, \ldots, N$. We suppose that $\Omega_1, \ldots, \Omega_l$ are "upstream" outlets, while $\Omega_{l+1}, \ldots, \Omega_N$ are "downstream" outlets. By incompressibility,

$$\sum_{i=1}^{l} \Phi_i = \sum_{i=l+1}^{N} \Phi_i .$$

The case of a piping system with three outlets is sketched in Figure 2.

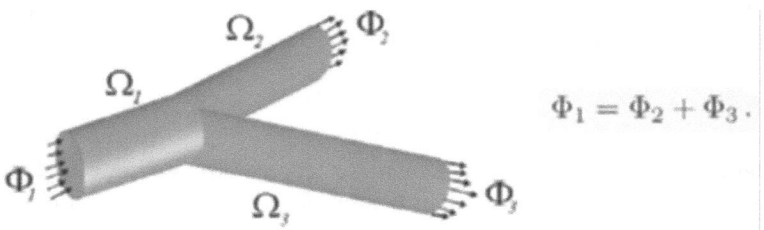

$$\Phi_1 = \Phi_2 + \Phi_3 .$$

FIGURE 2. Sketch of a piping system.

The first problem that one should address when modeling a piping system is how to prescribe boundary conditions at the open sections of the outlets. As we mentioned at the beginning of this Section 1, there are basically two ways of doing it. The first one assumes that each outlet, Ω_i, can be extended to a semi-infinite pipe of constant cross-section, possibly depending on i. The second one, instead, keeps the outlets of finite length and prescribes suitable boundary conditions at the open sections. Of course, these latter conditions should be chosen such as to meet a certain number of requirements that will be discussed in the next section.

In the current section we consider the unbounded domain approach. In this case, on the basis of what we have shown previously, it appears natural to assume that, eventually in each outlet, the flow becomes fully developed and, consequently, the velocity field approaches that of the Poiseuille flow corresponding to the flow-rate in that particular outlet. This situation is analogous to the more familiar one of a flow past an obstacle, where one assumes that, at large distance from the body, the flow velocity eventually reaches a constant value.

In order to not obscure the underlying ideas, we shall suppose that the system consists of only two (semi-infinite) outlets, Ω_1 and Ω_2, of cross-section S_1 and S_2, respectively. Therefore, the region of flow, Ω, can be written as follows:

$$\Omega = \Omega_1 \cup \Omega_0 \cup \Omega_2 \,,$$

where Ω_0 is a compact set of $\mathbb{R}^3$ and, in possibly different coordinate systems,

$$\Omega_1 = \{x \in \mathbb{R}^3 : \ x_1 < 0 \,, \ (x_2, x_3) \in S_1\} \,,$$

$$\Omega_2 = \{x \in \mathbb{R}^3 : \ x_1 > 0 \,, \ (x_2, x_3) \in S_2\} \,.$$

We shall also often use the following splitting of Ω:

$$\Omega = \widetilde{\Omega}_1 \cup \widetilde{\Omega}_0 \cup \widetilde{\Omega}_2 \,,$$

with

$$\widetilde{\Omega}_1 := \{x \in \Omega_1 : \ x_1 < -1\} \,, \quad \widetilde{\Omega}_2 := \{x \in \Omega_2 : \ x_1 > 1\} \quad \widetilde{\Omega}_0 := \Omega - (\widetilde{\Omega}_1 \cup \widetilde{\Omega}_2) \,. \tag{1.103}$$

In the following two subsections, we shall consider, separately, the two cases of constant flow-rate (steady-state motions) and time-periodic flow-rates (time-periodic motions). In performing this study, the next simple lemma is particularly useful.

Lemma 1.5. *Let Ω be Lipschitz, and let $v \in W^{1,2}(\Omega)$ and $u, w \in W_0^{1,2}(\Omega)$. Then the following inequalities hold:*

$$\|w\|_2 \leq C\|\nabla w\|_2,$$

$$|(u \cdot \nabla v, w)| \leq C\|\nabla u\|_2\|\nabla v\|_2\|\nabla w\|_2 \,,$$

where $C = C(\Omega) > 0$.

Proof. The first inequality follows by using (1.79) in each Ω_i along with the inequality

$$\|w\|_{2,\omega} \leq C_\omega \|\nabla w\|_2 \,, \tag{1.104}$$

holding for any function w in $W^{1,2}(\omega)$, ω a Lipschitz, bounded domain, and that vanishes on a subset of $\partial\omega$ having a non-zero (two-dimensional) Lebesgue measure; see [37, Exercise II.4.10]. The second inequality is a consequence of the first one, of the Hölder inequality and of the embedding $W^{1,2}(\Omega) \subset L^4(\Omega)$. $\square$

1.3.1. Steady-state case. Assuming that the flow-rate Φ is constant, the mathematical problem consists then in finding a pair $(\boldsymbol{v} = \boldsymbol{v}(x), p = p(x))$ such that

$$\left.\begin{array}{l} \Delta\boldsymbol{v} - \nabla p = \boldsymbol{v} \cdot \nabla\boldsymbol{v} \\[2mm] \nabla \cdot \boldsymbol{v} = 0 \end{array}\right\} \quad \text{in } \Omega,$$

$$\int_S \boldsymbol{v} \cdot \boldsymbol{n}\, dS = \Phi, \qquad\qquad (1.105)$$

$$\boldsymbol{v}|_{\partial\Omega} = \boldsymbol{0},$$

$$\lim_{|x|\to\infty,\ x\in\Omega_i} (\boldsymbol{v}(x) - \boldsymbol{v}_{Pi}(x)) = \boldsymbol{0}, \ i = 1, 2.$$

These equations are written in non-dimensional form, with the same scale quantities used for (1.77). Furthermore, S is a generic cross-section, that is, any intersection of a plane with Ω that reduces to S_i in Ω_i, and $\boldsymbol{n}$ is the unit normal to S oriented from Ω_1 toward Ω_2. Finally, $\boldsymbol{v}_{Pi}$, $i = 1, 2$, are the (non-dimensional) Poiseuille velocity fields corresponding to the flow-rate Φ.

In order to solve (1.105) it is useful to introduce the so-called (unit) *flow-rate carrier*. By this we mean a field $\boldsymbol{a}$ defined in Ω and satisfying the following properties.

(i) $\boldsymbol{a} \in W^{2,q}(\omega)$, for all $q \geq 1$ and for all bounded domains $\omega \subset \Omega$;

(ii) $\nabla \cdot \boldsymbol{a} = 0$ in Ω, $\boldsymbol{a}|_{\partial\Omega} = \boldsymbol{0}$;

(iii) $\displaystyle\int_S \boldsymbol{a} \cdot \boldsymbol{n}\, dS = 1$;

(iv) $\Phi\boldsymbol{a}(x) = \boldsymbol{v}_{P1}(x)$ for all $x \in \widetilde{\Omega}_1$ and $\Phi\boldsymbol{a}(x) = \boldsymbol{v}_{P2}(x)$ for all $x \in \widetilde{\Omega}_2$, where $\widetilde{\Omega}_1$ and $\widetilde{\Omega}_2$ are defined in (1.103);

(v) there is $C = C(\Omega_0, S_1, S_2) > 0$ such that $\max_{x\in\Omega} (|\boldsymbol{a}(x)| + |\nabla\boldsymbol{a}(x)|) \leq C$.

The existence of the field $\boldsymbol{a}$ is shown in [36, Section VI.1]. Here, we shall give a brief description of how to construct the field $\boldsymbol{a}$, referring to [36, Section VI.1] for details. Let $\zeta_1 = \zeta_1(x)$ and $\zeta_2 = \zeta_2(x)$ be two smooth functions satisfying the properties

$$\zeta_1(x) = \left\{\begin{array}{ll} 1 & x_1 \leq -1 \\[2mm] 0 & x_1 \geq -1/2 \end{array}\right. \qquad x \in \Omega_1,$$

$$\zeta_2(x) = \left\{\begin{array}{ll} 1 & x_1 \geq 1 \\[2mm] 0 & x_1 \leq 1/2 \end{array}\right. \qquad x \in \Omega_2,$$

and set

$$\boldsymbol{U} := \zeta_1\overline{\boldsymbol{v}}_{P1} + \zeta_2\overline{\boldsymbol{v}}_{P2}, \qquad\qquad (1.106)$$

where $\overline{\boldsymbol{v}}_{Pi} = \boldsymbol{v}_{Pi}/\Phi$. Clearly, the field $\boldsymbol{U}$ is smooth everywhere in Ω and reduces to $\overline{\boldsymbol{v}}_{Pi}$ in $\widetilde{\Omega}_i$. However, $\boldsymbol{U}$ is not solenoidal and so we need to add to it a suitable "correction". To this end, let $\widetilde{\Omega}_0 := \Omega - \overline{(\widetilde{\Omega}_1 \cup \widetilde{\Omega}_2)}$ and let $\boldsymbol{W}$ be a solution to the problem

$$\nabla \cdot \boldsymbol{W} = -\nabla \cdot \boldsymbol{U} \ \text{ in } \ \widetilde{\Omega}_0 \,,$$

$$\boldsymbol{W} \in W_0^{2,q}(\widetilde{\Omega}_0) \,, \quad \|\boldsymbol{W}\|_{1,q} \le C \,\|\nabla \cdot \boldsymbol{U}\|_q \,, \quad \|\boldsymbol{W}\|_{2,q} \le C \,\|\nabla \cdot \boldsymbol{U}\|_{1,q} \,, \tag{1.107}$$

where $C = C(\widetilde{\Omega}_0, q) > 0$. The existence of $\boldsymbol{W}$ is shown in [36, Section VI.1]. If we extend $\boldsymbol{W}$ to zero outside $\widetilde{\Omega}_0$ and continue to denote by $\boldsymbol{w}$ such an extension, we easily deduce that $\boldsymbol{a} := \boldsymbol{U} + \boldsymbol{W}$ satisfies all properties (i)–(v) listed above.

We now look for a solution to (1.105) of the form $(\boldsymbol{v} = \boldsymbol{u} + \boldsymbol{\alpha}, p)$ where $\boldsymbol{\alpha} := \Phi \, \boldsymbol{a}$ and $\boldsymbol{u}$ satisfies the problem

$$\left. \begin{aligned} \Delta \boldsymbol{u} - \nabla p &= \boldsymbol{u} \cdot \nabla \boldsymbol{u} + \boldsymbol{\alpha} \cdot \nabla \boldsymbol{u} + \boldsymbol{u} \cdot \nabla \boldsymbol{\alpha} + \boldsymbol{F} \\ \nabla \cdot \boldsymbol{u} &= 0 \end{aligned} \right\} \ \text{ in } \Omega,$$

$$\int_S \boldsymbol{u} \cdot \boldsymbol{n} \, dS = 0 \,, \quad \boldsymbol{u}|_{\partial\Omega} = \boldsymbol{0}, \tag{1.108}$$

$$\lim_{|x| \to \infty, \ x \in \Omega_i} \boldsymbol{u}(x) = \boldsymbol{0} \,, \ i = 1, 2 \,,$$

where

$$\boldsymbol{F} := -\Delta \boldsymbol{\alpha} + \boldsymbol{\alpha} \cdot \nabla \boldsymbol{\alpha} \,.$$

Notice that, by the property (iv) of $\boldsymbol{a}$, we have

$$\boldsymbol{F}(x) = G_i \boldsymbol{e}_1 \ \text{ for all } x \in \widetilde{\Omega}_i, \ i = 1, 2, \tag{1.109}$$

where G_i is the axial pressure gradient of the corresponding Poiseuille flow (see (1.6)). The existence of a solution to (1.108) can be shown by means of the classical Galerkin method in the space $\mathcal{D}_0^{1,2}(\Omega)$. We recall that, if $\partial\Omega$ is (locally) Lipschitz, then $\mathcal{D}_0^{1,2}(\Omega)$ consists of those solenoidal vector fields in Ω that belong to $W^{1,2}(\Omega)$ and have zero trace at the boundary. Let $\{\boldsymbol{\varphi}_k\} \subset \mathcal{D}(\Omega)$ be a base in $\mathcal{D}_0^{1,2}(\Omega)$ and set

$$\boldsymbol{u}_m = \sum_{k=1}^m c_{km} \boldsymbol{\varphi}_k \,,$$

where the coefficients c_{km} are determined from the system of equations

$$\int_\Omega \nabla \boldsymbol{u}_m : \nabla \boldsymbol{\varphi}_k = -\int_\Omega (\boldsymbol{u}_m \cdot \nabla \boldsymbol{u}_m + \boldsymbol{\alpha} \cdot \nabla \boldsymbol{u}_m + \boldsymbol{u}_m \cdot \nabla \boldsymbol{\alpha}) \cdot \boldsymbol{\varphi}_k - \int_\Omega \boldsymbol{F} \cdot \boldsymbol{\varphi}_k \,, \tag{1.110}$$

with $k = 1, 2, \ldots, m$. It is known, see [37, Section XI.3], that a solution to (1.110) exists if we can show that the Dirichlet norm of $\boldsymbol{u}_m$ is uniformly bounded. If we dot-multiply both sides of (1.110) by c_{km} and sum over k from 1 to m we obtain

$$\|\nabla \boldsymbol{u}_m\|_2^2 = \int_\Omega \boldsymbol{u}_m \cdot \nabla \boldsymbol{u}_m \cdot \boldsymbol{\alpha} - \int_\Omega \boldsymbol{F} \cdot \boldsymbol{u}_m \,. \tag{1.111}$$

By the property (v) of the field $\boldsymbol{a}$, by Schwarz inequality and by Lemma 1.5 we easily get

$$\int_\Omega \boldsymbol{u}_m \cdot \nabla \boldsymbol{u}_m \cdot \boldsymbol{\alpha} \le C_1 |\Phi| \, \|\nabla \boldsymbol{u}_m\|_2^2 \,, \tag{1.112}$$

where $C_1 = C_1(\Omega) > 0$. Furthermore, using the splitting $\Omega = \tilde{\Omega}_1 \cup \tilde{\Omega}_0 \cup \tilde{\Omega}_1$, we deduce

$$\int_\Omega \boldsymbol{F} \cdot \boldsymbol{u}_m = \int_{\tilde{\Omega}_0} \boldsymbol{F} \cdot \boldsymbol{u}_m + \sum_{i=1}^2 \int_{\tilde{\Omega}_i} \boldsymbol{F} \cdot \boldsymbol{u}_m$$

$$\leq C_2(|\Phi| + |\Phi|^2)\|\nabla \boldsymbol{u}_m\|_2 + \sum_{i=1}^2 \int_{\tilde{\Omega}_i} \boldsymbol{F} \cdot \boldsymbol{u}_m, \tag{1.113}$$

where $C_2 = C_2(\Omega) > 0$ and we have used property (i) of $\boldsymbol{a}$ together with (1.79). We next observe that, by (1.109),

$$\int_{\tilde{\Omega}_1} \boldsymbol{F} \cdot \boldsymbol{u}_m = G_1 \int_{-\infty}^{-1} \left(\int_S \boldsymbol{u}_m \cdot \boldsymbol{e}_1 \, dS \right) dx_1 = 0,$$

because, since $\boldsymbol{u}_m$ is of compact support and solenoidal, we have

$$\int_S \boldsymbol{u}_m \cdot \boldsymbol{e}_1 \, dS = 0.$$

Likewise,

$$\int_{\tilde{\Omega}_2} \boldsymbol{F} \cdot \boldsymbol{u}_m = G_1 \int_1^\infty \left(\int_S \boldsymbol{u}_m \cdot \boldsymbol{e}_1 \, dS \right) dx_1 = 0,$$

and so, from (1.111)–(1.113), it follows that

$$(1 - |\Phi| C_1)\|\nabla \boldsymbol{u}_m\| \leq C_2(|\Phi| + |\Phi|^2).$$

Therefore, if $|\Phi| < 1/C_1$, namely the magnitude of the flow-rate is suitably restricted, the previous inequality furnishes the desired bound for the Dirichlet norm of $\boldsymbol{u}_m$. Once this a-priori bound has been established, it is routine to show that we can select a subsequence $\{\boldsymbol{u}_{m'}\}$ and find a vector $\boldsymbol{u} \in \mathcal{D}_0^{1,2}(\Omega)$ such that $\boldsymbol{u}_{m'} \to \boldsymbol{u}$, as $m' \to \infty$, in suitable topologies, and that $\boldsymbol{u}$ satisfies the following "weak form" of (1.108):

$$\int_\Omega \nabla \boldsymbol{u} : \nabla \boldsymbol{\varphi} = -\int_\Omega (\boldsymbol{u} \cdot \nabla \boldsymbol{u} + \boldsymbol{a} \cdot \nabla \boldsymbol{u} + \boldsymbol{u} \cdot \nabla \boldsymbol{a}) \cdot \boldsymbol{\varphi} - \int_\Omega \boldsymbol{F} \cdot \boldsymbol{\varphi},$$

for all $\boldsymbol{\varphi} \in \mathcal{D}(\Omega)$. Furthermore, there exists a scalar field p such that the pair $(\boldsymbol{v} := \boldsymbol{u} + \boldsymbol{\alpha}, p)$ is of class $C^\infty(\Omega)$ and satisfies $(1.105)_{1,2,3}$. The boundary condition $(1.105)_4$ is also satisfied in the ordinary sense, provided $\partial\Omega$ is sufficiently smooth, e.g., of class C^2; see [37, Sections XI.1, XI.2 and XI.3]. Concerning the asymptotic condition $(1.105)_5$, from Theorem 1.4 we know that the decay rate is exponential, provided $\partial\Omega$ is sufficiently smooth. Our final consideration regards the uniqueness of these solutions. In [37, Theorem XI.3.2] it is shown that they are unique on condition that, again, the magnitude of the flow-rate is below a suitable constant depending only on Ω.

The results discussed so far can be then summarized in the following.

Theorem 1.5. *Assume* Ω *of class* C^2 *and let* $\Phi \in \mathbb{R}$. *Denote by* $\mathcal{C}_\Phi$ *the class of fields* $\boldsymbol{v}$ *such that*

$$\mathcal{C}_\Phi = \{(\boldsymbol{v}, p) \text{ solve } (1.105) \text{ and } \|\nabla(\boldsymbol{v} - \boldsymbol{v}_{Pi})\|_{2,\Omega_i} < \infty, \quad i = 1, 2\} \qquad (1.114)$$

where $\boldsymbol{v}_{Pi}$ *are the Poiseuille velocity in* Ω_i, $i = 1, 2$, *corresponding to the flow-rate* Φ. *There is a constant* $C = C(\Omega) > 0$ *such that if*

$$|\Phi| < C,$$

then problem (1.105) *has one and only one solution* $(\boldsymbol{v}, p)$ *with* $\boldsymbol{v}$ *in the class* $\mathcal{C}_\Phi$. *This solution satisfies the estimate*

$$\sum_{i=1}^{2} \|\nabla(\boldsymbol{v} - \boldsymbol{v}_{Pi})\|_2^2 + \|\nabla\boldsymbol{v}\|_{2,\Omega_0} \leq C_1 |\Phi|,$$

with $C_1 = C_1(\Omega) > 0$ *and, in addition, it is infinitely differentiable in* Ω. *Moreover, if* Ω *is of class* C^∞, $\boldsymbol{v} - \boldsymbol{v}_{Pi}$ *and* $\nabla p - G_i$, $i = 1, 2$, *together with their derivatives of arbitrary order, decay exponentially fast in the corresponding outlets* Ω_i.

Remark 1.6. The problem of existence of solutions to (1.105) was proposed by J. Leray to O.A. Ladyzhenskaya during his visit to St. Petersburg (Leningrad, at that time) in 1958. For this reason such an existence problem is often referred to as "Leray's Problem"; see [36, Chapter VI], [37, Chapter XI]. In 1959, Ladyzhenskaya published a paper (see *Dokl. Akad. Nauk. SSSR*, **124**, 551–553) claiming its resolution *without restrictions on the magnitude of the flow-rate*. However, her proof contains a major mistake, as pointed out by R. Finn in 1965 (see *Proc. Symp. Appl. Math.*, **17**, 121–153). Since then, many mathematicians, including C.J. Amick, O.A. Ladyzhenskaya, V.A. Solonnikov, K. Pileckas, H. Morimoto, H. Fujita and P.J. Rabier, have given several significant contributions to this intriguing problem without, however, being able to remove the restriction on the magnitude of the flow-rate. Clearly, the question appears to be extremely difficult and it is not even known if it admits a positive answer. In this regards, we wish to mention the result of Ladyzhenskaya and Solonnikov [71], who prove that problem (1.105) always has a solution for arbitrary flow-rate, provided we give up the asymptotic condition $(1.105)_5$. Moreover, their solutions satisfy also $(1.105)_5$ if the magnitude of the flow-rate is sufficiently "small". In this case, they are also unique.

Remark 1.7. In the case of $N > 2$ outlets, the results of Theorem 1.5 continue to hold, provided $\left(\sum_{i=1}^{N} |\Phi_i|\right) < C$, where Φ_i is the flow-rate in the outlet Ω_i and C is a suitable positive constant depending only on Ω.

1.3.2. Time-periodic case. The methods used for the case of a constant flow-rate can be suitably generalized to prove existence of time-periodic flows of period 2π (say), corresponding to a time-periodic flow-rate of the same period 2π. In the case of two outlets, the problem is formulated as follows. Given a (dimensionless) time-periodic function, $\Phi = \Phi(t)$, of period 2π, find a pair $(\boldsymbol{v}(x,t), p(x,t))$ 2π-periodic in t, such that

$$\left.\begin{aligned} \frac{\partial \boldsymbol{v}}{\partial t} + \boldsymbol{v} \cdot \nabla \boldsymbol{v} &= \Delta \boldsymbol{v} - \nabla p \\ \nabla \cdot \boldsymbol{v} &= 0 \end{aligned}\right\} \quad \text{in } \Omega \times (-\pi, \pi),$$

$$\int_S \boldsymbol{v} \cdot \boldsymbol{n} \, dS = \Phi(t), \qquad (1.115)$$

$$\boldsymbol{v}|_{\partial\Omega} = \boldsymbol{0},$$

$$\lim_{|x| \to \infty, \ x \in \Omega_i} (\boldsymbol{v}(x,t) - \boldsymbol{v}_{Pi}(x,t)) = \boldsymbol{0}, \quad i = 1, 2, \quad t \in (-\pi, \pi),$$

where $\boldsymbol{v}_{Pi}$ are time-periodic Poiseuille flows corresponding to the flow-rate Φ_i and whose existence has been determined in Theorem 1.2. Moreover, in the dimensionless equation (1.115) we have chosen the scale velocity $U = \nu/d$, where d is a characteristic length of the cross-section S.

The first result concerning the well-posedness of problem (1.115) is due to Beirão da Veiga [7], who proved existence of weak solutions (á la Leray–Hopf), provided the magnitude of the flow-rate and of its first derivative is suitably restricted. Under similar restrictions and by using the ideas of Galdi and Silvestre [45], we shall now prove existence *and uniqueness* of *strong* solutions.

To this end, we need some preparatory results. Hereafter we will assume that Ω is uniformly of class C^2. [9] We begin to introduce a 2π-periodic *flow-rate carrier* $\boldsymbol{A}(x,t)$ satisfying the following properties.

(i) $\boldsymbol{A} \in L^2(-\pi, \pi; W^{2,2}(\omega))$, $\dfrac{\partial \boldsymbol{A}}{\partial t} \in L^2(-\pi, \pi; L^2(\omega))$, for all bounded domains $\omega \subset \Omega$;

(ii) $\nabla \cdot \boldsymbol{A}(x,t) = 0$ for a.a. $(x,t) \in \Omega \times (-\pi, \pi)$, $\boldsymbol{A}(x,t)|_{\partial\Omega} = \boldsymbol{0}$, for a.a. $t \in (-\pi, \pi)$;

(iii) $\displaystyle\int_S \boldsymbol{A} \cdot \boldsymbol{n} \, dS = \Phi(t)$;

(iv) $\boldsymbol{A}(x,t) = \boldsymbol{v}_{P1}(x,t)$ for all $(x,t) \in \widetilde{\Omega}_1 \times (-\pi, \pi)$ and $\boldsymbol{A}(x,t) = \boldsymbol{v}_{P2}(x,t)$ for all $x \in \widetilde{\Omega}_2 \times (-\pi, \pi)$, where $\widetilde{\Omega}_i$, $i = 1, 2$, is defined in (1.103);

(v) for any bounded domain $\omega \subset \Omega$ there is $C = C(\omega) > 0$ such that

$$\max_{t \in [-\pi, \pi]} \|\boldsymbol{A}\|_{1,2,\omega}^2 + \int_{-\pi}^{\pi} \left(\left\| \frac{\partial \boldsymbol{A}}{\partial t} \right\|_{2,\omega}^2 + \|\boldsymbol{A}(t)\|_{2,2,\omega}^2 \right) dt \leq C \, \|\Phi\|_{W^{1,2}(-\pi,\pi)}^2;$$

---

[9] We recall that a domain D is said to be uniformly of class C^2 if D lies always on one part of its boundary ∂D and for every $y_0 \in \partial D$, there exists a ball B centered at y_0 and of radius ρ, independent of y_0, such that $\partial D \cap B$ admits a Cartesian representation of the form $y_3 = f(y_1, y_2)$ where f is a C^2-function in its domain such that f and all its derivatives up to the order 2 included are bounded by a constant M independent of y_0.

(vi) there is $K_i = K_i(S_i) > 0$, $i = 1, 2$ such that

$$\max_{t \in [-\pi, \pi]} \|\boldsymbol{A}\|^2_{W^{1,2}(S_i)} + \int_{-\pi}^{\pi} \left(\left\| \frac{\partial \boldsymbol{A}}{\partial t} \right\|^2_{L^2(S_i)} + \|\boldsymbol{A}(t)\|^2_{W^{2,2}(S_i)} \right) dt$$

$$\leq K_i^2 \|\Phi\|^2_{W^{1,2}(-\pi, \pi)}.$$

The construction of the field $\boldsymbol{A}$ makes use of Theorem 1.2 and is formally analogous to that of the field $\boldsymbol{a}$ in the steady-state case that we described in the previous section (with $q = 2$). More precisely, we set $\boldsymbol{A} := \boldsymbol{U} + \boldsymbol{W}$, where $\boldsymbol{U}$ and $\boldsymbol{W}$ are defined in (1.106) and (1.107) and where, this time, the Poiseuille velocity fields $\boldsymbol{v}_{P1}$ and $\boldsymbol{v}_{P2}$ are those constructed in Theorem 1.2. Then, taking into account of the properties of $\boldsymbol{W}$ and of the following, further one

$$\left\| \frac{\partial \boldsymbol{W}}{\partial t} \right\|_{1,2} \leq C \left(\left\| \frac{\partial}{\partial t} (\nabla \cdot \boldsymbol{U}) \right\|_2 \right), \tag{1.116}$$

with $C = C(\widetilde{\Omega}_0) > 0$, we easily show the validity of the above conditions (i)–(vi).

We next consider an increasing sequence of suitable bounded domains, $\{\Omega^{(k)}\}$, "invading" Ω. Set

$$\widetilde{\Omega}_{1k} = \{x \in \widetilde{\Omega}_1 : x_1 \geq -k\}, \quad \widetilde{\Omega}_{2k} = \{x \in \widetilde{\Omega}_2 : x_1 \leq k\};$$

then $\Omega^{(k)}$ is defined as

$$\Omega^{(k)} = \widetilde{\Omega}_0 \cup \widetilde{\Omega}_{1k} \cup \widetilde{\Omega}_{2k} \cup \omega_{1k} \cup \omega_{2k},$$

where the domains ω_{ik} satisfy the conditions

(A) $\omega_{1k} \subset \{x \in \Omega_1 : x \in (-k-1, -k)\}$, $\omega_{2k} \subset \{x \in \Omega_1 : x \in (k, k+1)\}$,

(B) $\omega_{1(k+1)} = \omega_{1k} + (-1, 0, 0)$, $\omega_{2(k+1)} = \omega_{1k} + (1, 0, 0)$.

Thus, recalling that Ω is uniformly of class C^2, also $\Omega^{(k)}$ will be of class C^2 uniformly with respect to $k \in \mathbb{N}$, namely, the constants ρ and M defining the C^2-regularity of $\Omega^{(k)}$ (see footnote [(9)]) are independent of k.

Clearly, $\Omega = \cup_{k=1}^{\infty} \Omega^{(k)}$. We will use the notation

$$\widetilde{\Omega}_1^{(k)} := \widetilde{\Omega}_{1k} \cup \omega_{1k}, \quad \widetilde{\Omega}_2^{(k)} := \widetilde{\Omega}_{2k} \cup \omega_{2k},$$

so that

$$\Omega^{(k)} = \widetilde{\Omega}_1^{(k)} \cup \widetilde{\Omega}_0 \cup \widetilde{\Omega}_2^{(k)}.$$

Let $P = P^{(k)}$ be the orthogonal projection operator of $L^2(\Omega^{(k)})$ onto its subspace $L^2_\sigma(\Omega^{(k)})$ of solenoidal functions having vanishing normal component at $\partial \Omega^{(k)}$ in the trace sense.

The following results hold.

Lemma 1.6. *Set*

$$(\boldsymbol{v}, \boldsymbol{w}) := (\boldsymbol{v}, \boldsymbol{w})_{\Omega^{(k)}}, \quad \|\cdot\|_s := \|\cdot\|_{s,\Omega^{(k)}}.$$

There exists a constant $C = C(\Omega) > 0$, independent of k, such that

$$| (\boldsymbol{u} \cdot \nabla \boldsymbol{A}, \boldsymbol{v}) | \leq C \|\Phi\|_{W^{1,2}(-\pi,\pi)} \|\nabla \boldsymbol{u}\|_2 \|\nabla \boldsymbol{v}\|_2 \,, \qquad (1.117)$$

for all $\boldsymbol{u}, \boldsymbol{v} \in W_0^{1,2}(\Omega^{(k)})$, and

$$| (\boldsymbol{u} \cdot \nabla \boldsymbol{A}, \boldsymbol{v}) | \leq C \|\Phi\|_{W^{1,2}(-\pi,\pi)} \|P\Delta \boldsymbol{u}\|_2 \|\boldsymbol{v}\|_2 \,,$$

$$| (\boldsymbol{A} \cdot \nabla \boldsymbol{u}, \boldsymbol{v}) | \leq C \|\Phi\|_{W^{1,2}(-\pi,\pi)} \|P\Delta \boldsymbol{u}\|_2 \|\boldsymbol{v}\|_2 \,,$$

$$(1.118)$$

for all $\boldsymbol{u} \in \mathcal{D}_0^{1,2}(\Omega^{(k)}) \cap W^{2,2}(\Omega^{(k)})$, $\boldsymbol{v} \in L^2(\Omega^{(k)})$.

Proof. In order to show the lemma, we need a number of inequalities that we collect here:

$$\|\boldsymbol{w}\|_{r,S} \leq K_1 \|\nabla \boldsymbol{w}\|_{2,S} \,, \ r \in [1,\infty), \ \ \boldsymbol{w} \in W_0^{1,2}(S) \,,$$

$$\|\boldsymbol{w}\|_{\infty,S} \leq K_2 \|\nabla \boldsymbol{w}\|_{3,S} \,, \ \ \boldsymbol{w} \in W_0^{1,3}(S) \,,$$

$$\|\nabla \boldsymbol{w}\|_{3,S} \leq K_2 \|\boldsymbol{w}\|_{2,2,S} \,, \ \ \boldsymbol{w} \in W^{2,2}(S) \,,$$

$$\|\boldsymbol{w}\|_{q,\widetilde{\Omega}_0} \leq K_3 \|\nabla \boldsymbol{w}\|_{2,\widetilde{\Omega}_0} \,, \ q \in [1,6], \ \ \boldsymbol{w} \in W^{1,2}(\widetilde{\Omega}_0), \ \boldsymbol{w}|_{\Gamma_0} = \boldsymbol{0},$$

$$\|\nabla \boldsymbol{w}\|_3 \leq K_4 \|P\Delta \boldsymbol{w}\|_2 \,, \ \ \boldsymbol{w} \in \mathcal{D}_0^{1,2}(\Omega^{(k)}) \cap W^{2,2}(\Omega^{(k)}) \,,$$

$$\|\boldsymbol{w}\|_{2,2} \leq K_4 \|P\Delta \boldsymbol{w}\|_2 \,, \ \ \boldsymbol{w} \in \mathcal{D}_0^{1,2}(\Omega^{(k)}) \cap W^{2,2}(\Omega^{(k)}) \,,$$

$$(1.119)$$

where S is any bounded domain in $\mathbb{R}^2$, $\Gamma_0 := \partial\Omega \cap \partial\widetilde{\Omega}_0$, and $K_1 = K_1(r,S) > 0$, $K_2 = K_2(S) > 0$, $K_3 = K_3(q,\Omega) > 0$ and $K_4 = K_4(\Omega) > 0$. Inequalities $(1.119)_{1,2,3}$ are classical, and we refer to [36, Chapter I] for a proof. We next observe that since

$$\|\boldsymbol{w}\|_{2,\widetilde{\Omega}_0} \leq K_5 \|\nabla \boldsymbol{w}\|_{2,\widetilde{\Omega}_0} \,, \ \ \boldsymbol{w} \in W^{1,2}(\widetilde{\Omega}_0), \ \boldsymbol{w}|_{\Gamma_0} = \boldsymbol{0} \,, \qquad (1.120)$$

with $K_5 = K_5(\widetilde{\Omega}_0) > 0$, see, e.g., [36, Exercise II.4.10], inequality $(1.119)_4$ follows from standard embedding theorems. Furthermore, by (1.79) we deduce, for any $\boldsymbol{w} \in C_0^\infty(\Omega^{(k)})$,

$$\|\boldsymbol{w}\|_{2,\widetilde{\Omega}_i^{(k)}} \leq c_i \|\nabla \boldsymbol{w}\|_{2,\widetilde{\Omega}_i^{(k)}} \,, \ i = 1,2, \qquad (1.121)$$

where $c_i = c_i(S_i) > 0$, $i = 1,2$. Therefore, from (1.120) and (1.121) we conclude

$$\|\boldsymbol{w}\|_2 \leq c \|\nabla \boldsymbol{w}\|_2 \,, \ \ \boldsymbol{w} \in W_0^{1,2}(\Omega^{(k)}) \,, \qquad (1.122)$$

where $c = c(\Omega) > 0$. Since we also have

$$(\boldsymbol{w}, P\Delta \boldsymbol{w}) = (\boldsymbol{w}, \Delta \boldsymbol{w}) = -\|\nabla \boldsymbol{w}\|_2 \,, \ \ \boldsymbol{w} \in \mathcal{D}_0^{1,2}(\Omega^{(k)}) \cap W^{2,2}(\Omega^{(k)}) \,,$$

by (1.122) and by Schwarz inequality we find

$$\|\nabla \boldsymbol{w}\|_2 \leq c \|P\Delta \boldsymbol{w}\|_2 \,, \ \ \boldsymbol{w} \in \mathcal{D}_0^{1,2}(\Omega^{(k)}) \cap W^{2,2}(\Omega^{(k)}) \,. \qquad (1.123)$$

We now recall the inequalities (see, e.g., [57, Lemma 1])

$$\|\nabla \boldsymbol{w}\|_3 \leq K_6 \left(\|\nabla \boldsymbol{w}\|_2^{\frac{1}{2}} \|D^2 \boldsymbol{w}\|_2^{\frac{1}{2}} + \|\nabla \boldsymbol{w}\|_2 \right), \ \ \boldsymbol{w} \in W^{2,2}(\Omega^{(k)}) \,, \qquad (1.124)$$

and

$$\|D^2 w\|_2 \le c_1 \left(\|P \Delta w\|_2 + \|\nabla w\|_2 \right), \tag{1.125}$$

where $D^2 w$ stands for an arbitrary second derivative of w and the constants K_6 and $c_1 > 0$ are independent of w and k. Combining (1.124) and (1.125) we find, in particular,

$$\|\nabla w\|_3 \le K_7 \left(\|\nabla w\|_2^{\frac{1}{2}} \|P \Delta w\|_2^{\frac{1}{2}} + \|\nabla w\|_2 \right), \quad w \in W^{2,2}(\Omega^{(k)}), \tag{1.126}$$

with $K_7 = K_7(\Omega) > 0$ independent of k. Consequently, $(1.119)_5$ follows from (1.122), (1.123) and (1.126), while $(1.119)_6$ follows from (1.122), (1.123) and (1.125). We now proceed to the proof of (1.117). By the Hölder inequality, by $(1.119)_4$ with $q = 4$, and by the properties (ii) and (v) of $\boldsymbol{A}$ we find

$$|(\boldsymbol{u} \cdot \nabla \boldsymbol{A}, \boldsymbol{v})| = |(\boldsymbol{u} \cdot \nabla \boldsymbol{v}, \boldsymbol{A})|$$

$$\le \|\boldsymbol{A}(t)\|_{4, \widetilde{\Omega}_0} \|\boldsymbol{u}\|_4 \|\nabla \boldsymbol{v}\|_2 + \sum_{i=1}^{2} \left| \int_{\widetilde{\Omega}_i^{(k)}} \boldsymbol{u} \cdot \nabla \boldsymbol{v} \cdot \boldsymbol{A} \right|$$

$$\le C_1 \|\Phi\|_{W^{1,2}(-\pi,\pi)} \|\nabla \boldsymbol{u}\|_2 \|\nabla \boldsymbol{v}\|_2 + \sum_{i=1}^{2} \left| \int_{\widetilde{\Omega}_i^{(k)}} \boldsymbol{u} \cdot \nabla \boldsymbol{v} \cdot \boldsymbol{A} \right|,$$

where $C_1 = C_1(\Omega) > 0$. Moreover, extending $\boldsymbol{u}$ and $\boldsymbol{v}$ to zero outside $\widetilde{\Omega}^{(k)}$, for all $k \in \mathbb{N}$, we have

$$\left| \int_{\widetilde{\Omega}_1^{(k)}} \boldsymbol{u} \cdot \nabla \boldsymbol{v} \cdot \boldsymbol{A} \right| = \left| \int_{-k-1}^{-1} \left(\int_{S_1} \boldsymbol{u} \cdot \nabla \boldsymbol{v} \cdot \boldsymbol{A} \, dS \right) dx_1 \right|.$$

By virtue of the Hölder inequality, by $(1.119)_1$ with $r = 4$, and by the properties (iv) and (vi) of $\boldsymbol{A}$, we obtain

$$\left| \int_{-k-1}^{-1} \int_{S_1} \boldsymbol{u} \cdot \nabla \boldsymbol{v} \cdot \boldsymbol{A} \, dS \, dx_1 \right| \le \int_{-k-1}^{-1} \|\boldsymbol{A}\|_{4,S_1} \|\boldsymbol{u}\|_{4,S_1} \|\nabla \boldsymbol{v}\|_{2,S_1} \, dx_1$$

$$\le k_1^2 \|\boldsymbol{A}(t)\|_{1,2,S_1} \int_{-k-1}^{-1} \|\nabla \boldsymbol{u}\|_{2,S_1} \|\nabla \boldsymbol{v}\|_{2,S_1} \, dx_1$$

$$\le k_1^2 K_1 \|\Phi\|_{W^{1,2}(-\pi,\pi)} \|\nabla \boldsymbol{u}\|_2 \|\nabla \boldsymbol{v}\|_2.$$

Likewise, we show

$$\left| \int_{\widetilde{\Omega}_2^{(k)}} \boldsymbol{u} \cdot \nabla \boldsymbol{v} \cdot \boldsymbol{A} \right| \le k_1^2 K_1 \|\Phi\|_{W^{1,2}(-\pi,\pi)} \|\nabla \boldsymbol{u}\|_2 \|\nabla \boldsymbol{v}\|_2,$$

and so we conclude the validity of (1.117). We now pass to the proof of $(1.118)_1$. We extend $\boldsymbol{v}$ to $\boldsymbol{0}$ outside $\Omega^{(k)}$ and continue to denote by $\boldsymbol{v}$ this extension. From the Hölder inequality and from $(1.119)_6$ and by the properties (ii) and (v) of $\boldsymbol{A}$

we find

$$|(\boldsymbol{u} \cdot \nabla \boldsymbol{A}, \boldsymbol{v})| \leq \|\nabla \boldsymbol{A}(t)\|_{2,\widetilde{\Omega}_0} \|\boldsymbol{u}\|_\infty \|\boldsymbol{v}\|_2 + \sum_{i=1}^{2} \left| \int_{\widetilde{\Omega}_i^{(k)}} \boldsymbol{u} \cdot \nabla \boldsymbol{A} \cdot \boldsymbol{v} \right|$$

$$\leq C_2 \|\Phi\|_{W^{1,2}(-\pi,\pi)} \|P\Delta \boldsymbol{u}\|_2 \|\boldsymbol{v}\|_2 + \sum_{i=1}^{2} \left| \int_{\widetilde{\Omega}_i^{(k)}} \boldsymbol{u} \cdot \nabla \boldsymbol{A} \cdot \boldsymbol{v} \right|,$$

$$(1.127)$$

where $C_2 = C_2(\Omega) > 0$. From the definition of $\widetilde{\Omega}_1^{(k)}$, we have

$$\left| \int_{\widetilde{\Omega}_1^{(k)}} \boldsymbol{u} \cdot \nabla \boldsymbol{A} \cdot \boldsymbol{v} \right| \leq \left| \int_{\widetilde{\Omega}_{1k}} \boldsymbol{u} \cdot \nabla \boldsymbol{A} \cdot \boldsymbol{v} \right| + \left| \int_{\omega_{1k}} \boldsymbol{u} \cdot \nabla \boldsymbol{A} \cdot \boldsymbol{v} \right| =: I_1 + I_2. \quad (1.128)$$

Recalling the definition of $\widetilde{\Omega}_{1k}$, the properties of $\boldsymbol{A}$ and $(1.119)_6$, we find

$$I_1 \leq \int_{-k}^{-1} \|\nabla \boldsymbol{A}\|_{2,S_1} \|\boldsymbol{u}\|_{\infty,S_1} \|\boldsymbol{v}\|_{2,S_1} \, dx_1$$

$$\leq C_3 \|\Phi\|_{W^{1,2}(-\pi,\pi)} \int_{-k}^{-1} \|\boldsymbol{u}\|_{2,2,S_1} \|\boldsymbol{v}\|_{2,S_1} \, dx_1 \qquad (1.129)$$

$$\leq C_4 \|\Phi\|_{W^{1,2}(-\pi,\pi)} \|P\Delta \boldsymbol{u}\|_2 \|\boldsymbol{v}\|_2,$$

where $C_4 = C_4(\Omega) > 0$. In order to evaluate I_2, we observe that, given the stated properties of $\boldsymbol{u}$, we can approximate $\boldsymbol{u}$ in $W^{1,s}(\Omega^{(k)})$, $s > 3$, with a sequence of functions $\{\boldsymbol{u}_m\} \subset C_0^\infty(\Omega^{(k)})$. If we use the embedding $W^{1,s}(\Omega^{(k)}) \subset L^\infty(\Omega^{(k)})$ with the embedding constant $C = C(\Omega^{(k)}, s) > 0$, we deduce

$$\left| \int_{\omega_{1k}} (\boldsymbol{u}_m - \boldsymbol{u}) \cdot \nabla \boldsymbol{A} \cdot \boldsymbol{v} \right| \leq \|\boldsymbol{u}_m - \boldsymbol{u}\|_\infty \|\nabla \boldsymbol{A}\|_2 \|\boldsymbol{v}\|_2 \leq C\|\boldsymbol{u} - \boldsymbol{u}_m\|_{1,s} \|\nabla \boldsymbol{A}\|_2 \|\boldsymbol{v}\|_2,$$

which, in turn, implies that

$$I_2 = \lim_{m\to\infty} \left| \int_{\omega_{1k}} \boldsymbol{u}_m \cdot \nabla \boldsymbol{A} \cdot \boldsymbol{v} \right|. \qquad (1.130)$$

However, by $(1.119)_2$, by the Hölder inequality and by the properties of $\boldsymbol{A}$, we have

$$\left| \int_{\omega_{1k}} \boldsymbol{u}_m \cdot \nabla \boldsymbol{A} \cdot \boldsymbol{v} \right| \leq \int_{-k-1}^{-k} \|\nabla \boldsymbol{A}\|_{2,S_1} \|\boldsymbol{u}_m\|_{\infty,S_1} \|\boldsymbol{v}\|_{2,S_1} \, dx_1$$

$$\leq C_5 \|\Phi\|_{W^{1,2}(-\pi,\pi)} \int_{-k-1}^{-k} \|\nabla \boldsymbol{u}_m\|_{3,S_1} \|\boldsymbol{v}\|_{2,S_1} \, dx_1$$

$$\leq C_5 \|\Phi\|_{W^{1,2}(-\pi,\pi)} \|\nabla \boldsymbol{u}_m\|_3 \|\boldsymbol{v}\|_2,$$

where $C_5 = C_5(\Omega) > 0$. From this relation, from (1.130) and from the property of the sequence $\{\boldsymbol{u}_m\}$ we recover

$$I_2 \leq C_5 \|\Phi\|_{W^{1,2}(-\pi,\pi)} \|\nabla \boldsymbol{u}\|_3 \|\boldsymbol{v}\|_2,$$

and so, by $(1.119)_5$,

$$I_2 \leq C_6 \|\Phi\|_{W^{1,2}(-\pi,\pi)} \|P\Delta u\|_2 \|v\|_2,$$

where $C_6 = C_6(\Omega) > 0$. From this latter equation and from (1.128) and (1.129), we find

$$\left| \int_{\widetilde{\Omega}_1^{(k)}} u \cdot \nabla A \cdot v \right| \leq C_6 \|\Phi\|_{W^{1,2}(-\pi,\pi)} \|P\Delta u\|_2 \|v\|_2, \qquad (1.131)$$

with $C_6 = C_6(\Omega) > 0$. Likewise, we show

$$\left| \int_{\widetilde{\Omega}_2^{(k)}} u \cdot \nabla A \cdot v \right| \leq C_7 \|\Phi\|_{W^{1,2}(-\pi,\pi)} \|P\Delta u\|_2 \|v\|_2, \qquad (1.132)$$

for another $C_7 = C_7(\Omega) > 0$, so that $(1.118)_1$ follows from (1.127), (1.131) and (1.132). In order to prove $(1.118)_2$, we again extend v to $\mathbf{0}$ outside $\Omega^{(k)}$ and continue to denote by v this extension. From the Hölder inequality and from $(1.119)_4$ with $q = 6$, $(1.119)_5$ and by the properties (ii) and (v) of A we obtain

$$|(A \cdot \nabla u, v)| \leq \|A(t)\|_{6,\widetilde{\Omega}_0} \|\nabla u\|_3 \|v\|_2 + \sum_{i=1}^{2} \left| \int_{\widetilde{\Omega}_i^{(k)}} A \cdot \nabla u \cdot v \right|$$

$$\leq C_7 \|\Phi\|_{W^{1,2}(-\pi,\pi)} \|P\Delta u\|_2 \|v\|_2 + \sum_{i=1}^{2} \left| \int_{\widetilde{\Omega}_i^{(k)}} A \cdot \nabla u \cdot v \right|, \qquad (1.133)$$

where $C_7 = C_7(\Omega) > 0$. As in the proof of $(1.118)_1$ we perform the splitting

$$\left| \int_{\widetilde{\Omega}_1^{(k)}} A \cdot \nabla u \cdot v \right| \leq \left| \int_{\widetilde{\Omega}_{1k}} A \cdot \nabla u \cdot v \right| + \left| \int_{\omega_{1k}} A \cdot \nabla u \cdot v \right| =: \mathcal{I}_1 + \mathcal{I}_2. \qquad (1.134)$$

By the properties of A, by $(1.119)_1$ with $r = 6$ and by $(1.119)_5$, we find

$$\mathcal{I}_1 \leq \int_{-k}^{-1} \|A\|_{6,S_1} \|\nabla u\|_{3,S_1} \|v\|_{2,S_1} dx_1 \leq C_8 \|\Phi\|_{W^{1,2}(-\pi,\pi)} \int_{-k}^{-1} \|u\|_{2,2,S_1} \|v\|_{2,S_1} dx_1$$

$$\leq C_9 \|\Phi\|_{W^{1,2}(-\pi,\pi)} \|P\Delta u\|_2 \|v\|_2, \qquad (1.135)$$

where $C_9 = C_9(\Omega) > 0$. Furthermore, using again the Hölder inequality along with $(1.119)_1$ with $r = 6$ and $(1.119)_5$, we show that

$$\mathcal{I}_2 \leq \int_{-k-1}^{-k} \|A\|_{6,S_1} \|\nabla u\|_{3,S_1} \|v\|_{2,S_1} dx_1$$

$$\leq C_{10} \|\Phi\|_{W^{1,2}(-\pi,\pi)} \int_{-k-1}^{-k} \|\nabla u\|_{3,S_1} \|v\|_{2,S_1} dx_1 \qquad (1.136)$$

$$\leq C_{10} \|\Phi\|_{W^{1,2}(-\pi,\pi)} \|\nabla u\|_3 \|v\|_2$$

$$\leq C_{11} \|\Phi\|_{W^{1,2}(-\pi,\pi)} \|P\Delta u\|_2 \|v\|_2,$$

where $C_{11} = C_{11}(\Omega) > 0$. From (1.134)–(1.136) it follows that

$$\left| \int_{\widetilde{\Omega}_1^{(k)}} \boldsymbol{A} \cdot \nabla \boldsymbol{u} \cdot \boldsymbol{v} \right| \leq C_{12} \|\Phi\|_{W^{1,2}(-\pi,\pi)} \|P\Delta\boldsymbol{u}\|_2 \|\boldsymbol{v}\|_2 , \qquad (1.137)$$

where $C_{12} = C_{12}(\Omega) > 0$. In a completely analogous manner we show that

$$\left| \int_{\widetilde{\Omega}_2^{(k)}} \boldsymbol{A} \cdot \nabla \boldsymbol{u} \cdot \boldsymbol{v} \right| \leq C_{13} \|\Phi\|_{W^{1,2}(-\pi,\pi)} \|P\Delta\boldsymbol{u}\|_2 \|\boldsymbol{v}\|_2 , \qquad (1.138)$$

with $C_{13} = C_{13}(\Omega) > 0$, and so $(1.118)_2$ follows from (1.133), (1.134), (1.137) and (1.138). $\qquad\square$

Remark 1.8. Inequalities $(1.119)_{5,6}$ continue to hold on the entire domain Ω. This can be easily proved by using the fact that the constants involved do not depend on k. Therefore, exactly by the same arguments used in the proof of Lemma 1.6, we can show that inequalities (1.117) and $(1.117)_1$ continue to hold if we replace $\Omega^{(k)}$ with the whole domain Ω.

We are now in a position to show the main result of this section.

Theorem 1.6. *Let Ω be uniformly of class C^2 and assume that $\Phi \in W^{1,2}(-\pi,\pi)$ with $\Phi(-\pi) = \Phi(\pi)$. Then, there exists a positive constant $C = C(\Omega)$ such that if*

$$\|\Phi\|_{W^{1,2}(-\pi,\pi)} < C , \qquad (1.139)$$

problem (1.115) has at least one time-periodic solution $(\boldsymbol{v}, p)$ of period 2π, with $\boldsymbol{v} = \boldsymbol{u} + \boldsymbol{A}$, such that

$$\boldsymbol{u} \in L^\infty(-\pi,\pi; \mathcal{D}_0^{1,2}(\Omega)) \cap L^2(-\pi,\pi; W^{2,2}(\Omega)), \quad \frac{\partial \boldsymbol{u}}{\partial t} \in L^2(-\pi,\pi; L^2(\Omega)),$$

$$\nabla p \in L^2(-\pi,\pi; L^2(\Omega)).$$

$$(1.140)$$

Moreover, this solution satisfies the estimate

$$\operatorname*{ess\,sup}_{t\in(-\pi,\pi)} \|\boldsymbol{u}(t)\|_{1,2}^2 + \int_{-\pi}^{\pi} \left(\left\| \frac{\partial \boldsymbol{u}}{\partial t} \right\|_2^2 + \|\boldsymbol{u}(t)\|_{2,2}^2 \right) dt \leq C_1 \|\Phi\|_{W^{1,2}(-\pi,\pi)}^2 , \quad (1.141)$$

where $C_1 = C_1(\Omega,) > 0$.

Concerning the asymptotic conditions $(1.115)_5$ they are satisfied in the following sense:

$$\lim_{r\to\infty} \max_{x\in\Omega_{ir}} |\boldsymbol{v}(x,t) - \boldsymbol{v}_{Pi}(x,t)| = 0 , \quad i = 1,2 \text{ for a.a. } t \in [-\pi,\pi] , \qquad (1.142)$$

where $\Omega_{1r} := \{x \in \Omega_i : x_1 < -r\}$ and $\Omega_{2r} := \{x \in \Omega_i : x_1 > r\}$.

Finally, if $(\widetilde{\boldsymbol{u}}, \widetilde{p})$ is any other solution in the class (1.140), then $\boldsymbol{u} \equiv \widetilde{\boldsymbol{u}}$, $\widetilde{p} \equiv p$.

Proof. We begin to explain the basic ideas. First of all, as in the steady-state case, we write $v = u + A$, where u satisfies the problem

$$\left.\begin{array}{l} \dfrac{\partial u}{\partial t} + u \cdot \nabla u + A \cdot \nabla u + u \cdot \nabla A = \Delta u - \nabla p + f \\[2mm] \nabla \cdot u = 0 \end{array}\right\} \quad \text{in } \Omega \times (-\pi, \pi),$$

$$\int_S u \cdot n \, dS = 0 \,, \quad u|_{\partial\Omega} = 0,$$

$$\lim_{|x| \to \infty,\ x \in \Omega_i} u(x) = 0 \,, \quad i = 1, 2,$$

(1.143)

with

$$f := -\frac{\partial A}{\partial t} + \Delta A - A \cdot \nabla A.$$

By the property (iv) of A, we have

$$f(x, t) = -G_i(t) e_1 \quad \text{for all } x \in \widetilde{\Omega}_i, \ i = 1, 2, \tag{1.144}$$

where $G_i(t)$ is the axial pressure gradient of the corresponding Poiseuille flow (see (1.4)). We next consider the increasing sequence of C^2-bounded domains, $\{\Omega^{(k)}\}$, introduced previously and show that, on each domain $\Omega^{(k)}$ problem (1.143) has a 2π-periodic solution $(u^{(k)}, p^{(k)})$ in the class (1.140). Successively, we prove that these "approximating solutions" can be bounded in suitable norms, uniformly in k, along a subsequence at least, and that in the limit of large k they converge, again in appropriate topologies, to a solution of problem (1.143) in the class (1.140). In order to find a solution in each $\Omega^{(k)}$, we shall use the Galerkin method with the special basis constituted by the eigenfunctions, $\{\psi_j^{(k)}\}$, of the Stokes problem:

$$P\Delta\psi_j^{(k)} = -\lambda^{(k)}\psi_j^{(k)}, \quad \psi_j^{(k)} \in L_\sigma^2(\Omega^{(k)}) \cap W_0^{1,2}(\Omega) \cap W^{2,2}(\Omega^{(k)}), \tag{1.145}$$

where P, we recall, is the orthogonal projection operator of $L^2(\Omega^{(k)})$ onto its subspace $L_\sigma^2(\Omega^{(k)})$ of solenoidal functions having vanishing normal component at $\partial\Omega^{(k)}$. The family $\{\psi_j^{(k)}\}$ is orthonormal in $L^2(\Omega^{(k)})$. We extend each $\psi_j^{(k)}$ to zero in Ω, outside $\Omega^{(k)}$ and continue to denote by $\psi_j^{(k)}$ this extension. It is easy to check that $\psi_j^{(k)}$ carries no flow-rate, that is,

$$\int_S \psi_j^{(k)} \cdot n \, dS = 0.$$

An approximating solution to problem (1.143) in $\Omega^{(k)}$ is sought of the form

$$u_m^{(k)}(x, t) = \sum_{i=1}^m c_{mi}^{(k)}(t) \psi_i^{(k)}(x)$$

where the coefficients $c_{mi}^{(k)}(t)$, $i = 1, \ldots, m$ are solutions to the following system of ordinary differential equations:

$$\frac{d}{dt}(\boldsymbol{u}_m^{(k)}, \boldsymbol{\psi}_i^{(k)}) + \left((\boldsymbol{u}_m^{(k)} \cdot \nabla \boldsymbol{u}_m^{(k)} + \boldsymbol{A} \cdot \nabla \boldsymbol{u}_m^{(k)} + \boldsymbol{u}_m^{(k)} \cdot \nabla \boldsymbol{A}), \boldsymbol{\psi}_i^{(k)} \right)$$
$$= (\Delta \boldsymbol{u}_m^{(k)}, \boldsymbol{\psi}_i^{(k)}) + (\boldsymbol{f}, \boldsymbol{\psi}_i^{(k)}), \quad i = 1, 2, \ldots, m, \tag{1.146}$$

and where $(\cdot, \cdot)$ denotes the $L^2(\Omega^{(k)})$ scalar product of vector functions. We shall now prove that (1.146) has a 2π-periodic solution, provided the norm of Φ in $W^{1,2}(-\pi, \pi)$ is suitably restricted. To this end, multiplying both sides of (1.146) by $c_{mi}^{(k)}(t)$ and summing over i from 1 to m, we get (superscript k omitted)

$$\frac{1}{2}\frac{d}{dt}\|\boldsymbol{u}_m\|_2^2 + \|\nabla \boldsymbol{u}_m\|_2^2 = (\boldsymbol{u}_m \cdot \nabla \boldsymbol{u}_m, \boldsymbol{A}) + (\boldsymbol{f}, \boldsymbol{u}_m). \tag{1.147}$$

In view of (1.117), we have

$$(\boldsymbol{u}_m \cdot \nabla \boldsymbol{u}_m, \boldsymbol{A}) \le C_1 \|\Phi\|_{W^{1,2}(-\pi,\pi)} \|\nabla \boldsymbol{u}\|_2^2, \tag{1.148}$$

where $C_1 = C_1(\Omega) > 0$. We observe next that, from (1.144) and for all $k \in \mathbb{N}$, we have

$$(\boldsymbol{f}, \boldsymbol{u}_m) = \int_{\tilde{\Omega}_0} \boldsymbol{f} \cdot \boldsymbol{u}_m - G_1(t) \int_{-k-1}^{-1} \int_{S_1} \boldsymbol{u}_m \cdot \boldsymbol{e}_1 - G_2(t) \int_1^{k+1} \int_{S_2} \boldsymbol{u}_m \cdot \boldsymbol{e}_1,$$

where we used the fact that $\boldsymbol{u}_m = \boldsymbol{0}$ outside $\Omega^{(k)}$. Thus, since $\boldsymbol{u}_m$ carries no flow-rate, from the previous relation and from the Schwarz inequality and inequality (1.120) we find

$$(\boldsymbol{f}, \boldsymbol{u}_m) \le \left(\int_{\tilde{\Omega}_0} |\boldsymbol{f}|^2 \right)^{1/2} \left(\int_{\tilde{\Omega}_0} |\boldsymbol{u}_m|^2 \right)^{1/2}$$
$$\le C_2 \left(\int_{\tilde{\Omega}_0} |\boldsymbol{f}|^2 \right)^{1/2} \|\nabla \boldsymbol{u}_m\|_2 \le C_3 \int_{\tilde{\Omega}_0} |\boldsymbol{f}|^2 + \tfrac{1}{4}\|\nabla \boldsymbol{u}_m\|_2^2, \tag{1.149}$$

with $C_3 = C_3(\Omega) > 0$. Collecting (1.147), (1.148) and (1.149) and imposing

$$\|\Phi\|_{W^{1,2}(-\pi,\pi)} < \frac{1}{4C_2}, \tag{1.150}$$

we thus deduce

$$\frac{d}{dt}\|\boldsymbol{u}_m\|_2^2 + \|\nabla \boldsymbol{u}_m\|_2^2 \le 2C_4 \int_{\tilde{\Omega}_0} |\boldsymbol{f}|^2. \tag{1.151}$$

We now notice that, from the definition of $\boldsymbol{f}$, from (1.150), from elementary embedding inequalities and from property (v) of the flow-rate carrier $\boldsymbol{A}$ it easily follows that

$$\int_{-\pi}^{\pi} \int_{\tilde{\Omega}_0} |\boldsymbol{f}|^2 dt \le C_5 \|\Phi\|_{W^{1,2}(-\pi,\pi)}^2, \tag{1.152}$$

with $C_5 = C_5(\Omega) > 0$. Consequently, integrating both sides of (1.151) from $-\pi$ to π, we obtain, on the one hand, that $\|\boldsymbol{u}_m(t)\|_2 = \sum_{i=1}^{m} |c_{mi}(t)|^2$ is uniformly bounded in $t \in [-\pi, \pi]$, provided $\|\boldsymbol{u}_m(-\pi)\|_2 = \sum_{i=1}^{m} |c_{mi}(-\pi)|^2 < \infty$, which implies that the solutions $c_{mi}(t)$ to the system (1.146) can be extended to the whole interval $[-\pi, \pi]$, and, on the other hand, that the following estimate holds:

$$\int_{-\pi}^{\pi} \|\nabla \boldsymbol{u}_m(t)\|_2^2 dt \leq \left(\|\boldsymbol{u}_m(-\pi)\|_2^2 - \|\boldsymbol{u}_m(\pi)\|_2^2 + C_6 \|\Phi\|_{W^{1,2}(-\pi,\pi)}^2 \right), \qquad (1.153)$$

with $C_6 = C_6(\Omega) > 0$. It is easy to prove that we can choose $\boldsymbol{u}_m(-\pi)$ in such a way that $\boldsymbol{u}_m(-\pi) = \boldsymbol{u}_m(\pi)$. Actually, by a routine calculation that we omit and refer to [90] for details, we prove that the map

$$T : c_{mi}(-\pi) \in \mathbb{R}^m \mapsto c_{mi}(\pi) \in \mathbb{R}^m$$

is continuous. It is also easy to prove that T transforms the ball of $\mathbb{R}^m$, B_R, of radius R into itself, for a suitable R. To this end, using (1.79) in the second term on the left-hand side of (1.151) along with Gronwall's lemma and (1.152) we obtain

$$\|\boldsymbol{u}_m(\pi)\|_2^2 \leq \|\boldsymbol{u}_m(-\pi)\|_2^2 e^{-C_7 \pi} + C_5 \|\Phi\|_{W^{1,2}(-\pi,\pi)}^2, \qquad (1.154)$$

with $C_7 = C_7(\Omega) > 0$. Thus, if we choose, for example,

$$R^2 = 2C_5 \frac{\|\Phi\|_{W^{1,2}(-\pi,\pi)}^2}{1 - e^{-C_7 \pi}},$$

T transforms B_R into itself and, therefore, by Brower's theorem, T has a fixed point in B_R, that is, $\boldsymbol{u}_m(-\pi) = \boldsymbol{u}_m(\pi)$. We may thus extend, by periodicity, $\boldsymbol{u}_m(t)$ to the whole real line. Furthermore, since $\|\boldsymbol{u}_m(-\pi)\|_2 \leq R$, from (1.151)–(1.153) we also have

$$\|\boldsymbol{u}_m(t)\|_2^2 + \int_{-\pi}^{\pi} \|\nabla \boldsymbol{u}_m(t)\|_2^2 dt \leq C_8 \|\Phi\|_{W^{1,2}(-\pi,\pi)}^2 \qquad (1.155)$$

where $C_8 = C_8(\Omega) > 0$ is independent of k. We now show further bounds for the sequence $\{\boldsymbol{u}_m\}$. If we dot-multiply both sides of (1.146) by $\lambda_i^{(k)} c_{mi}(t)$, sum over i from 1 to m and take into account (1.145), we find

$$\frac{1}{2} \frac{d}{dt} \|\nabla \boldsymbol{u}_m\|_2^2 + \|P \Delta \boldsymbol{u}_m\|_2^2 = ((\boldsymbol{u}_m \cdot \nabla \boldsymbol{u}_m + \boldsymbol{A} \cdot \nabla \boldsymbol{u}_m + \boldsymbol{u}_m \cdot \nabla \boldsymbol{A}), P \Delta \boldsymbol{u}_m)$$

$$+ (\boldsymbol{f}, P \Delta \boldsymbol{u}_m). \qquad (1.156)$$

In order to estimate the first term on the right-hand side of (1.156), we recall the classical Sobolev inequality

$$\|\boldsymbol{w}\|_{6,\Omega^{(k)}} \leq K \|\nabla \boldsymbol{w}\|_{2,\Omega^{(k)}},$$

where K is a positive universal constant independent of k. Thus, from the Hölder inequality, from (1.126), and from Young's inequality

$$ab \leq \varepsilon^{-q/q'} \frac{a^{q'}}{q'} + \varepsilon \frac{b^q}{q}, \quad a, b, \varepsilon > 0, \quad \frac{1}{q} + \frac{1}{q'} = 1, \quad q \in (1, \infty), \qquad (1.157)$$

we obtain

$$(\boldsymbol{u}_m \cdot \nabla \boldsymbol{u}_m, P\Delta \boldsymbol{u}_m) \leq C_9 \|\boldsymbol{u}_m\|_6^2 \|\nabla \boldsymbol{u}_m\|_3^2 + (1/4)\|P\Delta \boldsymbol{u}_m\|_2^2$$

$$\leq C_{10} \left(\|\nabla \boldsymbol{u}_m\|_2^4 + \|\nabla \boldsymbol{u}_m\|_2^6 \right) + (1/2)\|P\Delta \boldsymbol{u}_m\|_2^2, \quad (1.158)$$

where $C_{10} = C_{10}(\Omega) > 0$. The second and third term on the right-hand side of (1.156) are estimated by (1.118). Thus, setting

$$\|\Phi\|_{W^{1,2}(-\pi,\pi)} := D,$$

we have

$$(\boldsymbol{A} \cdot \nabla \boldsymbol{u}, P\Delta \boldsymbol{u}_m) + (\boldsymbol{u} \cdot \nabla \boldsymbol{A}, P\Delta \boldsymbol{u}_m) \leq C_{13} D \, \|P\Delta \boldsymbol{u}_m\|_2^2, \quad (1.159)$$

where $C_{13} = C_{13}(\Omega) > 0$. Moreover, by the same reasonings leading to (1.149), we find

$$(\boldsymbol{f}, P\Delta \boldsymbol{u}_m) \leq C_{16} \int_{\tilde{\Omega}_0} |\boldsymbol{f}|^2 + (1/4)\|P\Delta \boldsymbol{u}_m\|_2^2, \quad (1.160)$$

with $C_{14} = C_{14}(\Omega) > 0$. Then, from (1.156)–(1.160) it follows that there exists a constant $C_{15} = C_{15}(\Omega) > 0$ such that if $D < C_{15}$, the following inequality holds:

$$\frac{d}{dt}\|\nabla \boldsymbol{u}_m\|_2^2 + \|P\Delta \boldsymbol{u}_m\|_2^2 \leq C_{16} \left(\|\nabla \boldsymbol{u}_m\|_2^4 + \|\nabla \boldsymbol{u}_m\|_2^6 \right) + C_{17} \int_{\tilde{\Omega}_0} |\boldsymbol{f}|^2, \quad (1.161)$$

with $C_i = C_i(\Omega) > 0$, $i = 16, 17$. One consequence of (1.161) is obtained by using $(1.119)_6$ on the second term on its left-hand side. We get

$$\frac{d}{dt}\|\nabla \boldsymbol{u}_m\|_2^2 + \|\nabla \boldsymbol{u}_m\|_2^2 \leq C_{18} \left(\|\nabla \boldsymbol{u}_m\|_2^4 + \|\nabla \boldsymbol{u}_m\|_2^6 \right) + C_{19} \int_{\tilde{\Omega}_0} |\boldsymbol{f}|^2, \quad (1.162)$$

with $C_i = C_i(\Omega) > 0$, $i = 18, 19$. We then go back to (1.155) and observe that, by the mean-value theorem, there exists t_m (possibly depending on m) in $(-\pi, \pi)$ such that

$$\|\nabla \boldsymbol{u}_m(t_m)\|_2^2 \leq \frac{C_8}{2\pi} D^2.$$

So, using (1.152) and the 2π-periodicity of $\boldsymbol{u}_m$, we can readily show that, if D^2 is below a certain positive constant depending only on Ω, equation (1.162) implies that

$$\|\nabla \boldsymbol{u}_m(t)\|_2^2 \leq C_{20} D^2, \text{ for all } t \in [-\pi, \pi], \quad (1.163)$$

where, again, the constant depends only on Ω. Furthermore, if we integrate both sides of (1.161) over $(-\pi, \pi)$ and employ (1.152), (1.163) and the fact that D is below a suitable constant, we find, for some $C_{21} = C_{21}(\Omega) > 0$,

$$\int_{-\pi}^{\pi} \|P\Delta \boldsymbol{u}_m(t)\|_2^2 dt \leq C_{21} D, \quad (1.164)$$

namely, by $(1.119)_6$ and (1.163),

$$\int_{-\pi}^{\pi} \|\boldsymbol{u}_m(t)\|_{2,2}^2 dt \leq C_{22} D, \quad (1.165)$$

with $C_{22} = C_{22}(\Omega) > 0$. We next look for an estimate on the time derivative of $\boldsymbol{u}_m$. To this end, if we multiply both sides of (1.146) by $\dfrac{dc_{mi}}{dt}$ and sum over i from 1 to m we deduce

$$\frac{1}{2}\frac{d}{dt}\|\nabla\boldsymbol{u}_m\|_2^2 + \left\|\frac{\partial\boldsymbol{u}_m}{\partial t}\right\|_2^2 = -\left((\boldsymbol{u}_m\cdot\nabla\boldsymbol{u}_m + \boldsymbol{A}\cdot\nabla\boldsymbol{u}_m + \boldsymbol{u}_m\cdot\nabla\boldsymbol{A}), \frac{\partial\boldsymbol{u}_m}{\partial t}\right)$$

$$+ (\boldsymbol{f}, \frac{\partial\boldsymbol{u}_m}{\partial t}).$$

(1.166)

Again using the Hölder inequality, (1.157) and (1.119)$_6$, we find

$$\left|\left((\boldsymbol{u}_m\cdot\nabla\boldsymbol{u}_m, \frac{\partial\boldsymbol{u}_m}{\partial t}\right)\right| \leq \|\boldsymbol{u}_m\|_\infty^2\|\nabla\boldsymbol{u}_m\|_2^2 + (1/4)\left\|\frac{\partial\boldsymbol{u}_m}{\partial t}\right\|_2^2$$

$$\leq C_{23}\|P\Delta\boldsymbol{u}_m\|_2^2\|\nabla\boldsymbol{u}_m\|_2^2 + (1/4)\left\|\frac{\partial\boldsymbol{u}_m}{\partial t}\right\|_2^2,$$

(1.167)

with $C_{23} = C_{23}(\Omega) > 0$. The second and the third term on the right-hand side of (1.166) can be estimated by (1.118) to obtain

$$\left|\left(\boldsymbol{A}\cdot\nabla\boldsymbol{u}_m, \frac{\partial\boldsymbol{u}_m}{\partial t}\right) + \left(\boldsymbol{u}_m\cdot\nabla\boldsymbol{A}, \frac{\partial\boldsymbol{u}_m}{\partial t}\right)\right| \leq C_{24}D^2\|P\Delta\boldsymbol{u}_m\|_2^2 + (1/4)\left\|\frac{\partial\boldsymbol{u}_m}{\partial t}\right\|_2^2,$$

(1.168)

where $C_{24} = C_{24}(\Omega) > 0$. Finally, by an argument entirely analogous to that leading to (1.149), we get

$$|(\boldsymbol{f}, \frac{\partial\boldsymbol{u}_m}{\partial t})| \leq C_{25}\int_{\tilde{\Omega}_0}|\boldsymbol{f}|^2 + (1/4)\left\|\frac{\partial\boldsymbol{u}_m}{\partial t}\right\|_2^2,$$

(1.169)

where $C_{25} = C_{25}(\Omega) > 0$. From (1.166)–(1.169), and taking into account (1.152), (1.163) and (1.164), we conclude that there exists a constant $C_{26} = C_{26}(\Omega) > 0$ such that

$$\int_{-\pi}^{\pi}\left\|\frac{\partial\boldsymbol{u}_m}{\partial t}\right\|_2^2 dt \leq C_{26}D^2.$$

(1.170)

Once the uniform (in m and k) estimates (1.146), (1.155), (1.163), (1.165) and (1.170) have been obtained, it becomes routine to prove the existence of a 2π-periodic solution to (1.143). We will sketch the procedure here, referring the reader to [45] for details. Using the above estimates we can show that there exist a subsequence $\{\boldsymbol{u}_{m'}\}$ and a 2π-periodic in time vector field $\boldsymbol{u}^{(k)}(x, t)$ such that

$$\boldsymbol{u}_{m'} \to \boldsymbol{u}^{(k)} \quad \text{weakly in } L^2(-\pi, \pi; W^{2,2}(\Omega^{(k)})),$$

$$\frac{\partial\boldsymbol{u}_{m'}}{\partial t} \to \frac{\partial\boldsymbol{u}^{(k)}}{\partial t} \quad \text{weakly in } L^2(-\pi, \pi; L^2(\Omega^{(k)})),$$

$$\boldsymbol{u}_{m'}(t) \to \boldsymbol{u}^{(k)}(t) \quad \text{weakly in } W^{1,2}(\Omega^{(k)}), \text{ uniformly in } t \in (-\pi, \pi).$$

Furthermore, $\boldsymbol{u}^{(k)}$ satisfies the problem

$$
\left.\begin{aligned}
\frac{\partial \boldsymbol{u}^{(k)}}{\partial t} &+ \boldsymbol{u}^{(k)} \cdot \nabla \boldsymbol{u}^{(k)} + \boldsymbol{A} \cdot \nabla \boldsymbol{u}^{(k)} + \boldsymbol{u}^{(k)} \cdot \nabla \boldsymbol{A} \\
&= \Delta \boldsymbol{u}^{(k)} - \nabla p^{(k)} + \boldsymbol{f} \\
\nabla \cdot \boldsymbol{u}^{(k)} &= 0
\end{aligned} \right\} \quad \text{in } \Omega^{(k)} \times (-\pi, \pi), \tag{1.171}
$$

$$
\int_S \boldsymbol{u}^{(k)} \cdot \boldsymbol{n} \, dS = 0, \quad \boldsymbol{u}|_{\partial \Omega^{(k)}} = \boldsymbol{0},
$$

for a suitable $p \in L^2(-\pi, \pi; W^{1,2}(\Omega^{(k)}))$. In addition, the sequence of fields $\boldsymbol{u}^{(k)}$ continues to satisfy (1.146), (1.155), (1.163), (1.165) and (1.170) where, as we observed, the bounds do not depend on k. Thus, we can select a subsequence converging, in suitable topologies, to a vector field $\boldsymbol{u}$ belonging to the class $(1.140)_1$ and obeying (1.141). In addition, we can find a suitable scalar field p in the class $(1.140)_2$ which together with $\boldsymbol{u}$ satisfies (1.141) and $(1.143)_{1,2,3}$. As far as condition (1.142), we recall the inequality

$$
\max_{x \in \Sigma_r} |\boldsymbol{w}(x)| \leq C \|\boldsymbol{w}\|_{2,2,\Sigma_r}, \tag{1.172}
$$

where $\Sigma = \{x \in \Omega_1 : r < x_1 < r+1\}$ and where the constant C is independent of r; see, e.g., [36, Chapter II]. We now apply (1.172) to $\boldsymbol{u}(x,t)$ and notice that $\boldsymbol{u} = \boldsymbol{v} - \boldsymbol{v}_{P1}$ in $\widetilde{\Omega}_1$ and that $\boldsymbol{u} \in W^{2,2}(\Omega)$ for a.a. $t \in \mathbb{R}$. Thus, (1.142) in Ω_1 follows by letting $r \to \infty$ in (1.172) with $\boldsymbol{w} \equiv \boldsymbol{u}$. A similar proof holds for Ω_2. It remains to show uniqueness. Let $(\boldsymbol{u}_1, p_1)$ be the solution to (1.143) that we have just constructed, let $(\boldsymbol{u}_2, p_2)$ be another solution in the class (1.140), and set $\boldsymbol{u} := \boldsymbol{u}_2 - \boldsymbol{u}_1$, $p := p_2 - p_1$. From (1.143) we then find

$$
\left.\begin{aligned}
\frac{\partial \boldsymbol{u}}{\partial t} &+ \boldsymbol{u}_2 \cdot \nabla \boldsymbol{u} + \boldsymbol{u} \cdot \nabla \boldsymbol{u}_1 + \boldsymbol{A} \cdot \nabla \boldsymbol{u} + \boldsymbol{u} \cdot \nabla \boldsymbol{A} = \Delta \boldsymbol{u} - \nabla p \\
\nabla \cdot \boldsymbol{u} &= 0
\end{aligned} \right\} \quad \text{in } \Omega,
$$

$$
\int_S \boldsymbol{u} \cdot \boldsymbol{n} \, dS = 0, \quad \boldsymbol{u}|_{\partial \Omega} = \boldsymbol{0}, \tag{1.173}
$$

$$
\lim_{|x| \to \infty, \ x \in \Omega_i} \boldsymbol{u}(x) = \boldsymbol{0}, \quad i = 1, 2.
$$

If we dot-multiply both sides of $(1.173)_1$ by $\boldsymbol{u}$, integrate by parts over Ω, then integrate over the time-interval $(-\pi, \pi)$, and employ the 2π-periodicity of $\boldsymbol{u}$, we find

$$
\int_{-\pi}^{\pi} \|\nabla \boldsymbol{u}(t)\|_2^2 \, dt = - \int_{-\pi}^{\pi} [(\boldsymbol{u} \cdot \nabla \boldsymbol{u}_2, \boldsymbol{u}) + (\boldsymbol{u} \cdot \nabla \boldsymbol{A}, \boldsymbol{u})] \, dt. \tag{1.174}
$$

By Remark 1.8 and by (1.117), we have

$$
\int_{-\pi}^{\pi} |(\boldsymbol{u} \cdot \nabla \boldsymbol{A}, \boldsymbol{u})| \leq C \|\Phi\|_{W^{1,2}(-\pi,\pi)} \int_{-\pi}^{\pi} \|\nabla \boldsymbol{u}\|_2^2. \tag{1.175}
$$

Moreover, by the Hölder inequality,

$$\int_{-\pi}^{\pi} |(\boldsymbol{u} \cdot \nabla \boldsymbol{u}_2, \boldsymbol{u})| \, dt \leq \int_{-\pi}^{\pi} \|\boldsymbol{u}\|_4^2 \|\nabla \boldsymbol{u}_1\|_2 \, dt \, .$$

Now, by $(1.119)_4$ with $q = 4$ and by (1.79) we obtain

$$\|\boldsymbol{u}\|_2 \leq C_{27} \|\nabla \boldsymbol{u}\|_2 \, ,$$

with $C_{27} = C_{27}(\Omega) > 0$. So, using the embedding $W^{1,2}(\Omega) \subset L^4(\Omega)$, we find

$$\|\boldsymbol{u}\|_4 \leq C_{28} \|\nabla \boldsymbol{u}\|_2 \, , \tag{1.176}$$

with $C_{28} = C_{28}(\Omega) > 0$. Moreover, since $\boldsymbol{u}_1$ satisfies (1.141), we have

$$\operatorname*{ess\,sup}_{t \in (-\pi,\pi)} \|\nabla \boldsymbol{u}_1\|_2 \leq C_{29} \|\Phi\|_{W^{1,2}(-\pi,\pi)} \, , \tag{1.177}$$

where $C_{29} = C_{29}(\Omega) > 0$. From (1.174)–(1.177) it easily follows that, if a condition of the type (1.139) is satisfied for a suitable constant C, then $\boldsymbol{u}(x,t) = \boldsymbol{0}$, a.e. in $\Omega \times (-\pi, \pi)$. The proof of the theorem is thus completed. $\qquad \square$

Remark 1.9. As in the steady-state case, see Remark 1.7, the results of Theorem 1.5 can be directly extended to the situation where Ω has $N > 2$ outlets Ω_i, provided we require that the flow-rates Φ_i are time-periodic with the same period 2π and satisfy a condition of the type

$$\sum_{i=1}^{N} \|\Phi_i\|_{W^{1,2}(-\pi,\pi)} < C \, ,$$

for a suitable positive constant C.

1.3.3. Attainability of steady-state and time-periodic flow. In Section 1.1.3 we considered the problem of attainability of steady-state and time-periodic flow in the class of fully developed flow. The objective of this section is to study this problem in its complete generality. Thus, assume that the flow-rate is increased from zero to a given constant (respectively, time-periodic) function Φ. We wish to show that if the magnitude of Φ is suitably restricted, the corresponding unsteady flow will converge, as time goes to infinity, to the corresponding steady-state (respectively, time-periodic) flow.

Just for the sake of simplicity, we shall consider attainability in the time-periodic case, leaving the (simpler) steady-state case to the interested reader as an exercise. Furthermore, we will assume that the domain Ω satisfies the assumptions of Theorem 1.6.

Thus, let $\overline{\boldsymbol{v}} = \overline{\boldsymbol{v}}(x,t)$, $\overline{p} = \overline{p}(x,t)$ be the solution constructed in Theorem 1.6 corresponding to the flow-rate $\overline{\Phi}$ satisfying the restriction (1.139) and let $\overline{\boldsymbol{v}}_{Pi}$ be the corresponding Poiseuille flow in Ω_i, $i = 1, 2$. Moreover, let $\boldsymbol{v} = \boldsymbol{v}(x,t)$, $p = p(x,t)$ be a solution to (1.3) corresponding to the flow-rate $\psi\Phi$, where $\psi = \psi(t)$ is the "ramping" function introduced in Section 1.3.3. Thus, setting

$$\boldsymbol{w} := \boldsymbol{v} - \overline{\boldsymbol{v}}, \quad \phi := p - \overline{p},$$

we have

$$\left.\begin{array}{l} \dfrac{\partial \boldsymbol{w}}{\partial t} + \boldsymbol{w} \cdot \nabla \boldsymbol{w} + \overline{\boldsymbol{v}} \cdot \nabla \boldsymbol{w} + \boldsymbol{u} \cdot \nabla \overline{\boldsymbol{v}} = \Delta \boldsymbol{w} - \nabla \phi \\[2mm] \nabla \cdot \boldsymbol{w} = 0 \end{array}\right\} \quad \text{in } \Omega \times (0, \infty),$$

$$\int_S \boldsymbol{w}(x, t) \cdot \boldsymbol{n} \, dS = (\psi - 1)\overline{\Phi}(t), \quad t \in (0, \infty),$$

$$\boldsymbol{w}(x, t)|_{\partial\Omega} = \boldsymbol{0}, \quad t \in (0, \infty), \quad \boldsymbol{w}(x, 0) = -\overline{\boldsymbol{v}}(x, 0), \quad x \in \Omega.$$

$$(1.178)$$

The attainability problem consists then in finding a solution to (1.178) satisfying the condition

$$\lim_{t \to \infty} (\boldsymbol{w}(x, t) - \overline{\boldsymbol{v}}(x, t)) = \boldsymbol{0}, \quad x \in \Omega. \tag{1.179}$$

Similarly to the problems treated in the previous Sections 1.3.1 and 1.3.2, also for problem (1.178) an existence proof, in a suitable function class, can be safely accomplished by the construction of an appropriate flow-rate carrier, in conjuction with the Galerkin method and proper a-priori estimates. In the following, we shall limit ourselves to derive these estimates formally, leaving to the interested reader the task of combining them with the Galerkin method to produce the existence result. As a byproduct, our estimates will also show in which sense (1.179) is achieved.

We begin to introduce a flow-rate carrier, $\boldsymbol{B}$, defined as follows. Let

$$\widetilde{\boldsymbol{U}} = \zeta_1 \boldsymbol{U}_1 + \zeta_2 \boldsymbol{U}_2$$

where

$$\boldsymbol{U}_1(x, t) = (V_1(x, t)\boldsymbol{e}_1 - \overline{\boldsymbol{v}}_{P1}(x, t)), \quad x \in \Omega_1,$$

$$\boldsymbol{U}_2(x, t) = (V_2(x, t)\boldsymbol{e}_1 - \overline{\boldsymbol{v}}_{P2}(x, t)), \quad x \in \Omega_2,$$

and V_i, $i = 1, 2$, solves (1.30)–(1.31) in $S_i \times (0, \infty)$, $i = 1, 2$. We then set

$$\boldsymbol{B} = \widetilde{\boldsymbol{U}} + \widetilde{\boldsymbol{W}}$$

with $\widetilde{\boldsymbol{W}}$ given in (1.107) with $\boldsymbol{U} \equiv \widetilde{\boldsymbol{U}}$. Notice that, since $\widetilde{\boldsymbol{U}}(x, 0) = \boldsymbol{U}(x, 0)$, we may take $\widetilde{\boldsymbol{W}}(x, 0) = \boldsymbol{W}(x, 0)$ and so, recalling that $V_1(x, 0) = V_2(x, 0) = 0$, $x \in \Omega$, and the definition of $\boldsymbol{A}$ (see the beginning of Section I.3.2), we may deduce

$$\boldsymbol{A}(x, 0) + \boldsymbol{B}(x, 0) = \boldsymbol{0}, \quad x \in \Omega. \tag{1.180}$$

Using the properties of $\widetilde{\boldsymbol{W}}$ given in (1.107), with $q = 2$, and (1.116), and the results of Theorem 1.3, it becomes a straightforward procedure to establish that $\boldsymbol{B}$ satisfies, in particular, the following conditions.

(i) $\boldsymbol{B} \in L^2(0, \infty; W^{2,2}(\omega))$, $\dfrac{\partial \boldsymbol{B}}{\partial t} \in L^2(0, \infty; L^2(\omega))$, for all bounded domains $\omega \subset \Omega$;

(ii) $\nabla \cdot \boldsymbol{B}(x, t) = 0$ for a.a. $(x, t) \in \Omega \times (0, \infty)$, $\boldsymbol{B}(x, t)|_{\partial\Omega} = \boldsymbol{0}$, for a.a. $t \in (0, \infty)$;

(iii) $\displaystyle\int_S \boldsymbol{B} \cdot \boldsymbol{n} \, dS = (\psi - 1)\Phi(t)$;

(iv) for any bounded domain $\omega \subset \Omega$ there are $c_1 = c_1(\omega, \overline{\Phi}) > 0$ and $c_2 = c_2(\omega, T) > 0$ such that

$$\sup_{t \geq 0}(\|\boldsymbol{B}(t)\|_{1,2,\omega}^2 \, e^{\lambda_1 t}) + \int_0^\infty \left(\left\| \frac{\partial \boldsymbol{B}}{\partial t} \right\|_{2,\omega}^2 + \|\boldsymbol{B}(t)\|_{2,2,\omega}^2 \right) e^{\lambda_1 t} dt \leq C \,,$$

$$\max_{t \in [0,T]} \|\boldsymbol{B}(t)\|_{1,2,\omega}^2 \leq c_2 \, \|\overline{\Phi}\|_{W^{1,2}(-\pi,\pi)}^2 \,, \quad \text{for all } T > 0 \,;$$

(v) there are $C_i = C_i(S_i, \overline{\Phi}) > 0$ and $K_i = K_i(S_i, T) > 0$, $i = 1, 2$, such that

$$\sup_{t \geq 0}(\|\boldsymbol{B}(t)\|_{W^{1,2}(S_i)}^2 e^{\lambda_1 t}) + \int_0^\infty \left(\left\| \frac{\partial \boldsymbol{B}}{\partial t} \right\|_{L^2(S_i)}^2 + \|\boldsymbol{B}(t)\|_{W^{2,2}(S_i)}^2 \right) e^{\lambda_1 t} dt \leq C_i,$$

$$\max_{t \in [0,T]} \|\boldsymbol{B}(t)\|_{1,2,S_i}^2 \leq K_i \, \|\overline{\Phi}\|_{W^{1,2}(-\pi,\pi)}^2 \,, \quad \text{for all } T > 0 \,.$$

We next write $\boldsymbol{w} = \boldsymbol{u} + \boldsymbol{B}$ and obtain, from (1.178), (1.180) and from the properties of $\boldsymbol{B}$, that $\boldsymbol{u}$ satisfies the initial-boundary value problem

$$\left. \begin{aligned} \frac{\partial \boldsymbol{u}}{\partial t} &+ \boldsymbol{u} \cdot \nabla \boldsymbol{u} + \overline{\boldsymbol{v}} \cdot \nabla \boldsymbol{u} + \boldsymbol{u} \cdot \nabla \overline{\boldsymbol{v}} \\ &+ \boldsymbol{u} \cdot \nabla \boldsymbol{B} + \boldsymbol{B} \cdot \nabla \boldsymbol{u} = \Delta \boldsymbol{u} - \nabla \phi + \boldsymbol{F} \\ \nabla \cdot \boldsymbol{u} &= 0 \end{aligned} \right\} \quad \text{in } \Omega \times (0,\infty),$$

$$\int_S \boldsymbol{u}(x,t) \cdot \boldsymbol{n} \, dS = 0 \,, \quad t \in (0,\infty),$$

$$\boldsymbol{u}(x,t)|_{\partial\Omega} = \boldsymbol{0} \,, \quad t \in (0,\infty), \quad \boldsymbol{u}(x,0) = -\overline{\boldsymbol{v}}(x,0) + \boldsymbol{A}(x,0) := \boldsymbol{u}_0(x), \quad x \in \Omega,$$
$$\tag{1.181}$$

with

$$\boldsymbol{F} := -\frac{\partial \boldsymbol{B}}{\partial t} - \boldsymbol{B} \cdot \nabla \boldsymbol{B} - \overline{\boldsymbol{v}} \cdot \nabla \boldsymbol{B} - \boldsymbol{B} \cdot \nabla \overline{\boldsymbol{v}} + \Delta \boldsymbol{B} \,. \tag{1.182}$$

Notice that, because of (1.141), it follows that

$$\boldsymbol{u}_0 \in W^{1,2}(\Omega) \,. \tag{1.183}$$

We also notice that (1.3), (1.32) and the properties of $\boldsymbol{B}$ imply

$$\boldsymbol{F} = -G_i(t)\boldsymbol{e}_1 - (\overline{\boldsymbol{v}} - \overline{\boldsymbol{v}}_{Pi}) \cdot \nabla \boldsymbol{B} - \boldsymbol{B} \cdot \nabla(\overline{\boldsymbol{v}} - \overline{\boldsymbol{v}}_{Pi}), \quad x \in \widetilde{\Omega}_i, \quad i = 1, 2. \tag{1.184}$$

Dot-multiplying both sides of $(1.181)_1$ by $\boldsymbol{u}$, integrating by parts over Ω and taking into account $(1.181)_2$ and the solenoidality of $\boldsymbol{B}$, we get

$$\frac{1}{2}\frac{d}{dt}\|\boldsymbol{u}\|_2^2 + \|\nabla\boldsymbol{u}\|_2^2 = -(\boldsymbol{u} \cdot \nabla \boldsymbol{B}, \boldsymbol{u}) + (\boldsymbol{u} \cdot \nabla \boldsymbol{u}, \overline{\boldsymbol{u}}) - (\boldsymbol{u} \cdot \nabla \boldsymbol{A}, \boldsymbol{u}) + (\boldsymbol{F}, \boldsymbol{u}), \tag{1.185}$$

where $\overline{\boldsymbol{u}} := \overline{\boldsymbol{v}} - \boldsymbol{A}$. In view of Remark 1.8, we find

$$|(\boldsymbol{u} \cdot \nabla \boldsymbol{A}, \boldsymbol{u})| \leq C_1 \|\overline{\Phi}\|_{W^{1,2}(-\pi,\pi)} \|\nabla \boldsymbol{u}\|_2^2 \,, \tag{1.186}$$

with $C_1 = C_1(\Omega) > 0$, while, by the Hölder inequality, by the embedding $W^{1,2}(\Omega) \subset L^4(\Omega)$, by Lemma 1.5 and by (1.141) we get

$$|(\boldsymbol{u} \cdot \nabla \boldsymbol{u}, \overline{\boldsymbol{u}})| \leq C_2 \|\overline{\Phi}\|_{W^{1,2}(-\pi,\pi)} \|\nabla \boldsymbol{u}\|_2^2, \tag{1.187}$$

where $C_2 = C_2(\Omega) > 0$. Moreover, reasoning exactly as in the proof of (1.117), we show that

$$|(\boldsymbol{u} \cdot \nabla \boldsymbol{B}, \boldsymbol{u})| \leq C_3(\|\boldsymbol{B}\|_{1,2,\widetilde{\Omega}_0} + \sum_{i=1}^2 \|\boldsymbol{B}\|_{W^{1,2}(S_i)}) \|\nabla \boldsymbol{u}\|_2^2.$$

Thus, by properties (iv) and (v) of $\boldsymbol{B}$, we find

$$|(\boldsymbol{u} \cdot \nabla \boldsymbol{B}, \boldsymbol{u})| \leq \tfrac{1}{4} \|\nabla \boldsymbol{u}\|_2^2 \tag{1.188}$$

provided $\|\overline{\Phi}\|_{W^{1,2}(-\pi,\pi)} < C_4$ for some $C_4 = C_4(\Omega) > 0$. It remains to estimate the last term on the right-hand side of (1.185). Taking into account (1.184) and that $\boldsymbol{u}$ carries no flow-rate, with the help of Schwarz inequality and of Lemma 1.5 we find

$$(\boldsymbol{F}, \boldsymbol{u}) \leq C_5 \left(\int_{\widetilde{\Omega}_0} |\boldsymbol{F}|^2 + \sum_{i=1}^2 \left(\|\overline{\boldsymbol{v}} - \overline{\boldsymbol{v}}_{Pi}\|_4^2 \|\nabla \boldsymbol{B}\|_{4,S_i}^2 + \|\boldsymbol{B}\|_{\infty,S_i}^2 \|\overline{\boldsymbol{v}} - \overline{\boldsymbol{v}}_{Pi}\|_{1,2}^2 \right) \right)$$
$$+ \tfrac{1}{4} \|\nabla \boldsymbol{u}\|_2^2,$$

where $C_5 = C_5(\Omega) > 0$. Using in this latter relation $(1.119)_{1-3}$, the embedding $W^{1,2}(\Omega) \subset L^4(\Omega)$ and (1.141) we deduce

$$(\boldsymbol{F}, \boldsymbol{u}) \leq C_5 \int_{\widetilde{\Omega}_0} |\boldsymbol{F}|^2 + C_6 \sum_{i=1}^2 \|\boldsymbol{B}\|_{2,2,S_i}^2 + \tfrac{1}{4} \|\nabla \boldsymbol{u}\|_2^2, \tag{1.189}$$

with $C_6 = C_6(\Omega, \overline{\Phi}) > 0$. We next observe that from the Minkowski and Schwarz inequalities and from the embeddings $W^{2,2}(\widetilde{\Omega}_0) \subset W^{1,4}(\widetilde{\Omega}_0) \subset L^\infty(\widetilde{\Omega})$ we obtain

$$\|\overline{\boldsymbol{v}} \cdot \nabla \boldsymbol{B} + \boldsymbol{B} \cdot \nabla \overline{\boldsymbol{v}}\|_2 \leq C_7 \|\boldsymbol{B}\|_{2,2,\widetilde{\Omega}_0} \left(\|\overline{\boldsymbol{v}} - \boldsymbol{A}\|_{4,\widetilde{\Omega}_0} + \|\overline{\boldsymbol{v}} - \boldsymbol{A}\|_{1,2,,\widetilde{\Omega}_0} \right.$$
$$\left. + \|\boldsymbol{A}\|_{4,\widetilde{\Omega}_0} + \|\boldsymbol{A}\|_{1,2,,\widetilde{\Omega}_0} \right) \tag{1.190}$$

and

$$\|\boldsymbol{B} \cdot \nabla \boldsymbol{B}\|_{2,\widetilde{\Omega}_0} \leq C_7 \|\boldsymbol{B}\|_{1,2,\widetilde{\Omega}_0} \|\boldsymbol{B}\|_{2,2,\widetilde{\Omega}_0}, \tag{1.191}$$

with $C_7 = C_7(\Omega) > 0$. Thus, collecting (1.189)–(1.191) and taking into account (1.182), (1.141) and property (v) of the field $\boldsymbol{A}$ and property (iv) of the field $\boldsymbol{B}$, we conclude

$$(\boldsymbol{F}, \boldsymbol{u}) \leq \mathcal{F}(t) + \tfrac{1}{2} \|\nabla \boldsymbol{u}\|_2^2, \quad \mathcal{F} := C_8 \left(\|\frac{\partial \boldsymbol{B}}{\partial t}\|_{2,\widetilde{\Omega}_0}^2 + \|\boldsymbol{B}\|_{2,2,\widetilde{\Omega}_0}^2 + \sum_{i=1}^2 \|\boldsymbol{B}\|_{2,2,S_i}^2 \right),$$
$$\tag{1.192}$$

with $C_8 = C_8(\widetilde{\Omega}_0, \overline{\Phi}) > 0$. Consequently, from (1.184)–(1.188) and (1.192) we obtain that, if $\|\overline{\Phi}\|_{W^{1,2}(-\pi,\pi)}$ is below a suitable constant depending only on Ω, the following inequality holds:

$$\frac{d}{dt}\|\boldsymbol{u}\|_2^2 + \tfrac{1}{2}\|\nabla\boldsymbol{u}\|_2^2 \le 2\mathcal{F}.$$

If we use the first inequality of Lemma 1.6 in this relation, we obtain

$$\frac{d}{dt}\|\boldsymbol{u}\|_2^2 + C_9\|\boldsymbol{u}\|_2^2 \le 2\mathcal{F}.$$

with $C_9 = C_9(\Omega) > 0$. Integrating this relation and taking into account (1.183), we find

$$\|\boldsymbol{u}(t)\|_2^2 \le \|\boldsymbol{u}_0\|_2^2 e^{-C_9 t} + 2e^{-C_9 t}\int_0^t \mathcal{F}(s)e^{C_9 s}ds.$$

From this latter inequality, from (1.192) and from the properties (iv) and (v) of $\boldsymbol{B}$ we deduce

$$\|\boldsymbol{u}(t)\|_2^2 \le \|\boldsymbol{u}_0\|_2^2 e^{-C_9 t} + 2C_{10}e^{-C_9 t}\int_0^t e^{(C_9 - \lambda_1)s}ds, \qquad (1.193)$$

with $C_{10} = C_{10}(\Omega, \overline{\Phi}) > 0$, from which it follows, in particular, that $\|\boldsymbol{u}(t)\|_2$ is uniformly bounded in time and that, furthermore, it decays to zero exponentially fast. This latter shows the way in which condition (1.179) is satisfied.

Remark 1.10. (1) The boundedness and asymptotic properties just obtained can be shown also in other, stronger norms like, for example, the $W^{1,2}$-norm.
(2) As we observed earlier, the estimate (1.193), in conjunction with the Galerkin method, can be used to prove existence of (weak) solutions to problem (1.181). Strong solutions can be obtained by an "invading domain" technique completely analogous to that used in Section 1.3.2.
(3) By using more careful estimates, we can prove that, for $\|\overline{\Phi}\|_{W^{1,2}(-\pi,\pi)}$ sufficiently small, the quantity C_{10} in (1.193) is of the form $C_{11}\|\overline{\Phi}\|_{W^{1,2}(-\pi,\pi)}$, where C_{11} is a positive constant depending on Ω and on $|\psi'|$.

1.4. Mathematical modeling of a piping system. Bounded domain approach with "do-nothing" boundary conditions

In the numerical simulation of the flow of a liquid in a piping system it appears quite obvious that the system can not be modeled with the "unbounded outlets" method that we have introduced and discussed in the previous sections. In fact, the system has to be "truncated" at some point in the outlets, and appropriate "artificial" boundary conditions must be imposed at the open parts of the boundary, S_i, $i = 1, \ldots, N$; see Figure 3.

Even though a natural choice could be that of imposing the condition that the velocity of the liquid matches that of a (suitable) Poiseuille flow, we immediately realize that this choice need not be appropriate. The reason being that, as we have shown in Section 1.2, in a steady-state flow the entry length in each semi-infinite

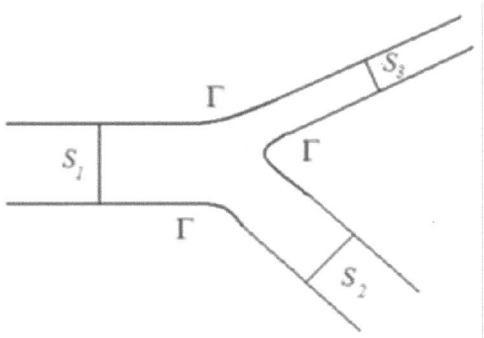

FIGURE 3. Truncated piping system.

outlet – that is, the distance from the inlet of each semi-infinite pipe at which the flow becomes fully developed – depends on the velocity distribution at the inlet. So, with this choice of boundary conditions, what could be a "reasonable" truncation at a certain distance in each semi-infinite outlet in some flow, in another flow it could show to be completely wrong.

In 1991, P.M. Gresho [48] proposed a type of boundary conditions that, as further shown by the systematic numerical investigation of J.G. Heywood, R. Rannacher and S. Turek [58], in several significant cases do not suffer from problems caused by where, in the outlets, the truncation occurs. These boundary conditions, often called "do-nothing" boundary conditions, are formulated as follows. Let Γ and S_i, $i = 1, \dots, N$, denote the "lateral" surface and the open parts, respectively, of the piping system; see Figure 3. The do-nothing boundary conditions are then expressed by the following (non-dimensional) requirements:

$$\boldsymbol{v}|_\Gamma = \mathbf{0}, \quad \left(\frac{\partial \boldsymbol{v}}{\partial \boldsymbol{n}} - p\boldsymbol{n}\right)\bigg|_{S_i} = P_i \boldsymbol{n}, \quad i = 1, \dots, N. \tag{1.194}$$

In (1.194), $\boldsymbol{v}$, p are velocity and pressure field of the fluid satisfying the Navier–Stokes equations (1.1), $\boldsymbol{n} = \boldsymbol{n}(S_i)$ is the outer unit normal to S_i, $i = 1, \dots, N$, while the quantities P_i are prescribed functions. Typically, for steady-state flow, P_i is a given constant, while, for time-dependent flow, P_i is a given function of time.

For the numerical investigation of system (1.194) with boundary conditions (1.194) and for more detailed information about the physical meaning of the quantities P_i, we refer the interested reader to the above cited papers of Gresho, of Heywood et al., and to the article of Rannacher in this volume [91]. In what follows, we shall be mostly focused on the study of several mathematical properties of the steady-state solutions to problem (1.1)–(1.194), see Section 1.4.1, while in Section 1.4.2 we shall investigate the (few) known properties of time-dependent solutions.

Besides the basic issues of existence and uniqueness for the boundary and the initial-boundary value problems, there is a number of significant questions that one can treat. They concern the relation between solutions to (1.1)–(1.194) and solutions to the problem in the whole unbounded piping system. In particular, we may address the following problems.

(a) Convergence of solutions to (1.199) on a sequence of "invading" domains, to solutions of (1.1) on the whole unbounded piping system.

(b) Conversely, given a solution corresponding to a given flow-rate in an unbounded piping system, construct a sequence of solutions on truncated domains approaching the given one.

(c) Control of the data of the sequence of solutions on the truncated domains by means of the flow-rate of the limiting solution and vice versa.

We will answer the above questions in the case of steady-state flow. The extension of our results to the time-dependent case appears to be a somewhat challenging task, mainly, due to the lack of a satisfactory existence theory; see Section 1.4.2.

As usual, we shall assume, for simplicity, that the piping system has only two outlets and that, on both open parts of the outlets, the do-nothing boundary conditions are prescribed. Since the pressure p may be defined up to a constant, we assume, without loss of generality, that $P_1 = 0$ and set $P_2 = P$. The case of more than two outlets, as well as the case of Dirichlet boundary conditions on some of the open parts and do-nothing boundary conditions on the remaining open parts present no relevant difficulty and are easily treated by exactly the same methods.

Throughout this section we shall use the following notation. As usual, by $\Omega = \Omega_1 \cup \Omega_0 \cup \Omega_2$ we indicate the unbounded piping system considered in the previous sections. By $\mathcal{V} \subset \Omega$ we denote the "truncated" region of flow, bounded by the lateral surface Γ and by the two surfaces S_i, $i = 1, 2$. These latter are taken to be flat and orthogonal to the axis of the i-th outlet, so that S_i coincides with the cross-section in each outlet. Even though many results that we will find do not require such a high degree of regularity, we shall assume that Γ is a surface of class C^2. Thus, $\mathcal{V}$ has the Lipschitz regularity.

We may split $\mathcal{V}$ as follows:

$$\mathcal{V} = \mathcal{V}_1 \cup \Omega_0 \cup \mathcal{V}_2, \tag{1.195}$$

where

$$\mathcal{V}_1 = \{x \in \Omega_1 : x_1 \in (-b_1, 0), \ x \in S_1\}, \quad \mathcal{V}_2 = \{x \in \Omega_2 : x_1 \in (0, b_2), \ x \in S_2\}, \tag{1.196}$$

with b_1 and b_2 positive real numbers which, without loss of generality, we may take such that $b_i \geq 2$, $i = 1, 2$. Sometimes, we shall also use the splitting

$$\mathcal{V} = \widetilde{\mathcal{V}}_1 \cup \widetilde{\Omega}_0 \cup \widetilde{\mathcal{V}}_2, \tag{1.197}$$

where

$$\widetilde{\mathcal{V}}_1 = \{x \in \mathcal{V}_1 : x_1 \in (-b_1, -1)\}, \quad \widetilde{\mathcal{V}}_2 = \{x \in \mathcal{V}_2 : x_1 \in (1, b_2)\}, \tag{1.198}$$

and $\widetilde{\Omega}_0$ is defined in (1.103).

1.4.1. Boundary-value problem. The objective of this section is to investigate the well-posedness of the boundary-value problem (in non-dimensional form)

$$
\left.\begin{aligned}
\Delta \boldsymbol{v} - \nabla p &= \boldsymbol{v} \cdot \nabla \boldsymbol{v} \\
\nabla \cdot \boldsymbol{v} &= 0
\end{aligned}\right\} \text{ in } \mathcal{V},
$$

$$
\boldsymbol{v}|_\Gamma = \left(\frac{\partial \boldsymbol{v}}{\partial \boldsymbol{n}} - p\boldsymbol{n}\right)\bigg|_{S_1} = \boldsymbol{0}, \quad \left(\frac{\partial \boldsymbol{v}}{\partial \boldsymbol{n}} - p\boldsymbol{n}\right)\bigg|_{S_2} = P\boldsymbol{n}.
$$

$$(1.199)$$

In these equations, $\boldsymbol{n} = \boldsymbol{n}(S_i)$ is the outer unit normal to S_i and P is a prescribed (non-dimensional) constant.

In order to reach our goal, we need some preliminary considerations. Let

$$
H(\mathcal{V}) := \{\boldsymbol{\varphi} \in W^{1,2}(\mathcal{V}) : \ \nabla \cdot \boldsymbol{\varphi} = 0, \ \boldsymbol{\varphi}|_\Gamma = \boldsymbol{0}\}. \tag{1.200}
$$

Using the properties of $W^{1,2}(\mathcal{V})$ along with inequality (1.104), it is easy to prove that H is a Hilbert space with associated scalar product

$$
(\nabla \boldsymbol{\varphi}_1, \nabla \boldsymbol{\varphi}_2) := \int_{\mathcal{V}} \nabla \boldsymbol{\varphi}_1 : \nabla \boldsymbol{\varphi}_2. \tag{1.201}
$$

The following lemma holds.

Lemma 1.7. *Set*

$$
(\cdot, \cdot) := (\cdot, \cdot)_{\mathcal{V}}.
$$

There exists a constant $C = C(S_1, S_2, \Omega_0) > 0$, independent of b_i, $i = 1, 2$, such that

$$
|(\boldsymbol{\varphi} \cdot \nabla \boldsymbol{\psi}, \boldsymbol{\chi})| \leq C \|\nabla \phi\|_2 \|\nabla \boldsymbol{\psi}\|_2 \|\nabla \boldsymbol{\chi}\|_2,
$$

for all $\boldsymbol{\varphi}, \boldsymbol{\psi}, \boldsymbol{\chi} \in H(\mathcal{V})$.

Proof. We recall the inequality [37, Lemma XI.2.1]

$$
\int_t^{t+1} \left(\int_{S_2} |\boldsymbol{\varphi}|^4 \, dS\right) dx \leq \kappa \left(\int_t^{t+1} \left(\int_{S_2} |\nabla \boldsymbol{\varphi}|^2 \, dS\right) dx\right)^2, \quad \boldsymbol{\varphi} \in H(\mathcal{V}), \tag{1.202}
$$

holding for arbitrary $t \in [0, b_2 - 1]$, with a positive constant κ depending only on S_2. Next, take $m \in \mathbb{N}$ such that $m \leq b_2 < m + 1$. If we set $\delta := b_2 - m \ (< 1)$, we then find

$$
|(\boldsymbol{\varphi} \cdot \nabla \boldsymbol{\psi}, \boldsymbol{\chi})_{\mathcal{V}_2}| \leq \int_0^\delta \int_{S_2} |\boldsymbol{\varphi} \cdot \nabla \boldsymbol{\psi} \cdot \boldsymbol{\chi}| \, dS dx_1 + \sum_{l=0}^{m-1} \int_{l+\delta}^{l+\delta+1} \int_{S_2} |\boldsymbol{\varphi} \cdot \nabla \boldsymbol{\psi} \cdot \boldsymbol{\chi}| \, dS \, dx_1
$$

$$
\leq \int_0^1 \int_{S_2} |\boldsymbol{\varphi} \cdot \nabla \boldsymbol{\psi} \cdot \boldsymbol{\chi}| \, dS dx_1 + \sum_{l=0}^{m-1} \int_{l+\delta}^{l+\delta+1} \int_{S_2} |\boldsymbol{\varphi} \cdot \nabla \boldsymbol{\psi} \cdot \boldsymbol{\chi}| \, dS \, dx_1.
$$

By the Hölder inequality we obtain

$$|(\boldsymbol{\varphi} \cdot \nabla \boldsymbol{\psi}, \boldsymbol{\chi})_{\mathcal{V}_2}|$$

$$\leq \left(\int_0^1 \int_{S_2} |\boldsymbol{\varphi}|^4 dS \, dx_1 \right)^{\frac{1}{4}} \left(\int_0^1 \int_{S_2} |\boldsymbol{\chi}|^4 dS \, dx_1 \right)^{\frac{1}{4}} \left(\int_0^1 \int_{S_2} |\nabla \boldsymbol{\psi}|^2 dS \, dx_1 \right)^{\frac{1}{2}}$$

$$+ \sum_{l=0}^{m-1} \left(\int_{l+\delta}^{l+\delta+1} \int_{S_2} |\boldsymbol{\varphi}|^4 dS \, dx_1 \right)^{\frac{1}{4}} \left(\int_{l+\delta}^{l+\delta+1} \int_{S_2} |\boldsymbol{\chi}|^4 dS \, dx_1 \right)^{\frac{1}{4}} \left(\int_{l+\delta}^{l+\delta+1} \int_{S_2} |\nabla \boldsymbol{\psi}|^2 dS \, dx_1 \right)^{\frac{1}{2}},$$

and so, with the help of (1.202), we find

$$|(\boldsymbol{\varphi} \cdot \nabla \boldsymbol{\psi}, \boldsymbol{\psi})_{\mathcal{V}_2}| \leq \kappa^2 \|\nabla \boldsymbol{\varphi}\|_2 \|\nabla \boldsymbol{\psi}\|_2 \|\nabla \boldsymbol{\chi}\|_2$$

$$+ \kappa^2 \|\nabla \boldsymbol{\varphi}\|_2 \sum_{l=0}^{m-1} \left(\int_{l+\delta}^{l+\delta+1} \int_{S_2} |\nabla \boldsymbol{\psi}|^2 dS \, dx_1 \right)^{\frac{1}{2}} \left(\int_{l+\delta}^{l+\delta+1} \int_{S_2} |\nabla \boldsymbol{\chi}|^2 dS \, dx_1 \right)^{\frac{1}{2}}$$

$$\leq 2\kappa^2 \|\nabla \boldsymbol{\varphi}\|_2 \|\nabla \boldsymbol{\psi}\|_2 \|\nabla \boldsymbol{\chi}\|_2 \,,$$

$$(1.203)$$

where, in the last step, we have used the Cauchy inequality. In a completely similar fashion, we show

$$|(\boldsymbol{\varphi} \cdot \nabla \boldsymbol{\psi}, \boldsymbol{\chi})_{\mathcal{V}_1}| \leq 2\kappa^2 \|\nabla \boldsymbol{\varphi}\|_2 \|\nabla \boldsymbol{\psi}\|_2 \|\nabla \boldsymbol{\chi}\|_2 \,. \qquad (1.204)$$

Furthermore, by the Hölder inequality and by $(1.119)_4$, we find

$$|(\boldsymbol{\varphi} \cdot \nabla \boldsymbol{\psi}, \boldsymbol{\psi})_{\Omega_0}| \leq \|\boldsymbol{\varphi}\|_4 \|\boldsymbol{\psi}\|_4 \|\nabla \boldsymbol{\chi}\|_2 \leq C_1 \|\nabla \boldsymbol{\varphi}\|_2 \|\nabla \boldsymbol{\psi}\|_2 \|\nabla \boldsymbol{\chi}\|_2 \,,$$

with $C = C(\Omega_0) > 0$. The lemma then follows from this latter displayed relation and from (1.203)–(1.204). □

We shall next provide a "weak" formulation of problem (1.199) as follows; see [58]. We dot-multiply both sides of $(1.199)_1$ by $\boldsymbol{\varphi} \in H(\mathcal{V})$, integrate by parts over $\mathcal{V}$ and use conditions $(1.199)_{2,3,4}$ to obtain (see (1.201) for notation)

$$(\nabla \boldsymbol{v}, \nabla \boldsymbol{\varphi}) = -(\boldsymbol{v} \cdot \nabla \boldsymbol{v}, \boldsymbol{\varphi}) + P \int_{S_2} \boldsymbol{\varphi} \cdot \boldsymbol{n} \, dS \,, \quad \text{for all } \boldsymbol{\varphi} \in H(\mathcal{V}). \qquad (1.205)$$

A *weak solution* to (1.199) is a field $\boldsymbol{v} \in H(\mathcal{V})$ satisfying (1.205).

Before we proceed to proving existence and uniqueness of weak solutions, we would like to make the following comments.

(a) The above definition of a weak solution is meaningful because, due to Lemma 1.7 and to the Schwarz inequality and (1.79), each term in (1.205) is finite.

(b) Consider the functional

$$\mathcal{F}(\boldsymbol{w}) := (\nabla \boldsymbol{v}, \nabla \boldsymbol{w}) + (\boldsymbol{v} \cdot \nabla \boldsymbol{v}, \boldsymbol{w}) \,, \quad \boldsymbol{w} \in W_0^{1,2}(\mathcal{V}) \,,$$

where $\boldsymbol{v}$ is a weak solution to (1.199). In view of Lemma 1.7, $\mathcal{F}$ is linear and bounded on $W_0^{1,2}(\mathcal{V})$. Moreover, if we take in (1.205) $\boldsymbol{\varphi} \in \mathcal{D}_0^{1,2}(\mathcal{V})$, we find

$$\mathcal{F}(\boldsymbol{\varphi}) = 0 \,, \quad \text{for all } \boldsymbol{\varphi} \in \mathcal{D}_0^{1,2}(\mathcal{V}).$$

Then, by [36, Corollary III.5.1] there exists a "pressure field" $p \in L^2(\mathcal{V})$ such that

$$(\nabla v, \nabla w) + (v \cdot \nabla v, w) = (p, \nabla \cdot w), \quad \text{for all } w \in W_0^{1,2}(\mathcal{V}). \tag{1.206}$$

(c) From (1.206) and from classical interior regularity results for the Navier–Stokes equations [37, Lemma VIII.1.2 and Theorem VIII.5.1], we obtain that every weak solution, v, and the corresponding pressure field, p, defined above in (b) are, in fact, in $C^\infty(\mathcal{V})$ and the pair (v, p) satisfies $(1.199)_1$ in the ordinary sense.

(d) So far as the regularity up to the boundary of weak solutions, we observe that, by well-known results [37, Theorem VIII.5.2], they are (at least) continuous up to boundary Γ. Concerning the regularity on the boundaries S_1 and S_2 we have the following. Setting $\mathbb{R}_+^3 = \{x \in \mathbb{R}^3 : \ x_1 > 0\}$, it can be shown that the linear system

$$\left. \begin{array}{l} \Delta u - \nabla \tau = G \\ \\ \nabla \cdot u = 0 \end{array} \right\} \quad \text{in } \mathbb{R}_+^3, \qquad \left(\frac{\partial u}{\partial n} - \tau n \right) \Bigg|_{x_1=0} = g,$$

with G and g prescribed, is of the Agmon–Douglas–Nirenberg-type. As a consequence, by using a standard procedure, and taking into account that S_i is flat, one can show that any weak solution v and corresponding pressure p is of class C^∞ in $B \cap \Omega := \beta$, where B is an arbitrary ball centered at a point of S_i, such that $\text{dist}(\beta, \Gamma) > 0$. It is readily seen, then, that a weak solution v and the associated pressure p assume the boundary condition $(1.199)_4$ at each point of S_i in the ordinary sense. In fact, if we integrate by parts (1.205) and use the fact that, by (c), (v, p) satisfy (1.199), we obtain

$$\sum_{i=1}^{2} \int_{S_i} \left(\frac{\partial v}{\partial n} - pn - P_i n \right) \cdot \varphi \, dS = 0, \quad \text{for all } \varphi \in H(\mathcal{V}), \tag{1.207}$$

where $P_1 := 0$ and $P_2 := P$. Let

$$\phi = \phi e_1 + \zeta_2 e_2 + \zeta_3 e_3 := \phi e_1 + \zeta,$$

where

$$\phi, \, \zeta_2, \, \zeta_3 \in C_0^\infty(S_2), \quad \int_{S_2} \phi \, dS = 0.$$

By [36, Exercise III.3.4], we may extend ϕ in $\mathcal{V}$ to a function, $\varphi \in H(\mathcal{V})$, such that, for some small $\varepsilon > 0$, $\varphi(x) = 0$ whenever $\text{dist}(x, \Gamma) < \varepsilon$ or $\text{dist}(x, S_1) < \varepsilon$. Replacing such a φ in (1.207) we obtain

$$\int_{S_2} \left(\frac{\partial v}{\partial n} - pn - Pn \right) \cdot \phi \, dS = 0,$$

that is, by recalling that $n(S_2) = e_1$ and by the arbitrariness of ϕ and ζ,

$$\int_{S_2} \left(\frac{\partial v_1}{\partial n} - p - P \right) \phi \, dS = 0, \quad \text{for all } \phi \in C_0^\infty(S_2) \text{ with } \int_{S_2} \phi \, dS = 0,$$

$$\int_{S_2} \frac{\partial v}{\partial n} \cdot \zeta \, dS = 0, \quad \text{with } \zeta = \zeta_2 e_2 + \zeta_3 e_3, \text{ any } \zeta_i \in C_0^\infty(S_2), \, i = 2, 3.$$

These two conditions then imply that there exists a constant, c_2, such that

$$\frac{\partial v}{\partial n} - pn - Pn \bigg|_{S_2} = c_2 n. \tag{1.208}$$

Likewise, we show

$$\frac{\partial v}{\partial n} - pn \bigg|_{S_1} = c_1 n. \tag{1.209}$$

We next replace (1.208) and (1.209) into (1.207) to get

$$c_1 \int_{S_1} \varphi \cdot n \, dS + c_2 \int_{S_2} \varphi \cdot n \, dS = 0, \quad \text{for all } \varphi \in H(\mathcal{V}). \tag{1.210}$$

So, if we choose φ such that

$$\int_{S_2} \varphi \cdot n = 1,$$

by the Gauss theorem and by the fact that $\nabla \cdot \varphi = 0$ in $\mathcal{V}$ we find

$$\int_{S_1} \varphi \cdot n = -1,$$

and from (1.210) we infer $c_1 = c_2 := -c$. Thus, if we modify p by the addition of the constant c, from (1.208) and (1.209) we obtain that v and p satisfy the desired boundary conditions on S_i, $i = 1, 2$.

In order to show existence and uniqueness for (1.199), in analogy with what we developed for the unbounded case, it is convenient to use a suitable flow-rate carrier, $\alpha := \Phi a$, with a of the type introduced in Section I.3.1 and Φ to be determined appropriately; see (1.212). Thus, if we write $v = u + \alpha$ and replace it into (1.205), we obtain

$$(\nabla u, \nabla \varphi) = -(u \cdot \nabla u, \varphi) - (\alpha \cdot \nabla u, \varphi) - (u \cdot \nabla \alpha, \varphi) - (\alpha \cdot \nabla \alpha, \varphi)$$
$$-(\nabla \alpha, \nabla \varphi) + P \int_{S_2} \varphi \cdot n \, dS, \tag{1.211}$$

for all $\varphi \in H(\mathcal{V})$. In view of the property (iv) of a, we have

$$(\alpha \cdot \nabla \alpha, \varphi) = (\alpha \cdot \nabla \alpha, \varphi)_{\widetilde{\Omega}_0}.$$

Moreover, again by the same property, we find

$$(\nabla \alpha, \nabla \varphi) = (\nabla \alpha, \nabla \varphi)_{\widetilde{\Omega}_0} + \sum_{i=1}^{2} (\nabla v_{Pi}, \nabla \varphi)_{\widetilde{\mathcal{V}}_i},$$

where each v_{Pi} carries the flow-rate Φ. Integrating by parts the last two terms on the right-hand side of this relation and taking into account (1.3) and (1.8)$_2$, we find

$$(\nabla v_{P1}, \nabla \varphi)_{\tilde{\mathcal{V}}_1} = (\Phi M_1(b_1 - 1) + c) \int_{S_1} \varphi \cdot n \, dS,$$

$$(\nabla v_{P2}, \nabla \varphi)_{\tilde{\mathcal{V}}_2} = \Phi M_2(b_2 - 1) \int_{S_2} \varphi \cdot n \, dS,$$

where c is an arbitrary constant and $M_i = M_i(S_i) > 0$, $i = 1, 2$. Thus, if we choose $c = -\Phi M_1(b_1 - 1)$ and

$$\Phi = \frac{P}{M_2(b_2 - 1)}, \tag{1.212}$$

from (1.211) we conclude, for all $\varphi \in H(\mathcal{V})$,

$$(\nabla u, \nabla \varphi) = -(u \cdot \nabla u, \varphi) - (\alpha \cdot \nabla u, \varphi) - (u \cdot \nabla \alpha, \varphi) - \mathcal{G}, \tag{1.213}$$

where

$$\mathcal{G} := -(\alpha \cdot \nabla \alpha, \varphi)_{\tilde{\Omega}_0} - (\nabla \alpha, \nabla \varphi)_{\tilde{\Omega}_0}. \tag{1.214}$$

Remark 1.11. The constant M_2 in (1.212) can be explicitly evaluated. In fact, we have

$$M_2 = M_2(S_2) = \left(\int_{S_2} \varphi \, dS \right)^{-1},$$

where φ is the solution to (1.7). For example, if S_2 is a circle of radius r, we find $M_2 = \pi R^4 / 8$.

The following result holds.

Theorem 1.7. *There are positive constants K_i, $i = 1, 2$, depending only on Ω_0, S_1 and S_2, but otherwise independent of b_1 and b_2, such that, if*

$$|P| < b_2 K_1, \tag{1.215}$$

there exists one and only one weak solution, v, to (1.199) that satisfies the inequality

$$\|\nabla(v - \alpha)\|_2 \le \frac{K_2}{b_2} |P|, \tag{1.216}$$

where $\alpha = \Phi a$, with a flow-rate carrier introduced in Section 1.3.1 and Φ given in (1.212).

Proof. Let us first consider the linearized problem

$$(\nabla u, \nabla \varphi) = \mathcal{F}(\varphi), \quad \text{for all } \varphi \in H(\mathcal{V}), \tag{1.217}$$

where $\mathcal{F}$ is a given, bounded linear functional on $H(\mathcal{V})$ and set

$$\|\mathcal{F}\|_{-1,2} := \sup_{\varphi \in H(\mathcal{V}), \varphi \neq 0} \frac{|\mathcal{F}(\varphi)|}{\|\nabla \varphi\|_2}.$$

Since $H(\mathcal{V})$ is a Hilbert space with scalar product (1.201), we deduce, by the Riesz theorem, the existence of one and only one $\boldsymbol{u} \in H(\mathcal{V})$ satisfying (1.217) along with the estimate

$$\|\nabla \boldsymbol{u}\|_2 = \|\mathcal{F}\|_{-1,2}. \tag{1.218}$$

Let us now consider the map

$$T : \boldsymbol{w} \in \mathcal{B}_\rho \mapsto \boldsymbol{u} \in H(\mathcal{V}),$$

where $\mathcal{B}_\rho$ is the ball of radius ρ in $H(\mathcal{V})$ centered at 0, while $\boldsymbol{u}$ satisfies the problem

$$(\nabla \boldsymbol{u}, \nabla \boldsymbol{\varphi}) = -(\boldsymbol{w} \cdot \nabla \boldsymbol{w}, \boldsymbol{\varphi}) - (\boldsymbol{\alpha} \cdot \nabla \boldsymbol{w}, \boldsymbol{\varphi}) - (\boldsymbol{w} \cdot \nabla \boldsymbol{\alpha}, \boldsymbol{\varphi}) - \mathcal{G}$$
$$\text{for all } \boldsymbol{\varphi} \in H(\mathcal{V}), \tag{1.219}$$

with $\mathcal{G} = \mathcal{G}(\boldsymbol{\varphi})$ given in (1.214). We shall show that, provided $|P|/b_2$ is below a suitable constant and ρ is chosen appropriately, the map T is a contraction of $\mathcal{B}_\rho$ into itself, thus recovering the existence part of the theorem. To prove the contracting property of T, we begin to observe that the functional

$$\mathcal{F}_1 : \boldsymbol{\varphi} \in H(\mathcal{V}) \mapsto (\boldsymbol{u}_1 \cdot \nabla \boldsymbol{u}_2, \boldsymbol{\varphi}) \in \mathbb{R}, \quad \boldsymbol{u}_1, \boldsymbol{u}_2 \in H(\mathcal{V}),$$

is linear and, by Lemma 1.7, it is bounded with

$$\|\mathcal{F}_1\|_{-1,2} \le C_1 \|\nabla \boldsymbol{u}_1\|_2 \|\nabla \boldsymbol{u}_2\|_2, \tag{1.220}$$

where $C_1 = C_1(\Omega_0, S_1, S_2) > 0$. We next observe that, by (1.79) and (1.119)$_4$ it easily follows that

$$\|\boldsymbol{\varphi}\|_2 \le C_2 \|\nabla \boldsymbol{\varphi}\|_2, \quad \boldsymbol{\varphi} \in H(\mathcal{V}) \tag{1.221}$$

with $C_2 = C_2(\Omega_0, S_1, S_2) > 0$. Consequently, since by the property (v) of $\boldsymbol{a}$ it is

$$\max_{x \in \mathcal{V}} (|\boldsymbol{\alpha}(x) + |\nabla \boldsymbol{\alpha}(x)|) \le C |\Phi|$$

with $C = C(\Omega_0, S_1, S_2) > 0$, by the Schwarz inequality and by (1.221) we find that the functional

$$\mathcal{F}_2 : \boldsymbol{\varphi} \in H(\mathcal{V}) \mapsto -(\boldsymbol{\alpha} \cdot \nabla \boldsymbol{w}, \boldsymbol{\varphi}) - (\boldsymbol{w} \cdot \nabla \boldsymbol{\alpha}, \boldsymbol{\varphi})$$

is also (linear) and bounded with

$$\|\mathcal{F}_2\|_{-1,2} \le C_3 |\Phi| \|\nabla \boldsymbol{w}\|_2 \tag{1.222}$$

where $C_3 = C_3(\widetilde{\Omega}_0, S_1, S_2) > 0$. Finally, it is obvious that $\mathcal{G}$ given in (1.214) is a linear functional on $H(\mathcal{V})$ and that, again from the Schwarz inequality and from (1.119)$_4$,

$$\|\mathcal{G}\|_{-1,2} \le C_4 \left(|\Phi| + |\Phi|^2\right), \tag{1.223}$$

where $C_4 = C_4(\Omega_0, S_1, S_2) > 0$. From the results shown for problem (1.217), and with the help of (1.218), (1.220), (1.222) and (1.223), we may then state that, for any $\boldsymbol{w} \in \mathcal{B}_\rho$, (1.219) has one and only one solution $\boldsymbol{u} \in H(\mathcal{V})$, which, in addition, satisfies the estimate

$$\|\nabla \boldsymbol{u}\|_2 \le C_1 \rho^2 + C_3 |\Phi| \rho + C_4(|\Phi| + |\Phi|^2). \tag{1.224}$$

Thus, if we take

$$|\Phi| \leq C_5 := \frac{1}{4C_1(1+C_4)^2 + 2C_3(1+C_4) + C_4}, \quad \rho = 2(1+C_4)|\Phi|, \quad (1.225)$$

from (1.224) it follows that

$$\|\nabla u\|_2 \leq \tfrac{1}{2}\rho, \quad (1.226)$$

that is, T transforms $\mathcal{B}_\rho$ in itself. Moreover, setting $u_i := T(w_i)$, $i = 1, 2$, $u := u_1 - u_2$, $w := w_1 - w_2$, from (1.219) we deduce, for all $\varphi \in H(\mathcal{V})$,

$$(\nabla v, \nabla \varphi) = -(w_1 \cdot \nabla w, \varphi) - (w \cdot \nabla w_2, \varphi) - (\alpha \cdot \nabla w, \varphi) - (w \cdot \nabla \alpha, \varphi). \quad (1.227)$$

Consequently, applying to (1.227) the results obtained for the linear problem (1.217) along with (1.220) and (1.222), we find

$$\|\nabla u\|_2 \leq \|\nabla w\|_2 \left[C_1 \left(\|\nabla w_1\|_2 + \|\nabla w_2\|_2 \right) + C_3 |\Phi| \right],$$

which, in turn, in view of (1.225), yields

$$\|\nabla u\|_2 \leq (C_1(2(1+C_4) + C_3) |\Phi| \|\nabla w\|_2 < \frac{1}{2}\|\nabla w\|_2. \quad (1.228)$$

The map T is then a contraction and the fixed point u satisfies (1.213)–(1.214). Furthermore, once we take into account that the flow-rate Φ is related to P by (1.212), from (1.225) and (1.226), we find that $v := u + \alpha$ satisfies also (1.216). Finally, the uniqueness part follows by setting $w \equiv u$ in (1.228). $\qquad\square$

Remark 1.12. It is worth emphasizing that the flow-rate, Φ, carried by the vector field α is, in general, distinct from the flow-rate, $\widetilde{\Phi}$, carried by the velocity field v. Now, in practical problems of blood flow modeling, the physical parameter that one prescribes is the flow-rate $\widetilde{\Phi}$. However, if one uses the "do nothing" boundary conditions one has to prescribe P. Thus, as pointed out to me by Anne Robertson, it would be extremely interesting to find some sort of relation between $\widetilde{\Phi}$ and P. This problem, even for small data, seems to be quite challenging because it definitely involves the "geometry" of the pipe. Moreover, it is not excluded that velocity fields corresponding to the different values of P may carry the same flow-rate.

Remark 1.13. Among other things, the previous Theorem 1.7 ensures that any weak solution to (1.199) is unique in the ball, $\mathcal{B}_R$, of $H(\mathcal{V})$ centered at the origin and of radius $R = K|P|$, for a suitable $K > 0$; see (1.216), (1.225) and (1.214). In other words, two weak solutions corresponding to the same data and lying in $\mathcal{B}_R$ must necessarily coincide. It then arises spontaneously the question of whether a given solution satisfying (1.216) is unique in the class of *all possible* weak solutions corresponding to the same data. This question, that admits a positive (and simple) answer in the case of the "classical" Dirichlet boundary conditions, is, to date, open, in the case of the do-nothing boundary conditions $(1.199)_{3,4,5}$. To see where and how the problem arises, let v_1 and v_2 be two weak solutions corresponding to the same data P. Setting $v = v_1 - v_2$, from (1.205) we find

$$(\nabla v, \nabla \varphi) = -(v \cdot \nabla v, \varphi) - (v_1 \cdot \nabla v, \varphi) - (v \cdot \nabla v_1, \varphi), \quad \text{for all } \varphi \in H(\mathcal{V}).$$

Setting $\boldsymbol{\varphi} = \boldsymbol{v}$ into this equation we thus get

$$\|\nabla\boldsymbol{v}\|_2^2 = (\boldsymbol{v}\cdot\nabla\boldsymbol{v},\boldsymbol{v}) + (\boldsymbol{v}_1\cdot\nabla\boldsymbol{v},\boldsymbol{v}) + (\boldsymbol{v}\cdot\nabla\boldsymbol{v}_1,\boldsymbol{v})\,. \qquad (1.229)$$

Applying Lemma 1.7 to the last two terms on the right-hand side of (1.229) and using (1.216), we get

$$(1 - C|P|)\|\nabla\boldsymbol{v}\|_2^2 \le (\boldsymbol{v}\cdot\nabla\boldsymbol{v},\boldsymbol{v})\,,$$

where $C = C(\mathcal{V}) > 0$, and so, if $|P| < 1/(2C)$, say, we obtain

$$\|\nabla\boldsymbol{v}\|_2^2 \le 2(\boldsymbol{v}\cdot\nabla\boldsymbol{v},\boldsymbol{v})\,. \qquad (1.230)$$

Now, if we were using Dirichlet boundary conditions on $\partial\mathcal{V}$, namely, $\boldsymbol{v} = \boldsymbol{0}$ on the *whole* $\Gamma \cup S_1 \cup S_2$, the term on the right-hand side of (1.230) would vanish and uniqueness would follow under the assumption that *only* the solution $\boldsymbol{v}_1$ is "small". However, with the do-nothing boundary conditions, at the surfaces S_1 and S_2 the normal component of $\boldsymbol{v}$ need not vanish, so that, the right-hand side of (1.230) is, in principle, not zero and it is given by

$$\tfrac{1}{2}\sum_{i=1}^{2}\int_{S_i}|\boldsymbol{v}|^2\boldsymbol{v}\cdot\boldsymbol{n}\,dS.$$

One very unpleasant consequence of this fact is that *we do not know, in general, if the only solution corresponding to zero data is the solution $\boldsymbol{v} = \boldsymbol{0}$*. We only know that it is unique in the ball $\mathcal{B}_\rho$.

1.4.2. Relation between steady-state flow in bounded and unbounded piping systems. Our next objective is to find the relation between solutions on the truncated domain determined in Theorem 1.7 and solutions in the whole unbounded piping system found in Theorem 1.5. To this end, we introduce the notation

$$\mathcal{V}^{(k)} = \mathcal{V}_1^{(k)} \cup \widetilde{\Omega}_0 \cup \mathcal{V}_1^{(k)}\,,$$

$$\mathcal{V}_1^{(k)} := \{x \in \Omega_1:\ -k-1 < x_1 < -1\}\,, \quad \mathcal{V}_2^{(k)} := \{x \in \Omega_2:\ 1 < x_1 < k+1\}\,,$$

$$S_1^{(k)} := \{x \in \Omega_1:\ x_1 = -k-1\}\,, \quad S_2^{(k)} := \{x \in \Omega_2:\ x_1 = k+1\}\,.$$

The following result holds.

Theorem 1.8. *Consider the sequence of problems*

$$\left.\begin{aligned}
\Delta\boldsymbol{v}^{(k)} - \nabla p^{(k)} &= \boldsymbol{v}^{(k)}\cdot\nabla\boldsymbol{v}^{(k)} \\
\nabla\cdot\boldsymbol{v}^{(k)} &= 0
\end{aligned}\right\} \ in\ \mathcal{V}^{(k)},$$

$$\boldsymbol{v}^{(k)}|_{\Gamma} = \left(\frac{\partial\boldsymbol{v}^{(k)}}{\partial\boldsymbol{n}} - p^{(k)}\boldsymbol{n}\right)\Big|_{S_1^{(k)}} = \boldsymbol{0},\quad \left(\frac{\partial\boldsymbol{v}^{(k)}}{\partial\boldsymbol{n}} - p^{(k)}\boldsymbol{n}\right)\Big|_{S_2^{(k)}} = P^{(k)}\boldsymbol{n}\,,$$

$$(1.231)$$

where $\{P^{(k)}\}$ is a sequence of given real numbers such that

$$\lim_{k\to\infty}\left(\frac{P^{(k)}}{k}\right) = P\,.$$

Moreover, set

$$\Phi^{(k)} := \frac{P^{(k)}}{kM}, \quad \Phi := \frac{P}{M}; \quad \boldsymbol{\alpha}^{(k)} := \Phi^{(k)}\boldsymbol{a}, \quad \boldsymbol{\alpha} := \Phi\boldsymbol{a},$$

where $M = M(S_2) > 0$ and $\boldsymbol{a}$ is the flow-rate carrier defined in Section I.3.1. Then, the following statements hold.

(a) *There exists a constant $K = K(S_1, S_2, \Omega_0) > 0$ such that if*

$$\left|\frac{P^{(k)}}{k}\right| < K,$$

it follows that

$$\|\nabla(\boldsymbol{u}^{(k)} - \boldsymbol{u})\|_{2,\mathcal{V}^{(k)}} \le C_0|\Phi^{(k)} - \Phi| + C_1\,e^{-k\,C_2}, \tag{1.232}$$

where $\boldsymbol{v}^{(k)} := \boldsymbol{u}^{(k)} + \boldsymbol{\alpha}^{(k)}$ and $\boldsymbol{v} := \boldsymbol{u} + \boldsymbol{\alpha}$ are the velocity fields associated to the uniquely determined solutions to problems (1.199) and (1.105) obtained in Theorem 1.7 and Theorem 1.5, respectively, while $C_0 = C_0(\Omega) > 0$ and $C_i = C_i(\Omega, \Phi, \|\nabla\boldsymbol{u}\|_2) > 0$, $i = 1, 2$.

(b) *Conversely, there is $K^* = K^*(\Omega) > 0$, such that, for any $\Phi \in \mathbb{R}$ with $|\Phi| < K^*$, the corresponding unique solution to (1.105), in the class $\mathcal{C}_\Phi$, obtained in Theorem 1.5 can be approximated, in the sense of (1.232), by a sequence of solutions $\{\boldsymbol{v}^{(k)} := \boldsymbol{u}^{(k)} + \boldsymbol{\alpha}^{(k)}, p^{(k)}\}$ to (1.231) corresponding to $P^{(k)} = k\,M\Phi$.*

Before we give the proof of the theorem, we wish to observe the following.

Remark 1.14. (1) The constant M in the theorem coincides with the constant M_2 given in (1.212) and can be explicitly computed; see Remark 1.11.
(2) In case (b), in Equation (1.232) we can drop the first term on the right-hand side and we can replace $\boldsymbol{u}^{(k)} - \boldsymbol{u}$ with $\boldsymbol{v}^{(k)} - \boldsymbol{v}$. In fact, in this case, we have $\Phi^{(k)} = \Phi$, and, consequently, $\boldsymbol{\alpha}^{(k)} = \boldsymbol{\alpha}$ which furnishes $\boldsymbol{v}^{(k)} - \boldsymbol{v} = \boldsymbol{u}^{(k)} - \boldsymbol{u}$. In case (a), by the triangle inequality and by (1.232), we have

$$\|\nabla(\boldsymbol{v}^{(k)} - \boldsymbol{v})\|_{2,\mathcal{V}^{(k)}} \le \|\nabla(\boldsymbol{u}^{(k)} - \boldsymbol{u})\|_{2,\mathcal{V}^{(k)}} + \|\nabla(\boldsymbol{\alpha}^{(k)} - \boldsymbol{\alpha})\|_{2,\mathcal{V}^{(k)}}$$

$$\le C_1\,e^{-k\,C_2} + C\,k|\Phi^{(k)} - \Phi|,$$

where $C = C(\Omega) > 0$. Therefore, the rate of convergence of $\boldsymbol{v}^{(k)}$ to $\boldsymbol{v}$ (and that of $\boldsymbol{u}^{(k)}$ to $\boldsymbol{u}$ as well) in the Dirichlet norm depends also on the (prescribed) rate of convergence of $\Phi^{(k)}$ to Φ.
(3) By using inequality (1.232) along with local estimates for the Stokes problem in the outlets Ω_i (see (1.96) and [37, Section XI.4]), we can show convergence of $(\boldsymbol{u}^{(k)}, p^{(k)})$ to $(\boldsymbol{u}, p)$, together with their derivatives up to the order N, also in pointwise norms. The number N depends only on the regularity of $\partial\Omega$. If, in particular, Ω is of class C^∞, then $N = \infty$.

Proof of Theorem 1.8. We recall that

$$\Delta\boldsymbol{\alpha} = \nabla p_{Pi}, \quad \boldsymbol{\alpha} \cdot \nabla\boldsymbol{\alpha} = \boldsymbol{0} \quad \text{in } \widetilde{\Omega}_i, \quad i = 1, 2,$$

where p_{Pi}, $i = 1, 2$, are the pressure fields associated to the Poiseuille flow in the outlet Ω_i. Thus, if we dot-multiply both sides of (1.108) by $\boldsymbol{\varphi} \in \mathcal{V}^{(k)}$ and integrate by parts over $\mathcal{V}^{(k)}$ we get (with $(\cdot, \cdot) \equiv (\cdot, \cdot)_{\mathcal{V}^{(k)}}$)

$$
\begin{aligned}
(\nabla \boldsymbol{u}, \nabla \boldsymbol{\varphi}) = {}& -(\boldsymbol{u} \cdot \nabla \boldsymbol{u}, \boldsymbol{\varphi}) - (\boldsymbol{\alpha} \cdot \nabla \boldsymbol{u}, \boldsymbol{\varphi}) - (\boldsymbol{u} \cdot \nabla \boldsymbol{\alpha}, \boldsymbol{\varphi}) \\
& -(\nabla \boldsymbol{\alpha}, \nabla \boldsymbol{\varphi})_{\widetilde{\Omega}_0} - (\boldsymbol{\alpha} \cdot \nabla \boldsymbol{\alpha}, \boldsymbol{\varphi})_{\widetilde{\Omega}_0} \\
& + \sum_{i=1}^{2} \int_{S_i^{(k)}} \left(\frac{\partial \boldsymbol{u}}{\partial \boldsymbol{n}} - (p - p_{Pi})\boldsymbol{n} \right) \cdot \boldsymbol{\varphi} \, dS \,,
\end{aligned}
\tag{1.233}
$$

where we have used the fact that

$$
\begin{aligned}
(\Delta \boldsymbol{\alpha}, \boldsymbol{\varphi})_{\widetilde{\Omega}_0} &= -(\nabla \boldsymbol{\alpha}, \nabla \boldsymbol{\varphi})_{\widetilde{\Omega}_0} + \sum_{i=1}^{2} \int_{S_i^{(1)}} \frac{\partial \boldsymbol{\alpha}}{\partial \boldsymbol{n}} \cdot \boldsymbol{\varphi} \, dS \\
&= -(\nabla \boldsymbol{\alpha}, \nabla \boldsymbol{\varphi})_{\widetilde{\Omega}_0} + \sum_{i=1}^{2} \int_{S_i^{(1)}} \frac{\partial \boldsymbol{v}_{Pi}}{\partial \boldsymbol{n}} \cdot \boldsymbol{\varphi} \, dS = -(\nabla \boldsymbol{\alpha}, \nabla \boldsymbol{\varphi})_{\widetilde{\Omega}_0} \,.
\end{aligned}
$$

Furthermore, by (1.213), we have

$$
\begin{aligned}
(\nabla \boldsymbol{u}^{(k)}, \nabla \boldsymbol{\varphi}) = {}& -(\boldsymbol{u}^{(k)} \cdot \nabla \boldsymbol{u}^{(k)}, \boldsymbol{\varphi}) - (\boldsymbol{\alpha}^{(k)} \cdot \nabla \boldsymbol{u}^{(k)}, \boldsymbol{\varphi}) - (\boldsymbol{u}^{(k)} \cdot \nabla \boldsymbol{\alpha}^{(k)}, \boldsymbol{\varphi}) \\
& -(\boldsymbol{\alpha}^{(k)} \cdot \nabla \boldsymbol{\alpha}^{(k)}, \boldsymbol{\varphi})_{\widetilde{\Omega}_0} - (\nabla \boldsymbol{\alpha}^{(k)}, \nabla \boldsymbol{\varphi})_{\widetilde{\Omega}_0} \,.
\end{aligned}
\tag{1.234}
$$

If we set $\boldsymbol{w}^{(k)} := \boldsymbol{u}^{(k)} - \boldsymbol{u}|_{\mathcal{V}^{(k)}}$ and $\boldsymbol{A}^{(k)} := \boldsymbol{\alpha}^{(k)} - \boldsymbol{\alpha}$, from (1.233)–(1.234) we find

$$
\begin{aligned}
(\nabla \boldsymbol{w}^{(k)}, \nabla \boldsymbol{\varphi}) = {}& -(\boldsymbol{u}^{(k)} \cdot \nabla \boldsymbol{w}^{(k)}, \boldsymbol{\varphi}) - (\boldsymbol{w}^{(k)} \cdot \nabla \boldsymbol{u}, \boldsymbol{\varphi}) - (\boldsymbol{A}^{(k)} \cdot \nabla \boldsymbol{u}^{(k)}, \boldsymbol{\varphi}) \\
& -(\boldsymbol{\alpha} \cdot \nabla \boldsymbol{w}^{(k)}, \boldsymbol{\varphi}) - (\boldsymbol{u}^{(k)} \cdot \nabla \boldsymbol{A}^{(k)}, \boldsymbol{\varphi}) - (\boldsymbol{w}^{(k)} \cdot \nabla \boldsymbol{\alpha}, \boldsymbol{\varphi}) \\
& -(\nabla \boldsymbol{A}^{(k)}, \nabla \boldsymbol{\varphi})_{\widetilde{\Omega}_0} - (\boldsymbol{\alpha}^{(k)} \cdot \nabla \boldsymbol{A}^{(k)}, \boldsymbol{\varphi})_{\widetilde{\Omega}_0} - (\boldsymbol{A}^{(k)} \cdot \nabla \boldsymbol{\alpha}, \boldsymbol{\varphi})_{\widetilde{\Omega}_0} \\
& - \sum_{i=1}^{2} \int_{S_i^{(k)}} \left(\frac{\partial \boldsymbol{u}}{\partial \boldsymbol{n}} - (p - p_{Pi})\boldsymbol{n} \right) \cdot \boldsymbol{\varphi} \, dS \,,
\end{aligned}
\tag{1.235}
$$

We now observe that, by the properties of $\boldsymbol{\alpha}$ and $\boldsymbol{\alpha}^{(k)}$, and by the assumption on the sequence $P^{(k)}/k$,

$$
\begin{aligned}
&\max_{x \in \Omega} \left(|\boldsymbol{A}^{(k)}(x)| + |\nabla \boldsymbol{A}^{(k)}(x)| \right) + \|\nabla \boldsymbol{A}^{(k)}\|_{2, \widetilde{\Omega}_0} \leq C_1 |\Phi^{(k)} - \Phi| , \\
&\max_{x \in \Omega} \left(|\boldsymbol{\alpha}(x)| + |\nabla \boldsymbol{\alpha}(x)| \right) + \|\nabla \boldsymbol{\alpha}\|_{2, \widetilde{\Omega}_0} \leq C_1 |\Phi| , \\
&\max_{x \in \Omega} |\boldsymbol{\alpha}^{(k)}| \leq C_1 \,,
\end{aligned}
\tag{1.236}
$$

with $C_1 = C_1(\Omega) > 0$. We next choose $\varphi \equiv w^{(k)}$ into (1.235), employ Lemma 1.7, (1.236), (1.220)$_1$ and the Schwarz inequality to obtain

$$
\begin{aligned}
\|\nabla w^{(k)}\|_2^2 \leq\ & C_2 \left(\|\nabla u^{(k)}\|_2 + \|\nabla u\|_2 + |\Phi|\right) \|\nabla w^{(k)}\|_2^2 \\
& + C_3 |\Phi^{(k)} - \Phi| \left(\|\nabla u^{(k)}\|_2 + |\Phi| + 1\right) \|\nabla w^{(k)}\|_2 \\
& + C_4 \left(\max_{x \in S_1^{(k)} \cup S_2^{(k)}} |\nabla u(x)| + \max_{x \in S_1^{(k)}} |p(x) - p_{P1}(x)| \right. \\
& \left. + \max_{x \in S_2^{(k)}} |p(x) - p_{P2}(x)| \right) \left(\int_{S_1^{(k)} \cup S_2^{(k)}} |w^{(k)}|^2 dS\right)^{1/2}
\end{aligned}
\tag{1.237}
$$

where $C_i = C_i(\Omega) > 0$, $i = 2, 3, 4$. From Theorem 1.5 and Theorem 1.7 and by assumption, we know that

$$
\|\nabla u^{(k)}\|_2 \leq C_5 |\Phi^{(k)}|, \quad \|\nabla u\|_2 \leq C_5 |\Phi|,
\tag{1.238}
$$

with $C_5 = C_5(\Omega) > 0$, provided $|\Phi^{(k)}|$ (in case (a)) or $|\Phi|$ (in case (b)) is below a suitable constant, K, depending only on Ω. Thus, by taking K sufficiently small (if necessary), and by observing that, from (1.79),

$$
\left(\int_{S_1^{(k)} \cup S_2^{(k)}} |w^{(k)}|^2 dS\right)^{1/2} \leq C_6 \|\nabla w^{(k)}\|_2
$$

with $C_6 = C_6(S_1, S_2) > 0$, from (1.237) we conclude

$$
\|\nabla w^{(k)}\|_2 \leq C_7 |\Phi^{(k)} - \Phi| + C_8 \left(\max_{x \in S_1^{(k)} \cup S_2^{(k)}} |\nabla u(x)| + \sum_{i=1}^{2} \max_{x \in S_i^{(k)}} |p(x) - p_{Pi}(x)|\right)
\tag{1.239}
$$

where the positive constants C_7 and C_8 depend only on Ω. However, from [37, Lemma XI.4.1, Lemma XI.4.4 and Remark XI.4.3] we know that there exist two positive constants C_9 and C_{10} depending only on Ω, Φ and $\|\nabla u\|_2^2$ such that

$$
\max_{x \in S_1^{(k)} \cup S_2^{(k)}} |\nabla u(x)| + \sum_{i=1}^{2} \max_{x \in S_i^{(k)}} |p(x) - p_{Pi}(x)| \leq C_9 e^{-C_{10}k},
$$

so that the theorem follows from this latter inequality and from (1.239). $\qquad\square$

1.4.3. Initial-boundary value problem. In this section we shall discuss the well-posedness of the following initial-boundary value problem (in non-dimensional form and with the same notation as Section 1.4.1):

$$\left.\begin{array}{l} \dfrac{\partial \boldsymbol{v}}{\partial t} + \boldsymbol{v} \cdot \nabla \boldsymbol{v} = \Delta \boldsymbol{v} - \nabla p \\[2mm] \nabla \cdot \boldsymbol{v} = 0 \end{array}\right\} \quad \text{in } \mathcal{V},$$

$$\boldsymbol{v}|_{\Gamma} = \left(\dfrac{\partial \boldsymbol{v}}{\partial \boldsymbol{n}} - p\boldsymbol{n}\right)\bigg|_{S_1} = \boldsymbol{0}, \quad \left(\dfrac{\partial \boldsymbol{v}}{\partial \boldsymbol{n}} - p\boldsymbol{n}\right)\bigg|_{S_2} = P\boldsymbol{n},$$

$$\boldsymbol{v}(x,0) = \boldsymbol{v}_0(x),$$

(1.240)

where $P = P(t)$, $t \geq 0$, and $\boldsymbol{v}_0 = \boldsymbol{v}_0(x)$, $x \in \mathcal{V}$, are prescribed functions.

The challenging feature of problem (1.240) consists in the fact that, because of the "do-nothing" boundary conditions – and unlike the classical "no-slip" conditions – we can *not* prove that $(\boldsymbol{v} \cdot \nabla \boldsymbol{v}, \boldsymbol{v}) = 0$ (see also Remark 1.13). The immediate effect of this observation is that one is no longer able to show that the kinetic energy of the fluid is controlled, at *all* times, *only* by the function P and by the initial data (of arbitrary size), due to the fact that, in principle, some uncontrolled "backflow" can take place at the open sections of the pipe, namely, at the surfaces S_i, $i = 1, 2$. The mathematical consequence of the lack of controllability of the kinetic energy at all times is that one is not able to prove global existence even of *weak solutions* in the three-dimensional case.

Actually, if $\mathcal{V}$ is two-dimensional, that is, in the case of a flow in a channel, one can still prove global existence of weak solutions (a la Leray–Hopf), on condition that the size of P and of the initial data is appropriately restricted. This result can be easily achieved by combining the Galerkin method together with a suitable a-priori estimate that we shall now derive, at least formally. If we dot-multiply both sides of (1.240)$_1$ by $\boldsymbol{v}$, integrate over $\mathcal{V}$ and use the boundary conditions (1.240)$_{3,4}$, we obtain

$$\tfrac{1}{2}\dfrac{d}{dt}\|\boldsymbol{v}\|_2^2 + \|\nabla \boldsymbol{v}\|_2^2 = -(\boldsymbol{v} \cdot \nabla \boldsymbol{v}, \boldsymbol{v}) + P\int_{S_2} \boldsymbol{v} \cdot \boldsymbol{n}\, dS.$$

(1.241)

We now recall the inequality

$$\|\boldsymbol{v}\|_{4,D} \leq C\,\|\boldsymbol{v}\|_{2,D}^{\frac{1}{2}}\|\nabla \boldsymbol{v}\|_{2,D}^{\frac{1}{2}},$$

(1.242)

where D is a bounded, Lipschitz domain of $\mathbb{R}^2$, $\boldsymbol{v}$ belongs to $W^{1,2}(D)$ and vanishes on a portion of ∂D having nonzero (one-dimensional) Lebesgue measure, and $C = C(D) > 0$; see [36, Exercise II.2.9 and Exercise II.4.10]. Thus, from this inequality and from that of Hölder, we obtain

$$|(\boldsymbol{v} \cdot \nabla \boldsymbol{v}, \boldsymbol{v})| \leq \|\boldsymbol{v}\|_4^2\|\nabla \boldsymbol{v}\|_2 \leq C^2\|\boldsymbol{v}\|_2\|\nabla \boldsymbol{v}\|_2^2.$$

(1.243)

Furthermore, let

$$\phi = \begin{cases} 0 & \text{if } x_1 \leq b_2 - 1, \\ 2x_1 - 2(b_2 - 1) & \text{if } x_1 \in (b_2 - 1, b_2 - \tfrac{1}{2}), \\ 1 & \text{if } x_1 \geq b_2 - \tfrac{1}{2}. \end{cases} \quad .$$

By taking into account that $\nabla \cdot \boldsymbol{v} = 0$ in $\mathcal{V}$, and that $\boldsymbol{v}|_\Gamma = \boldsymbol{0}$, we find

$$\left| \int_{S_2} \boldsymbol{v} \cdot \boldsymbol{n}\, dS \right| = \left| \int_{S_2} \phi \boldsymbol{v} \cdot \boldsymbol{n}\, dS \right| = \left| \int_{\mathcal{V}} \nabla \phi \cdot \boldsymbol{v} \right| \le C_1 \|\boldsymbol{v}\|_2\,, \quad \boldsymbol{v} \in H(\mathcal{V})\,, \qquad (1.244)$$

where $C_1 = C_1(S_2) > 0$. Thus, employing into (1.241) this latter relation together with (1.243), it follows that

$$\tfrac{1}{2}\frac{d}{dt}\|\boldsymbol{v}\|_2^2 + (1 - C^2\|\boldsymbol{v}\|_2)\|\nabla \boldsymbol{v}\|_2^2 \le C_1\,|P|\,\|\boldsymbol{v}\|_2\,.$$

Finally, if we use (1.221) and (1.157) on the right-hand side of this latter inequality, we conclude

$$\frac{d}{dt}\|\boldsymbol{v}\|_2^2 + (1 - 2C^2\|\boldsymbol{v}\|_2)\|\nabla \boldsymbol{v}\|_2^2 \le C_2\,|P|^2\,, \qquad (1.245)$$

where $C_2 = C_2(\mathcal{V}) > 0$. From this relation it readily follows that if $\|\boldsymbol{v}_0\|_2^2 \le 1/(8C^2)$, say, then

$$\|\boldsymbol{v}(t)\|_2^2 < 1/(4C^2)\,, \quad \text{for all } t > 0\,, \qquad (1.246)$$

provided $\|P\|_{L^\infty(0,\infty)}$ is suitably restricted. Actually, assume, by contradiction, that t^* is the first (finite) instant at which $\|\boldsymbol{v}(t^*)\|_2^2 = 1/(4C^2)$. With the help of (1.221) and (1.245) we then find

$$\|\boldsymbol{v}(t^*)\|_2^2 \le \|\boldsymbol{v}_0\|_2^2 e^{-\kappa t} + C_2\, e^{-\kappa t^*} \int_0^{t^*} e^{\kappa t^*} |P(s)|^2 ds\,,$$

where $\kappa = \kappa(\mathcal{V}) > 0$. Thus,

$$\|\boldsymbol{v}(t^*)\|_2^2 < \|\boldsymbol{v}_0\|_2^2 + C_2\|P\|_{L^\infty(0,\infty)}^2\,,$$

and so, if we choose $C_2\|P\|_{L^\infty(0,\infty)}^2 < 1/(8C^2)$, from the previous inequality and from the assumption on $\boldsymbol{v}_0$, we get $\|\boldsymbol{v}(t^*)\|_2^2 < 1/(4C^2)$ which gives a contradiction. Once (1.245) has been established, we go back to (1.245) and find that

$$\int_0^t \|\nabla \boldsymbol{v}(s)\|_2^2 ds \le 2\|\boldsymbol{v}_0\|_2^2 + 2C_2 \int_0^t |P(s)|^2 ds\,. \qquad (1.247)$$

Estimates (1.246) and (1.247), in conjunction with the Galerkin method, allow us then to prove the existence of weak solutions under the stated restrictions on the data.

Remark 1.15. Concerning the existence of two-dimensional global *strong* solutions for small data, the situation is not completely clear. By "strong", we mean that the velocity field $\boldsymbol{v} \in L^\infty(0,T; H(\mathcal{V})) \cap L^2(0,T; L^\infty(\mathcal{V}))$, while $\partial \boldsymbol{v}/\partial t$, and $\widetilde{\Delta}\boldsymbol{v}$ are in $L^2(\mathcal{V} \times [0,T])$, where $\widetilde{\Delta}$ is the (suitably defined) Stokes operator. In fact, the validity of a result of existence of strong solutions (for small data) in the two-dimensional case is claimed by Heywood et al. in [58, Section 6]. However, the argument given there is based on the validity of the second inequality below Eq. (48) of that paper, for which the authors refer to an article that does not appear in their list of references. Moreover, the proof of that inequality does not seem to be completely obvious. Nevertheless, by the methods employed in this section it is possible to show existence and uniqueness of global solutions in a suitable regularity class; see Remark 1.17.

If $\mathcal{V}$ is three-dimensional, it is at once established that the procedure presented above fails to produce global existence of weak solutions. In fact, in such a case, the inequality (1.242) is replaced by the following (see [36, loc. cit.] and the next Lemma 1.8):

$$\|\boldsymbol{v}\|_4 \leq C\|\boldsymbol{v}\|_2^{\frac{1}{4}}\|\nabla\boldsymbol{v}\|_2^{\frac{3}{4}},$$

which now, instead of (1.243), yields

$$|(\boldsymbol{v}\cdot\nabla\boldsymbol{v},\boldsymbol{v})| \leq C^2\|\boldsymbol{v}\|_2^{\frac{1}{2}}\|\nabla\boldsymbol{v}\|_2^{\frac{5}{2}},$$

and the argument employed previously for the two-dimensional case does not work. Consequently, *the existence of (even weak) global solutions to* (1.240) *remains open in the three-dimensional case.*

However, some *local* in time theorems of existence of solutions to (1.240) in suitable function class, with corresponding uniqueness results, are indeed available. In fact, in the paper [66], Kučera and Skalák show, by means of a fixed-point argument, existence of solutions such that (with $H(\mathcal{V})$ defined in (1.200))

$$\boldsymbol{v}\in L^2(0,T^*;L_\sigma^2(\mathcal{V})),\quad \frac{\partial\boldsymbol{v}}{\partial t}\in L^2(0,T^*;H(\mathcal{V})),\quad \frac{\partial^2\boldsymbol{v}}{\partial t^2}\in L^2(0,T^*;H^{-1}(\mathcal{V})),$$
$$(1.248)$$

provided $\boldsymbol{v}_0$ is sufficiently smooth and that P and $\boldsymbol{v}_0$ satisfy certain further restrictions, including the condition $P(0)=0$. [10] These solutions are still "weak" (in the sense that they do not possess second spatial derivatives, even in the interior of $\mathcal{V}$), but, nevertheless, they are proved to be unique.

In the remaining part of this section, we shall prove local existence solutions to (1.240) in a class somewhat stronger than (1.248). Our main tools are Lemma 1.7, the following Lemma 1.8 and the classical Galerkin method.

Remark 1.16. As in the steady-state case, all the estimates that we will derive are independent of the numbers b_1 and b_2 defining the region of flow $\mathcal{V}$; see (1.195). [11] Therefore, we may, in principle, apply the methods of the previous section to investigate if and how flows in the bounded region $\mathcal{V}$ converge to flow in the unbounded pipe Ω.

The following lemma holds.

Lemma 1.8. *There exists a constant* $C = C(S_1, S_2, \Omega_0) > 0$ *such that*

$$\|\boldsymbol{\varphi}\|_4 \leq C\,\|\boldsymbol{\varphi}\|_2^{1/4}\|\nabla\boldsymbol{\varphi}\|_2^{3/4},\quad \text{for all } \boldsymbol{\varphi}\in H(\mathcal{V}).$$

---

[10] As a matter of fact, the results of [66] cover more general situations than that described by problem (1.240). Actually, the domain $\mathcal{V}$ need not be "pipe-like" with cuts orthogonal to the axis in each outlet, whereas the boundary conditions include the case when P is prescribed as (suitable) function of space and time.

[11] The notation we will use throughout this section is the same as that of Section 1.4.1.

Proof. From a well-known embedding inequality (see, e.g., [36, Exercise II.2.9]) we have

$$\|\varphi\|_{4,\omega} \le C_1 \|\varphi\|_2^{1/4} \|\nabla\varphi\|_2^{3/4}, \tag{1.249}$$

with $C_1 = C_1(\omega) > 0$, where ω is any bounded, Lipschitz domain of Ω with $\partial\Omega \cap \partial\omega$ having non-zero (two-dimensional) Lebesgue measure. So, in particular,

$$\|\varphi\|_{4,\Omega_0} \le C_1 \|\varphi\|_2^{1/4} \|\nabla\varphi\|_2^{3/4}. \tag{1.250}$$

Moreover, setting

$$\|\cdot\|_{q,\mathcal{V}_{2;t,t+1}}^q := \int_t^{t+1} \int_{S_2} |\cdot|^q \, dS \, dx_1,$$

with t arbitrary in $[0, b_2 - 1]$, from the proof of Lemma XI.2.1 of [37] (see the procedure after formula (2.5)), it follows that

$$\|\varphi\|_{4,\mathcal{V}_{2;t,t+1}}^4 \le \kappa \int_t^{t+1} (\|\varphi\|_{2,\mathcal{V}_{2;t,t+1}}^2 + \|\varphi\|_{2,\mathcal{V}_{2;t,t+1}} \|\nabla\varphi\|_{2,\mathcal{V}_{2;t,t+1}}) \int_{S_2} |\nabla\varphi|^2 \, dS \, dx_1,$$

where $\kappa = \kappa(S_2) > 0$, which implies

$$\|\varphi\|_{4,\mathcal{V}_{2;t,t+1}}^4 \le \kappa \left(\|\varphi\|_2^2 \|\nabla\varphi\|_{2,\mathcal{V}_{2;t,t+1}}^2 + \|\varphi\|_{2,\mathcal{V}} \|\nabla\varphi\|_2 \|\nabla\varphi\|_{2,\mathcal{V}_{2;t,t+1}}^2 \right). \tag{1.251}$$

As in Lemma 1.7, we take $m \in \mathbb{N}$ such that $m \le b_2 < m+1$ and set $\delta := b_2 - m$ (<1). We then have

$$\begin{aligned}
\|\varphi\|_{4,\mathcal{V}_2}^4 &\le \int_0^\delta \int_{S_2} |\varphi|^4 dS \, dx_1 + \sum_{l=0}^{m-1} \|\varphi\|_{4,\mathcal{V}_{2;l+\delta,l+\delta+1}}^4 \\
&\le \int_0^1 \int_{S_2} |\varphi|^4 dS \, dx_1 + \sum_{l=0}^{m-1} \|\varphi\|_{4,\mathcal{V}_{2;l+\delta,l+\delta+1}}^4.
\end{aligned} \tag{1.252}$$

From (1.249) we get

$$\int_0^1 \int_{S_2} |\varphi|^4 dS \le C_2 \|\varphi\|_2 \|\nabla\varphi\|_2^3, \tag{1.253}$$

with $C_2 = C_2(S_2) > 0$. Furthermore, from (1.251) it follows that

$$\sum_{l=0}^{m-1} \|\varphi\|_{4,\mathcal{V}_{2;l+\delta,l+\delta+1}}^4 \le \kappa \left(\|\varphi\|_2^2 \|\nabla\varphi\|_2^2 + \|\varphi\|_2 \|\nabla\varphi\|_2^3 \right) \le C_3 \|\varphi\|_2 \|\nabla\varphi\|_2^3, \tag{1.254}$$

where $C_3 = C_3(S_2, \Omega_0) > 0$ and where, in the second inequality, we have used (1.221). Collecting (1.252)–(1.254) we deduce

$$\|\varphi\|_{4,\mathcal{V}_2}^4 \le C_4 \|\varphi\|_2 \|\nabla\varphi\|_2^3, \tag{1.255}$$

with $C_4 = C_4(S_2, \Omega_0) > 0$. In analogous fashion we can show

$$\|\varphi\|_{4,\mathcal{V}_1}^4 \le C_5 \|\varphi\|_2 \|\nabla\varphi\|_2^3, \tag{1.256}$$

with $C_5 = C_5(S_1, \Omega_0) > 0$. The lemma then follows from (1.250), (1.248) and (1.249). $\square$

We are now in a position to prove the main result of this section.

Theorem 1.9. *Let $v_0 \in W^{2,2}(\mathcal{V}) \cap H(\mathcal{V})$ and let $P \in W^{1,2}(0,T)$. Suppose, further, that v_0 satisfies at least one of the conditions:*

(a) $v_0 \in W_0^{2,2}(\mathcal{V})$,

(b) $\left.\dfrac{\partial v_0}{\partial n}\right|_{S_1} = 0$, $\left.\dfrac{\partial v_0}{\partial n}\right|_{S_2} = P(0)\, n$.

Then, there exists $T^* = T^*(P, v_0, \Omega_0, S_1, S_2) \in (0,T)$ and a uniquely determined field v in the class

$$v \in C(0,T^*; W^{1,2}(\mathcal{V})), \quad \frac{\partial v}{\partial t} \in L^{\infty}(0,T^*; L_{\sigma}^2(\mathcal{V})) \cap L^2(0,T^*; W^{1,2}(\mathcal{V})), \quad (1.257)$$

that satisfies (1.240) in the following generalized sense:

$$\left(\frac{\partial v}{\partial t}, \varphi\right) + (\nabla v, \nabla \varphi) = -(v \cdot \nabla v, \varphi) + P \int_{S_2} \varphi \cdot n \, dS, \quad \text{for all } \varphi \in H(\mathcal{V}). \quad (1.258)$$

Finally, the following estimate holds:

$$\max_{t \in [0,T^*]} \|v(t)\|_{1,2} + \operatorname*{ess\,sup}_{t \in [0,T^*]} \left\|\frac{\partial v}{\partial t}(t)\right\|_2 + \int_0^{T^*} \|\nabla v_t\|_2^2 \, dt$$

$$\leq C \left(\|v_0\|_{2,2}^2 + |\mathcal{P}|_{W^{1,2}(0,T)} + \beta |P(0)|\right), \quad (1.259)$$

where $C = C(S_1, S_2, \Omega_0) > 0$ and where β is 1 or 0 according to whether v_0 satisfies condition (a) *or* (b).

Proof. The proof is quite classical and makes use of the Galerkin method. Let $\{\psi_k\}$ be a basis of $H(\mathcal{V})$ with $\psi_1 = v_0$, and consider the following sequence of "approximating problems" (with $\dfrac{\partial(\cdot)}{\partial t} \equiv (\cdot)_t$):

$$(v_{mt}, \psi_k) + (\nabla v_m, \nabla \psi_k) = -(v_m \cdot \nabla v_m, \psi_k) + P \int_{S_2} \psi_k \cdot n \, dS, \quad k = 1, \ldots, m,$$

$$v_m(x,t) := \sum_{k=1}^{m} c_{km}(t) \psi_k(x).$$

$$(1.260)$$

The unknown coefficients c_{km} must be determined from the system of ODE's (1.260) with the initial conditions

$$c_{1m}(0) = 1, \quad c_{2m}(0) = c_{3m}(0) = \cdots = c_{mm}(0) = 0. \quad (1.261)$$

We now perform the following operations. (a) We multiply both sides of $(1.260)_1$ by c_{km} and sum over k from 0 to m; (b) we multiply both sides of $(1.260)_1$ by dc_{km}/dt and sum over k from 0 to m, and, finally, (c) we take the time derivative of $(1.260)_1$, then multiply both sides of the resulting equation by dc_{km}/dt and sum over k from 0 to m. We thus obtain the following three equations where, for simplicity, the subscript m has been omitted.

$$\frac{1}{2}\frac{d}{dt}\|\boldsymbol{v}\|_2^2 + \|\nabla\boldsymbol{v}\|_2^2 = -(\boldsymbol{v}\cdot\nabla\boldsymbol{v},\boldsymbol{v}) + P\int_{S_2}\boldsymbol{v}\cdot\boldsymbol{n}\,dS,$$

$$\frac{1}{2}\frac{d}{dt}\|\nabla\boldsymbol{v}\|_2^2 + \|\boldsymbol{v}_t\|_2^2 = -(\boldsymbol{v}\cdot\nabla\boldsymbol{v},\boldsymbol{v}_t) + P\int_{S_2}\boldsymbol{v}_t\cdot\boldsymbol{n}\,dS,\tag{1.262}$$

$$\frac{1}{2}\frac{d}{dt}\|\boldsymbol{v}_t\|_2^2 + \|\nabla\boldsymbol{v}_t\|_2^2 = -(\boldsymbol{v}_t\cdot\nabla\boldsymbol{v},\boldsymbol{v}_t) - (\boldsymbol{v}\cdot\nabla\boldsymbol{v}_t,\boldsymbol{v}_t) + P'\int_{S_2}\boldsymbol{v}_t\cdot\boldsymbol{n}\,dS\,.$$

In order to estimate the right-hand sides of the above equations, besides Lemma 1.7 and Lemma 1.8, we recall the inequality

$$\left|\int_{S_2}\boldsymbol{\varphi}\cdot\boldsymbol{n}\,dS\right| \le C_1(S_2)\|\nabla\boldsymbol{\varphi}\|_2\,,\quad \boldsymbol{\varphi}\in H(\mathcal{V})\,,\tag{1.263}$$

which is easily established with the help of the Schwarz inequality and of (1.79). Thus, using (1.263) together with (1.157), we find, for any $\varepsilon > 0$,

$$\left|P\int_{S_2}\boldsymbol{v}\cdot\boldsymbol{n}\,dS\right| + \left|P\int_{S_2}\boldsymbol{v}_t\cdot\boldsymbol{n}\,dS\right| + \left|P'\int_{S_2}\boldsymbol{v}_t\cdot\boldsymbol{n}\,dS\right|$$
$$\le C_2(|P|^2 + |P'|^2 + \|\nabla\boldsymbol{v}\|_2^2) + \varepsilon\|\nabla\boldsymbol{v}_t\|_2^2\,,\tag{1.264}$$

with $C_2 = C_2(S_2,\varepsilon) > 0$. Moreover, by Lemma 1.7, we have

$$|(\boldsymbol{v}\cdot\nabla\boldsymbol{v},\boldsymbol{v})| \le C_3\,\|\nabla\boldsymbol{v}\|_2^3\,,\tag{1.265}$$

where $C_3 = C_3(S_1,S_2,\Omega_0) > 0$. Furthermore, with the help of Hölder inequality, of Lemma 1.8 and of (1.157) we obtain

$$|(\boldsymbol{v}_t\cdot\nabla\boldsymbol{v},\boldsymbol{v}_t)| \le \|\boldsymbol{v}\|_4^2\|\nabla\boldsymbol{v}\|_2 \le C_4\|\boldsymbol{v}_t\|_2^{\frac{1}{2}}\|\nabla\boldsymbol{v}_t\|_2^{\frac{3}{2}}\|\nabla\boldsymbol{v}\|_2$$
$$\le C_5\left(\|\boldsymbol{v}_t\|_2^4 + \|\nabla\boldsymbol{v}\|_2^8\right) + \varepsilon\|\nabla\boldsymbol{v}_t\|_2^2\,,\tag{1.266}$$

with $C_5 = C_5(S_1,S_2,\Omega_0,\varepsilon) > 0$. In analogous way, we get

$$|(\boldsymbol{v}\cdot\nabla\boldsymbol{v}_t,\boldsymbol{v}_t)| \le \|\boldsymbol{v}\|_4\|\boldsymbol{v}_t\|_4\|\nabla\boldsymbol{v}_t\|_2 \le C_6\|\boldsymbol{v}\|_2^{\frac{1}{4}}\|\boldsymbol{v}_t\|_2^{\frac{1}{4}}\|\nabla\boldsymbol{v}\|_2^{\frac{3}{4}}\|\nabla\boldsymbol{v}_t\|_2^{\frac{7}{4}}$$
$$\le C_7\left(\|\boldsymbol{v}\|_2^8 + \|\boldsymbol{v}_t\|_2^8 + \|\nabla\boldsymbol{v}\|_2^8\right) + \varepsilon\|\nabla\boldsymbol{v}_t\|_2^2\,,\tag{1.267}$$

with $C_7 = C_7(S_1,S_2,\Omega_0,\varepsilon) > 0$. Finally, again by the same arguments, we find

$$|(\boldsymbol{v}\cdot\nabla\boldsymbol{v},\boldsymbol{v}_t)| \le \|\boldsymbol{v}\|_4\|\boldsymbol{v}_t\|_4\|\nabla\boldsymbol{v}\|_2 \le C_8\|\boldsymbol{v}\|_2^{\frac{1}{4}}\|\boldsymbol{v}_t\|_2^{\frac{1}{4}}\|\nabla\boldsymbol{v}_t\|_2^{\frac{3}{4}}\|\nabla\boldsymbol{v}\|_2^{\frac{7}{4}}$$
$$\le C_9\left(\|\boldsymbol{v}\|_2^8 + \|\boldsymbol{v}_t\|_2^8 + \|\nabla\boldsymbol{v}\|_2^{14}\right) + \varepsilon\|\nabla\boldsymbol{v}_t\|_2^2\,,\tag{1.268}$$

where $C_9 = C_9(S_1,S_2,\Omega_0,\varepsilon) > 0$. Thus, setting

$$Y := \|\boldsymbol{v}\|_2^2 + \|\nabla\boldsymbol{v}\|_2^2 + \|\boldsymbol{v}_t\|_2^2\,,\quad \mathcal{P} := |P|^2 + |P'|^2\,,$$

from (1.262) and (1.264)–(1.268), by taking ε sufficiently small, we deduce

$$\frac{dY}{dt} + \|\nabla v_t\|_2^2 \leq C_{10} \left(\mathcal{Q}(Y) + \mathcal{P} \right), \qquad (1.269)$$

where $C_{10} = C_{10}(S_1, S_2, \Omega_0) > 0$ and $\mathcal{Q}(z)$ is a suitable polynomial with $\mathcal{Q}(0) = 0$. We next observe that

$$Y(0) \leq C_{11}(\|v_0\|_{2,2}^2 + \beta |P(0)|), \qquad (1.270)$$

with $C_{11} = C_{11}(S_1, S_2, \Omega_0) > 0$, and where $\beta = 1$ or $\beta = 0$ according to whether v_0 satisfies the assumption (a) or (b) in the theorem. In fact, from $(1.260)_2$ and (1.261), and recalling that $\boldsymbol{\psi}_1 = v_0$, we find

$$Y(0) \leq \|v_0\|_{1,2}^2 + \|v_{mt}(0)\|_2^2.$$

Now, from $(1.260)_1$ it follows that

$$\|v_{mt}(0)\|_2^2 = -(\nabla v_0, \nabla v_{mt}(0)) - (v_0 \cdot \nabla v_0, v_{mt}(0)) + P(0) \int_{S_2} v_{mt}(0) \cdot \boldsymbol{n} \, dS. \qquad (1.271)$$

Under the assumption (a) of the theorem, since $v_0 \in W_0^{2,2}(\mathcal{V})$, we find

$$-(\nabla v_0, \nabla v_{mt}(0)) = (\Delta v_0, v_{mt}(0)) \leq \|\Delta v_0\|_2 \|v_{mt}(0)\|_2, \qquad (1.272)$$

while, by (1.244), we also get

$$P(0) \int_{S_2} v_{mt}(0) \cdot \boldsymbol{n} \, dS \leq C_{12}|P(0)| \|v_{mt}(0)\|_2, \qquad (1.273)$$

where $C_{12} = C_{12}(S_2) > 0$. Furthermore, by the Hölder inequality it follows that

$$|(v_0 \cdot \nabla v_0, v_{mt}(0))| \leq \|v_0\|_6 \|\nabla v_0\|_3 \|v_{mt}(0)\|_2.$$

Employing in this relation $(1.119)_{1,4}$ and (1.124) we easily obtain

$$|(v_0 \cdot \nabla v_0, v_{mt}(0))| \leq C_{13} \|v_0\|_2^2 \|v_{mt}(0)\|_2, \qquad (1.274)$$

where $C_{13} = C_{13}(S_1, S_2, \Omega_0) > 0$. Therefore, (1.270) follows from (1.271)–(1.274). Likewise, under the assumption (b) of the theorem, by integration by parts, we find

$$-(\nabla v_0, \nabla v_{mt}(0)) + P(0) \int_{S_2} v_{mt}(0) \cdot \boldsymbol{n} \, dS = (\Delta v_0, v_{mt}(0))$$

$$\leq \|\Delta v_0\|_2 \|v_{mt}(0)\|_2,$$

and so (1.270) follows again from this latter relation and from (1.274).

We now go back to (1.269). Disregarding, temporarily, the second term on its left-hand side, we easily show that there is $T^* = T^*(C_{10}, \|v_0\|_{2,2}, |\mathcal{P}|_{W^{1,2}(0,T)}) \in (0, T)$ such that

$$Y(t) \leq C_{14}(\|v_0\|_{2,2}^2 + |\mathcal{P}|_{W^{1,2}(0,T)} + \beta |P(0)|), \quad \text{for all } t \in [0, T^*], \qquad (1.275)$$

with $C_{14} = C_{14}(S_1, S_2, \Omega_0) > 0$. Using this information back into (1.269), we then conclude

$$\int_0^{T^*} \|\nabla \boldsymbol{v}_t\|_2^2 \, dt \le C_{15}(\|\boldsymbol{v}_0\|_{2,2}^2 + |\mathcal{P}|_{W^{1,2}(0,T)} + \beta |P(0)|), \qquad (1.276)$$

with $C_{15} = C_{15}(S_1, S_2, \Omega_0) > 0$. With the estimates (1.275) and (1.276) in our hand, it is then routine to prove that, as $m \to \infty$, we can select a subsequence, $\{\boldsymbol{v}_{m'}\}$, converging to a vector field $\boldsymbol{v}$ in the function class (1.257), which, further, obeys (1.259). Moreover, with the help of (1.260)$_1$, we also show that $\boldsymbol{v}$ satisfies (1.258). The proof of existence is therefore completed. In order to show uniqueness, let $\boldsymbol{v}_1$ and $\boldsymbol{v}_2$ two solutions to (1.258) in the class (1.257), corresponding to the same P and same $\boldsymbol{v}_0$. Setting $\boldsymbol{u} = \boldsymbol{v}_1 - \boldsymbol{v}_2$ we thus obtain

$$(\frac{\partial \boldsymbol{u}}{\partial t}, \boldsymbol{\varphi}) + (\nabla \boldsymbol{u}, \nabla \boldsymbol{\varphi}) = -(\boldsymbol{v}_1 \cdot \nabla \boldsymbol{u}, \boldsymbol{\varphi}) - (\boldsymbol{u} \cdot \nabla \boldsymbol{v}_2, \boldsymbol{\varphi}), \quad \text{for all } \boldsymbol{\varphi} \in H(\mathcal{V}). \quad (1.277)$$

Choosing $\boldsymbol{\varphi} = \boldsymbol{u}$ into (1.277) furnishes

$$\tfrac{1}{2}\frac{d}{dt}\|\boldsymbol{u}\|_2^2 + \|\nabla \boldsymbol{u}\|_2^2 = -(\boldsymbol{v}_1 \cdot \nabla \boldsymbol{u}, \boldsymbol{u}) - (\boldsymbol{u} \cdot \nabla \boldsymbol{v}_2, \boldsymbol{u}). \qquad (1.278)$$

By the Hölder inequality and by Lemma 1.8, we have

$$|(\boldsymbol{v}_1 \cdot \nabla \boldsymbol{u}, \boldsymbol{u})| + |(\boldsymbol{u} \cdot \nabla \boldsymbol{v}_2, \boldsymbol{u})| \le \|\boldsymbol{v}_1\|_4 \|\boldsymbol{u}\|_4 \|\nabla \boldsymbol{u}\|_2 + \|\boldsymbol{u}\|_4^2 \|\nabla \boldsymbol{v}_2\|_2$$

$$\le C_{15}\left(\|\boldsymbol{v}_1\|_2^{\frac{1}{4}}\|\boldsymbol{v}_1\|_2^{\frac{3}{4}} + \|\nabla \boldsymbol{v}_2\|_2\right)\left(\|\boldsymbol{u}\|_2^{\frac{1}{4}}\|\nabla \boldsymbol{u}\|_2^{\frac{7}{4}} + \|\boldsymbol{u}\|_2^{\frac{1}{2}}\|\nabla \boldsymbol{u}\|_2^{\frac{3}{2}}\right), \tag{1.279}$$

with $C_{15} = C_{15}(S_1, S_2, \Omega_0) > 0$. By assumption, we have

$$\|\boldsymbol{v}_1\|_2^{\frac{1}{4}}\|\boldsymbol{v}_1\|_2^{\frac{3}{4}} + \|\nabla \boldsymbol{v}_2\|_2 \le \kappa \mathcal{D},$$

where κ is a numerical constant and $\mathcal{D}$ is the right-hand side of (1.259). Replacing this information back into (1.279) and employing (1.157) we deduce

$$|(\boldsymbol{v}_1 \cdot \nabla \boldsymbol{u}, \boldsymbol{u})| + |(\boldsymbol{u} \cdot \nabla \boldsymbol{v}_2, \boldsymbol{u})| \le C_{16}\|\boldsymbol{u}\|_2^2 + \tfrac{1}{2}\|\nabla \boldsymbol{u}\|_2^2,$$

with $C_{16} = C_{16}(\mathcal{D}, S_1, S_2, \Omega_0) > 0$. Thus, from this latter relation and (1.278) we conclude

$$\frac{d}{dt}\|\boldsymbol{u}(t)\|_2^2 \le 2C_{16}\|\boldsymbol{u}(t)\|_2^2, \quad \boldsymbol{u}(0) = \boldsymbol{0},$$

which, in turn, by Gronwall's lemma, implies $\boldsymbol{u}(t) = \boldsymbol{0}$, for all $t \in [0, T^*]$. Uniqueness is then accomplished and the proof of the theorem is completed. $\qquad \square$

Remark 1.17. By the procedure employed in the proof of Theorem 1.8, it is not difficult to show that, if $\mathcal{V} \subset \mathbb{R}^2$, problem (1.240) admits a unique *global* solution in the class (1.257), provided the data $\boldsymbol{v}_0$ and P are, in addition, suitably restricted in size. The key role in this proof is played by (a) inequality (1.242) that, in two dimensions, replaces the inequality of Lemma 1.8, and (b) by the two global estimates given in (1.246) and (1.247). We leave the (few) details to the interested reader.

2. Problems in non-Newtonian fluid mechanics

As is well known, the constitutive equation that defines an incompressible, viscous Navier–Stokes fluid is characterized by the linear dependence of the Cauchy stress tensor, $\boldsymbol{T}$, on the stretching tensor, $\boldsymbol{D} = \boldsymbol{D}(\boldsymbol{v}) := \frac{1}{2}(\nabla\boldsymbol{v} + (\nabla\boldsymbol{v})^{\top})$, with $\boldsymbol{v}$ velocity field of the liquid. Specifically, we have

$$\boldsymbol{T} = -p\,\boldsymbol{I} + 2\mu\,\boldsymbol{D}\,, \tag{2.1}$$

where p is the pressure field, namely, the Lagrange multiplier associated to the incompressibility constraint, while μ is a positive *constant* called shear stress viscosity. Fluids obeying the constitutive law (2.1) are called *Newtonian*. Every other fluid is called *non-Newtonian*.

Objective of this chapter is to present a mathematical analysis of a number of significant problems associated to certain non-Newtonian models that are mostly used in the engineering community and that are specifically related to blood flow issues. Of course, since for a non-Newtonian fluid the relation between $\boldsymbol{T}$ and $\boldsymbol{D}$ is *nonlinear*, the corresponding governing equations are more complicated than the Navier–Stokes equations. Consequently, the mathematical analysis of non-Newtonian models is, in principle, much more complicated than that for a Newtonian fluid, which, as we know, is already very difficult. Therefore, before embarking ourselves on a potentially much more complex journey, we would like to present some (among many others!) significant arguments motivating a mathematical study of non-Newtonian models.

2.1. Why non-Newtonian models?

It is a well-established experimental fact that many real incompressible fluids, including blood, can not be described, under all circumstances, by the Navier–Stokes equations. Some of these experiments are very easy to reproduce, even in a home-made laboratory, and we would like to describe them here.

Probably, one of the most famous experiments is the *rod-climbing* or *Weissenberg effect*. Here a cylindrical container, $\mathcal{C}$, is filled with liquid, $\mathcal{L}$, and a cylindrical rod is immersed in $\mathcal{L}$ with its axis parallel to that of $\mathcal{C}$; see Figure 4. Next, the rod is rotated and kept at a constant angular velocity, Ω. Now the response of the free surface of $\mathcal{L}$ to the rotation of the rod is dramatically different depending on the physical characteristic of $\mathcal{L}$. In fact it will reach an equilibrium configuration which for a Newtonian liquid, like water, will be like the one sketched in Figure 4(a), while for certain non-Newtonian liquids, like ordinary hair shampoo, the equilibrium surface will "climb the rod", as shown in Figure 4(b). Moreover, if the angular velocity is further increased, then, in the non-Newtonian case, the free surface may start oscillating periodically, while climbing the rod; see [61, pp. 510–521].

Another not less remarkable and even simpler example of non-Newtonian effects is provided by the *sedimentation of symmetric particles in a liquid*. In this

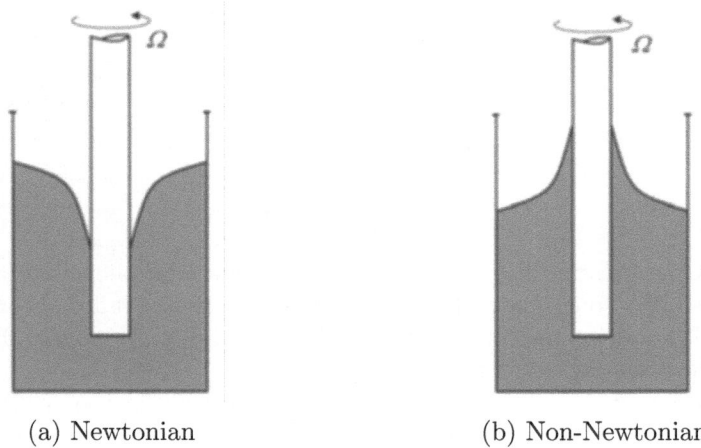

(a) Newtonian (b) Non-Newtonian

FIGURE 4. Rod-climbing effect.

experiment rigid particles of constant density, possessing rotational and fore-and-aft symmetry (cylinders of constant density, for instance) are dropped from rest in a vertical container filled with a liquid, $\mathcal{L}$. If the appropriate Reynolds number [12] is not "too large", the particle will reach a well-defined stable orientation that is completely independent of the initial orientation.

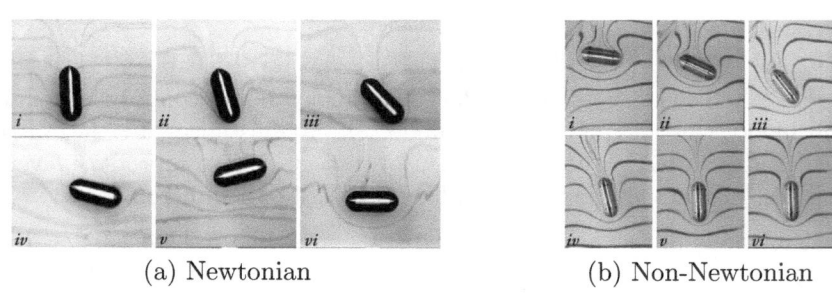

(a) Newtonian (b) Non-Newtonian

FIGURE 5. Sedimentation of symmetric particles in a liquid.

In particular, for a Navier–Stokes liquid like water, the particle will orient itself with its major axis of symmetry, $\mathbf{a}$, perpendicular to the direction of gravity, $\mathbf{g}$, while in a liquid like hair shampoo it will orient itself with $\mathbf{a}$ parallel to $\mathbf{g}$; see Figure 5. It is important to emphasize that the dimensionless numbers involved in these experiments may be very small. For example, for cylinders made of plastic, Teflon, aluminum and titanium, with length $\sim$ 2cm and diameter in the range

---

[12] Typically, the Reynolds number is defined as Ud/ν where U and d are the average speed and the hydraulic radius of the particle, respectively, while ν is the kinematic viscosity of the liquid.

0.25 ∼ 1cm, sedimenting in a 1.5 – 2% solution of Polyox in water, it is found that the Reynolds number varies from 0.016 to ∼ 5, while the Weissenberg number [13] ranges between 0.048 and ∼ 0.3; see [78].

As a final, simple example of different behavior of Newtonian and non-Newtonian liquids, we shall consider the *sedimentation of homogeneous spheres in the vicinity of a rigid wall.* In this case, a sphere of constant density is dropped in a vertical container, and "close enough" to one of its walls, so that "wall effects" become relevant. What is observed is that the sphere will move away from the wall or closer to it depending on whether the liquid is Newtonian (Figure 6(a)) or non-Newtonian (Figure 6(b)). Moreover, the equilibrium distance to the wall, h_e, will depend also on the physical characteristics of the sphere and of the liquid; see [5].

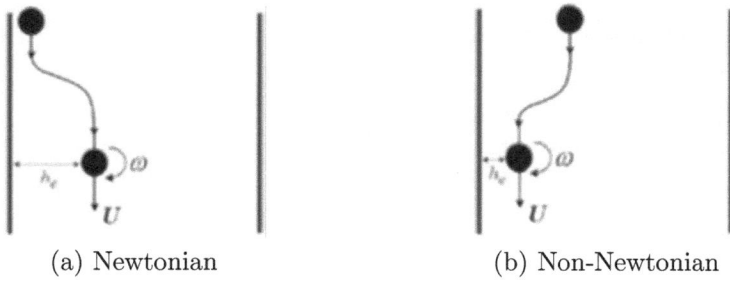

(a) Newtonian   (b) Non-Newtonian

FIGURE 6. Sedimentation of spheres nearby a vertical wall.

From the physical point of view, all the above experiments can be (qualitatively) explained by the fact that certain non-Newtonian liquids are able to exert a type of forces due to the so-called *normal stresses*, which are absent in the case of a Newtonian (Navier–Stokes) liquid; see Sections 2.3.1 and 2.3.2.

Besides the "normal stress effect", there is another important feature that is absent in a Navier–Stokes liquid and which may show up in more complex liquids like, for example, blood. In fact, in a Navier–Stokes liquid (e.g., water), the shear viscosity μ (see (2.1)) is constant for a reasonable wide range of shear rate, while in certain non-Newtonian liquids the viscosity may depend on the shear rate in a suitable range of shear rate. In particular, it may monotonically decrease or increase with increasing shear rate. Liquids showing the former behavior are referred to as *shear-thinning*, while liquids showing the latter are called *shear-thickening.*

A remarkable example of shear-thinning fluid is blood. In fact, in the range of low/mid shear rate (up to a few hundred $\sec^{-1}$) the blood shows shear-thinning properties (see Figure 7), while at higher shear rates the viscosity is practically constant. This phenomenon is commonly attributed to the properties of blood cells

---

[13] A dimensionless number that measures the "elasticity" of the fluid.

that change under increasing shear rate, and it is explained as follows [97]. In the low-shear region the red blood cells tend to aggregate with the effect of producing large clusters (so-called *rouleaux*). As the shear rate increases the shear stress begins to produce some disaggregation and overall reduces the size of aggregates. In the mid-shear region cell aggregates are largely destroyed by the stress levels present at these shear rates. Moreover, red cells orient and deform in order to pass adjacent cells. All the above changes in the aggregation of cells produces an overall variation in the blood viscosity. At high shear-rate they become strongly aligned to the direction of shear and form layers that slide on adjacent plasma layers. This structure is quite stable in this regime of shear-rate, with the consequence of keeping the overall viscosity constant.

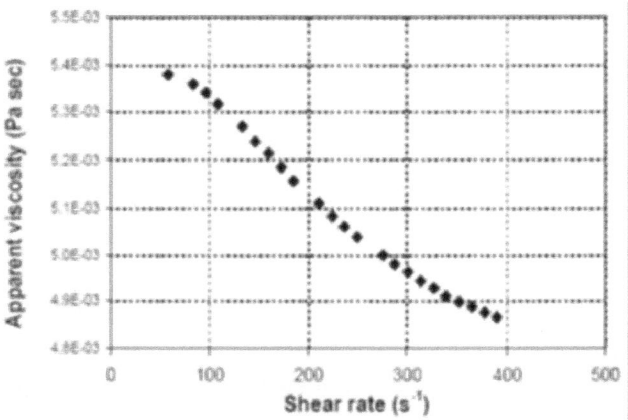

FIGURE 7. Dependence of blood viscosity on shear rate, showing the shear-thinning property of blood (after Chang et al. [21])

For further and more complete information about the physics and the appropriate modeling of non-Newtonian fluids, we refer the interested reader to [4] and also to the article of A.M. Robertson [94] in this volume.

2.2. Problems related to generalized Newtonian models

As mentioned previously, the two characteristic features of the Newtonian model are the linear dependence of the stress tensor T on D, and the fact that the shear viscosity μ is constant; see (2.1). Thus, a "natural" way of generalizing the constitutive equation (2.1) is to drop the linearity assumption, as well as to allow the dependence of μ on the stretching tensor D. Therefore, a simple generalization of (2.1) would lead to assume that T is a *quadratic polynomial* in D, with coefficients depending on D:

$$T = (-p + h_0)I + h_1 D + h_2 D^2 . \qquad (2.2)$$

As a matter of fact, it turns out that the constitutive equation (2.2) gives us the most general dependence of T upon D, that is compatible with basic physical requirements. Specifically, let us assume that

$$T = -p I + \sigma(D) \,, \tag{2.3}$$

where σ is a symmetric, [14] tensor-valued function of D. Then, a necessary and sufficient condition for the constitutive equation to be compatible with the *principle of material invariance* [15] is that T has to be of the form (2.2), where the coefficients h_i, $i = 0, 1, 2$, can only be functions of the principal invariants of D. Thus, taking into account that trace $D = \nabla \cdot v = 0$, it follows that h_i may depend only on $|D| := \sqrt{D : D}$ and on det D. Fluids obeying the constitutive law (2.2) with $h_0 = 0$, and with the mentioned restrictions on h_i, $i = 1, 2$, are called *Reiner–Rivlin fluids*.

It is important to observe that constitutive equations of the type (2.2), though describing fluids more general than those obeying the linear relation (2.1), do not seem to explain, even in a semi-quantitative way, the experiments we have mentioned in the previous section. Actually, in a little known paper of 1959, J. Serrin [102] used the Reiner–Rivlin model to give a mathematical explanation of the "rod-climbing" effect, which, as a by-product, would furnish a method for measuring some material parameters associated with the liquid model. His calculations assume that the coefficient h_2 (the so called "cross-viscosity") be a non-zero constant. However, it was later demonstrated by C. Leigh in 1961 [74], that the assumption $h_2 =$ constant$\neq 0$ is incompatible with the Second Principle of Thermodynamics and, consequently, Serrin's analysis is not applicable. Concerning the interaction liquid-particles and preferred orientation in sedimentation experiments, more recently, A. Vaidya [111] has proved, by elementary symmetry arguments, that models of the type (2.2) are unable to predict the orientation of symmetric cylinders in non-Newtonian liquids described in the preceding section. In other words, despite certain nonlinear effects are taken into account, the homogeneous cylinder will orient itself as in a Navier–Stokes liquid.

A special sub-class of the non-Newtonian constitutive equation (2.2) is represented by *generalized Newtonian models*, where $h_0 = h_2 = 0$ and h_1, the "generalized shear viscosity", depends only on the second invariant of D, namely, $|D|$. Thus, generalized Newtonian liquids are described by the constitutive equation

$$T = -p I + h_1 D \,, \quad h_1 = h_1(|D|) \,. \tag{2.4}$$

Among these models, a very popular one within the engineering community is the so-called "power law" model, introduced by W. Ostwald as early as 1925 [85], where

$$h_1(|D|) = \mu_1 |D|^{q-2} \tag{2.5}$$

with $\mu_1 > 0$ and $q > 1$. Notice that the classical Newtonian (Navier–Stokes) case corresponds to $q = 2$. The popularity of the empirical model (2.5) is essentially due to the fact that a wide class of flow problems can be solved analytically for it; see,

---

[14] The assumption of symmetry is necessary because T must be symmetric [94].

[15] Roughly speaking, this fundamental principle requires that the physical properties of a material must be independent of the observer. For more details, see the article [94] in this volume.

e.g., [13, Section 4.2]. Nevertheless, (2.5) may become completely unrealistic in the shear-thinning regime $q < 2$, in that the viscosity of the liquid grows unbounded ($h_1 \to \infty$) at small shear rates ($|\boldsymbol{D}| \to 0$), and the error between calculated values with (2.5) and experimental data can be quite large [13, Section 4.1]. For this and other reasons, in 1968 P.J. Carreau proposed, in his Ph.D. Thesis, a "more realistic" model given by

$$h_1(|\boldsymbol{D}|) = \mu_0 + \mu_1(\mu_2 + |\boldsymbol{D}|^2)^{\frac{q-2}{2}}, \tag{2.6}$$

where μ_i, $i = 0, 1, 2$, are positive material constants; see [19]. Obviously, (2.6) does not present the unbounded viscosity drawback of (2.5) and, moreover, it furnishes a very good curve-fitting of experimental data for a number of liquids and in a wide range of shear rate values; see [13, Section 4.1(a)] and the references therein. We shall not deal with more details of these and of other models of generalized Newtonian liquids, and refer the interested reader to [13, Chapter 4], and to the article of A.M. Robertson [94] in this volume.

In the mathematical community, generalized Newtonian models became particularly fashionable only in the late 1960s, when O.A. Ladyzhenskaya started a systematic investigation of the well-posedness of initial-boundary- and boundary-value problems associated to certain generalized Newtonian models [67, 68, 69]. The timing is not surprising, because, as we shall see later on, the mathematical analysis of generalized non-Newtonian models requires certain tools from the theory of monotone operators, pioneered by the work of G.J. Minty [82] and F.E. Browder [18] in the early 1960s. Ladyzhenskaya's investigation was basically focused on models where

$$h_1(|\boldsymbol{D}|) = \mu_0 + \mu_1|\boldsymbol{D}|^{q-2}, \tag{2.7}$$

and where $\mu_0, \mu_1 > 0$, and $q > 1$. This constitutive law is a particular case of (2.6) when we take, in this latter, $\mu_2 = 0$. Ladyzhenskaya's results were particularly interesting, for both initial-boundary- and boundary-value problems. [16] In fact, for the former, she was able to prove three-dimensional *global* existence *and uniqueness* in the *same* function class, provided $q \geq 9/4$. [17] For the latter, she showed existence of solutions in a possibly *multiply-connected* bounded domain, Ω, and corresponding to *non-homogeneous* boundary data $\boldsymbol{v}_*$, that satisfy *only* the compatibility condition

$$\int_{\partial\Omega} \boldsymbol{v}_* \cdot \boldsymbol{n} = 0,$$

where $\boldsymbol{n}$ is the unit outer normal to $\partial\Omega$, under the assumption that $q > 2$. It is worth emphasizing that either one of the above problems is an outstanding open question in the case of the Navier–Stokes equations.

---

[16] In her above-cited papers, Ladyzhenskaya considers also cases where $h_1 = \mu_0 + \mu_1|\nabla\boldsymbol{v}|^{q-2}$ or $h_1 = \mu_0 + \mu_1|\nabla \times \boldsymbol{v}|^{q-2}$. We would like to emphasize that these choices make no sense from the physical point of view, since the corresponding constitutive equations would not obey the principle of material invariance that we mentioned before.
[17] Actually, for existence she only required $q \geq 12/5$.

Over the years, Ladyzhenskaya's theorems were improved and/or extended in several directions by different authors, including J.-L. Lions, J. Nečas, J. Frehse and their collaborators, either with the objective of lowering the exponent q, or even by considering more general hypothesis on the function h_1 than those expressed by (2.7). In particular, great effort has been directed to prove well-posedness results for models where the constant μ_0 in the constitutive assumption (2.7) is zero, that is, for models such as (2.5). In this respect, we wish to emphasize one more time that, even though very challenging from a mathematical viewpoint, the power-law model (2.5) is physically not realistic for all values of $|\boldsymbol{D}|$.

In the following sections we shall present some basic results concerning the well-posedness of the boundary-value problem associated to certain generalized Newtonian models, in bounded domains as well as in a piping system.

For a complete and update list of results concerning the well-posedness of the *initial-boundary value problem*, we refer the interested reader to the recent paper of J. Wolf [115] and to the monograph [79].

Concerning regularity issues (a topic that is still unsettled, in many respects, and currently under deep investigation) we refer to the papers of C. Ebmeyer [30] and to the more recent ones of H. Beirão da Veiga [8, 9, 10, 11].

Finally, we would like to indicate a number of problems that, seemingly, have received little or no attention and that are, nevertheless, of great significance in several applications and, in particular, in blood flow problems.

(i) Existence, uniqueness and attainability of time-periodic motions in an unbounded piping system subject to a given time-periodic flow-rate (see Sections 1.3.2 and 1.3.3, for the analogous problem for a Navier–Stokes liquid).

(ii) Formulation of appropriate boundary conditions for a "truncated" bounded piping system and study of the corresponding well-posedness (see Section I.4, for the analogous problem for a Navier–Stokes liquid).

(iii) Well-posedness of the exterior boundary-value problem.

2.2.1. Boundary-value problem in bounded domain. In this and the next sections, we shall be interested in the unique solvability of the boundary-value problem

$$
\left.
\begin{aligned}
\nabla \cdot (h_1(|\boldsymbol{D}(\boldsymbol{v})|)\boldsymbol{D}(\boldsymbol{v})) &= \boldsymbol{v} \cdot \nabla\boldsymbol{v} + \nabla p + \boldsymbol{f} \\
\nabla \cdot \boldsymbol{v} &= 0
\end{aligned}
\right\} \quad \text{in } \Omega, \tag{2.8}
$$

$$
\boldsymbol{v}|_{\partial\Omega} = \boldsymbol{v}_*,
$$

where $\boldsymbol{f}$ and $\boldsymbol{v}_*$ are prescribed functions, Ω is a domain of $\mathbb{R}^3$, either bounded or unbounded, and, we recall, $\boldsymbol{D} = \boldsymbol{D}(\boldsymbol{w})$ is the symmetric part of $\nabla\boldsymbol{w}$.

We will assume that h_1 is of the form

$$
h_1 = \kappa_0 + h(|\boldsymbol{D}|), \tag{2.9}
$$

where h is a continuous scalar function defined on the linear subspace of $\mathbb{R}^{3\times3}$, $\mathcal{T}$, of second-order tensors, satisfying the following properties.

(i) *Coercivity*:
$$h(|\boldsymbol{A}|)\boldsymbol{A} : \boldsymbol{A} \geq \kappa_1 \boldsymbol{A}^q - \kappa_2 \,, \quad q > 1 \,; \tag{2.10}$$

(ii) *Growth*:
$$|h(|\boldsymbol{A}|)\boldsymbol{A}| \leq \kappa_3 \left(|\boldsymbol{A}|^{q-1} + 1 \right) \,, \quad q > 1 \,; \tag{2.11}$$

(iii) *Monotonicity*:
$$(h(|\boldsymbol{A}|)\boldsymbol{A} - h(|\boldsymbol{B}|)\boldsymbol{B}) : (\boldsymbol{A} - \boldsymbol{B}) > 0 \,, \text{ if } \boldsymbol{A} \neq \boldsymbol{B} \,. \tag{2.12}$$

Here, $\boldsymbol{A}$ and $\boldsymbol{B}$ are arbitrary elements from $\mathcal{T}$, and κ_i, $i = 0, 1, 2, 3$ are constants, depending at most on q, such that $\kappa_i > 0$, $i = 1, 3$, $\kappa_0 \geq 0$ and $\kappa_2 \in \mathbb{R}$.

Remark 2.1. The constitutive laws (2.5)–(2.7) satisfy (2.9)–(2.11). More specifically, the power-law (2.5) and the Ladyzhenskaya "model" (2.7) satisfy (2.9) with $\kappa_0 = 0$ and $\kappa_0 \equiv \mu_0$, respectively. Moreover, they both satisfy (2.10)–(2.11) with $\kappa_1 = \kappa_3 \equiv \mu_1$, $\kappa_2 = 0$. As far as the Carreau law (2.6) is concerned, it clearly satisfies (2.9) with $\kappa_0 \equiv \mu_0$. Furthermore, it is easy to show that it satisfies (2.10) [18], whereas the validity of (2.11) is trivially established if $q > 2$, and becomes a consequence of [96, Lemme 3.1, Eqs. (3.2) and (3.5)], if $1 < q < 2$. Finally, it can be shown that (2.5)–(2.7) also satisfy (2.12); see [96, Lemme 3.1].

We shall begin to furnish some existence and uniqueness results for the problem (2.8)–(2.12), when Ω is *bounded*. To this end, we shall follow some arguments due, basically, to O.A. Ladyzhenskaya [67, 68, 69] and J.-L. Lions [76, Chaptre 2, §5].

We begin to put (2.8) in a weak form. If we dot-multiply both sides of (2.8)$_1$ by $\boldsymbol{\varphi} \in \mathcal{D}(\Omega)$ and then integrate by parts over Ω, we have
$$(\kappa_0 \boldsymbol{D}(\boldsymbol{v}) + h(\boldsymbol{D}(\boldsymbol{v}))\boldsymbol{D}(\boldsymbol{v}), \boldsymbol{D}(\boldsymbol{\varphi})) = (\boldsymbol{v}\cdot\nabla\boldsymbol{\varphi}, \boldsymbol{v}) - (\boldsymbol{f}, \boldsymbol{\varphi}), \quad \text{for all } \boldsymbol{\varphi} \in \mathcal{D}(\Omega)\,. \tag{2.13}$$
Next, we set
$$\mathcal{V}^q = \mathcal{V}^q(\Omega) := \begin{cases} \mathcal{D}_0^{1,q}(\Omega) \cap \mathcal{D}_0^{1,2}(\Omega) & \text{if } \kappa_0 > 0\,, \\[2mm] \mathcal{D}_0^{1,q}(\Omega) & \text{if } \kappa_0 = 0\,, \end{cases}$$
$$\mathcal{W}^q = \mathcal{W}^q(\Omega) := \begin{cases} W^{1,q}(\Omega) \cap W^{1,2}(\Omega) & \text{if } \kappa_0 > 0\,, \\[2mm] W^{1,q}(\Omega) & \text{if } \kappa_0 = 0\,, \end{cases}$$
and define $\boldsymbol{v}$ to be a *weak solution* to the problem (2.8)–(2.12) if $\boldsymbol{v} = \boldsymbol{u} + \boldsymbol{V}$, where $\boldsymbol{u} \in \mathcal{V}^q(\Omega)$, and $\boldsymbol{V} \in \mathcal{W}^q(\Omega)$ with $\nabla \cdot \boldsymbol{V} = 0$ in Ω, and $\boldsymbol{V} = \boldsymbol{v}_*$ at $\partial\Omega$ (in the trace sense). [19]

---

[18] The case $q \geq 2$ is obvious. For the case $q < 2$, it is enough to show that there exist constants $c_i = c_i(q) > 0$, $i = 1, 2$, such that
$$\frac{x^2}{(1+x^2)^{(2-q)/2}} \geq c_1 x^q - c_2\,, \quad x \geq 0\,.$$

Now, if $1 \leq x$, the inequality follows at once with $c_1 = 2^{(q-2)/2}$ and arbitrary $c_2 \geq 0$. If $x < 1$, then the inequality again follows with $c_1 = c_2 = 1$. Thus, in conclusion, (2.6) satisfies (2.10) with $\kappa_1 = 2^{(q-2)/2}$ and $\kappa_3 = 1$.

[19] If $\kappa_0 = 0$, the above definition of weak solution is meaningful if $q \geq 6/5$; see Remark 2.1.

2.2.1(a) Existence results with $v_* \equiv 0$. In this section we shall prove the existence of a weak solution to (2.8)–(2.12) in the case when $v_* = 0$, referring to the next subsection for the case $v_* \not\equiv 0$. We shall achieve this goal under the hypothesis of the validity of the physically more realistic case $\kappa_0 > 0$ in (2.9). The case $\kappa_0 = 0$ which gives, in fact, more restrictive and less complete results, will be discussed and presented in Remark 2.1.

Specifically, we have the following.

Theorem 2.1. *Let Ω be a Lipschitz, bounded domain of $\mathbb{R}^3$. Assume that $f \in \mathcal{D}_0^{-1,q'}(\Omega) \cap \mathcal{D}_0^{-1,2}(\Omega)$, $q > 1$, and that h_1 satisfies (2.9)–(2.12) with $\kappa_0 > 0$. Then, for any given $q > 1$, problem (2.8)–(2.12), with $v_* \equiv 0$, has at least one corresponding weak solution $v \in \mathcal{D}_0^{1,2}(\Omega) \cap \mathcal{D}_0^{1,q}(\Omega)$.*

Proof. Let $\{\psi_k\} \subset \mathcal{D}(\Omega)$ be a basis in $\mathcal{D}_0^{1,2}(\Omega)$ whose linear hull can approximate any $\varphi \in \mathcal{D}(\Omega)$ in the $C^1(\overline{\Omega})$-norm. [20]. We then look for solutions to (2.8) of the form

$$v_m = \sum_{k=1}^{m} c_{km} \psi_k$$

$$(h_1(|D(v_m)|)D(v_m), D(\psi_k)) = (v_m \cdot \nabla \psi_k, v_m) - (f, \psi_k), \quad k = 1, \ldots, m.$$
(2.14)

Multiplying through both sides of $(2.14)_2$ by c_{km}, summing over k from 1 to m we get

$$\kappa_0 \|D(v_m)\|_2^2 + (h(D(v_m))D(v_m), D(v_m)) = -(f, v_m).$$
(2.15)

Thus, from (2.15) and (2.10) we deduce

$$\kappa_0 \|D(v_m)\|_2^2 + \kappa_1 \|D(v_m)\|_q^q \leq -(f, v_m) + \kappa_2 |\Omega|.$$
(2.16)

We then use the Korn inequality

$$\|\nabla v\|_r \leq C_1 \|D(v)\|_r, \quad \text{for all } v \in W_0^{1,r}(\Omega), \quad r > 1, \quad C_1 = C_1(\Omega, r) > 0, \quad (2.17)$$

into (2.16), to find that

$$\|\nabla v_m\|_2 \leq C_2 \left(\|f\|_{-1,2} + 1\right),$$

$$\|\nabla v_m\|_q \leq C_3 \left(\|f\|_{-1,q'}^{\frac{1}{q-1}} + 1\right),$$
(2.18)

where $C_2 = C_2(\kappa_2, \Omega) > 0$ and $C_3 = C_3(\kappa_2, \Omega, q) > 0$. The inequalities in (2.18) allow us, on the one hand, to prove that (2.14) has at least one solution $(c_{1m}, \ldots, c_{mm})$ for all $m \in \mathbb{N}$ (see [76]), and, on the other hand, to find a field v

---

[20] The sequence $\{\psi_k\}$ satisfying the mentioned properties can be easily constructed by the methods used in [36, Lemma VII.2.1].

and a subsequence, that we continue to denote by $\{\boldsymbol{v}_m\}$, such that

$$\boldsymbol{v}_m \to \boldsymbol{v} \quad \text{weakly in } W_0^{1,q}(\Omega) \text{ and } W_0^{1,2}(\Omega)\,,$$

$$\boldsymbol{v}_m \to \boldsymbol{v} \begin{cases} \text{strongly in } L^r(\Omega)\,, \text{ all } r \in [1, \dfrac{3q}{3-q})\,, \text{ if } q < 3\,, \\[2mm] \text{strongly in } L^r(\Omega)\,, \text{ all } r \in [1, \infty)\,, \text{ if } q = 3\,, \\[2mm] \text{strongly in } C^{0,\lambda}(\Omega)\,, \text{ all } \lambda \in (0, 1 - 3/q)\,, \text{ if } q > 3. \end{cases} \quad (2.19)$$

We shall now pass to the limit $m \to \infty$ in $(2.14)_2$ and, by using the information (2.19), will show that we can replace in it $\boldsymbol{v}_m$ with $\boldsymbol{v}$. It is simple to prove that, for any fixed k,

$$(\boldsymbol{v}_m \cdot \nabla \boldsymbol{\psi}_k, \boldsymbol{v}_m) \to (\boldsymbol{v} \cdot \nabla \boldsymbol{\psi}_k, \boldsymbol{v})\,. \quad (2.20)$$

In fact, by the Poincaré inequality, we have

$$\|\boldsymbol{v}_m\|_2 \le C_4 \|\nabla \boldsymbol{v}_m\|_2 \le C_4(\|\boldsymbol{f}\|_{-1,2} + 1)\,,$$

where $C_4 = C_4(\kappa_2, \Omega) > 0$. Moreover, by (2.19), $\boldsymbol{v}_m(x) \to \boldsymbol{v}(x)$ for almost all $x \in \Omega$, possibly along a subsequence. Thus, from [76, Lemme I.1.3] it follows that $v_{mi}v_{mj}$ converges weakly to $v_i v_j$, $i, j = 1, 2, 3$. Since $\nabla \boldsymbol{\psi}_k \in \mathcal{D}(\Omega)$, (2.20) follows. We shall next consider the convergence of the term on the left-hand side of (2.14). From (2.11) and (2.18) it follows that the sequence $\{h(|\boldsymbol{D}(\boldsymbol{v}_m)|)\boldsymbol{D}(\boldsymbol{v}_m)\}$ is bounded in $L^{q'}(\Omega)$, and so there exists a subsequence, still denoted by $\{\boldsymbol{v}_m\}$, and an element $\boldsymbol{G} \in L^{q'}(\Omega)$ such that

$$h(|\boldsymbol{D}(\boldsymbol{v}_m)|)\boldsymbol{D}(\boldsymbol{v}_m) \to \boldsymbol{G} \quad \text{weakly in } L^{q'}(\Omega)\,. \quad (2.21)$$

Therefore, from (2.14), $(2.19)_1$, (2.20) and (2.21) we conclude that

$$\kappa_0(\boldsymbol{D}(\boldsymbol{v}), \boldsymbol{D}(\boldsymbol{\psi}_k)) + (\boldsymbol{G}, \boldsymbol{D}(\boldsymbol{\psi}_k)) = (\boldsymbol{v} \cdot \nabla \boldsymbol{\psi}_k, \boldsymbol{v}) - (\boldsymbol{f}, \boldsymbol{\psi}_k)\,, \quad k = 1, \dots, m\,. \quad (2.22)$$

In order to show the existence of a weak solution it is enough to prove that

$$(\boldsymbol{G}, \boldsymbol{D}(\boldsymbol{\varphi})) = (h(|\boldsymbol{D}(\boldsymbol{v})|)\boldsymbol{D}(\boldsymbol{v}), \boldsymbol{D}(\boldsymbol{\varphi}))\,, \quad \text{for all } \boldsymbol{\varphi} \in \mathcal{D}(\Omega)\,. \quad (2.23)$$

In fact, if (2.23) holds, then , by (2.22), we obtain that (2.13) is satisfied for $\boldsymbol{\varphi} \equiv \boldsymbol{\psi}_k$, for all $k \in \mathbb{N}$ and, therefore, by the properties of $\{\boldsymbol{\psi}_k\}$, for all $\boldsymbol{\varphi} \in \mathcal{D}(\Omega)$. The proof of (2.23), will be achieved through a procedure often referred to as "Minty–Browder trick" [82], [18]. Multiplying both sides of (2.22) by c_{km} and summing over k from 1 to m we find

$$\kappa_0(\boldsymbol{D}(\boldsymbol{v}), \boldsymbol{D}(\boldsymbol{v}_m)) + (\boldsymbol{G}, \boldsymbol{D}(\boldsymbol{v}_m)) = (\boldsymbol{v} \cdot \nabla \boldsymbol{v}_m, \boldsymbol{v}) - (\boldsymbol{f}, \boldsymbol{v}_m)\,. \quad (2.24)$$

We wish to pass to the limit $m \to \infty$ into this relation. Given the convergence properties (2.19) and the fact that $\boldsymbol{G} \in L^{q'}(\Omega)$, the left-hand side and the second term on the right-hand side converge to $\kappa_0 \|\boldsymbol{D}(\boldsymbol{v})\|_2^2 + (\boldsymbol{G}, \boldsymbol{D}(\boldsymbol{v}))$ and $(\boldsymbol{f}, \boldsymbol{v})$, respectively. Concerning the convergence of the nonlinear term, we observe that, since $\boldsymbol{v} \in \mathcal{D}_0^{1,2}(\Omega)$, by the Sobolev embedding theorem we have $\boldsymbol{v} \in L^4(\Omega)$ and so, from this fact and from the weak convergence of ∇v_m to ∇v in $L^2(\Omega)$ (see $(2.19)_1$) we obtain

$$\lim_{m \to \infty} (v \cdot \nabla v_m, v) = (v \cdot \nabla v, v). \tag{2.25}$$

Now, again recalling that $v \in \mathcal{D}_0^{1,2}(\Omega)$, by a standard density argument, we can easily show that $(v \cdot \nabla v, v) = 0$. Consequently, from this observation and from (2.25), passing to the limit $m \to \infty$ in (2.24), we finally conclude

$$\kappa_0 \|D(v)\|_2^2 + (G, D(v)) = -(f, v). \tag{2.26}$$

We next observe that, setting

$$\sigma(v) := h(|D(v)|)D(v),$$

by $(2.14)_2$ it easily follows that

$$\kappa_0 \|D(v_m)\|_2^2 + (\sigma(v_m), D(v_m)) = -(f, v_m),$$

so that, letting $m \to \infty$ in this relation and using (2.26), we have

$$\lim_{m \to \infty} \left[\kappa_0 \|D(v_m)\|_2^2 + (\sigma(v_m), D(v_m)) \right] = \kappa_0 \|D(v)\|_2^2 + (G, D(v)). \tag{2.27}$$

Furthermore, by $(2.12)_2$, we find

$$(\kappa_0(D(v_m) - D(\Psi)) + (\sigma(v_m) - \sigma(\Psi)), D(v_m) - D(\Psi)) \geq 0,$$

for all $\Psi \in \mathcal{D}_0^{1,q}(\Omega) \cap \mathcal{D}_0^{1,2}(\Omega)$. Thus, passing to the limit $m \to \infty$ into this latter relation and using (2.27) we infer that

$$(\kappa_0(D(v) - D(\Psi)) + (G - \sigma(\Psi)), D(v) - D(\Psi)) \geq 0$$
$$\text{for all } \Psi \in \mathcal{D}_0^{1,q}(\Omega) \cap \mathcal{D}_0^{1,2}(\Omega). \tag{2.28}$$

If we now choose $\Psi = v - \varepsilon\varphi$, $\varepsilon > 0$, $\varphi \in \mathcal{D}(\Omega)$, from (2.28) we get

$$(\varepsilon\kappa_0 D(\varphi) + (G - \sigma(v - \varepsilon\varphi)), D(\varphi)) \geq 0 \quad \text{for all } \varphi \in \mathcal{D}(\Omega).$$

We now let $\varepsilon \to 0$ into this relation and use the continuity property of $h(D)$, the growth condition (2.11) and the Lebesgue dominated convergence theorem to show that

$$(G - \sigma(v), D(\varphi)) \geq 0 \quad \text{for all } \varphi \in \mathcal{D}(\Omega).$$

Repeating the above procedure with $-\varphi$ in place of φ, we then arrive at (2.23), which concludes the proof of the theorem. $\qquad \square$

Remark 2.2. If we assume that the material constant κ_0 in the constitutive hypothesis (2.13) vanishes (as in the case of the "power-law" model (2.5)), then the above argument furnishes existence of weak solutions provided we impose the restriction $q \geq 9/5$. Actually, if $\kappa_0 = 0$, the very definition of weak solution requires $q \geq 6/5$. In fact, taking into account that, this time, $v \in \mathcal{D}_0^{1,q}(\Omega)$ only, for the nonlinear term on the right-hand side of (2.13) to be convergent we need $v \in L^2(\Omega)$ and this, by the Sobolev embedding theorem, is ensured provided $q \geq 6/5$. Now, assuming $\boldsymbol{f} \in \mathcal{D}_0^{-1,q'}(\Omega)$, from (2.16) with $\kappa_0 = 0$, and (2.17), instead of (2.18) we only obtain the *weaker* estimate

$$\|\nabla \boldsymbol{v}_m\|_q \le C\left(\|\boldsymbol{f}\|_{-1,q'}^{\frac{1}{q'-1}} + 1\right), \tag{2.29}$$

where $C = C(q,\Omega,\kappa_1,\kappa_2) > 0$. The proof of the existence of weak solutions proceeds along the same lines of the proof when $\kappa_0 > 0$ without conceptual changes, till the passage to the limit given in (2.25). In fact, in order that this relation holds, we need a further restriction from below on the exponent q. This because for the convergence of the nonlinear term in (2.25) we need $\boldsymbol{v} \in L^{2q'}(\Omega)$, and since $\boldsymbol{v} \in \mathcal{D}_0^{1,q}(\Omega)$, by the Sobolev embedding theorem, this happens if $q \ge 9/5$. The rest of the proof remains virtually unchanged and so we may conclude that weak solutions exist also when $\kappa_0 = 0$, provided the material parameter q is not less than $9/5$. The case when $\kappa_0 = 0$ and $q \in (6/5, 9/5)$ has been considered by M. Růžička [95], and by J. Frehse et al. in [33, 34]. In particular, by using a completely different approach than the one presented here, these latter authors show that a weak solution does exist for all given q in the above range. Their method avoids the use of the Minty–Browder trick (which essentially requires the restriction $q \ge 9/5$) and replaces it by showing the existence of a suitable approximating sequence of solutions $\{\boldsymbol{v}_m\}$ (different than the Galerkin approximations) with the properties that $\nabla \boldsymbol{v}_m(x) \to \nabla \boldsymbol{v}(x)$ for a.a. x belonging to the compact subsets of Ω. For further and full details, we refer the reader to the paper [34].

The next question that we would like to address concerns the existence of a suitable pressure field associated to the weak solutions to problem (2.8)–(2.12). Actually, from the summability properties of the weak solution and from [36, Corollary III.5.1] we easily obtain that if Ω is Lipschitz and $\kappa_0 > 0$, there exists a scalar field $p \in L^s(\Omega)$, with $s = 2$ if $q < 2$, and $s = q'$ if $q > 2$, such that

$$(\kappa_0 \boldsymbol{D}(\boldsymbol{v}) + h(|\boldsymbol{D}(\boldsymbol{v})|)\boldsymbol{D}(\boldsymbol{v}), \boldsymbol{D}(\boldsymbol{\psi})) = (\boldsymbol{v} \cdot \nabla\boldsymbol{\psi}, \boldsymbol{v}) - (\boldsymbol{f}, \boldsymbol{\psi}) + (p, \nabla \cdot \boldsymbol{\psi})$$
$$\text{for all } \boldsymbol{\psi} \in C_0^\infty(\Omega). \tag{2.30}$$

However, following a procedure introduced in [33], we can give more detailed regularity properties of the pressure, provided we assume Ω more regular (of class C^2, for instance). Consider the Stokes problems

$$(\nabla \boldsymbol{w}_1, \nabla \boldsymbol{\varphi}) = (h(|\boldsymbol{D}(\boldsymbol{v})|)\boldsymbol{D}(\boldsymbol{v}), \nabla\boldsymbol{\varphi}), \ \boldsymbol{w}_1 \in \mathcal{D}_0^{1,q'}(\Omega),$$
$$(\nabla \boldsymbol{w}_2, \nabla \boldsymbol{\varphi}) = \kappa_0(\boldsymbol{D}(\boldsymbol{v}), \nabla\boldsymbol{\varphi}), \ \boldsymbol{w}_2 \in \mathcal{D}_0^{1,2}(\Omega), \tag{2.31}$$
$$(\nabla \boldsymbol{w}_3, \nabla \boldsymbol{\varphi}) = -(\boldsymbol{v} \cdot \nabla\boldsymbol{\varphi}, \boldsymbol{v}), \ \boldsymbol{w}_3 \in \mathcal{D}_0^{1,6}(\Omega) \cap W^{2,3/2}(\Omega), \ \text{all } r > 1,$$

where $\boldsymbol{\varphi}$ is an arbitrary element from $\mathcal{D}(\Omega)$ and $\boldsymbol{v}$ is the weak solution determined in Theorem 2.1. In view of the properties of the weak solution and of the embedding

$W^{1,2}(\Omega) \subset L^6(\Omega)$, we know that the fields $\boldsymbol{w}_i$, $i = 1, 2, 3$, exist; see [36, Chapter IV]. Moreover, we know that there are $p_1 \in L^{q'}(\Omega)$, $p_2 \in L^2(\Omega)$ and $p_3 \in W^{1,3/2}(\Omega)$ such that

$$(\nabla \boldsymbol{w}_1, \nabla \boldsymbol{\psi}) = (h(|\boldsymbol{D}(\boldsymbol{v})|)\boldsymbol{D}(\boldsymbol{v}), \nabla \boldsymbol{\psi}) - (p_1, \nabla \cdot \boldsymbol{\psi}),$$

$$(\nabla \boldsymbol{w}_2, \nabla \boldsymbol{\psi}) = \kappa_0(\boldsymbol{D}(\boldsymbol{v}), \nabla \boldsymbol{\psi}) - (p_2, \nabla \cdot \boldsymbol{\psi}), \qquad (2.32)$$

$$(\nabla \boldsymbol{w}_3, \nabla \boldsymbol{\psi}) = -(\boldsymbol{v} \cdot \nabla \boldsymbol{\psi}, \boldsymbol{v}) - (p_3, \nabla \cdot \boldsymbol{\psi}),$$

for all $\boldsymbol{\psi} \in C_0^\infty(\Omega)$. Moreover, adding the three equations in (2.31) side by side and taking into account that $\boldsymbol{v}$ satisfies (2.13), we obtain

$$(\nabla(\boldsymbol{w}_1 + \boldsymbol{w}_2 + \boldsymbol{w}_3), \nabla \boldsymbol{\varphi}) = -(\boldsymbol{f}, \boldsymbol{\varphi}).$$

If we assume, for example, $\boldsymbol{f} \in D_0^{1,2}(\Omega)$, then, again by known results on the Stokes problem (see [36, Theorem IV.1.1 and Theorem IV.6.1]), it follows that there exists $p^\star \in L^2(\Omega)$ such that

$$(\nabla(\boldsymbol{w}_1 + \boldsymbol{w}_2 + \boldsymbol{w}_3), \nabla \boldsymbol{\psi}) = -(\boldsymbol{f}, \boldsymbol{\psi}) + (p^\star, \nabla \cdot \boldsymbol{\psi}), \qquad (2.33)$$

for all $\boldsymbol{\psi} \in C_0^\infty(\Omega)$. Therefore, setting $p := p^\star + \sum_{i=1}^3 p_i$ from (2.32)–(2.33) we get that p satisfies (2.30). Summarizing, we have the following.

Theorem 2.2. *Let $\boldsymbol{v}$ be a weak solution to (2.8)–(2.12) with $\kappa_0 > 0$. Assume that Ω is of class C^2 and that $\boldsymbol{f} \in D_0^{-1,2}(\Omega)$. Then, there exists a scalar field p such that* [21]

$$p = p_1 + p_2 + p_3,$$

where $p_1 \in L^{q'}(\Omega)$, $p_2 \in L^2(\Omega)$ and $p_3 \in W^{1,3/2}(\Omega)$, and such that $(\boldsymbol{v}, p)$ satisfy (2.30). In particular, p has the following global summability properties: $p \in L^2(\Omega)$, if $q < 2$, while $p \in L^{q'}(\Omega)$, if $q > 2$. For this latter to hold it is sufficient to assume Ω Lipschitz.

Remark 2.3. If $\kappa_0 = 0$ (see Remark 2.1), we have, as expected, that p may have different regularity properties than in the case $\kappa_0 > 0$. Actually, under the assumption $\boldsymbol{f} \in D_0^{-1,q'}(\Omega)$ and that Ω is of class C^2, by the same argument used previously we show that $p = \tilde{p}_1 + \tilde{p}_2$, with $\tilde{p}_1 \in L^{q'}(\Omega)$ and $\tilde{p}_2 \in W^{1,s}(\Omega)$, where $s = 3q/(6-q)$ if $q < 3$, any $s < q$ if $q = 3$, and $s = q$ if $q > 3$. Therefore, in particular, by the Sobolev embedding theorem, $p \in L^{q'}(\Omega)$ if $q \geq 9/4$ and $p \in L^{3q/(6-q)}$ if $q \in [6/5, 9/4)$. This latter property can be directly proved by using [36, Corollary II.5.1], in which case it is sufficient to require Ω only Lipschitzian.

---

[21] In case $q > 2$, the part p_3 of the pressure can be shown to be even more regular. We leave the task of finding this extra regularity to the interested reader (see also Remark 2.2).

2.2.1(b) Existence results with $v_* \not\equiv 0$. Our next objective is the study of problem (2.8)–(2.12) in the case when $v_* \not\equiv 0$. Before doing this, we would like to recall that, in the Newtonian case, that is, in the case of the Navier–Stokes equations, the existence is known provided the datum v_* (belongs to a suitable function class and) satisfies the conditions

$$\int_{\Gamma_i} v_* \cdot n = 0, \quad i = 1, \ldots, N, \tag{2.34}$$

where Γ_i are the N connected components of the boundary $\partial\Omega$ of the region of flow [37, Theorem VII.4.1]. However, the necessary condition imposed on the datum v_* by the incompressibility property $\nabla \cdot v = 0$, only requires that

$$\int_{\partial\Omega} v_* \cdot n = 0. \tag{2.35}$$

Clearly, (2.34) is more restrictive than (2.35), unless $N = 1$. It is an outstanding open question in the mathematical theory of the Navier–Stokes equations to prove of disprove existence of solutions to the non-homogeneous boundary-value problem when only the natural compatibility condition (2.35) is satisfied. Nevertheless, if the liquid is non-Newtonian and obeys the constitutive law (2.4), (2.9)–(2.12), we shall show that the corresponding boundary-value problem (2.8) is solvable under the assumption (2.35), provided $q > 2$. This latter means, roughly speaking, that the liquid has to be "just a little bit" shear-thickening, that is, it is enough that its viscosity increases "only slightly" with increasing shear rate. It is interesting to observe that this result holds regardless of whether κ_0 is zero or not zero. If $q < 2$ and $\kappa_0 > 0$, it can be shown that existence holds under the same conditions of a Navier–Stokes liquid, that is, under the more restrictive assumption (2.34). Finally, if $q < 2$ and $\kappa = 0$, the situation is even worse and existence is known only if v_* (satisfies (2.35) and) is small in a suitable norm [16].

In what follows, we shall limit ourselves to prove existence of weak solutions to (2.8)–(2.12) when $q > 2$. More precisely, we wish to find a vector field $v \in W^{1,q}(\Omega)$, $q > 2$, with $\nabla \cdot v = 0$ in Ω, $v = v_*$ at $\partial\Omega$, and satisfying (2.13). To this end, assume that v_* is prescribed in $W^{1-1/q,q}(\partial\Omega)$ and satisfies condition (2.35). It is then known that, if Ω is Lipschitz, we can find a vector field $V \in W^{1,q}(\Omega)$ such that $\nabla \cdot V = 0$ in Ω and $V = v_*$ (in the trace sense) at $\partial\Omega$; see [36, Exercise III.3.4]. We then look for solutions v to (2.13) in the form $v := u + V$ where $u \in \mathcal{D}_0^{1,q}(\Omega)$. Then, as in the case $v_* = 0$, we use the Galerkin method to construct approximating solutions to (2.13) (see (2.14)), where now

$$v_m = V + \sum_{k=1}^{m} c_{km}\psi_k =: V + u_m, \tag{2.36}$$

$$(h_1(|D(v_m)|)D(v_m), D(\psi_k)) = (v_m \cdot \nabla\psi_k, v_m) - (f, \psi_k), \quad k = 1, \ldots, m.$$

The existence proof will then be identical to that of Theorem 2.1 provided we show that the approximating solutions satisfy a suitable uniform estimate analogous to

(2.18). In order to prove this latter, we multiply $(2.36)_2$ by c_{km} and sum over k from 1 to m. We thus get (the subscript "m" is, for simplicity, omitted)

$$\kappa_0(\boldsymbol{D}(\boldsymbol{v}), \boldsymbol{D}(\boldsymbol{u})) + (h(\boldsymbol{D}(\boldsymbol{v}))\boldsymbol{D}(\boldsymbol{v}), \boldsymbol{D}(\boldsymbol{u})) = (\boldsymbol{v} \cdot \nabla \boldsymbol{u}, \boldsymbol{v}) - (\boldsymbol{f}, \boldsymbol{u}). \tag{2.37}$$

Now, from $(2.36)_1$, (2.10)–(2.11) and from the Hölder inequality it follows that

$$\kappa_0(\boldsymbol{D}(\boldsymbol{v}), \boldsymbol{D}(\boldsymbol{u})) + (h(\boldsymbol{D}(\boldsymbol{v}))\boldsymbol{D}(\boldsymbol{v}), \boldsymbol{D}(\boldsymbol{u})) \geq \kappa_0\|\boldsymbol{D}(\boldsymbol{v})\|_2^2 + \kappa_1\|\boldsymbol{D}(\boldsymbol{v})\|_q^q$$

$$-\kappa_0(\boldsymbol{D}(\boldsymbol{v}), \boldsymbol{D}(\boldsymbol{V})) - (h(\boldsymbol{D}(\boldsymbol{v}))\boldsymbol{D}(\boldsymbol{v}), \boldsymbol{D}(\boldsymbol{V})) - \kappa_2|\Omega|$$

$$\geq \kappa_0\|\boldsymbol{D}(\boldsymbol{v})\|_2^2 + \kappa_1\|\boldsymbol{D}(\boldsymbol{v})\|_q^q$$

$$-\tfrac{1}{2}\kappa_0\|\boldsymbol{D}(\boldsymbol{v})\|_2^2 - \tfrac{1}{2}\kappa_1\|\boldsymbol{D}(\boldsymbol{v})\|_q^q - C_1(\mathcal{V}+1)$$

$$\geq \tfrac{1}{2}\kappa_0\|\boldsymbol{D}(\boldsymbol{v})\|_2^2 + \tfrac{1}{2}\kappa_1\|\boldsymbol{D}(\boldsymbol{v})\|_q^q - C_1(\mathcal{V}+1),$$

where $C_1 = C_1(q, \kappa_0, \kappa_1, \kappa_2, \kappa_3, \Omega)) > 0$, and

$$\mathcal{V} := \|\boldsymbol{D}(\boldsymbol{V})\|_2^2 + \|\boldsymbol{D}(\boldsymbol{V})\|_q^q. \tag{2.38}$$

Therefore, by the triangle inequality, we conclude

$$\kappa_0(\boldsymbol{D}(\boldsymbol{v}), \boldsymbol{D}(\boldsymbol{u})) + (h(\boldsymbol{D}(\boldsymbol{v}))\boldsymbol{D}(\boldsymbol{v}), \boldsymbol{D}(\boldsymbol{u}))$$
$$\geq \tfrac{1}{2}\kappa_0\|\boldsymbol{D}(\boldsymbol{u})\|_2^2 + \tfrac{1}{2}\kappa_1\|\boldsymbol{D}(\boldsymbol{u})\|_q^q - C_2(1+\mathcal{V}), \tag{2.39}$$

where $C_2 = C_2(q, \kappa_0, \kappa_1, \kappa_2, \kappa_3, \Omega) > 0$. Moreover, observing that $(\boldsymbol{v} \cdot \nabla \boldsymbol{u}, \boldsymbol{u}) = 0$, after an integration by parts, we find that

$$(\boldsymbol{v} \cdot \nabla \boldsymbol{u}, \boldsymbol{v}) = (\boldsymbol{V} \cdot \nabla \boldsymbol{u}, \boldsymbol{V}) - (\boldsymbol{u} \cdot \nabla \boldsymbol{V}, \boldsymbol{u}), \tag{2.40}$$

and so, by the Hölder inequality,

$$(\boldsymbol{v} \cdot \nabla \boldsymbol{u}, \boldsymbol{v}) \leq \|\nabla \boldsymbol{u}\|_q \|\boldsymbol{V}\|_{2q'}^2 + \|\boldsymbol{u}\|_{2q'}^2 \|\nabla \boldsymbol{V}\|_q.$$

Now, since $q > 2$, we have the embedding $W^{1,q}(\Omega) \subset L^{2q'}(\Omega)$. Therefore, using in this latter displayed relation Young's inequality (see (1.157) of Section 1) and Korn's inequality (2.17) we obtain

$$(\boldsymbol{v} \cdot \nabla \boldsymbol{u}, \boldsymbol{v}) \leq \tfrac{1}{4}\kappa_1\|\boldsymbol{D}(\boldsymbol{u})\|_q^q + C_3\|\boldsymbol{V}\|_{1,q}^{2q/(q-2)}, \tag{2.41}$$

where $C_3 = C_3(\kappa_0, \kappa_1, \Omega, q) > 0$. Finally, employing again Young's and Korn's inequalities, we increase the last term on the right-hand side of (2.37) as follows:

$$|(\boldsymbol{f}, \boldsymbol{u})| \leq \tfrac{1}{8}\kappa_1\|\boldsymbol{D}(\boldsymbol{u})\|_q^q + C_4\|\boldsymbol{f}\|_{-1,q'}^{q'}, \tag{2.42}$$

with $C_4 = C_4(\kappa_0, \kappa_1, \Omega, q) > 0$. Collecting (2.37)–(2.39) and (2.41)–(2.42) we then obtain the desired estimate

$$\|\boldsymbol{D}(\boldsymbol{u})\|_q^q \leq C_5\left(\|\boldsymbol{f}\|_{-1,q'}^{q'} + \|\boldsymbol{D}(\boldsymbol{V})\|_2^2 + \|\boldsymbol{D}(\boldsymbol{V})\|_q^q + \|\boldsymbol{V}\|_{1,q}^{2q/(q-2)} + 1\right), \tag{2.43}$$

with $C_5 = C_5(\kappa_0, \kappa_1, \kappa_2, \kappa_3, \Omega, q) > 0$, which in turn, by the triangle inequality and by Korn's inequality implies the following one on $\boldsymbol{v}$:

$$\|\nabla \boldsymbol{v}\|_q^q \leq C_6\left(\|\boldsymbol{f}\|_{-1,q'}^{q'} + \|\boldsymbol{D}(\boldsymbol{V})\|_2^2 + \|\boldsymbol{D}(\boldsymbol{V})\|_q^q + \|\boldsymbol{V}\|_{1,q}^{2q/(q-2)} + 1\right), \tag{2.44}$$

for another $C_6 = C_6(\kappa_0, \kappa_1, \kappa_2, \kappa_3, \Omega, q) > 0$.

Therefore, using (2.44) together with exactly the same procedure used in the proof of Theorem 2.1, we can show that from the sequence $\{\boldsymbol{v}_m\}$ we can select a subsequence $\{\boldsymbol{v}_k\}$ which, as $k \to \infty$, converges in the appropriate topologies to a solenoidal field $\boldsymbol{v} \in W^{1,q}(\Omega)$, $q > 2$, satisfying (2.13). Of course, by construction, $\boldsymbol{v} = \boldsymbol{v}_*$ at $\partial\Omega$.

We have thus proved the following result.

Theorem 2.3. *Let Ω be a Lipschitz, bounded domain of $\mathbb{R}^3$. Assume that $\boldsymbol{f} \in \mathcal{D}_0^{-1,q'}(\Omega)$, $q > 2$, that h_1 satisfies (2.9)–(2.12) with $\kappa_0 \geq 0$ and that $\boldsymbol{v}_* \in W^{1-1/q,q}(\partial\Omega)$ and satisfies the compatibility condition (2.35). Then, for any given $q > 2$, problem (2.8) has at least one corresponding weak solution. Specifically, there exists $\boldsymbol{v} \in W^{1,q}(\Omega)$ with $\nabla \cdot \boldsymbol{v} = 0$ in Ω, $\boldsymbol{v} = \boldsymbol{v}_*$ at $\partial\Omega$, and such that (2.13) is satisfied.*

Remark 2.4. We wish to emphasize the following two facts. In the first place, the assumption $q > 2$ is crucial for the estimate of the second term on the right-hand side of (2.40). In the second place, the condition $\kappa_0 > 0$ is not needed in deriving (2.43), which therefore continues to hold also for $\kappa_0 = 0$. Of course, since Ω is bounded, (2.43) furnishes an estimate also for the L^2-norm of $\boldsymbol{D}(\boldsymbol{u})$.

2.2.1(c) Uniqueness results. Our next task is the investigation of the uniqueness of weak solutions. Specifically, we shall prove a uniqueness result under the assumption that the function h_1 satisfies the conditions (2.9), (2.11) and (2.12), with $\kappa_0 > 0$. An analogous result when $\kappa_0 = 0$ will be discussed in Remark 2.4.

Theorem 2.4. *Let Ω be a Lipschitz, bounded domain of $\mathbb{R}^3$ and let $\boldsymbol{v}_1, \boldsymbol{v}_2$ be two weak solutions to problem (2.8), where h_1 satisfies (2.9), (2.11) and (2.12) with $\kappa_0 > 0$ and a given $q > 1$. Then, if*

$$\|\boldsymbol{v}_1\|_3 < \frac{\sqrt{3}\,\kappa_0}{2}\,,$$

necessarily $\boldsymbol{v}_1 = \boldsymbol{v}_2$.

Proof. Set $\boldsymbol{u} := \boldsymbol{v}_1 - \boldsymbol{v}_2$. From (2.13) we then find

$$\kappa_0(\boldsymbol{D}(\boldsymbol{u}), \boldsymbol{D}(\boldsymbol{\varphi})) + (h(\boldsymbol{D}(\boldsymbol{v}_1))\boldsymbol{D}(\boldsymbol{v}_1) - h(\boldsymbol{D}(\boldsymbol{v}_2))\boldsymbol{D}(\boldsymbol{v}_2), \boldsymbol{D}(\boldsymbol{\varphi}))$$

$$= (\boldsymbol{u} \cdot \nabla\boldsymbol{\varphi}, \boldsymbol{u}) + (\boldsymbol{u} \cdot \nabla\boldsymbol{\varphi}, \boldsymbol{v}_1) + (\boldsymbol{v}_1 \cdot \nabla\boldsymbol{\varphi}, \boldsymbol{u}) \qquad (2.45)$$

$$\text{for all } \boldsymbol{\varphi} \in \mathcal{D}(\Omega)\,.$$

Let $\{\boldsymbol{u}_k\}$ be a sequence from $\mathcal{D}(\Omega)$ converging to $\boldsymbol{u}$ in the norm of $\mathcal{D}_0^{1,2}(\Omega) \cap \mathcal{D}_0^{1,q}(\Omega)$. By [36, Theorem III.6.1], this sequence exists. We replace $\boldsymbol{u}_k$ for $\boldsymbol{\varphi}$ in (2.45) and pass to the limit $k \to \infty$ on both sides of the resulting equation. In view of the summability properties of weak solutions and of the assumption (2.11) made on h_1, it is easy to show the validity of the relation

$$\kappa_0\|\boldsymbol{D}(\boldsymbol{u})\|^2 + (h(\boldsymbol{D}(\boldsymbol{v}_1))\boldsymbol{D}(\boldsymbol{v}_1) - h(\boldsymbol{D}(\boldsymbol{v}_2))\boldsymbol{D}(\boldsymbol{v}_2), \boldsymbol{D}(\boldsymbol{u})) = (\boldsymbol{u} \cdot \nabla\boldsymbol{u}, \boldsymbol{v}_1)\,. \quad (2.46)$$

Now, by the Hölder inequality and by the Sobolev inequality

$$\|u\|_6 \leq \frac{2}{\sqrt{3}} \|\nabla u\|_2 \,,$$

see, e.g., [36, p. 31], we find

$$|(u \cdot \nabla u, v_1)| \leq \|u\|_6 \|\nabla u\|_2 \|v_1\|_3 \leq \frac{2}{\sqrt{3}} \|v_1\|_3 \|\nabla u\|_2^2 \,. \tag{2.47}$$

Therefore, inserting this latter inequality back into (2.46) and recalling the assumption (2.12), we find

$$(\kappa_0 - \frac{2}{\sqrt{3}} \|v_1\|_3) \|\nabla u\|_2^2 \leq 0 \,,$$

from which the result follows. □

Remark 2.5. If $\kappa_0 = 0$, uniqueness results can still be recovered, provided we strengthen the monotonicity assumption on the function h ($\equiv h_1$). For example, suppose that for all symmetric tensors A, B, the function h satisfies the property

$$(h(A)A - h(B)B) : (A - B) \geq \kappa_4 |A - B|^2 (|A| + |B|)^{q-2} \quad \text{if } q > 2 \tag{2.48}$$

where $\kappa_4 = \kappa_4(q) > 0$. Then, from (2.46) with $\kappa_0 = 0$, and from (2.48) and (2.17), we readily find that

$$\|\nabla u\|_2^2 \leq C(u \cdot \nabla u, v_1) \,,$$

where $C = C(\Omega, \kappa_4) > 0$, from which, with the help of (2.47), we again obtain uniqueness, provided the norm $\|v_1\|_3$ is sufficiently small. [22] It must be emphasized that all the constitutive relations (2.5)–(2.7) satisfy the property (2.48) for any $q > 2$; see [96, Lemme 3.1]. If $q \in (1, 2)$ it can be shown that the constitutive equations (2.5)-(2.7) satisfy, instead, the inequality

$$(h(A)A - h(B)B) : (A - B) \geq \kappa_5 \frac{|A - B|^2}{(\delta + |A|^{2-q} + |B|^{2-q})} \quad \text{if } q \in (1, 2) \,, \tag{2.49}$$

where $\kappa_5 = \kappa_5(q) > 0$ and $\delta \geq 0$; see [96, Lemme 3.1]. In such a circumstance the proof of uniqueness requires a little more work than the case $q > 2$. In addition, and most important, it shows a drawback that is not present in the case $q > 2$. Specifically, besides the condition $q \geq 9/5$, the known uniqueness results require that *both* solutions, v_1 and v_2, be sufficiently "small" in suitable norms (that is, we have just *local* uniqueness); see [16]. More precisely, given two weak solutions v_1 and v_2 to (2.8) with h_1 satisfying (2.9) with $\kappa_0 = 0$, (2.11) and (2.49) for some $q \in [9/5, 2)$, there exists a positive constant C depending only on Ω and on the material parameters, such that if $\|\nabla v_i\|_{1,q} < C$, $i = 1, 2$, then $v_1 = v_2$. Also in this case, Ω is requested to be Lipschitz. For the proof of this theorem we refer to [16]. To date, no uniqueness result for weak solutions is known if h ($\equiv h_1$) satisfies (2.11), (2.49) and $q \in [6/5, 9/5)$.

---

[22] Actually, in this case, we can also find uniqueness imposing restrictions on the L^s-norm of v_1, with $s < 3$. We leave the simple proof of this statement to the interested reader.

2.2.2. Boundary-value problem in unbounded piping system. In this section we shall investigate the existence and uniqueness of steady flow of generalized non-Newtonian liquids in an unbounded system of pipes with prescribed flow-rate, Φ. As in the previous section, we shall use the liquid model given by (2.4) where h_1 satisfies (2.9)–(2.12) with $\kappa_0 > 0$. The case $\kappa_0 = 0$ will be discussed in Remark 2.10.

For the sake of simplicity, we shall assume that the piping system consists of a "distorted" pipe $\Omega = \Omega_0 \cup \Omega_1 \cup \Omega_2$, where Ω_i, $i = 0, 1, 2$, are defined at the beginning of Section 1.3. [23] Therefore, the relevant boundary-value problem is described by the equations

$$\left. \begin{array}{l} \nabla \cdot (h_1(|\boldsymbol{D}(\boldsymbol{v})|)\boldsymbol{D}(\boldsymbol{v})) = \boldsymbol{v} \cdot \nabla \boldsymbol{v} + \nabla p \\[2mm] \nabla \cdot \boldsymbol{v} = 0 \end{array} \right\} \quad \text{in } \Omega,$$

$$\int_S \boldsymbol{v} \cdot \boldsymbol{n} \, dS = \Phi, \quad \boldsymbol{v}|_{\partial\Omega} = \boldsymbol{0}. \tag{2.50}$$

As in the Newtonian (Navier–Stokes) case, in order to solve this problem and to find the asymptotics of the velocity field, we shall introduce, for the case at hand, the analogue of Hagen–Poiseuille flow in an infinite straight pipe (see Section 1.1). This will be the object of the next section.

2.2.2(a) Fully developed steady-state flows. Let us consider a generalized Newtonian liquid (described by (2.4), (2.9)–(2.12)) in an infinite, straight pipe, Ω, of constant cross-section S (bounded domain of $\mathbb{R}^2$). In what follows, we suppose S of class C^2. Assuming x_1 parallel to the axis of the pipe, it is then easy to check that $(2.50)_{1,2,4}$ admits a solution of the type

$$\boldsymbol{v}_P(x) = U(x_2, x_3)\boldsymbol{e}_1, \quad p_P(x) = -\Gamma x_1, \tag{2.51}$$

with $\Gamma \in \mathbb{R}$, provided the pair (U, Γ) satisfies the boundary-value problem

$$\kappa_0 \Delta' U + \nabla' \cdot (h(|\nabla' U|)\nabla' U) = -\Gamma, \quad U(x_2, x_3)|_S = 0, \tag{2.52}$$

where the prime means differentiation only with respect to x_2 and x_3. In obtaining (2.52) we have noticed that, for the velocity field given in (2.51), the only nonzero components of $\boldsymbol{D}$ are $D_{12} = D_{21} = \dfrac{\partial U}{\partial x_2}$ and $D_{13} = D_{31} = \dfrac{\partial U}{\partial x_3}$, so that $2|\boldsymbol{D}| = |\nabla' U|$. [24] We recall that, from the physical point of view, Γ is the (constant) axial gradient of pressure. In analogy with the Newtonian case, we call the flow (2.51)–(2.52) the *Hagen–Poiseuille* or, more simply, the *Poiseuille* flow of the generalized Newtonian model.

If the constant Γ is prescribed, then it is easy to show the existence of a weak solution to (2.52). By this latter, we mean a scalar field U such that (primes

---

[23] Throughout this section, the notation will be the same as that introduced in Section 1.3.

[24] Rigorously, we should have used another symbol $\tilde{h}$, say, for the function h in (2.52) where $\tilde{h}(|\nabla' U|) = h(\frac{1}{2}|\nabla' U|)$. However, in order to avoid to introduce further notation, we set $\tilde{h} \equiv h$.

omitted)

$$\kappa_0 \left(\nabla U, \nabla \phi \right) + \left(h(|\nabla U|) \nabla U, \nabla \phi \right) = (\Gamma, \phi) \quad \text{for all } \phi \in V^q(S),$$ (2.53)

where

$$V^q = V^q(S) := \begin{cases} W_0^{1,q}(S) \cap W_0^{1,2}(S) & \text{if } \kappa_0 > 0, \\ W_0^{1,q}(S) & \text{if } \kappa_0 = 0. \end{cases}$$

In fact, by formally multiplying both sides of $(2.52)_1$ by U and by integrating by parts, in view of the properties of the function h we find (with $\| \cdot \|_{r,S} \equiv \| \cdot \|_r$)

$$\kappa_0 \|\nabla U\|_2 + \|\nabla U\|_q \leq C,$$ (2.54)

where $C = C(q, S, \Gamma) > 0$. Thus, using the Galerkin method along with exactly the same procedure adopted in the proof of Theorem 2.1, we can show that, for any given $q > 1$, there exists U satisfying (2.53)–(2.54).

If, however, the flow-rate,

$$\Phi = \int_S U(x_2, x_3) \, dS,$$ (2.55)

is prescribed, then the task is slightly less obvious, since, in this case, we have to find U *and* Γ satisfying (2.53), which thus becomes an *inverse problem*. To show existence in this case, we set

$$\widehat{C}_0^\infty(S) = \{\varphi \in C_0^\infty(S) : \int_S \varphi(x_2, x_3) \, dS = 0\},$$

and denote by $\widehat{W}_0^{1,r}(S)$, $1 < r < \infty$, the completion of $\widehat{C}_0^\infty(S)$ in the $W^{1,r}$-norm. We also set

$$\widehat{V}^q = \widehat{V}^q(S) := \begin{cases} \widehat{W}_0^{1,q}(S) \cap \widehat{W}_0^{1,2}(S) & \text{if } \kappa_0 > 0, \\ \widehat{W}_0^{1,q}(S) & \text{if } \kappa_0 = 0. \end{cases}$$

We then define the pair (U, Γ) to be a weak solution to (2.52) corresponding to the flow-rate $\Phi \in \mathbb{R}$, if $U \in V^q(S)$, and (a) U satisfies (2.55), and (b) the pair (U, Γ) obeys (2.53).

The following result holds.

Proposition 2.1. *Let h satisfy the assumptions (2.10)–(2.12) for some $q > 1$, and let $\kappa_0 \geq 0$. Then, for any $\Phi \in \mathbb{R}$ there exists one and only one weak solution (U, Γ) to (2.52). Assume, further, that either $q > 2$ or $\kappa_0 > 0$. Then the following two properties hold.*

 (i) *$U \in C^0(\overline{S})$ and there is a constant $C > 0$ depending at most on q, S, Φ and κ_0 such that*

$$\|U\|_{C^0(\overline{S})} \leq C.$$

(ii) *The map*

$$M : \Phi \in \mathbb{R} \mapsto (U, \Gamma) \in C^0(\overline{S}) \times \mathbb{R}$$

 is continuous.

Proof. Proving the existence of a weak solution, (U, Γ), to (2.52) is equivalent to proving the existence of $U \in V^q(S)$ satisfying (2.55) along with the condition

$$\kappa_0 \left(\nabla U, \nabla \varphi\right) + (h(|\nabla U|)\nabla U, \nabla \varphi) = 0 \quad \text{for all } \varphi \in \widehat{V}^q(S). \tag{2.56}$$

Actually, it is obvious that if (U, Γ) is a weak solution to (2.52), then U satisfies (2.56). Conversely, assume that $U \in V^q(\Omega)$ satisfies (2.56), and let ϕ be an arbitrary element of $V^q(S)$, and $\chi \in V^q(S)$ with $\int_S \chi \, dS = 1$. We then choose in (2.56)

$$\varphi = \phi - \chi \int_S \phi(x_2, x_3) \, dS \,.$$

Clearly, $\varphi \in \widehat{V}^q(S)$. Thus, setting

$$\boldsymbol{h} := \kappa_0 \nabla U + h(|\nabla U|)\nabla U \,, \tag{2.57}$$

we find

$$(\boldsymbol{h}, \nabla \phi) = (\Gamma, \phi) \,, \quad \text{for all } \phi \in V^q(S) \,,$$

where

$$\Gamma := \int_S \boldsymbol{h} \cdot \nabla \chi \, dS \,, \tag{2.58}$$

which shows that the pair (U, Γ) satisfies (2.53). Notice that the constant Γ does not depend on the particular choice of χ. In fact, if χ_1 is another function from $V^q(S)$ with $\int_S \chi \, dS = 1$, we deduce that $\chi - \chi_1 \in \widehat{V}^q(S)$ and so, by (2.56),

$$\int_S \boldsymbol{h} \cdot \nabla \chi \, dS = \int_S \boldsymbol{h} \cdot \nabla \chi_1 \, dS \,.$$

Furthermore, from (2.57)–(2.58) it follows that

$$|\Gamma| \leq C \left(\kappa_0 \|\nabla U\|_1 + \|h(|\nabla U|)\nabla U\|_1\right) \,, \tag{2.59}$$

with $C = C(S) > 0$. Now, the existence of a function $U \in V^q(S)$ satisfying (2.55) and (2.56) can be proved by the same arguments used in the proof of Theorem 2.1, provided we show an a-priori estimate for U, of the type given in (2.54). In order to obtain this latter, we look for a solution U of the form $U = u + a$, where $a = \Phi A$, and $A \in C_0^\infty(S)$, with

$$\int_S A(x_2, x_3) \, dS = 1 \,.$$

Clearly, u has zero flow-rate through S. Thus, by formally replacing u for φ in (2.56) we get

$$\kappa_0 \|\nabla U\|_2^2 + (h(|\nabla U|)\nabla U, \nabla u) = 0 \,. \tag{2.60}$$

Moreover, by the properties of the function h, we also have

$$(h(|\nabla U|)\nabla U, \nabla u) \geq C_1 \|\nabla U\|_q^q - (h(|\nabla U|)\nabla U, \nabla a) - C_2|S|$$

$$\geq \tfrac{1}{2} C_1 \|\nabla U\|_q^q - C_3 \left(\|\nabla a\|_q^q + 1\right) \,,$$

where $C_i = C_i(\kappa_i) > 0$, $i = 1, 2$, and $C_3 = C_3(\kappa_1, \kappa_2, q, S) > 0$. Substituting this latter inequality in (2.60) proves the following one:

$$\kappa_0 \|\nabla U\|_2^2 + \|\nabla U\|_q^q \leq C_4 \left(|\Phi|^q + 1\right) \qquad (2.61)$$

with $C_4 = C_4(\kappa_1, \kappa_2, q, S) > 0$, which furnishes the desired estimate. Concerning uniqueness, let (U_1, Γ_1) and (U_2, Γ_2) be two weak solutions corresponding to the same flow-rate Φ, and set $U := U_1 - U_2$, $\Gamma := \Gamma_1 - \Gamma_2$. From (2.53) we find

$$\kappa_0(\nabla U, \nabla \phi) + (h(|\nabla U_1|)\nabla U_1 - h(|\nabla U_2|)\nabla U_2, \nabla \phi) = (\Gamma, \phi) \quad \text{for all } \phi \in V^q(S). \qquad (2.62)$$

Since $\widehat{V}^q(S) \subset V^q(S)$ and since $U \in \widehat{V}^q(\Omega)$ [25] from (2.62) we get

$$\kappa_0 \|\nabla U\|_2^2 + (h(|\nabla U_1|)\nabla U_1 - h(|\nabla U_2|)\nabla U_2, \nabla U) = 0,$$

which, in particular, implies

$$(h(|\nabla U_1|)\nabla U_1 - h(|\nabla U_2|)\nabla U_2, \nabla U) \leq 0.$$

However, by the assumption (2.12) and by the fact that $U \in V^q(S)$, from this latter inequality we infer $U = 0$ a.e. in S, which, once replaced in (2.62), furnishes $\Gamma = 0$, and uniqueness follows. Let us now show that $U \in C^0(\overline{S})$, under the stated assumptions. If $q > 2$, then, by the Sobolev embedding theorem and by (2.61), the claim is obvious. If $q < 2$ and $k_0 > 0$, we notice that, from the assumption (2.11) on h, we find that

$$\phi \in V^q(S) \mapsto -(h(|\nabla U|)\nabla U, \nabla \phi) + (\Gamma, \phi)$$

defines a bounded linear functional and, consequently, by well-known results on elliptic equations, from (2.53) we obtain that $U \in V^{q'}(S)$ together with the estimate

$$\|\nabla U\|_{V^{q'}(S)} \leq C_5 \left(\|\nabla U\|_{V^q(S)} + |\Gamma| \right), \qquad (2.63)$$

where $C_5 = C_5(\kappa_0, S, q) > 0$. Thus, the required property again follows from the Sobolev embedding theorem and (2.59) and (2.61). It remains to show the continuity of the map M, namely, for any given Φ_0 and $\varepsilon > 0$, there exists $\delta = \delta(\Phi_0, \varepsilon) > 0$ such that

$$|\Phi - \Phi_0| < \delta \quad \Longrightarrow \quad \|U - U_0\|_{C^0(\overline{S})} + |\Gamma - \Gamma_0| < \varepsilon, \qquad (2.64)$$

---

[25] It is easy to show that $\widehat{W}_0^{1,r}(S) = \{\varphi \in W_0^{1,r}(S) : \int_S \varphi \, dS = 0\} \equiv V_0^{1,r}(S)$. In fact, clearly, $\widehat{W}_0^{1,r}(S) \subset V_0^{1,r}(S)$. Conversely, let $u \in V_0^{1,r}(S)$. We have to prove that there exists a sequence $\{u_k\} \subset \widehat{C}_0^\infty(S)$ converging to u in the $W^{1,r}$-norm. Since $V_0^{1,r}(S) \subset W_0^{1,r}(S)$, there exists a sequence $\{\psi_k\} \subset C_0^\infty(S)$ converging to u in $W^{1,r}(S)$. Let $\zeta \in C_0^\infty(S)$ with $\int_S \zeta \, dS = 1$. Then, the sequence

$$u_k := \psi_k - \zeta \int_S \psi_k \, dS, \quad k \in \mathbb{N},$$

belongs to $\widehat{C}_0^\infty(S)$ and converges to u in $W^{1,r}(S)$.

where (U, Γ) and (U_0, Γ_0) are weak solutions corresponding to Φ and Φ_0, respectively. In fact, assume that (2.64) does not hold. Then we can find sequences $\{\Phi_n\}$, $\{U_n, \Gamma_n\}$ and a number $\varepsilon_0 > 0$, independent of n, such that

$$\lim_{n \to \infty} \Phi_n = \Phi_0, \quad \text{and} \quad \|U_n - U_0\|_{C^0(\overline{S})} + |\Gamma_n - \Gamma_0| \geq \varepsilon_0 \quad \text{for all } n \in \mathbb{N}. \quad (2.65)$$

We recall that, by (2.58), U_n and Γ_n are related by the equation

$$\Gamma_n = \int_S (\kappa_0 \nabla U_n + h(|\nabla U_n|)\nabla U_n) \cdot \nabla \chi. \quad (2.66)$$

Now, from (2.61), if $q > 2$, and from (2.59), (2.61) and (2.63), if $q < 2$ and $\kappa_0 > 0$, and from (2.66), we may find subsequences $\{U_{n_k}, \Gamma_{n_k}\}$, such that

$$U_{n_k} \to \widetilde{U} \quad \text{weakly in } V^s(S), \quad s = q, q',$$
$$\lim_{n_k \to \infty} \Gamma_{n_k} = \widetilde{\Gamma}, \quad (2.67)$$

for some $\widetilde{U} \in V^s(S)$ and $\widetilde{\Gamma} \in \mathbb{R}$. By using the monotonicity property of h along with (2.67), it is a simple task to prove that there is $\widetilde{\Gamma} \in \mathbb{R}$ such that $(\widetilde{U}, \widetilde{\Gamma})$, is a weak solution to (2.52) corresponding to the flow-rate Φ_0. Therefore, in view of uniqueness, $\widetilde{U} = U_0$, $\widetilde{\Gamma} = \Gamma_0$ and so, by taking into account that the embedding $V^s(S) \subset C^0(\overline{S})$, $s > 2$, is compact, conditions (2.67) contradict (2.65) and the proof of the proposition is completed. □

Remark 2.6. Unlike the Navier–Stokes case (see Section I.1.1), for the generalized Newtonian models considered here, in general, it is not immediate to find the relation between the flow-rate Φ and axial pressure gradient Γ. However, in the case of the power-law model (2.5), due to the homogeneity of the constitutive equation, it is possible to find a simple relation. Specifically, set

$$U = \frac{1}{\mu_1^{\frac{1}{q-1}}} \frac{\Gamma}{|\Gamma|^{\frac{q-2}{q-1}}} u, \quad (2.68)$$

where u solves the problem

$$\nabla \cdot (|\nabla u|^{q-2} \nabla u) = -1, \quad u|_{\partial S} = 0. \quad (2.69)$$

Then, it is immediate to check that U is a solution to (2.52) with $\kappa_0 = 0$ and $h = \mu_1 |\nabla U|^{q-2} \nabla U$. Therefore, from (2.68) we obtain the relation

$$\Phi = M \frac{1}{\mu_1^{\frac{1}{q-1}}} \frac{\Gamma}{|\Gamma|^{\frac{q-2}{q-1}}},$$

with $M := \int_S u \, dS$, or, equivalently,

$$\Gamma = \mu_1 \frac{\Phi}{M} \left| \frac{\Phi}{M} \right|^{q-2}.$$

Observe that M depends only on S. In the case when S is a circle of radius R, the solution to (2.68) is easily found to be

$$u = C(R,q) \left[1 - \left(\frac{r}{R} \right)^{\frac{q}{q-1}} \right],$$

where $C(R,q) > 0$ depends only on R and q; see, e.g., the article of A.M. Robertson [94] in this volume. It is clear that the above type of scaling does not work any more if $\kappa_0 > 0$, or even if $\kappa_0 = 0$ and $h \equiv h_1$ is given by (2.6) with $\mu_0 = 0$.

Our next objective is to establish some regularity results for the solution determined in Proposition 2.1. To this end we need to impose some further restrictions on the function h. Specifically, setting

$$a_{ij}(p) := \frac{\partial}{\partial p_j} \left(h(|p|)p_i \right), \quad i,j = 1,2,$$

we shall assume the ellipticity and growth properties

$$\begin{aligned} a_{ij}(p)\,\xi_i\,\xi_j &\geq \lambda \left(\kappa + |p| \right)^{q-2} |\xi|^2, \\ |a_{ij}(p)| &\leq \Lambda \left(\kappa + |p| \right)^{q-2}, \end{aligned} \tag{2.70}$$

where $\lambda > 0$, $\Lambda > 0$ and $\kappa \geq 0$ are constants, possibly depending on q.

By a straightforward calculation, one shows that the function h associated to each constitutive relation (2.5)–(2.7) satisfies (2.70). The following result holds.

Proposition 2.2. *Let S be of class $C^{1,\beta}$, for some $\beta > 0$ and let h satisfy (2.70). Then, if either $q > 2$ or $\kappa_0 > 0$, the weak solution (U,Γ) to (2.52) corresponding to a given $\Phi \in \mathbb{R}$, satisfies $U \in C^{1,\gamma}(\overline{S})$, for some $\gamma > 0$. Moreover, there exists a constant $C > 0$ depending, at most, on q, S, κ_0 and Φ), such that*

$$\|U\|_{C^{1,\gamma}(S)} \leq C.$$

Proof. Under the given assumptions on q and κ_0, by Proposition 2.1, we know that U is bounded in S by some constant depending, at most, on q, S κ_0 and Φ. The result is then an immediate consequence of [75, Theorem 1]. $\square$

Remark 2.7. (1) In the case of the power-law model, it can be shown that u, and hence U, is in $C^0(\overline{S})$, for *all* $q \in (1,\infty)$, despite the fact that $\kappa_0 = 0$ (see Remark 2.1). In fact, one can prove the stronger result that, for any given $q > 1$, the corresponding weak solution u to (2.69) is in $W^{2,2}(\Omega)$; see [77]. Then, by [75, Theorem 1], it follows that $u \in C^{1,\gamma}(\overline{S})$, for some $\gamma > 0$.

(2) In the case of the Carreau model (2.6), the regularity of the weak solution to (2.52), corresponding to a given Φ, follows also when $\mu_0 \equiv \kappa_0 = 0$ (see Remark 2.1). In fact, by classical results on non-degenerate quasi-linear elliptic equations in divergence form, it follows that if S is of class $C^{2,\beta}$, for some $\beta > 0$, then $U \in C^{2,\beta}(\overline{S})$; see, e.g., [72, Theorem 8.3 in Chapter 4].

2.2.2(b) Steady-state flow in a distorted pipe. The main result of this section will concern existence and uniqueness for problem (2.50), under suitable assumptions on the function h_1. In order to achieve this goal, as in the Newtonian case, it is appropriate to furnish an equivalent formulation of (2.50), and to put this latter in a weak form. To this end, we begin to introduce the *flow-rate carrier* $\gamma = \gamma(x)$ satisfying the following properties.

(i) $\gamma \in C^1(\overline{\omega})$, for all bounded domains $\omega \subset \Omega$;

(ii) $\nabla \cdot \gamma = 0$ in Ω, $\gamma|_{\partial\Omega} = \mathbf{0}$;

(iii) $\displaystyle\int_S \gamma \cdot \mathbf{n}\, dS = \Phi$;

(iv) $\gamma(x) = \mathbf{v}_{P1}(x)$ for all $x \in \widetilde{\Omega}_1$ and $\gamma(x) = \mathbf{v}_{P2}(x)$ for all $x \in \widetilde{\Omega}_2$, where, we recall, $\widetilde{\Omega}_1$ and $\widetilde{\Omega}_2$ are defined in (1.103), and $\mathbf{v}_{Pi}$, $i = 1, 2$, are the Poiseuille velocity fields in Ω_i defined in (2.51) and corresponding to the flow-rate Φ;

(v) for any $\varepsilon > 0$, there is $\delta = \delta(\varepsilon) > 0$ such that $\max\limits_{x \in \Omega}|\gamma(x)| < \varepsilon$, if $|\Phi| < \delta$.

Taking into account Proposition 2.1, the existence of a field γ satisfying (i)–(iv) can be proven exactly by the same methods used to construct the flow-rate carrier $\mathbf{a}$ of Section I.3.1. Moreover, the property (v) follows from Proposition 2.1(ii).

We next notice that, in view of the properties of the fields γ and $\mathbf{v}_{Pi}(x)$, we find that

$$\left.\begin{array}{l} \nabla \cdot (\kappa_0 + h(|\mathbf{D}(\gamma)|))\,\mathbf{D}(\gamma) - \gamma \cdot \nabla\gamma = -\Gamma_i e_1 \\[2mm] \gamma \cdot \nabla\gamma = \mathbf{0} \end{array}\right\} \quad \text{in } \Omega_i\,, \ i = 1, 2\,. \qquad (2.71)$$

Therefore, since

$$\int_{S_i} \varphi \cdot \mathbf{n}\, dS = 0 \quad \text{for all } \varphi \in \mathcal{D}_0^{1,r}(\Omega)\,, \ i = 1, 2\,, \ 1 < r < \infty\,,$$

from the preceding relation we deduce

$$\int_\Omega (\kappa_0 + h(|\mathbf{D}(\gamma)|))\,\mathbf{D}(\gamma) : \mathbf{D}(\varphi) = -\int_{\widetilde{\Omega}_0} \gamma \cdot \nabla\gamma \cdot \varphi$$
$$-\int_{\widetilde{\Omega}_0} (\kappa_0 + h(|\mathbf{D}(\gamma)|))\,\mathbf{D}(\gamma) : \mathbf{D}(\varphi)\,, \qquad (2.72)$$

for all such φ. We shall now give a suitable generalized formulation of problem (2.50). We set

$$\mathcal{V}^q(\Omega) := \mathcal{D}_0^{1,q}(\Omega) \cap \mathcal{D}_0^{1,2}(\Omega)\,, \quad q \in (1, \infty)\,, \ q \neq 2\,.$$

Clearly, $\mathcal{V}^q(\Omega)$, endowed with the natural norm $\|\mathbf{u}\|_{\mathcal{V}^q(\Omega)} \equiv \|\nabla\mathbf{u}\|_q + \|\nabla\mathbf{u}\|_2$, is a reflexive, separable Banach space. We next set $\mathbf{v} = \mathbf{u} + \gamma$, multiply both sides of $(2.50)_1$ by $\varphi \in \mathcal{V}^q(\Omega)$, integrate by parts over Ω and use (2.72). We thus obtain the equation

$$((\kappa_0 + h(|\boldsymbol{D}(\boldsymbol{u}+\boldsymbol{\gamma})|)) \ \boldsymbol{D}(\boldsymbol{u}+\boldsymbol{\gamma}) - (\kappa_0 + h(|\boldsymbol{D}(\boldsymbol{\gamma})|)) \ \boldsymbol{D}(\boldsymbol{\gamma}), \boldsymbol{D}(\boldsymbol{\varphi}))$$
$$= (\boldsymbol{u} \cdot \nabla\boldsymbol{\varphi}, \boldsymbol{u}) + (\boldsymbol{u} \cdot \nabla\boldsymbol{\varphi}, \boldsymbol{\gamma}) + (\boldsymbol{\gamma} \cdot \nabla\boldsymbol{\varphi}, \boldsymbol{u}) + \boldsymbol{F}(\boldsymbol{\varphi})$$
$$\text{for all } \boldsymbol{\varphi} \in V^q(\Omega) \,,$$

(2.73)

where

$$\boldsymbol{F}(\boldsymbol{\varphi}) := \int_{\widetilde{\Omega}_0} \boldsymbol{\gamma} \cdot \nabla\boldsymbol{\gamma} \cdot \boldsymbol{\varphi} + \int_{\widetilde{\Omega}_0} (\kappa_0 + h(|\boldsymbol{D}(\boldsymbol{\gamma})|)) \, \boldsymbol{D}(\boldsymbol{\gamma}) : \boldsymbol{D}(\boldsymbol{\varphi}) \,. \qquad (2.74)$$

We shall say that $\boldsymbol{u} : \Omega \mapsto \mathbb{R}^3$ is a *weak solution to problem* (2.50) if and only if
(a) $\boldsymbol{u} \in \mathcal{V}^q(\Omega)$, and (b) $\boldsymbol{u}$ satisfies (2.73)–(2.74).

The proof of existence and uniqueness of weak solutions requires the following
more stringent assumptions on the function h,

$$|h(|\boldsymbol{A}|)\boldsymbol{A} - h(|\boldsymbol{B}|)\boldsymbol{B}| \le \kappa_4(\kappa_5 + |\boldsymbol{A}|^{q-2} + |\boldsymbol{B}|^{q-2})|\boldsymbol{A} - \boldsymbol{B}|,$$
$$(h(|\boldsymbol{A}|)\boldsymbol{A} - h(|\boldsymbol{B}|)\boldsymbol{B}) : (\boldsymbol{A} - \boldsymbol{B}) \ge \kappa_6|\boldsymbol{A} - \boldsymbol{B}|^q \,,$$

(2.75)

for arbitrary second-order symmetric tensors $\boldsymbol{A}$, $\boldsymbol{B}$ and $q > 2$, with $\kappa_4, \kappa_6 > 0$ and
$\kappa_5 \ge 0$ material constants.

Remark 2.8. Conditions (2.75) imply (2.10)–(2.12). Moreover, all constitutive
equations (2.5)–(2.7) satisfy the properties (2.75); see [96, Lemme 3.1].

Theorem 2.5. *Let h_1 satisfy the assumptions* (2.9), (2.75) *with $\kappa_0 > 0$ and for
some some $q > 2$, and assumptions* (2.70) *as well. Then, given $\Phi \in \mathbb{R}$, there exists
a constant $C = C(\Omega, q, \kappa_0, \kappa_4, \kappa_5, \kappa_6) > 0$ such that, if $|\Phi| < C$, problem* (2.50)
*has at leat one one weak solution $\boldsymbol{u}$. Furthermore, there exists a positive constant
$M = M(\Omega, q, \kappa_0, \kappa_4, \kappa_5, \kappa_6)$, such that if $\|\nabla\boldsymbol{u}\|_2 < M$, then $\boldsymbol{u}$ is the only weak
solution corresponding to Φ.*

Proof. It is convenient to write (2.73)–(2.74) as a nonlinear equation in the space
$\mathcal{V}^\star := [\mathcal{V}^q(\Omega)]'$. Throughout the proof, we shall denote by $\langle \cdot, \cdot \rangle$ the duality pairing
between $\mathcal{V}^\star$ and $\mathcal{V}^q(\Omega)$. In view of the assumption (2.75)$_1$, from the properties
of the function $\boldsymbol{\gamma}$ and from the Hölder inequality, we easily obtain that, for any
$\boldsymbol{u} \in \mathcal{V}^q(\Omega)$, the map

$$\boldsymbol{\varphi} \in \mathcal{V}^q(\Omega) \mapsto$$
$$((\kappa_0 + h(|\boldsymbol{D}(\boldsymbol{u}+\boldsymbol{\gamma})|)) \, \boldsymbol{D}(\boldsymbol{u}+\boldsymbol{\gamma}) - (\kappa_0 + h(|\boldsymbol{D}(\boldsymbol{\gamma})|)) \, \boldsymbol{D}(\boldsymbol{\gamma}), \boldsymbol{D}(\boldsymbol{\varphi})) \in \mathbb{R}$$

defines an element $A(\boldsymbol{u})$ of $\mathcal{V}^\star$, such that

$$\langle A(\boldsymbol{u}), \boldsymbol{\varphi} \rangle = ((\kappa_0 + h(|\boldsymbol{D}(\boldsymbol{u}+\boldsymbol{\gamma})|))\boldsymbol{D}(\boldsymbol{u}+\boldsymbol{\gamma})$$
$$- (\kappa_0 + h(|\boldsymbol{D}(\boldsymbol{\gamma})|)) \, \boldsymbol{D}(\boldsymbol{\gamma}), \boldsymbol{D}(\boldsymbol{\varphi}))$$

(2.76)

for all $\boldsymbol{\varphi} \in \mathcal{V}^q(\Omega)$, with

$$\|A(\boldsymbol{u})\|_{\mathcal{V}^\star} \le C_1 \left(\|\nabla\boldsymbol{u}\|_2 + \|\nabla\boldsymbol{u}\|_q^{q-1} \right), \qquad (2.77)$$

where C_1 is a positive constant independent of $\boldsymbol{u}$. Moreover, on the one hand, in view of the embedding $\mathcal{V}^q(\Omega) \subset L^4(\Omega)$, we have

$$|(\boldsymbol{u} \cdot \nabla \boldsymbol{\varphi}, \boldsymbol{u})| \leq \|\boldsymbol{u}\|_4^2 \|\boldsymbol{\varphi}\|_{\mathcal{V}^q(\Omega)} \leq C_2 \|\boldsymbol{u}\|_{\mathcal{V}^q(\Omega)}^2 \|\boldsymbol{\varphi}\|_{\mathcal{V}^q(\Omega)} \,,$$

and, on the other hand, by the Poincaré inequality (see (1.79) in Section 1), the Schwarz inequality and the properties of the function $\boldsymbol{\gamma}$ it follows that

$$|(\boldsymbol{u} \cdot \nabla \boldsymbol{\varphi}, \boldsymbol{\gamma}) + (\boldsymbol{\gamma} \cdot \nabla \boldsymbol{\varphi}, \boldsymbol{u})| \leq C_3 \|\boldsymbol{u}\|_{\mathcal{V}^q(\Omega)} \|\boldsymbol{\varphi}\|_{\mathcal{V}^q(\Omega)} \,,$$

where C_i, $i = 2, 3$, are positive constants independent of $\boldsymbol{u}$ and $\boldsymbol{\varphi}$. Therefore, taking into account the linear dependence on $\boldsymbol{\varphi}$ of the above quantities, we can find a uniquely determined $B(\boldsymbol{u}) \in \mathcal{V}^\star$ such that

$$\langle B(\boldsymbol{u}), \boldsymbol{\varphi} \rangle = (\boldsymbol{u} \cdot \nabla \boldsymbol{\varphi}, \boldsymbol{u}) + (\boldsymbol{u} \cdot \nabla \boldsymbol{\varphi}, \boldsymbol{\gamma}) + (\boldsymbol{\gamma} \cdot \nabla \boldsymbol{\varphi}, \boldsymbol{u}) \,, \tag{2.78}$$

for all $\boldsymbol{\varphi} \in \mathcal{V}^q(\Omega)$. Finally, again by the properties of the function $\boldsymbol{\gamma}$, it is immediate to prove that there exists a constant $C_4 > 0$ independent of $\boldsymbol{\varphi}$ such that $|\boldsymbol{F}(\boldsymbol{\varphi})| \leq C_4 \|\boldsymbol{\varphi}\|_{\mathcal{V}^q(\Omega)}$, and so there exists $f \in \mathcal{V}^\star$ such that

$$\langle f, \boldsymbol{\varphi} \rangle = \boldsymbol{F}(\boldsymbol{\varphi}) \,, \tag{2.79}$$

for all $\boldsymbol{\varphi} \in \mathcal{V}^q(\Omega)$. From (2.76), (2.78)–(2.79), we find that the existence of a weak solution to (2.50) is equivalent to finding $\boldsymbol{u} \in \mathcal{V}^q(\Omega)$ such that

$$\langle A(\boldsymbol{u}), \boldsymbol{\varphi} \rangle = \langle B(\boldsymbol{u}), \boldsymbol{\varphi} \rangle + \langle f, \boldsymbol{\varphi} \rangle \quad \text{for all } \boldsymbol{\varphi} \in \mathcal{V}^q(\Omega). \tag{2.80}$$

The proof of existence can be now handled by means of the classical theory of monotone operators. Let $\{\boldsymbol{\varphi}_k\}$ be a sequence from $\mathcal{D}(\Omega)$ which is an orthogonal basis in $\mathcal{D}_0^{1,2}(\Omega)$ and whose linear hull is dense in $\mathcal{V}^q(\Omega)$. We look for an approximated solution to (2.80) of the form

$$\boldsymbol{u}_m := \sum_{k=1}^{m} c_{km} \boldsymbol{\varphi}_k \,, \tag{2.81}$$

$$\langle A(\boldsymbol{u}_m), \boldsymbol{\varphi}_k \rangle = \langle B(\boldsymbol{u}_m), \boldsymbol{\varphi}_k \rangle + \langle f, \boldsymbol{\varphi}_k \rangle, \quad k = 1, \dots, m \,.$$

As in the case of the bounded domain treated in the previous section, the existence of a solution to (2.81) is obtained provided we show a uniform estimate for $\boldsymbol{u}_m$, $m \in \mathbb{N}$. We thus multiply both sides of (2.81) by c_{km} and sum over k from 1 to m. Taking into account that $\langle B(\boldsymbol{u}_m), \boldsymbol{u}_m \rangle = (\boldsymbol{u}_m \cdot \nabla \boldsymbol{u}_m, \boldsymbol{\gamma})$, from (2.75), (2.76), and from the Korn inequality, we find

$$\|\nabla \boldsymbol{u}_m\|_2^2 + \|\nabla \boldsymbol{u}_m\|_q^q \leq C_5 (\boldsymbol{u}_m \cdot \nabla \boldsymbol{u}_m, \boldsymbol{\gamma}) + \|f\|_{\mathcal{V}^\star} \|\boldsymbol{u}\|_{\mathcal{V}^q(\Omega)} \,, \tag{2.82}$$

where $C_5 > 0$ is independent of m. By the Poincaré inequality we find that

$$|(\boldsymbol{u}_m \cdot \nabla \boldsymbol{u}_m, \boldsymbol{\gamma})| \leq C_6 \max_{x \in \Omega} |\boldsymbol{\gamma}(x)| \|\nabla \boldsymbol{u}_m\|_2^2 \,,$$

with $C_6 = C_6(\Omega) > 0$, and so, by the property (v) of $\boldsymbol{\gamma}$, we can choose $|\Phi|$ so small that

$$C_6 \max_{x \in \Omega} |\boldsymbol{\gamma}(x)| < \frac{1}{2} \,.$$

From (2.82) we thus find

$$\|\nabla \boldsymbol{u}_m\|_2^2 + \|\nabla \boldsymbol{u}_m\|_q^q \leq 2\|f\|_{\mathcal{V}^\star}\|\boldsymbol{u}\|_{\mathcal{V}^q(\Omega)},$$

which, with the help of (1.157) in Section 1, in turn, implies

$$\|\nabla \boldsymbol{u}_m\|_2^2 + \|\nabla \boldsymbol{u}_m\|_q^q \leq C_7 \left(\|f\|_{\mathcal{V}^\star}^2 + \|f\|_{\mathcal{V}^\star}^{\frac{q}{q-1}}\right), \tag{2.83}$$

with $C_7 = C_7(q) > 0$. This is the desired estimate that proves the existence of a solution to (2.81), for all $m \in \mathbb{N}$. From (2.83) we may select a subsequence, that we continue to denote by $\{\boldsymbol{u}_m\}$ and $\boldsymbol{u} \in \mathcal{V}^q(\Omega)$ such that

$$\boldsymbol{u}_m \to \boldsymbol{u} \quad \text{weakly in } \mathcal{V}^q(\Omega).$$

Moreover, from (2.77) and (2.83) it follows that the sequence $\{A(\boldsymbol{u}_m)\}$ is uniformly bounded in $\mathcal{V}^\star$ and, consequently, there exists $G \in \mathcal{V}^\star$ such that

$$\lim_{m \to \infty} \langle A(\boldsymbol{u}_m), \boldsymbol{\varphi} \rangle = \langle G, \boldsymbol{\varphi} \rangle, \quad \text{for all } \boldsymbol{\varphi} \in \mathcal{V}^q(\Omega). \tag{2.84}$$

Having established these properties, the existence proof becomes completely analogous to that of Theorem 2.1 and it will only be sketched here. Since

$$\lim_{m \to \infty} \langle B(\boldsymbol{u}_m), \boldsymbol{\varphi}_k \rangle = \langle B(\boldsymbol{u}), \boldsymbol{\varphi}_k \rangle,$$

in view of the properties of the sequence $\{\boldsymbol{\varphi}_k\}$, it is enough to show that

$$\langle A(\boldsymbol{u}), \boldsymbol{\varphi} \rangle = \langle G, \boldsymbol{\varphi} \rangle, \quad \text{for all } \boldsymbol{\varphi} \in \mathcal{V}^q(\Omega). \tag{2.85}$$

From $(2.81)_2$ and (2.84) we can show that

$$\lim_{m \to \infty} \langle A(\boldsymbol{u}_m), \boldsymbol{u}_m \rangle = \langle G, \boldsymbol{u} \rangle. \tag{2.86}$$

Now, from the monotonicity properties of the function h, $(2.75)_2$, we obtain, for all $\boldsymbol{\psi} \in \mathcal{V}^q(\Omega)$,

$$\langle A(\boldsymbol{u}_m) - A(\boldsymbol{\psi}), \boldsymbol{u}_m - \boldsymbol{\psi} \rangle \geq 0,$$

and so, passing to the limit $m \to \infty$ and taking into account (2.86), we get

$$\langle G - A(\boldsymbol{\psi}), \boldsymbol{u} - \boldsymbol{\psi} \rangle \geq 0 \quad \text{for all } \boldsymbol{\psi} \in \mathcal{V}^q(\Omega).$$

Therefore, choosing $\boldsymbol{\psi} = \boldsymbol{u} + \lambda\boldsymbol{\varphi}$, $\lambda > 0$ and letting $\lambda \to 0$, by the continuity property of h we find

$$\langle G - A(\boldsymbol{u}), \boldsymbol{\varphi} \rangle \geq 0 \quad \text{for all } \boldsymbol{\varphi} \in \mathcal{V}^q(\Omega),$$

which proves (2.85). The proof of existence is then accomplished. Let now, $\boldsymbol{u}_1$ be another weak solution corresponding to Φ, and let $\boldsymbol{w} := \boldsymbol{u} - \boldsymbol{u}_1$. From (2.80) it follows that

$$\langle A(\boldsymbol{u}) - A(\boldsymbol{u}_1), \boldsymbol{w} \rangle = \langle B(\boldsymbol{u}) - B(\boldsymbol{u}_1), \boldsymbol{w} \rangle,$$

which, taking into account $(2.75)_2$ and (2.78), in particular, implies

$$\|\nabla \boldsymbol{w}\|_2^2 \leq C_7 \left((\boldsymbol{w} \cdot \nabla \boldsymbol{w}, \boldsymbol{u}) + (\boldsymbol{w} \cdot \nabla \boldsymbol{w}, \boldsymbol{\gamma})\right) \leq C_8 \left(\|\nabla \boldsymbol{u}\|_2 + \max_{x \in \Omega}|\boldsymbol{\gamma}(x)|\right)\|\nabla \boldsymbol{w}\|_2^2, \tag{2.87}$$

where we have used the embedding $W^{1,2}(\Omega) \subset L^4(\Omega)$ and where $C_i > 0$, $i = 7, 8$, are independent of $\boldsymbol{u}$ and $\boldsymbol{u}_1$. By the property (v) of $\boldsymbol{\gamma}$, there exists $\delta > 0$ such that

$C_8 \max_{x \in \Omega} |\boldsymbol{\gamma}(x)| < \frac{1}{2}$, provided $|\Phi| < \delta$. Under the sated assumptions, the uniqueness result is then an immediate consequence of (2.87). □

Remark 2.9. If $\kappa_0 > 0$ and $1 < q < 2$, unique solvability still holds, provided h satisfies appropriate assumptions. More precisely, suppose that h is a continuous function satisfying (2.12) along with the following one:

$$|h(|\boldsymbol{A}|)\boldsymbol{A} - h(|\boldsymbol{B}|)\boldsymbol{B}| \le \kappa_7 |\boldsymbol{A} - \boldsymbol{B}|, \qquad (2.88)$$

for all symmetric tensors $\boldsymbol{A}$ and $\boldsymbol{B}$ and for some constant $\kappa_7 > 0$ independent of $\boldsymbol{A}$ and $\boldsymbol{B}$. We then define a weak solution to problem (2.50) to be a vector field $\boldsymbol{u} \in \mathcal{D}_0^{1,2}(\Omega)$ satisfying (2.73)–(2.74) with $\mathcal{V}^q(\Omega)$ replaced by $\mathcal{D}_0^{1,2}(\Omega)$. Because of (2.88) (and of the properties of $\boldsymbol{\gamma}$), it is readily checked that every term in (2.73) is meaningful. Thus, by the same arguments used in the proof of Theorem 2.5, we can show the existence of a weak solution on the condition that the magnitude of the flow-rate does not exceed a certain constant. This solution is also unique if its Dirichlet norm is sufficiently "small". Notice that the Carreau law (2.6) satisfies (2.88); see [96, Lemme 3.1, Eqs. (3.4) and (3.5)].

Remark 2.10. If $\kappa_0 = 0$, as in the case of the power-law model (see Remark 2.1), the procedure employed in the proof of Theorem 2.5 fails. However, by using a "weighted" approach, with weight related to the Hagen–Poiseuille velocity fields, $\boldsymbol{v}_{Pi}$, in Ω_i, $i = 1, 2$, E. Marušić-Paloka has shown in [80] some existence and uniqueness results of a weak solution for the power-law model (see (2.5)), with $q > 2$ for "small" flow-rate. This approach requires a detailed information about the dependence of $\boldsymbol{v}_{Pi}$ on the cross-sectional coordinates and, consequently, the proof is given only in the case of a circular cross section, where an explicit solution is known (see Remark 2.6). The case $q < 2$ remains open, even in the case of circular cross-sections.

2.3. Problems related to viscoelastic liquid models

As we observed in Section 2.2, there are two main properties of many real incompressible fluids that can not be explained by the Newtonian (Navier–Stokes) model. They are (a) the dependence of the viscosity on the shear, and (b) the normal-stress effects, where the fluid shows a stress distribution normal to the direction of shear, *other than* the one due to the pressure. However, whereas the generalized Newtonian model and, more generally, the Reiner–Rivlin model (namely, (2.2) with $h_0 = 0$ and h_i, $i = 1, 2$, functions of the principal invariants), are able to furnish results in quite a good agreement with experiments in case (a), there appears to be no obvious way that they could explain experiments where normal-stress effects play a fundamental role, such as "rod-climbing" and orientation of free-falling symmetric particles. In this respect, Clifford Truesdell makes the following witty comments [109, p. 116].

> It was natural to expect that if one constant, μ, could specify a Navier–Stokes fluid completely, the two functions [h_1] and [h_2] of the Reiner–Rivlin theory ought to be more than enough to explain any deviation from classical

theory, so various rheologists suggested particular functions $[h_1]$ and $[h_2]$ as being more intuitive or physical than others. As the normal-stress effects came to be understood better, more precise experiments were performed, which made it clear that no choice whatever of $[h_1]$ and $[h_2]$ could fit the data. This experimental fact surprised such empiricists as believed that if one disposable constant was good, half a dozen would surely be better. Indeed, this example shows that while adjustment of one constant may provide a splendid first approximation for all or nearly all incompressible fluids, two disposable functions of two variables may fail to improve that approximation quantitatively.

In the late 1950s W. Noll [83] proposed another, very general model of a fluid, called *simple fluid*, [26] that, among other things, could give a very good account of normal-stress effects. The main characteristic of this model, roughly speaking, is that the response of the material does not depend on the *current* status of the material but, rather on its *hystory*. In more precise words, the constitutive equation of the fluid does not depend on the deformation at the current time t, but on the whole "hystory" of the deformation, from $-\infty$ to t. It goes without saying that this class of constitutive equations is quite general and, as such, of possibly little practical use. Therefore, in a series of papers, B.D. Coleman and W. Noll [24, 25, 26] introduced and analyzed the concept of "fading memory" which essentially describes "how far back" a fluid may remember its past hystory. Mathematically, this translates into a suitable expansion of the constitutive equation (which, in the case of simple fluids becomes a *functional*) in terms of a "retardation parameter", r, so that the constitutive equation is represented by a power series in r. If the series is truncated after the linear term, we obtain the Navier–Stokes model (2.1), whereas if it is truncated after the quadratic term, cubic term, etc., we obtain the *second-order, third order, etc.*, approximation of the simple fluid. A popular truncation in the engineering community is the second-order one, that furnishes the following relation between stress and deformation:

$$\boldsymbol{T} = -p\boldsymbol{I} + \alpha_0 \boldsymbol{A}_1 + \alpha_1 \boldsymbol{A}_2 + \alpha_2 \boldsymbol{A}_1^2, \qquad (2.89)$$

where $\boldsymbol{A}_1 = 2\boldsymbol{D}$ ($\boldsymbol{D}$ stretching tensor of the fluid),

$$\boldsymbol{A}_2 := \frac{\partial \boldsymbol{A}_1}{\partial t} + \boldsymbol{v} \cdot \nabla \boldsymbol{A}_1 + \boldsymbol{A}_1 \cdot (\nabla \boldsymbol{v})^\top + \nabla \boldsymbol{v} \cdot \boldsymbol{A}_1, \qquad (2.90)$$

($\boldsymbol{v}$ velocity field of the fluid), and where the coefficients α_i, $i = 0, 1, 2$, are, at most, functions of the principal invariants of $\boldsymbol{D}$. The fluid whose stress tensor is given by (2.89)–(2.90) with α_i=const., $i = 0, 1, 2$, and $\alpha_0 = \mu$ (shear viscosity) is called *second-order* or *grade* 2 fluid. It is worth emphasizing that, in this framework, the relations (2.89)–(2.90) do *not* represent a constitutive equation, but, rather, only a suitable approximation of it. It is, therefore, not surprising that, as we shall see, the initial-boundary value problem for the equations of motion associated to (2.89)–(2.90) presents some incongruities. Among these, the most striking one is that, if one accepts the values of the constant α_1 measured by rheologists, which

---

[26] Actually, Noll proposed the more general model of *simple material*.

gives $\alpha_1 < 0$, the rest state is *nonlinearly unstable* in the sense of Liapounov; see Section 2.4.2. In fact, if we restrict ourselves to *linear* instability of the rest state, then D.D. Joseph has shown that, if $\alpha_1 < 0$, all truncations of the simple fluid of *arbitrary* order, n, will lead to equations of motion where the rest state is linearly unstable [60]. In this respect, Joseph observes [61]

> ... it is wrong to study stability u sing constitutive expressions for "fluids" [27] of grade n. These expressions arise in response to slow deformations. In fact, there is no such a thing as a constitutive equation without prior specification of the domain of deformations in which the constitutive equation lives. However good rigid body mechanics is for some problems, it is obviously no good for studying deformation of strained bodies.

In principle, these incongruities would disappear, had we considered the *complete* constitutive equations, without any approximation. Actually, again in the words of Clifford Truesdell [109, p. 132],

> [It is not proven] any relation at all between the *solutions* [28] of the differential equations of motion for the fluid of grade 2 and the *solutions* [29] of the equations of motion of the general simple fluid that the particular fluid of grade 2 approximates.

Nevertheless, all inconsistencies vanish if we restrict our analysis to steady-state (or quasi-steady state) motions of the fluid, where the values of the parameters α_1 and α_2 are, in fact, measured. As a matter of fact, as we shall see, in this case the second-order fluid gives qualitative and rigorous predictions which are in a very good agreement with experiments, in a significant range of physical parameters.

Other, more complicated, popular models in the engineering community that do not suffer from the drawbacks of the second-order fluid are the so-called viscoelastic fluids of *rate-type* or *differential type*. The main characteristic of these models is that the non-Newtonian part of the Cauchy stress tensor obeys a constitutive equation expressed by first-order differential equation in time that should be studied in conjunction with the linear momentum equation (and conservation of mass). Referring to [13, Chapter 7], and to the article of A.M. Robertson [94] in this volume for further information regarding this type of models, here we shall recall one of the most widely used, the *Oldroyd*-B fluid. In this case, the Cauchy stress tensor $\boldsymbol{T}$ is given by

$$\boldsymbol{T} = -p\boldsymbol{I} + 2\mu_\infty \boldsymbol{D} + \boldsymbol{\tau}\,, \tag{2.91}$$

where $\boldsymbol{\tau}$ satisfies the constitutive equation

$$\lambda_1 \left(\frac{\partial \boldsymbol{\tau}}{\partial t} + \boldsymbol{v} \cdot \nabla \boldsymbol{\tau} - \boldsymbol{\tau} \cdot (\nabla \boldsymbol{v})^\top - \nabla \boldsymbol{v} \cdot \boldsymbol{\tau} \right) + \boldsymbol{\tau} = 2\left(\mu_0 - \mu_\infty \right) \boldsymbol{D}\,, \tag{2.92}$$

---

[27] In quotes in the original text.
[28] Emphasized in the original text.
[29] Emphasized in the original text.

with λ_1, μ_0 and μ_∞ are positive material constants, with $\mu_\infty < \mu_0$. Notice that if $\lambda_1 = 0$, then (2.89)–(2.92) reduce to the Navier–Stokes constitutive equation (2.1), with $\mu \equiv \mu_0$. The constitutive equation (2.92) is the prototype of much more general models, including upper and lower convected Maxwell, Johnson-Segalman, generalized Jeffreys and others which, from the mathematical viewpoint, amount to add a certain number of nonlinear terms into (2.92). The mathematical problems related to these more general models present, more or less, the same degree of difficulty as those related to (2.92) and, consequently, can be treated by the same arguments.

The objective of this section is to provide some results and methods related to the mathematical theory of viscoelastic fluids, modeled by second-order and Oldroyd-B fluids. Further results associated to more general models will also be considered.

The main idea behind the mathematical (and numerical) study of the problems related to the above viscoelastic models (and to their generalizations) was introduced by M. Renardy [92] and, successively and independently, by G.P. Galdi et al. [39], in a different context. It consists in splitting the original equations of motion into a Stokes-like system and into a transport-like vector equation. This decomposition is, in fact, quite natural if one looks at the models from a physical point of view. Actually, they have to combine two different and coexisting features of the liquid, namely, diffusion (typical of the parabolic character of the Navier–Stokes equations) and transport (typical of the hyperbolic character of an elastic material).

Before performing the mathematical study in the following Sections 2.4 and 2.5, we would like to show how the normal stress-effects, inherent in these models, are able to account for a qualitative or even semi-quantitative explanation of the "rod-climbing" and "particle orientation" experiments we mentioned in Section 2.1 and that the Reiner–Rivlin model was not able to predict. A completely rigorous proof of the "particle orientation" phenomenon is deferred to Section 3.

2.3.1. A semi-quantitative explanation of the "rod-climbing" effect.

A viscoelastic fluid partially fills the region between two infinite, vertical cylinders, C_1 and C_2, of radii r_1 and r_2, respectively, with C_2 at rest and C_1 rotating with a constant angular velocity, $\boldsymbol{\Omega} = \Omega e_3$, parallel to its axis; see Figure 8.

We suppose that the motion of the liquid is steady. Thus, if we assume the second-order fluid model, the motion of the liquid is governed by the equations

$$\rho \boldsymbol{v} \cdot \nabla \boldsymbol{v} = \nabla \cdot \boldsymbol{T},$$
$$\nabla \cdot \boldsymbol{v} = 0,$$

(2.93)

where ρ is the (constant) density of the liquid, $\boldsymbol{T}$ is given in (2.89)–(2.90), and $\boldsymbol{v} = \boldsymbol{v}(x)$. By a direct calculation, we show that the Couette velocity field

$$\boldsymbol{v}_c = \left(\frac{A}{r} + B\,r \right) e_\theta\,, \quad A := \Omega r_1\,, \quad B := \frac{\Omega(r_1/r_2)}{r_2[(r_1/r_2)^2 - 1]}$$

(2.94)

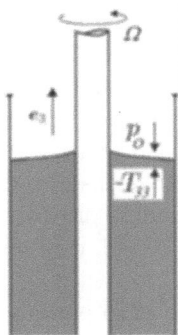

along with the corresponding pressure field

$$p_c = \rho \, g x_3 - \rho \int (A + Br^2)^2 r^{-3} \, dr \qquad (2.95)$$

is a solution to (2.93), satisfying the boundary conditions

$$\boldsymbol{v}_c(r_1) = \Omega \, r_1 \boldsymbol{e}_\theta \,, \quad \boldsymbol{v}_c(r_2) = \boldsymbol{0} \,.$$

In the above equations, g is the gravity, (r, θ) is a polar coordinate system in the plane $x_3 = \text{const.}$, and $\{\boldsymbol{e}_r, \boldsymbol{e}_\theta\}$ is the associated canonical base. By another (tedious but) straightforward calculation, from (2.89)–(2.90) we find that the component T_{33} of the stress tensor, evaluated along $(\boldsymbol{v}_c, p_c)$, is given by

$$T_{33} = -p_c - \alpha_1 \boldsymbol{v}_c \cdot \Delta \boldsymbol{v}_c - \tfrac{1}{4}(2\alpha_1 + \alpha_2)|\boldsymbol{A}_1(\boldsymbol{v}_c)|^2 \,. \qquad (2.96)$$

Moreover, by direct inspection, from (2.94) we find

$$\boldsymbol{v}_c \cdot \Delta \boldsymbol{v}_c = \boldsymbol{0} \,, \quad |\boldsymbol{A}_1(\boldsymbol{v}_c)|^2 = 8 \frac{A^2}{r^4} \,,$$

so that (2.96) furnishes

$$T_{33} = -p_c - 2(2\alpha_1 + \alpha_2)\frac{A^2}{r^4} \,. \qquad (2.97)$$

If the free surface, Γ, of the liquid does not greatly vary from the horizontal, it is reasonable to disregard there the contribution of the component $T_{r\theta}$ of the stress tensor, so that the equation of Γ is found simply by equating the component $-T_{33}$ to the atmospheric pressure p_{atm}. Thus, from (2.95) and (2.96) we deduce that the equation of Γ is given by

$$x_3 = -\frac{p_{\text{atm}}}{\rho g} - \frac{\gamma}{\rho g \, r^4} + \frac{1}{g} \int (A + Br^2)^2 r^{-3} \, dr \,, \quad \gamma := 2(2\alpha_1 + \alpha_2) \,. \qquad (2.98)$$

From this equation, we can easily show the "rod-climbing" effect. Actually, differentiating both sides of (2.98) with respect to r and evaluating this derivative at

r_1 (at the surface of the inner cylinder, that is), we obtain

$$\chi := \left.\frac{dx_3}{dr}\right|_{r=r_1} = \frac{A^2}{gr_1^5}\left\{\left[1 - \left(\frac{r_1}{r_2}\right)^2\right]^2 r_1^2 + \frac{4\gamma}{\rho}\right\}. \tag{2.99}$$

Therefore, we will have that the liquid "climbs the rod" if and only if $\chi < 0$, and so, a sufficient condition for this to happen is that

$$r_1^2 < -\frac{4\gamma}{\rho}. \tag{2.100}$$

Since the observed value of $\gamma := 2\alpha_1 + \alpha_2$ is $\simeq -21\mathrm{gr/cm}$ [61, Section 17.11], condition (2.99) gives the qualitative information that rod-climbing can occur if the radius of the inner cylinder is not too large. In addition, it is interesting to observe that, from (2.99) it also follows that, in the case of a Newtonian fluid ($\gamma \equiv 0$), the slope of the surface in the vicinity of the inner cylinder is always increasing.

Remark 2.11. A study of the "rod-climbing" effect for a second-order liquid has been performed by D.D. Joseph and R.L. Fosdick [64, 65], by means of a *formal* series expansion in the angular velocity of the rod. However, we wish to emphasize that a *rigorous* proof of rod-climbing is an outstanding *open question*.

2.3.2. A qualitative explanation of particle orientation. The phenomenon of particle orientation in a liquid is obviously related to the *torque* that the liquid is able to exert on the body. In particular, the possible equilibrium configurations of the body (either stable or unstable) will be those at which the torque is zero. Following [106] and [63], we shall qualitatively evaluate the torque exerted by a Navier–Stokes fluid and by normal stress in a second-order fluid, and show that they are completely different and, in fact, competing. In order to make things simpler, we shall consider a two-dimensional version of the phenomenon consisting of an ellipse translating with a constant velocity, U, through an incompressible liquid. We assume that the viscous effects of the liquid are negligible. [30] The streamline distribution is then of the type described in Figure 9(a).

We now recall that, for a second-order liquid the stress vector, t, at a surface Σ with unit outer normal n, is given by (disregarding shear viscosity terms)

$$t = -pn + \tfrac{1}{2}(2\alpha_1 + \alpha_2)|A_1|^2 n; \tag{2.101}$$

see [12]. Now, if $\alpha_1 = \alpha_2 = 0$, namely, in the Newtonian case, the only stress on the surface is due to the pressure, p, which is a maximum at the two stagnation points S_1 and S_2 and is directed toward the body, as indicated in Figure 9(b). It is then clear that these stresses will generate a couple that tends to orient the ellipse with its major axis perpendicular to U (see Figure 5(a)). At the other extreme, let us consider the stress distribution generated by the purely viscoelastic term. From (2.101) it follows that this latter is higher at those points of the surface where the

---

[30] As we shall see in Chapter 3, this assumption can be rigorously justified, provided the Reynolds number of the particle is not too "large".

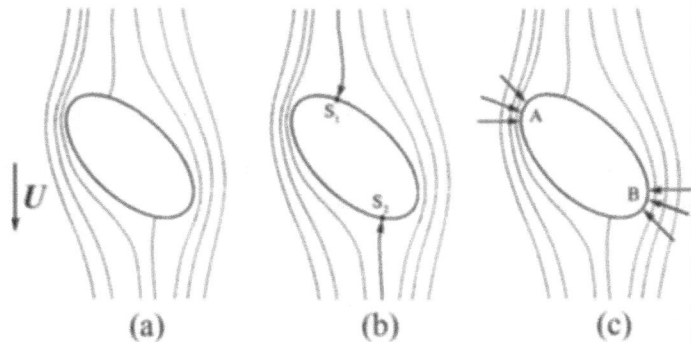

FIGURE 9. (a) Streamline distribution in an inviscid liquid around an
ellipse translating with velocity $\boldsymbol{U}$; (b) S_1 and S_2 are the two stagnation
points of the flow, where the pressure is a maximum; (c) The purely
viscoelastic normal stress is maximal at the points A and B where the
streamlines are denser.

velocity gradient of the liquid is larger, that is, where the streamlines are denser.
With a view to Figure 9(c), we see that the streamlines crowd around the points
A and B of the surface. Therefore, since $2\alpha_1 + \alpha_2 \sim -23\,\mathrm{cm/gr} < 0$, these stresses
are directed toward the body and will generate a couple that tends to orient the
ellipse with its major axis parallel to $\boldsymbol{U}$, in agreement with the experiment (see
Figure 5(b)).

2.4. Some results in the mathematical theory of second-order fluids

In this section we shall concentrate on certain mathematical topics related to
second-order fluids. To this end, we recall that the linear momentum equation
$\rho\, d\boldsymbol{v}/dt = \nabla \cdot \boldsymbol{T} + \rho \boldsymbol{f}$, where $\boldsymbol{T}$ is given in (2.89)–(2.90) (with $\alpha_0 \equiv \mu$), and $\boldsymbol{f}$ is
the body force acting on the fluid, along with the incompressibility condition, can
be written as (see [35])

$$\left.\begin{aligned}
\frac{\partial}{\partial t}\left(\boldsymbol{v} - \alpha_1 \Delta \boldsymbol{v}\right) + \boldsymbol{\omega} \times \boldsymbol{v} - \alpha_1 (\Delta \boldsymbol{\omega} \times \boldsymbol{v}) \\
= \nu \Delta \boldsymbol{v} - \nabla P + \boldsymbol{N}(\boldsymbol{v}) + \boldsymbol{f} \\
\nabla \cdot \boldsymbol{v} = 0
\end{aligned}\right\} \quad \text{in } \Omega \times (0, T), \quad (2.102)$$

In these equations, Ω is the region of flow, $T > 0$, $\boldsymbol{\omega} := \nabla \times \boldsymbol{v}$, and

$$\begin{aligned}
\boldsymbol{N}(\boldsymbol{v}) &:= (\alpha_1 + \alpha_2)\left\{\Delta \boldsymbol{v} \cdot \boldsymbol{A}_1 + 2\nabla \cdot \left[\nabla \boldsymbol{v} \cdot (\nabla \boldsymbol{v})^\top\right]\right\}, \\
P &:= \frac{1}{\rho}\left[p - \alpha_1 \boldsymbol{v} \cdot \Delta \boldsymbol{v} + \frac{\alpha_1}{4}|\boldsymbol{A}_1|^2 + \frac{1}{2}\rho|\boldsymbol{v}|^2\right].
\end{aligned} \qquad (2.103)$$

Furthermore, we set $\nu := \mu/\rho$ and we continue to denote by α_1 and α_2 the ratios
α_1/ρ and α_2/ρ, respectively. To (2.102) we shall append the following adherence

boundary conditions:

$$\boldsymbol{v}(x,t) = \boldsymbol{0}, \quad \text{for all } (x,t) \in \partial\Omega \times (0,T). \tag{2.104}$$

In the following subsections we shall furnish some results concerning well-posedness of the initial-boundary value problem and of the boundary-value problem associated to (2.102)–(2.103). In particular, we shall show that in the experimentally observed range of physical parameters the rest state is always unstable. This is no paradox since, as we noted in Section 2.3, the second-order fluid is only an approximation of the simple fluid and, as such, it may give erroneous predictions. Actually, it should be expected that erroneous predictions occur in time-dependent situations, given that the second-order fluid comes from an approximation that requires "sufficiently slow flow" of the fluid. As we mentioned previously, these inconsistencies disappear in a steady-state flow.

2.4.1. Well-posedness of the boundary-value problem. In this section we shall be interested to the steady-state problem associated to (2.102)–(2.104), obtained by formally neglecting the dependence on t of the fields $\boldsymbol{v}$ and P. This leads to the boundary-value problem

$$\left. \begin{aligned} \boldsymbol{\omega} \times \boldsymbol{v} - \alpha_1(\Delta\boldsymbol{\omega} \times \boldsymbol{v}) &= \nu\Delta\boldsymbol{v} - \nabla P + \boldsymbol{N}(\boldsymbol{v}) + \boldsymbol{f} \\ \nabla \cdot \boldsymbol{v} &= 0 \end{aligned} \right\} \quad \text{in } \Omega, \tag{2.105}$$
$$\boldsymbol{v}|_{\partial\Omega} = \boldsymbol{0},$$

where $\boldsymbol{N}$ is defined in (2.103)$_1$. As we mentioned previously, unique solvability of (2.105) can be obtained by the following "splitting method", originally introduced by V. Coscia and G.P. Galdi [27]; see also [35]. By a formal Helmholtz-like decomposition of $\Delta\boldsymbol{v}$ we have

$$\left. \begin{aligned} \Delta\boldsymbol{v} &= \boldsymbol{u} + \nabla\pi \\ \nabla \cdot \boldsymbol{v} &= 0 \end{aligned} \right\} \quad \text{in } \Omega, \tag{2.106}$$
$$\boldsymbol{v}_{\partial\Omega} = \boldsymbol{0},$$

where, from (2.105), the field $\boldsymbol{u}$ satisfies the problem

$$\left. \begin{aligned} \nu\boldsymbol{u} - \alpha_1\boldsymbol{v} \times \nabla \times \boldsymbol{u} &= \nabla \times \boldsymbol{v} \times \boldsymbol{v} - \boldsymbol{N}(\boldsymbol{v}) + \nabla\phi - \boldsymbol{f} \\ \nabla \cdot \boldsymbol{u} &= 0 \end{aligned} \right\} \quad \text{in } \Omega, \tag{2.107}$$
$$\boldsymbol{u} \cdot \boldsymbol{n}|_{\partial\Omega} = 0,$$

with $\phi := P - \nu\pi$. It is easy to check that (2.105) and (2.106)–(2.107) are equivalent problems, provided $\boldsymbol{u}$ and $\boldsymbol{v}$ are sufficiently regular. Moreover, for a given $\boldsymbol{u}$, (2.106), is the well-known Stokes problem for $(\boldsymbol{v}, \pi)$ with Dirichlet boundary data in the domain Ω, whereas, for a given $\boldsymbol{v}$, (2.107) is a "vector" transport problem for $(\boldsymbol{u}, \phi)$. Now, consider the map $M : \boldsymbol{\varphi} \mapsto \boldsymbol{u}$, defined (formally) as the composition of the map $M_1 : \boldsymbol{\varphi} \mapsto \boldsymbol{v}$ defined by

$$\left.\begin{array}{l} \Delta \boldsymbol{v} = \boldsymbol{\varphi} + \nabla \pi \\[2mm] \nabla \cdot \boldsymbol{v} = 0 \end{array}\right\} \quad \text{in } \Omega \tag{2.108}$$

$$\boldsymbol{v}_{\partial \Omega} = \boldsymbol{0} \,,$$

with the map $M_2 : \boldsymbol{v} \mapsto \boldsymbol{u}$ defined by (2.107). The existence of a solution to (2.106)–(2.107) (and so, equivalently, to (2.105)) will be established if we show that the map M, suitably defined, has a fixed point. This latter is indeed shown in the paper of V. Coscia and G.P. Galdi [27]. Specifically, setting

$$X_m(\Omega) = \{ \boldsymbol{v} \in W^{m,2}(\Omega) : \ \nabla \cdot \boldsymbol{v} = 0 \text{ in } \Omega \,, \ \boldsymbol{v} \cdot \boldsymbol{n}|_{\partial \Omega} = 0 \}$$

with $\boldsymbol{n}$ unit outer normal to $\partial \Omega$, the following result holds; see [27] and [35, Theorem 5.1].

Theorem 2.6. *Let Ω be a bounded domain of class C^{m+2}, $\boldsymbol{f} \in W^{m,2}(\Omega)$, $m \geq 2$, and $\alpha_1, \alpha_2 \in \mathbb{R}$. Then, there exists a constant $C = C(\Omega, m, |\alpha_1 + \alpha_2|, \alpha_1, \nu) > 0$ such that, if $\|\boldsymbol{f}\|_{m,2} \leq C$, problem (2.105) has a unique solution $\boldsymbol{v} \in X_{m+2}(\Omega)$, $P \in W^{m+1}(\Omega)$. Thus, if, in particular, $m = 3$, then $\boldsymbol{v} \in C^3(\overline{\Omega})$, $P \in C^1(\overline{\Omega})$.*

Proof. We shall *sketch a proof of* Theorem 2.6, referring the reader to the papers [27], [35] for full details. We propose the following two preparatory lemmas. The first one is a classical result on the Stokes problem and its proof can be found, for example, in [36, Theorem IV.6.1]. The proof of the second one is given in [35, Lemma 3.4].

Lemma 2.1. *Let Ω be of class C^m, for some $m \geq 0$. Then, given $\boldsymbol{\varphi} \in W^{m,2}(\Omega)$, there exists one and only one solution to (2.108) $(\boldsymbol{v}, \pi) \in [W^{m+2,2}(\Omega) \cap \mathcal{D}_0^{1,2}(\Omega)] \times W^{m+1,2}(\Omega)$. In particular, this solution satisfies the estimate*

$$\|\boldsymbol{v}\|_{m+2,2} + \|p\|_{m+1,2} \leq C\|\boldsymbol{\varphi}\|_m \,,$$

where $C = C(m, \Omega) > 0$.

Lemma 2.2. *Let Ω be of class C^m, $m \geq 2$, and let*

$$\boldsymbol{v} \in X_{m+2}(\Omega), \quad \text{with } \|\boldsymbol{v}\|_{m+2,2} \leq C D \,,$$

$$\boldsymbol{f} \in W^{m,2}(\Omega), \quad \text{with } \|\boldsymbol{f}\|_{m,2} \leq \beta D \,,$$

for some $C > 0$. Then, there exists a constant $c = c(\Omega, m, |\alpha_1 + \alpha_2|, \alpha_1, \nu) > 0$ such that if $D < c$, $\beta < c$, problem (2.107) admits a unique solution, $\boldsymbol{u}$, ϕ, such that $(\boldsymbol{u}, \phi) \in X_m(\Omega) \times W^{m+1,2}(\Omega)$. Moreover $\|\boldsymbol{u}\|_m \leq D$.

Finally, we recall the following Schauder fixed point theorem.

Lemma 2.3. *A compact mapping M of a closed bounded, convex set G of a Banach space Y into itself has a fixed point.*

We now choose M as described previously, $Y := X_{m+1}(\Omega)$, $m \geq 1$, and, for $D > 0$, we define

$$G := \{\varphi \in Y : \|\varphi\|_{m,2} \leq D\} \,.$$

Recalling the definition of M, from Lemma 2.1 and Lemma 2.2 we obtain that M is well defined and that, if D is sufficiently small, namely, if $\|f\|_{m,2}$ is sufficiently small, we have $M(G) \subset G$. Moreover, the compact embedding $X_m \subset X_{m-1}$ implies the compactness of M. Then it remains to check the continuity of M in the Y-norm. To this end, it can be readily shown that, in view of the compactness of M, it is enough to check the continuity of M in the L^2-norm. To prove this latter, let v_n and v be the solutions to (2.108) corresponding to φ_n and φ, respectively, and set $u_n := M(\varphi_n)$, $u := M(\varphi)$. We now subtract to (2.107) the same equation written with $u \equiv u_n$ and multiply the resulting equation by $u - u_n$. Integrating by parts over Ω, we find

$$\begin{aligned}
&\nu\|u - u_n\|_2^2 + \alpha_1(\nabla \times (u - u_n) \times v, u - u_n) \\
&+ \alpha_1(\nabla \times u_n \times (v - v_n), u - u_n) - ((\omega - \omega_n) \times v, u - u_n) \\
&- (\omega_n \times (v - v_n), u - u_n) + (N(v) - N(v_n), u - u_n) = 0 \,,
\end{aligned} \qquad (2.109)$$

where $\omega_n := \nabla \times v_n$. By the Schwarz inequality, by (2.103) and by the Sobolev embedding theorem one can show that (see [35, Lemma 3.1])

$$\begin{aligned}
&|((\omega - \omega_n) \times v, u - u_n) + (\omega_n \times (v - v_n), u - u_n) \\
&\qquad + (N(v) - N(v_n), u - u_n)| \\
&\leq C_1(\|v\|_{m+2,2} + \|v_n\|_{m+2,2})\|v - v_n\|_{m+2,2}\|u - u_n\|_2 \,,
\end{aligned} \qquad (2.110)$$

with $C_1 = C_1(\Omega, m, |\alpha_1 + \alpha_2|) > 0$. Furthermore, again by the Schwarz inequality and the Sobolev theorem, we find

$$\begin{aligned}
|(\nabla \times (u - u_n) \times v, u - u_n)| &= |((u - u_n) \cdot \nabla v, u - u_n) \\
&+ ((u - u_n) \times \nabla \times v, u - u_n)| \\
&\leq C_2\|v\|_{m+2,2}\|u - u_n\|_2^2 \,,
\end{aligned} \qquad (2.111)$$

with $C_2 = C_2(\Omega, m) > 0$. From Lemma 2.1 and from the fact that $\varphi, \varphi_n \in G$, we deduce

$$\|v\|_{m+2,2}, \ \|v_n\|_{m+2,2} \leq C\,D, \quad \|v - v_n\|_{m+2,2} \leq C\,\|\varphi - \varphi_n\|_{m,2} \,,$$

and so, by choosing D sufficiently small, from (2.109)–(2.111) we conclude

$$\nu\|u - u_n\|_2 \leq C_3\|\varphi - \varphi_n\|_2 \,,$$

with $C_3 > 0$ independent of φ, u and n. We have thus shown the continuity of the map M which, therefore, by Schauder's theorem Lemma 2.3, possesses a fixed point in G. This concludes the existence proof. Concerning uniqueness, let (v_1, p_1) and (v_2, p_2) be two solutions of (2.105) corresponding to the same data and belonging to the functional class stated in the theorem. Moreover, denote by u_1 and u_2 the Helmholtz projections of Δv_1 and Δv_2 on $X_0(\Omega)$. We recall that both $\boldsymbol{u}_i$, $i = 1, 2$, satisfy (2.106). Thus, letting $\boldsymbol{u} := \boldsymbol{u}_1 - \boldsymbol{u}_2$ and using a procedure similar to that used to prove the continuity of the map M, we can show that

$$\nu\|\boldsymbol{u}\|_{m,2} \leq C_4 \left(\|\boldsymbol{v}_1\|_{m+2,2} + \|\boldsymbol{v}_2\|_{m+2,2}\right) \|\boldsymbol{u}\|_{m,2}\,,$$

where $C_4 = C_4(\Omega, m, \alpha_1, \alpha_2)$; see [27] for details. Since

$$\|\boldsymbol{v}_1\|_{m+2,2} + \|\boldsymbol{v}_2\|_{m+2,2} \leq C_5\, D = C_6 \|\boldsymbol{f}\|_{m,2}\,,$$

with $C_i > 0$, $i = 5, 6$, independent of $\boldsymbol{v}_k$, $k = 1, 2$. These latter two displayed equations then prove uniqueness if $\|\boldsymbol{f}\|_{m,2}$ is sufficiently small. $\square$

The method used in the proof of Theorem 2.6 can be applied to more general situations like, for example, steady motion of a second-order fluid in a piping system under the action of a constant flow-rate, Φ. This amounts to solve problem (2.105) with $\boldsymbol{f} = \boldsymbol{0}$ and with the side condition

$$\int_S \boldsymbol{v} \cdot \boldsymbol{n}\, dS = \Phi\,,$$

where S is the cross-section of the "distorted pipe" Ω (see Sections 1.3 and 1.3.1). In fact, one can show that the Hagen–Poiseuille flow given in Section 1.1.1 is also a solution to (2.105) (in an infinite straight pipe). Thus, similarly to the Navier–Stokes case, one can look for a solution to the problem in the form $\boldsymbol{v} = \boldsymbol{w} + \Phi\boldsymbol{a}$, where $\boldsymbol{a}$ is the "flow-rate carrier" constructed in Section 1.3.1. To the resulting equation for $\boldsymbol{w}$ we then apply the "splitting method" adopted in the proof of Theorem 2.6. This procedure, ultimately furnishes existence and uniqueness to the problem, at least if $|\Phi|$ does not exceed a given constant. The details of this proof can be found in the paper of K. Pileckas et al. [87].

The same method can be suitably adapted to consider well-posedness in exterior domains. The interested reader is referred to the papers of G.P. Galdi et al. [44], A. Novotný et al. [84].

We finally observe that this "splitting method" has also been successfully applied to numerical studies of viscoelastic fluids by A. Sequeira and her associates; see, e.g., [100], [101].

2.4.2. Instability of the rest state and well-posedness of the initial-boundary value problem. Detailed experiments performed by rheologists consistently show a value of the parameter α_1 in the range $-100 \div -1\,\mathrm{gr/cm}$; see, e.g., [61, Section 17.11]. However, it is easy to see that, if $\alpha_1 < 0$, the rest state of a second-order fluid is nonlinearly unstable in the sense of Liapounov. We shall suppose, for simplicity, that Ω is bounded. Define the norm

$$\|\boldsymbol{v}\| := \max_{x \in \Omega} |\nabla\boldsymbol{v}(x)| + \|\nabla\boldsymbol{v}\|_2$$

and assume, *per absurdum*, that the rest state is stable in the norm $\|\cdot\|$ in the Liapunov sense, that is,

$$\forall \varepsilon > 0, \ \exists \delta(\varepsilon) > 0 \ : \ \|\boldsymbol{v}(0)\| < \delta \implies \sup_{t \geq 0} \|\boldsymbol{v}(t)\| < \varepsilon\,. \tag{2.112}$$

By dot-multiplying both sides of $(2.102)_1$ by v, by integrating by parts over Ω and by taking into account $(2.102)_2$–(2.104), we obtain

$$\frac{dN}{dt} = \nu\|\nabla v\|_2^2 + (N(v), v) \,, \qquad (2.113)$$

where

$$N(t) = \tfrac{1}{2}\int_\Omega \left(|\alpha_1|\,|\nabla v|^2 - |v|^2\right) \,.$$

By a straightforward calculation, we show that

$$(N(v), v) = 4(\alpha_1 + \alpha_2)\int_\Omega [\nabla v \cdot (\nabla v)^\top] \cdot \nabla v \geq -4|\alpha_1 + \alpha_2|\,\|v\|\,\|\nabla v\|^2 \,.$$

Therefore, from (2.112)–(1.207) we may choose $\varepsilon > 0$ such that

$$\frac{dN}{dt} \geq \lambda N(t) \,, \quad \lambda := \frac{\nu - 4\varepsilon|\alpha_1 + \alpha_2|}{|\alpha_1|} > 0 \,.$$

As a consequence, we get

$$N(t) \geq N(0)\exp(\lambda t) \,, \quad \text{for all } t > 0 \,, \qquad (2.114)$$

provided $\|v(0)\| < \delta$. Let u_n be the normalized eigenfunction of the Stokes operator corresponding to the eigenvalue σ_n, namely,

$$\left.\begin{array}{l} -\Delta u_n = \sigma_n u_n + \nabla p_n \\[4pt] \nabla \cdot u_n = 0 \end{array}\right\} \text{ in } \Omega,$$

$$u_n = 0 \ \text{ at } \partial\Omega \,.$$

It is well known that $\sigma_n \to \infty$ as $n \to \infty$. We choose as initial data $v(0) = A_n\,u_n$ where $A_n > 0$ and u_n satisfy the conditions

$$\frac{\|\nabla u_n\|^2}{\|u_n\|_2^2} \equiv \sigma_n > \frac{1}{|\alpha_1|} \,, \quad A_n\|u_n\| < \delta \,.$$

From (2.114) it then follows that

$$\|v(t)\|^2 \geq \frac{(|\alpha_1|\,\sigma_n - 1)}{|\alpha_1|}\|v(0)\|_2^2 \exp(\lambda t) \,, \quad \text{for all } t \geq 0 \,,$$

and hence

$$\|v(t)\| \to \infty \ \text{ as } t \to \infty \,,$$

contradicting the hypothesis of stability.

However, if α_1 is negative, but not too large in magnitude, one can still prove that the initial-boundary value problem associated to (2.102)–(2.103) has a unique, classical solution, at least for a short time interval. Specifically, we have the following result; see [35, Theorem 4.1].

Theorem 2.7. *Let Ω be a bounded domain of $\mathbb{R}^3$ of class C^{m+2}, $m \geq 2$, and let $\alpha_1 > -\lambda_1$, where λ_1 is the smallest eigenvalue of the Stokes operator in Ω with Dirichlet boundary conditions. Then, for any $t_0 > 0$, $\boldsymbol{v}(\cdot, t_0) \equiv \boldsymbol{v}_0 \in X_{m+2}$ and $\boldsymbol{f} \in L_{\mathrm{loc}}^\infty(\overline{\mathbb{R}}_+; W^{m,2}(\Omega))$, there exists $T = T(\alpha_1, \alpha_2, \boldsymbol{v}_0, \Omega, m) > 0$ such that problem (2.102)–(2.103) has a unique solution in $I := [t_0, t_0 + T]$ satisfying*

$$\boldsymbol{v} \in C^0(I; X_{m+1}(\Omega)) \cap L^\infty(I, W^{m+2}(\Omega)), \quad \nabla P \in L^\infty(I, W^{m+1}(\Omega)),$$

$$\frac{\partial \boldsymbol{v}}{\partial t} \in L^\infty(I, W^{m+1}(\Omega)).$$

In particular, if $m = 3$ and if, in addition, $\partial \boldsymbol{f}/\partial t \in L_{\mathrm{loc}}^\infty(\overline{\mathbb{R}}_+ W^{3,2}(\Omega))$, then

$$\boldsymbol{v} \in C^1(I; C^3(\overline{\Omega})), \quad P \in C^0(I; C^1(\overline{\Omega})).$$

We shall not include here the proof of this theorem. Rather, we shall limit ourselves to observe that its method of proof is again based on a "splitting" argument similar to that used in the solvability of the boundary-value problem. Actually, in this case, the solution is sought as a fixed point of the map $\widetilde{M} : \boldsymbol{\varphi} \mapsto \boldsymbol{u}$, defined as the composition of the map $\widetilde{M_1} : \boldsymbol{\varphi} \mapsto \boldsymbol{v}$, defined as

$$\left. \begin{array}{c} \boldsymbol{v} - \alpha_1 \Delta \boldsymbol{v} = \boldsymbol{\varphi} + \nabla \pi \\[2mm] \nabla \cdot \boldsymbol{v} = 0 \end{array} \right\} \quad \text{in } \Omega \times I,$$

$$\boldsymbol{v}_{\partial\Omega \times I} = \boldsymbol{0},$$

and of the map $\widetilde{M_2} : \boldsymbol{v} \mapsto \boldsymbol{u}$, where $\boldsymbol{u}$ solves the time-dependent vector transport problem

$$\left. \begin{array}{c} \dfrac{\partial \boldsymbol{u}}{\partial t} + \sigma \left(\boldsymbol{u} - \boldsymbol{v} \right) - \boldsymbol{v} \times \nabla \times \boldsymbol{u} = \boldsymbol{N}(\boldsymbol{v}) - \nabla \phi + \boldsymbol{f} \\[2mm] \nabla \cdot \boldsymbol{u} = 0 \end{array} \right\} \quad \text{in } \Omega \times I,$$

$$\boldsymbol{u} \cdot \boldsymbol{n}|_{\partial\Omega \times I} = 0, \quad \boldsymbol{u}(\cdot, t_0) = \boldsymbol{u}(t_0),$$

with $\sigma := \nu/\alpha_1$ and $\phi := P + \pi/\alpha_1 + \partial\pi/\partial t$. Notice that if we formally take $\boldsymbol{\varphi} = \boldsymbol{u}$ in the above equations, we obtain the initial-boundary value problem associated to (2.102)–(2.103).

2.5. Some results in the mathematical theory of Oldroyd-B fluids and related models

In this section we will be interested in well-posedness questions related to the equations

$$\left.\begin{array}{l} \dfrac{\partial \boldsymbol{v}}{\partial t} + \boldsymbol{v} \cdot \nabla \boldsymbol{v} = \mu_\infty \Delta \boldsymbol{v} - \nabla p + \nabla \cdot \boldsymbol{\tau} + \boldsymbol{f} \\[2mm] \nabla \cdot \boldsymbol{v} = 0 \\[2mm] \lambda_1 \left(\dfrac{\partial \boldsymbol{\tau}}{\partial t} + \boldsymbol{v} \cdot \nabla \boldsymbol{\tau} - \boldsymbol{\tau} \cdot (\nabla \boldsymbol{v})^\top - \nabla \boldsymbol{v} \cdot \boldsymbol{\tau} \right) + \boldsymbol{\tau} \\[4mm] \qquad\qquad\qquad = 2\left(\mu_0 - \mu_\infty\right) \boldsymbol{D}(\boldsymbol{v}) \end{array}\right\} \ \text{in } \Omega \times (0,T),$$

$$\boldsymbol{v}(x,t)|_{\partial\Omega \times (0,T)} = \boldsymbol{0},$$

(2.115)

and to their steady-state counterpart, obtained by suppressing the time-dependence on the unknowns $\boldsymbol{v}$, p and $\boldsymbol{\tau}$. In these equations λ_1 and λ are constant, positive material parameters, while ν is the coefficient of kinematic viscosity. As we mentioned in Section 2.3, (2.115) describes the motion of a viscoelastic Oldroyd-B fluid with associated boundary condition. Notice that, if $\lambda_1 = 0$, then $(2.115)_{1,2,3}$ reduce to the Navier–Stokes equations.

It is not hard to show – as noticed for the first time by M. Renardy [92] in the steady-state case and then, successively, further developed in the general case, steady or unsteady, by C. Guillopé and J.-C. Saut [51, 52] – that (2.115) can be split into a Stokes-like problem and into a transport-like problem of the same types considered for the second-order fluid in the previous sections. Similarly to Theorem 2.1 and Theorem 2.2, the existence proof is based upon finding the fixed point of a suitable composite map, M, involving both problems. Actually, in the case at hand, the splitting is even more "natural" than that for a fluid of grade 2. In fact, we define $M := M_2 \circ M_1$, and $M_1 : (\boldsymbol{\varphi}, \boldsymbol{\sigma}) \mapsto \boldsymbol{v}$ where $\boldsymbol{v}$ is the solution to the Stokes problem

$$\left.\begin{array}{l} \dfrac{\partial \boldsymbol{v}}{\partial t} = \mu_\infty \Delta \boldsymbol{v} - \nabla p + \boldsymbol{f}_1 \\[3mm] \nabla \cdot \boldsymbol{v} = 0 \end{array}\right\} \ \text{in } \Omega \times (0,T),$$

(2.116)

$$\boldsymbol{v}(x,t)|_{\partial\Omega \times (0,T)} = \boldsymbol{0}, \quad \boldsymbol{v}(x,0) = \boldsymbol{v}_0(x), \ x \in \Omega,$$

with

$$\boldsymbol{f}_1 := -\boldsymbol{\varphi} \cdot \nabla \boldsymbol{\varphi} + \nabla \cdot \boldsymbol{\sigma} + \boldsymbol{f},$$

and $M_2 : \boldsymbol{v} \mapsto \boldsymbol{\tau}$ where $\boldsymbol{\tau}$ satisfies the transport equation

$$\lambda_1 \left(\dfrac{\partial \boldsymbol{\tau}}{\partial t} + \boldsymbol{v} \cdot \nabla \boldsymbol{\tau} \right) + \boldsymbol{\tau} = \boldsymbol{f}_2 \ \text{in } \Omega \times (0,T), \quad \boldsymbol{\tau}(x,0) = \boldsymbol{\tau}_0(x), \ x \in \Omega, \quad (2.117)$$

with

$$\boldsymbol{f}_2 := \lambda_1 (\boldsymbol{\tau} \cdot (\nabla \boldsymbol{v})^\top + \nabla \boldsymbol{v} \cdot \boldsymbol{\tau}) + 2\left(\mu_0 - \mu_\infty\right) \boldsymbol{D}(\boldsymbol{v}).$$

In (2.116)–(2.117), $\boldsymbol{v}_0$ and $\boldsymbol{\tau}_0$ are prescribed vector and second-order tensor, respectively. It is clear that the existence of a solution to the initial-boundary value problem associated to (2.115) is established if one shows that M has a fixed point.

Classical results on the Stokes problem and on the time-dependent transport equation allow C. Guillopé and J.-C. Saut [51, 52] to establish that the map M

has a fixed point in a suitable function class, at least for T sufficiently "small". However, if the size of v_0, τ_0, and f in appropriate norms, is suitably restricted and if *some further restriction* on the magnitude of the dimensionless quantity $\omega := 1 - \mu_\infty/\mu_0$ is imposed, then one can take $T = \infty$. More precisely, we have the following result for whose proof we refer to the cited papers of C. Guillopé and J.-C. Saut.

Theorem 2.8. *Let Ω be a bounded domain of $\mathbb{R}^3$ of class C^3, and assume that*

$$f \in L^2_{\mathrm{loc}}(\mathbb{R}_+; W^{1,2}(\Omega))\,, \quad \frac{\partial f}{\partial t} \in L^2_{\mathrm{loc}}(\mathbb{R}_+; W^{-1,2}_0(\Omega))\,,$$

$$v_0 \in \mathcal{D}^{1,2}_0(\Omega) \cap W^{2,2}(\Omega)\,, \quad \tau_0 \in W^{2,2}(\Omega)\,.$$

Then, there exists $T > 0$, such that problem (2.115) with initial conditions $v(x,0) = v_0(x)$ and $\tau(x,0) = \tau_0(x)$ has one and only one solution (v,p,τ) in the time interval $(0,T)$, in the class

$$v \in L^2(0,T; W^{3,2}(\Omega)) \cap C^0([0,T]; \mathcal{D}^{1,2}_0(\Omega) \cap W^{2,2}(\Omega))\,,$$

$$\frac{\partial v}{\partial t} \in L^2(0,T; \mathcal{D}^{1,2}_0(\Omega)) \cap C^0([0,T]; L^2_\sigma(\Omega))\,,$$

$$p \in L^2(0,T; W^{2,2}(\Omega))\,, \quad \tau \in C^0([0,T]; W^{2,2}(\Omega))\,.$$

Assume, in addition, that Ω is of class C^4. Then, there exist constants $\omega_0 > 0$ and $C > 0$ such that if $0 < \omega < \omega_0$ and

$$\|v_0\|_{2,2} + \|\tau_0\|_{2,2} + \operatorname*{ess\,sup}_{t \in \mathbb{R}_+} \left(\|f(t)\|_{1,2} + \left\| \frac{\partial f}{\partial t}(t) \right\|_{-1,2} \right) < C\,,$$

we may take $T = \infty$.

Remark 2.12. An interesting generalization of Theorem 2.8 can be found in the paper of A. Hakim [56]. Besides considering more general models of the type mentioned in Section 2.3, in this paper the author allows the viscosity to be shear-dependent. In other words, the assumption that the coefficient $\tilde{\mu} := \mu_0 - \mu_\infty$ in (2.92) is independent of $|D|$ is relaxed to include the more physical relevant case of when $\tilde{\mu}$ is a function of $|D|$. Particular cases include the "power-law" model (see (2.5)) and the Carreau model (see (2.6)). Hakim's results hold when $\Omega \subset \mathbb{R}^2$.

The global existence result of Theorem 2.8, suitably elaborated and under the assumption that f does not depend on time, leads to the existence of steady-state solutions for problem (2.115), namely, to a triple $v = v(x)$, $p = p(x)$, and $\tau = \tau(x)$ such that

$$\left. \begin{aligned} &\varphi \cdot \nabla\varphi = \mu_\infty \Delta v - \nabla p + \nabla \cdot \tau + f \\ &\nabla \cdot v = 0 \\ &\lambda_1 \left(v \cdot \nabla\tau - \tau \cdot (\nabla v)^\top - \nabla v \cdot \tau \right) + \tau = 2 \left(\mu_0 - \mu_\infty \right) D(v) \end{aligned} \right\} \quad \text{in } \Omega, \tag{2.118}$$

$$v(x)|_{\partial\Omega} = 0\,.$$

However, this result, besides the requirement of "smallness" of $\boldsymbol{f}$ in a suitable norm, necessitates the further condition $0 < \omega < \omega_0$, for some $\omega_0 > 0$. We recall that the only physical condition imposed on ω is that it is positive and strictly less than one. We shall now sketch a simple argument that shows why this extra condition is needed for the above mentioned fixed point proof to work. Let $M : (\boldsymbol{\varphi}, \boldsymbol{\sigma}) \mapsto (\boldsymbol{v}, \boldsymbol{\tau})$ be the composite map defined through the maps $M_1 : (\boldsymbol{\varphi}, \boldsymbol{\sigma}) \mapsto \boldsymbol{v}$ with

$$\left.\begin{aligned} \boldsymbol{v} \cdot \nabla \boldsymbol{v} &= \mu_\infty \Delta \boldsymbol{v} - \nabla p + \nabla \cdot \boldsymbol{\sigma} + \boldsymbol{f} \\[2mm] \nabla \cdot \boldsymbol{v} &= 0 \end{aligned}\right\} \ \text{in } \Omega, \tag{2.119}$$

$$\boldsymbol{v}(x)|_{\partial\Omega} = \boldsymbol{0}\,,$$

and $M_2 : \boldsymbol{v} \mapsto \boldsymbol{\tau}$ where

$$\lambda_1 \left(\boldsymbol{v} \cdot \nabla \boldsymbol{\tau} - \boldsymbol{\tau} \cdot (\nabla \boldsymbol{v})^\top - \nabla \boldsymbol{v} \cdot \boldsymbol{\tau}\right) + \boldsymbol{\tau} = 2\left(\mu_0 - \mu_\infty\right) \boldsymbol{D}(\boldsymbol{v}) \ \text{ in } \Omega. \tag{2.120}$$

We assume, for example, that M is defined on S_D, $D > 0$, where

$$S_D := \{(\boldsymbol{\varphi}, \boldsymbol{\sigma}) \in [\mathcal{D}_0^{1,2}(\Omega) \cap W^{m+2,2}(\Omega)] \times W^{m+1,2}(\Omega) :$$

$$\|\boldsymbol{\varphi}\|_{m+2,2} + \|\boldsymbol{\sigma}\|_{m+1,2} \le D\} \subset W^{m+1,2}(\Omega) \times W^{m,2}(\Omega)$$

and $m \ge 2$. The first thing to show is that, under suitable assumptions on $\boldsymbol{f}$, M maps S_D into itself for an appropriate choice of D. To this end, by classical results concerning the Stokes operator in a bounded domain (see Lemma 2.1), and taking into account that $W^{m,2}(\Omega)$ is an algebra for $m \ge 2$, from (2.119) we get

$$\mu_\infty \|\boldsymbol{v}\|_{m+2,2} \le C_1 \left(\|\boldsymbol{\varphi}\|_{m+2}^2 + \|\boldsymbol{\sigma}\|_{m+1,2} + \|\boldsymbol{f}\|_m\right), \tag{2.121}$$

where $C_1 = C_1(\Omega, m) > 0$. Furthermore, by known results on the steady-state transport equation (see also Lemma 2.2) we can prove the following estimate for (2.120):

$$\|\boldsymbol{\tau}\|_{m+1,2} \le C_2 \left((\mu_0 - \mu_\infty)\|\boldsymbol{v}\|_{m+2,2} + \|\boldsymbol{v}\|_{m+2,2}\|\boldsymbol{\tau}\|_{m+1,2}\right), \tag{2.122}$$

with $C_2 = C_2(\Omega, m, \lambda_1) > 0$. Recalling that $(\boldsymbol{\varphi}, \boldsymbol{\sigma}) \in S_D$ and choosing $D = \boldsymbol{f}/\beta$, $\beta > 0$, from (2.121)–(2.122) it follows, in particular, that

$$\|\boldsymbol{\tau}\|_{m+1} \le C_3 \left((\mu_0 - \mu_\infty)\|\boldsymbol{v}\|_{m+2,2} + (D^2 + D + \beta\,D)\|\boldsymbol{\tau}\|_{m+1,2}\right),$$

with $C_3 = C_3(\Omega, m, \lambda_1, \mu_\infty) > 0$. Thus, choosing D sufficiently small, from this latter relation we get

$$\|\boldsymbol{\tau}\|_{m+1} \le 2C_3 \left(\mu_0 - \mu_\infty\right)\|\boldsymbol{v}\|_{m+2,2}\,. \tag{2.123}$$

Employing again (2.121) into (2.123), we get, for some $C_4 = C_4(\Omega, m, \lambda_1, \mu_\infty) > 0$,

$$\|\boldsymbol{\tau}\|_{m+1} \le 2C_4\,\omega(D^2 + D + \beta\,D)$$

from which it is obvious that if we want $\|\boldsymbol{\tau}\|_{m+1} \le D$, we have to impose that ω is less than a suitable quantity.

It is of some interest to investigate if the above restriction on ω can be eliminated. This problem has been addressed by R. Talhouk [104] who was able to remove this restriction, by recurring to an idea originally due to M. Renardy

[92]. Talhouk's method of proof is again based on a fixed-point argument, but, as expected, the map involved is different than the one we have just considered. Referring the reader to [104] for a proof, here we shall limit ourselves to state the main result.

Theorem 2.9. *Let Ω be a bounded domain in $\mathbb{R}^3$ of class C^{m+2}, $m \geq 1$, $\boldsymbol{f} \in W^{m,2}(\Omega)$ and $\omega := (1 - \mu_\infty/\mu_0) \in (0,1)$. Then, there exists a constant $C = C(\Omega, m, \lambda_1, \mu_\infty, \mu_0) > 0$ such that, if $\|\boldsymbol{f}\|_{m,2} \leq C$, problem (2.118) has at least one solution*

$$(\boldsymbol{v}, p, \boldsymbol{\tau}) \in [\mathcal{D}_0^{1,2}(\Omega) \cap W^{m+1,2}(\Omega)] \times W^{m,2}(\Omega) \times W^{m,2}(\Omega) := Y^m(\Omega).$$

Moreover, there exists $C_0 > 0$, independent of the particular solution, such that if two solutions, corresponding to the same $\boldsymbol{f}$, have their $Y^m(\Omega)$-norm less than C_0, they must coincide.

The results of Theorem 2.9 have been generalized and/or extended in several directions. Generalizations and extensions include the cases when the viscoelastic model is "shear thinning" or "shear-thickening" as in Remark 2.12, and the flow occurs in an exterior domain. These results were found by C. Guillopé and her collaborators [53, 50] and, independently, by A. Sequeira and N. Arada [2, 3].

For well-posedness questions of the boundary-value problem associated to fluids of rate-type in piping system, we refer, among others, to the paper by M.A. Fontelos and A. Friedman [32], and that of K. Pileckas et al. [87].

Remark 2.13. Seemingly, an area that has not been contributed yet, even though very significant from the point of view of applications, is that of existence, uniqueness and stability of time-periodic flow in an unbounded piping system for fluids of rate-type. Moreover, to my knowledge, there is no result concerning mathematical modelling and corresponding well-posedness results for a *bounded* (truncated) system of pipes, for both second-order and rate-type fluid models.

3. Problems in liquid-particle interaction

The motion of small particles in a viscous liquid represents one of the main focuses of applied research; see, e.g., [93, 62] and the references cited therein. Actually, studies on liquid-particle interaction cover a wide range of applications, and can be ubiquitously found in the engineering literature, from manufacturing of short-fiber composites [1, 73] to separation of macromolecules by electrophoresis, [49, 55, 107, 105], from flow-induced microstructures [62] to blood flow [97, 15, 110] and particle-laden materials [22]. The presence of the particles affects the flow of the liquid, and this, in turn, affects the motion of the particles, so that the problem of determining the flow characteristics for the combined system liquid-particle is highly coupled. It is just this latter feature that makes any fundamental problem related to liquid-particle interaction a particularly challenging one.

In Sections 2.1 and 2.3.2 we have given several examples of particle-liquid interaction and, more important, we have emphasized how this interaction, for a fixed type of particles, may dramatically depend on the physical properties of the liquid. For example, we have noticed that a homogeneous particle in the shape of an ellipsoid, dropped from rest in a viscous liquid, will orient itself with its broadside perpendicular or parallel to the gravity according to whether the liquid is Newtonian or viscoelastic, no matter what the initial orientation.

Another type of problem that requires separate attention is the motion of spherical particles in the shear flow of a viscous liquid. This problem drew the attention of many researchers, from both theoretical and experimental viewpoints, after the famous series of experiments performed by G. Segrè and A. Silberberg [98, 99] on the inertial migration of neutrally buoyant spheres in a the shear flow of a Navier–Stokes liquid in a pipe, at low Reynolds number. In this experiment, a number of spheres of same density as the liquid, is initially at rest in a pipe of circular cross-section of radius R. The flow-rate of the liquid is then raised to a fixed constant value and then, in a certain range of (particle) Reynolds number, it is observed that, eventually, the spheres will chain to form a stable ring located at a distance from the axis of the pipe of approximately $0.6\,R$; see Figure 10.

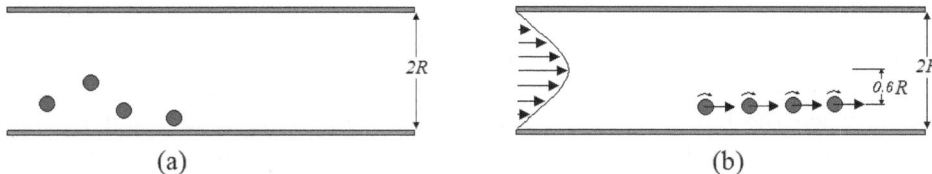

FIGURE 10. Sketch of the Segrè-Silberberg experiment; (a) floating spheres (zero flow-rate); (b) chaining spheres (non-zero, constant flow-rate)

A most striking application of the migration of "spherical" particles in shear flow can be found in the lateral migration of erythrocytes (red blood cells) in Poiseuille flow in a channel, under a sufficiently large shear rate. A detailed experimental analysis of this phenomenon was performed by W.S.J. Uijttwaal et al. in [110], where the migration of blood cells was studied in the shear flow of a phosphate buffer solution in a duct. The geometry of the flow chamber and of the particles make the setting, basically, two-dimensional. In Figure 11 are reported some of the findings of [110], concerning the concentration of erythrocytes versus the width, w, of the duct, from which it appears that the highest concentration is at about $0.6\,w$ from the center of the channel. The wall shear rate is 2800 sec^{-1}, which makes erythrocites behave, basically, as rigid particles [97].

Objective of this chapter is to provide a mathematical analysis of (a) orientation of symmetric particles sedimenting in a viscous liquid, and (b) migration of spherical particles in a channel under shear flow.

In case (a), we shall show, in particular, that the competition between "inertial effects" (purely Newtonian) and the "normal-stress effects" (purely viscoelastic) is responsible for the orientation of a homogeneous symmetric and *sufficiently*

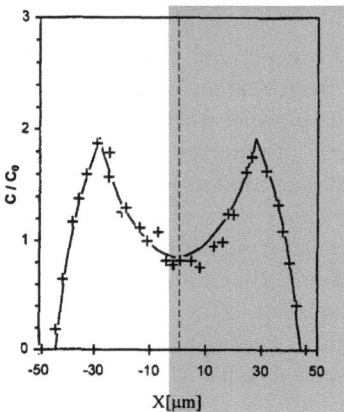

FIGURE 11. Erythrocyte concentration versus channel width in Poiseuille flow, with particle Reynolds number 10^{-3} and wall shear rate 2800 sec^{-1} (after Uijttwaal et al. [110]).

smooth particle in a viscoelastic liquid: whenever the former is predominant, the particle will orient with its broadside, b, perpendicular to the gravity, g, and, vice versa, when the latter prevails, then b and g are parallel. This result is shown in a small range of the relevant dimensionless parameters, the Reynolds and the Weissenberg numbers, in agreement with the experimental findings. Moreover, we shall also present some experimental work which shows that cylindrical particles free-falling in a viscoelastic liquid may show an equilibrium angle between b and g somewhere between 0 and $\pi/2$, depending on whether they have sharp flat ends or smooth round ends (*"shape-tilting"* phenomenon).

Concerning problem (b), we shall consider the two-dimensional version of it, where the pipe is modeled as a channel and the sphere becomes a "disk". Thus, not only shall we provide a mathematical analysis of the Segrè–Silberberg experiment but, in addition, we shall also consider the case of non-buoyant particles and the influence of viscoelasticity on the migration. In both cases, we shall use the second-order fluid to model the viscoelastic liquid. Our presentation of the results will closely follow the papers of G.P. Galdi et al. [42] and of G.P. Galdi and V. Heuveline [40]. We also refer to these papers for the many further references on the subject.

3.1. Sedimentation of symmetric particles in viscoelastic liquid

Suppose that a rigid body, $\mathcal{B}$, is moving in a viscoelastic liquid, $\mathcal{L}$, filling the whole space, [31] under the sole action of gravity g, namely, $\mathcal{B}$ is *sedimenting in* $\mathcal{L}$. We

---

[31] We assume that the liquid occupies the entire space exterior to the body. From the physical point of view, this means that we neglect "wall effects". In fact, as observed experimentally, these latter are completely irrelevant for the orientation phenomenon we are interested in.

assume that $\mathcal{B}$ is a homogeneous body of revolution around an axis, a, with fore-and-aft symmetry and that its translational velocity, $\boldsymbol{U}$, and angular velocity, $\boldsymbol{\omega}$, are constant in time with respect to an inertial frame. Our objective in this section is to find the possible equilibrium configurations of $\mathcal{B}$, determined by the angle, θ, between a and $\boldsymbol{g}$ at *small and nonzero* Reynolds and Weissenberg numbers. In order to solve this problem, we make the following observations.

(a) As observed experimentally, in the above range of dimensionless numbers the motion of $\mathcal{B}$ is purely translational, that is, $\boldsymbol{\omega} = \boldsymbol{0}$.

(b) At all possible equilibrium configurations, determined by θ, the total torque, $\mathcal{M}$, exerted by the liquid on the body must vanish.

On the basis of these two observations we shall restrict ourselves to translational motions of $\mathcal{B}$, and will evaluate $\mathcal{M}$ along these motions. With this in mind, we begin to write the Cauchy stress, $\boldsymbol{T}$, of $\mathcal{L}$, in dimensionless form, as

$$\boldsymbol{T} = \boldsymbol{T}(\boldsymbol{w}, \phi) := \boldsymbol{T}_N(\boldsymbol{w}, \phi) - \lambda\,\boldsymbol{S}(\boldsymbol{w})\,, \qquad (3.1)$$

where

$$\boldsymbol{T}_N(\boldsymbol{w}, \phi) := -\phi\boldsymbol{I} + 2\boldsymbol{D}(\boldsymbol{w}),$$
$$\boldsymbol{S}(\boldsymbol{w}) := (\boldsymbol{A}_2(\boldsymbol{w}) + \varepsilon\boldsymbol{A}_1(\boldsymbol{w}) \cdot \boldsymbol{A}_1(\boldsymbol{w}))\,, \qquad (3.2)$$

$\lambda = -\alpha_1 V/d\mu$ (Weissenberg number), $\varepsilon = \alpha_2/\alpha_1$, d is the diameter of $\mathcal{B}$, V is a scaling speed, and μ is the (constant) shear viscosity coefficient of $\mathcal{L}$. Finally, the constants α_1 and α_2 and the tensor fields $\boldsymbol{A}_1$ and $\boldsymbol{A}_2$ are defined in Section 2.3. Notice that, since $\alpha_1 < 0$ in a realistic liquid, we have $\lambda > 0$. It is convenient to refer the equation of motion of the system $\{\mathcal{B}, \mathcal{L}\}$ to a frame attached to $\mathcal{B}$ with the origin at the center of mass of $\mathcal{B}$. Therefore, the relevant equations become [42]

$$\left.\begin{array}{l} \mathrm{Re}\,\boldsymbol{u} \cdot \nabla\boldsymbol{u} = \Delta\boldsymbol{u} - \nabla p - \lambda\nabla \cdot \boldsymbol{S}(\boldsymbol{u}) + \boldsymbol{g} \\[2mm] \nabla \cdot \boldsymbol{u} = 0 \end{array}\right\} \text{ in } \Omega,$$

$$\lim_{|x|\to\infty} \boldsymbol{u}(y) = -\boldsymbol{U}, \qquad (3.3)$$

$$\boldsymbol{u}(x) = \boldsymbol{0}, \quad \boldsymbol{y} \in \Sigma,$$

$$m\boldsymbol{g} = \int_\Sigma \boldsymbol{T}(\boldsymbol{u}, p) \cdot \boldsymbol{n}, \qquad (3.4)$$

$$\int_\Sigma \boldsymbol{x} \times \boldsymbol{T}(\boldsymbol{u}, p) \cdot \boldsymbol{n} = \boldsymbol{0}. \qquad (3.5)$$

In (3.3)–(3.5), Ω is the (time-independent) region exterior to $\mathcal{B}$ occupied by $\mathcal{L}$, $\boldsymbol{u}$ is the velocity field of $\mathcal{L}$ relative to $\mathcal{B}$, $\mathrm{Re} := Vd/\mu$ is the Reynolds number, m is the effective mass of $\mathcal{B}$, [32] $\boldsymbol{g}$ is a unit vector in the direction of gravity, and $\Sigma := \partial\Omega$ (the bounding surface of $\mathcal{B}$). We recall that a solution to this problem is given by the quadruple $\{\boldsymbol{u}, p, \boldsymbol{U}, \boldsymbol{g}\}$, which we will call *translational steady fall.* It

---

[32] Namely, $m = (\rho_s/\rho_f - 1)|\mathcal{B}|/d^3$, where ρ_s and ρ_f denote densities of $\mathcal{B}$ and $\mathcal{L}$, respectively, and $|\mathcal{B}|$ is the volume of $\mathcal{B}$. In what follows, we shall assume that $m > 0$.

is worth emphasizing that also $\boldsymbol{g}$ is an unknown, since the motion of the system is referred to a frame attached to $\mathcal{B}$, whose orientation with respect to gravity is part of the problem.

Concerning problem (3.3)–(3.5), one can establish a preliminary existence result which, roughly speaking, states that there are at least two possible main classes of solutions. In order to state this result, we introduce the functional class $\mathcal{A}_C$ as follows. For a given $C > 0$, we shall say that a solution $\{\boldsymbol{u}, p, \boldsymbol{U}, \boldsymbol{g}\}$ to (3.3)–(3.5) *belongs to the class* $\mathcal{A}_C$ if and only if, for some $q \in (1, 2)$ and $t > 1$,

$$\operatorname{Re}^{\frac{1}{2}} \|\boldsymbol{u} + \boldsymbol{U}\|_{\frac{2q}{2-q}} + \operatorname{Re}^{\frac{1}{4}} \|\nabla \boldsymbol{u}\|_{\frac{4q}{4-q}} + \|D^2 \boldsymbol{u}\|_{1,q} + \|D^2 \boldsymbol{u}\|_{1,t}$$

$$+ \|\nabla p\|_q + \|\nabla p\|_t \leq C \,.$$

We have the following result, for whose proof we refer to Theorem 3.1 and Theorem 3.2 of [42].

Theorem 3.1. *Let $\mathcal{B}$ be a homogeneous body of revolution around an axis* a, *of class C^3 and possessing fore-and-aft symmetry. Then, there exist* Re_0, λ_0, $C > 0$ *depending only on $\mathcal{B}$ and ε such that, for* $\operatorname{Re} < \operatorname{Re}_0$, $\lambda < \lambda_0$, *there are at least two types of translational steady falls* $\{\boldsymbol{u}, p, \boldsymbol{U}, \boldsymbol{g}\} \in \mathcal{A}_C$, *and they are determined by the following directions of the acceleration of gravity $\boldsymbol{g}$:*

(a) $\boldsymbol{g}$ *is parallel to* a;
(b) $\boldsymbol{g}$ *is orthogonal to* a.

In both cases, $\boldsymbol{g}$ is parallel to $\boldsymbol{U}$, with $\boldsymbol{U} \cdot \boldsymbol{g} > 0$. Moreover, if $\{\boldsymbol{u}_1, p_1, \boldsymbol{U}, \boldsymbol{g}_1\} \in \mathcal{A}_C$ is another translational steady fall corresponding to the same velocity $\boldsymbol{U}$, there exist Re_1, $\lambda_1 > 0$ *depending only on $\mathcal{B}$, ε, and C such that, for* $\operatorname{Re} < \operatorname{Re}_1$, $\lambda < \lambda_1$, *we have $\boldsymbol{u} \equiv \boldsymbol{u}_1$, $p \equiv p_1$, and $\boldsymbol{g} = \boldsymbol{g}_1$.*

In other words, this theorem states, in particular, that in a suitable functional class and at small Reynolds and Weissenberg numbers, the body $\mathcal{B}$ can orient itself with its axis of revolution either parallel or orthogonal to gravity. Our next task is to show that, at small Reynolds and Weissenberg numbers, these are the only possible orientations that $\mathcal{B}$ can achieve. In order to reach this goal, we will evaluate the torque $\mathcal{M}$ exerted by the fluid on $\mathcal{B}$, when $\mathcal{B}$ is in generic a translational motion and show that $\mathcal{M}$ is zero if and only if either condition (a) or (b) holds. We shall sketch the proof of this result here, referring to [42] for full details. We recall that

$$\mathcal{M} = -\int_\Sigma \boldsymbol{x} \times \boldsymbol{T} \cdot \boldsymbol{n} \,. \tag{3.6}$$

In order to evaluate $\mathcal{M}$, we begin to introduce the following *auxiliary fields* $(h^{(i)}, p^{(i)})$, and $(H^{(i)}, P^{(i)})$, $i = 1, 2, 3$, satisfying the boundary value problems

$$\left. \begin{array}{c} \Delta h^{(i)} = \nabla p^{(i)} \\ \nabla \cdot h^{(i)} = 0 \end{array} \right\} \text{ in } \Omega,$$

$$h^{(i)}(x) = e_i, \quad x \in \Sigma,$$

$$\lim_{|x| \to \infty} h^{(i)}(x) = 0,$$

(3.7)

and

$$\left. \begin{array}{c} \Delta H^{(i)} = \nabla P^{(i)} \\ \nabla \cdot H^{(i)} = 0 \end{array} \right\} \text{ in } \Omega,$$

$$H^{(i)}(x) = e_i \times x, \quad x \in \Sigma,$$

$$\lim_{|x| \to \infty} H^{(i)}(x) = 0,$$

(3.8)

where $\{e_i\}$ is the canonical basis in $\mathbb{R}^3$. The fields $(h^{(i)}, p^{(i)})$ [respectively, the fields $(H^{(i)}, P^{(i)})$] are velocity and pressure fields of $\mathcal{L}$ when $\mathcal{B}$ is translating [respectively, $\mathcal{B}$ is rotating] in $\mathcal{L}$ along three orthogonal directions. It is evident that the auxiliary fields depend *only* on geometric properties of $\mathcal{B}$ such as size, shape, symmetry, etc. Existence and uniqueness of $(h^{(i)}, p^{(i)})$, and $(H^{(i)}, P^{(i)})$ is well known [36, Chapter V]. Multiplying $(3.3)_1$ by $H^{(i)}$, integrating by parts over Ω, using $(3.8)_{2,3,4}$ and the fact that (u, p) are in the class $\mathcal{A}_C$, we find

$$-\mathcal{M}_i = 2 \int_\Omega D(u) : D(H^{(i)}) - \lambda \int_\Omega S : D(H^{(i)}) + \operatorname{Re} \int_\Omega u \cdot \nabla u \cdot H^{(i)}. \quad (3.9)$$

The first integral on the right-hand side of this relation can be evaluated by multiplying $(3.8)_1$ by $u + U$ and integrating by parts over Ω . We get

$$2 \int_\Omega D(u) : D(H^{(i)}) = U \cdot \int_\Sigma T_N(H^{(i)}, P^{(i)}) \cdot n. \quad (3.10)$$

From (3.9), and (3.10) we thus obtain

$$\mathcal{M} = \mathcal{M}^S + \operatorname{Re} \mathcal{M}^I + \lambda \mathcal{M}^V \quad (3.11)$$

where $(i = 1, 2, 3)$

$$\mathcal{M}_i^S := -U \cdot \int_\Sigma T_N(H^{(i)}, P^{(i)}) \cdot n,$$

$$\mathcal{M}_i^I := -\int_\Omega u \cdot \nabla u \cdot H^{(i)}, \quad (3.12)$$

$$\mathcal{M}_i^V := \int_\Omega S(u) : D(H^{(i)})$$

are the torque in the Stokes approximation (*i.e.* Re $= \lambda = 0$), the torque due to inertia (*i.e.* $\lambda = 0$), and the torque due to the non-Newtonian character of $\mathcal{L}$ (*i.e.* Re $= 0$), respectively.

We now denote by $(\boldsymbol{u}_S, p_S)$ the solution to $(3.3)_{1,\dots,4}$, and set

$$
\mathcal{M}_i^{0,I} = -\int_{\mathcal{D}} \boldsymbol{u}_S \cdot \nabla \boldsymbol{u}_S \cdot \boldsymbol{H}^{(i)}
$$
$$
\mathcal{M}_i^{0,V} = \int_{\mathcal{D}} \boldsymbol{S}(\boldsymbol{u}_S) : \boldsymbol{D}(\boldsymbol{H}^{(i)}).
$$
(3.13)

From (3.11)–(3.13) we thus get

$$
\boldsymbol{\mathcal{M}} = \boldsymbol{\mathcal{M}}^S + \mathrm{Re}\,\boldsymbol{\mathcal{M}}^{0,I} + \lambda \boldsymbol{\mathcal{M}}^{0,V} + \boldsymbol{\mathcal{N}}
$$
(3.14)

where

$$
\boldsymbol{\mathcal{N}} = \mathrm{Re}\left(\boldsymbol{\mathcal{M}}^I - \boldsymbol{\mathcal{M}}^{0,I}\right) + \lambda\left(\boldsymbol{\mathcal{M}}^{NN} - \boldsymbol{\mathcal{M}}^{0,NN}\right)
$$
$$
\equiv \mathrm{Re}\,\boldsymbol{\mathcal{N}}_1 + \lambda \boldsymbol{\mathcal{N}}_2.
$$

The expression (3.14) for $\boldsymbol{\mathcal{M}}$ applies to a generic body translating with velocity $\boldsymbol{U}$. We now specialize it to the case when $\mathcal{B}$ has the afore-mentioned symmetry properties. Without loss of generality, we take the x_2-axis of a frame attached to $\mathcal{B}$ coinciding with the axis of revolution $\mathbf{a}$ of $\mathcal{B}$, and assume the translational velocity $\boldsymbol{U}$ contained in the plane x_1, x_2 with $U_1, U_2 > 0$; see Figure 12.

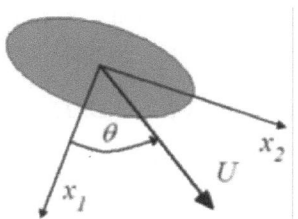

FIGURE 12. Choice of the axes.

With these choices, one can show the following results, for whose proof we refer again to Section 3 of [42]:

$$
\boldsymbol{\mathcal{M}}^S = \mathbf{0},
$$
$$
\mathcal{M}_1^{0,I} = \mathcal{M}_2^{0,I} = \mathcal{M}_1^{0,V} = \mathcal{M}_2^{0,V} = 0,
$$

and

$$
\mathcal{M}_3^{0,I} = U_1 U_2 \mathcal{G}_I, \qquad \mathcal{M}_3^{0,V} = U_1 U_2 \mathcal{G}_{\mathcal{V},\varepsilon}
$$
(3.15)

where

$$
\mathcal{G}_I := -\int_{\mathcal{D}} \left(\boldsymbol{h}^{(1)} \cdot \nabla \boldsymbol{h}^{(2)} + \boldsymbol{h}^{(2)} \cdot \nabla \boldsymbol{h}^{(1)}\right) \cdot \boldsymbol{H}^{(3)}
$$
(3.16)

and

$$\mathcal{G}_{\mathcal{V},\varepsilon} := -\int_{\mathcal{D}} \Big(\boldsymbol{h}^{(1)} \cdot \nabla \boldsymbol{A}_1(\boldsymbol{h}^{(2)}) + (\nabla \boldsymbol{h}^{(1)})^{\mathsf{T}} \cdot \boldsymbol{A}_1(\boldsymbol{h}^{(2)}) + \boldsymbol{A}_1(\boldsymbol{h}^{(1)}) \cdot \nabla \boldsymbol{h}^{(2)}$$

$$+ \boldsymbol{h}^{(2)} \cdot \nabla \boldsymbol{A}_1(\boldsymbol{h}^{(1)}) + (\nabla \boldsymbol{h}^{(2)})^{\mathsf{T}} \cdot \boldsymbol{A}_1(\boldsymbol{h}^{(1)}) + \boldsymbol{A}_1(\boldsymbol{h}^{(2)}) \cdot \nabla \boldsymbol{h}^{(1)}$$

$$+ 2\varepsilon \boldsymbol{A}_1(\boldsymbol{h}^{(1)}) \cdot \boldsymbol{A}_1(\boldsymbol{h}^{(2)}) \Big) : \boldsymbol{D}(\boldsymbol{H}^{(3)}) .$$

(3.17)

Notice that the torque in the Stokes approximation, $\mathcal{M}^S$, is identically zero, showing that "purely viscous effects" play no role in the orientation of the symmetric body. Furthermore, $\mathcal{G}_{\mathcal{I}}$ and, for a fixed ε, $\mathcal{G}_{\mathcal{V},\varepsilon}$ depend only on the geometric properties of $\mathcal{B}$, such as size or shape, but they are otherwise independent of the orientation of $\mathcal{B}$ and of the properties of the liquid. The quantities $\mathcal{G}_{\mathcal{I}}$ and $\mathcal{G}_{\mathcal{V},\varepsilon}$ are referred to as the *inertial torque coefficient* and the *visco-elastic torque coefficient*, respectively. From (3.14)–(3.17) we deduce that

$$\mathcal{M} = (\mathrm{Re}\,\mathcal{G}_{\mathcal{I}} + \lambda \mathcal{G}_{\mathcal{V},\varepsilon})\, U_1 U_2 \boldsymbol{e}_3 + \mathcal{N} .$$

The following fundamental result follows from [46] and [42].

Theorem 3.2. *Let $\mathcal{B}$ satisfy the assumptions of* Theorem 3.1 *and let $\{\boldsymbol{u}, p, \boldsymbol{U}, \boldsymbol{g}\}$ be in the class $\mathcal{A}_C$. Suppose* [33]

$$\mathrm{Re}\,\mathcal{G}_{\mathcal{I}} + \lambda \mathcal{G}_{\mathcal{V},\varepsilon} \neq 0. \qquad (3.18)$$

Then, there exists a positive number c_0, depending only on $\mathcal{B}$ and ε, such that for $0 < \mathrm{Re}\,, \lambda < c_0$, we have [34]

$$\tfrac{1}{2}\,|\mathrm{Re}\,\mathcal{G}_{\mathcal{I}} + \lambda \mathcal{G}_{\mathcal{V},\varepsilon}|\, U_1 U_2 \leq |\mathcal{M}| \leq \tfrac{3}{2}\,|\mathrm{Re}\,\mathcal{G}_{\mathcal{I}} + \lambda \mathcal{G}_{\mathcal{V},\varepsilon}|\, U_1 U_2 . \qquad (3.19)$$

Moreover, there exist positive numbers c_1, c_2, and γ, depending only on $\mathcal{B}$ and ε, such that

$$\mathcal{M} = (\mathrm{Re}\,\mathcal{G}_{\mathcal{I}} + \lambda \mathcal{G}_{\mathcal{V},\varepsilon})\, U_1 U_2 \boldsymbol{e}_3 + \mathcal{N} \qquad (3.20)$$

where $|\mathcal{N}| \leq c_2 \left(\mathrm{Re}^{\,1+\gamma} + \lambda^{1+\gamma} \right)$.

Let us analyze some consequences of Theorem 3.2. In the first place, we know that the equilibrium positions for $\mathcal{B}$ correspond to $\mathcal{M} = \boldsymbol{0}$ and so, provided (3.18) holds, this can happen only if $\boldsymbol{U}$ is either directed along the axis of revolution a of $\mathcal{B}$ or it is perpendicular to it. From Theorem 3.1 it then follows that $\boldsymbol{U}$ has the same orientation as $\boldsymbol{g}$ and so we conclude that *provided (3.18) holds, the only possible orientations of $\mathcal{B}$ at small Reynolds and Weissenberg numbers are with a either parallel or perpendicular to $\boldsymbol{g}$.*

---

[33] Notice that if $\mathrm{Re}\,\mathcal{G}_{\mathcal{I}} + \lambda \mathcal{G}_{\mathcal{V},\varepsilon} = 0$, all orientations are allowed (at small Re and λ). In a real experiment, however, this vanishing condition is practically unattainable.

[34] The numbers $\frac{1}{2}$ and $\frac{3}{2}$, in (3.19) can be replaced by $1-\eta$ and $1+\eta$, respectively, $0 < \eta < 1$, in which case the constant c_0 in the statement of the theorem will depend also on η.

Secondly, let us now consider the stability of such orientations. Since $U_1 = |U| \cos\theta$, $U_2 = |U| \sin\theta$ (see Figure 12), Equation (3.20), at *first order* in Re and λ furnishes

$$\mathcal{M} = |U|^2 (\operatorname{Re}\mathcal{G}_\mathcal{I} + \lambda\,\mathcal{G}_\mathcal{V}) \sin\theta \cos\theta e_3. \tag{3.21}$$

Thus, if we limit ourselves to perturbations in the form of infinitesimal disorientations of a with respect to g, of the type $\delta\theta e_3$, we have

$$\left.\frac{d(\mathcal{M}\cdot e_3)}{d\theta}\right|_{\theta=\theta_0} < 0 \Longrightarrow \quad \text{stability},$$

$$\left.\frac{d(\mathcal{M}\cdot e_3)}{d\theta}\right|_{\theta=\theta_0} > 0 \Longrightarrow \quad \text{instability},$$

where θ_0 denotes the equilibrium configuration (that is, θ_0 is either 0 or $\pi/2$). Consequently, we obtain

$$\theta = 0 \quad \begin{cases} \text{stable if} & \operatorname{Re}\mathcal{G}_\mathcal{I} < -\lambda\,\mathcal{G}_\mathcal{V}, \\ \text{unstable if} & \operatorname{Re}\mathcal{G}_\mathcal{I} > -\lambda\,\mathcal{G}_\mathcal{V}, \end{cases}$$

$$\theta = \frac{\pi}{2} \quad \begin{cases} \text{stable if} & \operatorname{Re}\mathcal{G}_\mathcal{I} > -\lambda\,\mathcal{G}_\mathcal{V}, \\ \text{unstable if} & \operatorname{Re}\mathcal{G}_\mathcal{I} < -\lambda\,\mathcal{G}_\mathcal{V}. \end{cases}$$

From this we see that the competition between the inertial torque (= inertial effects) and viscoelastic torque (= normal-stress effects) is responsible for the stability/instability of the configurations $\theta = 0, \pi/2$.

The above results are better stated in terms of the *elasticity number* $E := \lambda/\operatorname{Re}$. Actually, set

$$E_c := -\frac{\mathcal{G}_\mathcal{I}}{\mathcal{G}_{\mathcal{V},\varepsilon}}.$$

Clearly, E_c depends only on the geometric properties of $\mathcal{B}$ and on the material constant ε. The previous results can then be summarized in the following theorem.

Theorem 3.3. *Let $\mathcal{B}$ be a body of revolution around an axis a, possessing fore-and-aft symmetry, and of class C^3. Then, there exists a positive number c_0, depending only on $\mathcal{B}$ and ε, such that for $0 < \operatorname{Re}, \lambda < c_0$ the following properties hold.*

(a) *If $E \neq E_c$, there exists two and only two classes of translational steady falls that $\mathcal{B}$ can perform, namely, those characterized by a being either parallel (class Pa) or perpendicular (class Pe) to gravity.*

(b) *At first order in Re and λ, if $E < E_c$, then the falls in the class Pe are stable and those in the class Pa are unstable, whereas, if $E > E_c$, the reverse conclusion holds.*

The mathematical predictions of Theorem 3.3 have been analyzed in detail in [42], when $\mathcal{B}$ is a prolate spheroid of eccentricity e. In this case, the "critical" elasticity number E_c, for fixed ε, is a function of e only. It is found that $\mathcal{G}_\mathcal{I} < 0$, for all $e \in (0,1)$, while, for physically realistic values of ε, $\mathcal{G}_{\mathcal{V},\varepsilon} > 0$ for all $e \in (0,1)$, so that $E_c > 0$ for all these values of e as well. The results are summarized in

Figure 13. A comparison of Theorem 3.3 with the experiments of Joseph and Liu [78] is reported in Figure 14. In agreement with [78], the value of ε is chosen to be -1.8.

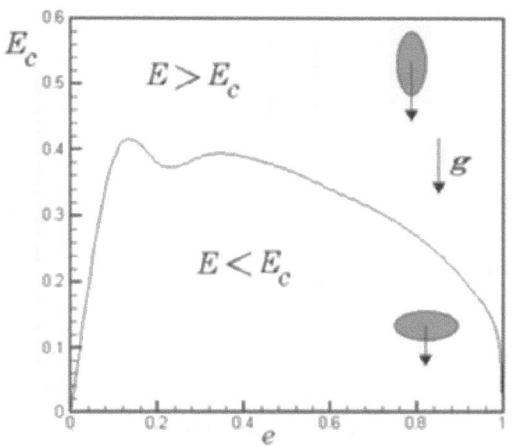

FIGURE 13. Critical elasticity number E_c versus eccentricity e, for $\varepsilon = -1.8$. According to Theorem 3.3, if $E > E_c$, the ellipsoid falls with its major axis a parallel to g, while if $E < E_c$, the fall with a parallel to g is stable.

3.2. Shape-tilting phenomenon

In this section we wish to present and briefly discuss another interesting and puzzling property of sedimentation of symmetric bodies in a viscoelastic liquid, the so-called "shape-tilting" phenomenon. This curious phenomenon was systematically studied from the experimental viewpoint by J. Wang et al. in [114], and still lacks of any mathematical explanation. Specifically, a homogeneous cylindrical particle with round ends is dropped in a polymeric liquid. Let us call a the axis of the cylinder. Then, as expected, the particle will eventually reach an equilibrium configuration with its broadside parallel to the direction of gravity. Now, if the ends of the same particle are made flat, the particle will eventually find an equilibrium configuration with an angle θ somewhere between $0°$ and $90°$; see Figure 15. It has been experimentally found that the equilibrium angle θ, for a fixed material, is a monotonically increasing function of the aspect ratio of the cylinder defined as the ratio of its length L to its diameter d. Some significant findings are given in Figure 16.

Probably, the shape-tilting phenomenon can be qualitatively interpreted by the fact that the presence of sharp edges enhances the normal-stress effects, but no definitive conclusion can be drawn. However, if this is the case, a rigorous interpretation appears to be rather difficult, in that the mathematical analysis of particle

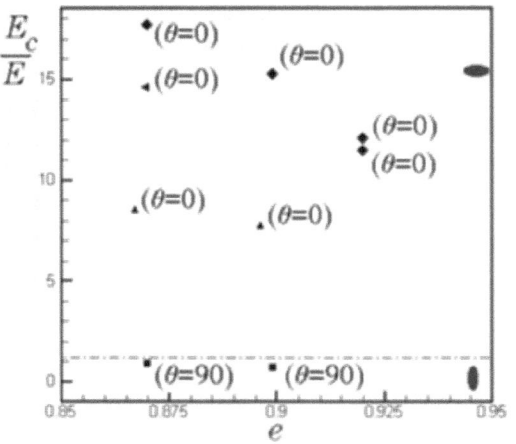

FIGURE 14. Results of Theorem 3.3 versus experimental results of Liu and Joseph [78]. The different symbols refer to the different materials used in the experiments. ♦ Brass; ▲ Aluminum; ■ Plastic; ◄ Tin. The observed equilibrium angles are mentioned besides the plotted points. The dashed line indicates the critical ratio $E/E_c = 1$. Theorem 3.3 correctly predicts that all data point lying above the line should have $\theta = 0°$, otherwise $\theta = 90°$.

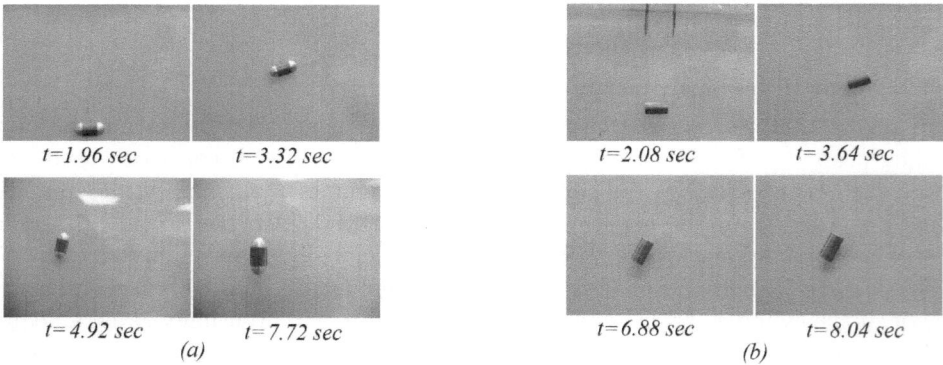

FIGURE 15. (a) Snapshots of sedimentation of an aluminum cylinder with round ends in a 0.75% aqueous polythylene oxide solution; (b) Same particle as in (a) with flat ends, and same polymeric solution.

orientation discussed in the previous section requires very smooth particles, of class C^3 at least.

Another intriguing aspect to this problem has been very recently added by the experimental findings of A. Vaidya [112]. The kinematic viscosity coefficient (at zero shear) of the polymeric solution used in his experiments is 100 times greater than that of the solutions used in [114]. As a result, the sedimentation of a cylinder in a 4-feet high tank can take up to several hours. In such a case, Vaidya finds that there is no shape-tilting and that all particles eventually orient themeselvs with their broad side parallel to gravity, for aspect ratios ranging between 0.75 and 2.00.

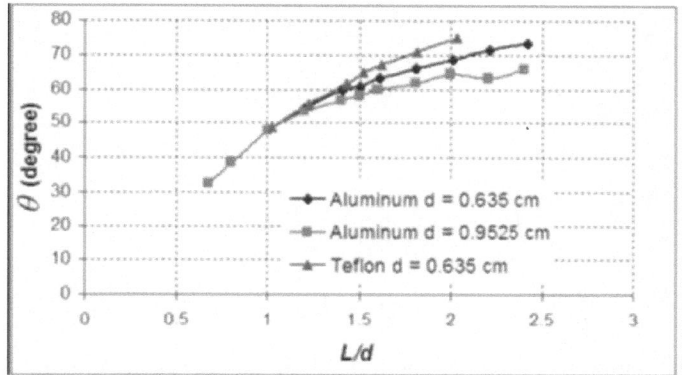

FIGURE 16. Dependence of the equilibrium angle θ on the aspect ratio for different materials (from J. Wang et al. [114]). When L/d becomes "sufficiently large", a cylinder with flat ends will orient itself the same way as a cylinder with round ends, namely, with its broad side parallel to gravity. The polymeric solution is the same as that of Figure 1.5.

3.3. Motion of a disk in the shear flow of a liquid in a horizontal channel

Our objective in this and the next sections is to investigate equilibrium positions and velocity of a disk moving in the shear (Poiseuille) flow of a viscoelastic liquid in a horizontal channel. As usual, the liquid is modeled by a second-order liquid. We are interested in *steady motions* of the system liquid-disk, that is, the translational velocity U and the angular velocity ω of the disk Σ are constant in time, and the motion of the fluid as seen from a frame $\mathcal{I}$ attached to Σ and moving with velocity U is *independent of time*. Thus, for a steady motion to occur, it is clear that the (unknown) velocity U of Σ must be directed along the channel walls which we will take, without loss of generality, parallel to the x_1-axis of the frame $\mathcal{I}$. We also take the origin of $\mathcal{I}$ coinciding with the center of Σ and use the thickness d of the channel as a length scale. It is convenient to define $V \equiv \sqrt{gd}$ as velocity scale, with g acceleration of gravity. We then find that the Poiseuille flow (v_0, p_0), as seen from $\mathcal{I}$ assumes the following dimensionless form (see Section 1.1.1):

$$v_0(x_2; h) = -6Fr[h^2 + h(2x_2 - 1) + x_2(x_2 - 1)]e_1 \equiv Fr f(x_2, h)e_1,$$

$$p_0(x_1) = -12\frac{\Phi}{\mu d^2}x_1, \tag{3.22}$$

where $Fr = \frac{\Phi}{Vd}$, is the Froude number, Φ is the given flow rate (which, without loss, we assume to be positive), μ is the shear viscosity of the liquid and $-h$ and $1 - h$ are the x_2-coordinates of the walls Γ_1 and Γ_2 of the channel; see Figure 17.

Thus, mathematically, our problem can be formulated as follows.

Find $\{v, p, \omega, U, h\}$ satisfying the dimensionless equations

$$\left. \begin{array}{c} \nabla \cdot \boldsymbol{T}(v, P) = \mathcal{R} v \cdot \nabla v \\ \\ \nabla \cdot v = 0 \end{array} \right\} \quad \text{in } \Omega,$$

$$v \big|_S = \omega e_3 \times x, \quad v \big|_{\Gamma_1} = v \big|_{\Gamma_2} = -U,$$

$$\lim_{|x_1| \to \infty} (v(x_1, x_2) - v_0(x_2; h) + U) = 0, \tag{3.23}$$

$$\int_{-h}^{1-h} v_1(x_1, x_2) dx_2 = Fr - U \cdot e_1,$$

$$\int_S \boldsymbol{T}(v, P) \cdot n = \boldsymbol{G}, \quad \int_S x \times \boldsymbol{T}(v, P) \cdot n = \boldsymbol{0}.$$

Here Ω is the region occupied by the fluid, S is the surface of Σ, while $\boldsymbol{T} = \boldsymbol{T}(v, P)$ is the Cauchy stress tensor given by

$$\boldsymbol{T}(v, P) = \boldsymbol{T}_N(v, P) - \lambda \boldsymbol{S}(v) \tag{3.24}$$

with $\boldsymbol{T}_N$ and $\boldsymbol{S}$ defined in (3.2). Moreover, $P = p - \mathcal{R} x_2$, $\mathcal{R} = \frac{\rho V d}{\mu}$ (the disk Reynolds number), ρ is the fluid density, $\lambda = \frac{-\alpha_1 V}{d\mu}$ (the Weissenberg number) and $\boldsymbol{G} = -\mathcal{R}\alpha e_2$. Finally,

$$\alpha = \pi (R/d)^2 (\rho_s/\rho - 1), \tag{3.25}$$

where R, ρ_s are the radius and the density of the disk, respectively. [35]

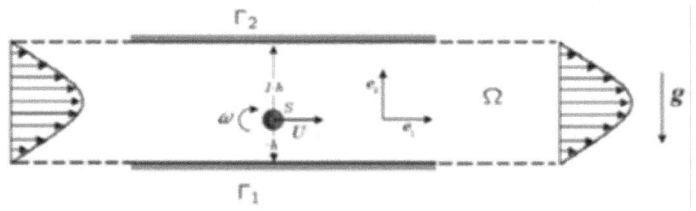

FIGURE 17. Schematic view of the system.

Our strategy in solving the problem develops according to the following steps.

Step 1. We prove the existence and uniqueness of a solution to problem (3.23) for any *given* $h \in (a, 1 - a)$, where $a := R/d$ ($< 1/2$). The corresponding solution $\{v, P, U, \omega\}$ will have $U = U_1 e_1 + U_2 e_2$ where, U_2, in general, need not be zero. However, by virtue of the existence and uniqueness result, we find that there is a

----

[35] We shall assume throughout that $\rho_s \geq \rho$ so that $\alpha \geq 0$.

map $\mathcal{M} : h \in (a, 1 - a) \mapsto U_2 \in \mathbb{R}$. Therefore, all possible equilibrium positions will be given by the zeros of the map $\mathcal{M}$.

Step 2. Our next objective is the evaluation of these zeros. Following [40], we shall evaluate the zeros of $\mathcal{M}$ at first order in $\mathcal{R}$ and λ. However, as shown in [17], the zeros of $\mathcal{M}$ can be precisely computed also to higher orders in $\mathcal{R}$ and λ.

However, both steps require a preliminary study of the associated *Stokes problem*, formally obtained by setting $\lambda = \mathcal{R} = 0$ in (3.22)–(3.24). This preparatory study, very interesting on its own, will be performed in the following Section 3.3.1. Successively, in Section 3.3.2 we shall prove the result stated in Step 1, while in Section 3.3.3 we will give a complete description of the equilibrium positions according to Step 2. In order to simplify the presentation, we shall give a detailed proof of these results only in the case of a Newtonian (Navier–Stokes) liquid. Their modifications due to viscoelastic effects will be summarized in Section 3.3.3. For details about these latter, we refer the reader to the paper [40].

3.3.1. Stokes approximation. Let us consider problem (3.24) in absence of viscoelastic effects and in the limit of vanishing Reynolds number. We thus get the following *Stokes approximation* of the original problem.

$$\left.\begin{array}{l} \nabla \cdot \boldsymbol{T}(\boldsymbol{v}_s, P_s) = \boldsymbol{0} \\ \nabla \cdot \boldsymbol{v}_s = 0 \end{array}\right\} \quad \text{in } \Omega,$$

$$\boldsymbol{v}_s\big|_S = \omega_s \boldsymbol{e}_3 \times \boldsymbol{x}, \quad \boldsymbol{v}_s\big|_{\Gamma_1} = \boldsymbol{v}_s\big|_{\Gamma_2} = -\boldsymbol{U}_s,$$

$$\lim_{|x_1| \to \infty} \left(\boldsymbol{v}_s(x_1, x_2) - \boldsymbol{v}_0(x_2; h) + \boldsymbol{U}_s\right) = 0, \tag{3.26}$$

$$\int_{-h}^{1-h} v_{s1}(x_1, x_2)dx_2 = Fr - \boldsymbol{U}_s \cdot \boldsymbol{e}_1,$$

$$\int_S \boldsymbol{T}(\boldsymbol{v}_s, P) \cdot \boldsymbol{n} = \boldsymbol{0}, \quad \int_S \boldsymbol{x} \times \boldsymbol{T}(\boldsymbol{v}_s, P_s) \cdot \boldsymbol{n} = \boldsymbol{0},$$

with $\boldsymbol{T}(\boldsymbol{v}_s, P_s) := \boldsymbol{T}_N(\boldsymbol{v}_s, P_s)$ and where the unknowns are $\{\boldsymbol{v}_s, P_s, \boldsymbol{U}_s, \omega_s, h\}$. Our next goal is to decouple the above system, that is, we shall separate the equations of the body from the equations of the liquid. To this end, we begin to rewrite the last two equations in (3.26) in an equivalent form. Let us introduce the *auxiliary fields* $\{\boldsymbol{w}^{(i)}, \pi^{(i)}\}$ defined as solutions to the linear problems $(i = 1, 2, 3)$

$$\left.\begin{array}{l} \nabla \cdot \boldsymbol{T}_N(\boldsymbol{w}^{(i)}, \pi^{(i)}) = \boldsymbol{0} \\ \nabla \cdot \boldsymbol{w}^{(i)} = 0 \end{array}\right\} \quad \text{in } \Omega,$$

$$\boldsymbol{w}^{(i)}\big|_S = \boldsymbol{\beta}_i, \quad \boldsymbol{w}^{(i)}\big|_{\Gamma_1} = \boldsymbol{w}^{(i)}\big|_{\Gamma_2} = 0, \tag{3.27}$$

$$\lim_{|x_1| \to \infty} \boldsymbol{w}^{(i)}(x_1, x_2) = 0,$$

where $\boldsymbol{\beta}_i = \boldsymbol{e}_i$ for $i = 1, 2$, and $\boldsymbol{\beta}_3 = \boldsymbol{e}_3 \times \boldsymbol{x}$. Notice that all fields $\boldsymbol{w}, \pi$ depend only on h. Furthermore, $\boldsymbol{w}^{(i)}, p^{(i)}$ and all corresponding derivatives decay exponentially fast to zero as $|x| \to \infty$ [37, Chapter XI]. In view of the symmetry $x_1 \to -x_1$ of problems (3.27), it is easily shown that the following relations hold:

$$\int_S \boldsymbol{T}_N(\boldsymbol{w}^{(2)}, \pi^{(2)}) \cdot \boldsymbol{n} = \mathcal{T}_2(h)\boldsymbol{e}_2\,, \quad \int_S \boldsymbol{x} \times \boldsymbol{T}_N(\boldsymbol{w}^{(2)}, \pi^{(2)}) \cdot \boldsymbol{n} = 0\,,$$

$$\int_S \boldsymbol{T}_N(\boldsymbol{w}^{(3)}, \pi^{(3)}) \cdot \boldsymbol{n} = \mathcal{T}_3(h)\boldsymbol{e}_1\,, \quad \int_S \boldsymbol{T}_N(\boldsymbol{w}^{(1)}, \pi^{(1)}) \cdot \boldsymbol{n} = \mathcal{T}_1(h)\boldsymbol{e}_1,$$

$$\int_S \boldsymbol{v}_0 \cdot \boldsymbol{T}_N(\boldsymbol{w}^{(1)}, \pi^{(1)}) \cdot \boldsymbol{n} = Fr\mathcal{F}_1(h)\boldsymbol{e}_1\,, \quad \int_S \boldsymbol{v}_0 \cdot \boldsymbol{T}_N(\boldsymbol{w}^{(2)}, \pi^{(2)}) \cdot \boldsymbol{n} = 0\,,$$

$$\int_S \boldsymbol{v}_0 \cdot \boldsymbol{T}_N(\boldsymbol{w}^{(3)}, \pi^{(3)}) \cdot \boldsymbol{n} = Fr\mathcal{F}_3(h)\boldsymbol{e}_1,$$

$$(3.28)$$

where $\mathcal{T}_i(h)$, $i = 1, 2, 3$, and $\mathcal{F}_i(h)$, $i = 1, 3$, depend only on h. Moreover, obviously,

$$\int_S \boldsymbol{x} \times \boldsymbol{T}_N(\boldsymbol{w}^{(1)}, \pi^{(1)}) \cdot \boldsymbol{n} = \mathcal{R}_1(h)\boldsymbol{e}_3\,, \quad \int_S \boldsymbol{x} \times \boldsymbol{T}_N(\boldsymbol{w}^{(3)}, \pi^{(3)}) \cdot \boldsymbol{n} = \mathcal{R}_2(h)\boldsymbol{e}_3, \quad (3.29)$$

where, again, $\mathcal{R}_i(h)$, $i = 1, 3$, depend only on the location of the disk. It is easy to see that

$$\mathcal{T}_3(h) = \mathcal{R}_1(h)\,, \quad \text{for all } > 1 - a > h > a\,. \tag{3.30}$$

Actually, if we multiply $(3.27)_1$ with $i = 3$ by $\boldsymbol{w}^{(1)}$, integrate by parts over Ω and take into account (3.28), we obtain

$$\mathcal{T}_3(h) = 2\int_\Omega \boldsymbol{D}(\boldsymbol{w}^{(1)}) : \boldsymbol{D}(\boldsymbol{w}^{(3)})\,.$$

Likewise, if we multiply $(3.27)_1$, with $i = 1$, by $\boldsymbol{w}^{(3)}$, integrate by parts over Ω and take into account (3.28), we obtain

$$\mathcal{R}_1(h) = 2\int_\Omega \boldsymbol{D}(\boldsymbol{w}^{(3)}) : \boldsymbol{D}(\boldsymbol{w}^{(1)})\,,$$

and (3.30) follows. Moreover, we have that

$$\mathcal{T}_1(h) > 0\,, \mathcal{T}_2(h) > 0\,, \mathcal{R}_2(h) > 0\,, \mathcal{T}_1(h)\mathcal{R}_2(h) - \mathcal{R}_1(h)\mathcal{T}_3(h) > 0\,, \tag{3.31}$$

for all $1 - a > h > a$. In fact, let $\boldsymbol{w} = \sum_{i=1}^3 \lambda_i \boldsymbol{w}^{(i)}$, $\Pi = \sum_{i=1}^3 \lambda_i \pi^{(i)}$, $\lambda_i \in \mathbb{R}$, $i = 1, 2, 3$. From (3.27) we find

$$\left.\begin{array}{c} \nabla \cdot \boldsymbol{T}_N(\boldsymbol{w}, \Pi) = 0 \\[2mm] \nabla \cdot \boldsymbol{w} = 0 \end{array}\right\} \quad \text{in } \Omega,$$

$$\boldsymbol{w}\,|_S = \sum_{i=1}^3 \lambda_i \boldsymbol{\beta}_i\,, \quad \boldsymbol{w}\,|_{\Gamma_1} = \boldsymbol{w}\,|_{\Gamma_2} = 0, \tag{3.32}$$

$$\lim_{|x_1| \to \infty} \boldsymbol{w}(x_1, x_2) = 0.$$

Multiplying $(3.32)_1$ by $\boldsymbol{w}$, integrating by parts over Ω and taking into account (3.30) and (3.28) we get

$$\lambda_1^2 \mathcal{T}_1 + \lambda_2^2 \mathcal{T}_2 + \lambda_3^2 \mathcal{R}_2 + 2\lambda_1 \lambda_3 \mathcal{T}_3 = 2 \int_\Omega |\boldsymbol{D}(\boldsymbol{w})|^2 .$$

Since the right-hand side of this equation is always positive, unless $\lambda_1 = \lambda_2 = \lambda_3 = 0$, and the λ's are arbitrary, the property (3.31) follows. We now multiply $(3.26)_1$ by $\boldsymbol{w}^{(i)}$, $i = 1, 2, 3$, integrate by parts over Ω and use the asympotic properties of $\boldsymbol{w}^{(i)}$ to obtain ($i = 1, 2, 3$)

$$\int_S \boldsymbol{\beta}_i \cdot \boldsymbol{T}(\boldsymbol{v}_s, P_s) \cdot \boldsymbol{n} = 2 \int_\Omega \boldsymbol{D}(\boldsymbol{v}_s) : \boldsymbol{D}(\boldsymbol{w}^{(i)}) . \tag{3.33}$$

Likewise, multiplying $(3.27)_1$ by $\boldsymbol{v}_s - \boldsymbol{v}_0 + \boldsymbol{U}_s$ and integrating by parts over Ω we find ($i = 1, 2, 3$)

$$\int_S (\boldsymbol{U}_s + \omega_s \boldsymbol{e}_3 \times \boldsymbol{x} - \boldsymbol{v}_0) \cdot \boldsymbol{T}_N(\boldsymbol{w}^{(i)}, \pi^{(i)}) \cdot \boldsymbol{n} = 2 \int_\Omega \boldsymbol{D}(\boldsymbol{v}_s) : \boldsymbol{D}(\boldsymbol{w}^{(i)}) . \tag{3.34}$$

From (3.28), (3.29), (3.33) and (3.34), one deduces that the last two equations in (3.26) are equivalent to the following ones:

$$\omega_s \mathcal{R}_1(h) + U_{s1} \mathcal{T}_1(h) = Fr \mathcal{F}_1(h),$$

$$U_{s2} \mathcal{T}_2(h) = 0, \tag{3.35}$$

$$\omega_s \mathcal{R}_2(h) + U_{s1} \mathcal{T}_3(h) = Fr \mathcal{F}_2(h).$$

Thanks to (3.35), we are in a position to give a full answer to the equilibrium problem for the disk in the Stokes approximation.

Proposition 3.1. *For any given $Fr > 0$ and $h \in (a, 1 - a)$, problem (3.26) has one and only one solution $\{\boldsymbol{v}_s, P_s, \boldsymbol{U}_s, \omega_s\}$, where*

$$\boldsymbol{v}_s = Fr \left[A(h)(\boldsymbol{w}^{(1)} - \boldsymbol{e}_1) + B(h)\boldsymbol{w}^{(3)} + \boldsymbol{w}^{(4)} \right] := Fr \, \overline{\boldsymbol{v}}_s ,$$

$$P_s = U_s \pi^{(1)} + \omega_s \pi^{(3)} + Fr \, w^{(4)},$$

$$\boldsymbol{U}_s = U_s \boldsymbol{e}_1 := Fr \, A(h) \, \boldsymbol{e}_1, \quad \omega_s = Fr \, B(h),$$

and where A and B are functions of h only, and

$$\left. \begin{aligned} \nabla \cdot \boldsymbol{T}(\boldsymbol{w}^{(4)}, \pi^{(4)}) = 0 \\ \nabla \cdot \boldsymbol{w}^{(4)} = 0 \end{aligned} \right\} \quad in \ \Omega,$$

$$\boldsymbol{w}^{(4)} \big|_S = 0, \quad \boldsymbol{w}^{(4)} \big|_{\Gamma_1} = \boldsymbol{w}^{(4)} \big|_{\Gamma_2} = 0, \tag{3.36}$$

$$\lim_{|x_1| \to \infty} \left(\boldsymbol{w}^{(4)}(x_1, x_2) - f(x_2, h)\boldsymbol{e}_1 \right) = 0 .$$

Moreover, for each fixed h and Fr, U_s and ω_s are the solutions to the linear system

$$\omega_s \mathcal{R}_1(h) + U_s \mathcal{T}_1(h) = Fr \mathcal{F}_1(h),$$

$$\omega_s \mathcal{R}_2(h) + U_s \mathcal{T}_3(h) = Fr \mathcal{F}_2(h). \tag{3.37}$$

Proof. For any given Fr and h in their respective ranges, in view of (3.31) we can find a unique solution, $\{U = U_s e_1, \omega_s\}$, to (3.35) satisfying the properties stated in the proposition. Clearly, by construction, with this choice of U_s and ω_s, equations $(3.26)_{6,7}$ are automatically satisfied. Moreover, by the properties of the fields w and π, It is immediately checked that

$$v_s := U_s(w^{(1)} - e_1) + \omega_s w^{(3)} + Fr w^{(4)},$$

$$P_s := U_s \pi^{(1)} + \omega_s \pi^{(3)} + Fr w^{(4)}$$

solve $(3.26)_{1,\dots,5}$. The proof of the proposition is thus completed. $\square$

Remark 3.1. One interesting consequence of Proposition 3.1 is that *in the Stokes approximation all admissible values of h are equilibrium heights.* Consequently, the migration of the disk must come from the inertial-effects of the liquid (the nonlinear terms), a fact first discovered by F.P. Bretherton in 1962 [14].

The nonzero component, U_s, of the translational velocity and the angular velocity, ω_s, as functions of h and for fixed Fr can be computed from (3.37). This has been done in [40] and the results are reported in Figure 18.

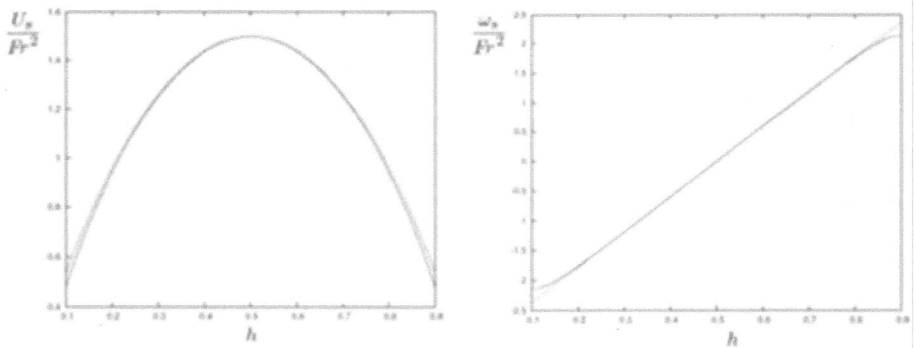

FIGURE 18. Graphs of U_s/Fr^2 and ω_s/Fr^2 versus h computed from (3.37) (solid line). The dotted line corresponds to the velocity and vorticity, respectively, in the Poiseuille flow (3.22).

3.3.2. A fundamental existence and uniqueness theorem. The goal of this section is to show the following result.

Theorem 3.4. *Let $h \in (a, 1 - a)$, $Fr \geq 0$ and $G : \mathbb{R}_+ \mapsto \mathbb{R}^3$ be given, where G is a (real) analytic function of $\mathcal{R}$. Then there exists $\mathcal{R}_0 > 0$ such that, if $\mathcal{R} < \mathcal{R}_0$, problem (3.23)–(3.24) with $\lambda = 0$ has one and only one solution $\{v, P, U, \omega\}$ such that*

$$(v - v_0 + U) \in W^{2,2}(\Omega), \quad P \in W^{1,2}(\Omega).$$

Moreover, the solution is (real) analytic in $\mathcal{R}$ and, denoted by $\{v_s, P_s, U_s, \omega_s\}$ the solution to (3.23) corresponding to $\mathcal{R} = \lambda = 0$, the series

$$v = v_s + \sum_{n=1}^{\infty} v_n \mathcal{R}^n, \quad P = P_s + \sum_{n=1}^{\infty} P_n \mathcal{R}^n,$$

$$U = U_s + \sum_{n=1}^{\infty} U_n \mathcal{R}^n, \quad \omega = \omega_s + \sum_{n=1}^{\infty} \omega_n \mathcal{R}^n,$$

are absolutely convergent in the norms of $W^{2,2}(\Omega)$, $W^{1,2}(\Omega)$, $\mathbb{R}^2$ and $\mathbb{R}$, respectively.

In order to prove this result, we need some preliminary considerations and preparatory lemmas. We first put problem (3.23)–(3.24) in an equivalent form. To this end, we begin to construct a suitable extension of v_0, along the lines of the argument of Section I.3.1. Let $\zeta = \zeta(x_1, x_2)$ be a smooth function such that (with $r = \sqrt{x_1^2 + x_2^2}$)

$$\zeta(x_1, x_2) = \begin{cases} 1 & \text{if } r > 2\delta \\ 0 & \text{if } r < \delta \end{cases}$$

and set

$$a_h = (a_{h1}(x_1, x_2), a_{h2}(x_1, x_2))$$

where

$$a_{h1}(x_1, x_2) = \frac{\partial \zeta}{\partial x_2} \int_{-h}^{x_2} f(\eta, h) d\eta + \zeta(x_1, x_2) f(x_2, h),$$

$$a_{2h}(x_1, x_2) = -\frac{\partial \zeta}{\partial x_1} \int_{-h}^{x_2} f(\eta, h) d\eta,$$

where $f(x_2, h)$ is defined in (3.22). Taking into account that $\max f(x_2, h) = 3/2$, and that $\max |f'(x_2, h)| = 6$, by direct inspection one shows that a_h satisfies the following properties:

1. $a_h \in C^{\infty}(\Omega)$;
2. $\nabla \cdot a_h = 0$ in Ω;
3. $a_h(x_1, x_2) = v_0(x_2; h), \ |x_1| > 2\delta$;
4. $|\nabla a_h(x)| \le M, \ x \in \Omega$;
5. $\|a \cdot \nabla a\|_q \le M, \ 1 \le q \le \infty$;
6. $\int_{-h}^{1-h} a_{h1}(x_1, x_2)(\eta) d\eta = \int_{-h}^{1-h} f(\eta, h) d\eta = 1$;

where M is independent of h. We next set

$$u = v - Fr a_h + U, \quad \Pi = P - \zeta p_0.$$

Using the properties of $\boldsymbol{a}_h$ we then obtain that the field $\boldsymbol{u}$ satisfies the problem

$$\left.\begin{aligned} \nabla \cdot \boldsymbol{T}(\boldsymbol{u}, \Pi) &= \mathcal{R}\left(B(\boldsymbol{u}, \boldsymbol{u}) - B(\boldsymbol{U}, \boldsymbol{u}) + Fr(B(\boldsymbol{a}_h, \boldsymbol{u}) \right. \\ &\quad \left. +B(\boldsymbol{u}, \boldsymbol{a}_h) - B(\boldsymbol{U}, \boldsymbol{a}_h))\right) + \boldsymbol{F} \\ \nabla \cdot \boldsymbol{u} &= 0 \end{aligned}\right\} \quad \text{in } \Omega,$$

$$\boldsymbol{u}\,|_S = \omega e_3 \times \boldsymbol{x} + \boldsymbol{U}, \quad \boldsymbol{u}\,|_{\Gamma_1} = \boldsymbol{u}\,|_{\Gamma_2} = \boldsymbol{0},$$

$$\lim_{|x_1| \to \infty} \boldsymbol{u}(x_1, x_2) = \boldsymbol{0},\qquad\qquad (3.38)$$

$$\int_{-h}^{1-h} u_1(x, x_2) dx_2 = 0,$$

$$\int_S \boldsymbol{T}(\boldsymbol{u}, p) \cdot \boldsymbol{n} = \boldsymbol{G}, \quad \int_S \boldsymbol{x} \times \boldsymbol{T}(\boldsymbol{u}, p) \cdot \boldsymbol{n} = \boldsymbol{0}\,,$$

where $\boldsymbol{T} \equiv \boldsymbol{T}_N$, $B(\boldsymbol{a}, \boldsymbol{b}) = \boldsymbol{a} \cdot \nabla \boldsymbol{b}$ and $\boldsymbol{F} = \mathcal{R}Fr\, B(\boldsymbol{a}_h, \boldsymbol{a}_h) + Fr^2 \boldsymbol{g}$, with $\boldsymbol{g}$ a function of bounded support. We shall now look (formally) for a solution to (3.38) of the form

$$\boldsymbol{u} = \sum_{n=0}^{\infty} \boldsymbol{u}_n \mathcal{R}^n\,, \quad \Pi = \sum_{n=0}^{\infty} \Pi_n \mathcal{R}^n\,, \quad \boldsymbol{U} = \sum_{n=0}^{\infty} \boldsymbol{U}_n \mathcal{R}^n\,, \quad \omega = \sum_{n=0}^{\infty} \omega_n \mathcal{R}^n\,,$$

where the zero-th order terms are the solution to the following Stokes problem (determined in Proposition 3.1):

$$\left.\begin{aligned} \nabla \cdot \boldsymbol{T}(\boldsymbol{u}_0, \Pi_0) &= Fr\boldsymbol{g} \\ \nabla \cdot \boldsymbol{u}_0 &= 0 \end{aligned}\right\} \quad \text{in } \Omega,$$

$$\boldsymbol{u}_0\,|_S = \omega_0 e_3 \times \boldsymbol{x} + \boldsymbol{U}_0, \quad \boldsymbol{u}_0\,|_{\Gamma_1} = \boldsymbol{u}_0\,|_{\Gamma_2} = \boldsymbol{0},$$

$$\lim_{|x_1| \to \infty} \boldsymbol{u}_0(x_1, x_2) = \boldsymbol{0},\qquad\qquad (3.39)$$

$$\int_{-h}^{1-h} u_{01}(x, x_2) dx_2 = 0,$$

$$\int_S \boldsymbol{T}(\boldsymbol{u}_0, \Pi_0) \cdot \boldsymbol{n} = \boldsymbol{G}_0, \quad \int_S \boldsymbol{x} \times \boldsymbol{T}(\boldsymbol{u}_0, \Pi_0) \cdot \boldsymbol{n} = \boldsymbol{0}\,,$$

while, for $n \geq 1$,

$$\left.\begin{aligned}\nabla \cdot \boldsymbol{T}(\boldsymbol{u}_{n+1}, \Pi_{n+1}) &= \sum_{k=0}^{n} \left(B(\boldsymbol{u}_{n-k}, \boldsymbol{u}_k) - B(\boldsymbol{U}_{n-k}, \boldsymbol{u}_k) \right) \\ &\quad + Fr\left(B(\boldsymbol{a}_h, \boldsymbol{u}_n) + B(\boldsymbol{u}_n, \boldsymbol{a}_h) - B(\boldsymbol{U}_n, \boldsymbol{a}_h) \right) + \widetilde{\boldsymbol{F}} \\ \nabla \cdot \boldsymbol{u}_{n+1} &= 0 \end{aligned}\right\} \text{ in } \Omega,$$

$$\boldsymbol{u}_{n+1} \big|_S = \omega_{n+1} \boldsymbol{e}_3 \times \boldsymbol{x} + \boldsymbol{U}_{n+1}, \tag{3.40}$$

$$\boldsymbol{u}_{n+1} \big|_{\Gamma_1} = \boldsymbol{u}_{n+1} \big|_{\Gamma_2} = 0,$$

$$\lim_{|x_1| \to \infty} \boldsymbol{u}(x_1, x_2) = 0,$$

$$\int_{-h}^{1-h} \boldsymbol{u}_{n+1} \cdot \boldsymbol{e}_1(x, x_2) dx_2 = 0,$$

$$\int_S \boldsymbol{T}(\boldsymbol{u}_{n+1}, \Pi_{n+1}) \cdot \boldsymbol{n} = \boldsymbol{G}_{n+1}, \quad \int_S \boldsymbol{x} \times \boldsymbol{T}(\boldsymbol{u}_{n+1}, \Pi_{n+1}) \cdot \boldsymbol{n} = \boldsymbol{0}.$$

In (3.40) $\widetilde{\boldsymbol{F}} = Fr^2 B(\boldsymbol{h}, \boldsymbol{h})$ if $n = 0$, and $\widetilde{\boldsymbol{F}} = \boldsymbol{0}$ otherwise. Moreover, $\boldsymbol{G}_k$ are the coefficients of the power series of $\boldsymbol{G}$. We want to show that problems (3.39) and (3.40) are solvable for all $n \geq 0$ with corresponding estimates. Let $\mathcal{H}(\Omega)$ be the class of functions $\boldsymbol{\varphi}$ such that

1. $\boldsymbol{\varphi} \in C_0^\infty(\overline{\Omega})$;
2. $\nabla \cdot \boldsymbol{\varphi} = 0$ in Ω; $\qquad\qquad\qquad\qquad\qquad\qquad\qquad$ (3.41)
3. $\boldsymbol{\varphi} \equiv 0$ in a neighborhood of Γ_1 and Γ_2;
4. $\boldsymbol{\varphi} = \overline{\boldsymbol{\varphi}} \equiv \boldsymbol{\Phi}_1 + \Phi_2 \boldsymbol{e}_3 \times \boldsymbol{y}$, for some $\boldsymbol{\Phi}_1 \in \mathbb{R}^2, \Phi_2 \in \mathbb{R}$, in a neighborhood of S.

Reasoning as in [38], one can show the validity of the Poincaré inequality

$$\|\boldsymbol{\varphi}\|_2 \leq \gamma_0 \|\boldsymbol{D}(\boldsymbol{\varphi})\|_2, \tag{3.42}$$

where γ_0 is a constant independent of h. Furthermore, one shows that the "translational velocity" $\boldsymbol{\Phi}_1$ and the "spin" Φ_2 of a generic $\boldsymbol{\varphi} \in \mathcal{H}(\Omega)$ can be controlled by the L^2-norm of $\boldsymbol{D}$. Specifically, we have

$$|\boldsymbol{\Phi}_1| + |\Phi_2| \leq \gamma \|\boldsymbol{D}(\boldsymbol{\varphi})\|_2 \tag{3.43}$$

where γ depends only on R. We shall denote by $H(\Omega)$ the completion of $\mathcal{H}(\Omega)$ in the norm $\|\boldsymbol{D}(\cdot)\|_2$.

We have the following lemma.

Lemma 3.1. *Let* $\boldsymbol{F}_1 \in L^2(\Omega)$, $\boldsymbol{F}_2 \in \mathbb{R}^2$ *be given. Then, the problem*

$$\left.\begin{array}{r}\nabla \cdot \boldsymbol{T}(\boldsymbol{u}, \Pi) = \boldsymbol{F}_1 \\[2mm] \nabla \cdot \boldsymbol{u} = 0\end{array}\right\} \ in\ \Omega,$$

$$\boldsymbol{u}\,|_S = \omega \boldsymbol{e}_3 \times \boldsymbol{x} + \boldsymbol{U}, \quad \boldsymbol{u}\,|_{\Gamma_1} = \boldsymbol{u}\,|_{\Gamma_2} = \boldsymbol{0},$$

$$\lim_{|x_1|\to\infty} \boldsymbol{u}(x_1, x_2) = \boldsymbol{0}, \tag{3.44}$$

$$\int_{-h}^{1-h} u_1(x, x_2)dx_2 = 0,$$

$$\int_S \boldsymbol{T}(\boldsymbol{u}, \Pi) \cdot \boldsymbol{n} = \boldsymbol{F}_2, \quad \int_S \boldsymbol{x} \times \boldsymbol{T}(\boldsymbol{u}, \Pi) \cdot \boldsymbol{n} = \boldsymbol{0},$$

has one and only one solution $\{\boldsymbol{u} \in W^{2,2}(\Omega), \Pi \in W^{1,2}(\Omega), \boldsymbol{U}, \omega\}$. *Moreover, this solution satisfies the estimate*

$$|\boldsymbol{U}| + |\omega| + \|\boldsymbol{u}\|_{2,2} + \|\Pi\|_{1,2} \le C \left(\|\boldsymbol{F}_1\|_2 + |\boldsymbol{F}_2|\right), \tag{3.45}$$

where $C = C(\Omega) > 0$.

Proof. We give a weak formulation of the problem. Thus, multiplying $(3.44)_1$ by $\boldsymbol{\varphi} \in \mathcal{H}(\Omega)$, and integrating by parts over Ω, we find

$$\boldsymbol{\Phi}_1 \cdot \int_S \boldsymbol{T}(\boldsymbol{u}, p) \cdot \boldsymbol{n} + \Phi_2 \boldsymbol{e}_3 \cdot \int_S \boldsymbol{x} \times \boldsymbol{T}(\boldsymbol{u}, p) \cdot \boldsymbol{n} - \int_\Omega \boldsymbol{D}(\boldsymbol{u}) : \boldsymbol{D}(\boldsymbol{\varphi}) = \int_\Omega \boldsymbol{F}_1 \cdot \boldsymbol{\varphi},$$

where $\boldsymbol{\Phi}_1 + \Phi_2 \boldsymbol{e}_3 \times \boldsymbol{x}$ is the trace of $\boldsymbol{\varphi}$ at S. Therefore, using the last two equations in $(3.44)_2$ it follows that

$$\int_\Omega \boldsymbol{D}(\boldsymbol{u}) : \boldsymbol{D}(\boldsymbol{\varphi}) = -\int_\Omega \boldsymbol{F}_1 \cdot \boldsymbol{\varphi} + \boldsymbol{\Phi}_1 \cdot \boldsymbol{F}_2. \tag{3.46}$$

We shall say that $\{\boldsymbol{u}, h, \omega, \boldsymbol{U}\}$ is a *weak solution* to problem (3.44) if and only if: (i) $\boldsymbol{u} \in H(\Omega)$, (ii) $\boldsymbol{u} = \omega \boldsymbol{e}_3 \times \boldsymbol{x} + U\boldsymbol{e}_1$, $\boldsymbol{x} \in S$, and (iii) $\boldsymbol{u}$ satisfies (3.46) for all $\boldsymbol{\varphi} \in \mathcal{H}(\Omega)$.

The existence of a field $\boldsymbol{u}$ satisfying requirements (i) and (iii) can be proved by the classical Galerkin method. As is known, the method furnishes existence provided we obtain a suitable a-priori bound on the solution. This latter can be obtained as follows. Replacing, formally, $\boldsymbol{\varphi}$ in (3.46) with $\boldsymbol{u}$, we get

$$\|\boldsymbol{D}(\boldsymbol{u})\|_2^2 = -\int_\Omega \boldsymbol{F}_1 \cdot \boldsymbol{u} + \boldsymbol{U} \cdot \boldsymbol{F}_2. \tag{3.47}$$

Using in (3.47) the Schwarz inequality along with (3.42) and (3.43) we find

$$\|\boldsymbol{D}(\boldsymbol{u})\|_2 \le \gamma_0 \|\boldsymbol{F}_1\|_2 + \gamma\,|\boldsymbol{F}_2|, \tag{3.48}$$

which furnishes the desired a-priori estimate. By means of (3.48) and of the Galerkin method we thus establish the existence of a weak solution that, in addition, satisfies (3.47). From standard regularity theory, see, e.g., [36, Lemma VI.1.2], we have that $u \in W^{2,2}(\Omega)$ and that it satisfies $(3.44)_1$ for some $\Pi \in W^{1,2}(\Omega)$. Moreover, the following estimate holds:

$$\|u\|_{2,2} + \|\Pi\|_{1,2} \le c \left(\|F_1\|_2 + \|D(u)\|_2 + |U| + |\omega| \right), \qquad (3.49)$$

where we used the inequality

$$\|\nabla u\|_2 \le \sqrt{2} \|D(u)\|_2 ; \qquad (3.50)$$

see [38]. The lemma then follows from (3.48), (3.49) and (3.43). $\qquad \square$

Lemma 3.2. *Let $v, w \in W^{1,2}(\Omega)$. Then*

$$\|B(v, w)\|_2 \le c \|D(v)\|_2 \|D(w)\|_2 ,$$

where $c = c(\Omega) > 0$.

Proof. By the Schwarz inequality, we have

$$\|B(v, w)\|_2 \le \|v\|_4 \|w\|_4 .$$

Since

$$\|u\|_4 \le c \|\nabla u\|_2 \quad u \in W^{1,2}(\Omega) ,$$

see, e.g., [36, Lemma IX.2.1], the lemma follows from these last two displayed inequalities and from (3.50). $\qquad \square$

Set

$$V_n := \|u_n\|_{2,2} + \|\Pi\|_{1,2} + |U_n| + |\omega_n|,$$

$$H_n := \sum_{k=0}^{n} \left(B(u_{n-k}, u_k) - B(U_{n-k}, u_k) \right) + Fr(B(a_h, u_n)$$

$$+ B(u_n, a_h) - B(U_n, a_h)) + \widetilde{F_n} ,$$

where $\widetilde{F_n} = \widetilde{F}$ if $n = 1$ and $\widetilde{F_n} = 0$ otherwise. From Lemma 3.2 and from the properties 1–6 of the function a_h it easily follows that

$$\|H_n\|_2 \le c \left(\sum_{k=0}^{n} V_{n-k} V_k + Fr \, V_n + Fr^2 \delta_{n1} \right), \quad n \ge 0. \qquad (3.51)$$

We now apply the results of Lemma 3.1 and Lemma 3.2 to problems (3.39) and (3.40) and use (3.51). We thus get

$$V_0 \le c \, (Fr + |G_0|),$$

$$V_{n+1} \le c \left(\sum_{k=0}^{n} V_{n-k} V_k + Fr \, V_n + Fr^2 \delta_{n1} + |G_{n+1}| \right), \quad n \ge 0. \qquad (3.52)$$

Let A_n be defined through the recurrent relations

$$A_0 = c \, (Fr + |G_0|),$$

$$A_{n+1} = c \left(\sum_{k=0}^{n} A_{n-k} A_k + Fr \, A_n + Fr^2 \delta_{n1} + |G_{n+1}| \right), \quad n \ge 0. \qquad (3.53)$$

Clearly, $V_k \leq A_k$, for all $k \geq 0$. We shall now show that, provided $\mathcal{R}$ is sufficiently restricted, the series $Z := \sum_{n=0}^{\infty} A_n \mathcal{R}^n$ is converging, thus implying the convergence of $\sum_{n=0}^{\infty} V_n \mathcal{R}^n$ which will complete the existence part of the theorem. To reach our goal we observe that, multiplying both sides of $(3.53)_2$ by $\mathcal{R}^n$, using the Cauchy product formula and summing over n from 0 to ∞, we obtain

$$Z - A_0 = c \left[\mathcal{R}(Z^2 + Fr\, Z + Fr^2) + \mathcal{S} \right] \tag{3.54}$$

where $\mathcal{S} = \sum_{n=0}^{\infty} |G_n| \mathcal{R}^n - |G_0|$. The solution to (3.54) that reduces to a_0 at $x = 0$ is given by

$$Z(x) = \frac{1}{2c\mathcal{R}} \left[(1 - c\mathcal{R}Fr) - \sqrt{(1 - c\mathcal{R}Fr)^2 - 4(A_0 + cFr^2 + \mathcal{S})\, c\mathcal{R}} \right]$$

which is positive and has an analytic branch provided

$$1 > c\mathcal{R}Fr, \quad (1 - c\mathcal{R}Fr)^2 > 4(A_0 + cFr^2 + \mathcal{S})\, c\mathcal{R}.$$

Let $B > 0$ be such that the series $\mathcal{S} + |G_0|$ converges for all $\mathcal{R} \in (0, B]$ and let $G_M = G_M(B)$ denote an upper bound for $\mathcal{S} + |G_0|$, uniformly in $\mathcal{R} \in (0, B]$. Taking into account that $A_0 = V_0$ and inequality $(3.52)_1$, we thus deduce that this latter condition is satisfied provided we choose $\mathcal{R} < \min\{B, C(Fr + Fr^2 + G_M)^{-1}\}$, where $C > 0$ depends only on Ω. The theorem is therefore proved. $\square$

3.3.3. Evaluation of equilibrium heights. It must be emphasized that the "translational velocity" U of solutions given in Theorem 3.4 need *not* be directed along the walls Γ_1, Γ_2, that is, these solutions may have $U_2 \neq 0$. However, from the same theorem, we know that for any given h there exists one and only one corresponding $U_2 \in \mathbb{R}$. Our next objective is to find for which values, h_{eq}, of h we have that $U_2 = 0$. Such values h_{eq} will give precisely the possible equilibrium values for the heights h.

To reach this goal, we rewrite the last two equations in (3.23) in an equivalent form. To this end, we dot-multiply both sides of $(3.23)_1$ by $\boldsymbol{w}^{(i)}$, $i = 1, 2, 3$, integrate by parts over Ω and use the asympotic properties of $\boldsymbol{w}^{(i)}$ to obtain $(i = 1, 2, 3)$

$$\int_S \boldsymbol{\beta}_i \cdot \boldsymbol{T}(\boldsymbol{v}, P) \cdot \boldsymbol{n} = \mathcal{R} \int_\Omega \boldsymbol{v} \cdot \nabla \boldsymbol{v} \cdot \boldsymbol{w}^{(i)} + 2 \int_\Omega \boldsymbol{D}(\boldsymbol{v}) : \boldsymbol{D}(\boldsymbol{w}^{(i)}). \tag{3.55}$$

From (3.28), (3.29), (3.34) and (3.55), one deduces that the last two equations in (3.23) are equivalent to the following ones:

$$\omega \mathcal{R}_1(h) + U_1 \mathcal{T}_1(h) = Fr \mathcal{F}_1(h) - \mathcal{R} \int_\Omega \boldsymbol{v} \cdot \nabla \boldsymbol{v} \cdot \boldsymbol{w}^{(1)},$$

$$U_2 \mathcal{T}_2(h) = -\mathcal{R} \int_\Omega \boldsymbol{v} \cdot \nabla \boldsymbol{v} \cdot \boldsymbol{w}^{(2)} - \mathcal{R}\alpha, \tag{3.56}$$

$$\omega \mathcal{R}_2(h) + U_1 \mathcal{T}_3(h) = Fr \mathcal{F}_2(h) - \mathcal{R} \int_D \boldsymbol{v} \cdot \nabla \boldsymbol{v} \cdot \boldsymbol{w}^{(3)},$$

where, we recall, α is defined in (3.25). From these equations it is possible to draw a number of consequences. Actually, in view of Theorem 3.4, from $(3.56)_2$ it follows that

$$\mathcal{R} U_2^{(1)} \mathcal{T}_2(h) = Fr^2 \mathcal{R} \mathcal{G}(h) - \mathcal{R}\alpha + \Lambda, \tag{3.57}$$

where

$$\mathcal{G}(h) := \int_\Omega \overline{\boldsymbol{v}}_s \cdot \nabla \overline{\boldsymbol{v}}_s \cdot \boldsymbol{w}^{(2)}, \tag{3.58}$$

and

$$|\Lambda| \leq C \mathcal{R}^2, \quad C = C(\Omega, Fr) > 0.$$

Thus, at first order in $\mathcal{R}$ and λ, we find that U_2 is given by

$$U_2^{(1)} = \frac{Fr^2}{\mathcal{T}_2(h)} [\mathcal{G}(h) - K] \tag{3.59}$$

where $K = \frac{\alpha}{Fr^2}$. Since the equilibrium heights h_{eq} are those at which $U_2 = 0$, from (3.59) we deduce that, at first order in $\mathcal{R}$ and λ, h_{eq} is the solution to the equation

$$\mathcal{G}(h) = K. \tag{3.60}$$

Notice that the quantity $\mathcal{R} Fr^2 \mathcal{G}(h)$ represents, at first order in $\mathcal{R}$ the component of the force exerted by the liquid in the direction orthogonal to the translational velocity of the disk (lift). Solutions to the equation (3.60) were computed numerically in [40], and will be presently discussed. We observe that, once the values for h_{eq} have been obtained, from $(3.56)_{1,3}$ and with the help of Theorem 3.4, we may calculate also the translational and angular velocities of the disk, $U_1^{(1)}$ and $\omega^{(1)}$, at first order in $\mathcal{R}$. In fact, we have

$$\omega^{(1)} \mathcal{R}_1(h) + U_1^{(1)} \mathcal{T}_1(h) = -Fr^2 \int_\Omega \overline{\boldsymbol{v}}_s \cdot \nabla \overline{\boldsymbol{v}}_s \cdot \boldsymbol{w}^{(1)},$$

$$\omega^{(1)} \mathcal{R}_2(h) + U_1^{(1)} \mathcal{T}_3(h) = -Fr^2 \int_D \overline{\boldsymbol{v}}_s \cdot \nabla \overline{\boldsymbol{v}}_s \cdot \boldsymbol{w}^{(3)}. \tag{3.61}$$

However, the numerical computation of [40] gives that the two integrals on the right-hand side of (3.61) are zero, so that, at first order in $\mathcal{R}$ the translational and angular velocity of the disk coincides with that evaluated in the Stokes approximation and given in Proposition 3.1 (see also Figure 18). In fact, as shown in [17], the first further non-zero contributions to U_1 and ω are at the second order in $\mathcal{R}$.

Let us now study the equilibrium equation (3.60) in some detail, starting with the case $\alpha = 0$. From the physical point of view, this means that the disk has zero buoyancy. With a view to Figure 19, we then find three possible solutions

$$h^{(1)} = 0.261, \quad h^{(2)} = 0, \quad h^{(3)} = 0.738.$$

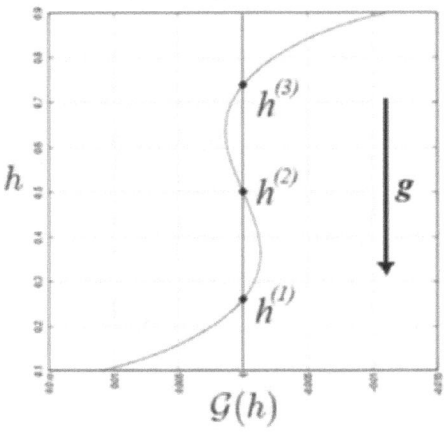

FIGURE 19. Equilibrium heights for zero buoyancy.

Notice that $h^{(1)}$ and $h^{(3)}$ are *stable*, while $h^{(2)}$ is *unstable*. These conclusions come from the following argument. Consider a small variation in h at the equilibrium position $h^{(1)}$. If it is at the left of $h^{(1)}$, that is, the sphere is pushed downward in the channel, $\mathcal{G}(h)$ is positive there and, therefore, the lift is positive. This means that the fluid will exert a force in the upward direction that will try to bring the particle back to $h^{(1)}$. Analogously, if the perturbation is at the right of $h^{(1)}$, that is, the sphere is pushed upward in the channel, $\mathcal{G}(h)$ is negative there and the lift is negative. This means that the liquid will exert a force in the downward direction that will bring the particle back to $h^{(1)}$. For the same reasons, $h^{(3)}$ is stable and $h^{(2)}$ is unstable. We may summarize these results by saying that a given equilibrium position h_{eq} is *stable* if the slope of $\mathcal{G}(h)$ is *negative* at h_{eq} and it is *unstable* if it is *positive*.

Let us next consider the case of a *negative buoyancy*, that is, the density of the particle is larger than that of the liquid. We then have $\alpha > 0$. From (3.60) we thus deduce that the equilibrium heights are given by the intersection of the straight line α/Fr^2 (represented by the solid straight line in Figure 20) with the curve $\mathcal{G}(h)$.

With a view to Figure 19, wee see that all equilibrium heights move downward in the channel. In particular, the stable ones, $h^{(1)}$ and $h^{(3)}$, become closer to the bottom plate. We also have the following interesting "jump phenomenon" in the equilibrium height. Consider a particle in the top half of the channel and set

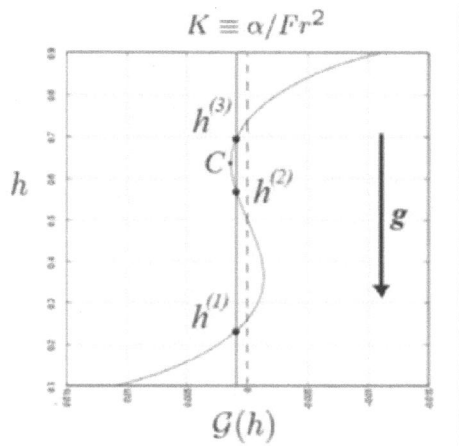

FIGURE 20. Equilibrium for a nonzero buoyancy.

$\delta = \mathcal{G}(h_c)$, where h_c is the coordinate of the point C, the local maximum of $\mathcal{G}(h)$; see Figure 20. In [40] the value for δ was computed and found $\delta = 0.00134$. Therefore, if

$$\alpha/Fr^2 < 0.00134\,, \tag{3.62}$$

the position $h^{(3)}$ exists and it is locally stable. However, if

$$\alpha/Fr^2 > 0.00134\,, \tag{3.63}$$

the particle will jump to the position $h^{(1)}$ on the lower half of the channel, that always exists and is always locally stable. This fact has a very simple interpretation from the physical point of view. In fact, for a given flow rate (Fr), a particle can stay in equilibrium in the top half of the channel if it is not "too heavy", that is, if there is enough lift. Otherwise, the particle will fall down. Notice that, for a given α, we can always increase Fr (that is we can always increase the flow rate) in such a way that the particle stays in the equilibrium position $h^{(3)}$. The above result can also be interpreted in a different way. We *fix* the buoyancy, α, of the particle and take Fr sufficiently small in such a way that (3.63) holds, namely, there is only one stable equilibrium height $(h^{(1)})$ located on the branch of the curve $\mathcal{G}(h)$ close to the bottom plate. If we increase Fr, we will reach a critical value at which (3.62) is valid, and another stable equilibrium height $h^{(3)}$ appears.

Using (3.59) and taking into account that $U_2^{(1)} = dh/dt$, in [40] the trajectories of disks, starting at different heights in the channel, have been computed. The results for $\alpha = 1$ and $Fr = 2.58, 2.77$ and 10 are shown in Figure 21, 22 and 23, respectively.

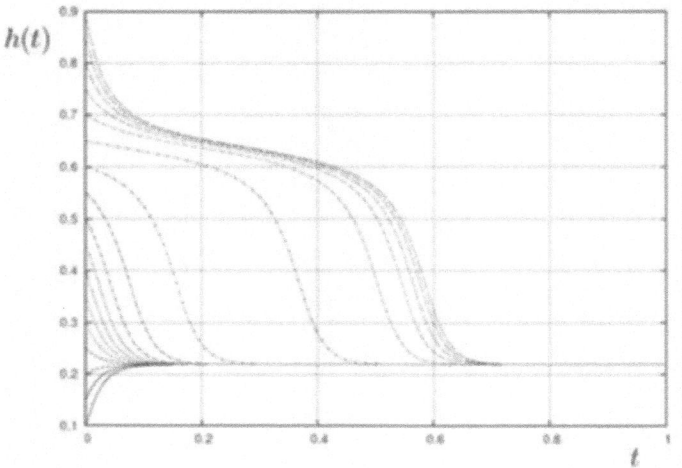

FIGURE 21. Trajectories of the disk, at first order in $\mathcal{R}$, for $\alpha = 1$ and $Fr = 2.58$. The magnitude of the flow-rate is not large enough to generate an equilibrium position in the upper part of the channel. See (3.62).

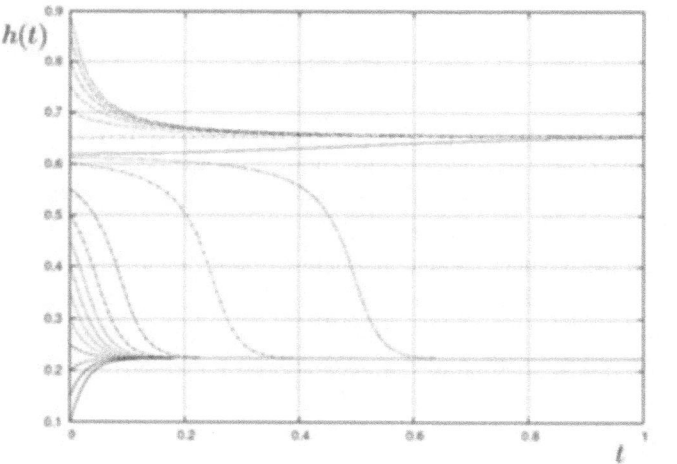

FIGURE 22. Trajectories of the disk, at first order in $\mathcal{R}$, for $\alpha = 1$ and $Fr = 2.77$. The magnitude of the flow-rate is large enough to generate an equilibrium position in the upper part of the channel. See (3.63). However, not every particle initially in the upper part of the channel will reach the upper equilibrium position.

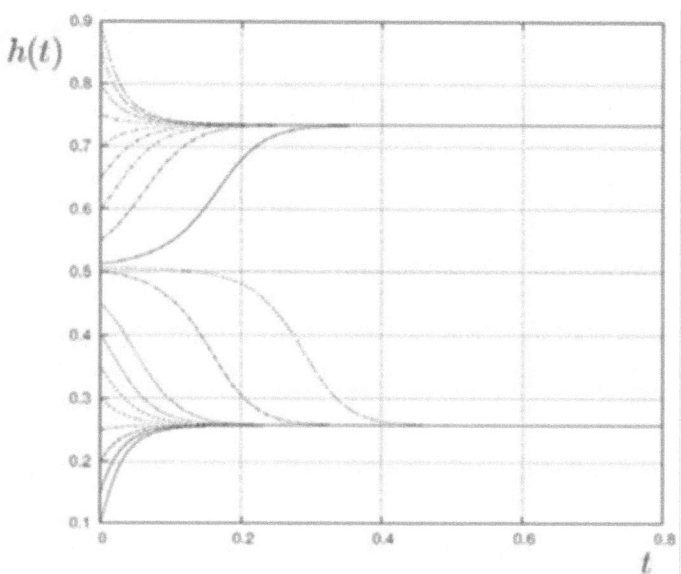

FIGURE 23. Trajectories of the disk, at first order in $\mathcal{R}$, for $\alpha = 1$ and $Fr = 10.0$. The magnitude of the flow-rate is now large enough, and every particle initially in the upper part of the channel will reach the upper equilibrium position.

We wish to end this section with a final remark. In view of the analyticity of the solution established in Theorem 3.4, it is clear that, in principle, from (3.35), we can evaluate U_2, ω and h to any fixed order in the Reynolds number $\mathcal{R}$. Of course, the effective computation may become more and more complicated at higher order. Evaluation of the above quantities up to the $5th$ order included has been done in [17].

3.3.4. Evaluation of equilibrium heights. Viscoelastic case. In this section we shall analyze how the viscoelastic properties of the liquid can modify the equilibrium heights determined for the purely Newtonian case in Section 3.3.4. As usual, we shall employ the second-order fluid model. Therefore, the non-Newtonian characteristic will reduce to normal-stress effects. Also, as mentioned previously, we shall limit ourselves to describe the main ideas and the main results, referring the reader to [40] for further details.

The starting point of the analysis is the replacement of (3.56) with one that takes into account the non-Newtonian character of the liquid, expressed by the tensor S defined in (3.2). Now, the following relations can be proven:

$$\omega \mathcal{R}_1(h) + U_1 \mathcal{T}_1(h) = Fr \mathcal{F}_1(h) - \mathcal{R} \int_\Omega \boldsymbol{v} \cdot \nabla \boldsymbol{v} \cdot \boldsymbol{w}^{(1)} + \lambda \int_\Omega \boldsymbol{S}(\boldsymbol{v}) : \boldsymbol{D}(\boldsymbol{w}^{(1)}),$$

$$U_2 \mathcal{T}_2(h) = -\mathcal{R} \int_\Omega \boldsymbol{v} \cdot \nabla \boldsymbol{v} \cdot \boldsymbol{w}^{(2)} + \lambda \int_\Omega \boldsymbol{S}(\boldsymbol{v}) : \boldsymbol{D}(\boldsymbol{w}^{(2)}) - \mathcal{R}\,\alpha, \quad (3.64)$$

$$\omega \mathcal{R}_2(h) + U_1 \mathcal{T}_3(h) = Fr \mathcal{F}_2(h) - \mathcal{R} \int_D \boldsymbol{v} \cdot \nabla \boldsymbol{v} \cdot \boldsymbol{w}^{(3)} + \lambda \int_\Omega \boldsymbol{S}(\boldsymbol{v}) : \boldsymbol{D}(\boldsymbol{w}^{(3)}).$$

As in the Newtonian case, the next step is to "expand" $\{\boldsymbol{v}, p, \boldsymbol{U}, \omega\}$ around the Stokes solution given in Proposition 3.1. In fact, one can show that, in particular, $(3.64)_2$ furnishes

$$\mathcal{R}(1 + E)U_2^{(1)} \mathcal{T}_2(h) = \mathcal{R}\left(Fr^2 \mathcal{G}(h) + E\,Fr^2 \mathcal{G}_V(h) - \alpha\right) + \Lambda \qquad (3.65)$$

where $E := \lambda/\mathcal{R}$ is the elasticity number, $\mathcal{G}$ is defined in (3.58),

$$\mathcal{G}_V(h) := \int_\Omega \boldsymbol{S}(\overline{\boldsymbol{v}}_s) : \boldsymbol{D}(\boldsymbol{w}^{(2)}) \qquad (3.66)$$

and

$$|\Lambda| \le C\,\mathcal{R}^2, \quad C = C(\Omega, Fr, E) > 0\,.$$

Thus, at first order in $\mathcal{R}$, we find that U_2 is given by

$$U_2^{(1)} = \frac{Fr^2}{(1 + E)\mathcal{T}_2(h)}\left[\mathcal{G}(h) + E\mathcal{G}_V(h) - K\right] \qquad (3.67)$$

where K is defined after (3.59). From (3.67) it follows that the equilibrium heights are now determined by the equation

$$\mathcal{G}(h) + E\mathcal{G}_V(h) = K\,, \qquad (3.68)$$

which is solved numerically in [40], for different values of E and K. Since the viscoelastic effect is measured through the parameter E, we are interested in how solutions $h = h_{eq}$ of (3.68) change by increasing E. In [40] it is found that the overall effect of viscoelasticity is to move the disk toward the middle of the channel and not away from it, as it happens in the purely Newtonian case. In Figures 24 and 25 is reported the variation of h_{eq} with E for $K = 0$, that is, when the disk and the liquid have the same density. We see that the two stable equilibrium heights for $E = 0$ eventually merge in only one situated in the middle of the channel.

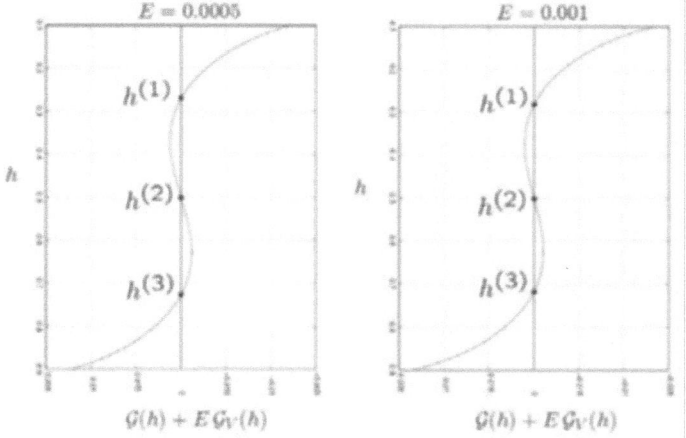

FIGURE 24. Variation of the equilibrium heights with E for $K = 0$. It is seen that the equilibrium heights move toward the middle of the channel. This property is enhanced at higher values of E, as shown in Figure 25.

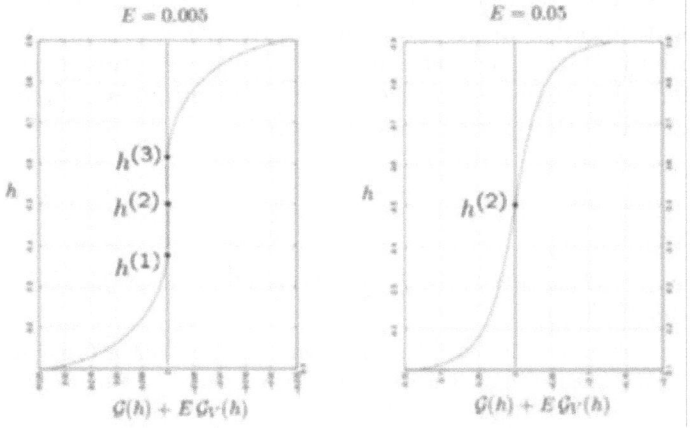

FIGURE 25. Variation of the equilibrium heights with E for $K = 0$. The two distinct stable equilibrium heights merge into the (stable) one located in the middle of the channel.

In Figure 26 it is reported the computation of the equilibria for fixed E and increasing $K > 0$. Here, the situation is qualitatively analogous to the Newtonian case. In fact, the upper equilibrium position tends to disappear, whereas the lower one tends to move toward the bottom wall.

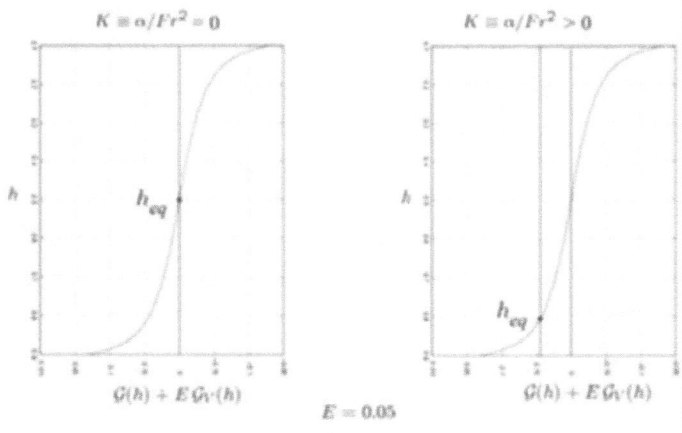

FIGURE 26. Variation of the equilibrium heights with K for $E = 0.05$. For this value of the elasticity the only equilibrium height is in the middle of the channel. As K increases, it moves continuously toward the lower wall.

We end with a final remark concerning the translational and angular velocity of the disk. As a matter of fact, it was found in [40] that the integrals appearing on the right-hand side of $(3.64)_{1,3}$ are, effectively, zero. As a consequence, the first-order contributions to the translational and angular velocities of the disk are zero and, therefore, they coincide with the analogous quantities evaluated in the Stokes approximation (see Proposition 3.1 and Figure 18).

References

[1] Advani, A.S., *Flow and Rheology in Polymer Composites Manufacturing*, Elsevier, Amsterdam (1994).

[2] Arada, N. and Sequeira, A., Strong Steady Solutions for a Generalized Oldroy-B Model with Shear-Dependent Viscosity in a Bounded Domain, *Math. Mod. and Meth. in Appl. Sci.* **13** (2003), 1303–1323.

[3] Arada, N. and Sequeira, A., Steady Flows of Shear-Dependent Oldroyd-B Fluids around an Obstacle, *J. Math. Fluid Mech.* **7** (2005), 451-483.

[4] Astarita, G. and Marucci, G., *Principles of Non-Newtonian Fluid Mechanics*, McGraw-Hill (1974).

[5] Becker, L.E., McKinley, G.H., and Stone, H.A., Sedimentation of a Sphere Near a Plane Wall: Weak Non-Newtonian and Inertial Effects, *J. Non-Newtonian Fluid Mech.* **63** (1996), 201–233.

[6] Beirão da Veiga, H., On the Existence of Strong Solutions to a Coupled Fluid-Structure Evolution Problem, *J. Math. Fluid Mech.* **6** (2004), 21–52.

[7] Beirão da Veiga, H., Time periodic solutions of the Navier–Stokes equations in unbounded cylindrical domains—Leray's problem for periodic flows. *Arch. Ration. Mech. Anal.* **178** (2005), 301–325.

[8] Beirão da Veiga, H., On the Regularity of Flows with Ladyzhenskaya Shear-Dependent Viscosity and Slip or Nonslip Boundary Conditions, *Comm. Pure Appl. Math.* **58** (2005), 552–577.

[9] Beirão da Veiga, H., On Some Boundary Value Problems for Incompressible Viscous Flows with Shear Dependent Viscosity, *Elliptic and Parabolic Problems*, Progr. Nonlinear Differential Equations Appl., Vol. 63, Birkhäuser, Basel, 2005, 23–32.

[10] Beirão da Veiga, H., On Some Boundary Value Problems for Flows with Shear Dependent Viscosity, *Variational Analysis and Applications* Nonconvex Optim. Appl., Vol. 79, Springer, New York, 2005, 161–172.

[11] Beirão da Veiga, H., Navier–Stokes Equations with Shear-Dependent Viscosity. Regularity up to the Boundary, *J. Math. Fluid Mech.*, in press.

[12] Berker, R., 1964, Contrainte sur un Paroi en Contact avec un Fluide Visqueux Classique, un Fluide de Stokes, un Fluide de Coleman-Noll, *C.R. Acad. Sci. Paris*, **285**, 5144–5147.

[13] R. B. Bird, R. C. Armstrong, and O. Hassager, *Dynamics of Polymeric Liquids*, Volume I, John Wiley & Sons, second ed. (1987).

[14] Bretherton, F.P., The motion of a Rigid Particle in a Shear Flow at Low Reynolds Number. *J. Fluid Mech.* **14** (1962), 284–304.

[15] Bitbol, M., Red Blood Cell Orientation in Orbit $C = 0$, *Biophys. J.* **49** (1986), 1055–1068.

[16] Blavier, E. and Mikelić, A., On the Stationary Quasi-Newtonian Flow Obeying a Power Law, *Math. Meth. Appl. Sci.* **18** (1995), 927–948.

[17] Bönisch, S. and Galdi, G.P., Lift and Migration of Spheres in a Two-Dimensional Channel, in progress.

[18] Browder, F.E., Existence and Uniqueness Theorems for Solutions of Nonlinear Boundary Value Problems, *Proc. Sympos. Appl. Math.* **17** Amer. Math. Soc.. Providence, R.I., 1965, 24–49.

[19] Carreau, P.J., *Rheological equations from molecular network theories*, Ph.D. thesis, University of. Wisconsin (1968).

[20] Chambolle, A., Desjardins, B., Esteban, M.J., Grandmont, C., Existence of Weak Solutions for an Unsteady Fluid-Plate Interaction Problem, *J. Math. Fluid Mech.* **7** (2005), 368–404.

[21] Chang, W., Trebotich, D., Lee, L.P., and Liepmann, D., Blood Flow in Simple Microchannels, *Proceedings of the 1st Annual International IEEE-EMBS Special Topic Conference on Microtechnologies in Medicine & Biology, Lyon*, France (2000).

[22] Chhabra R.P., *Bubbles, Drops and Particles in Non-Newtonian Fluids*, CRC Press (1993).

[23] Cheng, C.H.A, Cutand, D., and Shkoller, D., Navier–Stokes Equations Interacting with a Nonlinear Elastic Shell (2006), preprint.

[24] Coleman, B.D. and Noll, W., On Certain Steady Flows of General Fluids, *Arch. Rational Mech. Anal.* **3** (1959), 289–303.

[25] Coleman, B.D. and Noll, W., An Approximation Theorem for Functionals with Applications in Continuum Mechanics, *Arch. Rational Mech. Anal.* **6** (1960), 55–70.

[26] Coleman, B.D. and Noll, W., Simple Fluids with Fading Memory, *Second-Order Effects in Elasticity, Plasticity and Fluid Dynamics*, Oxford, Pergamon Press, 1962, 530–552.

[27] Coscia, V., and Galdi, G.P., Existence, Uniqueness and Stability of Regular Steady Motions of a Second Grade Fluid, *Int. J. Nonl. Mech.* **29** (1994), 493–516.

[28] Coutand, D., and Shkoller, S., Motion of an Elastic Solid Inside of an Incompressible Viscous Fluid, *Arch. Rational Mech. Anal.* **176** (2005), 25–102.

[29] Coutand, D., and Shkoller, S., On the Interaction Between Quasilinear Elastodynamics and the Navier–Stokes Equations, *Arch. Rational Mech. Anal.* **179** (2006), 303–352.

[30] Ebmeyer, C., Steady Flow of Fluids with Shear-Dependent Viscosity under Mixed Boundary Value Conditions in Polyhedral Domains, *Math. Models Methods Appl. Sci.* **10** (2000), 629–650.

[31] Feireisl, E., On the Motion of Rigid Bodies in a Viscous Incompressible Fluid, *J. Evol. Equ.* **3** (2003), 419–441.

[32] Fontelos, M.A. and Friedman, A., Stationary non-Newtonian Fluid Flows in Channel-Like and Pipe-Like Domains, *Arch. Ration. Mech. Anal.* **151** (2000), 1–43.

[33] Frehse, J., Málek,J., and Steinhauer, M., An Existence Result for Fluids with Shear Dependent Viscosity–Steady Flows, *Nonlinear Anal. Th. Methods Appl.* **30** (1997), 3041–3049.

[34] Frehse, J., Málek,J., and Steinhauer, M., On Analysis of Steady Flows of Fluids with Shear-Dependent Viscosity Based on the Lipschitz Truncation Method, *SIAM J. Math. Anal.* **34** (2003), 1064–1083.

[35] Galdi, G.P., Mathematical Theory of Second-Grade Fluids, *Stability and Wave Propagation in Fluids and Solids*, G.P. Galdi ed., Springer-Verlag, Berlin, 1995, 67–104.

[36] Galdi, G.P., *An Introduction to the Mathematical Theory of the Navier–Stokes Equations: Linearised Steady Problems*, Springer Tracts in Natural Philosophy, Vol. 38, Springer-Verlag, 2nd Corrected Edition (1998).

[37] Galdi, G.P., *An Introduction to the Mathematical Theory of the Navier–Stokes Equations: Nonlinear Steady Problems*, Springer Tracts in Natural Philosophy, Vol. 39, Springer-Verlag, 2nd Corrected Edition (1998).

[38] Galdi, G.P., 2002, On the Motion of a Rigid Body in a Viscous Liquid: A Mathematical Analysis with Applications, *Handbook of Mathematical Fluid Mechanics*, North-Holland Elsevier Science, Vol. 1 (2002), 653–792.

[39] Galdi, G.P., Grobbelaar, M. and Sauer, N., Existence and Uniqueness of Classical Solutions of the Equations of Motion for Second-Grade Fluids, *Arch. Rational Mech. Anal.* **124** (1993), 221–237.

[40] Galdi, G.P, and Heuveline, V., Lift and Sedimentation of Particles in the Flow of a Viscoelastic Liquid in a Channel, *Free and Moving Boundarie Analysis, Simulation and Control*, R. Glowinski and J.-P. Zolesio Eds, CRC Publ., in press.

[41] Galdi, G.P., Pileckas, K. and Silvestre, A.L, Relation Between Pressure-Drop and Flow Rate in Unsteady Poiseuille Flow, *Zeitschrift für Angewandte Mathematik und Physik* (ZAMP), submitted.

[42] Galdi, G.P., Pokorný, M., Vaidya, A., Joseph, D.D., and Feng, J., Orientation of Symmetric Bodies Falling in a Second-Order Liquid at Non-Zero Reynolds Number, *Math. Models Methods Appl. Sci.* **12** (2002), 1653–1690.

[43] Galdi, G.P., and Robertson, A.M., The Relation Between Flow Rate and Axial Pressure Gradient for Time-Periodic Poiseuille Flow in a Pipe, *J. Math. Fluid Mech.* **7** suppl. 2 (2005), 215–223.

[44] Galdi, G.P., Sequeira, A. and Videman, J., Steady Motions of a Second-Grade Fluid in an Exterior Domain, *Adv. Math. Sci. Appl.* **7** (1997), 977–995.

[45] Galdi, G.P. and Silvestre, A.L., Existence of Time-Periodic Solutions to the Navier-Stokes Equations Around a Moving Body *Pacific J. Math.* **223** (2006), 251–268.

[46] Galdi G.P., and Vaidya A., Translational Steady Fall of Symmetric Bodies in a Navier–Stokes Liquid, with Application to Particle Sedimentation, *J. Math. Fluid Mech.* **3** (2001), 183–211.

[47] Grandmont, C., Existence for a Three-Dimensional Steady State Fluid-Structure Interaction Problem, *J. Math. Fluid Mech* **4** (2002), 76–94.

[48] Gresho, P.M., Some Current CFD Issues Relevant to the Incompressible Navier–Stokes Equations, *Comput. Methods Appl. Mech. Eng.* **87** (1991), 201–252.

[49] Grossman, P.D., and Soane, D.S., Orientation Effects on the Electrophoretic Mobility of Rod-Shaped Molecules in Free Solution, *Anal. Chem.* **62**, (1990), 1592–1596.

[50] Guillopé, G., Hakim, A., and Talhouk, R., Existence of Steady Flows of Slightly Compressible Viscoelastic Fluids of White-Metzner Type Around an Obstacle, *Comm. Pure Appl. Anal.* **4** (2005), 23–43.

[51] Guillopé C. and J.-C. Saut, Existence Results for the Flow of Viscoelastic Fluids with a Differential Constitutive Law, *Nonlinear Anal. Th. Methods Appl.* **15** (1990), 849–869.

[52] Guillopé, C. and J.-C. Saut, Existence and Stability of Steady Flows of Weakly Viscoelastic Fluids, *Proc. Roy. Soc. Edinburgh* A119 (1991), 137–158.

[53] Guillopé, G. and Talhouk, R., Steady Flows of Slightly Compressible Viscoelastic Fluids of Jeffreys' Type Around an Obstacle, *Diff. Int. Eq.* **16** (2003), 1293–1320.

[54] Hagen, G. On the Motion of Water in Narrow Cylindrical Tubes, *Pogg. Ann.*, **46** (1839), 423–442.

[55] Hames, B.D., and Rickwood, D., Eds., *Gel Electrophoresis of Proteins*, IRL Press, Washington, D.C. (1984).

[56] Hakim, A., Mathematical Analysis of Viscoelastic Fluids of White-Metzner Type, *J. Math. Anal. Appl.* **185** (1994), 675–705.

[57] Heywood, J.G., The Navier–Stokes Equations: On the Existence, Regularity and Decay of Solutions, *Indiana U. Math. J.*, **29** (1980), 639–681.

[58] Heywood, J.G., Rannacher, R., and Turek, S., Artificial Boundaries and Flux and Pressure Conditions for the Incompressible Navier–Stokes Equations, *Int. J. Numer. Meth. in Fluids* **22** (1996), 325–352.

[59] Horgan, C.O., and Wheeler, L.T., Spatial Decay Estimates for the Navier–Stokes Equations with Application to the Problem of Entry Flow, *SIAM J. Appl. Math.* **35** (1978), 97–116.

[60] Joseph, D.D., Instability of the Rest State of Fluids of Arbitrary Grade Larger than One, *Arch. Rational Mech. Anal.* **75** (1980), 251–256.

[61] Joseph, D.D., *Fluid Dynamics of Viscoelastic Liquids*, Applied Mathematical Sciences, **84**, Springer-Verlag (1990).

[62] Joseph, D.D., 2000, Interrogations of Direct Numerical Simulation of Solid-Liquid Flow, Web Site:
http://www.aem.umn.edu/people/faculty/joseph/interrogation.html

[63] Joseph, D.D., and Feng, J., A Note on the Forces that Move Particles in a Second-Order Fluid, *J. Non-Newtonian Fluid Mech.* **64** (1996), 299–302.

[64] Joseph, D.D., and Fosdick, R.L., The Free Surface on a Liquid Between Cylinders Rotating at Different Speeds. Part I, *Arch. Rational Mech. Anal.* **49** (1973), 321–380.

[65] Joseph, D.D., Beavers, G.S., and and Fosdick, R.L., The Free Surface on a Liquid Between Cylinders Rotating at Different Speeds. Part II, *Arch. Rational Mech. Anal.* **49** (1973), 381–401.

[66] Kučera, P. and Skalák, Z., Local Solutions to the Navier–Stokes Equations with Mixed Boundary Conditions, *Acta Appl. Math.* **54** (1998), 275–288.

[67] Ladyzhenskaya, O.A., On Some New Equations Describing Dynamics of Incompressible Fluids and on Global Solvability of Boundary Value Problems to These Equations, *Trudy Steklov Math. Inst.* **102** (1967), 85-104.

[68] Ladyzhenskaya, O.A., On Some Modifications of the Navier–Stokes Equations for Large Gradients of Velocity, *Zap. Nauchn. Sem. Leningrad. Otdel. Mat. Inst. Steklov* (LOMI) **7** (1968), 126-154.

[69] Ladyzhenskaya, O.A., *The Mathematical Theory of Viscous Incompressible Flow*, Gordon and Breach, New York (1969).

[70] Ladyzhenskaya, O.A., *Boundary Value Problems of Mathematical Physics*, Springer-Verlag (1985).

[71] Ladyzhenskaya, O.A., and Solonnikov, V.A., Determination of Solutions of Boundary Value Problems for Steady-State Stokes and Navier–Stokes Equations in Domains Having an Unbounded Dirichlet Integral, *Zap. Nauchn. Sem. Leningrad Otdel. Mat. Inst. Steklov* (LOMI) **96** (1980), 117–160; English Transl.: *J.Soviet Math.*, **2** no. 1 (1983), 728–761.

[72] Ladyzhenskaya, O.A. and Ural'ceva, N.N., *Linear and Quasilinear Elliptic Equations*, Academic Press, New York, (1968).

[73] Lee, S.C., Yang, D.Y., Ko, J., and You, J.R., Effect of compressibility on flow field and fiber orientation during the filling stage of injection molding *J Mater. Process. Tech.* **70** (1997), 83–92.

[74] Leigh, D.C., Non-Newtonian Fluids and the Second Law of Thermodynamics, *Phys. Fluids* **5** (1962), 501–502.

[75] Lieberman, G.M., Boundary Regularity for Solutions of Degenerate Elliptic Equations, *Nonlinear Anal. Th. Methods Appl.* **12** (1988), 1203–1219.

[76] Lions, J.-L., *Quelques Methodes de Résolution des Problèmes aux Limites Non Linéaires,* Dunod, Paris (1969).

[77] Liu, W.B. and Barrett, W., A Remark on the Regularity of the Solutions of the p–Laplacian and its Application to their Finite Element Approximation, *J. Math. Anal. Appl.* **178** (1993), 470–487.

[78] Liu, Y.J., and Joseph, D.D., Sedimentation of Particles in Polymer Solutions, *J. Fluid Mech.* **255** (1993), 565–595.

[79] Malek, J., Nečas, J., Rokyta, M., and Růžička, M., *Weak and Measure-Valued Solutions to Evolutionary PDEs,* Vol. 13 Chapman & Hall , London (1996).

[80] Marušić-Paloka, E., Steady Flow of a Non-Newtonian Fluid in Unbounded Channels and Pipes, *Math. Mod. Meth. Appl. Sci.* **10** (2000), 1425–1445.

[81] Miller, R.K., Feldstein, A., Smoothness of solutions of Volterra integral equations with weakly singular kernels, *SIAM J. Math. Anal.* **2** (1971), 242–258.

[82] Minty, G.J., On a 'Monotonicity' Method for the Solution of Nonlinear Equations in Banach Spaces, *Proc. Nat. Acad. Sci. U.S.A.* **50** (1963), 1038–1041.

[83] Noll, W., A Mathematical Theory of the Mechanical Behavior of Continuous Media, *Arch. Rational Mech. Anal.* **2** (1958), 197–226.

[84] Novotný, A. Sequeira, A. Videman, J.H., Steady Motions of Viscoelastic Fluids in Three-Dimensional Exterior Domains. Existence, Uniqueness and Asymptotic Behaviour, *Arch. Ration. Mech. Anal.* **149** (1999), 49–67.

[85] Ostwald, K., Über die Geschwindigkeitsfunktion der Viskosität disperser Systeme I, *Kolloid Zeit.* **36** (1925), 99–117.

[86] Pileckas, K., Navier–Stokes System in Domains with Cylindrical Outlets to Infnity. Leray's Problem, *Handbook of Mathematical Fluid Mechanics,* North-Holland Elsevier Science, in press.

[87] Pileckas, K., Sequeira, A., and Videman, J.H., Steady Flows of Viscoelastic Fluids in Domains with Outlets to Infinity *J. Math. Fluid Mech.* **2** (2000), 185–218.

[88] Poiseuille, J.L.M., Recherches Experimentales sur le Mouvement des Liquides dans les Tubes de Tres Petits Diameters, *C. R. Acad. Sci. Paris* **11** (1840), 961–967.

[89] Prilepko, A.I., Orlovsky, D.G., Vasin, I.A., *Methods for Solving Inverse Problems in Mathematical Physics,* Marcel Dekker, New York, Basel (1999).

[90] Prouse, G., Soluzioni Periodiche delle Equazioni di Navier–Stokes, *Rend. Atti. Accad. Naz. Lincei* **35** (1963), 403–409.

[91] Rannacher, R., Methods for Numerical Flow Simulation, article in this volume.

[92] Renardy, M., Existence of Slow Steady Flows of Viscoelastic Fluids with Differential Constitutive Equations *Z. Angew. Math. Mech.* **65** (1985), 449–451.

[93] Roco, M.C., (Ed.), *Particulate Two-Phase Flow,* Butterworth-Heinemann Publ., Series in Chemical Engineering (1993).

[94] Robertson, A.M., Review of Relevant Continuum Mechanics, article in this volume.

[95] Růžička, M., A Note on Steady Flow of Fluids with Shear Dependent Viscosity, *Nonlinear Anal. Th. Methods Appl.* **30** (1997), 3029–3039.

[96] Sandri, D., Sur L'Approximation Numérique des Écoulements Quasi-Newtoniens dont la Viscosité suit la loi Puissance ou la loi de Carreau, *Math. Mod. Numer. Anal.* **27** (1993), 131–155.

[97] Schmid-Schonbein, H., and Wells,R., Fluid Drop-Like Transition of Erythrocytes under shear, *Science* **165** (1969), 288–291.

[98] Segrè, G., and Silberberg, A., Radial Poiseuille Flow of Suspensions, *Nature* **189** (1961), 209–210.

[99] Segrè G., and Silberberg, A., Behaviour of Macroscopic Rigid Spheres in Poiseuille Flow, Part I, *J. Fluid Mech.* **14** (1962), 115–135.

[100] Sequeira, A., and Baía, M., A Finite Element Approximation for the Steady Solution of a Second-Grade Fluid Model, *J. Comput. Appl. Math.* **111** (1999), 281–295.

[101] Sequeira, A., and Videman, J.-H. Mathematical Results and Numerical Methods for Steady Incompressible Viscoelastic Fluid Flows, *Math. Appl.* **528** (2001), 339–365.

[102] Serrin, J.B, Poiseuille and Couette Flow of Non-Newtonian Fluids, *Z. Angew. Math. Mech.* **39** (1959), 295–299.

[103] Simader, C.G., and Sohr, H., 1997, *The Dirichlet Problem for the Laplacian in Bounded and Unbounded Domains*, Pitman Research Notes in Mathematics Series, Longman Scientific & Technical, Vol. 360.

[104] Talhouk, R., Existence Results for Steady Flow of Weakly Compressible Viscoelastic Fluids with Differential Constitutive Law, *Diff. and Integral Eq.* **12** (1999), 741–722.

[105] Tinland, B., Meistermann, L., Weill, G., Simultaneous Measurements of Mobility, Dispersion, and Orientation of DNA During Steady-Field Gel Electrophoresis Coupling a Fluorescence Recovery After Photobleaching Apparatus with a Fluorescence Detected Linear Dichroism Setup, *Phys. Rev. E* **61** (2000), 6993–6998.

[106] Thomson, W., and Tait, P.G., *Natural Philosophy*, Vols 1, 2, Cambridge University Press (1879).

[107] Trainor, G.L., DNA Sequencing, Automation and Human Genome, *Anal. Chem.* **62** (1990), 418–426.

[108] Tricomi, F.G., *Integral Equations*, Intersience, New York (1957).

[109] Truesdell, C.A., The Meaning of Viscometry in Fluid Dynamics, *Ann. Rev. Fluid Mech.* **6** (1974), 111–146.

[110] Uijttewaal, W.S.J., Nijhof, E.-J., and Heethaar R.M., Lateral Migration of Blood Cells and Microspheres in Two-Dimensional Poiseuille Flow : a Laser-Doppler Study, *J. Biomech.* **27** (1994), 35–42.

[111] Vaidya, A., A Note on the Orientation of Symmetric Rigid Bodies Sedimenting in a Power-Law Fluid, *Appl. Math. Letters* **18** (2005), 1332–1338.

[112] Vaidya, A., Observations on the Transient Nature of Shape-Tilting Bodies Sedimenting in Polymeric Liquids, *J. Fluids and Struct.* **22** (2006), 253–259.

[113] Vejvoda, O., Herrmann, L., Lovicar, V., Sova, M.; Straškraba, I., and Štědrý, M., *Partial Differential Equations: Time-Periodic Solutions,* Martinus Nijhoff Publishers, The Hague (1981).

[114] Wang, J., Bail, R.-Y., Lewandowski, C., Galdi, G.P. and Joseph, D.D., Sedimntation of Cylindrical Particles in a Viscoelastic Liquid: Shape-Tilting, *China Particuology* **2** (2004), 13–18.

[115] Wolf, J., Existence of Weak Solutions to the Equations of Non-Stationary Motion of Non-Newtonian Fluids with Shear-Rate Dependent Viscosity, *J. Math. Fluid Mech.*, in press.

[116] Womersley, J.R., Method for the Calculation of Velocity, Rate of Flow and Viscous Drag in Arteries when the Pressure Gradient is Known, *J. Physiol.* **127** (1955), 553-556.

Giovanni P. Galdi
Department of Mechanical Engineering and Materials Science
University of Pittsburgh
648 Benedum Hall, 3700 O'Hara Street
Pittsburgh, PA 15261
USA
e-mail: `galdi@engr.pitt.edu`

Hemodynamical Flows. Modeling, Analysis and Simulation
Oberwolfach Seminars, Vol. 37, 275–332
© 2008 Birkhäuser Verlag Basel/Switzerland

Methods for Numerical Flow Simulation

Rolf Rannacher

The author acknowledges the contributions by several former and present members of his work group and the support by the German Research Foundation (DFG).

Introduction

This chapter introduces into computational methods for the simulation of PDE-based models of laminar hemodynamical flows. We discuss space and time discretization with emphasis on operator-splitting and finite-element Galerkin methods because of their flexibility and rigorous mathematical basis. Special attention is paid to the simulation of pipe flow and the related question of artificial outflow boundary conditions. Further topics are efficient methods for the solution of the resulting algebraic problems, techniques of sensitivity-based error control and mesh adaptation, as well as flow control and model calibration. We concentrate on *laminar* flows in which all relevant spatial and temporal scales can be resolved and no additional modeling of turbulence effects is required. This covers most of the relevant situations of hemodynamical flows. The numerical solution of the corresponding systems is complicated mainly because of the incompressibility constraint which enforces the use of implicit methods and its essentially parabolic or elliptic character which requires the prescription of boundary conditions along the whole boundary of the computational domain.

The material of this article is based on long-standing joint work of several former and present members of the Numerical Methods Group at the Institute of Applied Mathematics, University of Heidelberg. For more details the following publications may be consulted: Rannacher [65, 66], Rannacher/Turek [67, 48], Becker [3, 4, 5], Becker/Rannacher [13, 14], Bangerth/Rannacher [2], Braack/Richter [20], Becker/Vexler [15], Heuveline/Rannacher [47], Most of the numerical examples have been obtained using the software tools 'Gascoigne' [34], 'Hiflow' [45], and the graphics packages 'VisuSimple' [77, 8] and 'HiVision' [46, 18]. More references to the relevant literature will be given at the respective places in the text, below. The contents of this article are organized as follows:

Contents

1. Finite-element methods for the simulation of viscous flow

1.1. The Navier–Stokes equations

The continuum-mechanical model of the flow of a viscous Newtonian fluid is the system of conservation equations for mass, momentum and energy:

$$\partial_t \rho + \nabla \cdot (\rho v) = 0, \tag{1.1}$$

$$\partial_t(\rho v) + \rho v \cdot \nabla v - \nabla \cdot (\mu \nabla v + \tfrac{1}{3}\mu \nabla \cdot v I) + \nabla p_{\text{tot}} = \rho f, \tag{1.2}$$

$$\partial_t(c_p \rho T) + c_p \rho v \cdot \nabla T - \nabla \cdot (\lambda \nabla T) = h. \tag{1.3}$$

In the following, the fluid is assumed as *incompressible* and the density as homogeneous, i.e., $\rho \equiv \text{const.}$, so that (1.1) reduces to the constraint $\nabla \cdot v = 0$. Further, in the isothermal case, the energy equation decouples from the momentum and continuity equations and, setting $\rho \equiv 1$, the Navier–Stokes system can be written in short as

$$\partial_t v + v \cdot \nabla v - \nu \Delta v + \nabla p = f, \tag{1.4}$$

$$\nabla \cdot v = 0, \tag{1.5}$$

with the kinematic viscosity $\nu > 0$. This system is supplemented by appropriate initial and boundary conditions,

$$v_{|t=0} = v^0, \quad v_{|\Gamma_{\text{rigid}}} = 0, \quad v_{|\Gamma_{\text{in}}} = v^{\text{in}}, \quad (\mu \partial_n v - pn)_{|\Gamma_{\text{out}}} = qn, \qquad (1.6)$$

where Γ_{rigid}, Γ_{in}, and Γ_{out} are the rigid part, the inflow part and the outflow part, respectively, of the flow domain's boundary $\partial\Omega$. The role of the natural outflow boundary condition on Γ_{out} will be discussed in greater detail below. In this formulation the flow domain may be two- or three-dimensional. This model is made dimensionless through a scaling transformation with the Reynolds number $Re = UL/\nu$ as the characteristic parameter, where U is the reference velocity and L the characteristic length, e.g., $U \approx \max|v^{\text{in}}|$ and $L \approx \text{diam}(\Omega)$.

1.1.1. Variational formulation. The finite-element discretization of the Navier–Stokes problem (1.4–1.5) is based on its variational formulation. In order to understand the behavior of this method, we need a certain amount of mathematical formalism. We will use the following subspaces of the usual Lebesgue function space $L^2(\Omega)$ of square-integrable functions on a domain $\Omega \subset \mathbb{R}^d$ ($d = 2$ or 3):

$$L^2_0(\Omega) = \left\{ \varphi \in L^2(\Omega) : (v, 1) = 0 \right\}, \quad H^1(\Omega) = \left\{ v \in L^2(\Omega), \nabla v \in L^2(\Omega)^d \right\},$$

and $H^1_0(\Gamma; \Omega) = \left\{ v \in H^1(\Omega), v_{|\Gamma} = 0 \right\}$, for some (non-trivial) part Γ of the boundary $\partial\Omega$, as well as the corresponding inner products

$$(u, v) = \int_\Omega uv \, dx, \quad (\nabla u, \nabla v) = \int_\Omega \nabla u \cdot \nabla v \, dx,$$

and norms $\|v\| = (v, v)^{1/2}$ and $\|\nabla v\| = (\nabla v, \nabla v)^{1/2}$. These are all spaces of $\mathbb{R}$-valued functions. Spaces of $\mathbb{R}^d$-valued functions $v = (v_1, \ldots, v_d)$ are denoted by boldface-type, but no distinction is made in the notation of norms and inner products; thus, $\mathbf{H}^1_0(\Gamma; \Omega) = H^1_0(\Gamma; \Omega)^d$ has norm $\|\nabla v\| = (\sum_{i=1}^d \|\nabla v_i\|^2)^{1/2}$, etc. All the other notation is self-explaining:

$$\partial_t u = \frac{\partial u}{\partial t}, \quad \partial_i u = \frac{\partial u}{\partial x_i}, \ i = 1, 2, 3, \quad \partial_n v = n \cdot \nabla v, \quad \partial_\tau = \tau \cdot \nabla v,$$

where n and τ are the normal and tangential unit vectors along the boundary $\partial\Omega$, respectively, and the dot '$\cdot$' indicates the inner product of vector quantities.

The pressure p in the Navier–Stokes equations is uniquely (possibly up to a constant) determined by the velocity field v. There holds the stability estimate ('inf-sup' stability)

$$\inf_{q \in L^2(\Omega)} \left\{ \sup_{\varphi \in \mathbf{H}^1_0(\Gamma; \Omega)} \frac{(q, \nabla \cdot \varphi)}{\|q\| \, \|\nabla \varphi\|} \right\} \geq \gamma_0 > 0, \qquad (1.7)$$

where $L^2(\Omega)$ has to be replaced by $L^2_0(\Omega)$ in the case $\Gamma = \partial\Omega$. Finally, we introduce the bilinear and trilinear forms

$$a(u, v) := \nu(\nabla u, \nabla v), \quad b(p, v) := -(p, \nabla \cdot v), \quad n(u, v, w) := (u \cdot \nabla v, w),$$

and the abbreviations

$$\mathbf{H} := \mathbf{H}_0^1(\Gamma_D; \Omega), \quad L := L^2(\Omega) \quad \left(L := L_0^2(\Omega) \text{ in the case } \Gamma_D = \partial\Omega \right),$$

where $\Gamma_D = \Gamma_{\text{in}} \cup \Gamma_{\text{rigid}}$. With this notation the variational formulation of the Navier–Stokes problem (1.4,1.5) reads as follows:

Problem 1.1. Find functions $v(\cdot, t) \in v^{\text{in}} + \mathbf{H}$ and $p(\cdot, t) \in L$, where $v(\cdot, t)$ is continuous on $[0, T]$ and differentiable on $(0, T]$, such that $v_{|t=0} = v^0$, and

$$(\partial_t v, \varphi) + a(v, \varphi) + n(v, v, \varphi) + b(p, \varphi) = (f, \varphi) \quad \forall \varphi \in \mathbf{H}, \tag{1.8}$$

$$b(\chi, v) = 0 \quad \forall \chi \in L. \tag{1.9}$$

The corresponding stationary problem reads

Problem 1.2. Find $v \in v^{\text{in}} + \mathbf{H}$ and $p \in L$, such that

$$a(v, \varphi) + n(v, v, \varphi) + b(p, \varphi) = (f, \varphi) \quad \forall \varphi \in \mathbf{H}, \tag{1.10}$$

$$b(\chi, v) = 0 \quad \forall \chi \in L. \tag{1.11}$$

It is well known that, for $\Gamma_{\text{out}} = \emptyset$, the stationary problem (1.10,1.11) possesses a unique solution, which is also a classical solutions if the data of the problems are smooth enough. However, for small viscosity, i.e., large Reynolds number, these solutions may be unstable. For the non-stationary problem (1.8,1.9) the existance of a unique solution is known for general data in two dimensions, but in tree dimensions only for sufficiently small data, e.g., $\|\nabla v^0\| \approx \nu$, or on sufficiently short intervals of time, $0 \leq t \leq T$, with $T \approx \nu$. For more details on the mathematical theory of the Navier–Stokes equations the reader may consult, e.g., [32] and the article Galdi [33] in this volume.

1.2. Discretization of space

For the discretization in space the finite-element Galerkin method is considered. Let $\mathbb{T}_h$ be decompositions of $\overline{\Omega}$ into (closed) cells K (triangles or quadrilaterals in 2d, and tetrahedra or hexahedra in 3d) such that:

- $\overline{\Omega} = \cup\{K \in \mathbb{T}_h\}$.
- Any two cells K, K' only intersect in common faces, edges or vertices.
- The decomposition $\mathbb{T}_h$ matches the decomposition $\partial\Omega = \Gamma_{\text{in}} \cup \Gamma_{\text{rigid}} \cup \Gamma_{\text{out}}$.

In the following, we will also allow decompositions with a certain limited number of 'hanging nodes' in order to ease local mesh refinement and coarsening. For the cells $K \in \mathbb{T}_h$, let $h_K := \text{diam}(K)$, ρ_K the radius of maximal ball contained in K, and $h := \max_{K \in \mathbb{T}_h} h_K$. In the following, we will assume *shape-regularity*,

$$ch_K \leq \rho_K \leq h_K, \tag{1.12}$$

but *quasi-uniformity*, $\max_{K \in \mathbb{T}_h} h_K \leq c \min_{K \in \mathbb{T}_h} h_K$, is not required.

The 'finite-element' spaces $H_h \subset H$ corresponding to the decompositions $\mathbb{T}_h$ are defined by

$$H_h := \{v_h \in H, \ v_{h|K} \in P(K), \ K \in \mathbb{T}_h\},$$

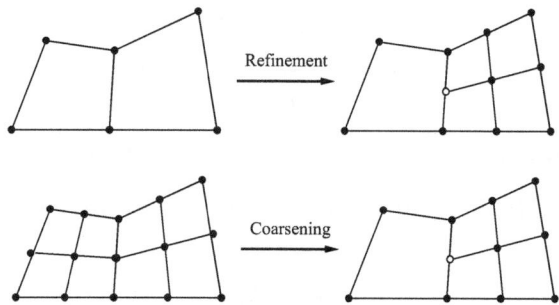

FIGURE 1. Refinement and coarsening in quadrilateral meshes using 'hanging' nodes.

where $P(K)$ are certain spaces of elementary functions on the cells K. In the simplest case, $P(K)$ are full polynom spaces, $P(K) = P_r(K)$, for some degree $r \geq 1$. On general quadrilateral or hexahedral cells, one uses 'parametric' elements, i.e., the local shape functions are constructed by using transformations $\psi_K : \hat{K} \to K$ between the 'physical' cell K and a fixed 'reference unit-cell' $\hat{K}$, i.e., $v_{h|K}(\psi_K(\cdot)) \in P_r(\hat{K})$. This construction is necessary, in general, in order to preserve 'conformity' (i.e., global continuity) of the cell-wise defined functions $v_h \in H_h$. For example, the use of *bilinear* shape functions $\varphi \in \text{span}\{1, x_1, x_2, x_1 x_2\}$ on a quadrilateral mesh in 2D employs likewise *bilinear* transformations $\psi_K : \hat{K} \to K$.

In a finite-element discretization *'consistency'* is expressed in terms of local approximation properties of the shape functions used. For example, in the case of a second-order approximation using *linear* or *d-linear* shape functions, there holds locally on each cell K:

$$\|v - I_h v\|_K + h_K \|\nabla(v - I_h v)\|_K \leq c_I h_K^2 \|\nabla^2 v\|_K. \tag{1.13}$$

Here, $I_h v \in H_h$ is the natural 'nodal interpolation' of a function $v \in H \cap H^2(\Omega)$, i.e., $I_h v$ coincides with v with respect to certain 'nodal functionals' (e.g., point values at vertices, mean values over edges or faces, etc.). The 'interpolation constant' is usually of size $c_I \sim 0.1-1$.

On a finite-element mesh $\mathbb{T}_h$ of Ω, one defines spaces of 'discrete' trial and test functions, $\mathbf{H}_h \subset' \mathbf{H}$ and $L_h \subset L$. The notation $\mathbf{H}_h \subset' \mathbf{H}$ indicates that in this discretization the spaces $\mathbf{H}_h$ may be 'non-conforming', i.e., the discrete velocities v_h are continuous across the interelement boundaries and zero along the rigid boundaries only in an approximate sense, with the cell-wise defined forms

$$a_h(\varphi, \psi) := \sum_{K \in \mathbb{T}_h} \nu(\nabla \varphi, \nabla \psi)_K, \quad b_h(\chi, \varphi) := -\sum_{K \in \mathbb{T}_h} (\chi, \nabla \cdot \varphi)_K,$$

with analogous definitions for $n_h(\varphi, \psi, \xi)$ and $\|\nabla \varphi\|_h$. With this notation the discrete versions of the stationary Navier–Stokes problem (1.10,1.11) read as follows:

Problem 1.3. Find $v_h \in v_h^{\text{in}} + \mathbf{H}_h$ and $p_h \in L_h$, such that

$$a_h(v_h, \varphi_h) + n_h(v_h, v_h, \varphi_h) + b_h(p_h, \varphi_h) = (f, \varphi_h) \quad \forall \, \varphi_h \in \mathbf{H}_h,$$

$$b_h(\chi_h, v_h) = 0 \quad \forall \, \chi_h \in L_h,$$

where v_h^{in} is a suitable approximation of the inflow data v^{in}.

The spaces $\mathbf{H}_h \times L_h$ are required to satisfy a discrete (uniform) 'inf-sup' condition:

$$\inf_{q_h \in L_h} \left\{ \sup_{\varphi_h \in \mathbf{H}_h} \frac{b_h(q_h, \varphi_h)}{\|q_h\| \|\nabla \varphi_h\|_h} \right\} \geq \gamma > 0. \tag{1.14}$$

This ensures that the discrete problems possess solutions which are uniquely determined in $\mathbf{H}_h \times L_h$ and stable. Under certain smoothness conditions on the solution and for sufficiently good boundary approximation, there holds the a priori estimate

$$\|\nabla(v - v_h)\|_h + \|p - p_h\| \leq ch^r \{ \|v\|_{H^{r+1}} + \|p\|_{H^r} \}, \tag{1.15}$$

if trial functions of degree $r \in \{1, 2\}$ for the velocity and at least of degree $r - 1$ for the pressure are used.

1.2.1. Examples of Stokes elements. We recall some standard finite-element spaces, which are frequently used in the spatial discretization of the Navier–Stokes equations.

(1) *The non-conforming 'rotated' d-linear $\widetilde{Q}_1^{nc}/P_0$ Stokes element:* This is the natural quadrilateral analogue of the well-known triangular *non-conforming* finite element of Crouzeix/Raviart (see Girault/Raviart [36]) and is sometimes referred to as 'Rannacher/Turek element' (see Rannacher/Turek [67]). It works well in two- as well as in three dimensions and is implemented in the FEATFLOW code; see Turek [72, 74] and URL: http://www.featflow.de. The reference velocity shape functions are 'rotated' d-linear with piecewise constant pressures:

$$\widetilde{Q}_1^{nc}(K) = \begin{cases} \{q \circ \psi_T^{-1} : q \in P_1 + \text{span}\{x_1^2 - x_2^2\}\}, & \text{for } d = 2, \\ \{q \circ \psi_T^{-1} : q \in P_1 + \text{span}\{x_1^2 - x_2^2, x_2^2 - x_3^2\}\}, & \text{for } d = 3. \end{cases}$$

The nodal functionals are the mean values over edges (in 2D) or faces (in 3D), $F_S(v_h) = |S|^{-1} \int_S v_h \, do$, for the velocity, and the mean value over the cell, $F_K(p_h) = |K|^{-1} \int_K p_h \, dx$, for the pressure. The corresponding finite-element spaces are defined by

$$\mathbf{H}_h := \left\{ \begin{array}{l} v_h \in \mathbf{L}^2(\Omega) : v_{h|K} \in Q_1^{nc}(K)^d, \ K \in \mathbb{T}_h, \ \text{continuous w.r.t.} \\ \text{the nodal functionals } F_S(\cdot), \ \text{and } F_S(v_h) = 0 \ \text{for } S \subset \Gamma \end{array} \right\},$$

$$L_h := \{q_h \in L^2(\Omega) : q_{h|K} \in P_0(K), \ K \in \mathbb{T}_h\}.$$

(2) *The conforming d-linear Q_1^c/Q_1^c Stokes element with pressure stabilization:* This Stokes element uses continuous isoparametric d-linear shape functions for both the velocity and the pressure approximations. The nodal values are just the

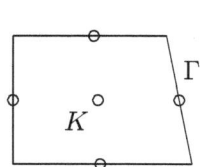
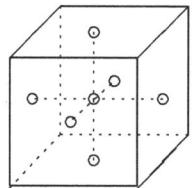

function values of the velocity and the pressure at the vertices of the cells, making this approximation particularly attractive in three dimensions:

$$Q_1^c(K) = \begin{cases} \{q \circ \psi_T^{-1} : q \in P_1 + \mathrm{span}\{x_1 x_2\}\}, & \text{for } d = 2, \\ \{q \circ \psi_T^{-1} : q \in P_1 + \mathrm{span}\{x_1 x_2, x_1 x_3, x_2 x_3, x_1 x_2 x_3\}\}, & \text{for } d = 3. \end{cases}$$

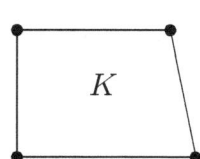
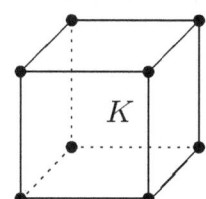

The corresponding finite-element spaces are defined by

$$\mathbf{H}_h := \{v_h \in \mathbf{H}_0^1(\Gamma; \Omega) : v_{h|K} \in Q_1^c(K)^d, \ K \in \mathbb{T}_h\},$$
$$L_h := \{q_h \in H^1(\Omega) : q_{h|K} \in Q_1^c(K), \ K \in \mathbb{T}_h\},$$

with the nodal functionals $F_a(v_h) = v_h(a)$ and $F_a(p_h) = p_h(a)$. This combination of spaces, however, would be unstable, i.e., it would fail to satisfy the inf-sup condition, if used together with the standard variational formulation of the Navier–Stokes system. Following Hughes et al. [51], for stabilization one may add certain least squares terms in the continuity equation (so-called 'pressure stabilization method'),

$$b(\chi_h, v_h) + c_h(\chi_h, p_h) = g_h(v_h; \chi_h),$$

where

$$c_h(\chi_h, p_h) = \frac{\alpha}{\nu} \sum_{K \in \mathbb{T}_h} h_K^2 (\nabla \chi_h, \nabla p_h)_K,$$

$$g_h(v_h; \chi_h) = \frac{\alpha}{\nu} \sum_{K \in \mathbb{T}_h} h_K^2 (\nabla \chi_h, f + \nu \Delta v_h - v_h \cdot \nabla v_h)_K.$$

The correction terms on the right-hand side have the effect that this modification (even for higher-order polynomial approximation) is fully consistent, since the additional terms cancel out if the exact solution $\{v, p\}$ is inserted. On fairly general meshes, one obtains a stable and consistent approximation of the Navier–Stokes problem with optimal-order accuracy. However, from a practical point of view, the above least-squares stabilization has several short-comings. First, being used together with the Neumann-type outflow boundary condition (1.6), it induces a non-physical numerical boundary layer along the outflow boundary. Second, the evaluation of the various terms in the stabilization forms $c_h(\chi_h, p_h)$ and $g_h(v_h; \chi_h)$ is very expensive, particularly in 3D. These problems can be resolved by using instead the so-called 'local pressure stabilization' (LPS) of Becker/Braack [7]. Here, the stabilization forms are $g_h = 0$ and

$$c_h(\chi_h, p_h) := (\nabla(\chi_h - \pi_{2h}\chi_h), \nabla(p_h - \pi_{2h}p_h)),$$

where π_{2h} denotes the projection into a space L_{2h} defined on a coarser mesh $\mathbb{T}_{2h}$. The resulting scheme is of second-order accurate, the evaluation of the system matrices is cheap and the consistency defect at the outflow boundary is avoided.

(3) *Higher-order Stokes elements:* One of the main advantages of the finite-element Galerkin method is that it provides systematic ways of dealing with complex geometries and of constructing higher-order approximation. Popular examples are:

- The triangular $P_2^\#/P_1^{dc}$ element with *continuous* piecewise quadratic velocity (augmented by a cubic 'bulb function') and *discontinuous* piecewise linear pressure approximation.
- The quadrilateral Q_2^c/Q_2^c element with *continuous* piecewise biquadratic velocity as well as pressure approximation and pressure stabilization.
- The triangular P_2^c/P_1^c or the quadrilateral Q_2^c/Q_1^c Taylor-Hood element with *continuous* piecewise quadratic or biquadratic velocity and piecewise linear or bilinear pressure approximation.

All these Stokes elements are inf-sup stable and third-order accurate for the velocity and second-order for the pressure, the errors measured in the respective L^2-norms; see Girault/Raviart [36] and Brezzi/Falk [22]. Practical experience shows that they yield much better approximations than the lower-order elements described above; see John [52] and Braack/Richter [20].

1.2.2. Treating dominant transport. In the case of higher Reynolds number (e.g., Re > 500 for the driven cavity, and Re > 50 for pipe flow around a cylinder) the standard finite-element models become unstable since they essentially use central-differences-like discretization of the advection term. For curing this, additional stabilization is needed:

- 'Upwind techniques' as common in FDM can also be used for FEM, but have only first-order accuracy and usually impose too much damping; see Tobiska/Schieweck [70].
- The Galerkin structure of the FEM is fully respected by the so-called 'streamline diffusion method' in form of an SUPG ('Streamline Upwinding Petrov Galerkin') method or LSSD ('Least-Squares Streamline Diffusion') method. The idea of streamline diffusion is to introduce artificial diffusion acting only in the transport direction while maintaining the second-order consistency of the scheme; see Hughes and Brooks [50].
- A new development for transport stabilization is the use of 'local-projection stabilization' (LPS) in the spirit of a 'multiscale' approach, such as already

described above for pressure stabilization. This has proven to be a very economical and sufficiently consistent method; see Braack/Burmann [19].

We describe the LSSD method for the Navier–Stokes system in greater detail. To this end, we introduce the product Hilbert-spaces $\mathbf{V} := \mathbf{H} \times L$ of pairs $u := \{v, p\}$ and $\varphi = \{\psi, \chi\}$ as well as their discrete analogues $\mathbf{V}_h := \mathbf{H}_h \times L_h$ of pairs $u_h := \{v_h, p_h\}$ and $\varphi_h = \{\psi_h, \chi_h\}$. On these spaces the semi-linear form

$$A(u)(\varphi) := a_h(v, \psi) + n_h(v, v, \psi) + b_h(p, \psi) - b(\chi, v) - (f, \psi)$$

is defined. Then, the variational formulation of the stationary Navier–Stokes equations is written in the following compact form:

Problem 1.4. Find $u \in (v_h^{\text{in}}, 0)^T + \mathbf{V}$, such that

$$A(u)(\varphi) = 0 \quad \forall \varphi \in \mathbf{V}.$$

For defining the stabilization, we introduce the differential operator $S(u) := -\nu \Delta + v \cdot \nabla$. Then, with the weighted L^2-bilinear form

$$(v, w)_\delta := \sum_{K \in \mathbb{T}_h} \delta_K (v, w)_K,$$

the LSSD stabilized finite-element approximation reads:

Problem 1.5. Find $u_h \in (v^{\text{in}}, 0)^T + \mathbf{V}_h$, such that

$$A(u_h)(\varphi_h) + (S(u_h)u_h - f, S(u_h)\varphi_h)_\delta = 0 \quad \forall \varphi_h \in \mathbf{V}_h.$$

The stabilization form contains the terms

$$\sum_{K \in \mathbb{T}_h} \delta_K (\nabla p_h, \nabla \chi_h)_K, \quad \sum_{K \in \mathbb{T}_h} \delta_K (v_h \cdot \nabla v_h, v_h \cdot \nabla \psi_h)_K, \quad \sum_{K \in \mathbb{T}_h} \delta_K (\nabla \cdot v_h, \nabla \cdot \psi_h)_K,$$

where the first term stabilizes the pressure-velocity coupling for the conforming Q_1^c/Q_1^c Stokes element, the second term stabilizes the transport operator, and the third term enhances mass conservation. The other terms introduced in the stabilization are correction terms which guarantee second-order consistency for the stabilized scheme. Practical experience and analysis suggest the following choice of the stabilization parameters:

$$\delta_K = \alpha \Big(\frac{\nu}{h_K^2} + \frac{\beta |v_h|_{K;\infty}}{h_K} \Big)^{-1},$$

with values $\alpha \approx \frac{1}{12}$ and $\beta \approx \frac{1}{6}$. The terms $-\nu \Delta v_h$ as well as $-\nu \Delta \psi_h$ in the stabilization are usually dropped, since they vanish or almost vanish on the low-order elements considered. Theoretical analysis shows that this kind of Galerkin stabilization actually leads to an improvement over the standard upwinding scheme.

1.3. The stationary algebraic problems

The discrete Navier–Stokes problem including simultaneously pressure and stream-line diffusion stabilization has to be converted into an algebraic system which can be solved on a computer. To this end, we choose appropriate local 'nodal bases' $\{\psi_h^i, i = 1, \ldots, N_v\}$ of the 'velocity space' $\mathbf{H}_h$, and $\{\chi_h^i, i = 1, \ldots, N_p\}$ of the 'pressure space' L_h, and expand the unknown solution $\{v_h, p_h\}$ in the form $v_h - v_h^{\text{in}} = \sum_{j=1}^{N_v} x_j \psi_h^j$ and $p_h = \sum_{j=1}^{N_p} y_j \chi_h^j$. Accordingly, we introduce the matrices

$$A = \big(a_h(\psi_h^j, \psi_h^i)\big)_{i,j=1}^{N_v}, \quad B = \big(b_h(\chi_h^j, \psi_h^i)\big)_{i,j=1}^{N_v, N_p},$$

$$N(x) = \big(n_h(v_h, \psi_h^j, \psi_h^i) + n_h(\psi_h^j, v_h^{\text{in}}, \psi_h^i)\big)_{i,j=1}^{N_v},$$

$$S(x) = \big((-\nu\Delta\psi_h^j + v_h \cdot \nabla\psi_h^j, -\nu\Delta\psi_h^i + v_h \cdot \nabla\psi_h^i)_\delta + (\nabla \cdot \psi_h^j, \nabla \cdot \psi_h^i)_\delta\big)_{i,j=1}^{N_v},$$

$$T(x) = \big((\nabla\chi_h^j, -\nu\Delta\psi_h^i + v_h \cdot \nabla\psi_h^i)_\delta\big)_{i,j=1}^{N_v, N_p}, \quad C = \big((\nabla\chi_h^j, \nabla\chi_h^i)_\delta\big)_{i,j=1}^{N_p},$$

and vectors

$$b = \big((f, \psi_h^i) - a(v_h^{\text{in}}, \psi_h^i) - n_h(v_h^{\text{in}}, v_h^{\text{in}}, \psi_h^i) + (f, v_h \cdot \nabla\psi_h^i)_\delta\big)_{i=1}^{N_v},$$

$$c = \big((f, \nabla\chi_h^i)_\delta\big)_{i=1}^{N_p}.$$

Notice that the non-homogeneous inflow data $v_{h|\Gamma_{\text{in}}} = v_h^{\text{in}}$ is implicitly incorporated into the system, and the stabilization only acts on velocity basis functions corresponding to interior nodes. With this notation the original variational formulation can equivalently be written in form of an algebraic system for the vectors $x \in \mathbb{R}^{N_v}$ and $y \in \mathbb{R}^{N_p}$ of expansion coefficients:

$$\begin{bmatrix} A + N(x) + S(x) & B + T(x) \\ -B^T + T(x)^T & C \end{bmatrix} \begin{bmatrix} x \\ y \end{bmatrix} = \begin{bmatrix} b \\ c \end{bmatrix}. \tag{1.16}$$

This system has essentially the features of a saddle-point problem (since C is small of size h^2) and is generically non-symmetric. This poses a series of problems for its iterative solution.

1.4. Discretization of time

We now consider the *non-stationary* Navier–Stokes system (Problem 1.1)

$$(\partial_t v, \varphi) + a(v, \varphi) + n(v, v, \varphi) + b(p, \varphi) = (f, \varphi) \quad \forall \varphi \in \mathbf{H},$$
$$b(\chi, v) = 0 \quad \forall \chi \in L, \tag{1.17}$$

where it is implicitly assumed that v as a function of time is continuous on $[0, \infty)$ and differentiable on $(0, \infty)$. The choice of the function spaces $\mathbf{H} \subset \mathbf{H}^1(\Omega)$ and $L \subset L^2(\Omega)$ depends again on the specific boundary conditions chosen for the problem to be solved. Due to the incompressibility constraint the non-stationary Navier–Stokes system has the character of a 'differential-algebraic equation' (in short 'DAE') of 'index two', in the language of ODE theory. This requires an implicit treatment of the pressure within the time-stepping process, while the other flow quantities may, in principle, be treated more explicitly.

1.4.1. The 'Rothe Method'. In the 'Rothe method', at first, the time variable is discretized by one of the common time-differencing schemes. For example, the backward Euler scheme leads to a sequence of stationary Navier–Stokes-like problems, starting from the given initial value v^0,

$$k_n^{-1}(v^n - v^{n-1}, \varphi) + a(v^n, \varphi) + n(v^n, v^n, \varphi) + b(p^n, \varphi) = (f^n, \varphi),$$
$$b(\chi, v^n) = 0,$$
(1.18)

for all $\{\varphi, \chi\} \in \mathbf{H} \times L$, where $k_n = t_n - t_{n-1}$ is the time step. Each of these problems is then solved by a spatial discretization method as described in the preceding section. This provides the flexibility to vary the spatial discretization, i.e., the mesh or the type of trial functions in the finite-element method, during the time stepping process. In the classical Rothe method the time-discretization scheme is kept fixed and only the size of the time step may change.

1.4.2. The 'Method of Lines'. The more traditional approach to solving time-dependent problems is the 'method of lines'. At first, the spatial variable is discrete, e.g., by a finite-element method as described in the preceding section, leading to a DAE system of the form

$$\begin{bmatrix} M & 0 \\ 0 & 0 \end{bmatrix} \begin{bmatrix} \dot{x}(t) \\ \dot{y}(t) \end{bmatrix} + \begin{bmatrix} A + N(x(t)) & B \\ -B^T & C \end{bmatrix} \begin{bmatrix} x(t) \\ y(t) \end{bmatrix} = \begin{bmatrix} b(t) \\ c(t) \end{bmatrix},$$

for $t \geq 0$, with the initial value $x(0) = x^0$ defined by $u_h^0 - u_h^{\mathrm{in}}(0) = \sum_{i=1}^{N_v} x_i^0 \psi_h^i$. The mass matrix M, the stiffness matrix A and the gradient matrix B are as defined above. The matrix C and the right-hand side c come from the pressure stabilization when using the conforming Q_1^c/Q_1^c Stokes element. Further, to simplify notation, the matrices $N(\cdot)$ and B as well as the vectors b and c are thought to contain also all further terms arising from the transport stabilization. This DAE system is now discretized with respect to time. Let k denote the time-step size. Frequently used schemes are the simple 'one-step-θ schemes', in which the time step $t_{n-1} \to t_n$ reads

$$\begin{bmatrix} M + \theta k A^n & \theta k B \\ -B^T & C \end{bmatrix} \begin{bmatrix} x^n \\ y^n \end{bmatrix} = \begin{bmatrix} [M - (1-\theta)k A^{n-1}] x^{n-1} + \theta k b^n + (1-\theta) k b^{n-1} \\ c^n \end{bmatrix},$$

where $x^n \approx x(t_n)$ and $A^n := A(x^n)$. Special cases are the *'forward Euler scheme'* for $\theta = 0$ (first-order explicit), the most popular *'Crank–Nicolson scheme'* for $\theta = 1/2$ (second-order implicit, A-stable), and the the *'backward Euler scheme'* for $\theta = 1$ (first-order implicit, strongly A-stable). These properties can be seen by applying the schemes to the scalar model equation $\dot{x} = \lambda x$. In this context it is related to a rational approximation of the exponential of the form

$$R_\theta(-\lambda) = \frac{1 - (\theta - \frac{1}{2})\lambda}{1 + \theta\lambda} = e^{-\lambda} + \mathbf{O}\left((\theta - \tfrac{1}{2})|\lambda|^2 + |\lambda|^3\right), \quad |\lambda| \leq 1.$$

The most robust implicit Euler scheme is very dissipative and therefore not suitable for the time-accurate computation of non-stationary flow. In contrast to that, the Crank–Nicolson scheme has only very little dissipation but occasionally suffers from instabilities caused by the possible occurrence of rough perturbations in the data which are not damped out due to the only weak stability properties of this scheme (not *strongly* A-stable). This defect can in principle be cured by an adaptive step-size selection but this may enforce the use of an unreasonably small time step, thereby increasing the computational cost. Alternative schemes of higher order are based on the (diagonally) implicit Runge–Kutta formulas or the backward differencing multi-step formulas, both being well known from the ODE literature. These schemes, however, have not yet found wide applications in practical flow computations, mainly because of their higher complexity and storage requirements compared with the simple Crank–Nicolson scheme. Another alternative to the Crank–Nicolson scheme is the so-called 'fractional-step-θ scheme'.

1.4.3. The Fractional-Step-θ Scheme. Each time step is split into three substeps $t_{n-1} \to t_{n-1+\theta} \to t_{n-\theta} \to t_n$, which read

$$\begin{bmatrix} M+\alpha\theta k A^{n-1+\theta} & \theta k B \\ -B^T & C \end{bmatrix} \begin{bmatrix} x^{n-1+\theta} \\ y^{n-1+\theta} \end{bmatrix} = \begin{bmatrix} [M-\beta\theta k A^{n-1}]x^{n-1} + \theta k b^{n-1} \\ c^{n-1+\theta} \end{bmatrix},$$

$$\begin{bmatrix} M+\beta\theta' k A^{n-\theta} & \theta' k B y^{n-\theta} \\ -B^T & C \end{bmatrix} \begin{bmatrix} x^{n-\theta} \\ y^{n-\theta} \end{bmatrix} = \begin{bmatrix} [M-\alpha\theta' k A^{n-1+\theta}]x^{n-1+\theta} + \theta' k b^{n-\theta} \\ c^{n-\theta} \end{bmatrix},$$

$$\begin{bmatrix} M+\alpha\theta k A^n & \theta k B \\ -B^T & C \end{bmatrix} \begin{bmatrix} x^n \\ y^n \end{bmatrix} = \begin{bmatrix} [M-\beta\theta k A^{n-\theta}]x^{n-\theta} + \theta k b_h^{n-\theta} \\ c^n \end{bmatrix}.$$

Here, $\theta = 1-1/\sqrt{2} = 0.292893\ldots$, $\theta' = 1-2\theta$, $\alpha \in (1/2, 1]$, and $\beta = 1-\alpha$, in order to ensure second-order accuracy, and strong A-stability. In the context of the scalar model equation this scheme reduces to a rational approximation of the exponential of the form

$$R_\theta(-\lambda) = \frac{(1 - \alpha\theta'\lambda)(1 - \beta\theta\lambda)^2}{(1 + \alpha\theta\lambda)^2(1 + \beta\theta'\lambda)} = e^{-\lambda} + \mathbf{O}(|\lambda|^3), \quad |\lambda| \leq 1.$$

1.4.4. Splitting and projection schemes. The fractional-step-θ scheme was originally introduced as an operator-splitting scheme in order to separate the two main difficulties in solving the problem, namely the nonlinearity causing asymmetry and the incompressibility constraint causing indefiniteness; see the survey article Glowinski [37] and the literature cited therein. However, since the efficient numerical handling of this kind of indefinite nonlinear problems is not much of a problem anymore today, we will not discuss this variant of the fractional-step-θ scheme. We rather briefly introduce the so-called 'projection methods' which are particularly efficient for solving non-stationary problems in certain situations. The 'classical' Chorin projection method has originally been designed in order to overcome the problem with the incompressibility constraint $\nabla \cdot v = 0$. The continuity equation is decoupled from the momentum equation through an iterative process (again 'operator splitting'). There are various schemes of this kind in the literature referred to as 'projection method', 'quasi-compressibility method', 'SIMPLE method', etc.

All these methods are based on the same principle idea. The continuity equation $\nabla \cdot v = 0$ is supplemented by certain stabilizing terms involving the pressure, e.g.,

$$(1) \quad \nabla \cdot v + \epsilon p = 0,$$

$$(2) \quad \nabla \cdot v - \epsilon \Delta p = 0, \quad \partial_n p_{|\partial\Omega} = 0,$$

$$(3) \quad \nabla \cdot v + \epsilon \partial_t p = 0, \quad p_{|t=0} = 0,$$

$$(4) \quad \nabla \cdot v - \epsilon \partial_t \Delta p = 0, \quad \partial_n p_{|\partial\Omega} = 0, \ p_{|t=0} = 0,$$

where the small parameter ϵ is usually taken as $\epsilon \approx h^\alpha$, or $\epsilon \approx k^\beta$, depending on the purpose of the procedure. For example, (1) corresponds to the classical 'penalty method', and (2) is the simplest form of the 'least squares pressure stabilization' scheme described above, with $\epsilon \approx h^2$ in both cases. Further, (3) corresponds to the 'quasi-compressibility method' with $\epsilon \approx k$, while (4) occurs in the context of Van Kan's second-order projection method with $\epsilon \approx k^2$. These approaches are closely related to the Chorin projection method. Since this method used to be particularly attractive for computing non-stationary incompressible flow, we will discuss it in some greater detail. For a comprehensive discussion of such types of schemes, we refer to the book Gresho/Sani [38].

For simplicity consider the case of pure homogeneous Dirichlet boundary conditions, $v_{|\partial\Omega} = 0$. Then, the projection method reads as follows:

Chorin Projection Method: For an admissible initial value v^0, solve for $n \geq 1$:
(i) Implicit 'Burgers step' for $\tilde{v}^n \in \mathbf{H}$:

$$k^{-1}(\tilde{v}^n - v^{n-1}) - \nu\Delta\tilde{v}^n + \tilde{v}^n \cdot \nabla\tilde{v}^n = f^n, \text{ in } \Omega. \qquad (1.19)$$

(ii) 'Projection step' for $v^n := \tilde{v}^n + k\nabla\tilde{p}^n$, where $\tilde{p}^n \in H^1(\Omega)$ is determined by

$$\Delta\tilde{p}^n = k^{-1}\nabla \cdot \tilde{v}^n, \text{ in } \Omega, \quad \partial_n\tilde{p}^n_{|\partial\Omega} = 0. \qquad (1.20)$$

Substep (ii) amounts to a Poisson equation for $\tilde{p}^n$ with homogeneous Neumann boundary conditions. It is this non-physical boundary condition, $\partial_n\tilde{p}^n_{|\partial\Omega} = 0$, which has caused a lot of controversial discussion about the value of the projection method. Nevertheless, the method has proven to work well for representing the velocity field in many flow problems of physical interest. It is very economical as it requires in each time step only the solution of a (nonlinear) advection-diffusion system and a scalar Neumann problem.

For the projection methods rigorous convergence results are available showing that the quantities $\tilde{p}^n$ are indeed reasonable approximations to the pressure $p(t_n)$. This may be understood by re-interpreting the projection method in the context of 'pressure stabilization'. To this end the quantity $v^{n-1} = \tilde{v}^{n-1} - k\nabla\tilde{p}^{n-1}$ is inserted into the momentum equation yielding

$$k^{-1}(\tilde{v}^n - \tilde{v}^{n-1}) - \nu\Delta\tilde{v}^n + (\tilde{v}^n \cdot \nabla)\tilde{v}^n + \nabla\tilde{p}^{n-1} = f^n, \quad \tilde{v}^n_{|\partial\Omega} = 0, \qquad (1.21)$$

$$\nabla \cdot \tilde{v}^n - k\Delta\tilde{p}^n = 0, \quad \partial_n\tilde{p}^n_{|\partial\Omega} = 0. \qquad (1.22)$$

This appears like an approximation of the Navier–Stokes equations involving a first-order (in time) 'pressure stabilization' term, i.e., the projection method can be viewed as a pressure stabilization method with a global stabilization parameter $\epsilon = k$, and an explicit treatment of the pressure term. As a byproduct, this also explains the success of the not inf-sup-stable Q_1^c/Q_1^c Stokes element in the context of non-stationary computations. The pressure error is actually confined to a small boundary strip of width $\delta \approx \sqrt{\nu k}$ and decays exponentially into the interior of Ω.

The projection approach can be extended to formally higher order. The most popular example is the so-called 'Van Kan method':

Van Kan Method: For admissible starting values v^0 and p^0 compute, for $n \geq 1$ and some $\alpha \geq \frac{1}{2}$:
(i) Second-order implicit 'Burgers step' for $\tilde{v}^n \in \mathbf{H}$:

$$k^{-1}(\tilde{v}^n - v^{n-1}) - \tfrac{1}{2}\nu\Delta(\tilde{v}^n + v^{n-1}) + \tilde{v}^n \cdot \nabla\tilde{v}^n + \nabla p^{n-1} = f^{n-1/2}. \qquad (1.23)$$

(ii) Pressure Poisson problem for $q^n \in H^1(\Omega)$:

$$\Delta q^n = \alpha^{-1}k^{-1}\nabla \cdot \tilde{v}^n, \quad \text{in } \Omega, \quad q^n_{|\partial\Omega} = 0. \qquad (1.24)$$

(iii) Pressure and velocity update:

$$v^n = \tilde{v}^n - \alpha k\nabla q^n, \quad p^n = p^{n-1} + q^n, \quad \text{in } \Omega. \qquad (1.25)$$

This scheme can also be interpreted in the context of pressure stabilization methods using a stabilization of the form

$$\nabla \cdot v - \alpha k^2 \partial_t \Delta p = 0, \quad \text{in } \Omega, \quad \partial_n p_{|\partial\Omega} = 0, \qquad (1.26)$$

i.e., this method appears like a quasi-compressibility method of the form (4) with $\epsilon \approx k^2$.

1.5. The quasi-stationary algebraic problems

As for the stationary Navier–Stokes problem the space-time discrete variational problems (1.18) are converted into algebraic systems. We continue using the notation from above for the nodal bases $\{\psi_h^i,\ i=1,\ldots,N_v\}$ of the velocity space $\mathbf{H}_h$ and $\{\chi_h^i,\ i=1,\ldots,N_p\}$ of the pressure space L_h, and the associated system matrices $A, B, N(x), S(x), T(x)$, and C. The corresponding velocity 'mass matrix' is

$$M := \left((\psi_h^i, \psi_h^j)\right)_{i,j=1}^{N_v}.$$

Then, the algebraic equations for the nodal vectors $x \in \mathbb{R}^{N_v}$ and $y \in \mathbb{R}^{N_p}$ read

$$\begin{bmatrix} M + kA + kN(x) + kS(x) & kB + kT(x) \\ -kB^T + kT(x)^T & C \end{bmatrix} \begin{bmatrix} x \\ y \end{bmatrix} = \begin{bmatrix} b \\ c \end{bmatrix}. \qquad (1.27)$$

Special multigrid methods for solving this particular problem efficiently are described in the article Turek [74] in this volume.

2. Numerical simulation of pipe flow

A typical setting of flow in hemodynamics is that of flow through a pipe or through
a system of pipes. Usually one is only interested in the flow properties in a smaller
section of the global system, for example in bypass simulation. Therefore, the
global flow region has to be truncated to a smaller domain in order to limit the
computational work or since the characteristics are not accessible outside that
region. This requires the use of artificial boundary conditions at the cut boundaries.
This section is concerned with the definition of such conditions and the theoretical
as well as practical difficulties going along with them. The material presented in
this section is mainly based on Heywood et al. [48]

2.1. Variational 'open' boundary conditions

We consider a prototypical flow configuration, a bifurcating pipe in two dimensions,
as shown in Figure 2. The underlying mathematical model is again the system of

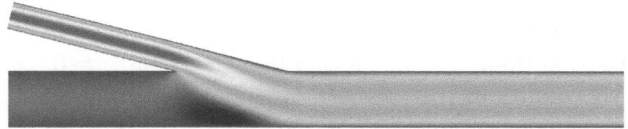

FIGURE 2.   Configuration of flow through a bifurcating pipe.

the incompressible Navier–Stokes equations with the natural no-slip boundary
conditions along the rigid walls Γ_{rigid},

$$\partial_t v - \nu \Delta v + v \cdot \nabla v + \nabla p = 0 \quad \text{in } \Omega,$$
$$\nabla \cdot v = 0 \quad \text{in } \Omega, \tag{2.1}$$
$$v = 0 \quad \text{on } \Gamma_{\text{rigid}}.$$

The question is that of how to deal with the 'open' boundaries Γ_{out} and $\Gamma_{\text{in}} = \Gamma_{\text{in}}^{(1)} \cup \Gamma_{\text{in}}^{(2)}$, which originate from truncating a larger flow region. In this case, we
may assume that these cut boundaries are straight and form right angles with the
rigid walls. Other situations require certain modifications in the following argument. Depending on the physical situation, which is to be modeled, the following
boundary conditions may be used:

- flow driven by prescribed inflow profile:

$$v = v^{\text{in}} \quad \text{on } \Gamma_{\text{in}},$$

- flow driven by prescribed mean flux:

$$\int_{\Gamma_{\text{in}}} v \cdot n \, do = F^{\text{in}},$$

- flow driven by prescribed pressure profile:

$$p = p^{\text{in}} \quad \text{on } \Gamma_{\text{in}},$$

- flow driven by prescribed mean pressure drop:

$$\int_{\Gamma_{\text{out}}} p \, do - \int_{\Gamma_{\text{in}}} p \, do = P.$$

All these boundary conditions contain flow quantities such as velocity profiles, fluxes, mean pressure drops, etc., which need to be prescribed in a particular situation. Hence its relevance depends on whether these quantities are available from measurements or can actively be enforced. In this view, prescribing mean quantities such as fluxes and mean pressure drops seems more natural (and physically meaningful) than prescribing velocity or pressure profiles.

If nothing particular is known about the flow behavior outside the computational domain, it is most natural to assume that the inflow and outflow pipe segments extend as straight pipes such that the main flow behaves like parallel pipe flow (Poiseuille flow) beyond the artificial boundaries. A further requirement is that of as little upstream effect as possible. We note that the Poiseuille flow satisfies several types of boundary conditions:

- Dirichlet condition: $\quad v_{|\Gamma_{\text{out}}} = v^{\text{Poiseuille}}$,
- Neumann condition: $\quad \partial_n v_{|\Gamma_{\text{out}}} = 0, \quad (v \times n)_{|\Gamma_{\text{out}}} = 0$,
- Periodicity condition: $\quad v_{|\Gamma_{\text{in}}} = v_{|\Gamma_{\text{out}}}$,
- Free outstream condition: $\quad (\nu \partial_n v - pn)_{|\Gamma_{\text{out}}} = 0$,

and mixtures of these. Some of these conditions are *essential* and have to be incorporated into the solution space $\mathbf{H}$, and others are *natural* and result from the variation principle.

We prefer the most simple outflow boundary condition, called the '(variational) do-nothing (or free-stream) condition', which seems most natural since it does not prescribe anything at Γ_{out}, i.e., it uses the solution space

$$\mathbf{H} := \{v \in H^1(\Omega)^d, \ v = 0 \ \text{on} \ \Gamma_{\text{rigid}} \cup \Gamma_{\text{in}}\},$$

in the variational formulation of problem (2.1).

Problem 2.1. Determine $v(\cdot, t) \in v^{\text{in}} + \mathbf{H}$, $p(\cdot, t) \in L$, such that $v(0) = v^0$ and

$$(\partial_t v, \psi) + a(v, \psi) + b(v, v, \psi) + b(p, \psi) = 0 \quad \forall \psi \in \mathbf{H},$$
$$b(\chi, v) = 0 \quad \forall \chi \in L. \tag{2.2}$$

Assuming the existence of a sufficiently smooth solution, the variation principle

$$0 = \nu(\nabla v \nabla \varphi) + (v \cdot \nabla v, \varphi) - (p, \nabla \cdot \varphi) = (-\nu \Delta v + v \cdot \nabla v + \nabla p, \varphi)$$
$$+ (\nu \partial_n v - pn, \varphi)_{\Gamma_{\text{out}}} \quad \forall \varphi \in \mathbf{H}$$

implies the *natural* outflow boundary condition

$$\nu \partial_n v_n - p = 0, \quad \nu \partial_n v_\tau = 0, \quad \text{on} \ \Gamma_{\text{out}}. \tag{2.3}$$

Further using the incompressibility constraint, we have

$$0 = \int_{\Gamma_{\text{out}}} \{\nu \partial_n v_n - p\} \, do = \int_{\Gamma_{\text{out}}} \{-\nu(n \times \nabla)v_n - p\} \, do = -\int_{\Gamma_{\text{out}}} p \, do,$$

i.e., the 'do-nothing' outflow boundary condition implies as *hidden* condition that the pressure has mean value zero at the outflow boundary,

$$\int_{\Gamma_{\text{out}}} p \, do = 0. \tag{2.4}$$

The performance of the 'do-nothing (free-stream)' outflow boundary condition is demonstrated in Figure 3 by showing pressure isolines for unsteady flow around an inclined ellipse at Re = 500 in a channel compared to the result in another channel of half its size. Notice that the stream of vortices seems to be almost undisturbed by the 'do-nothing' outflow boundary condition, even though the tangential flow is not zero. In this case the commonly used 'zero-flux' boundary condition,

$$\partial_n v_{|\Gamma_{\text{out}}} = 0, \quad (v \times n)_{|\Gamma_{\text{out}}} = 0, \tag{2.5}$$

is not satisfied.

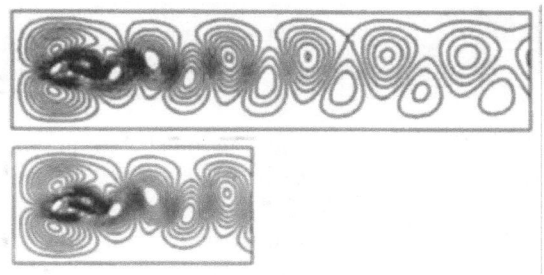

FIGURE 3. Vorticity patterns of channel flow around an inclined ellipse at Re = 500 compared to the result in another channel of half its size.

Next, we consider the case that the inflow velocity v^{in} is not known but rather a mean pressure drop P between inflow and outflow boundary is prescribed. This can be modeled by using the solution space

$$\mathbf{H} := \{ v \in H^1(\Omega)^d, \ v = 0 \ \text{on} \ \Gamma_{\text{rigid}} \}$$

in the variational formulation of problem (2.1).

Problem 2.2. Determine $v(\cdot, t) \in v^{\text{in}} + \mathbf{H}$, $p(\cdot, t) \in L$, such that $v(0) = v^0$ and

$$(\partial_t v, \psi) + \nu(\nabla v, \nabla \psi) + (v \cdot \nabla v, \psi) - (p, \nabla \cdot \psi) = -(P, n \cdot \psi)_{\Gamma_{\text{out}}} \quad \forall \psi \in \mathbf{H},$$
$$(\chi, \nabla \cdot v) = 0 \quad \forall \chi \in L. \tag{2.6}$$

This formulation contains the following natural boundary conditions:

$$\nu \partial_n v - pn = 0 \ \text{on} \ \Gamma_{\text{in}}, \quad \nu \partial_n v - pn = Pn \ \text{on} \ \Gamma_{\text{out}},$$
$$|\Gamma_{\text{in}}|^{-1} \int_{\Gamma_{\text{in}}} p \, do - |\Gamma_{\text{out}}|^{-1} \int_{\Gamma_{\text{out}}} p \, do = P. \tag{2.7}$$

2.2. Problems with the 'do-nothing' boundary condition

The naive use of the 'do-nothing' outflow boundary condition can result in undesirable behavior of the computed flow. In the following, we discuss some of these problems.

(i) Problem of multiple outlets. The use of the 'do-nothing' boundary condition at several outlets $\Gamma_i \subset \partial\Omega$ results in boundary conditions

$$\nu\partial_n v - pn = P_i n \quad \text{on } \Gamma_i, \qquad |\Gamma_i|^{-1}\int_{\Gamma_i} p\, do = P_i.$$

The question is that of the appropriate choice of the mean pressures P_i. Figure 4 shows the effect of the 'do-nothing' outflow boundary condition for flow through bifurcating channels of different lengths for $\mathrm{Re} = 20$.

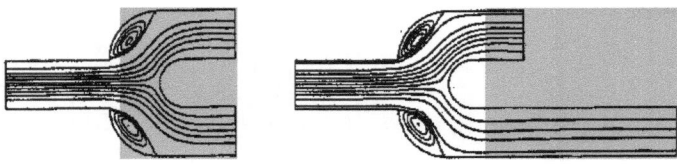

FIGURE 4. Effect of the 'do-nothing' outflow boundary condition for flow through a bifurcating channel.

(ii) Modifications of the variational formulation.

a) In order to preserve energy conservation in the case of velocity approximation, which is not exactly divergence-free, a common remedy is the use of a *symmetrized transport formulation*

$$(v \cdot \nabla v, \varphi) \;\rightarrow\; \tfrac{1}{2}(v \cdot \nabla v, \varphi) - \tfrac{1}{2}(v \cdot \nabla \varphi, v).$$

This change has no effect in the case of pure Dirichlet boundary conditions. But using the 'do-nothing' approach this modification leads to the outflow condition

$$\nu\partial_n v - pn - \tfrac{1}{2}|v_n|^2 n = 0, \quad \text{on } \Gamma_{\text{out}}, \tag{2.8}$$

which is not satisfied by the Poiseuille flow and consequently induces an undesirable flow behavior across the outflow boundary.

b) Recalling the physical origin of the Navier–Stokes equations one is motivated to use the *strain tensor formulation*

$$\nu(\nabla v, \nabla \varphi) \;\rightarrow\; \nu(D[v], D[\varphi]), \quad D[v] = \tfrac{1}{2}(\nabla v + \nabla v^T),$$

for the viscous term. Again this change has no effect in the case of pure Dirichlet boundary conditions. But using the 'do-nothing' approach this modification leads to the outflow boundary condition

$$n\cdot D[v] - pn = 0, \quad \text{on } \Gamma_{\text{out}}, \tag{2.9}$$

which is not satisfied by the Poiseuille flow, either, and induces an undesirable flow behavior across the outflow boundary.

Figure 5 shows the effect of using the 'do-nothing' concept for outflow prescription together with the symmetrized transport formulations and the deformation tensor formulation. In both cases non-physical behavior is observed, with streamlines bending either inwards or outwards at the outflow boundary. For comparison the correct streamline pattern of the Poiseuille flow is shown.

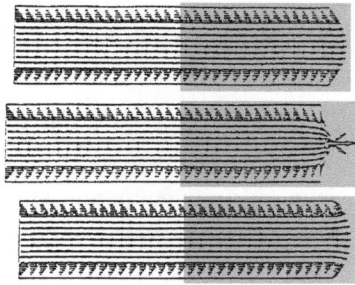

FIGURE 5. Effect of the 'do-nothing' outflow boundary condition in connection with the standard variational formulation (top), the symmetrized transport formulations (middle), and the deformation tensor formulation (bottom).

(iii) Further problems. The 'do-nothing' outflow boundary condition also works in the non-stationary case, e.g., in simulating von Karman vortex shedding, as demonstrated in Figure 3. However problems may occur in the following cases:

- acting non-zero forcing f, such as in heat-driven flow,
- if outflow boundary and rigid boundary do not form a right angle,
- if boundaries are moving by enforcement or in the context of fluid-structure interaction.

All these situations share the property that, for physical reasons, the flow cannot be expected to be essentially parallel across the outflow boundary.

2.3. The problem of well-posedness

Finally, we want to address the question of well-posedness of the variational formulation involving the 'do-nothing' outflow boundary condition. We will see that there are problems even in the case of stationary plain pipe flow.

(i) First, we consider the Stokes equations with the variational formulation

$$\nu(\nabla v, \nabla \psi) - (p, \nabla \cdot \psi) - (\chi, \nabla \cdot v) = -\sum_{i=1}^{n} (q^i, v \cdot n)_{\Gamma_i}, \qquad (2.10)$$

for all $\{\psi, \chi\} \in \mathbf{H} \times L$. Taking $\psi := v$ and $\chi := 0$, we obtain the 'energy relation'

$$\nu \|\nabla v\|^2 = -\sum_{i=1}^{n} (P_i, v \cdot n)_{\Gamma_i}. \qquad (2.11)$$

From this, we easily conclude existence, uniqueness and stability of solutions by standard arguments; see Heywood et al. [48].

(ii) Next, we turn to the Navier–Stokes equations with the variational formulation

$$\nu(\nabla v, \nabla \psi) + (v \cdot \nabla v, \psi) - (p, \nabla \cdot \psi) - (\chi, \nabla \cdot v) = -\sum_{i=1}^{n}(P_i, v \cdot n)_{\Gamma_i}, \qquad (2.12)$$

for all $\{\psi, \chi\} \in \mathbf{H} \times L$. Again taking $\psi := v$ and $\chi := 0$, we obtain the *nonlinear* 'energy relation'

$$\nu\|\nabla v\|^2 = -(v \cdot \nabla v, v) - \sum_{i=1}^{n}(P_i, v \cdot n)_{\Gamma_i} = \sum_{i=1}^{n}\int_{\Gamma_i}(\tfrac{1}{2}|v|^2 - P_i)v \cdot n\, do. \qquad (2.13)$$

From this, we conclude existence, uniqueness and stability of solutions for sufficiently small data $|P_i| \ll 1$ in a small $\mathbf{H}$-ball

$$B_\rho := \{w \in \mathbf{H},\ \|\nabla w\| < \rho\}.$$

We remark that in the case of pure Dirichlet boundary conditions, we can show an a priori bound of the solution in terms of data,

$$\|\nabla v\| \le C(\nu, v^{\mathrm{in}}, f). \qquad (2.14)$$

This in turn leads us to the result that the problem is well posed for sufficiently small data. Unfortunately, in the case of open outlets such an a priori bound is not known. This is reflected by the fact that not even the global uniqueness of the zero-solution has been proven yet,

$$\nu(\nabla v, \nabla \psi) + (v \cdot \nabla v, \psi) = 0 \quad \forall \psi \in \mathbf{H} \quad \Rightarrow \quad v = 0\,?$$

However this possible non-uniqueness could not be confirmed by numerical experiments.

2.4. The closure problem

The simulation of the flow through a section of a pipe or a pipe system requires the prescription of boundary conditions at the artificial cut boundaries. As seen above, these may be given in terms of mean fluxes or mean pressure drops. The determination of such conditions is related to a 'closure problem' since they are supposed to model the global behavior of the flow in the whole system and its interaction with the local flow behavior in the pipe section considered.

Closure by global lower-dimensional models. One attractive approach to solve the closure problem is the embedding of the full 3D model of a portion of a pipe into a lower-dimensional, e.g., 1D model, of the closed blood flow circuit. For the systematic development of this method, we refer to the series of papers Formaggia et al. [31], Quarteroni et al. [63], Quarteroni/Veneziani [64], Milisic/Quarteroni [58], and Fernandez et al. [30]. This approach allows one to model the qualitative behavior of the system, but does not lead to *quantitative* information, which would be needed, for instance, for assisting the operation of a particular patient.

Closure by measurement-based model calibration. An alternative approach to the closure problem uses model calibration based on local flow measurements. Suppose that the goal of the simulation is to predict the mean shear stress at the middle part $\Gamma \subset \Gamma_1$ of the bottom wall in Figure 6, which is relevant, for instance, in planing of bypass operations,

$$E(u) := \int_\Gamma n \cdot \sigma \cdot e_x \, do.$$

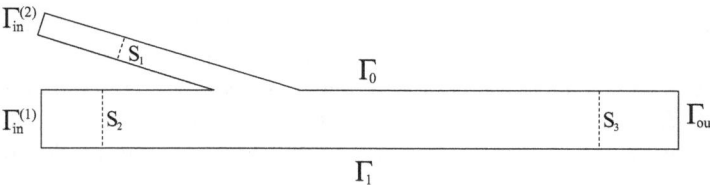

FIGURE 6. Configuration of the bypass setting.

In this case the mathematical model reads in strong form as follows:

$$-\nu\Delta v + v \cdot \nabla v + \nabla p = 0, \quad \nabla \cdot v = 0 \quad \text{in } \Omega,$$
$$v = \hat{v} \quad \text{on } \Gamma_{\text{in}}, \quad v = 0 \quad \text{on } \Gamma_0 \cup \Gamma_1,$$
$$\nu\partial_n v - p \cdot n = q_1 \cdot n \quad \text{on } \Gamma_{\text{in}}^{(1)}, \quad \nu\partial_n v - p \cdot n = q_2 \cdot n \quad \text{on } \Gamma_{\text{in}}^{(2)}, \tag{2.15}$$
$$\nu\partial_n v - p \cdot n = 0 \quad \text{on } \Gamma_{\text{out}}.$$

The unknown pressure mean values q_i at the openings $\Gamma_{\text{in}}^{(i)}$ are to be determined by *parameter estimation* from given measurements, for example, of the mean fluxes

$$C_j(u) := \int_{S_j} v \cdot n \, do,$$

along certain interior cross sections of the flow domain or other measurable local flow quantities. This approach will be described in the context of general flow optimization problems in the next section; for details see Vexler [75].

3. Mesh adaptation and model calibration

This section introduces into methods for mesh adaptivity and model calibration in numerical flow simulation. The emphasis is on viscous incompressible flows governed by the Navier–Stokes equations. The finite-element Galerkin method provides the basis for a common rigorous a posteriori error analysis.

A large part of the existing work on a posteriori error analysis deals with error estimation in global norms such as the 'energy norm' involving usually unknown stability constants. However, in most CFD applications, the error in a global norm does not provide useful bounds for the errors in the quantities of real physical interest. Such 'goal-oriented' error bounds can be derived by duality arguments borrowed from optimal control theory. These a posteriori error estimates provide the basis of a feedback process for successively constructing economical meshes and corresponding error bounds tailored to the particular goal of the computation. This approach, called the 'dual weighted residual method' (DWR method), is developed within an abstract functional analytic setting, thus providing the general guideline for applications to various kinds of flow models including also aspects of flow control and hydrodynamic stability.

The 'residual-based' error indicators largely exploit the structure of the underlying differential equations. This requires an appropriate discretization which inherits as much as possible of the structure and properties of the continuous model. Here, the method of choice is the *'continuous' finite-element Galerkin method (cG-FEM)* which is particularly suited for approximating models governed by viscous terms, such as the Navier–Stokes equations for moderate Reynolds numbers. For inviscid models or those with dominant transport, such as the Euler equations, the *'discontinuous' finite-element Galerkin method (dG-FEM)* shares most of the features of the traditional *finite volume method (FVM)* but is potentially more flexible with respect to mesh geometry and order of approximation. Both methods are based on variational formulations of the differential equations to be solved and allow for the rigorous derivation of a priori as well as a posteriori error estimates.

3.1. Principles of a posteriori error estimation

We begin with a brief discussion of the philosophy underlying the approaches to adaptivity which will be the subject of the following discussion. Let the goal of a simulation be the accurate and efficient computation of the value of a functional $J(u)$, the 'target quantity', with accuracy TOL from the solution u of a continuous model

$$\mathcal{A}(u) = F, \tag{3.1}$$

by using a discrete model $\mathcal{A}_h(u_h) = F_h$ of dimension N. The evaluation of the solution by the output functional $J(\cdot)$ represents what we exactly want to know of a solution. This may be, for instance, the stress or pressure near a critical point, certain local mean values of species concentrations, the drag and lift coefficient of a body in a viscous liquid, etc. Then, the goal of adaptivity is the 'optimal' use of computing resources to achieve either *minimal work for prescribed accuracy*, or *maximal accuracy for prescribed work*. These goals are approached by automatic mesh adaptation on the basis of local 'error indicators' taken from the computed solution u_h on the current mesh $\mathbb{T}_h = \{K\}$. Examples of such error indicators are:

- error indicators based on pure 'regularity' information,

$$\eta_K^{\mathrm{reg}} := h_K \|D_h^2 u_h\|_K,$$

where $D_h^2 u_h$ are certain second-order difference quotients,

- error indicators based on local gradient recovery such as the well-known 'Zienkiewicz–Zhu indicator' (see Ainsworth/Oden [1]),

$$\eta_K^{ZZ} := \|M_h(\nabla u_h) - \nabla u_h\|_K,$$

where $M_h(\nabla u_h)$ is obtained by locally averaging function values of ∇u_h,
- error indicators based on 'residual' information,

$$\eta_K^{\text{res}} := h_K \|R(u_h)\|_K,$$

where $R(u_h)_{|K}$ are certain 'residuals' of the computed solution.

According to the size of the error indicators the current mesh may be locally refined or coarsened, or a full remeshing may be performed. Although remeshing is very popular in CFD applications since it allows to use commercial mesh generators and to maintain certain mesh qualities, it also has some disadvantages. The most severe one is that remeshing destroys the regular hierarchical structure of successively refined meshes which makes the use of efficient multigrid solvers difficult. Therefore, all examples presented in this article employ hierarchical mesh adaptation, which allows for optimally efficient geometrical multigrid solution; see Becker/Braack [6].

The traditional approach to adaptivity aims at estimating the error with respect to the generic 'energy norm' of the problem in terms of the computable 'residual' $R(u_h) = 'F - \mathcal{A}(u_h)'$, which is well defined in the context of a finite-element Galerkin method,

$$\|u - u_h\|_E \le c\Big(\sum_{K \in \mathbb{T}_h} h_K^2 \rho_K^2 \Big)^{1/2}, \tag{3.2}$$

where $\rho_K := \|R(u_h)\|_K$, and the sum extends over all cells of the mesh $\mathbb{T}_h$. For references see the survey articles by Ainsworth/Oden [1] and Verfürth [76]. This approach seems rather generic as it is directly based on the variational formulation of the problem and allows to exploit its inherent coercivity properties. However, in most applications the error in the energy norm does not provide a useful bound on the error in the quantities of real physical interest. A more versatile method for a posteriori error estimation with respect to relevant error measures such as point values, line averages, etc., is obtained by using duality arguments as common in the a priori error analysis of finite-element methods. This approach has been systematically developed by C. Johnson and his co-workers [54, 29], and was then extended by the author and his group to a practical feedback method for mesh optimization termed 'dual weighted residual method' (DWR method); see Becker/Rannacher [12, 13, 14]. A general introduction to the DWR method and a variety of applications can be found in the survey article Becker/Rannacher [14] and the book Bangerth/Rannacher [2]. Variants of this approach have also been developed in the groups of A.T. Patera [57, 62], and J.T. Oden [60, 61].

The DWR method yields *weighted* a posteriori error bounds with respect to prescribed 'output functionals' $J(u)$ of the solution, of the form

$$|J(u) - J(u_h)| \leq \sum_{K \in \mathbb{T}_h} \rho_K \omega_K, \tag{3.3}$$

where the weights ω_K are obtained by approximately solving a linearized *dual problem*, $\mathcal{A}'(u)^* z = J$. The *dual solution* z may be viewed as a generalized Green function with respect to the output functional $J(\cdot)$, and accordingly the weight ω describes the effect of variations of the residual $R(u_h)$ on the error $J(u) - J(u_h)$, as consequence of mesh adaptation. This accomplishes control of

- error propagation in space (global pollution effect),
- interaction of various physical error sources (cross sensitivities).

In practice it is mostly impossible to determine the complex error interaction by analytical means, it rather has to be detected by computation. This automatically leads to a feed-back process in which error estimation and mesh adaptation go hand-in-hand leading to economical discretization for computing the quantities of interest.

3.2. The dual weighted residual (DWR) method

The theoretical basis of the DWR method is laid within the abstract framework of Galerkin approximation of variational equations in Hilbert space. The following presentation is adopted from Becker/Rannacher [14].

3.2.1. Approximation of stationary points. Let X be a Hilbert space and $L(\cdot)$ a differentiable functional on X. Its first-, second-, and third-order derivatives at some $x \in X$ are denoted by $L'(x)(\cdot)$, $L''(x)(\cdot, \cdot)$, and $L'''(x)(\cdot, \cdot, \cdot)$, respectively. Suppose that $x \in X$ is a stationary point of $L(\cdot)$, i.e.,

$$L'(x)(y) = 0 \quad \forall y \in X. \tag{3.4}$$

This equation is approximately solved by a Galerkin method using finite-dimensional subspaces $X_h \subset X$, parametrized by $h \in \mathbb{R}_+$. We seek $x_h \in X_h$ satisfying

$$L'(x_h)(y_h) = 0 \quad \forall y_h \in X_h. \tag{3.5}$$

For estimating the difference $L(x) - L(x_h)$, we start from the trivial identity

$$L(x) - L(x_h) = \int_0^1 L'(x_h + se)(e) \, ds,$$

where $e := x - x_h$. Approximating the integral by the trapezoidal rule yields

$$L(x) - L(x_h) = \tfrac{1}{2} \{ L'(x_h)(e) + L'(x)(e) \} + \tfrac{1}{2} \int_0^1 L'''(x_h + se)(e, e, e) \, s(s-1) \, ds.$$

Thus, observing (3.4) and (3.5), we obtain the following result.

Proposition 3.1. *For any solutions of the problems* (3.4) *and* (3.5), *we have the a posteriori error representation*

$$L(x) - L(x_h) = \tfrac{1}{2}L'(x_h)(x - y_h) + \mathcal{R}_h, \tag{3.6}$$

for arbitrary $y_h \in X_h$. *The remainder term* $\mathcal{R}_h$ *is cubic in the error* e,

$$\mathcal{R}_h := \tfrac{1}{2}\int_0^1 L'''(x_h + se)(e, e, e)\, s(s-1)\, ds.$$

Remark 3.2. In view of the possible non-uniqueness of the solutions x and x_h, the formulated goal of estimating the error quantity $L(x) - L(x_h)$ needs some explanation. The error representation (3.6) does not explicitly require that the approximation x_h is close to x. However, since it contains a remainder term in which the difference $x - x_h$ occurs, the result is useful only under the assumption that the convergence $x_h \to x$, as $h \to 0$, is known by *a priori* arguments.

3.2.2. Approximation of variational equations. Let $A(\cdot)(\cdot)$ be a differentiable semi-linear form and $F(\cdot)$ a linear functional defined on some Hilbert space V. We seek a solution $u \in V$ to the variational equation

$$A(u)(\varphi) = F(\varphi) \quad \forall \varphi \in V. \tag{3.7}$$

For a finite-dimensional subspace $V_h \subset V$, again parametrized by $h \in \mathbb{R}_+$, the corresponding Galerkin approximation $u_h \in V_h$ is determined by

$$A(u_h)(\varphi_h) = F(\varphi_h) \quad \forall \varphi_h \in V_h. \tag{3.8}$$

We assume that equations (3.7) and (3.8) possess solutions (not necessarily unique). Let the goal of the computation be the evaluation $J(u)$, where $J(\cdot)$ is a given differentiable functional. We want to embed this situation into the general setting of Proposition 3.1. To this end, we note that computing $J(u)$ from the solution of (3.7) can be interpreted as computing a stationary point $\{u, z\} \in V \times V$ of the Lagrangian

$$L(u, z) := J(u) - A(u)(z) + F(z),$$

with the dual variable $z \in V$, that is solving

$$A(u)(\psi) = F(\psi) \quad \forall \psi \in V, \tag{3.9}$$
$$A'(u)(\varphi, z) = J'(u)(\varphi) \quad \forall \varphi \in V. \tag{3.10}$$

In order to obtain a discretization of the system (3.9–3.10), in addition to (3.8), we solve the discrete adjoint equation

$$A'(u_h)(\varphi_h, z_h) = J'(u_h)(\varphi_h), \quad \varphi_h \in V_h. \tag{3.11}$$

We suppose that the dual problems also possess solutions $z \in V$ and $z_h \in V_h$, respectively. Notice that at the solutions $x = \{u, z\} \in X := V \times V$ and $x_h = \{u_h, z_h\} \in X_h := V_h \times V_h$, there holds

$$L(u, z) - L(u_h, z_h) = J(u) - J(u_h).$$

Hence, Proposition 3.1, applied to the Lagrangian $L(\cdot)(\cdot)$ on X yields a representation for the error $J(u) - J(u_h)$ in terms of the residuals

$$\rho(u_h)(\psi) := F(\psi) - A(u_h)(\psi),$$
$$\rho^*(z_h)(\varphi) := J'(u_h)(\varphi) - A'(u_h)(\varphi, z_h).$$

Since $L(u)(z)$ is linear in z, the remainder $\mathcal{R}_h$ only consists of the following three terms:

$$J'''(u_h + se)(e, e, e) - A'''(u_h + se)(e, e, e, z_h + se^*) - 3A''(u_h + se)(e, e, e^*),$$

where $e := u - u_h$ and $e^* := z - z_h$. This leads us to the following result.

Proposition 3.3. *For any solutions of the Euler–Lagrange systems* (3.9), (3.10) *and* (3.8), (3.11), *we have the* a posteriori *error representation*

$$J(u) - J(u_h) = \tfrac{1}{2}\rho(u_h)(z - \psi_h) + \tfrac{1}{2}\rho^*(z_h)(u - \varphi_h) + \mathcal{R}_h, \qquad (3.12)$$

for arbitrary $\psi_h, \varphi_h \in V_h$. *The remainder term* $\mathcal{R}_h$ *is cubic in the errors* $e := u - u_h$ *and* $e^* := z - z_h$,

$$\mathcal{R}_h := \tfrac{1}{2}\int_0^1 \big\{ J'''(u_h + se)(e, e, e) - A'''(u_h + se)(e, e, e, z_h + se^*)$$
$$- 3A''(u_h + se)(e, e, e^*) \big\}\, s(s - 1)\, ds.$$

The remainder term $\mathcal{R}_h$ in (3.12) is usually neglected. The evaluation of the resulting error estimator

$$\eta(u_h) := \tfrac{1}{2}\rho(u_h)(z - \psi_h) + \tfrac{1}{2}\rho^*(z_h)(u - \varphi_h),$$

for arbitrary $\psi_h, \varphi_h \in V_h$, requires the determination of approximations to the exact primal and dual solutions u and z, respectively.

Remark 3.4. We note that the error representation (3.12) is the nonlinear analogue of the trivial identity

$$J(e) = \rho(u_h)(z - \psi_h) = \rho^*(z_h)(u - \varphi_h) = F(e^*), \qquad (3.13)$$

in the linear case, for arbitrary $\varphi_h, \psi_h \in V_h$.

Integrating by parts in (3.12), we can derive a simpler error representation that does not contain the unknown primal solution u,

$$J(u) - J(u_h) = \rho(u_h)(z - \psi_h) + \widetilde{\mathcal{R}}_h, \qquad (3.14)$$

for arbitrary $\psi_h \in V_h$, with the remainder term

$$\widetilde{\mathcal{R}}_h = \int_0^1 \big\{ A''(u_h + se)(e, e, z) - J''(u_h + se)(e, e) \big\} s\, ds.$$

The evaluation of (3.14) only requires a guess for the dual solution z, but the remainder term $\widetilde{\mathcal{R}}_h$ is only quadratic in the error e.

3.2.3. Approximation of optimization problems. We continue using the notation from above. A differentiable 'cost-functional' $J(u, q)$ is now to be minimized under the equation constraint (3.7),

$$J(u, q) \to \min, \qquad A(u, q)(\varphi) = F(\varphi) \quad \forall \varphi \in V, \tag{3.15}$$

where q is the control from the 'control space' Q, and $A(\cdot, \cdot)(\cdot)$ is a differentiable semi-linear form on $V \times Q \times V$. On the space $X := V \times Q \times V$, we introduce the Lagrangian

$$L(u, q, \lambda) := J(u, q) - A(u, q)(\lambda) + F(\lambda),$$

with the adjoint variable $\lambda \in V$. We want to compute stationary points $x = \{u, q, \lambda\} \in X$ of L, that is, solutions of the variational equation

$$L'(x)(y) = 0 \quad \forall y \in X. \tag{3.16}$$

This is equivalent to the saddle-point system

$$A'_u(u, q)(\varphi, \lambda) = J'_u(u, q)(\varphi) \quad \forall \varphi \in V, \tag{3.17}$$

$$A(u, q)(\psi) = F(\psi) \quad \forall \psi \in V, \tag{3.18}$$

$$A'_q(u, q)(\chi, \lambda) = J'_q(u, q)(\chi) \quad \forall \chi \in Q. \tag{3.19}$$

We refer to the book Tröltzsch [71] for a general discussion of the Euler–Lagrange approach and the derivation of the corresponding 'KKT system' (3.17–3.19) of first-order necessary optimality conditions in the formulation of optimal control problems with PDEs.

For discretization of Equation (3.16), we introduce finite-dimensional subspaces $V_h \subset V$ and $Q_h \subset Q$, parametrized by $h \in \mathbb{R}_+$, and set $X_h := V_h \times Q_h \times V_h \subset X$. Then, approximations $x_h = \{u_h, q_h, \lambda_h\} \in X_h$ are determined by

$$L'(x_h)(y_h) = 0 \quad \forall y_h \in X_h, \tag{3.20}$$

which is equivalent to the discrete saddle-point problem

$$A'_u(u_h, q_h)(\varphi_h, \lambda_h) = J'_u(u_h, q_h)(\varphi_h) \quad \forall \varphi_h \in V, \tag{3.21}$$

$$A(u_h, q_h)(\psi_h) = F(\psi_h) \quad \forall \psi_h \in V_h, \tag{3.22}$$

$$A'_q(u_h, q_h)(\chi_h, \lambda_h) = J'_q(u_h, q_h)(\chi_h) \quad \forall \chi_h \in Q_h. \tag{3.23}$$

The residuals of these equations are defined by

$$\rho^*(\lambda_h)(\cdot) := J'_u(u_h, q_h)(\varphi_h) - A'_u(u_h, q_h)(\varphi_h, \lambda_h),$$

$$\rho(u_h)(\cdot) := F(\psi_h) - A(u_h, q_h)(\psi_h),$$

$$\rho^{**}(q_h)(\cdot) := J'_q(u_h, q_h)(\chi_h) - A'_q(u_h, q_h)(\chi_h, \lambda_h).$$

Again, since the pairs $\{u, q\}$ and $\{u_h, q_h\}$ satisfy the state equations, we have

$$L(u, q, \lambda) - L(u_h, q_h, \lambda_h) = J(u) - J(u_h).$$

Then, as before, we obtain from Proposition 3.1 the following result.

Proposition 3.5. *For any solutions of the saddle point problems* (3.17)–(3.19) *and* (3.21)–(3.23), *we have the a posteriori error representation*

$$J(u) - J(u_h) = \tfrac{1}{2}\rho(u_h)(\lambda - \psi_h) + \tfrac{1}{2}\rho^*(\lambda_h)(u - \varphi_h)$$
$$+ \tfrac{1}{2}\rho^{**}(q_h)(q - \chi_h) + \mathcal{R}_h, \tag{3.24}$$

for arbitrary $\varphi_h, \psi_h \in V_h$ *and* $\chi_h \in Q_h$. *The remainder term* $\mathcal{R}_h$ *is cubic in the errors* $e^u := u - u_h$, $e^\lambda := \lambda - \lambda_h$, *and* $e^q := q - q_h$.

For examples of the use of the error representation (3.24) for mesh adaptation in numerical optimal control, we refer to the 'landmark paper' Becker et al. [11] and to the forthcoming book Becker/Vexler [16].

3.2.4. Approximation of stability eigenvalue problems. Let $\hat{u} \in V$ and $\hat{u}_h \in V_h$ be a base solution and its Galerkin approximation determined by semilinear equations

$$a(\hat{u})(\hat{\psi}) = F(\hat{\psi}) \quad \forall \hat{\psi} \in V, \tag{3.25}$$

and

$$a(\hat{u}_h)(\hat{\psi}_h) = F(\hat{\psi}_h) \quad \forall \hat{\psi}_h \in V_h, \tag{3.26}$$

such as (3.7) and (3.8), respectively, with slightly changed notation for technical reasons. We consider the eigenvalue problem associated with the linearization of the semi-linear form $a(\cdot)(\cdot)$ about $\hat{u}$,

$$a'(\hat{u})(u, \varphi) = \lambda \, m(u, \varphi) \quad \forall \varphi \in V, \tag{3.27}$$

and its discrete analogues,

$$a'(\hat{u}_h)(u_h, \varphi_h) = \lambda_h \, m(u_h, \varphi_h) \quad \forall \varphi_h \in V_h. \tag{3.28}$$

Here, $m(\cdot, \cdot)$ is a symmetric, semidefinite bilinear form on V. We assume that $a'(\hat{u}_h)(\cdot, \cdot)$ is coercive on V and that $m(\cdot, \cdot)$ is compact, such that the Fredholm theory applies to this eigenvalue problem. From the eigenvalues $\lambda \in \mathbb{C}$ of (3.27) one can obtain information about the (dynamic) stability of the base solution $\hat{u}$. For $\mathrm{Re}\lambda > 0$ it is said to be 'linearly stable' and for $\mathrm{Re}\lambda < 0$ 'linearly unstable'. The related aspects of 'linear (hydrodynamic) stability theory' will be discussed in greater detail below.

In order to derive an *a posteriori* estimate for the eigenvalue error $\lambda - \lambda_h$, we introduce the spaces $\mathcal{V} := V \times V \times \mathbb{C}$ and $\mathcal{V}_h := V_h \times V_h \times \mathbb{C}$, and denote their elements by $U := \{\hat{u}, u, \lambda\}$ and $U_h := \{\hat{u}_h, u_h, \lambda_h\}$, respectively. Further, for $\Phi = \{\hat{\varphi}, \varphi, \mu\} \in \mathcal{V}$, we introduce a semi-linear form $A(\cdot)(\cdot)$ by

$$A(U)(\Phi) := f(\hat{\varphi}) - a(\hat{u})(\hat{\varphi}) - a'(\hat{u})(u, \varphi) + \lambda \, m(u, \varphi) + \overline{\mu}\{m(u, u) - 1\}.$$

With this notation the sets of equations (3.25,3.27) and (3.26,3.28) can be written in compact form as follows:

$$A(U)(\Phi) = 0 \quad \forall \Phi \in \mathcal{V}, \tag{3.29}$$
$$A(U_h)(\Phi_h) = 0 \quad \forall \Phi_h \in \mathcal{V}_h. \tag{3.30}$$

For controlling the error of this approximation, we choose the functional

$$J(\Phi) := \mu\, m(\varphi, \varphi),$$

for $\Phi = \{\hat{\varphi}, \varphi, \mu\} \in \mathcal{V}$, which is motivated by the fact that $J(U) = \lambda$, since $m(u, u) = 1$. In order to apply the general result of Theorem 3.3 to this situation, we have to identify the dual problems corresponding to (3.29) and (3.30). The dual solutions $Z = \{\hat{z}, z, \pi\} \in \mathcal{V}$ and $Z_h = \{\hat{z}_h, z_h, \pi_h\} \in \mathcal{V}_h$ are determined by the equation

$$A'(U)(\Phi, Z) = J'(U)(\Phi) \quad \forall \Phi \in \mathcal{V}, \tag{3.31}$$

and its discrete analogue

$$A'(U_h)(\Phi_h, Z_h) = J'(U_h)(\Phi_h) \quad \forall \Phi_h \in \mathcal{V}_h, \tag{3.32}$$

respectively. By a straightforward calculation (for the details see Heuveline and Rannacher [47]), we find that the dual solution $Z = \{\hat{z}, z, \pi\} \in \mathcal{V}$ is given by $z = u^*$ and $\pi = \lambda$, while $\hat{z} = \hat{u}^*$ is determined as solution of

$$a'(\hat{u})(\psi, \hat{u}^*) = -a''(\hat{u})(\psi, u, u^*) \quad \forall \psi \in V. \tag{3.33}$$

The corresponding residuals are defined by

$$\rho(\hat{u}_h)(\psi) := f(\psi) - a(\hat{u}_h; \psi),$$
$$\rho^*(\hat{u}_h^*)(\psi) := -a''(\hat{u})(\psi, u_h, u_h^*) - a'(\hat{u}_h)(\psi, \hat{u}_h^*),$$
$$\rho(u_h, \lambda_h)(\psi) := \lambda_h\, m(u_h, \psi) - a'(\hat{u}_h)(u_h, \psi),$$
$$\rho^*(u_h^*, \lambda_h)(\psi) := \lambda_h\, m(\psi, u_h^*) - a'(\hat{u}_h)(\psi, u_h^*).$$

Then, from Theorem 3.3, we obtain the following result.

Proposition 3.6. *For the eigenvalue problems (3.27) and (3.28), we have the error representation*

$$\begin{aligned} \lambda - \lambda_h &= \tfrac{1}{2}\big\{\rho(\hat{u}_h)(\hat{u}^* - \hat{\psi}_h) + \rho^*(\hat{u}_h^*; \hat{u} - \hat{\varphi}_h)\big\} \\ &\quad + \tfrac{1}{2}\big\{\rho(u_h, \lambda_h)(u^* - \psi_h) + \rho^*(u_h^*, \lambda_h)(u - \varphi_h)\big\} - \mathcal{R}_h, \end{aligned} \tag{3.34}$$

with arbitrary $\hat{\psi}_h$, ψ_h, $\hat{\varphi}_h$, $\varphi_h \in V_h$. The cubic remainder $\mathcal{R}_h$ is given by

$$\mathcal{R}_h = \tfrac{1}{2}(\lambda - \lambda_h)(e^v, e^{v*}) - \tfrac{1}{12}a''(\hat{u})(\hat{e}, \hat{e}, \hat{e}^*) - \tfrac{1}{12}a''(\hat{u})(\hat{e}, e, e^*),$$

where $\hat{e}^v := \hat{v} - \hat{v}_h$, $\hat{e}^{v*} := \hat{v}^* - \hat{v}_h^*$, $e^v := v - v_h$, and $e^{v*} := v^* - v_h^*$.

Remark 3.7. The result of Theorem 3.5 does not require the eigenvalue λ to be simple or non-degenerate. However, this is the generic case in most practical applications. The test for $m(v_h^*, v_h^*) \to \infty$ or $m(v_h^*, v_h^*) \gg 1$ can be used to detect either the degeneracy of the eigenvalue λ or, in the case $0 < \mathrm{Re}\lambda \ll 1$, the extension of the corresponding 'pseudo-spectrum' into the negative complex half-plane, which indicates possible dynamic instability of the base flow. For a more detailed discussion of this point, we refer to Heuveline and Rannacher [47].

3.3. Model problems and practical aspects

In this section, we describe the application of the foregoing abstract theory to prototypical model situations which usually occur as components of flow models. In this context, we also discuss the practical evaluation of the a posteriori error representations and their use for automatic mesh adaptation. The first model case is the *elliptic* Poisson equation, the second one a purely *hyperbolic* transport problem, and the third one the *parabolic* heat equation.

3.3.1. Elliptic model case: Poisson equation.

We consider the model problem

$$-\Delta u = f \quad \text{in } \Omega, \quad u = 0 \text{ on } \partial\Omega, \tag{3.35}$$

on a polygonal domain $\Omega \subset \mathbb{R}^2$. The natural solution space for the boundary value problem (3.35) is the Sobolev space $V = H_0^1(\Omega)$. The variational formulation of (3.35) seeks $u \in V$, such that

$$(\nabla u, \nabla \varphi) = (f, \varphi) \quad \forall \varphi \in V. \tag{3.36}$$

The finite-element approximation of (3.36) uses finite-dimensional subspaces

$$V_h = \{v \in V : v_{|K} \in P(K), \ K \in \mathbb{T}_h\},$$

defined on decompositions $\mathbb{T}_h$ of $\overline{\Omega}$ into triangles or quadrilaterals K (*cells*) of width $h_K = \operatorname{diam}(K)$; we write $h = \max_{K \in \mathbb{T}_h} h_K$ for the *global* mesh width. Here, $P(K)$ denotes a suitable space of polynomial-like functions defined on the cell $K \in \mathbb{T}_h$. We will mainly consider low-order finite elements on quadrilateral meshes where $P(K) = \widetilde{Q}_1(K)$ consists of shape functions which are obtained as usual via a bilinear transformation from the space of bilinear functions $Q_1(\hat{K}) = \operatorname{span}\{1, x_1, x_2, x_1 x_2\}$ on the reference cell $\hat{K} = [0,1]^2$ (*isoparametric bilinears*). Local mesh refinement or coarsening is realized by using *hanging nodes*. The variable corresponding to such a hanging node is eliminated from the system by linear interpolation of neighboring variables in order to preserve the conformity of the global ansatz, i.e., $V_h \subset V$ (for more details we refer to Carey/Oden [23]). The discretization of (3.36) determines $u_h \in V_h$ by

$$(\nabla u_h, \nabla \varphi_h) = (f, \varphi_h) \quad \forall \varphi_h \in V_h. \tag{3.37}$$

The essential feature of this approximation is the *Galerkin orthogonality* of the error $e = u - u_h$,

$$(\nabla e, \nabla \varphi_h) = 0, \quad \varphi_h \in V_h. \tag{3.38}$$

A priori error analysis. We begin with a brief discussion of the a priori error analysis for the scheme (3.37). By $i_h u \in V_h$, we denote the natural nodal interpolant of $u \in C(\overline{\Omega})$ satisfying $i_h u(a) = u(a)$ at all nodal points a. There holds (see, e.g., Brenner/Scott [21]):

$$\|u - i_h u\|_K + h_K^{1/2}\|u - i_h u\|_{\partial K} + h_K\|\nabla(u - i_h u)\|_K \le c_i h_K^2 \|\nabla^2 u\|_K, \tag{3.39}$$

with some *interpolation constant* c_i, usually $0.1 \leq c_i \leq 1.0$. By the projection property of the Galerkin finite-element scheme the interpolation estimate (3.39) directly implies the a priori *energy-norm* error estimate

$$\|\nabla e\| = \inf_{\varphi_h \in V_h} \|\nabla(u - \varphi_h)\| \leq c_i h^2 \|\nabla^2 u\|. \tag{3.40}$$

Further, employing a duality argument (so-called *Aubin–Nitsche trick*),

$$-\Delta z = \|e\|^{-1} e \quad \text{in } \Omega, \quad z = 0 \text{ on } \partial\Omega, \tag{3.41}$$

we obtain

$$\|e\| = (e, -\Delta z) = (\nabla e, \nabla z) = (\nabla e, \nabla(z - i_h z) \leq c_i c_s h \|\nabla e\|, \tag{3.42}$$

where the *stability constant* c_s is defined by the a priori bound $\|\nabla^2 z\| \leq c_s$. Together with the energy-error estimate (3.40), this implies the improved a priori L^2-norm error estimate

$$\|e\| \leq c_i^2 c_s h^2 \|\nabla^2 u\| \leq c_i^2 c_s^2 h^2 \|f\|. \tag{3.43}$$

A posteriori error analysis. Next, we seek to derive a posteriori error estimates. Let $J(\cdot)$ be an arbitrary (linear) *error functional* defined on V, and $z \in V$ the solution of the corresponding 'dual problem'

$$(\nabla\varphi, \nabla z) = J(\varphi) \quad \forall \varphi \in V. \tag{3.44}$$

Taking $\varphi = e$ in (3.44) and using the Galerkin orthogonality (3.38), in accordance with the general result of Theorem 3.3 and the relation (3.13), we obtain after cell-wise integration by parts the error representation

$$J(e) = \rho(u_h)(z - \varphi_h) = (\nabla e, \nabla(z - \varphi_h))$$
$$= \sum_{K \in \mathbb{T}_h} \left\{ (-\Delta u + \Delta u_h, z - \varphi_h)_K - (\partial_n u_h, z - \varphi_h)_{\partial K} \right\},$$

with an arbitrary $\varphi_h \in V_h$. This can be rewritten as

$$J(e) = \sum_{K \in \mathbb{T}_h} \left\{ (R_h, z - \varphi_h)_K + (r_h, z - \varphi_h)_{\partial K} \right\}, \tag{3.45}$$

with the 'cell-' and 'edge-residuals' defined by

$$R_{h|K} = f + \Delta u_h, \qquad r_{h|\Gamma} := \begin{cases} -\frac{1}{2}[\partial_n u_h], & \text{if } \Gamma \subset \partial K \setminus \partial\Omega, \\ 0, & \text{if } \Gamma \subset \partial\Omega, \end{cases}$$

where $[\nabla u_h]$ denotes the jump of ∇u_h across the cell edges Γ. From the error identity (3.45), we can infer an a posteriori error estimate of the form

$$|J(e)| \leq \eta(u_h) := \sum_{K \in \mathbb{T}_h} \eta_K(u_h), \tag{3.46}$$

with the cell-wise error indicators

$$\eta_K(u_h) := |(R_h, z - \varphi_h)_K + (r_h, z - \varphi_h)_{\partial K}|.$$

These indicators are 'consistent' in the sense that they vanish at the exact solution, $\eta_K(u) = 0$.

Proposition 3.8. *For the finite-element approximation of the Poisson equation (3.35), there holds the 'goal-oriented' a posteriori error estimate*

$$|J(e_h)| \leq \sum_{K \in \mathbb{T}_h} \rho_K \, \omega_K, \tag{3.47}$$

with the cell residuals ρ_K and weights ω_K defined by

$$\rho_K := \left(\|R_h\|_K^2 + h_K^{-1}\|r_h\|_{\partial K}^2 \right)^{1/2}, \quad \omega_K := \left(\|z - i_h z\|_K^2 + h_K\|z - i_h z\|_{\partial K}^2 \right)^{1/2}.$$

Remark 3.9. In general, the transition from the error identity (3.45) to the error estimate (3.46) and further to (3.47) causes significant over-estimation of the true error. The weights ω_K describe the dependence of the error $J(e_h)$ on variations of the cell residuals ρ_K. In practice they have to be determined computationally.

Remark 3.10. Since the relation (3.45) is an identity, any reformulation of it can be used for deriving estimates for the error $J(e)$. However, one has to be careful in extracting local refinement indicators. For example, one may prefer the form

$$J(e) = \sum_{K \in \mathbb{T}_h} \left\{ (f, z - \varphi_h)_K - (\nabla u_h, \nabla(z - \varphi_h))_K \right\},$$

which does not require the evaluation of normal derivatives across the interelement boundaries. But the corresponding local error indicators

$$\eta_K(u_h) := |(f, z - \varphi_h)_K - (\nabla u_h, \nabla(z - \varphi_h))_K|$$

are not consistent, i.e., $\eta_K(u) \neq 0$. Mesh adaptation based on these inconsistent error indicators generally results in unnecessary over-refinement.

Remark 3.11. Another approach to goal-oriented a posteriori error estimation uses so-called 'gradient recovery techniques' in the spirit of the ZZ approach. From the dual equation (3.44) employing Galerkin orthogonality, we obtain the error identity

$$J(e) = (\nabla e, \nabla e^*), \tag{3.48}$$

with the primal and dual errors $e = u - u_h$ and $e^* = z - z_h$, respectively. Let $M_K(\nabla u_h)$ be an approximation obtained by local averaging, satisfying

$$\|\nabla u - M_K(\nabla u_h)\|_K \ll \|\nabla u - \nabla u_h\|_{\tilde{K}},$$

where $\tilde{K}$ denotes an h_K-neighborhood of cell K. Then,

$$J(e) \approx \sum_{K \in \mathbb{T}_h} (M_K(\nabla u_h) - \nabla u_h, M_K(\nabla z) - \nabla z_h)_K,$$

and the mesh adaptation may be based on the local error indicators

$$\eta_K(u_h) := |(M_K(\nabla u_h) - \nabla u_h, M_K(\nabla z) - \nabla z_h)_K|.$$

For more details on this method see Korotov, Neitaanmäki and Repin [56]. Its possible success depends on the reliability of the approximation $M_K(\nabla u_h) \approx \nabla u$ which is to be expected only for isotropic elliptic problems.

Remark 3.12. The error representation (3.45) seems to suggest that the use of differently refined meshes for u and z may be advisable according to their mutual singularities. However, this is a misconception as it does not observe the special role of the *multiplicative* interaction between primal residuals and dual weights. *Primal and dual solutions do not need to be computed on different meshes if their singularities are located at different places.* This rule is well confirmed by numerical tests even for hyperbolic problems.

A posteriori energy-norm error bound. By the same type of argument as used above, we can also derive the traditional global energy-norm error estimates. To this end, we choose the functional

$$J(\varphi) := \|\nabla e\|^{-1}(\nabla e, \nabla \varphi)$$

in the dual problem (3.44). Its solution $z \in V$ satisfies $\|\nabla z\| \leq 1$. Applying Theorem 3.8, we obtain the estimate

$$\|\nabla e\| \leq \sum_{K \in \mathbb{T}_h} \rho_K \, \omega_K \leq \Big(\sum_{K \in \mathbb{T}_h} h_K^2 \rho_K^2 \Big)^{1/2} \Big(\sum_{K \in \mathbb{T}_h} h_K^{-2} \omega_K^2 \Big)^{1/2},$$

with residual terms and weights as defined above. Now, we use the approximation estimate

$$\Big(\sum_{K \in \mathbb{T}_h} \Big\{ h_K^{-2} \|z - i_h^* z\|_K^2 + h_K^{-1} \|z - i_h^* z\|_{\partial K}^2 \Big\} \Big)^{1/2} \leq c_i^* \, \|\nabla z\|, \tag{3.49}$$

where $i_h^* z \in V_h$ is a modified nodal interpolation which is defined and stable on $H^1(\Omega)$ (for the construction of such an operator see Brenner/Scott [21]). Using this, we easily deduce the a posteriori *energy-norm* error estimate

$$\|\nabla e\| \leq \eta_E(u_h) := c_i^* \Big(\sum_{K \in \mathbb{T}_h} h_K^2 \rho_K^2 \Big)^{1/2}. \tag{3.50}$$

An analogous argument also yields the usual a posteriori L^2-norm error bound

$$\|e\| \leq \eta_{L^2}(u_h) := c_i c_s \Big(\sum_{K \in \mathbb{T}_h} h_K^4 \rho_K^2 \Big)^{1/2}. \tag{3.51}$$

Evaluation of error estimates. From the a posteriori error estimate (3.46), we want to deduce criteria for local mesh adaptation and for the final stopping of the adaptation process. To this end, we have to evaluate the local cell error indicators

$$\eta_K(u_h) := |(R_h, z - \varphi_h)_K + (r_h, z - \varphi_h)_{\partial K}|,$$

for arbitrary $\varphi_h \in V_h$, and the global error estimator

$$\eta(u_h) = \sum_{K \in \mathbb{T}_h} \eta_K(u_h).$$

This requires the construction of an approximation $\tilde{z} \approx z \in V$, such that $\tilde{z} - i_h\tilde{z}$ can substitute $z - i_h z$, resulting in approximate cell error indicators

$$\tilde{\eta}_K(u_h) := |(R_h, \tilde{z} - i_h\tilde{z})_K + (r_h, \tilde{z} - i_h\tilde{z})_{\partial K}|$$

and the corresponding approximate error estimator

$$|J(e)| \approx \tilde{\eta}(u_h) := \sum_{K \in \mathbb{T}_h} \tilde{\eta}_K(u_h).$$

The goal is to achieve an optimal *effectivity index* for the error estimator $\tilde{\eta}(u_h)$,

$$I_{\text{eff}} := \lim_{TOL \to 0} \frac{\tilde{\eta}(u_h)}{|J(e)|} = 1.$$

Computational experience shows that asymptotic sharpness does not seem to be achievable with acceptable effort (see Becker/Rannacher [13]). With all the cheaper methods considered the effectivity index I_{eff} never really tends to 1, but in most relevant cases stays well below 2, what may actually be considered as good enough. There are two separate aspects to be considered: the sharpness of the global error bound $\tilde{\eta}(u_h)$ and the effectivity of the local error indicators $\tilde{\eta}_K$ which are used in the mesh refinement process.

Accurate a posteriori error estimation is a delicate matter. Already by once applying the triangle inequality,

$$|J(e)| \leq \eta(u_h) = \sum_{K \in \mathbb{T}_h} \eta_K(u_h), \tag{3.52}$$

asymptotic sharpness may be lost. This is seen, for instance, in the case $J(u) = u(0)$ when the exact as well as the approximate solution are anti-symmetric with respect to the x-axis meaning that $e(0) = 0$, but $\eta(u_h) \neq 0$.

Most practical ways of generating approximation $\tilde{z}$ are based on solving the dual problem numerically. Let $z_h \in V_h$ denote the approximation to z obtained on the current mesh by the same finite-element method as used for computing u_h.

1. *Approximation by higher-order methods:* The dual problem is solved by using *biquadratic* finite elements on the current mesh yielding an approximation $z_h^{(2)}$ to z. The resulting error estimator is denoted by

$$\eta^{(1)}(u_h) := \sum_{K \in \mathbb{T}_h} \left| (R_h, z_h^{(2)} - i_h z_h^{(2)})_K + (r_h, z_h^{(2)} - i_h z_h^{(2)})_{\partial K} \right|.$$

It is seen by theoretical analysis as well as by numerical experiments that in 'good' cases $\eta^{(1)}(u_h)$ has optimal effectivity, $I_{\text{eff}} \approx 1$. This rather expensive way of evaluating the error estimator is useful only in certain circumstances, e.g., for very irregular dual problems such as occurring in the solution of the Euler equations.

2. *Approximation by higher-order interpolation:* A cheaper strategy uses patchwise *biquadratic* interpolation of the bilinear approximation z_h on the current mesh yielding an approximation $i_{2h}^{(2)} z_h$ to z. This construction requires some special care for elements with hanging nodes, in order to preserve the higher-order accuracy of the interpolation process. The resulting global error estimator is denoted by

$$\eta^{(2)}(u_h) := \sum_{K \in \mathbb{T}_h} \left| (R_h, i_{2h}^{(2)} z_h - z_h)_K + (r_h, i_{2h}^{(2)} z_h - z_h)_{\partial K} \right|.$$

This rather simple strategy turns out to be surprisingly effective in many different situations. It is actually used in most of the computational examples discussed below. For the Poisson test problems considered there holds $I_{\mathrm{eff}} \approx 1.5 - 2$. For more details and for other strategies for evaluating the error estimators we refer to Becker/Rannacher [13, 14].

Remark 3.13. If on the basis of a numerical approximation to the dual solution z an approximate error representation $\tilde{\eta}(u_h)$ has been generated, one may hope to obtain an improved approximation to the target quantity by setting

$$\tilde{J}(u_h) := J(u_h) + \tilde{\eta}(u_h) \approx J(u).$$

Such a 'post-processing' can significantly improve the accuracy in computing $J(u)$. This idea has been systematically developed in Giles/Süli [35], particularly for finite-volume approximations of flow problems.

Strategies for mesh adaptation. We want to discuss some strategies for organizing local mesh adaptation on the basis of the a posteriori error estimates derived above. Suppose that we have an a posteriori error estimate of the form

$$|J(u) - J(u_h)| \approx \tilde{\eta}(u_h) \leq \sum_{K \in \mathbb{T}_h} \tilde{\eta}_K, \tag{3.53}$$

with local *cell-error indicators* $\tilde{\eta}_K = \tilde{\eta}_K(u_h)$. The prescribed error tolerance is TOL and the maximum number of mesh cells $N_{\max}$.

(i) *Error-balancing strategy:* Cycle through the mesh and seek to equilibrate the local error indicators according to

$$\tilde{\eta}_K \approx \frac{TOL}{N}, \quad N = \#\{K \in \mathbb{T}_h\}. \tag{3.54}$$

This process requires iteration with respect to the number of mesh cells N and eventually leads to $\tilde{\eta}(u_h) \approx TOL$.

(ii) *Fixed-fraction strategy:* Order cells according to the size of $\tilde{\eta}_K$,

$$\tilde{\eta}_{K_N} \geq \cdots \geq \tilde{\eta}_{K_i} \cdots \geq \tilde{\eta}_{K_1},$$

and refine 20% of cells with largest $\tilde{\eta}_K$ (or those which make up 20% of the estimator value) and coarsen 10% of those cells with smallest $\tilde{\eta}_K$. By this strategy, we may achieve a prescribed rate of increase of N (or keep it constant in solving non-stationary problems). The fixed fraction strategy is very robust and economical, and is therefore used in most of the examples discussed below.

(iii) *Mesh-optimization strategy:* Use the (heuristic) representation

$$\eta(u_h) = \sum_{K \in \mathbb{T}_h} \eta_K(u_h) \approx \int_\Omega h(x)^2 \Psi(x) \, dx, \tag{3.55}$$

directly for deriving a formula for an optimal mesh-size distribution $h_{\mathrm{opt}}(x)$, for details see Bangerth/Rannacher [2]. Corresponding 'optimal' meshes may be constructed by successive hierarchical refinement of an initial coarse mesh or by a sequence of complete remeshings.

3.4. Hyperbolic model case: transport problem

As a simple model case, we consider the scalar transport equation

$$\beta \cdot \nabla u = f \quad \text{in } \Omega, \quad u = g \text{ on } \partial\Omega_-, \tag{3.56}$$

on a domain $\Omega \subset \mathbb{R}^2$ with *inflow boundary* $\partial\Omega_- = \{x \in \partial\Omega, \, n \cdot \beta < 0\}$. Accordingly, $\partial\Omega_+ = \partial\Omega \backslash \partial\Omega_-$ is the *outflow boundary*. The transport vector β is assumed as constant for simplicity. Then, the natural solution space is $V = \{v \in L^2(\Omega),$ $\beta \cdot \nabla v \in L^2(\Omega)\}$. This problem is discretized using the Galerkin finite-element method with *streamline diffusion* stabilization (see Hansbo/Johnson [39] and also Johnson [53]). On quadrilateral meshes $\mathbb{T}_h$, we define subspaces

$$V_h = \{v \in H^1(\Omega), \, v_{|T} \in \widetilde{Q}_1(K), \, K \in \mathbb{T}_h\}$$

again consisting of (isoparametric) *bilinear* finite elements. The discrete solution $u_h \in V_h$ is defined by

$$(\beta \cdot \nabla u_h - f, \varphi_h + \delta\beta \cdot \nabla\varphi_h) + (n \cdot \beta(g - u_h), \varphi_h)_{\partial\Omega_-} = 0 \quad \forall \varphi_h \in V_h, \tag{3.57}$$

where the stabilization parameter is locally determined by $\delta_K = \alpha h_K$. In this formulation the inflow boundary condition is imposed in the weak sense. This facilitates the use of a duality argument in generating a posteriori error estimates. Let $J(\cdot)$ be a given functional for controlling the error $e = u - u_h$. Following the DWR approach, we consider the corresponding dual problem

$$A_h(\varphi, z) = (\beta \cdot \nabla\varphi, z + \delta\beta \cdot \nabla z) - (n \cdot \beta\varphi, z)_{\partial\Omega_-} = J(\varphi) \quad \forall \varphi \in V, \tag{3.58}$$

which is a transport problem with transport in the negative β-direction. We note that the stabilized bilinear form $A_h(\cdot, \cdot)$ is used in the duality argument in order to achieve an optimal treatment of the stabilization terms; for a detailed discussion of this point see Houston et al. [49]. The resulting error representation reads

$$J(e) = (\beta \cdot \nabla e, z - \varphi_h + \delta\beta \cdot \nabla(z - \varphi_h)) - (n \cdot \beta e, z - z_h)_{\partial\Omega_-},$$

for arbitrary $\varphi_h \in V_h$. This leads us in the following result.

Proposition 3.14. *For the approximation of the transport problem* (3.56) *by the finite-element scheme* (3.57), *there holds the a posteriori error estimate*

$$|J(e)| \leq \sum_{K \in \mathbb{T}_h} \rho_K \omega_K, \tag{3.59}$$

with the cell residuals ρ_K and weights ω_K defined by

$$\rho_K := \left(\|f - \beta \cdot \nabla u_h\|_K^2 + h_K^{-1}\|n \cdot \beta(u_h - g)\|_{\partial K \cap \partial\Omega_-}^2\right)^{1/2},$$

$$\omega_K := \left(\|z - \varphi_h\|_K^2 + \delta_K^2\|\beta \cdot \nabla(z - \varphi_h)\|_K^2 + h_K\|z - \varphi_h\|_{\partial K \cap \partial\Omega_-}^2\right)^{1/2}.$$

Remark 3.15. We note that the a posteriori error bound (3.59) explicitly contains the mesh size h_K and the stabilization parameter δ_K as well. This gives us the possibility to simultaneously adapt both parameters, which is particularly advantageous in capturing sharp layers in the solution.

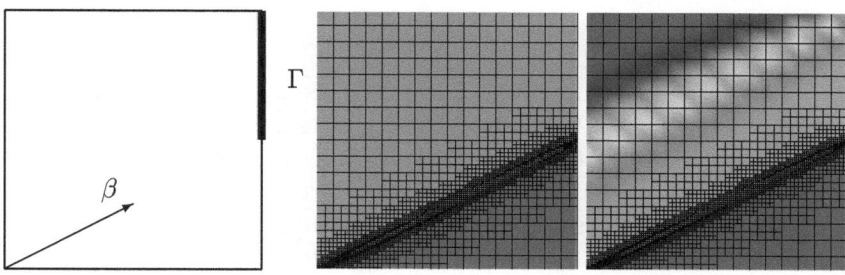

FIGURE 7. *Configuration and grids of the test computation for the model transport problem (3.56) (left), primal solution (middle) and dual solution (right) on an adaptively refined mesh.*

A simple thought experiment helps to understand the features of the error estimate (3.59). Let $\Omega = (0,1)^2$ and $f = 0$. We take the functional $J(u) = (1, n \cdot \beta u)_{\partial\Omega_+}$. The corresponding dual solution is $z \equiv 1$, so that $J(e) = 0$. Hence

$$(1, n \cdot \beta u_h)_{\partial\Omega_+} = (1, n \cdot \beta u)_{\partial\Omega_+} = -(1, n \cdot \beta g)_{\partial\Omega_-},$$

which recovers the well-known global conservation property of the scheme.

Next, we take again the unit square $\Omega = (0,1)^2$ and $f = 0$, and consider the case of constant transport $\beta = (1, 0.5)^T$ and inflow data $g(x,0) = 0$, $g(0,y) = 1$. The quantity to be computed is part of the outflow as indicated in Figure 7:

$$J(u) = \int_\Gamma \beta \cdot nu \, ds \, .$$

The mesh refinement is organized according to the *fixed-fraction strategy* described above. Table 1 shows results for this test computation (see Hartmann [41]). The corresponding meshes and the primal as well as the dual solution are presented in Figure 7. Notice that there is no mesh refinement enforced of the dual solution along the upper line of discontinuity since here the residual of the primal solution is almost zero. Apparently, this has not much effect on the accuracy of the error estimator.

Remark 3.16. The results of this simple test show a somewhat counter-intuitive feature of error estimation using the DWR method. The evaluation of the a posteriori error estimator for a functional output $J(u)$ does not require extra mesh refinement in approximating the dual solution z. It is most economical and sufficiently accurate to compute both approximations u_h as well as z_h on the same (adapted) mesh. This is due to the multiplicative occurrence of residual ρ_K and weight ω_K in the error representation formulas. In areas where the primal solution u is smooth, and therefore the residual of u_h small, the error in approximating

TABLE 1. *Convergence results of the test computation for the model transport problem* (3.56).

L	N	$J(e)$	η	I_{eff}
0	256	2.01e-2	2.38e-2	1.18
2	634	1.09e-2	1.21e-2	1.11
4	1315	6.25e-3	7.88e-3	1.26
6	2050	4.21e-3	5.37e-3	1.27
8	2566	3.90e-3	5.01e-3	1.28
10	3094	3.41e-3	4.71e-3	1.38

the weight may be large, due to irregularities in z, without significantly affecting the accuracy of the error representation.

Remark 3.17. The simple transport problem (3.56) is the prototype of the Euler equations for modeling inviscid compressible flow. Adaptive finite-elemente methods for the Euler equations using the DWR method have been developed in a series of papers (Hartmann [41] and Hartmann/Houston [42, 43, 44]).

3.4.1. Parabolic model case: heat equation. We consider the heat-conduction problem

$$\partial_t u - \Delta u = f \quad \text{in } Q_T, \quad u_{|t=0} = u^0 \quad \text{in } \Omega, \quad u_{|\partial\Omega} = 0 \quad \text{on } I, \tag{3.60}$$

on a space-time region $Q_T = \Omega \times I$, where $\Omega \subset \mathbb{R}^d$, $d \geq 1$, and $I = [0, T]$; the coefficient a may vary in space. This model is used to describe diffusive transport of energy or certain species concentrations.

The discretization of problem (3.60) uses a Galerkin method in space-time. We split the time interval $[0, T]$ into subintervals $I_n = (t_{n-1}, t_n]$ according to

$$0 = t_0 < \cdots < t_n < \cdots < t_N = T, \qquad k_n := t_n - t_{n-1}.$$

At each time level t_n, let $\mathbb{T}_h^n$ be a regular finite-element mesh as defined above with local mesh width $h_K = \text{diam}(K)$, and let $V_h^n \subset H_0^1(\Omega)$ be the corresponding finite-element subspace with d-linear shape functions. Extending the spatial mesh to the corresponding space-time slab $\Omega \times I_n$, we obtain a global space-time mesh consisting of $(d+1)$-dimensional cubes $Q_K^n := K \times I_n$. On this mesh, we define the global finite-element space

$$V_h^k = \big\{ v \in W, \; v(\cdot, t)_{|Q_K^n} \in \widetilde{Q}_1(K), \; v(x, \cdot)_{|Q_K^n} \in P_r(I_n) \; \forall \, Q_K^n \big\},$$

where $W = L^2((0, T); H_0^1(\Omega))$ and $r \geq 0$. For functions from this space and their time-continuous analogues, we use the notation

$$v^{n+} = \lim_{t \to t_n + 0} v(t), \quad v^{n-} = \lim_{t \to t_n - 0} v(t), \quad [v]^n = v^{n+} - v^{n-}.$$

The discretization of problem (3.60) is based on a variational formulation which allows the use of functions, which are discontinuous in time at the time instants t_n. This method, termed *dG(r) method* (*discontinuous* Galerkin method in time), determines approximations $U \in V_h^k$ by requiring

$$A(U, \varphi) = 0 \quad \forall \varphi \in V_h^k, \tag{3.61}$$

with the semi-linear form

$$A(u, \varphi) = \sum_{n=1}^{N} \int_{I_n} \left\{ (\partial_t u, \varphi) + (\nabla u, \nabla \varphi) - (f, \varphi) \right\} dt + \sum_{n=1}^{N} ([u]_{n-1}, \varphi_{n-1}^+),$$

where $u_0^- = u_0$. We note that the continuous solution u also satisfies equation (3.61) which again implies Galerkin orthogonality for the error $e = u - U$ with respect to the bilinear form $A(\cdot, \cdot)$. Since the test functions $\varphi \in V_h^k$ may be discontinuous at times t_n, the global system (3.61) decouples and can be written in form of a time-stepping scheme,

$$\int_{I_n} \left\{ (\partial_t U, \varphi) + (\nabla U, \nabla \varphi) \right\} dt + ([U]^{n-1}, \varphi^{(n-1)+}) = \int_{I_n} (f, \varphi) \, dt, \quad n = 1, \dots, N,$$

for all $\varphi \in V_h^n$. In the following, we consider only the lowest-order case $r = 0$, the so-called 'dG(0) method', which is closely related to the backward Euler scheme. For explaining the application of the DWR approach to this situation, we concentrate on the control of the spatial L^2-norm error $\|e^{N-}\|$ at the end time $T = t_N$, corresponding to the error functional

$$J(\varphi) := (\varphi^{N-}, e^{N-}) \|e^{N-}\|^{-1}.$$

The corresponding dual problem in space-time reads as

$$\begin{aligned} \partial_t z - \Delta z &= 0 \quad \text{in } \Omega \times I, \\ z_{|t=T} &= \|e^{N-}\|^{-1} e^{N-} \text{ in } \Omega, \quad z_{|\partial\Omega} = 0 \text{ on } I. \end{aligned} \tag{3.62}$$

In this situation the abstract error representations (3.12) or (3.13) take the form

$$\begin{aligned} J(e) = \sum_{n=1}^{N} \sum_{K \in \mathbb{T}_h^n} \Big\{ &(R_h^k, z - I_h^k z)_{K \times I_n} + (r_h^k, z - I_h^k z)_{\partial K \times I_n} \\ &- ([U]^{n-1}, (z - I_h^k z)^{(n-1)+})_K \Big\}, \end{aligned} \tag{3.63}$$

with an appropriate approximation $I_h^k z \in V_h^k$ and the local residuals

$$R_{h|K}^k := f - \partial_t U + \Delta U, \qquad r_{h|\Gamma}^k := \begin{cases} -\frac{1}{2}[\partial_n U], & \text{if } \Gamma \subset \partial T \backslash \partial\Omega, \\ 0, & \text{if } \Gamma \subset \partial\Omega. \end{cases}$$

Here, we use the natural interpolation $I_h^k z \in V_h^k$ which is defined by

$$\int_{I_n} I_h^k z(a, t) \, dt = \bar{z}(a), \qquad \bar{z}(x) := \int_{I_n} z(x, t) \, dt, \quad x \in \overline{\Omega},$$

for all nodal points a of the mesh $\mathbb{T}_h^n$. Observing that the time-integrated equation residual $\overline{R_h^k} = \overline{f - \partial_t U + \Delta U}$ as well as the jump-residual r_h^k are constant in time, the a posteriori error representation can be rewritten in the form

$$J(e) = \sum_{n=1}^{N} \sum_{K \in \mathbb{T}_h^n} \left\{ (f - \overline{f}, z - I_h^k z)_{Q_K^n} + (\overline{R_h^k}, \overline{z} - I_h^k z)_{Q_K^n} \right.$$
$$\left. + (r_h^k, \overline{z} - I_h^k z)_{\partial K \times I_n} - ([U]^{n-1}, (z - I_h^k z)^{(n-1)+})_K \right\}. \tag{3.64}$$

From this error representation we conclude the following result:

Proposition 3.18. *For the approximation of the heat conduction problem by the dG(0)-FEM, there holds the a posteriori error estimate*

$$|J(e)| \leq \sum_{n=1}^{N} \sum_{K \in \mathbb{T}_h^n} \left\{ \rho_K^{h,n} \, \omega_K^{h,n} + \rho_K^{k,n} \, \omega_K^{k,n} \right\}, \tag{3.65}$$

where the cell residuals and weights can be grouped as follows:

(i) *spatial terms:*

$$\rho_K^{h,n} := \left(\|\overline{R_h^k}\|_{K \times I_n}^2 + h_K^{-1} \|r_h^k\|_{\partial K \times I_n}^2 \right)^{1/2},$$
$$\omega_K^{h,n} := \left(\|\overline{z} - I_h^k z\|_{K \times I_n}^2 + h_K \|\overline{z} - I_h^k z\|_{\partial K \times I_n}^2 \right)^{1/2},$$

(ii) *temporal terms:*

$$\rho_K^{k,n} := \left(\|f - \overline{f}\|_{Q_K^n}^2 + k_n^{-1} \|[U]^{n-1}\|_K^2 \right)^{1/2},$$
$$\omega_K^{k,n} := \left(\|z - I_h^k z\|_{K \times I_n} + k_n \|(z - I_h^k z)^{(n-1)+}\|_K^2 \right)^{1/2}.$$

In the error estimator (3.65) the effect of the space discretization is separated from that of the time discretization. On each space-time cell Q_K^n the indicator $\eta_{K,h}^n := \rho_{K,h}^n \omega_{K,h}^n$ can be used for controlling the spatial mesh width h_K and the indicator $\eta_{K,k}^n := \rho_{K,k}^n \omega_{K,k}^n$ for the time step k_n, i.e., spatial mesh size and time step can be adapted independently. The weights $\omega_{K,h}^n$ and $\omega_{K,k}^n$ are evaluated in the same way as described for the stationary case by post-processing a computed approximation $z_h \in V_h^k$ to the dual solution z. An analogous a posteriori error estimator can also be derived for higher-order time stepping schemes, such as the dG(1) scheme and the cG(1) scheme, the latter being closely related to the popular Crank–Nicolson scheme. For more details, we refer to Hartmann [40], Bangerth/Rannacher [2], and the literature cited therein. The first complete a posteriori error analysis of dG methods for parabolic problems has been given in a sequence of papers Eriksson et al. [25], and Eriksson/Johnson [26, 27, 28].

Numerical test. The performance of mesh adaptation based on the error identity (3.64) is illustrated by a simple test in two space dimensions where the constructed exact solution represents a smooth rotating bump on the unit square. Figure 9 shows a sequence of adapted meshes at successive times obtained by controlling the spatial L^2-norm error at the end time $t_N = 0.5$ (see Hartmann [40]). We clearly see the effect of the weights in the error estimator which suppress the influence of the residuals during the initial period. Accordingly, the time step is kept coarse at the beginning and is successively refined when approaching the end time t_N.

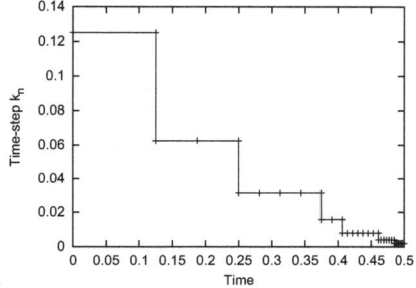 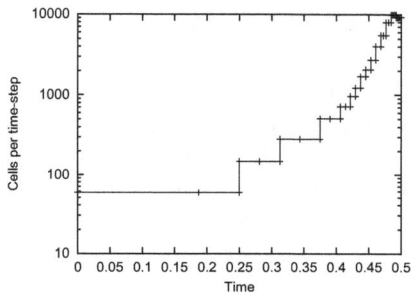

FIGURE 8. *Development of the time-step size (left) and the number N_n of mesh cells (right) over the time interval $I = [0, 0.5]$.*

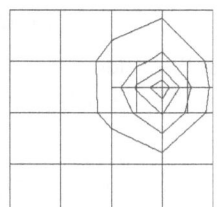 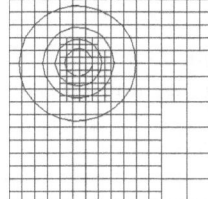 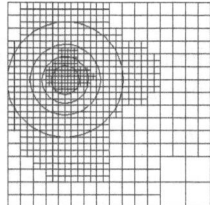 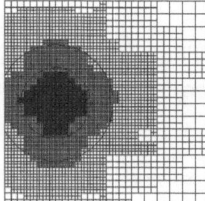

FIGURE 9. *Sequence of refined meshes for controlling the end-time error $\|e^{N-}\|$ shown at four consecutive times levels $t_n = 0.125000, \ldots, 0.5$.*

3.5. Application to flow models

We consider the *stationary* Navier–Stokes system for pairs $u := \{v, p\}$,

$$\mathcal{A}(u) := \left\{ \begin{matrix} -\nu \Delta v + v \cdot \nabla v + \nabla p - f \\ \nabla \cdot v \end{matrix} \right\} = 0, \tag{3.66}$$

with the usual boundary conditions

$$v|_{\Gamma_{\text{rigid}}} = 0, \quad v|_{\Gamma_{\text{in}}} = v^{\text{in}}, \quad \nu \partial_n v - n p|_{\Gamma_{\text{out}}} = 0.$$

where Γ_{in}, Γ_{out}, and Γ_{rgid} are the 'inflow', the 'outflow' and the 'rigid' part of the boundary. For this problem, we will consider the full cycle of numerical simulation (see Becker et al. [9]):

- computation of a target quantity $J(u)$ from the solution of
$$\mathcal{A}(u) = 0,$$

- minimization of $J(u)$ w.r.t. some control q quantity under the equation constraint
$$\mathcal{A}(u) + Bq = 0,$$

- investigation of the stability of the optimum state $\hat{u}$ by solving the stability eigenvalue problem
$$\mathcal{A}'(\hat{u})u = \lambda \mathcal{M}u.$$

The use of adaptive finite-element methods for all three problems can be treated within the same general framework laid out above.

The finite-element approximation of the Navier–Stokes system is based on its variational formulation. To this end, we recall some of the notation from the preeeding sections, i.e., the function spaces

$$L := L^2(\Omega), \quad \mathbf{H} := \{v \in H^1(\Omega)^d : v_{|\Gamma_{\text{in}} \cup \Gamma_{\text{rigid}}} = 0\}, \quad V := \mathbf{H} \times L,$$

and the semi-linear form ('energy form')

$$a(u)(\varphi) := (\nabla v, \nabla \varphi^v) + (v \cdot \nabla v - f, \varphi^v) - (p, \nabla \cdot \varphi^v) + (\varphi^p, \nabla \cdot v),$$

for arguments $u = \{v, p\}, \varphi = \{\varphi^v, \varphi^p\}$. The variational Navier–Stokes problem then seeks $u \in (v^{\text{in}}, 0) + V$, such that

$$a(u)(\varphi) = 0 \quad \forall \varphi \in V. \tag{3.67}$$

The Galerkin finite-element discretization uses the Q_1/Q_1-Stokes element, i.e., equal-order d-linear approximation of velocity and pressure (see Figure 10),

$$L_h \subset L, \quad \mathbf{H}_h \subset H, \quad V_h := \mathbf{H}_h \times L_h,$$

defined on quadrilateral or hexahedral meshes. As mentioned above, the Q_1/Q_1-Stokes element does not satisfy the usual 'inf-sup' stability condition. Following Hughes et al. [50, 51], this discretization is supplemented by 'least-squares' stabilization of pressure-velocity coupling and advection. The stabilized discrete problems seek $u_h \in (v_h^{\text{in}}, 0) + V_h$, such that

$$a_h(u_h)(\varphi_h) := a(u_h)(\varphi_h) + (\mathcal{A}(u_h), \mathcal{S}(u_h)\varphi_h)_\delta = 0 \quad \forall \varphi_h \in V_h, \tag{3.68}$$

where

$$\mathcal{S}(u)\varphi := \left\{ \begin{matrix} v \cdot \nabla \varphi^v + \nabla \varphi^p \\ \nabla \cdot \varphi^v \end{matrix} \right\}, \quad (\varphi, \psi)_\delta := \sum_{K \in \mathbb{T}_h} \delta_K (\varphi, \psi)_K.$$

The stabilization parameter δ_K is chosen adaptively as described above by

$$\delta_K = \alpha \left(\frac{\nu}{h_K^2} + \frac{\beta |v_h|_{K;\infty}}{h_K} \right)^{-1}, \tag{3.69}$$

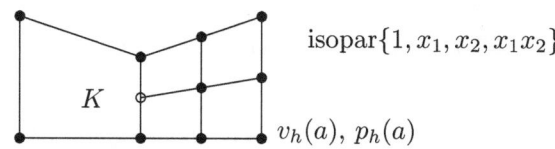

FIGURE 10. *Quadrilateral mesh patch with a 'hanging' node.*

with values $\alpha \approx \frac{1}{12}$ and $\beta \approx \frac{1}{6}$.

3.5.1. A posteriori error analysis. Let $J(\cdot)$ be a prescribed (linear) 'error functional' and $z = (z^v, z^p) \in V$ the associated dual solution determined by

$$\nu(\nabla\psi, \nabla z^v) - (\psi, v \cdot \nabla z^v) + (\psi, n \cdot v z^v)_{\Gamma_{\text{out}}} + (\psi, \nabla v\, v) = J(\psi), \qquad (3.70)$$

for all $\psi = (\psi^v, \psi^p) \in V$. The corresponding 'out-flow' boundary condition is of Robin-type, $\{\nu\partial_n z^v + n \cdot \hat{v} z^v - z^p n\}_{|\Gamma_{\text{out}}} = 0$. From the general Theorem 3.3, we can infer the following a posteriori error representation (Becker/Rannacher [14]):

$$J(u - u_h) = \tfrac{1}{2}\rho(u_h)(z - i_h z) + \tfrac{1}{2}\rho^*(u_h, z_h)(u - i_h u) + \mathcal{R}_h, \qquad (3.71)$$

where in this case the cubic remainder $\mathcal{R}_h$ can be bounded as follows:

$$|\mathcal{R}_h| \leq \|e^v\|\, \|\nabla e^v\|\, \|e^{v*}\|_\infty + \mathcal{O}(\delta\|e^v\|),$$

with the errors $e^v := v - v_h$, $e^{v*} := z^v - z^v_h$. Here, the primal residual is given by

$$\rho(u_h)(z - i_h z) := \sum_{K \in \mathbb{T}_h} \big\{ (R_h, z^v - i_h z^v)_K + (r_h, z^v - i_h z^v)_{\partial K}$$
$$+ (z^p - i_h z^p, \nabla \cdot v_h)_K + \ldots \big\},$$

with the cell and edge residuals ($[\ldots]$ denoting jumps across the cell edges or faces)

$$R_{h|K} := f - \nu\Delta v_h + v_h \cdot \nabla v_h + \nabla p_h,$$

$$r_{h|\Gamma} := \begin{cases} -\frac{1}{2}[\nu\partial_n v_h - n p_h], & \text{if } \Gamma \not\subset \partial\Omega, \\ 0, & \text{if } \Gamma \subset \Gamma_{\text{rigid}} \cup \Gamma_{\text{in}}, \\ -\nu\partial_n v_h + n p_h, & \text{if } \Gamma \subset \Gamma_{\text{out}}. \end{cases}$$

The corresponding dual residual has the form

$$\rho^*(u_h, z_h)(u - i_h u) := \sum_{K \in \mathbb{T}_h} \big\{ (R^*_h, v - i_h v)_K + (r^*_h, v - i_h v)_{\partial K}$$
$$+ (p - i_h p, \nabla \cdot z_h)_K + \ldots \big\},$$

with cell and edge residuals

$$R^*_{h|K} := j - \nu\Delta z^v_h - v_h \cdot \nabla z^v_h + \nabla v^T_h z^v_h - \nabla \cdot v_h z^v_h + \nabla z^p_h,$$

$$r^*_{h|\Gamma} := \begin{cases} -\frac{1}{2}[\nu\partial_n z^v_h + n \cdot v_h z^v_h - z^p_h n], & \text{if } \Gamma \not\subset \partial\Omega, \\ 0, & \text{if } \Gamma \subset \Gamma_{\text{rigid}} \cup \Gamma_{\text{in}}, \\ -\nu\partial_n z^v_h - n \cdot v_h z^v_h + z^p_h n, & \text{if } \Gamma \subset \Gamma_{\text{out}}. \end{cases}$$

From (3.71), we obtain the practical error estimator

$$\tilde{\eta}_\omega(u_h) := \tfrac{1}{2}\rho(u_h)(\tilde{z}_h - z_h) + \tfrac{1}{2}\rho^*(u_h, z_h)(\tilde{u}_h - u_h), \tag{3.72}$$

where $\tilde{u}_h$ and $\tilde{z}_h$ are approximations to u and z, respectively, obtained by post-processing the Galerkin solutions u_h and z_h, as described above.

A first example: 2D flow around a circular cylinder. We consider the laminar flow around the cross section of a cylinder in a 2D channel (with slightly displaced vertical position) as shown in Figure 11. This is a standard benchmark problem for which reference solutions are available (Schäfer/Turek [69]).

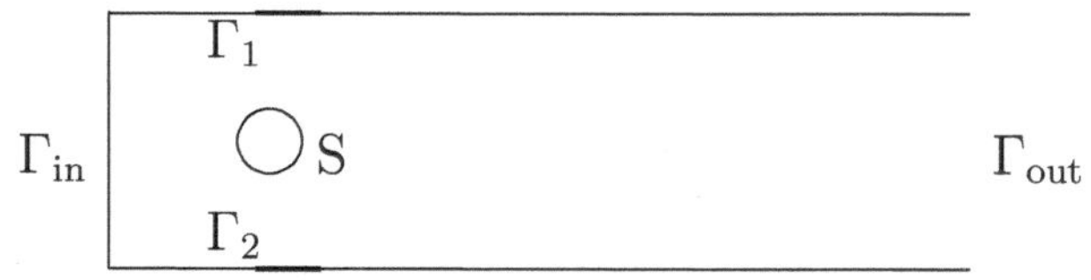

FIGURE 11. *Configuration of the benchmark problem 'viscous flow around a circular cylinder' with outlets $\Gamma_{1,2}$ for boundary control by pressure variation.*

One of the quantities of physical interest is the drag coefficient defined by

$$J_{\text{drag}}(u) = c_{\text{drag}} := \frac{2}{\bar{U}^2 D} \int_S n^T \sigma(v, p) e_1 \, ds,$$

where S is the surface of the cylinder, D its diameter, $\bar{U}$ the reference inflow velocity, $\sigma(v, p) = \frac{1}{2}\nu(\nabla v + \nabla v^T) + pI$ the stress force acting on S, and e_1 the unit vector in the main flow direction. In our example, the Reynolds number is $Re = \bar{U}^2 D/\nu = 20$, such that the flow is stationary. For evaluating the drag coefficient, one usually uses an equivalent volume formula,

$$J_{\text{drag}}(u) = \frac{2}{\bar{U}^2 D} \int_\Omega \{\sigma(v, p)\nabla\bar{e}_1 + \nabla \cdot \sigma(v, p)\bar{e}_1\} \, dx,$$

where $\bar{e}_1$ is an extension of e_1 to the interior of Ω with support along S. Notice that on the discrete level the two formulas differ. Theory and computation show that the volume formula yields more accurate and robust approximations of the drag coefficient (see Becker [3]).

Table 2 shows the results of the drag computation, where the effectivity index is again defined by $I_{\text{eff}} := \tilde{\eta}_\omega(u_h)/|J(e)|$ (see Becker [3]). Figure 12 shows refined meshes generated by the 'weighted' error estimator $\tilde{\eta}_{\text{weight}}(u_h)$ and by a (heuristic) 'energy norm' error indicator using only the primal cell residuals,

$$\eta^{\text{res}}(u_h) := \Big(\sum_{K \in \mathbb{T}_h} \big\{ h_K^2 \|R_h\|_K^2 + h_K \|r_h\|_{\partial K}^2 + h_K^2 \|\nabla \cdot v_h\|_K^2 \big\} \Big)^{1/2}.$$

A posteriori error estimates based on this type of residual error indicators have been derived by Oden et al. [59], Verfürth [76], and Bernardi et al. [17]. In Johnson et al. [55] duality arguments are used to obtain long-term error bounds for the non-stationary Navier–Stokes equations.

TABLE 2. *Results for drag and lift on adaptively refined meshes, error level of 1% indicated by bold face.*

Computation of drag				
L	N	c_{drag}	$\tilde{\eta}_{\text{drag}}$	I_{eff}
4	984	5.66058	$1.1e{-}1$	0.76
5	2244	5.59431	$3.1e{-}2$	0.47
6	**4368**	**5.58980**	$1.8e{-}2$	0.58
6	7680	5.58507	$8.0e{-}3$	0.69
7	9444	5.58309	$6.3e{-}3$	0.55
8	22548	5.58151	$2.5e{-}3$	0.77
9	41952	5.58051	$1.2e{-}3$	0.76
	∞	$5.579535\ldots$		

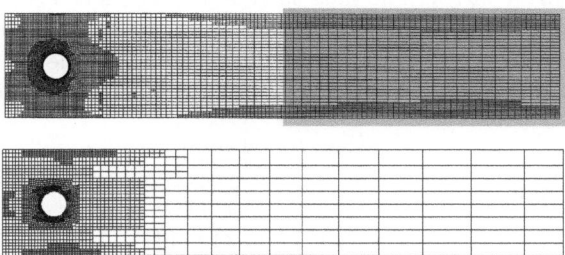

FIGURE 12. *Refined meshes generated by the 'residual error' estimator (top) and by the 'weighted' error estimator (bottom).*

A second example: 3D flow around a square cylinder. Next, we consider a 3D version of the above example, namely stationary channel flow around a cylinder with square cross-section. Again the target quantity of the computation is the drag coefficient c_{drag}. Table 3 shows the corresponding results compared to those obtained by mesh adaptation based on the residual error indicators η_K^{res} (see

Braack/Richter [20]). The superiority of goal-oriented mesh adaptation is clearly seen. However, one should not forget that these specially tuned meshes are not necessarily also appropriate for computing other flow quantities such as the global vortex structure of the flow or the pressure along the shear forces along the walls. For computing these quantities the mesh adaptation has to utilize the corresponding dual solutions.

TABLE 3. *Results of the drag computation:* (left) *with mesh adaptation by 'residual' error indicator,* (right) *with mesh adaptation by 'weighted' error indicator.*

N_{res}	c_d	N_{weight}	c_d
$3,696$	12.7888	$3,696$	12.7888
$21,512$	8.7117	$8,456$	9.8262
$80,864$	7.9505	$15,768$	8.1147
$182,352$	7.9142	$30,224$	8.1848
$473,000$	7.8635	$84,832$	7.8282
$-$	$-$	$162,680$	7.7788
$-$	$-$	$367,040$	7.7784
$-$	$-$	$700,904$	7.7769
∞	7.7730	∞	7.7730

3.6. Application in optimal flow control

Next, we present some results obtained for the minimization of the drag coefficients by boundary control. The data is chosen such that $\text{Re} = \overline{\text{U}}^2\text{D}/\nu = 40$ for the uncontrolled flow. The drag coefficient c_d is to be minimized by optimally adjusting the pressure prescription q at the secondary outlets $\Gamma_Q = \Gamma_1 \cup \Gamma_2$ (see Figure 11). This means that a state $u \in (v^{\text{in}}, 0) + V$ is sought, such that

$$J_{\text{drag}}(u) \rightarrow \min,$$

under the constraint

$$a(u, q)(\varphi) := a(u)(\varphi) + b(q, \varphi) = (f, \varphi^v) \quad \forall \varphi \in V, \tag{3.73}$$

where the control form is given by

$$b(q, \varphi) := -(q, n \cdot \varphi^v)_{\Gamma_Q}.$$

In Table 4, the values of the drag coefficient on optimized meshes, as shown in Figure 14, are compared with results obtained on globally refined meshes (see Becker [4, 5]). It is clear from these numbers that a significant reduction in the dimension of the discrete model is possible by using appropriately refined meshes. Figure 14 shows streamline plots of the uncontrolled ($q = 0$) and the controlled ($q = q^{\text{opt}}$) solution and a corresponding 'optimal' mesh.

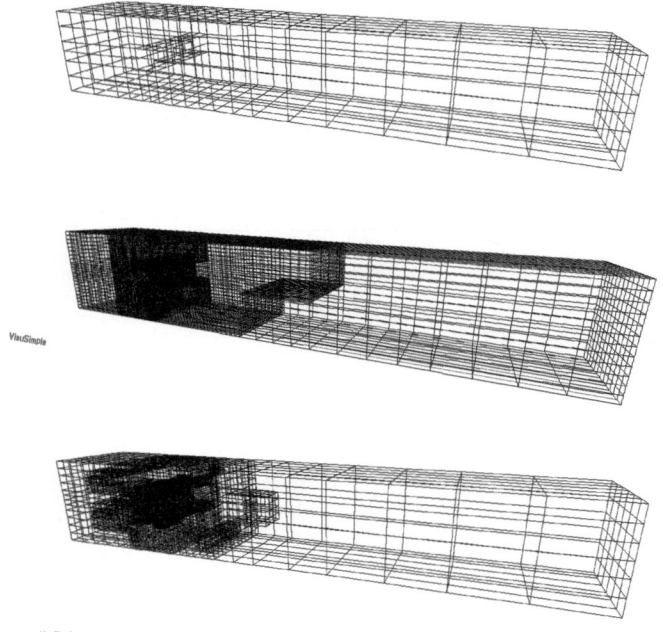

FIGURE 13. *Geometry-adapted coarse mesh (top) and refined meshes obtained by the 'energy-norm' (middle) and the 'weighted' error estimator (bottom).*

The locally refined mesh produced by the adaptive algorithm seems to contradict intuition since the recirculation behind the cylinder is not so well resolved. However, due to the particular structure of the optimal velocity field (most of the flow leaves the domain at the control boundary), it might be clear that this recirculation does not significantly influence the cost functional. Instead, a strong local refinement is produced near the cylinder, where the cost functional is evaluated, as well as near the control boundary. It remains the question whether the generated stationary 'optimal' flow is dynamically stable, i.e., can actually be realized in practice.

3.7. Application in hydrodynamic stability analysis

For investigating the stability of the stationary optimal state $\hat{u} = \{\hat{v}, \hat{p}\}$ obtained above by the linear stability theory, we have to solve the following non-symmetric eigenvalue problem for $u := \{v, p\} \in V$ and $\lambda \in \mathbb{C}$:

$$-\nu\Delta v + \hat{v} \cdot \nabla v + v \cdot \nabla \hat{v} + \nabla p = \lambda v, \quad \nabla \cdot v = 0, \tag{3.74}$$

TABLE 4. *Uniform refinement (left) versus adaptive refinement (right) in the drag minimization.*

Uniform refinement		Adaptive refinement	
N	$J_{\mathrm{drag}}^{\min}$	N	$J_{\mathrm{drag}}^{\min}$
10512	3.31321	1572	3.28625
41504	3.21096	4264	3.16723
164928	3.11800	11146	3.11972

FIGURE 14. *Velocity of the uncontrolled flow (top), the controlled flow (middle) and the corresponding adapted mesh (bottom).*

with homogeneous boundary conditions

$$v_{|\Gamma_{\mathrm{in}}\cup\Gamma_{\mathrm{rigid}}} = 0, \quad (\nu\partial_n v - pn)_{|\Gamma_{\mathrm{out}}\cup\Gamma_Q} = 0.$$

Its variational form reads

$$a'(\hat{u})(u,\varphi) = \lambda m(u,\varphi) \quad \forall\varphi \in V, \tag{3.75}$$

where $m(u,\varphi) := (v,\varphi^v)$. The associated 'adjoint' eigenvalue problem determines $v^* \in V$ and $\lambda^* = \bar{\lambda} \in \mathbb{C}$, such that

$$\nu(\nabla\psi, \nabla v^*) - (\psi, \hat{v}\cdot\nabla v^*) + (\psi, n\cdot\hat{v}v^*)_{\Gamma_{\mathrm{out}}} + (\psi, \nabla\hat{v}\,v) = \lambda^*(\psi, v^*), \tag{3.76}$$

for all $\psi \in V$. The dual eigenpair has to satisfy Robin-type outflow boundary conditions, $(\nu\partial_n v^* + n\cdot\hat{v}v^* - p^*n)_{|\Gamma_{\mathrm{out}}\cup\Gamma_Q} = 0$. From the general Theorem 3.6, we obtain the eigenvalue error estimator

$$|\lambda^{\mathrm{crit}} - \lambda_h^{\mathrm{crit}}| \leq \sum_{K \in \mathbb{T}_h} \{\hat{\eta}_K + \eta_K^\lambda\} + \mathcal{R}_h. \qquad (3.77)$$

where the cell error indicators $\hat{\eta}_K = \hat{\rho}_K(\hat{u}_h)\hat{\omega}_K$ and $\eta_K^\lambda = \rho_K(u_h)\omega_K$ represent the errors due to the approximation of the optimal base flow $\hat{u} = \{\hat{v}, \hat{p}\}$ and the approximation of the corresponding eigenpair $\{u, \lambda\}$, respectively. For instance, the primal eigenvalue error indicators are obtained from the residual term

$$\rho_K(u_h)(u^* - \psi_h) := \sum_{K \in \mathbb{T}_h} \{(R_h, \hat{v}^* - \psi_h)_K + (r_h, \hat{v}^* - \psi_h)_{\partial K}$$
$$+ (\hat{p}^* - \chi_h, \nabla \cdot \hat{v}_h)_K + \dots\},$$

with the cell and edge residuals defined by

$$R_{h|K} := \lambda v_h + \nu \Delta v_h - v_h \cdot \nabla v_h - \nabla p_h,$$

$$r_{h|\Gamma} := \begin{cases} -\frac{1}{2}[\nu \partial_n v_h - p_h n], & \text{if } \Gamma \not\subset \partial\Omega, \\ 0, & \text{if } \Gamma \subset \Gamma_{\mathrm{rigid}} \cup \Gamma_{\mathrm{in}}, \\ \nu \partial v_h - p_h n, & \text{if } \Gamma \subset \Gamma_{\mathrm{out}}. \end{cases}$$

This leads us to the following criterion for balancing linearization and discretization error:

$$\sum_{K \in \mathbb{T}_h} \hat{\eta}_K \leq \sum_{K \in \mathbb{T}_h} \eta_K^\lambda. \qquad (3.78)$$

This criterion together with the fixed-fraction strategy described above has been used in the computation of the critical eigenvalues of the optimum state $\hat{u}$. Figure 15 shows adapted meshes for computing the optimum stationary state and the corresponding critical eigenvalue λ^{crit}. We see that the eigenvalue computation requires more global mesh refinement. This is reflected by the results shown in Figure 16. On coarser meshes, such as used in the optimization process, the error due to the linearization about the wrong base solution $\hat{u}_h$ dominates the error due to the discretization of the eigenvalue problem, but on more globally refined meshes the picture changes and the linearization error falls below the discretization error. Still the meshes obtained by this adaptation process are more economical than simply using the meshes generated by the plain residual-based error estimator.

3.8. Calibration of flow models

In the following, we present a general approach to the numerical solution of (discrete) parameter identification problems based on the least-squares method. This can be used for parameter calibration in flow models such as in the closure problem discussed in Section 2.4. For more details, we refer to Vexler [75] and Becker/Vexler [15], see also Becker et al. [10], and the literature cited therein.

Let V denote the state space, $Q = \mathbb{R}^{n_p}$ the control space, $Z = \mathbb{R}^{n_m}$ the observation space with observation operator $C : V \to Z$, and observation $\bar{c} \in Z$.

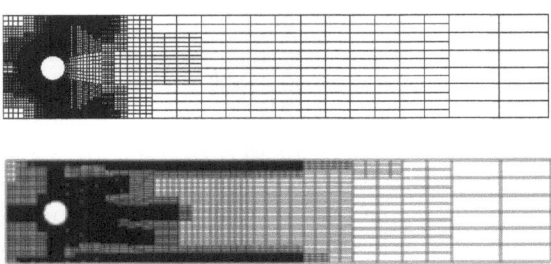

FIGURE 15. *Meshes obtained by the error estimators for the drag minimization (top) and the eigenvalue computation (bottom).*

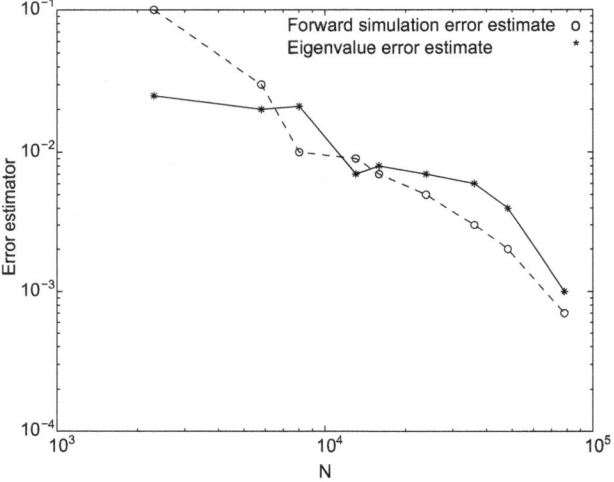

FIGURE 16. *The size of the two components of the error estimator* $\eta_h^\lambda(\hat{u}_h, \hat{u}_h^*, u_h, u_h^*, \lambda_h)$, *i.e., the errors in the base solution and the eigenvalue approximation.*

Then, the corresponding optimization problem seeks $u \in V$ and $q \in Q = \mathbb{R}^{n_p}$, such that

$$J(u) := \tfrac{1}{2}\|C(u) - \bar{c}\|_Z^2 \to \min, \qquad a(u,q)(\varphi) = 0 \quad \forall \varphi \in V, \qquad (3.79)$$

with an energy-semi-linear from $a(\cdot,\cdot)(\cdot)$ as described above. The corresponding necessary first-order optimality condition (analogous to (3.17-3.19) reads

$$a_u'(u,q)(\varphi,\lambda) = J_u'(u,q)(\varphi) \quad \forall \varphi \in V, \qquad (3.80)$$

$$a_q'(u,q)(\chi,\lambda) = J_q'(u,q)(\chi) \quad \forall \chi \in Q, \qquad (3.81)$$

$$a(u,q)(\psi) = 0 \quad \forall \psi \in V. \qquad (3.82)$$

The discretization of (3.79) uses finite-element spaces $V_h \subset V$ and determines $\{u_h, q_h\} \in V_h \times Q$, such that

$$J(u_h) = \tfrac{1}{2}\|C(u_h) - \bar{c}\|_Z^2 \to \min, \quad a(u_h, q_h)(\varphi_h) = 0 \quad \forall \varphi_h \in V_h. \tag{3.83}$$

If the form $a(\cdot, q)(\cdot)$ is regular for any $q \in Q$, we can define the solution operator $S : Q \to V$ and, setting $u = S(q)$, obtain the following unconstrained equivalent of (3.79) posed in the finite-dimensional space $Q = \mathbb{R}^{n_p}$:

$$j(q) := \tfrac{1}{2}\|c(q) - \bar{c}\|_Z^2 \to \min, \quad c(q) := C(S(q)), \ q \in Q.$$

The derivatives

$$G_{ij} := \partial_{q_j} c_i(q) = C_i'(u)(w_j), \quad G = (G_{ij})_{i,j=1}^{n_p},$$

are determined by the solutions $w_j \in V$ of the auxiliary equations

$$a_u'(u, q)(w_j, \varphi) = -a_{q_j}'(u, q)(1, \varphi) \quad \forall \varphi \in V.$$

Using this notation the necessary first-order optimality condition is

$$j'(q) = 0 \quad \Leftrightarrow \quad G^*(c(q) - \bar{c}) = 0.$$

This equation may be solved by a fixed-point iteration of the form

$$q_{k+1} = q_k + \delta q, \quad H_k \delta q = G_k^*(\bar{c} - c(q_k)), \quad G_k = c'(q_k),$$

with a suitably chosen preconditioning matrix H_k. Popular choices are:

- The full 'Newton algorithm', $H_k := G_k^* G_k + \langle c(q_k) - \bar{c}, c''(q_k) \rangle_Z$.
- The 'Gauß-Newton algorithm', $H_k := G_k^* G_k$.
- The 'update method', $H_k := G_k^* G_k + M_k$ with certain corrections M_k.

However, the full Newton algorithm is rarely used in this context since it involves the evaluation of the term $\langle c(q_h^k) - \bar{c}, c''(q_h^k) \rangle_Z$, particularly the second derivative $c_h''(q_h^k)$. This is rather expensive since it requires the solution of several auxiliary problems, depending on the dimension of Q. Since in the limit $k \to \infty$ the deviation $c_h(q_h) - \bar{c}$ is expected to be small, the Gauß–Newton algorithm is justified and its convergence is sometimes even super-linear.

3.8.1. A posteriori error estimation.
Controlling the error in the discretization of a parameter identification problem based on the control functional $J(\cdot)$ may be inappropriate for guiding mesh adaptation. Hence, in this situation, one follows another approach by choosing an error control functional $E(\cdot)$ which addresses the error in the controls more directly, e.g.,

$$E(u_h, q_h) := \tfrac{1}{2}\|q - q_h\|_Q^2.$$

Then, the systematic error control by the general approach described above applied to the Galerkin approximation of the saddle-point system (3.80–3.82) has to use an extra 'outer' dual problem with solution $z = \{z^u, z^q, z^\lambda\} \in V \times Q \times V$,

$$A'(u, q, \lambda)(\varphi, \chi, \psi, z^u, z^q, z^\lambda) = E'(u, q, \lambda)(\varphi, \chi, \psi), \tag{3.84}$$

for all $\{\varphi, \chi, \psi\} \in V \times Q \times V$, where

$$A(u, q, \lambda)(\varphi, \chi, \psi) := a'_u(u, q)(\varphi, \lambda) - J'_u(u, q)(\varphi)$$
$$+ a'_q(u, q)(\chi, \lambda) - J'_q(u, q)(\chi) + a(u, q)(\psi).$$

A careful analysis of this setting results in an a posteriori error representation of the following form (Vexler [75] and Becker/Vexler [15]):

$$E(u, q) - E(u_h, q_h) = \eta_h + \mathcal{R}_h + \mathcal{P}_h. \tag{3.85}$$

Here, the main part η_h of the estimator has the usual form

$$\eta_h = \tfrac{1}{2}\rho(u_h, q_h)(z - i_h z) + \tfrac{1}{2}\rho^*(u_h, q_h, z_h)(u - i_h u),$$

with the 'dual solution' $z \in V$ determined by the dual problem

$$a'_u(u, q)(\varphi, z) = -\langle G(G^*G)^{-1}\nabla E(q), C'(u)(\varphi)\rangle_Z \quad \forall \varphi \in V,$$

and the residuals

$$\rho(u_h, q_h)(\psi) := -a(u_h, q_h)(\psi),$$
$$\rho^*(u_h, q_h, z_h)(\varphi) := \langle G_h(G_h^*G_h)^{-1}\nabla E(q_h), C'(u_h)(\varphi)\rangle - a'_u(u_h, q_h)(\varphi, z_h).$$

The remainder $\mathcal{R}_h$ due to linearization is again cubic in the errors $u - u_h$, $q - q_h$, and $z - z_h$, and the additional error term $\mathcal{P}_h$ is bounded like

$$|\mathcal{P}_h| \leq \tilde{C}\, \|e\|_V \, \|C(u) - \bar{c}\|_Z. \tag{3.86}$$

Due to the particular features of the parameter identification problem, we can expect that $\|C(u) - \bar{c}\|_Z \ll 1$ for the optimal state. Hence, this term is neglected compared to the leading term η. Based on the a posteriori error bound (3.85) the mesh adaptation is organized as described above.

3.8.2. Application to the closure problem. The method described above is applied to the closure problem discussed in Section 2.4. The underlying model is

$$-\nu\Delta v + v \cdot \nabla v + \nabla p = 0, \quad \nabla \cdot v = 0 \quad \text{in } \Omega,$$
$$v = 0 \text{ on } \Gamma_{\text{rigid}}, \quad \nu\partial_n v - p \cdot n = 0 \text{ on } \Gamma_{\text{out}}, \tag{3.87}$$
$$\nu\partial_n v - p \cdot n = q_1 \cdot n \text{ on } \Gamma_{\text{in}}^{(1)}, \quad \nu\partial_n v - p \cdot n = q_2 \cdot n \text{ on } \Gamma_{\text{in}}^{(2)}.$$

Here, the unknown pressure mean values q_i at the openings $\Gamma_{\text{in}}^{(i)}$ are to be determined by *parameter estimation* from given measurements, for example, of the mean fluxes

$$C_j(u) := \int_{S_j} v \cdot n \, do,$$

along certain interior cross sections of the flow domain or other measurable local flow quantities as depicted in Figure 17.

Using the a posteriori error representation (3.85) in the parameter identification process described in the preceding section, adapted meshes are generated as shown in Figure 18. The resulting solution efficiency obtained on these meshes compared to uniformly or accordingly to the 'energy-norm' error estimator refined

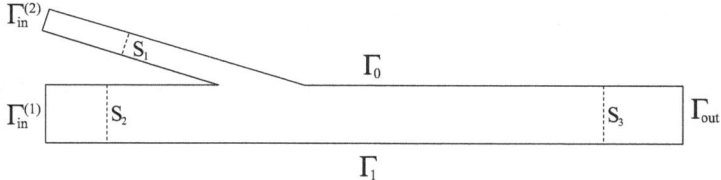

FIGURE 17. Configuration of the bypass problem for model calibration.

meshes are presented in Figure 19. The superiority of the sensitivity-driven mesh adaptation is clearly seen.

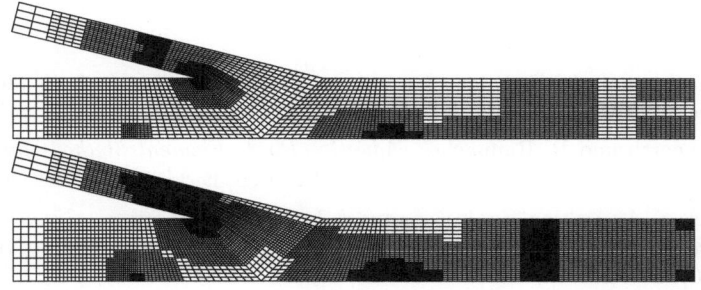

FIGURE 18. Adapted meshes in the model calibration process.

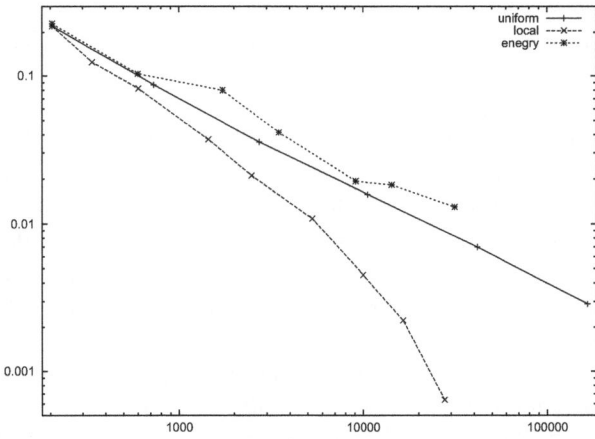

FIGURE 19. Efficiency of mesh adaptation in the model calibration process.

3.9. Current work and further development

Systematic mesh adaptation in optimal flow control by the DWR approach has great potential for significantly reducing the computational work. Current and future developments are on the following topics:

- Combination of spatial mesh size and time-step adaptation for solving the space-time KKT system in non-stationary flow control.
- Incorporation of control and state constraints, such as lower bounds on the lift coefficient, $c_{\text{lift}} \geq c_0$, or the suppression of local recirculation, $v_1 \geq 0$.
- Optimization in fluid-structure interaction problems, as discussed in the article Bönisch et al. [24] in this volume.

References

[1] M. Ainsworth and J.T. Oden. *A posteriori error estimation in finite-element analysis.* Comput. Methods Appl. Mech. Enrg. **142** (1997), 1–88.

[2] W. Bangerth and R. Rannacher, *Adaptive Finite-Element Methods for Differential Equations.* Lectures in Mathematics, ETH Zürich, Birkhäuser, Basel 2003.

[3] R. Becker, *An optimal control approach to a posteriori error estimation for finite-element discretizations of the Navier–Stokes equations.* East-West J. Numer. Math. **8** (2000), 257–274.

[4] R. Becker, *Mesh adaptation for stationary flow control.* J. Math. Fluid Mech. **3** (2001), 317–341.

[5] R. Becker, *Adaptive Finite Elements for Optimal Control Problems.* Habilitation thesis, Institute of Applied Mathematics, Univ. of Heidelberg, 2001, `http://numerik.iwr.uni-heidelberg.de/`.

[6] R. Becker and M. Braack, *Multigrid techniques for finite elements on locally refined meshes.* Numer. Linear Algebra Appl. **7** (2000), 363–379.

[7] R. Becker and M. Braack, *A finite-element pressure gradient stabilization for the Stokes equations based on local projections.* Calcolo **38** (2001), 173–199.

[8] R. Becker and Th. Dunne, *VisuSimple: An interactive visualization utility for scientific computing.* Abschlußband SFB 359, Reactive Flows, Diffusion and Transport (W. Jäger et al., eds.), Springer, Berlin-Heidelberg New York, 2007.

[9] R. Becker, V. Heuveline, and R. Rannacher, *An optimal control approach to adaptivity in computational fluid mechanics.* Int. J. Numer. Meth. Fluids. **40** (2002), 105–120.

[10] R. Becker, D. Meidner, R. Rannacher, and B. Vexler, *Adaptive finite-element methods for PDE-constrained optimal control problems.* In 'Reactive Flows, Diffusion and Transport', Abschlußband SFB 359, University of Heidelberg, (W. Jäger et al., eds.), Springer, Heidelberg, 2007.

[11] R. Becker, H. Kapp, and R. Rannacher, *Adaptive finite-element methods for optimal control of partial differential equations: basic concepts.* SIAM J. Optimization Control **39** (2000), 113–132.

[12] R. Becker and R. Rannacher, *Weighted a posteriori error estimates in FE methods.* Lecture ENUMATH-95, Paris, Sept. 18-22, 1995. In Proc. 'ENUMATH-97' (H.G. Bock, et al., eds.), pp. 621-637, World Scientific Publ., Singapore, 1998.

[13] R. Becker and R. Rannacher, *A feed-back approach to error control in finite-element methods: Basic analysis and examples.* East-West J. Numer. Math. **4** (1996), 237–264.

[14] R. Becker and R. Rannacher, *An optimal control approach to error estimation and mesh adaptation in finite-element methods.* Acta Numerica 2000 (A. Iserles, ed.), pp. 1-102, Cambridge University Press, 2001.

[15] R. Becker and B. Vexler, *Mesh refinement and numerical sensitivity analysis for parameter calibration of partial differential equations.* J. Comput. Phys. **206** (2005), 95-110.

[16] R. Becker and B. Vexler. *Optimal Flow Control by Adaptive Finite-Element Methods.* Advances in Mathematical Fluid Mechanics, Birkhäuser, Basel-Boston-Berlin, in preparation.

[17] C. Bernardi, O. Bonnon, C. Langouët, and B. Métivet. *Residual error indicators for linear problems: Extension to the Navier–Stokes equations.* In Proc. 9th Int. Conf. 'Finite Elements in Fluids', 1995.

[18] S. Bönisch and V. Heuveline, *Advanced flow visualization with HiVision.* Abschlußband SFB 359, Reactive Flows, Diffusion and Transport (W. Jäger et al., eds.), Springer, Berlin-Heidelberg New York, 2007.

[19] M. Braack and E. Burman: *Local projection stabilization for the Oseen problem and its interpretation as a variational multiscale method,* SIAM J. Numer. Anal. **43** (2006), 2544–2566.

[20] M. Braack and T. Richter, *Solutions of 3D Navier-Stokes benchmark problems with adaptive finite elements.* Computers and Fluids **35** (2006), 372–392.

[21] S. Brenner and R.L. Scott, *The Mathematical Theory of the Finite-Element Method.* Springer, Berlin Heidelberg New York, 1994.

[22] S. Brezzi and R. Falk, *Stability of higher-order Hood-Taylor methods.* J. Numer. Anal. **28** (1991), 581–590.

[23] G. Carey and J. Oden, *Finite Elements, Computational Aspects.* Volume III. Prentice-Hall, 1984.

[24] Th. Dunne and R. Rannacher. *Numerics of Fluid-Structure Interaction.* In 'Fluid-Structure Interaction: Modelling, Simulation, Optimisation' (H.-J. Bungartz and M. Schäfer, eds.), in Springer's LNCSE-Series, 2006.

[25] K. Eriksson, C. Johnson, and V. Thomée, *Time discretization of parabolic problems by the discontinuous Galerkin method.* RAIRO Model. Math. Anal. Numer. **19** (1985), 611-643.

[26] K. Eriksson and C. Johnson. *Adaptive finite-element methods for parabolic problems, I: A linear model problem.* SIAM J. Numer. Anal. **28** (1991), 43–77.

[27] K. Eriksson and C. Johnson. *Adaptive finite-element methods for parabolic problems, IV: Nonlinear problems.* SIAM J. Numer. Anal. **32** (1995), 1729–1749.

[28] K. Eriksson and C. Johnson. *Adaptive finite-element methods for parabolic problems, V: Long-time integration.* SIAM J. Numer. Anal. **32** (1995), 1750–1763.

[29] K. Eriksson, D. Estep, P. Hansbo, and C. Johnson, *Introduction to adaptive methods for differential equations.* Acta Numerica 1995 (A. Iserles, ed.), pp. 105–158, Cambridge University Press, 1995.

[30] M. Fernandez, V. Milisic, and A. Quarteroni, *Analysis of a geometrical multiscale blood flow model based on the coupling of ODE's and hyperbolic PDE's.* SIAM J. MMS **4** (2005), 215-236.

[31] L. Formaggia, J.F. Gerbeau, F. Nobile, and A. Quarteroni, *On the coupling of 3D and 1D Navier–Stokes equations for flow problems in compliant vessels.* Comp. Methods Appl. Mech. Engnrg. **191** (2001), 561-582.

[32] G.P. Galdi, *An Introduction to the Mathematical Theory of the Navier–Stokes Equations.* Vol. 1: Linearized Steady problems, Vol. 2: Nonlinear Steady Problems, Springer: Berlin-Heidelberg-New York, 1998.

[33] G.P. Galdi, *An introduction to the Navier–Stokes initial-boundary value problem.* In this volume.

[34] R. Becker and M. Braack, *Gascoigne: A C++ numerics library for scientific computing.* Institute of Applied Mathematics, University of Heidelberg, URL `http://www.gascoigne.uni-hd.de/`, 2005.

[35] M.B. Giles and E. Süli, *Adjoint methods for PDEs: a posteriori error analysis and postprocessing by duality.* Acta Numerica 2002 (A. Iserles, ed.), pp. 145–236, Cambridge University Press, 2002.

[36] V. Girault and P.-A. Raviart, *Finite-Element Methods for the Navier–Stokes Equations.* Springer: Berlin-Heidelberg-New York, 1986.

[37] R. Glowinski, *Finite-element methods for incompressible viscous flow.* In 'Handbook of Numerical Analysis' (P.G. Ciarlet and J.-L. Lions, eds.), Volume IX: Numerical Methods for Fluids (Part 3), North-Holland: Amsterdam, 2003.

[38] P.M. Gresho and R.L. Sani, *Incompressible Flow and the Finite-Element Method.* John Wiley & Sons: Chichester, 1998.

[39] P. Hansbo and C. Johnson. *Adaptive streamline diffusion finite-element methods for compressible flow using conservative variables.* Comput. Methods Appl. Mech. Engrg **87** (1991), 267–280.

[40] R. Hartmann, *A posteriori Fehlerschätzung und adaptive Schrittweiten- und Ortsgittersteuerung bei Galerkin-Verfahren für die Wärmeleitungsgleichung.* Diploma thesis, Institute of Applied Mathematics, University of Heidelberg, 1998.

[41] R. Hartmann, *Adaptive Finite-Element Methods for the Compressible Euler Equations.* Dissertation, Institute of Applied Mathematics, Universität Heidelberg, 2002.

[42] R. Hartmann and P. Houston, *Adaptive discontinuous Galerkin finite-element methods for the compressible Euler equations.* J. Comput. Phys. **183** (2002), 508–532.

[43] R. Hartmann and P. Houston, *Symmetric interior penalty DG methods for the compressible Navier–Stokes equations I: method formulation.* Int. J. Numer. Anal. Model. **3** (2006), 1–20.

[44] R. Hartmann and P. Houston, *Symmetric interior penalty DG methods for the compressible Navier–Stokes equations II: goal-oriented a posteriori error estimation.* Int. J. Numer. Anal. Model. **3** (2006), 141–162.

[45] V. Heuveline, *HiFlow: A multi-purpose finite-element package*, Rechenzentrum, Universität Karlsruhe, URL `http://hiflow.de/` .

[46] V. Heuveline, *HiVision: A visualization platform*, Rechenzentrum, Universität Karlsruhe, URL http://hiflow.de/.

[47] V. Heuveline and R. Rannacher, *Adaptive FEM for eigenvalue problems with application in hydrodynamic stability analysis*. Proc. Int. Conf. 'Advances in Numerical Mathematics', Moscow, Sept. 16–17, 2005 (W. Fitzgibbon et al., eds.), pp. 109–140, Institute of Numerical Mathematics RAS, Moscow, 2006.

[48] J. Heywood, R. Rannacher, and S. Turek, *Artificial boundaries and flux and pressure conditions for the incompressible Navier–Stokes equations*. Int. J. Numer. Math. Fluids **22** (1992), 325–352.

[49] P. Houston, R. Rannacher, and E. Süli. *A posteriori error analysis for stabilized finite-element approximation of transport problems*. Comput. Methods Appl. Mech. Enrg. **190** (2000), 1483–1508.

[50] T.J.R. Hughes and A.N. Brooks, *Streamline upwind/Petrov–Galerkin formulations for convection dominated flows with particular emphasis on the incompressible Navier–Stokes equation*. Comput. Meth. Appl. Mech. Engrg. **32** (1982), 199–259.

[51] T.J.R. Hughes, L.P. Franc, and M. Balestra, *A new finite-element formulation for computational fluid mechanics: V. Circumventing the Babuska-Brezzi condition: A stable Petrov-Galerkin formulation of the Stokes problem accommodating equal order interpolation*. Comput. Meth. Appl. Mech. Engrg. **59** (1986), 85–99.

[52] V. John, *Higher order finite-element methods and multigrid solvers in a benchmark problem for the 3D Navier–Stokes equations*. Int. J. Numer. Meth. Fluids **40** (2002), 775–798.

[53] C. Johnson, *Numerical Solution of Partial Differential Equations by the Finite-Element Method*. Cambridge University Press, Cambridge, 1987.

[54] C. Johnson. *Adaptive finite-element methods for diffusion and convection problems*. Comput. Methods Appl. Mech. Eng. **82** (1990), 301–322.

[55] C. Johnson, R. Rannacher, and M. Boman. *Numerics and hydrodynamic stability: Towards error control in CFD*. SIAM J. Numer. Anal. **32** (1995), 1058–1079.

[56] S. Korotov, P. Neitaanmäki and S. Repin. *A posteriori error estimation of goal-oriented quantities by the superconvergence patch recovery*, J. Numer. Math. **11** (2003), 33-53.

[57] L. Machiels, A.T. Patera, and J. Peraire, *Output bound approximation for partial differential equations; application to the incompressible Navier-Stokes equations*. Industrial and Environmental Applications of Direct and Large Eddy Numerical Simulation (S. Biringen, ed.), Springer, Berlin Heidelberg New York, 1998.

[58] V. Milisic and A. Quarteroni, *Analysis of lumped parameter models for blood flow simulations and their relation with 1D models*. M2AN, Vol.IV (2004), 613–632.

[59] J.T. Oden, W. Wu, and M. Ainsworth, *An a posteriori error estimate for finite-element approximations of the Navier–Stokes equations*. Comput. Methods Appl. Mech. Eng. **111** (1993), 185–202.

[60] J.T. Oden and S. Prudhomme. *On goal-oriented error estimation for elliptic problems: Application to the control of pointwise errors*. Comput. Methods Appl. Mech. Eng. **176** (1999), 313–331.

[61] J.T. Oden and S. Prudhomme. *Estimation of modeling error in computational mechanics*. Preprint, TICAM, The University of Texas at Austin, 2002.

[62] M. Paraschivoiu and A.T. Patera. *Hierarchical duality approach to bounds for the outputs of partial differential equations.* Comput. Methods Appl. Mech. Enrg. **158** (1998), 389–407.

[63] A. Quarteroni, S. Ragni, and A. Veneziani, *Coupling between lumped and distributed models for blood flow problems.* Computing and Visualization in Science **4** (2001) 2, 111–124.

[64] A. Quarteroni and A. Veneziani, *Analysis of a geometrical multiscale model based on the coupling of PDE's and ODE's for blood flow simulations.* SIAM J. MMS. **1** (2003), 173–195,

[65] R. Rannacher, *Finite-element methods for the incompressible Navier–Stokes equations.* In 'Fundamental Directions in Mathematical Fluid Mechanics' (G.P. Galdi et al., eds.), pp. 191–293, Birkhäuser, Basel, 2000.

[66] R. Rannacher, *Incompressible viscous flow.* In 'Encyclopedia of Computational Mechanics' (E. Stein, et al., eds.), Volume 3 'Fluids', John Wiley, Chichester, 2004.

[67] R. Rannacher and S. Turek, *Simple nonconforming quadrilateral Stokes element.* Numer. Meth. Part. Diff. Equ. **8** (1992), 97–111.

[68] R. Becker, D. Meidner, and B. Vexler, *RoDoBo: A C++ library for optimization with stationary and nonstationary PDEs.* Institute of Applied Mathematics, University of Heidelberg, URL `http://www.rodobo.uni-hd.de/`, 2005.

[69] M. Schäfer and S. Turek. *Benchmark computations of laminar flow around a cylinder.* In Flow Simulation with High-Performance Computers II (E. H. Hirschel, ed.), pp. 547–566, DFG priority research program results 1993-1995, vol. 52 of Notes Numer. Fluid Mech., Vieweg, Wiesbaden, 1996.

[70] L. Tobiska and F. Schieweck, *A nonconforming finite-element method of up-stream type applied to the stationary Navier–Stokes equation.* M^2AN **23** (1989), 627–647.

[71] F. Tröltzsch, *Optimale Steuerung partieller Differentialgleichungen - Theorie, Verfahren und Anwendungen.* Vieweg, Braunschweig, 2005.

[72] S. Turek, *Efficient solvers for incompressible flow problems: an algorithmic and computational approach.* Springer, Heidelberg-Berlin-New York, 1999.

[73] S. Turek and M. Schäfer, *Benchmark computations of laminar flow around a cylinder.* In 'Flow Simulation with High-Performance Computers II' (E.H. Hirschel, ed.), Volume 52 of Notes on Numerical Fluid Mechanics, Vieweg, Braunschweig, 1996.

[74] S. Turek, *Efficient solvers for incompressible flow problems.* In this volume.

[75] B. Vexler, *Adaptive Finite-Element Methods for Parameter Identification Problems.* Dissertation, Institute of Applied Mathematics, University of Heidelberg, 2004.

[76] R. Verfürth, *A Review of A Posteriori Error Estimation and Adaptive Mesh-Refinement Techniques.* Wiley/Teubner, New York Stuttgart, 1996.

[77] R. Becker and Th. Dunne, *VisuSimple: An open source interactive visualization utility for scientific computing.* Institute of Applied Mathematics, University of Heidelberg, 2005.

Rolf Rannacher
Institute of Applied Mathematics, University of Heidelberg, Im Neuenheier Feld 293/294
D-69120 Heidelberg, Germany
e-mail: `rannacher@iwr.uni-heidelberg.de`

Hemodynamical Flows. Modeling, Analysis and Simulation
Oberwolfach Seminars, Vol. 37, 333–378
© 2008 Birkhäuser Verlag Basel/Switzerland

Numerics of Fluid-Structure Interaction

Sebastian Bönisch, Thomas Dunne and Rolf Rannacher

The authors acknowledge the support by the German Research Foundation (DFG) through the International Research Training Group "Complex Processes: Modeling, Simulation and Optimization" (IGK 710) and the Research Unit "Fluid-Structure Interaction: Modelling Simulation, Optimization" (FOR 493).

Introduction

This chapter describes numerical methods for simulating the interaction of viscous liquids with rigid or elastic bodies.

General examples of fluid-solid/structure interaction (FSI) problems are flow transporting rigid or elastic particles (particulate flow), flow around elastic structures (airplanes, submarines) and flow in elastic structures (hemodynamics, transport of fluids in closed containers). In all these settings the dilemma in modeling the coupled dynamics is that the fluid model is normally based on an Eulerian perspective in contrast to the usual Lagrangian formulation of the solid model. This makes the setup of a common variational description difficult. However, such a variational formulation of FSI is needed as the basis of a consistent Galerkin discretization with a posteriori error control and mesh adaptation, as well as the solution of optimal control problems based on the Euler–Lagrange approach.

In this article, we describe variational methods for simulating two prototypical types of fluid-solid/structure interaction phenomena: the movement of rigid bodies in a viscous fluid and the interaction of a viscous fluid with an elastic structure. Figure 1 shows the gravity-driven motion of a non-symmetric body in a Newtonian fluid as the solution of the Navier–Stokes equations. Figure 2 shows the instability-induced oscillation of a thin elastic plate in a viscous fluid, computed by the usual Arbitrary Lagrangian–Eulerian ('ALE') method.

The material presented in this article is mainly based on the doctoral theses of the first two authors, Bönisch [9] and Dunne [20]. For more details the reader may also consult the articles Bönisch/Heuveline [10, 11], Dunne/Rannacher [21], and Bönisch et al. [13]. Most of the numerical examples have been provided using the software tools 'Gascoigne' [4], 'Hiflow' [31], and the graphics packages 'VisuSimple'

[55, 5] and 'HiVision' [32, 12]. More references to the relevant literature will be given at the respective places in the text, below.

Contents

1. Fluid-'single rigid body' interaction

We present a finite element Galerkin scheme for the detailed simulation of the free fall of a single rigid body in a viscous fluid filling all space. By residual-controlled adaptation of the mesh and the size of the computational domain the method achieves a high degree of flexibility and accuracy. Numerical results are presented for two-dimensional configurations which allow for stable and unstable quasi-stationary as well as non-stationary solutions. For example, in a stationary situation in two dimensions more than 10^7 unknowns would be necessary with a uniform mesh on a truncated domain Ω_h with $\operatorname{diam}(\Omega_h) = 2500$ to achieve an error of less than 10% in the fall velocity. Employing local mesh adaptation the same accuracy can be achieved with only $1.5 \cdot 10^6$ unknowns. Combining this with

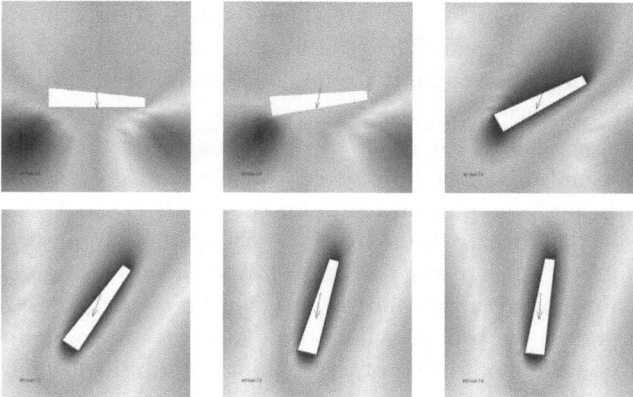

FIGURE 1. Fall of a non-symmetric body in a viscous liquid under gravity.

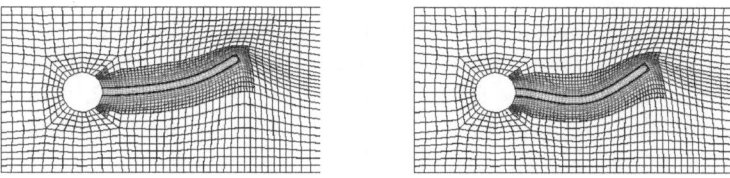

FIGURE 2.   Oscillating thin elastic plate in a viscous fluid, computed by the ALE approach.

more sophisticated artificial boundary conditions the domain can be reduced to $\mathrm{diam}(\Omega_h) = 100$ resulting in only $7.5 \cdot 10^5$ unknowns.

1.1. Model setup in the body frame

The free fall of a rigid body in a viscous fluid is one of the simplest examples of fluid-structure interaction. Particularly for rotationally symmetric bodies several theoretical results concerning the existence of quasi-stationary states and their stability are available; see Weinberger [58], Hu et al. [36, 37], Unverdi/Tryggvason [54], Desjardin/Esteban [19], Glowinski et al. [28], Burger et al. [17], and Conca et al. [18], Hoffmann/Starovoitov [33], Gunzburger et al. [29], Galdi [23, 24], and the references cited therein. For general non-symmetric and non-convex bodies these questions seem to be largely open. Here, numerical experiments may be able to guide theoretical analysis. Despite its simplicity the numerical solution of the underlying model equations poses severe difficulties. The conceptionally unbounded domain has to be truncated and appropriate artificial boundary conditions are needed. The dynamic behavior of the orientation of the body and the speed and direction of fall is determined by the quantities drag, lift, and torque, the accurate computation of which is rather delicate. The coupling of the relevant quantities in the model is highly nonlinear. The reliable computation of drag, lift, and torque requires a sufficiently large computational domain as well as local mesh refinement along the body's surface.

The numerical approach is based on the solution of the 'incompressible' Navier–Stokes equations in body-fixed coordinates and uses a finite-element discretization with mesh adaptation based on the DWR (Dual Weighted Residual) concept described in Becker/Rannacher [7]. By systematic mesh refinement the quantities of interest, such as the free fall velocity, the orientation of the body and the acting hydrodynamic force and torque, can be computed to any prescribed accuracy. This works for two as well as three-dimensional configurations. An additional feature is the monitoring of the stability of quasi-stationary solutions by computing critical eigenvalues. The performance of the current implementation of this method will be illustrated by the results of some simulations for two-dimensional configurations with stationary as well as non-stationary solutions. We investigate the existence of quasi-stationary states for symmetric bodies and the dynamic stability of these solutions. More details on the numerical methodology and the computational examples can be found in Bönisch et al. [13]. These 'experimental' observations may serve as stimulus for further theoretical analysis.

We consider the free fall of a solid body $\mathcal{S} \subset \mathbb{R}^d$ $(d = 2, 3)$ in an incompressible liquid $\mathcal{L}$ filling the whole space $\mathcal{D} := \mathbb{R}^d \setminus \mathcal{S}$. The solid body $\mathcal{S}$ is assumed to be a bounded domain and the velocity of its mass center C (resp. its angular velocity) are denoted by $\mathcal{V}_C$ (resp. Ω) in the inertial frame $\mathcal{F}$. The region occupied by $\mathcal{S}$ at time t is described by $S(t)$ and the corresponding attached frame is denoted by $\mathcal{R}(t)$. The fluid-body coupling occurs through Dirichlet boundary conditions. In the inertial frame $\mathcal{F}$ the equations of conservation of momentum and mass of the fluid as well as of linear and angular momentum of the body in their non-conservative form together with their natural boundary conditions are given by

$$
\text{Fluid} \quad
\begin{cases}
\rho \partial_t \mathrm{v} + \rho \mathrm{v} \cdot \nabla \mathrm{v} = \rho g + \nabla \cdot \sigma(\mathrm{v}, \mathrm{p}), \\
\nabla \cdot \mathrm{v} = 0, \quad \text{for } (x, t) \in \bigcup_{t>0} S(t)^c \times \{t\}, \\
\mathrm{v}(x, 0) = 0, \quad \lim_{|x| \to \infty} \mathrm{v}(x, t) = 0, \\
\mathrm{v}(x, t) = \mathcal{V}_C(t) + \Omega(t) \times (x - x_C(t)), \quad \text{for } x \in \partial S(t).
\end{cases}
\tag{1.1}
$$

$$
\text{Body} \quad
\begin{cases}
d_t(m_S \mathcal{V}_C) = m_S g - \int_{\partial S(t)} \sigma(\mathrm{v}, \mathrm{p}) \cdot N \, do, \\
d_t(J_{S(t)} \cdot \Omega) = - \int_{\partial S(t)} (x - x_C) \times [\sigma(\mathrm{v}, \mathrm{p}) \cdot N] \, do.
\end{cases}
\tag{1.2}
$$

Here, ρ is the constant density of $\mathcal{L}$, v and p are the Eulerian velocity field and pressure associated with $\mathcal{L}$, σ is the Cauchy stress tensor and ρg is the force of gravity which is assumed to be the only external force. We assume further a Navier–Stokes liquid model for which the Cauchy stress tensor is given by $\sigma(\mathrm{v}, \mathrm{p}) = -\mathrm{p}\mathbf{1} + \mu(\nabla \mathrm{v} + (\nabla \mathrm{v})^T)$, where μ is the shear viscosity. Further, m_S is the mass of the body, N is the unit normal to $\partial S(t)$ oriented toward the body and J_S

the inertia tensor with respect to the mass center C. We assume $\mathcal{V}_C(0) = 0$ and $\Omega(0) = 0$.

The straightforward formulation (1.1–1.2) has the disadvantage that the region $S(t)$ occupied by the liquid is time-dependent. This can be avoided by reformulating these equations in the body-attached frame $\mathcal{R}(t)$. If y denotes the position of a point P in the frame $\mathcal{R}(t)$ and x is the position of the same point in $\mathcal{F}$, we have

$$x = Q(t)y + x_C(t), \quad Q(0) = I, \quad x_C(0) = 0, \tag{1.3}$$

with Q an orthogonal linear transformation. In addition, we introduce the following transformed fields:

$$v(y,t) := Q^T \mathrm{v}(Qy + x_C, t), \quad p(y,t) := \mathrm{p}(Qy + x_C, t), \quad G := Q^T g, \tag{1.4}$$

$$V(y,t) := Q^T(\mathcal{V}_C + \Omega \times Qy), \quad \sigma(v,p) := Q^T \sigma(Q\mathrm{v}, \mathrm{p})Q, \quad \omega := Q^T \Omega, \tag{1.5}$$

and

$$V_C := Q^T \cdot \mathcal{V}_C, \quad n := Q^T \cdot N, \quad I_S := Q^T \cdot J_S \cdot Q, \quad \partial S := \partial S(0). \tag{1.6}$$

Using the transformations (1.3–1.6), we can reformulate the system of equations (1.1–1.2) in the following form:

$$\text{Fluid} \quad \begin{cases} \rho \partial_t v + \rho((v - V) \cdot \nabla)v + \rho \omega \times v = \nabla \cdot \sigma(v,p) + \rho G(t), \\ \nabla \cdot v = 0, \quad \text{for } (y,t) \in S(0)^c \times (0,\infty), \\ v(y,0) = 0, \quad \lim_{|y| \to \infty} v(y,t) = 0, \\ v(y,t) = V_C(t) + \omega(t) \times y, \quad \text{for } x \in \partial S(t). \end{cases} \tag{1.7}$$

$$\text{Body} \quad \begin{cases} m_S \dot{V}_C + m_S(\omega \times V_C) = m_S G(t) - \int_{\partial S} \sigma(v,p) \cdot n \, do, \\ I_S \cdot \dot{\omega} + \omega \times (I_S \cdot \omega) = - \int_{\partial S} y \times [\sigma(v,p) \cdot n] \, do, \\ d_t G - G \times \omega = 0. \end{cases} \tag{1.8}$$

In order to keep compatible notations for both the two- and three-dimensional case, we assume for $d = 2$ that $\omega := (0, 0, \omega)$ and similarly $y \times [\sigma \cdot n] = (0, 0, -y_2(\sigma \cdot n)_1 + y_1(\sigma \cdot n)_2)$. For $d = 2$, the second equation in (1.8) reduces to a scalar equation. The additional term $\omega \times v$ in the momentum equation in (1.7) corresponds to the *Coriolis force* induced by the frame transformation (1.3). In the body frame $\mathcal{R}(t)$ the direction of the gravitational force G depends on time t and therefore becomes an unknown to be computed. The third additional equation of (1.8) provides the equation needed for describing its variation. Its derivation relies on simple calculus related to the transformation (1.3). For more details regarding the derivation of these equations, we refer to Galdi [24], or Bönisch et al. [13].

1.2. The stationary free-fall problem

The solid body $\mathcal{S}$ is said to undergo a *free steady fall* if the translational and angular velocity V_C and ω are constant and if the motion of the liquid $\mathcal{L}$ is stationary in the frame $\mathcal{R}(t)$. The study of such a configuration is of great interest since it corresponds to so called *terminate state* motions of sedimenting particles for which many questions still remain open, e.g., the number of possible terminal states for a given body geometry, the orientation of the solid body, the stability of the corresponding solution (see Galdi [24] and references therein).

The free-steady-fall equations are obtained by requiring that v, p, V_C, ω, and G are time-independent. Comparing with (1.7–1.8), this leads us to the following system of equations:

$$\text{Fluid} \quad \begin{cases} \rho(v - V) \cdot \nabla v + \rho\omega \times v = \nabla \cdot \sigma(v, p) + \rho G, \\ \nabla \cdot v = 0, \quad \text{for } y \in \mathbb{R}^d \setminus S, \\ v(y, 0) = 0, \quad \lim_{|y| \to \infty} v(y) = 0, \\ v(y) = V(y) := V_C + \omega \times y \quad \text{for } y \in \partial S. \end{cases} \quad (1.9)$$

$$\text{Body} \quad \begin{cases} m_S(\omega \times V_C) = m_S G - \int_{\partial S} \sigma(v, p) \cdot n \, do, \\ \omega \times (I_S \cdot \omega) = -\int_{\partial S} y \times [\sigma(v, p) \cdot n] \, do, \\ G \times \omega = 0. \end{cases} \quad (1.10)$$

For the most general setup, we assume $\omega \neq 0$. Due to the third equation in (1.10), this configuration can be attained only for $d = 3$. Further it imposes G parallel to ω. The free-steady-fall problem can then be stated as follows:

Problem 1.1 (*Stationary fall in 3D*). Assume $d = 3$. Given ρ, $\sigma = \sigma(v, p)$, $|G| = |g|$, I_S, and m_S, find v, p, V_C, ω, and G, where $G = |g||\omega|^{-1}\omega$ if $\omega \neq 0$ (see Table 1), such that equations (1.9–1.10) hold.

An important subclass of free-steady-fall problems is given by the case $\omega = 0$ describing a solid body S, which falls in a purely *translational* motion. The problem formulation for this case is subtle since it depends not only on the dimension d of the problem but also on the geometrical properties of the solid.

At first, we assume that the equation $G \times \omega = 0$ has to be enforced and cannot be eliminated by means of any special geometrical properties of the solid S or of the flow configuration. For $d = 3$ such a translational problem is overdetermined and will therefore not be further considered (see Table 1). For $d = 2$ however this problem is well formulated in the sense that it involves six unknowns associated to six scalar equations. It can be stated as follows:

Problem 1.2 (*Stationary fall in 2D*). Assume $d = 2$. Given ρ, $\sigma = \sigma(v, p)$, $|G| = |g|$, I_S, m_S, and $\omega := 0$, find v, p, V_C, and the direction $\hat{G}$ of $G := |g|\hat{G}$, such that equations (1.9–1.10) hold.

The system of equations (1.9–1.10) describes different classes of free-fall regimes and configurations which are outlined on Table 1.

From the physical point of view, the reason of the overdetermination of the translational free steady fall for $d = 3$ can be interpreted by the fact that additional geometric properties of the solid body S have to prevent it from rotating (see Galdi/Vaidya [25]). Following Galdi [24], we consider now translational free-steady-fall problems for solid bodies with symmetry properties. Let $\{e_1, e_2, e_3\}$ be the

TABLE 1. Number of physical unknowns in the fluid-rigid body interaction problem depending on the setup.

Dimension	ω	Body-fluid setup	Number of	
			unknowns	scalar equations
3	$\neq 0$	general	10	10
2	$\neq 0$	not possible	-	-
3	$= 0$	overdetermined	9	10
2	$= 0$	general	6	6
3	$= 0$	symmetric	5	5
2	$= 0$	symmetric	4	4

canonical basis of $\mathbb{R}^3$. Assume that the solid body is homogeneous and symmetric with respect to the axis e_2. Further, the velocity field v and the pressure p describing the *terminal state* of the fluid $\mathcal{L}$ are assumed to be symmetric with respect to the axis e_2. One can show (see Galdi/Vaidya [25]) that every sufficiently smooth pair $\{v, p\}$ satisfies the following equations:

$$\int_{\partial S} \sigma(v, p) \cdot n = \eta \, e_2, \quad \eta \in \mathbb{R}, \tag{1.11}$$

$$\int_{\partial S} y \times [\sigma(v, p) \cdot n] = 0, \tag{1.12}$$

$$V = \alpha_V e_2, \ \alpha_V \in \mathbb{R}. \tag{1.13}$$

Therefore for the symmetric case, the equations (1.10) reduce to the following scalar equation:

$$-\left\{ \int_{\partial S} \sigma(v, p) \cdot n \, ds \right\}_2 + m_S |g| = 0, \tag{1.14}$$

since comparing the first equation of (1.10) with (1.11) implies $G = \pm |g| e_2$. We choose the orientation $G = -|g| e_2$ for the force of gravity. Under these assumptions of symmetry, the steady-free-fall problem can be formulated as follows:

Problem 1.3 (Symmetric steady fall in 2D). Given ρ, $\sigma = \sigma(v, p)$, $G = -|g| e_2$, I_S, m_S, and $\omega := 0$, find v, p, and the scalar quantity α_V defining $V := \alpha_V e_2$, such that equations (1.9) and (1.14) hold.

Problem 1.3 is well formulated for both three or two dimensions.

1.3. Numerical approximation

We begin with some standard notation. For a domain $\Omega \subset \mathbb{R}^d$, let $L^2(\Omega)$ denote the Lebesgue space of square-integrable functions on Ω equipped with the inner product and norm

$$(f, g)_\Omega := \int_\Omega f g \, dx, \quad \|f\|_\Omega := (f, f)^{1/2}.$$

Analogously, $L^2(\partial\Omega)$ denotes the space of square-integrable functions defined on the boundary $\partial\Omega$. The L^2-functions with generalized (in the sense of distributions) first-order derivatives in $L^2(\Omega)$ form the Sobolev space H^1, while $H_0^1 = \{v \in H^1(\Omega), v_{|\partial\Omega} = 0\}$ and $L_0^2(D) := \{q \in L^2(D) : (q, 1)_D = 0\}$.

1.3.1. The general non-stationary case. We first consider the setting of formulation (1.7–1.8) for the solution of the general free-fall problem. The unbounded domain $D := \mathbb{R}^d \backslash S$ filled by the liquid $\mathcal{L}$ is replaced by a bounded domain $\Omega \subset \mathbb{R}^d \backslash S$ which is chosen to be large enough in order that the liquid may be assumed to be at rest on Γ which denotes the boundary of Ω without ∂S. In the remainder of this paper, Ω is chosen such that the impact of this simplification on the quantities of interest is smaller than the discretization error. We refer to Bönisch et al. [14, 15] for a detailed discussion on this issue.

The key ingredient for the derivation of a weak form of the equations is an adequate choice of the velocity space allowing to eliminate the explicit formulation of the hydrodynamic force and torque on the solid body needed for the kinematic equations (1.10). This can be obtained by including the no-slip Dirichlet condition in the velocity space:

$$\mathcal{H}_1(D) := \left\{(v, V, \omega) : v \in [H_{loc}^1(\overline{D})]^d, V, \omega \in \mathbb{R}^d, v_{|\partial S} = V + \omega \times y\right\},$$

where $D := \mathbb{R}^d \backslash S$. The pressure p, defined only modulo constants, is assumed to lie in the space $L_0^2(D)$. For $U := \{(v, V_C, \omega), p, G\}$, $\Phi := \{(\varphi, \varphi_1, \varphi_2), q, \gamma\} \in \mathcal{H}_1(D) \times L_0^2(D) \times \mathbb{R}^d$, we define the semi-linear form

$$
\begin{aligned}
A(U)(\Phi) := {} & \rho(((v - (V_C + \omega \times y)) \cdot \nabla)v, \varphi)_D + (\omega \times v, \varphi)_D \\
& - (p, \nabla \cdot \varphi)_D + 2\mu(D(v), D(\varphi))_D - (\rho G, \varphi)_D \\
& - \varphi_1 \cdot [m_S(G - \omega \times V_C)] + \varphi_2 \cdot [\omega \times (I_S \cdot \omega)] - (\nabla \cdot v, q)_D,
\end{aligned}
$$

which is obtained by testing the equations (1.9–1.10) with $\Phi \in \mathcal{H}_1(D) \times L_0^2(D) \times \mathbb{R}^d$ and by integration by parts of the diffusive terms and the pressure gradient. Here, $D(v) := \frac{1}{2}(\nabla v + (\nabla v)^T)$ is the deformation tensor. The equations (1.9) modeling the balance of the linear and angular momentum can obviously be recovered by testing $A(U)(\Phi)$ with the functions $\Phi = \{(0, \varphi_1, 0), 0, 0\}$ and $\Phi = \{(0, 0, \varphi_2), 0, 0\}$), respectively. Further, we will use the bilinear form

$$m(\partial_t U, \Phi) := (\rho \partial_t v, \varphi) + (m_S \dot{V}_C, \varphi_1) + (\dot{I}_S \cdot \omega, \varphi_2) + (\dot{G}, \gamma).$$

Then, the variational formulation of the general non-stationary problem reads as follows:

Problem 1.4. Find a time-differentiable field $U(t) = \{(v(t), V_C(t), \omega(t)), p(t), G(t)\} \in \mathcal{H}_1(D) \times L_0^2(D) \times \mathbb{R}^d$, such that

$$m(\partial_t U, \Phi) + A(U)(\Phi) = 0 \quad \forall \Phi \in \mathcal{H}_1(D) \times L_0^2(D) \times \mathbb{R}^d. \tag{1.15}$$

Remark 1.5. The advantages of the formulation (1.15) rely on the fact that the force and torque on the solid body do not need to be computed explicitly. Numerical instabilities arising for the computation of these lower-dimensional integrals can therefore be avoided (see Hu et al. [37]).

Problem 1.4 is discretized in time by the so-called fractional-step-θ scheme described, e.g., in the article Rannacher [47] in this volume; see also Rannacher [46] and Turek [53]. The fluid-body interaction is handled by operator-splitting leading to the following time-stepping process:

1. Extrapolate V_C^{n-1}, ω^{n-1} to $\tilde{V}_C^n, \tilde{\omega}^n$ (explicit).
2. Compute v^n, p^n from the Navier–Stokes system by solving the three time substeps using the Newton method with geometric multigrid solution of the linear subsystems (implicit).
3. Compute the hydrodynamic forces acting on the body.
4. Update V_C^n, ω^n, G^n by applying one higher-order explicit time step to the corresponding ODEs (explicit).

The stationary subproblems within this scheme are discretized in space by the finite-element method on a truncated bounded domain $D_h = \cup\{K \in \mathbb{T}_h\}$, using the Q_2/Q_1 Taylor–Hood element on a quadrilateral mesh with hanging nodes for local mesh refinement (see Girault/Raviart [27] and also the article Rannacher [47] in this volume). The finite-element spaces are given by

$$W_1^h := \big\{((v, V, \omega), p) \in ([C(\overline{D_h})]^d \times \mathbb{R}^d \times \mathbb{R}^d) \times C(\overline{D_h}),$$
$$v|_K \in [Q_2]^d, \, p|_K \in Q_1, v|_{\partial S} = V + \omega \times y\big\},$$

in which the no-slip Dirichlet condition is included, in order to avoid the explicit formulation of the hydrodynamic force and torque on the solid body needed for the kinematic equations. Here the polynom spaces Q_k are to be understood in the isoparametric sense. This approximation is consistent with third order, i.e., there holds $\|v - I_h v\| = \mathcal{O}(h^3)$ for the nodal interpolant $I_h v$ of a sufficiently smooth velocity field.

1.3.2. The special stationary cases. The particular features of the various formulations of the *stationary* free-fall problem result in correspondingly simplified approximate problems. We first consider the most general setup of Problem 1.1, i. e., $\omega \neq 0$ and the related equations (1.9-1.10). For $U := \{(v, V_C, \omega), p\}$, $\Phi := \{(\varphi, \varphi_1, \varphi_2), q\} \in \mathcal{H}_1(D) \times L_0^2(D)$, we define the semi-linear form

$$A_1(U)(\Phi) := \rho(((v - (V_C + \omega \times y)) \cdot \nabla)v, \varphi)_D + (\omega \times v, \varphi)_D$$
$$- (p, \nabla \cdot \varphi)_D + 2\mu(D(v), D(\varphi))_D - (\rho|g||\omega|^{-1}\omega, \varphi)_D$$
$$- \varphi_1 \cdot [m_S(|g||\omega|^{-1}\omega - \omega \times V_C)] + \varphi_2 \cdot [\omega \times (I_S \cdot \omega)] - (\nabla \cdot v, q)_D.$$

A weak form of Problem 1.1 reads as follows:

Problem 1.6. Find $U := \{(v, V_C, \omega), p\} \in \mathcal{H}_1(D) \times L_0^2(D)$, such that

$$A_1(U)(\Phi) = 0 \quad \forall \Phi \in \mathcal{H}_1(D) \times L_0^2(D). \tag{1.16}$$

For the weak formulation of Problem 1.2 and Problem 1.3, the formulation (1.16) simplifies greatly since the free steady fall is then assumed to be translational. For the velocity field we define

$$\mathcal{H}_2(D) := \left\{ (v, V) : v \in [H^1_{loc}(\overline{D})]^d, \, V \in \mathbb{R}^d, \, v = V \text{ on } \partial S \right\}.$$

For $U := \{(v, V_C), p, \theta\} \in \mathcal{H}_2(D) \times L^2_0(D) \times \mathbb{R}$ and $\Phi := \{(\varphi, \varphi_1), q, \varphi_2\} \in \mathcal{H}_2(D) \times L^2_0(D) \times \mathbb{R}$, we define the semi-linear form

$$
\begin{aligned}
A_2(U)(\Phi) := {} & \rho(((v - V_C) \cdot \nabla)v, \varphi)_D - (p, \nabla \cdot \varphi)_D + 2\mu(D(v), D(\varphi))_D \\
& - (\nabla \cdot v, q)_D - \rho(G, \varphi) - m_S G \cdot \varphi_1 \\
& + (-y_2 \left\{\sigma(v, p) \cdot n\right\}_1 + y_1 \left\{\sigma(v, p) \cdot n\right\}_2, \varphi_2)_{\partial S},
\end{aligned}
$$

where G is assumed to be $G := |g| \binom{\cos \theta}{\sin \theta}$. The weak formulation of Problem 1.2 may then be stated as follows:

Problem 1.7. Find $U := \{(v, V_C), p, \theta\} \in \mathcal{H}_2(D) \times L^2_0(D) \times \mathbb{R}$, such that

$$A_2(U)(\Phi) = 0 \quad \forall \Phi \in \mathcal{H}_2(D) \times L^2_0(D) \times \mathbb{R}. \tag{1.17}$$

For Problem 1.3 the direction of the gravitation force G is not a variable anymore. Further due to equation (1.13) the direction of V_C is known to be collinear to e_2. For this configuration, we therefore define the space

$$\mathcal{H}_3(D) := \left\{ (v, \alpha_V) : v \in [H^1_{loc}(\overline{D})]^d, \, \alpha_V \in \mathbb{R}, \, v = \alpha_V e_2 \text{ on } \partial S \right\}, \tag{1.18}$$

for the velocity field. For $U := \{(v, \alpha_V), p\}$, $\Phi := \{(\varphi, \varphi_1), q\} \in \mathcal{H}_3(D) \times L^2_0(D)$, we define the semi-linear form

$$
\begin{aligned}
A_3(U)(\Phi) := {} & \rho(((v - \alpha_V e_2) \cdot \nabla)v, \varphi)_D - (p, \nabla \cdot \varphi)_D + 2\mu(D(v), D(\varphi))_D \\
& - (\nabla \cdot v, q)_D - (\rho G, \varphi)_D - m_S \varphi_1 e_2 \cdot G.
\end{aligned}
$$

The weak formulation for Problem 1.3 may then be stated as follows:

Problem 1.8. Find $U := \{(v, \alpha_V), p\} \in \mathcal{H}_3(D) \times L^2_0(D)$, such that

$$A_3(U)(\Phi) = 0 \quad \forall \Phi \in \mathcal{H}_3(D) \times L^2_0(D). \tag{1.19}$$

Using the finite-element spaces W^h_1 defined above, the discrete counterpart of Problem (1.6) reads as follows:

Problem 1.9. Find $U_h \in W^h_1$, such that

$$A_1(U_h)(\Phi_h) = 0 \quad \forall \Phi_h \in W^h_1. \tag{1.20}$$

Analogously, we define for Problem 1.7 and Problem 1.8, respectively, the following finite-element spaces:

$$W^h_2 := \left\{ (v, V, p, \theta) \in [C(\overline{\Omega})]^d \times \mathbb{R}^d \times C(\overline{\Omega}) \times \mathbb{R}, \, v_{|K} \in [Q_2]^d, p_{|K} \in Q_1, v_{|\partial S} = V \right\},$$
$$W^h_3 := \left\{ (v, \alpha_V, p) \in [C(\overline{\Omega})]^d \times \mathbb{R} \times C(\overline{\Omega}), \, v_{|K} \in [Q_2]^d, p_{|K} \in Q_1, v_{|\partial S} = \alpha_V e_2 \right\}.$$

The discrete counterpart of Problem (1.7) reads

Problem 1.10. Find $U_h \in W_2^h$, such that

$$A_2(U_h)(\Phi_h) = 0 \quad \forall \Phi_h \in W_2^h. \tag{1.21}$$

Analogously, the discrete counterpart of Problem (1.8) reads

Problem 1.11. Find $U_h \in W_3^h$, such that

$$A_3(U_h)(\Phi_h) = 0 \quad \forall \Phi_h \in W_3^h. \tag{1.22}$$

1.4. The issue of domain truncation

The truncation of the unbounded exterior domain to a bounded computational domain with artificial (homogeneous) 'outflow' boundary conditions creates errors which may be of significant size. To illustrate this point, we consider the particular situation with the parameters body length $l = 6$, body width $w = 1$, shear viscosity $\mu = 0.1$, and density $\rho = 1$. Figure 3 shows the dependence of the free-fall velocity on the diameter d_D of the truncated computational domain. We see that satisfactory accuracy is achieved only for $d_D \geq 400$ units.

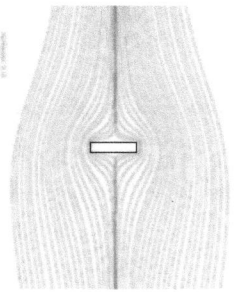
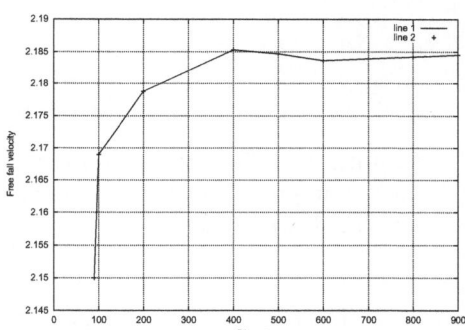

FIGURE 3. Effect of truncated domain (diameter) on the free-fall velocity of steady motion.

Asymptotic analysis shows that a 'parabolic wake' develops upstream and in the crosswind direction the flow behaves like potential flow. Hence for modeling the farfield behavior of the flow, one may use 'outflow' boundary conditions governed by the Gaussian (heat) kernel

$$v \sim \left(\frac{1}{2} C' x_1 x_2^{-3/2} e^{-\frac{U x_1^2}{4\nu x_2}}, C' x_2^{-1/2} e^{-\frac{U x_1^2}{4\nu x_2}} \right)^T, \quad C' := -\frac{Q\sqrt{U}}{2\sqrt{\pi\nu}},$$

and by the derivative Green function

$$v(x) \sim \frac{Q}{4\pi} \frac{x}{|x|^2}.$$

On this basis improved artificial boundary conditions can be derived for the truncated computational domain, which allow for a significant reduction of its size without sacrificing accuracy. This is demonstrated in Figure 4. In this approach the choice of the truncation diameter d_D can be adaptively controlled by residual-based a posteriori error estimates and the resulting errors may be balanced with the discretization error. For details, we refer to Wittwer [59], Bönisch et al. [14, 15]). Other approaches for treating unbounded domains are discussed in Tsynkov [51, 52].

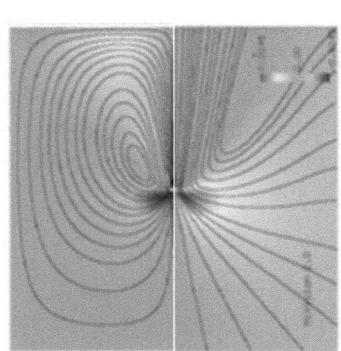
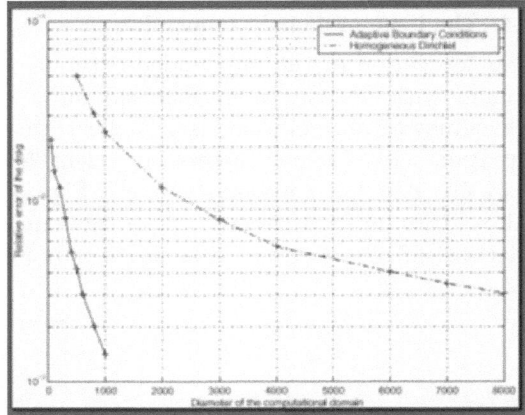

FIGURE 4. Effect of domain truncation (left) and gain in accuracy by improved 'outflow' boundary conditions based on asymptotic analysis (right).

Remark 1.12. The effect of truncating exterior domains to bounded computational domains has been analyzed in Bänsch/Dörfler [1] in the context of a posteriori energy-norm error estimation.

1.5. Toward economical meshes

Our goal in this section is to derive an a posteriori error estimator to control the accuracy of the most important output quantities in the free-fall problem, namely the velocity and the orientation of the falling body. At the same time this gives us strategies for an adequate mesh adaptation in order to obtain economical meshes. The derivation of a posteriori error estimates for the approximation of the continuous equation (1.16) by means of equation (1.20) relies on their interpretation as an optimal control problem and their embedding into the framework described in the article Rannacher [47] in this volume.

At first we recall from Becker/Rannacher [7] an abstract framework for the a posteriori error analysis of Galerkin approximation of general nonlinear variational equations; see also Becker et al. [6] and Bangerth/Rannacher [2]. Let $A(\cdot)(\cdot)$ be a differentiable semi-linear form defined on a function space W. The derivatives of $A(\cdot)(\cdot)$ at a point u in direction $\delta u, \delta v, \delta w$ are denoted by $A'(u)(\delta u, \cdot)$,

$A''(u)(\delta u, \delta v, \cdot)$, and $A'''(u)(\delta u, \delta v, \delta w, \cdot)$, e.g.,

$$A'(u)(\delta u, \varphi) := \lim_{\tau \to 0} \frac{1}{\tau} \{ A(u + \tau \delta u)(\varphi) - A(u)(\varphi) \}, \quad \varphi \in W.$$

Here, we use the convention that the dependence on the variables in the second round brackets is always linear while that with respect to the variable in the first brackets may be nonlinear. We assume that the variational equation

$$A(u)(\varphi) = 0 \quad \forall \varphi \in W, \tag{1.23}$$

has a solution $u \in W$. Suppose that the goal is to compute a certain physical quantity related to u by a differentiable functional $J(\cdot)$ with derivatives denoted by $J'(u)(\delta u)$, $J''(u)(\delta u, \delta v)$, and $J'''(u)(\delta u, \delta v, \delta w)$. Problem (1.23) is thought to be approximated by a Galerkin method using finite-dimensional subspaces $W_h \subset W$ parametrized by $h \in \mathbb{R}_+$. We assume that the associated discrete problems

$$A(u_h)(\varphi_h) = 0 \quad \forall \varphi_h \in W_h, \tag{1.24}$$

also possess solutions $u_h \in W_h$ with $J(u_h)$ being the approximation to the target quantity $J(u)$.

The aim is now to derive a posteriori estimates for the error $J(u) - J(u_h)$. To this end, we employ the Euler–Lagrange approach of optimal control theory. The problem of computing $J(u)$ from the solution of (1.23) can be equivalently formulated as computing stationary points $\{u, z\} \in W \times W$ of the Lagrangian functional

$$L(u)(z) := J(u) - A(u)(z), \tag{1.25}$$

with the adjoint variable $z \in W$. Hence we seek solutions $\{u, z\} \in W \times W$ to the Euler–Lagrange system

$$A(u)(\varphi) = 0 \quad \forall \varphi \in W, \tag{1.26}$$

$$A'(u)(z)(\varphi) = J'(u)(\varphi) \quad \forall \varphi \in W. \tag{1.27}$$

Notice that the first equation of this system is just the considered variational equation (1.23). The Galerkin approximation of system (1.26–1.27) in the subspace $W_h \subset W$ seeks pairs $\{u_h, z_h\} \in W_h \times W_h$ satisfying

$$A(u_h)(\varphi_h) = 0 \quad \forall \varphi_h \in W_h, \tag{1.28}$$

$$A'(u_h)(z_h)(\varphi_h) = J'(u_h)(\varphi_h) \quad \forall \varphi_h \in W_h. \tag{1.29}$$

To the approximate solutions $u_h \in W_h$, we associate the residual

$$\rho(u_h)(\cdot) := -A(u_h)(\cdot), \tag{1.30}$$

which is defined on all of W.

Proposition 1.13. *For the Galerkin approximation* (1.28)–(1.29) *of the Euler–Lagrange system* (1.26)–(1.27), *we have the a posteriori error representation*

$$J(u) - J(u_h) = \rho(u_h)(z - \varphi_h) + R_h \tag{1.31}$$

for arbitrary elements $\varphi_h \in W_h$. The remainder R_h is quadratic in the error $e := u - u_h$ and given by

$$R_h = \int_0^1 \left\{ A''(u_h + se)(e, e, z) - J''(u_h + se)(e, e) \right\} s \, ds. \tag{1.32}$$

Proof. The proof which relies on standard differential calculus can be found in Becker/Rannacher [7]. $\qquad\square$

This general approach will now be applied to the (steady) free-fall problem. We recall the governing semi-linear form

$$
\begin{aligned}
A_1(U)(\Phi) := {} & \rho(((v - (V_C + \omega \times y)) \cdot \nabla)v, \varphi)_D + (\omega \times v, \varphi)_D \\
& - (p, \nabla \cdot \varphi)_D + 2\mu(D(v), D(\varphi))_D - (\rho|g||\omega|^{-1}\omega, \varphi)_D \\
& - \varphi_1 \cdot [m_S(|g||\omega|^{-1}\omega - \omega \times V_C)] + \varphi_2 \cdot [\omega \times (I_S \cdot \omega)] - (\nabla \cdot v, q)_D,
\end{aligned}
$$

for arguments $U = \{(v, V_C, \omega), p\}$ and $\Phi = \{(\varphi, \varphi_1, \varphi_2), q\}$. The corresponding derivative which occurs in the dual problem has the form

$$
\begin{aligned}
A_1'(U)(\Psi, \Phi) := {} & \rho(((\psi - (\psi_1 + \psi_2 \times y)) \cdot \nabla)v, \varphi)) \\
& + \rho(((v - (V_C + \omega \times y)) \cdot \nabla)\psi, \varphi) + (\omega \times \psi, \varphi) + (\psi_2 \times v, \varphi) \\
& - (r, \nabla \cdot \varphi) + 2\mu(D(\psi), D(\varphi)) + \varphi_1 \cdot (\psi_2 \times V_C + \omega \times \psi_1) \\
& + \varphi_2 \cdot (\psi_2 \times (I_S \cdot \omega) + \omega \times (I_S \cdot \psi_2)) - (\nabla \cdot \psi, q),
\end{aligned}
$$

for arguments $U = \{(v, V_C, \omega), p\}$, $\Phi = \{(\varphi, \varphi_1, \varphi_2), q\}$, and $\Psi = \{(\psi, \psi_1, \psi_2), r\}$. At first, in order to avoid an overload of technicalities for the derivation, we consider the setup of the simplest Problem 1.3 with the governing semi-linear form

$$
\begin{aligned}
A_3(U)(\Phi) := {} & \rho(((v - \alpha_V e_2) \cdot \nabla)v, \varphi)_D - (p, \nabla \cdot \varphi)_D + 2\mu(D(v), D(\varphi))_D \\
& - (\nabla \cdot v, q)_D - (\rho G, \varphi)_D - m_S \varphi_1 e_2 \cdot G.
\end{aligned}
$$

For $U := \{(v, \alpha_V), p\} \in \mathcal{H}_3(D) \times L_0^2(D)$, the target functional for the control of the fall velocity of the body $\mathcal{S}$ is chosen as

$$J_3(U) := \alpha_V, \quad U \in \mathcal{H}_3(D) \times L_0^2(D). \tag{1.33}$$

The associated dual problem is given by

$$A_3'(U)(\Phi, Z) = J_3'(U)(\Phi) \quad \forall \Phi \in \mathcal{H}_3(D) \times L_0^2(D), \tag{1.34}$$

with the discrete analogue

$$A_3'(U_h)(\Phi_h, Z_h) = J_3'(U_h)(\Phi_h) \quad \forall \Phi_h \in W_3^h. \tag{1.35}$$

To the approximate solution $U_h \in W_3^h$ of the discrete Problem 1.11, we associate the residual

$$\rho_3(U_h)(\cdot) := -A_3(U_h)(\cdot). \tag{1.36}$$

Then, Proposition 1.13 gives us the following result:

Proposition 1.14. *Let* $U := \{(v, \alpha_V), p\} \in \mathcal{H}_3(D) \times L_0^2(D)$ *and* $Z := \{(z^v, z^\alpha), z^p\} \in \mathcal{H}_3(D) \times L_0^2(D)$ *be the solutions of* (1.19) *and* (1.34), *respectively. Further, let* U_h *and* Z_h *be their discrete counterparts, i.e., the solutions of* (1.22) *and* (1.35), *respectively. Then, there holds the error representation*

$$\alpha_V - \alpha_V^h = \rho_3(U_h)(Z - Z_h) + R_3, \tag{1.37}$$

where the remainder R_3 *is quadratic in the errors* $e^v := v - v_h$ *and* $e^\alpha := \alpha_V - \alpha_V^h$,

$$R_3 := \rho((e^v \cdot \nabla)e^v, z^v)_D - \rho e^\alpha((e_2 \cdot \nabla)e^v, z^v)_D.$$

Proof. The identity (1.37) is a direct consequence of the general error representation (1.31) of Proposition 1.13. To identify the remainder R_3, we note that

$$A_3''(U_h + sE)(E, E, Z) = 2\rho(((e^v - e^\alpha e_2) \cdot \nabla)e^v, z^v)_D$$
$$J_3''(U_h + sE)(E, E) = 0.$$

This completes the proof. $\qquad\square$

Remark 1.15. The dual problem associated to equation (1.34) possesses, despite its linear character, a structure similar to the primal Problem 1.8. The natural boundary condition of (1.34) is indeed

$$\int_{\partial S} [\sigma(z^v, z^p) \cdot n] \cdot e_2 \, d\sigma = 1. \tag{1.38}$$

which should be compared to (1.14).

For the more complex setup of Problem 1.6, one can derive an error representation similar to (1.37). In that context, however, due to the existence of additional nonlinear terms for the description of the gravitation force $G := |g||\omega|^{-1}\omega$, the remainder becomes much more complicated. In order to control the fall velocity of the solid body $\mathcal{S}$, we choose the functional

$$J_1(U) := \tfrac{1}{2}|V_C|^2, \quad U := \{(v, V_C, \omega), p\} \in \mathcal{H}_1(D) \times L_0^2(D).$$

The associated dual problem is defined as

$$A_1'(U)(\Phi, Z) = J_1'(U)(\Phi) \quad \forall \Phi \in \mathcal{H}_1(D) \times L_0^2(D), \tag{1.39}$$

with its discrete analogue

$$A_1'(U_h)(\Phi_h, Z_h) = J_1'(U_h)(\Phi_h) \quad \forall \Phi_h \in W_1^h. \tag{1.40}$$

To the approximate solution $u_h \in W_1^h$ of the discrete Problem 1.9, we associate the residual

$$\rho_1(U_h)(\cdot) := -A_1(U_h)(\cdot).$$

Analogously to Proposition 1.14, we obtain the following result.

Proposition 1.16. *Let* $U := \{(v, V_C, \omega), p\}$, $Z := \{(z^v, z^{V_C}, z^\omega), z^p\} \in \mathcal{H}_1(D) \times L_0^2(D)$ *be the solutions of* (1.16) *and* (1.39), *respectively. Further, let* U_h *and* Z_h *be their discrete counterparts in* W_1^h, *i.e., the solutions of* (1.20) *and* (1.40), *respectively. Then, we have the error representation*

$$J_1(U) - J_1(U_h) = \rho_1(U_h)(Z - Z_h) + R_1, \tag{1.41}$$

with a remainder R_1 quadratic in the errors $e^v := v - v_h$, $e^{V_C} := V_C - V_C^h$, and $e^\omega := \omega - \omega_h$,

$$R_1 := \rho((e^v \cdot \nabla)e^v, z^v)_D - \rho((e^{V_C} \cdot \nabla)e^v, z^v)_D - \rho(((e^\omega \times y) \cdot \nabla)e^v, z^v)_D$$
$$+ (e^\omega \times e^v, z^v)_D - z^{V_C} \cdot [e^\omega \times e^{V_C}] + z^\omega \cdot [e^\omega \times (I_S \cdot e^\omega)]$$
$$- \tfrac{1}{2}|e^{V_C}|^2 + O(|e^\omega|^2).$$

The term $O(|e^\omega|^2)$ is due to the unknown direction of the gravitational force.

Proof. The error representation (1.41) is derived in the same way as that of Proposition 1.14. It follows from the error representation derived in Proposition 1.13. The expression of the remainder R_1 follows from

$$A_1''(u_h + se)(e, e, z) = \rho(([e^v - (e^{V_C} + e^\omega \times y)] \cdot \nabla)e^v, z^v)_D$$
$$- 2z^{V_C} \cdot [e^\omega \times e^{V_C}] + 2z^\omega \cdot [e^\omega \times (I_S \cdot e^\omega)] + O(|e^\omega|^2),$$
$$J_1''(u_h + se)(e, e) = |e^{V_C}|^2.$$

This completes the proof.   □

Notice that Proposition 1.16 can be trivially extended to the configuration of Problem 1.7. In that context however, especially for stability analysis of the terminal state, the error control of the orientation of the solid body may be of great interest. Our proposed approach allows us indeed to control the orientation of the solid body by means of the functional

$$J_2(U) := \theta, \quad U := \{(v, V_C), p, \theta\} \in \mathcal{H}_2(D) \times L_0^2(D) \times \mathbb{R}.$$

The associated dual problem is defined as

$$A_2'(U)(\Phi, Z) = J_2'(U)(\Phi) \quad \forall \Phi \in \mathcal{H}_2(D) \times L_0^2(D) \times \mathbb{R}, \tag{1.42}$$

as well as its discrete analogue

$$A_2'(U_h)(\Phi_h, Z_h) = J_2'(U_h)(\Phi_h) \quad \forall \Phi_h \in W_2^h. \tag{1.43}$$

To the approximate solution $U_h \in W_2^h$ of the discrete problem 1.10 we associate the residual

$$\rho_2(U_h)(\cdot) := -A_2(U_h)(\cdot). \tag{1.44}$$

The discretization error on the orientation of the solid body $\mathcal{S}$ can be estimated by means of the following proposition:

Proposition 1.17. *Let* $U := \{(v, V_C), p, \theta\}$, $Z := \{(z^v, z^{V_C}), z^p, z^\theta\} \in \mathcal{H}_2(D) \times L_0^2(D) \times \mathbb{R}$ *be the solutions of (1.17) and (1.42), respectively. Further, let* U_h *and* Z_h *be their discrete counterparts, i.e., the solutions of (1.21) and (1.43), respectively. Then, there holds*

$$\theta - \theta_h = \rho_2(U_h)(Z - Z_h) + R_2, \tag{1.45}$$

with a remainder R_2 quadratic in the errors $e^v := v - v_h$, $e^{V_C} := V_C - V_C^h$, *and* $e^\theta := \theta - \theta_h$,

$$R_2 := \rho(((e^v - e^{V_C}) \cdot \nabla)e^v, z^v)_D$$

$$+ \tfrac{1}{2}|g|\left\{\rho\left(\begin{pmatrix}\cos\theta \\ \sin\theta\end{pmatrix}, z^v\right)_D + m_S\left(\begin{pmatrix}\cos\theta \\ \sin\theta\end{pmatrix} \cdot z^{V_C}\right)\right\}|e^\theta|^2.$$

Proof. The error representation (1.45) is a direct consequence of Proposition 1.13. To identify the remainder R_2, we note that

$$A_2''(U_h + sE)(E, E, Z) = 2\rho(((e^v - e^{V_C}) \cdot \nabla)e^v, z^v)_D$$

$$+ \rho|g|\left\{\left(\begin{pmatrix}\cos\theta \\ \sin\theta\end{pmatrix}, z^v\right)_D + \rho^{-1}m_S\begin{pmatrix}\cos\theta \\ \sin\theta\end{pmatrix} \cdot z^{V_C}\right\}|e^\theta|^2,$$

$$J_2''(U_h + sE)(E, E) = 0.$$

This completes the proof. □

Remark 1.18. We note that by the same approach as used above, goal-oriented a posteriori error estimates can also be derived for the hydrodynamical force and torque acting on the solid body $\mathcal{S}$,

$$J_{\psi_1}(U) := \int_{\partial S} [\sigma(v, p) \cdot n] \cdot \psi \, do, \qquad J_{\psi_2 \times y}(U) := \int_{\partial S} y \times [\sigma(v, p) \cdot n] \cdot \psi \, do.$$

This allows one to control any weighted combination of both quantities. This can be done by an adequate definition of the weights ψ_1 and ψ_2 of the trace $\psi = \psi_1 + \psi_2 \times y$, which determines the Dirichlet boundary condition for the corresponding dual solution, $z^v|_{\partial S} = \psi$.

Figure 5 shows adapted meshes obtained by using the DWR approach in the simulation of the 'steady fall problem'.

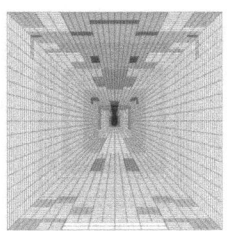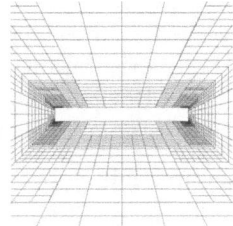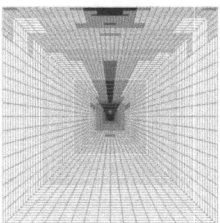

FIGURE 5. Adapted meshes – horizontal fall (left: D=800, right: D=100).

1.6. Hydrodynamic stability

The stability of the steady-fall solutions $\hat{u} = \{\hat{v}, \hat{p}\}$ shown in Figure 6 has been investigated by the linearized stability theory, i.e., by checking the eigenvalues of the corresponding linearized stability eigenvalue problem,

$$A_1'(\hat{U})(U, \Phi) = \lambda(U, \Phi) \quad \forall \Phi \in \mathcal{H}_1(D) \times L_0^2(D). \tag{1.46}$$

If one eigenvalue has real part $\mathrm{Re}\lambda \leq 0$, then the steady-state solution is (dynamically) unstable, i.e., it will not persist under arbitrarily small perturbations. Otherwise, if all eigenvalues have real parts $\mathrm{Re}\lambda > 0$, then the steady-state solution is called 'linearly stable'; see Heuveline/Rannacher [30] The results for the 'free-fall problem' are shown in Table 2.

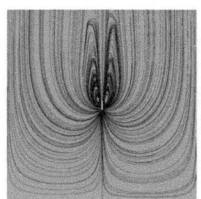 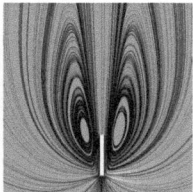 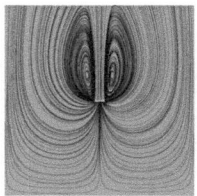 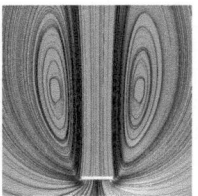

FIGURE 6. Stationary motion of a falling body in a Newtonian fluid: unstable vertical orientation (left) and stable horizontal orientation (right).

TABLE 2. Results of the stability analysis of the steady-free-fall problem.

2D case: $\mu = 0.1$	Real part of the critical eigenvalue	
domain diameter	vertical orientation	horizontal orientation
200	-0.82	0.81
800	-1.91	0.61
1000	-1.94	0.84

We see that in the 2D symmetric case for moderate Reynolds number ($\mu = 0.1$), there is one stable (horizontal body orientation) and one unstable (vertical body orientation) solution. For very small Reynolds number, i.e., for $\mu \geq 10^6$, all orientations correspond to (numerically) stable solutions.

1.7. Dynamics of non-stationary free fall

Finally, we report on some results obtained by the non-stationary version of the numerical method described above. In a two-dimensional setting, we simulate the free fall of a (symmetric) rod in a viscous liquid. We are interested in the different types of steady and unsteady fall patterns for varying Reynolds numbers, a question which has been experimentally studied in the literature, see Field et al. [22]

and Belmonte et al. [8]. The experimental studies identify four different regimes
of "free fall":

- Quasi-steady motion (low Re and moment of inertia).
- Oscillatory motion (higher Re and low moment of inertia).
- Tumbling motion (moderately large Re and very large moment of inertia).
- Chaotic motion (moderately large Re and moment of inertia).

All four types of motion could be realized by the numerical simulation, see Figure 7
(see [10] for more details).

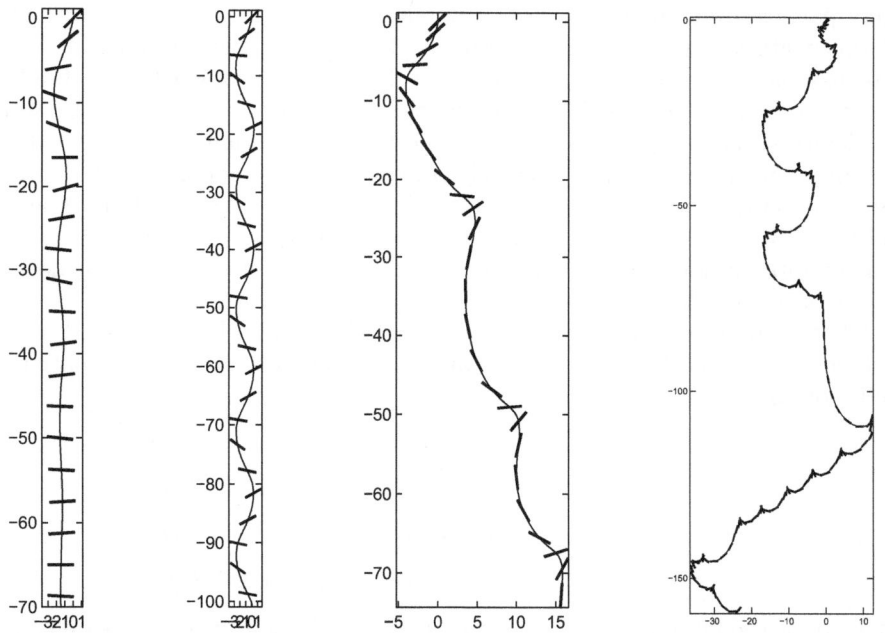

FIGURE 7. 'Free-fall' patterns of symmetric bodies (from left to
right): quasi-steady, oscillatory, tumbling, and 'chaotic' motion.

1.8. Open problems and further development

The free fall of a rigid body in a viscous fluid is a subject rich of theoretical as well
as computational and experimental problems. Though being relatively elementary
in its mathematical formulation this physical process poses many difficult questions, for example, for stationary and non-stationary fall patterns of non-symmetric
bodies, for criteria of the stability of corresponding quasi-stationary fall, for the
number of stable states depending on the geometry, and so on. Further, the study
of theses questions for various models of *non-Newtonian* fluids opens a whole new
field of interesting problems for research. An example is the question of the possible 'tilt angle' of a falling body depending on its shape and the nature of the

liquid, which is currently being investigated theoretically and numerically as well as experimentally, see Galdi et al. [26, 57].

2. Fluid-'many rigid bodies-wall' interaction

2.1. The stress-DLM method

Next, we describe the so-called 'stress-DLM' (Lagrange-multiplier-based fictitious-domain method) formulation of Patankar at al. [45, 43, 44, 42] for the direct numerical simulation of rigid particulate flows, particularly of the behavior of many particles and the interaction of particles with rigid walls. For more details on this material, we refer to Bönisch [9]. The idea of the stress-DLM method is to assume that the entire fluid-particle domain is occupied by the fluid and then to constrain the fluid inside the particle domain as a rigid body by setting the deformation tensor equal to zero. The latter constraint is represented by a Lagrange multiplier field in the particle domain, which can be interpreted as being the displacement field of a linear elastic body. The velocity and the Lagrange multiplier can be represented by an equal-order interpolation scheme in a finite-element formulation, which unlike as in the pressure-velocity coupling does not require extra stabilization.

Let Ω be the computational domain which includes both the fluid and the particle domain and let $P(t)$ be the particle domain. The governing equations for fluid motion are given by:

$$
\begin{aligned}
\rho_f \left(\partial_t v + v \cdot \nabla v \right) + \nabla p - \mu \Delta v &= \rho_f g, \quad \text{in } \Omega \setminus \overline{P(t)}, \\
\nabla \cdot v &= 0 \quad \text{in } \Omega \setminus \overline{P(t)}, \\
v = v^\partial(t) \quad \text{on } \partial \Omega(t), \quad v &= v^i \quad \text{on } \partial P(t), \\
v_{|t=0} &= v^0 \quad \text{in } \Omega \setminus \overline{P(0)},
\end{aligned}
\tag{2.1}
$$

where ρ_f is the fluid density, $U = \{v, p\}$ is the fluid velocity and pressure pair, v^i is the velocity of the fluid-particle interfaces $\partial P(t)$, and v^0 is the initial velocity. In the stress-DLM formulation the particles are treated as a fluid with an additional constraint to impose the rigidity. Accordingly, the governing equations for particle motion are:

$$
\begin{aligned}
\rho_s \left(\partial_t v + v \cdot \nabla v \right) + \nabla p - \mu \Delta v &= \rho_s g \quad \text{in } P(t), \\
\nabla \cdot v &= 0 \quad \text{in } P(t), \\
\nabla \cdot D[v] = 0 \quad \text{in } P(t), \quad n \cdot D[v] &= 0 \quad \text{on } \partial P(t), \\
v &= v^i \quad \text{on } \partial P(t), \\
v_{|t=0} &= v^0 \quad \text{in } P(0),
\end{aligned}
\tag{2.2}
$$

where ρ_s is the particle density. The third equation in (2.2) represents the rigidity constraint, that sets the deformation tensor, $D[v] := (\nabla v + \nabla v^T)/2$, in the particle domain equal to zero. Then, a combined weak formulation of the fluid-particle equations (2.1–2.2) can be derived by introducing a *distributed Lagrange multiplier* (DLM) Λ, which can be interpreted as an extra-stress inside the particle. With the spaces

$$V_0 := H_0^1(\Omega)^2, \quad L_0 := \{q \in L^2(\Omega), (q, 1)_\Omega = 0\},$$

the combined weak formulation reads as follows:

Problem 2.1. For $t > 0$, find $v \in v^\partial + V_0$, $p \in L_0$, $\Lambda \in H^1(P(t))^2$ satisfying

$$
\begin{aligned}
\big(\rho_f(\partial_t v + v \cdot \nabla v - g), \varphi\big)_\Omega &- (p, \nabla \cdot \varphi)_\Omega + (\chi, \nabla \cdot v)_\Omega + \mu(\nabla v, \nabla \varphi)_\Omega \\
&+ \big((\rho_s - \rho_f)(\partial_t v + v \cdot \nabla v - g), \varphi\big)_{P(t)} \\
&+ \big(D[\Lambda], D[\varphi]\big)_{P(t)} + \big(D[\psi], D[v]\big)_{P(t)} = 0,
\end{aligned}
\tag{2.3}
$$

for all $\varphi \in V_0, \chi \in L_0(\Omega), \psi \in H^1(P(t))^2$.

The fluid-particle interface condition is internal to the combined system (2.3). Hence the particle translational and angular velocities are not present in the combined form.

2.2. The fractional-step scheme

Equation (2.3) is solved by means of an operator-splitting time-stepping scheme. The algorithm is a variant of the scheme presented in Patankar et al. [45]:

Step 1: Calculation of the particle velocity. Given an approximation v^n of $v(t^n)$, find the translational velocity U^n and the angular velocity Ω^n of the particle,

$$MU^n = \int_{P(t^n)} \rho_s v^n \, dx, \quad I\Omega^n = \int_{P(t^n)} r \times \rho_s v^n \, dx, \tag{2.4}$$

where M is the mass of the particle, and I denotes the moment of inertia.

Step 2: Explicit update of particle position/orientation. This is achieved by the following subcycling procedure: Set $X^{n+1,0} := X^n$. For $k = 1, \ldots, K$:

$$
\begin{aligned}
X^{*n+1,k} &= X^{n+1,k-1} + \frac{\Delta t}{2K}(U^n + U^{n-1}), \\
X^{n+1} &= X^{*n+1,k} + \left(\frac{\Delta t}{K}\right)^2 \frac{1}{2M}\big(F(X^{n+1,k-1}) + F(X^{*n+1,k})\big).
\end{aligned}
\tag{2.5}
$$

Set $X^{n+1} := X^{n+1,K}$, and

$$A_c^{n+1} := \frac{1}{\Delta t^2}\left(X^{n+1} - X^n - \frac{1}{2}(U^n + U^{n-1})\right).$$

Here, F denotes the collision force acting on the particles to prevent them from penetrating each other or the walls of the domain. A_c is the acceleration of the particle due to collision. This term provides an additional body force acting on the particle and is included in the combined momentum equation to be solved in the subsequent steps.

Step 3: Solve the flow equations on Ω. Find $v^{n+1/2} \in v^\partial + V_0(t^{n+1})$ and $p^{n+1/2} \in L_0$, such that

$$
\begin{aligned}
\left(\rho_f(\Delta t^{-1}\{v^{n+1/2} - v^n\} + v^{n+1/2} \cdot \nabla v^{n+1/2} - g), \varphi\right)_\Omega \\
-\left(p^{n+1/2}, \nabla \cdot \varphi)\right)_\Omega + \left(\chi, \nabla \cdot v^{n+1/2}\right)_\Omega + \mu\left(\nabla v^{n+1/2}, \nabla\varphi\right)_\Omega = 0,
\end{aligned} \tag{2.6}
$$

for all $\{\varphi, \chi\} \in V_0 \times L_0^2(\Omega)$.

Step 4: Correct velocity in the particle domain. Find $v^{n+1} \in v^\partial + V_0(t^{n+1})$ and $\Lambda^{n+1} \in H^1(P(t^{n+1}))^2$, such that

$$
\begin{aligned}
\left(\rho_f \Delta t^{-1}(v^{n+1} - v^{n+1/2}), \varphi\right)_\Omega - \left(\rho_s A_c^{n+1}, \varphi\right)_{P(t^{n+1})} \\
+ \left(D[\Lambda^{n+1}], D[\varphi]\right)_{P(t^{n+1})} + \left(D[\psi], D[v^{n+1}]\right)_{P(t^{n+1})} \\
+ \left((\rho_s - \rho_f)(\Delta t^{-1}\{v^{n+1} - v^n\} + v^{n+1/2} \cdot \nabla v^{n+1/2} - g), \psi\right)_{P(t^{n+1})} = 0,
\end{aligned} \tag{2.7}
$$

for all $\{\varphi, \psi\} \in V_0 \times H^1(P(t^{n+1}))^2$. The last fractional step, Equation (2.7), adds computational cost to the solution procedure. To avoid this, Patankar et al. [41, 49]) proposed a fast projection scheme that eliminates the need to solve (2.7) by means of an iterative procedure.

2.2.1. Mesh adaptation in the stress-DLM method. In its original form the stress-DLM method has been described on uniform meshes in order to facilitate the use of special fast-solution algorithms and to keep the cost of the solution process low. However, this may either lead to insufficient resolution of the flow near the particle boundary and therefore to inaccurate representation of the particle interaction or to enormous costs when handling many particles on a very fine mesh. The alternative proposed in this article is to use locally adapted meshes, which still have a good degree of regularity. The refinement zones of the mesh are attached to the particles following purely geometrical criteria and are moved according to the movement of the particles. A further step of adaptation is applied in the numerical integration of bilinear forms along the particle-liquid interface. Here, the ususal 4-point Gauß rule is replaced by a summed Newton–Cotes rule, in order to cope with the discontinuity in the density across this interface, see Figure 8. This simple approach has proven to be very effective in several numerical tests.

2.2.2. Test case 'Dancing of Two Particles'. For validation the stress-DLM method has been applied for simulating the 2D motion of two rigid bodies subject to gravity in a vertical channel. Figure 9 shows the result where 'drafting', 'kissing' and 'tumbling' are seen as can be observed in corresponding experiments.

2.2.3. Test case 'Fall of Many Particles Through a Hole'. Next, the case of groups of particles is considered. Figure 10 shows the 2D motion of a number of circular particles subject to gravity in a vertical channel with a constriction. The initial 'pyramid-like' positioning of the particles is depicted on the left-hand side of Figure 10. The setup is slightly asymmetric (One particle on the right is missing.) in order to avoid symmetry-breaking being triggered by numerical instabilities. The

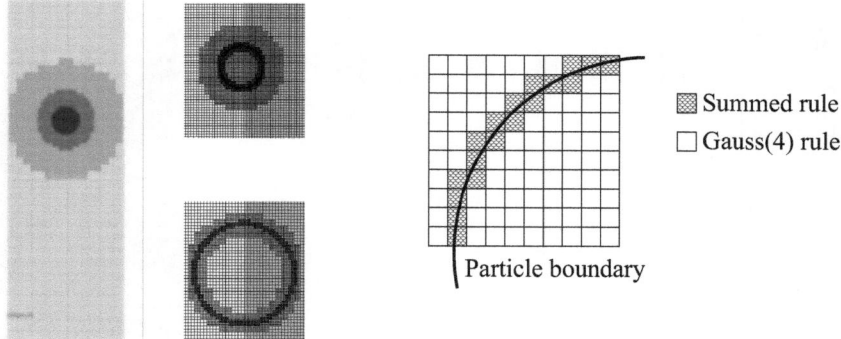

FIGURE 8. Mesh adaptivity in the stress-DLM method for a single particle and selective numerical integration along the fluid-particle interfaces.

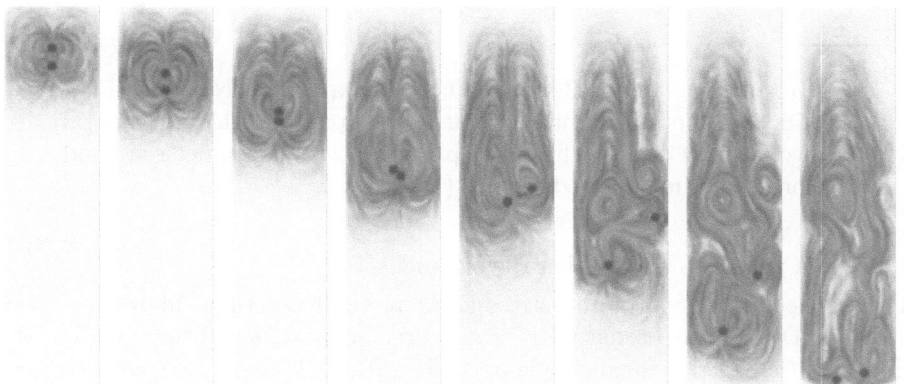

FIGURE 9. Free fall of two bodies in a viscous Newtonian liquid: 'drafting', 'kissing' and 'tumbling' phenomena are observed.

simulation was done with 41 particles of diameter 0.05 on dynamically adapted meshes with about $20,000 - 40,000$ cells and minimal mesh size $h_{\min} \approx 0.005$. The finest mesh was obtained by 4 global and 3 additional local refinement steps. This finest mesh would correspond to a globally refined mesh with about $440,000$ cells. The (uniform) time step was $\Delta t = 0.005$, i.e., 2000 time steps were needed for the computation over the relevant time interval $[0, 10]$. The whole simulation took about 1 day on an AMD Athlon64 3500+ computer. This time could be significantly reduced by optimizing the components of the multigrid solver used within each time step.

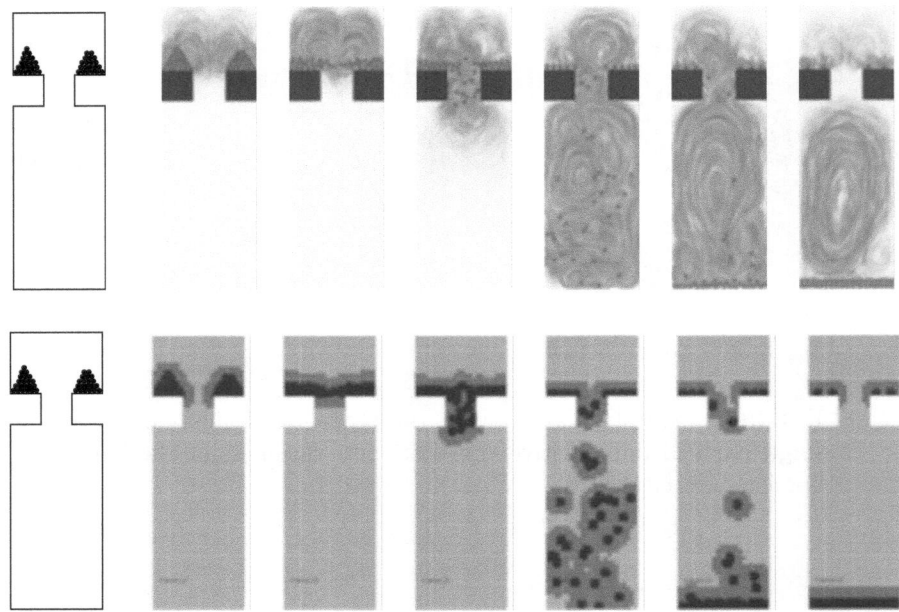

FIGURE 10. Free fall of many bodies in a viscous Newtonian liquid: sketch of the initial positions of the particles (left), temporal evolution of the flow field and particle positions (upper row) and corresponding adapted meshes (lower row).

2.3. Open problems and further development

The particular version of the stress-DLM method described above can also be used in 3D and can be naturally generalized to particles of any general shape. However, the restriction to simple particles with high degree of symmetry results in a drastic reduction of computational work, which allows for treating a larger number ($\geq 10^3$) of particles; see Tezduyar et al. [50] and Wan/Turek [56].

3. Fluid-'elastic structure' interaction

As a prototypical example of a fluid-structure interaction (FSI) problem, we consider the benchmark 'Vibrating Thin Plate', the configuration of which is shown in Figure 11; see Hron/Turek [34] and Dunne/Rannacher [21].

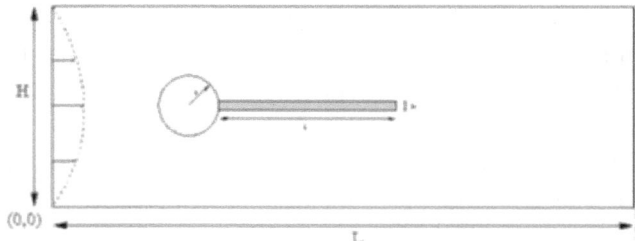

FIGURE 11. Configuration of the FSI benchmark 'Vibrating Thin Plate'.

The model characteristics used in this FSI example are that of an *incompressible Newtonian* fluid and of a *compressible St. Venant–Kirchhoff (STVK)* or an *incompressible neo-Hookean (INH)* material for the structure. The numerical approaches used for the simulation are summarized as follows:

- 'Monolithical' *arbitrary Lagrangian–Eulerian (ALE)* or *Eulerian–Eulerian* variational formulation.
- Galerkin finite-element method for fluid and structure.
- Mesh adaptation by the DWR approach based on numerically computed model sensitivities for the Galerkin residual as 'model perturbation'.
- Time stepping by the 'fractional-step-θ' scheme.
- Linearization by Newton-type iteration for fluid and structure part and functional iteration for 'interface'-capturing.
- GMRES with multigrid-preconditioning for the algebraic linear subproblems.

3.1. Solution methods for FSI problems

We briefly recall the common approaches for solving FSI problems.

(I) Combining the *Eulerian* and the *Lagrangian* setting involves conceptional difficulties. The time-varying fluid domain depends on the deformation of the structure domain. In turn, for determining the deformation of the structure the fluid boundary values (velocity and normal stress) are needed. In the *partitioned approach* each of the two subproblems is solved separately (using standard methods/codes), and so iterated to a solution of the coupled system.

(II) The *'arbitrary Lagrangian–Eulerian' (ALE)* method is based on a 'monolithical' variational formulation of the FSI problem. For representing the fluid-structure interface an auxiliary coordinate transformation ζ_f is introduced, which is determined by a variational equation. With its help the fluid problem is rewritten as one on the reference domain which is fixed in time. All computations are done on

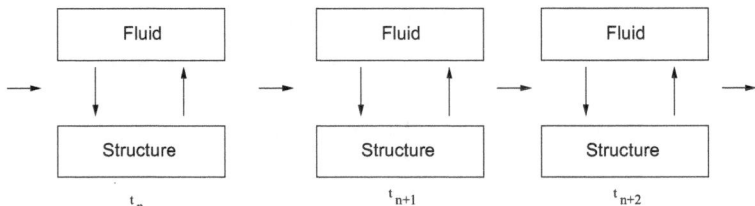

FIGURE 12.   Partitioned approach: Lagrangian and Eulerian frameworks iteratively coupled.

the reference domain/mesh and as a part of the computation the function ζ_f has to be determined at each time step; see Hron/Turek [35].

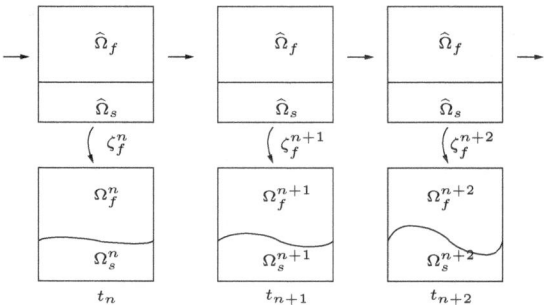

FIGURE 13.   Transformation approach: both frameworks Lagrangian.

(III) Both, the *partitioned* and the *transformation approach* to overcome the Euler-Lagrange discrepancy, explicitly track the fluid-structure interface by mesh adjustment and are referred to as *'interface tracking'* methods. The structure problem is left in its natural Lagrangian setting. An alternative is to treat the FSI problem in a purely Eulerian setting such as commonly used for describing two-phase flows; see Lui/Walkington [40]. A phase variable is employed on the fixed mesh to distinguish between the different phases, liquid and solid. This approach to identifying the fluid-structure interface is referred to as *'interface capturing'*. This approach is similar to the Level Set (LS) method but is realized in a form which avoids the need for reinitialization due to smearing effects. We emphasize that both approaches, the ALE and the Eulerian method, are based on 'monolithical' variational formulations of the FSI problem.

Typical results of a simulation based either on the ALE or the purely Eulerian approach are shown in Figure 14.

3.2. Variational formulation

We begin with introducing some notation, which slightly differs from the one occurring above and will be used throughout the following presentation. By $\Omega \subset$

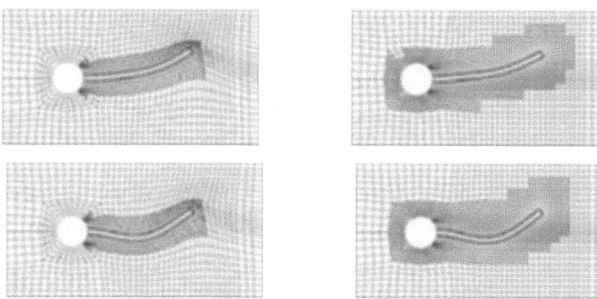

FIGURE 14. Results for the FSI benchmark 'Vibrating Thin Plate' obtained by the ALE (left) and the Eulerian (right) approaches.

$\mathbb{R}^d$ ($d = 2$ or $d = 3$), we denote the domain of definition of the FSI problem. The domain Ω is supposed to be *time-independent* but to consist of two possibly time-dependent subdomains, the fluid domain $\Omega_f(t)$ and the structure domain $\Omega_s(t)$. Unless needed, the explicit time-dependency will be skipped in this notation. The boundaries of Ω, Ω_f, and Ω_s are denoted by $\partial\Omega$, $\partial\Omega_f$, and $\partial\Omega_s$, respectively. The common interface between Ω_f and Ω_s is $\Gamma_i(t)$, or simply Γ_i. The initial structure domain is denoted by $\widehat{\Omega}_s$. Spaces, domains, coordinates, values (such as pressure, displacement, velocity) and operators associated to $\widehat{\Omega}_s$ (or $\widehat{\Omega}_f$) will likewise be indicated by a 'hat'.

Partial derivatives of a function f with respect to the i-th coordinate are denoted by $\partial_i f$, and the total time-derivative by $d_t f$. The divergences of vectors and tensors are written as $\mathrm{div} f = \sum_i \partial_i f_i$ and $(\mathrm{div} F)_i = \sum_j \partial_j F_{ij}$. The gradient of a vector-valued function v is the tensor $(\nabla v)_{ij} = \partial_j v_i$. By $[f]$, we denote the jump of a (possibly discontinuous) function f across an interior boundary, where n is always the unit vector n at points on that boundary.

For a set X, we denote by $L^2(X)$ the Lebesque space of square-integrable functions on X equipped with the usual inner product and norm

$$(f, g)_X := \int_X fg \, dx, \quad \|f\|_X = (f, f)_X^{1/2},$$

respectively, and correspondingly for vector- and matrix-valued functions. Mostly the domain X will be Ω, in which case we will skip the domain index in products and norms. For Ω_f and Ω_s, we similarly indicate the associated spaces, products, and norms by a corresponding index 'f' or 's'. Let $L_X := L^2(X)$ and $L_X^0 := L^2(X)/\mathbb{R}$. The functions in L_X (with $X = \Omega$, $X = \Omega_f(t)$, or $X = \Omega_s(t)$) with first-order distributional derivatives in L_X make up the Sobolev space $H^1(X)$. Further, $H_0^1(\partial X_D, X) = \{v \in H^1(X) : v_{|\partial X_D} = 0\}$, where ∂X_D is that part of the boundary ∂X at which Dirichlet boundary conditions are imposed. Further, we will use the function spaces $V_X := H^1(X)^d$, $V_X^0 := H_0^1(X)^d$, and for time-dependent functions

$$\mathcal{L}_X := L^2[0,T;L_X], \quad \mathcal{V}_X := L^2[0,T;V_X] \cap H^1[0,T;V_X^*],$$
$$\mathcal{L}_X^0 := L^2[0,T;L_X^0], \quad \mathcal{V}_X^0 := L^2[0,T;V_X^0] \cap H^1[0,T;V_X^*],$$

where V_X^* is the dual of V_X^0. Again, the X-index will be skipped in the case of $X = \Omega$, and for $X = \Omega_f$ and $X = \Omega_s$ a corresponding index 'f' or 's' will be used.

The fluid part is described by the usual 'incompressible' Navier–Stokes equations naturally written in the Eulerian frame using a time-dependent flow domain $\Omega_f(t)$. The unknowns are the pressure field $p_f \in \mathcal{L}_f$ and the velocity field $v_f \in v_f^D + \mathcal{V}_f$.

Problem 3.1 (Fluid model). Find $\{v_f, p_f\} \in \{v_f^D + \mathcal{V}_f^0\} \times \mathcal{L}_f$, such that $v_f(0) = v_f^0$, and

$$(\rho_f(\partial_t + v_f \cdot \nabla)v_f, \psi^v)_f + (\sigma_f, \epsilon(\psi^v))_f + (\nabla \cdot v_f, \psi^p)_f = f_f(\psi^v), \qquad (3.1)$$

for all $\{\psi^v, \psi^p\} \in V_f^0 \times L_f$, where

$$\sigma_f := -p_f I + 2\rho_f \nu_f \epsilon(v_f), \quad \epsilon(v) := \tfrac{1}{2}(\nabla v + \nabla v^T).$$

The structure part is described by the Lamé–Navier equations naturally written in the Lagrangian frame. We consider an 'incompressible neo-Hookean' (INH) and a compressible 'St. Venant–Kirchhoff' (STVK) material. The density of the structure is ρ_s. The material elasticity is usually described by the Lamé coefficients λ_s, μ_s. The model is formulated in Lagrangian coordinates in the domain $\hat{\Omega}_s$ with the scalar pressure field $\hat{p}_s \in \hat{\mathcal{L}}_f$ and the vector displacement and velocity fields $\hat{u}_s \in \hat{u}_s^D + \hat{\mathcal{V}}_s^0$, $\hat{v}_s \in \hat{v}_s^D + \hat{\mathcal{V}}_s^0$.

Problem 3.2 (INH structure model). Find $\{\hat{u}_s, \hat{v}_s, \hat{p}_s\} \in \{\hat{u}_s^D + \hat{\mathcal{V}}_s^0\} \times \{\hat{v}_s^D + \hat{\mathcal{V}}_s^0\} \times \hat{\mathcal{L}}_s$, s.t. $\hat{u}_s(0) = \hat{u}_s^0$, $\hat{v}_s(0) = \hat{v}_s^0$, and

$$(\rho_s d_t \hat{v}_s, \hat{\psi}^u)_{\hat{s}} + (\hat{\sigma}_s \hat{F}^{-T}, \hat{\epsilon}(\hat{\psi}^u))_{\hat{s}} = \hat{f}_s(\hat{\psi}^u),$$
$$(d_t \hat{u}_s - \hat{v}_s, \hat{\psi}^v)_{\hat{s}} = 0, \qquad (3.2)$$
$$(\det \hat{F}, \hat{\psi}^p)_{\hat{s}} = (1, \hat{\psi}^p)_{\hat{s}},$$

for all $\{\hat{\psi}^u, \hat{\psi}^v, \hat{\psi}^p\} \in \hat{V}_s^0 \times \hat{V}_s^0 \times \hat{L}_s$, where

$$\hat{F} := I + \hat{\nabla}\hat{u}_s, \quad \hat{\sigma}_s := -\hat{p}_s I + \mu_s(\hat{F}\hat{F}^T - I), \quad \hat{\epsilon}(\hat{\psi}^u) := \tfrac{1}{2}(\hat{\nabla}\hat{\psi}^u + \hat{\nabla}\hat{\psi}^{uT}).$$

Problem 3.3 (STVK structure model). Find $\{\hat{u}_s, \hat{v}_s\} \in \{\hat{u}_s^D + \hat{\mathcal{V}}_s^0\} \times \{\hat{v}_s^D + \hat{\mathcal{V}}_s^0\}$, such that $\hat{u}_s(0) = \hat{u}_s^0$, $\hat{v}_s(0) = \hat{v}_s^0$, and

$$(\rho_s d_t \hat{v}_s, \hat{\psi}^u)_{\hat{s}} + (\hat{J}\,\hat{\sigma}_s\,\hat{F}^{-T}, \hat{\epsilon}(\hat{\psi}^u))_{\hat{s}} = \hat{f}_s(\hat{\psi}^u),$$
$$(d_t \hat{u}_s - \hat{v}_s, \hat{\psi}^v)_{\hat{s}} = 0, \qquad (3.3)$$

for all $\{\hat{\psi}^u, \hat{\psi}^v\} \in \hat{V}_s^0 \times \hat{V}_s^0$, where

$$\hat{F} := I + \hat{\nabla}\hat{u}_s, \quad \hat{J} := \det\hat{F}, \quad \hat{E} := \tfrac{1}{2}(\hat{F}^T\hat{F} - I),$$

$$\hat{\epsilon}(\hat{\psi}^u) := \tfrac{1}{2}(\hat{\nabla}\hat{\psi}^u + \hat{\nabla}\hat{\psi}^{uT}), \quad \hat{\sigma}_s := \hat{J}^{-1}\hat{F}(\lambda_s(\mathrm{tr}\hat{E})I + 2\mu_s\hat{E})\hat{F}^T.$$

For writing the structure equations in the Eulerian frame, we use the (inverse) displacement function $D(x)$ of points in the deformed domain Ω_s back to points in the initial domain $\hat{\Omega}_s$,

$$\hat{D} : \hat{\Omega}_s \to \Omega_s, \quad \hat{D}(\hat{x}) = \hat{x} + \hat{u}_s = x,$$

$$D : \Omega_s \to \hat{\Omega}_s, \quad D(x) = x - u_s = \hat{x}.$$

Since $\det\hat{\nabla}\hat{D} = \det\hat{F} \neq 0$, the displacements D and $\hat{D}$ are well defined. With this notation, we introduce the Eulerian pressure p_s, displacement u_s,

$$p_s(x) = \hat{p}_s(D(x)) = \hat{p}_s(\hat{x}), \quad u_s(x) = \hat{u}_s(D(x)) = \hat{u}_s(\hat{x}).$$

Here, the difficulty is that u_s is only implicitly determined by $\hat{u}_s$, since $D(x)$ also depends on u_s. This leads us to introduce the 'set of initial positions' (IP set) $\Phi(t, \Omega)$ of all points of Ω at time t. If we look at a given 'material' point at the position $x \in \Omega$ and the time $t \in (0, T]$, then the value $\Phi(t, x)$ will tell us what the initial position of this point was at time $t = 0$. These points are transported in the full domain with a certain velocity w. The convection velocity in the structure will be the structure velocity itself, $w_{|\Omega_s} = v_s$, but in the fluid domain a harmonic (or biharmonic) extension is used for stability reasons. With this notation, the mapping Φ is determined by the following transport equation for the IP set:

Problem 3.4 (IP set equation). Find $\Phi \in \Phi_0 + \mathcal{V}^0$, such that

$$(\partial_t\Phi + w \cdot \nabla\Phi, \psi) = 0 \quad \forall\psi \in V^0, \tag{3.4}$$

where Φ_0 is a suitable extension of the Dirichlet data along the boundaries,

$$\Phi(0, x) = x, \quad x \in \Omega, \quad \Phi(t, x) = x, \quad \{t, x\} \in (0, T] \times \partial\Omega.$$

This means that $\hat{x} + \hat{u}(t, \hat{x}) = x$, for any point with the initial position $\hat{x}$ and the position x later at time t. Since $\hat{x} = \Phi(0, \hat{x}) = \Phi(t, x)$ and $\hat{u}(t, \hat{x}) = u(t, x)$, it follows that $\Phi = x - u$. Using this in the IP set transport equation yields

$$(\partial_t u - w + w \cdot \nabla u, \psi) = 0 \quad \forall\psi \in V^0. \tag{3.5}$$

Differentiating the identity $D(\hat{D}(\hat{x})) = \hat{x}$ yields

$$(I - \nabla u)(I + \hat{\nabla}\hat{u}) = I \quad \Leftrightarrow \quad \hat{\nabla}\hat{u} = (I - \nabla u)^{-1} - I.$$

Thus, the Cauchy stress tensor σ_s can be written for INH and STVK materials in the Eulerian framework as follows:

$$\sigma_s = \begin{cases} -p_s I + \mu_s(FF^T - I) & \text{(INH material)}, \\ J^{-1}F(\lambda_s(\mathrm{tr}E)I + 2\mu_s E)F^T & \text{(STVK material)}, \end{cases}$$

$$F = I + \hat{\nabla}\hat{u} = (I - \nabla u)^{-1}, \quad J = \det F, \quad E = \tfrac{1}{2}(F^T F - I).$$

This leads us to the following Eulerian formulation of the structure equations:

Problem 3.5 (Eulerian INH structure model). Find $\{u_s, v_s, p_s\} \in \{u_s^D + \mathcal{V}_s^0\} \times \{v_s^D + \mathcal{V}_s^0\} \times \mathcal{L}_s$, such that $u_s(0) = u_s^0$, $v_s(0) = v_s^0$,

$$
\begin{aligned}
(\rho_s(\partial_t v_s + v_s \cdot \nabla v_s), \psi^u)_s + (\sigma_s, \epsilon(\psi^u))_s &= f_s(\psi^u), \\
(\partial_t u_s - v_s + v_s \cdot \nabla u_s, \psi^v)_s + (\nabla \cdot v_s, \psi^p)_s &= 0,
\end{aligned}
\tag{3.6}
$$

for all $\{\psi^u, \psi^v, \psi^p\} \in V_s^0 \times V_s^0 \times L_s$ where $\sigma_s := -p_s I + \mu_s(FF^T - I)$,

$$
F := (I - \nabla u)^{-1}, \quad \epsilon(\psi^u) = \tfrac{1}{2}(\nabla \psi^u + \nabla \psi^{uT}).
$$

Problem 3.6 (Eulerian STVK structure model). Find $\{u_s, v_s\} \in \{u_s^D + \mathcal{V}_s^0\} \times \{v_s^D + \mathcal{V}_s^0\}$, such that $u_s(0) = u_s^0$, $v_s(0) = v_s^0$,

$$
\begin{aligned}
(\rho_s(\partial_t v_s + v_s \cdot \nabla v_s), \psi^u)_s + (\sigma_s, \epsilon(\psi^u))_s &= f_s(\psi^u), \\
(\partial_t u_s - v_s + v_s \cdot \nabla u_s, \psi^v)_s &= 0,
\end{aligned}
\tag{3.7}
$$

for all $\{\psi^u, \psi^v\} \in V_s^0 \times V_s^0$, where $\sigma_s := J^{-1}F(\lambda_s(\text{tr}E)I + 2\mu_s E)F^T$,

$$
F := (I - \nabla u)^{-1}, \quad J := \det F, \quad E := \tfrac{1}{2}(F^T F - I).
$$

Now, we can combine the Eulerian formulations of the fluid and structure part to a monolithical variational formulation of the coupled FSI problem in Eulerian frame. The IP set function Φ allows us to determine the characteristic functions χ_s and $\chi_f := 1 - \chi_s$ of structure and fluid domain, respectively, by

$$
\chi_f(t, x) = \begin{cases} 1, & \Phi(t, x) = x - u(t, x) \in \hat{\Omega}_f \setminus \hat{\Gamma}_i. \\ 0, & \Phi(t, x) = x - u(t, x) \in \hat{\Omega}_s. \end{cases}
$$

The fluid-structure interface is implicitly determined by

$$
\Gamma_i(t) = \{x \in \Omega, \ x - u(x, t) \in \hat{\Gamma}_i\}.
\tag{3.8}
$$

The evaluation of Φ requires a continuation of the displacement and its gradient from the structure domain into the fluid domain. The value of u in the fluid domain is determined by the choice of the convection velocity w, for which we use the *harmonic continuation* of the structure velocity to Ω. In the case of the compressible STVK material, we also use the harmonic continuation of the fluid's pressure into the structure domain.

Problem 3.7 (Eulerian formulation of FSI problem). Find $\{v, w, u, p\} \in \{v^D + \mathcal{V}^0\} \times \mathcal{V}^0 \times \mathcal{V}^0 \times \mathcal{L}$, such that $v(0) = v^0$, $u(0) = u^0$, and

$$
\begin{aligned}
(\rho(\partial_t v + v \cdot \nabla v), \psi) + (\sigma, \epsilon(\psi)) &= f(\psi) \quad \forall \psi \in V^0, \\
(\nabla \cdot v, \chi) &= 0 \quad \forall \chi \in L \quad \text{(INH material)}, \\
(\chi_f \nabla \cdot v, \chi) + (\chi_s \alpha_p \nabla p, \nabla \chi) &= 0 \quad \forall \chi \in L \quad \text{(STVK material)}, \\
(\partial_t u - w + w \cdot \nabla u, \psi) &= 0 \quad \forall \psi \in V^0, \\
(\chi_s(w - v), \psi) + (\chi_f \alpha_w \nabla w, \nabla \psi) &= 0 \quad \forall \psi \in V^0,
\end{aligned}
\tag{3.9}
$$

where α_p, α_w are positive constants, $\chi_s = 1 - \chi_f$, and

$$\chi_f := \begin{cases} 1, & x - u \in \hat{\Omega}_f \setminus \hat{\Gamma}_i, \\ 0, & x - u \in \hat{\Omega}_s, \end{cases} \qquad \rho/\sigma := \begin{cases} \rho_f/\sigma_f & \text{in } \Omega_f, \\ \rho_s/\sigma_s & \text{in } \Omega_s, \end{cases}$$

$$\sigma_f := -pI + 2\rho_f \nu_f \epsilon(v),$$

$$\sigma_s := \begin{cases} -pI + \mu_s(FF^T - I) & \text{(INH material)}, \\ J^{-1}F(\lambda_s(\operatorname{tr}E)I + 2\mu_s E)F^T & \text{(STVK material)}, \end{cases}$$

$$F := (I - \nabla u)^{-1}, \quad J := \det F, \quad E := \tfrac{1}{2}(F^T F - I).$$

In some situations the solution of an FSI problem may tend to a 'steady state' $\{v^*, u^*, w^*\}$ as $t \to \infty$ (e.g., driven cavity)

$$v_f^* := \lim_{t \to \infty} v_{|\Omega_f}, \quad v_s^* \equiv 0, \quad w^* \equiv 0,$$

$$u_s^* := \lim_{t \to \infty} u_{|\Omega_s}, \quad u_f^* = u_f^{\lim} := \lim_{t \to \infty} u_{|\Omega_f}.$$

Here, u_f^* depends on $w_{|\Omega_f}$ as harmonic extension of $w_{|\Omega_s}$.

Problem 3.8 (Eulerian formulation of the 'stationary' FSI problem). Find $\{u, v, p\} \in \{u^D + V^0\} \times \{v^D + V^0\} \times L$, such that

$$(\rho v \cdot \nabla v, \psi) + (\sigma, \epsilon(\psi)) = 0 \quad \forall \psi \in V^0,$$

$$(\nabla \cdot v, \chi) = 0 \quad \forall \chi \in L \quad \text{(INH material)},$$

$$(\chi_f \nabla \cdot v, \chi) + (\chi_s \alpha_p \nabla p, \nabla \chi) = 0 \quad \forall \chi \in L \quad \text{(STVK material)},$$

$$(\chi_f(u - u_f^{\lim}), \psi) + (\chi_s v, \psi) = 0 \quad \forall \psi \in V^0.$$

In the following Table 3, we summarize the two monolithical variational formulations of the FSI problem, the (arbitrary) Lagrangian–Eulerian (ALE) and the Eulerian–Eulerian formulation.

3.3. Numerical approximation

The Galerkin discretization is based on a variational formulation. For arguments $U = \{v, w, u, p\}$ and $\Psi = \{\psi^v, \psi^w, \psi^u, \psi^p\} \in \mathcal{W} := [\mathcal{V}]^4$, we introduce the space-time semi-linear form

$$A(U)(\Psi) := \int_0^T \Big\{ (\rho(\partial_t v + v \cdot \nabla v), \psi^v) + (\sigma(v), \epsilon(\psi^v))$$

$$+ \begin{cases} (\nabla \cdot v, \psi^p) & \text{(INH material)} \\ (\chi_f \nabla \cdot v, \psi^p) + (\chi_s \alpha_p \nabla p, \nabla \psi^p) & \text{(STVK material)} \end{cases}$$

$$- (f, \psi^v) + (\partial_t u - w + w \cdot \nabla u, \psi^u)$$

$$+ (\chi_s(w - v), \psi^w) + (\chi_f \alpha_w \nabla w, \nabla \psi^w) \Big\} dt.$$

Then, the variational formulation of the FSI problem in the Eulerian frame reads:

TABLE 3. Variational formulations of the FSI problem.

I) *ALE formulation* (INH):

$$\hat{\Omega} = \hat{\Omega}_f \cup \hat{\Gamma}_i \cup \hat{\Omega}_s$$

Find $\{\hat{v}, \hat{u}, \hat{p}\}$ *for* $\hat{v}(0)$, $\hat{u}(0)$ *with*

$$(\rho \hat{J} \partial_t \hat{v} + \hat{\chi}_f \rho \hat{J} (\partial_t \hat{u} - \hat{v}) \cdot \widehat{\nabla} \hat{v} \hat{F}^{-1}, \psi)$$
$$+ (\hat{J} \hat{\sigma} \hat{F}^{-T}, \hat{\epsilon}(\psi)) = \hat{f}(\psi)$$

$$(\hat{\chi}_s (\hat{J} - 1) + \hat{\chi}_f \widehat{\nabla} \cdot (\hat{J} \hat{v} \cdot \hat{F}^{-T}), \xi) = 0$$

$$(\partial_t \hat{u} - \hat{\chi}_s \hat{v}, \psi) + (\hat{\chi}_f \widehat{\nabla} \hat{u}, \widehat{\nabla} \psi) = 0$$

(harmonically continued $\hat{u}_s$ *into* Ω_f*)*

for all test fields $\{\psi, \xi, \psi, \psi\}$, *where*

$$\hat{\chi}_s(\hat{x}) := \begin{cases} 0, & \hat{x} \in \hat{\Omega}_f \\ 1, & \hat{x} \in \hat{\Omega}_s \cup \hat{\Gamma}_i \end{cases}$$

$$\hat{\chi}_f := 1 - \hat{\chi}_s$$

$$\hat{\sigma} := \begin{cases} -\hat{p}_f + \mu_f (\widehat{\nabla} \hat{v}_f \hat{F}^{-1} + \hat{F}^{-T} \widehat{\nabla} \hat{v}_f^T) \\ -\hat{p}_s + \mu_s (\hat{F} \hat{F}^T - I) \quad in \ \hat{\Omega}_s \cup \hat{\Gamma}_i \end{cases}$$

$$\hat{F} := I + \widehat{\nabla} \hat{u}, \quad \hat{J} := \det \hat{F}$$

II) *Eulerian formulation* (INH):

$$\Omega = \Omega_f \cup \Gamma_i \cup \Omega_s$$

Find $\{v, w, u, p\}$ *for* $v(0)$, $u(0)$ *with*

$$(\rho \partial_t v + \rho v \cdot \nabla v, \psi) + (\sigma, \epsilon(\psi)) = f(\psi)$$

$$(\nabla \cdot v, \xi) = 0$$

$$(\partial_t u - w + w \cdot \nabla u, \psi) = 0$$

$$(\chi_s (w - v), \psi) + (\chi_f \nabla w, \nabla \psi) = 0$$

(harmonically continued w_s *into* Ω_f*)*

for all test fields $\{\psi, \xi, \psi, \psi\}$, *where*

$$\chi_s(x) := \begin{cases} 0, & x - u \in \hat{\Omega}_f \\ 1, & x - u \in \hat{\Omega}_s \cup \hat{\Gamma}_i \end{cases}$$

$$\chi_s := 1 - \chi_f$$

$$\sigma := \begin{cases} -p_f I + 2\mu_f \epsilon(v_f) \quad in \ \Omega_f \\ -p_s I + \mu_s (FF^T - I) \quad in \ \Omega_s \end{cases}$$

$$F := (I - \nabla u)^{-1}$$

Problem 3.9. Find $U \in U^D + \mathcal{W}^0$, such that

$$A(U)(\Psi) = 0 \quad \forall \Psi \in \mathcal{W}^0, \tag{3.10}$$

where U^D is an appropriate extension of the Dirichlet boundary and initial data and the space $\mathcal{W}^0$ is defined by

$$\mathcal{W}^0 := \{\Psi \in [\mathcal{V}^0]^4, \ \psi^u(0) = \psi^v(0) = 0\}.$$

The spatial discretization is by 'equal-order' Q_1 finite elements (d-linear shape functions) for all unknowns and the time discretization by the cG(r) ('continuous' Galerkin or Crank–Nicolson scheme) method, with $r = 1$. The corresponding finite-element spaces for velocity and pressure on meshes $\mathbb{T}_h$ are denoted by V_h. This spatial discretization needs stabilization in order to compensate for the lacking 'inf-sup stability'. As an alternative to the usual SDLS (Streamline Diffusion Least-Squares) or SUPG (Streamline Upwinded Petrov Galerkin) stabilization of Hughes et al. [38, 39], we suggest the LPS (Local Projection Stabilization) method proposed in Becker/Braack [3] and Braack/Burman [16], which has several advantages over the other stabilization methods. For formulating this method, we use the cellwise weighted L^2-scalar product

$$(\varphi, \psi)_\delta := \sum_{K \in \mathbb{T}_h} \delta_K (\varphi, \psi)_K,$$

$$\delta_K := \alpha \big(\chi_f \rho_f \nu_f h_K^{-2} + \chi_s \mu_s h_K^{-2} + \beta \rho |v_h|_{\infty;K} h_K^{-1} + \gamma |w_h|_{\infty;K} h_K^{-1} \big)^{-1}.$$

Further, we define the 'fluctuation operator' $\pi_h : V_h \to V_{2h}$ on the finest mesh level $\mathbb{T}_h$ by $\pi_h = I - P_{2h}$, where $P_{2h} : V_h \to V_{2h}$ is the L^2-projection. For this construction it is assumed that the meshes $\mathbb{T}_h$ always consist of regular 2×2-patches of cells of width h_T, such that the spaces V_h and V_{2h} are well defined. With this notation, we define the stabilization form

$$S_\delta(U_h)(\Psi_h) := \int_0^T \big\{ (\nabla \pi_h p_h, \nabla \pi_h \psi_h^p)_\delta + (\rho v_h \cdot \nabla \pi_h v_h, v_h \cdot \nabla \pi_h \psi_h^v)_\delta$$

$$+ (w_h \cdot \nabla \pi_h u_h, w_h \cdot \nabla \pi_h \psi_h^u)_\delta \big\} \, dt.$$

In this form the first term stabilizes the fluid pressure, the second term the structure pressure, the third term the transport in the flow model, and the fourth term the transport of the displacement u_h. For more details see also the article Rannacher [47] in this volume.

Now, the finite-element approximation of the variational problem (3.10) is formulated with discrete analogues $\mathcal{W}_h^0$ of the spaces $\mathcal{W}^0$ as follows:

Problem 3.10 (Finite-element approximation of the FSI problem). Find $U_h \in U_h^D + \mathcal{W}_h^0$, such that

$$A_\delta(U_h)(\Psi_h) := A(U_h, \Psi_h) + S_\delta(U_h)(\Psi_h) = 0 \quad \forall \Psi_h \in \mathcal{W}_h^0. \tag{3.11}$$

For computing steady-state solutions, we use a pseudo-time stepping techniques based on the simple (first-order) backward Euler scheme. The solution of the nonlinear algebraic systems in each of the (implicit) time steps is achieved by a standard Newton method with adaptive step-length selection. The stabilization terms and the terms involving the characteristic function χ_f, determining the position of the interface, are treated by a simple functional iteration. The resulting linear subproblems are solved by GMRES with preconditioning by a geometric multigrid method with block-ILU smoothing.

3.4. Mesh adaptation

Since the finite-element approximation (3.11) of the FSI problem is based on a 'monolithical' variational formulation, we can apply the concept of the DWR (Dual Weighted Residual) method described above for organizing automatic mesh adaptation. Below, we will compare the performance of three different ways of mesh refinement:

- *uniform mesh refinement* using several steps of uniform (edge) bisection of a coarse initial mesh,

- *zonal mesh refinement* using a purely geometry-based criterion by marking all cells for refinement which have certain prescribed distances from the fluid-structure interface,
- *local mesh refinement* driven by a systematic residual-based criterion (DWR method) by marking all cells for refinement which have error indicators above a certain threshold.

We recall the basics of the DWR method in a simplified form applied to the present situation of finite-element approximation of the FSI problem. For details, see Becker/Rannacher [7] and the article Rannacher [47] in this volume.

If the value $J(U)$ for some (linear) functional $J(\cdot)$ is be computed, the approximation is to be controlled in terms of the error

$$|J(U) - J(U_h)| = |J(U - U_h)| \leq TOL.$$

We assume the existence of the directional derivative

$$A'(U)(\Phi, \Psi) := \lim_{\tau \to 0} \frac{1}{\tau} \{ A(U + \tau\Phi)(\Psi) - A(U)(\Psi) \}, \quad \Phi, \Psi \in \mathcal{W}^0,$$

and introduce the bilinear form

$$L(U, U_h)(\Phi, \Psi) := \int_0^1 A'(U_h + s(U - U_h))(\Phi, \Psi) \, ds.$$

Let $Z \in \mathcal{W}^0$ be the solution of the *'dual problem'*

$$L(U, U_h)(\Phi, Z) = J(\Phi) \quad \forall \Phi \in \mathcal{W}^0. \tag{3.12}$$

the existence of which is assumed. Then, taking $\Phi = U - U_h \in \mathcal{W}^0$ yields the error representation

$$J(U - U_h) = L(U, U_h)(U - U_h, Z)$$
$$= \int_0^1 A'(U_h + s(U - U_h))(U - U_h, Z) \, ds = A(U)(Z) - A(U_h)(Z).$$

Further, using the perturbed Galerkin orthogonality property

$$A(U)(\Psi_h) - A(U_h)(\Psi_h) = \text{'}STAB\text{'}, \quad \Psi \in \mathcal{W}_h^0,$$

we obtain

$$J(U - U_h) = A(U)(Z - \Psi_h) - A(U_h)(Z - \Psi_h) - S_\delta(U_h)(\Psi_h)$$
$$= F(Z - \Psi_h) - A(U_h)(Z - \Psi_h) - S_\delta(U_h)(\Psi_h)$$
$$=: \rho(U_h)(Z - \Psi_h) + \text{'}STAB\text{'}$$
$$= \sum_{K \in \mathbb{T}_h} \rho_K(U_h)(Z - \Psi_h) + \text{'}STAB\text{'},$$

where $\Psi_h \in \mathcal{W}^0$ is an arbitrary element. By cellwise integration by parts and rearranging boundary terms, we obtain

$$|J(U - U_h)|$$

$$\approx \sum_{K \in \tilde{\mathbb{T}}_h} \Big\{ (\nabla \cdot \sigma(U_h) - \rho v_h \cdot \nabla v_h, z^v - z_h^v)_K - (\tfrac{1}{2}[\sigma(U_h) \cdot n], z^v - z_h^v)_{\partial K}$$

$$- (\chi_f \nabla \cdot v_h + \chi_s \nabla \cdot u_h, z^p - z_h^p)_K - (\chi_f(u_h - u_f) + \chi_s v_h, z^u - z_h^u)_K \Big\},$$

and estimating further

$$|J(U - U_h)| \approx \sum_{K \in \tilde{\mathbb{T}}_h} \eta_K, \qquad \eta_K := \sum_{i=1}^{4} \rho_K^{(i)} \omega_K^{(i)}, \tag{3.13}$$

with the residual terms and weights

$$\rho_K^{(1)} := \|\nabla \cdot \sigma(U_h) - \rho v_h \cdot \nabla v_h\|_K, \quad \omega_K^{(1)} := \|Z - Z_h\|_K,$$

$$\rho_K^{(2)} := \tfrac{1}{2} h_K^{-1/2} \|[\sigma(U_h) \cdot n]\|_{\partial K}, \quad \omega_K^{(2)} := h_K^{1/2} \|z^v - z_h^v\|_{\partial K},$$

$$\rho_K^{(3)} := \|\chi_f \nabla \cdot v_h + \chi_s \nabla \cdot u_h\|_K, \quad \omega_K^{(3)} := \|z^p - z_h^p\|_K,$$

$$\rho_K^{(4)} := \|\chi_f(u_h - u_f) + \chi_s v_h\|_K, \quad \omega_K^{(4)} := \|z^u - z_h^u\|_K.$$

The steps from this 'theoretical' to a 'practical' error estimate are as follows:

- Linearization of the dual problem:

$$L(U, U_h)(\Phi, \Psi) \approx L(U_h, U_h)(\Phi, \Psi) = A'(U_h)(\Phi, \Psi).$$

This linearization causes an error, which is quadratic in the error $E := U - U_h$ and can be safely neglected in most cases (compare with the general error representation in Proposition 1.13).

- Solution of the 'discrete' dual problem for $Z_h \in \mathcal{W}_h^0$:

$$A'(U_h)(\Phi, Z_h) = J(\Phi_h) \quad \forall \Phi_h \in \mathcal{W}_h^0.$$

- Approximation of dual solution: From Z_h, we generate improved approximations to Z in a post-processing step by patchwise higher-order interpolation. On 2×2-patches of cells in $\mathbb{T}_h$ the 9 nodal values of the piecewise bilinear Z_h are used to construct a patchwise biquadratic function $\tilde{Z}$.

The result is a 'practical' a posteriori error estimate:

$$|J(U - U_h)| \approx \eta := \sum_{K \in \mathbb{T}_h} \rho_K(U_h) \omega_K(\tilde{Z}). \tag{3.14}$$

On the basis of this error estimate the automatic mesh adaptation is organized along the following lines. Let an error tolerance TOL be given.

1. Compute the primal solution U_h on the current mesh, starting from some initial state, e.g., that with zero deformation.
2. Compute the solution $\tilde{Z}_h$ of the approximate discrete dual problem.
3. Evaluate the cell-error indicators $\eta_K := \rho_K(U_h) \omega_K(\tilde{Z}_h)$.

4. If $\eta < TOL$ (the given tolerance), then accept U_h and evaluate $J(U_h)$, otherwise proceed to the next step.

5. Determine the 25% cells with largest and the 10% cells with smallest values of η_K. The cells of the first group are refined and those of the second group coarsened. Then, continue with Step 1. By this strategy it can be achieved that the number of cells stays about constant during the adaptation process within a time-stepping procedure.

Remark 3.11. The solvability of the dual problem (3.12) is not for granted and has to be established by exploiting the particular properties of the primal problem (3.11). This is a difficult task in view of the rather few existence results in the literature for general FSI problems. However, in our test calculation, we have never encountered difficulties in obtaining the 'discrete' dual solution. In the test examples presented below, the discrete dual problem has been set up simply by transposing the Newton matrix at the final solution.

Remark 3.12. The above assumption of differentiability of the semi-linear form $A(U)(\cdot)$ may cause concerns in treating the FSI problem since the dependence of the characteristic function $\chi_f(u)$ on the deflection u is generically not differentiable (only Lipschitzian). However, this non-differentiability is confined to the interface between fluid and structure which can be assumed to form a lower-dimensional manifold. Hence, for practical applications, after discretization along the interface, the directional derivative can safely be replaced by a mesh-size dependent difference quotient. This pragmatic approach has proven very successful in similar situations, e.g., for Hencky elasto-plasticity Rannacher/Suttmeier [48].

For illustrating the application of the DWR method to the FSI problem, particularly the computation of the linearized dual problem, we use the following simplified model situation formulated with the notation from above:

$$\Omega = \Omega_s \cup \Gamma_i \cup \Omega_f,$$

$$A(U)(\Psi) := \big(a(U)\epsilon[U], \epsilon[\Psi]\big) = (f, \Psi) \quad \forall \Psi \in V(\Omega),$$

where

$$a(U)(x) := \begin{cases} C_s, & x - U(x) \in \hat{\Omega}_s, \\ C_f, & x - U(x) \notin \hat{\Omega}_s. \end{cases}$$

The corresponding finite-element approximation is

$$A(U_h)(\Psi_h) = (f, \Psi_h) \quad \forall \Psi_h \in V_h,$$

with the residual functional

$$\rho(U_h)(\Psi) := (f, \Psi) - A(U_h)(\Psi), \quad \Psi \in V(\Omega).$$

Using the formal definition of the directional derivative $A'(U)(\Phi, \Psi)$, we find for sufficiently smooth interface Γ_i and Φ that

$$A'(U)(\Phi, \Psi) = \big(a(U)\epsilon[\Phi], \epsilon[\Psi]\big) + (n \cdot [C_s - C_f]\epsilon[\Phi], \Psi)_{\Gamma_i}.$$

Then, using the solution $Z \in V(\Omega)$ of the corresponding dual problem

$$A'(U)(\Phi, Z) = J(\Phi) \quad \forall \Phi \in V(\Omega),$$

we obtain the a posteriori error representation

$$J(U - U_h) \approx \rho(U_h)(Z - I_h Z).$$

The realization of this methods for the full FSI problem proceeds analogously but involves many more terms to evaluate. For the details, we refer to Dunne [20]. In the test examples considered below, the discrete dual problem has been set up simply by transposing the Newton matrix at the final solution.

3.5. Stationary test case 'Elastic Flow Cavity'

At first, we present a stationary test case. The fluid is modeled by the linear Stokes equations and the structure material is incompressible neo-Hookean. The flow and solid deformation velocity are small enough to allow for a stationary solution of the coupled linear systems. This solution is computed by a pseudo-time stepping method employing the implicit Euler scheme. A steady state is reached once the kinetic energy of the structure is below the tolerance $\|v_s\|^2 \leq 10^{-8}$.

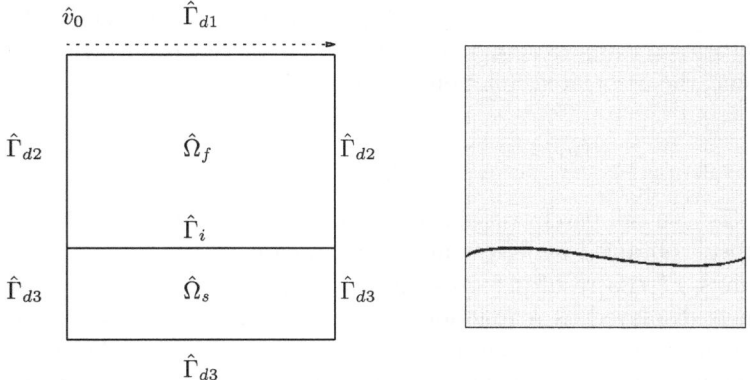

FIGURE 15. Configuration of the 'elastic flow cavity' problem (left) and final position of the interface (right).

The cavity has a size of 2×2, and its elastic part has a height of 0.5. The material constants are $\rho_f = \rho_s = 1$, $\nu_f = 0.2$, and $\mu_s = 2.0$. The tangential flow profile at the top boundary is given by $v_0(x) = 0.54x$, $x \in [0.0, 0.25]$, $v_0(x) = 1$, $x \in (0.25, 1.75)$, and $v_0(x) = 4(2 - x)$, $x \in [1.75, 2.0]$.

Figure 16 shows the development of $\|v_s\|^2$ during the pseudo-time-stepping process depending on the number of cells of the mesh and the mass error of the structure at the stationary state. As expected the kinetic energy tends to zero. The multiple 'bumps' in the time development of the kinetic energy occur due to the way the elastic structure reaches its stationary state by 'swinging' back and forth a few times. At the extreme point of each swing the kinetic energy has a local minimum. The mass error of the structure at the stationary state is of the expected order $O(h^2)$.

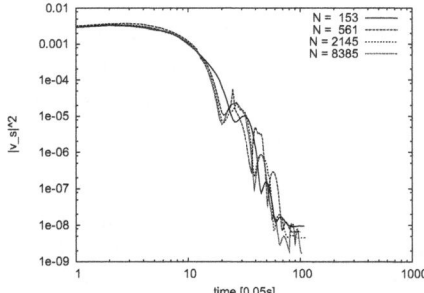 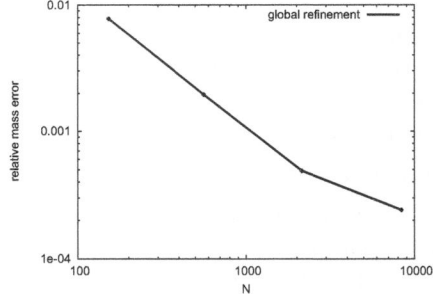

FIGURE 16. Time variation of $\|v_s\|^2$ for different N (left) and mass error in the steady state (right) on globally refined meshes.

For testing the 'goal-oriented' mesh adaptation, we take the value of the pressure at the point $A = (0.5, 1.0)^T$ which is located in the flow region. To avoid sharp singularities in the corresponding dual solution and the resulting unnecessary overrefinement, the associated functional is regularized to

$$J(u, p) := |S_A|^{-1} \int_{S_A} p \, dx \approx p(A),$$

where $S_A \in \mathbb{T}_h$ is a cell patch containing the point A. As reference value of $p(A)$, we use the result obtained on a very fine uniform mesh.

In Figures 17 the resulting pressure error and the relative error in mass conservation is displayed as a function of the number of mesh cells. Figure 18 shows a sequence of adapted meshes. As expected two effects can be seen. There is local refinement around the point of interest and since the position of the fluid-structure interface is a decisive factor for the pressure field, local refinement also occurs along the interface.

It may seem surprising that in Figure 17 there is no reduction of the mass error in the last iteration. This is due to the approach we are using here. After each step of mesh adaption a new primal solution is calculated, starting with the initial state of no deformation. The sensitivity analysis though does not take the initial state into account. Mesh adaption takes place around the final state of the interface, it does not reflect its initial state. An easy way of alleviating the mass error problem is to explicitly move a certain amount of local refinements with the interface from one time step to the next. Doing that though in this example would have made it unclear if the local refinement at the final interface position was due to the sensitivity analysis or the explicit movement of interface-bound refinement.

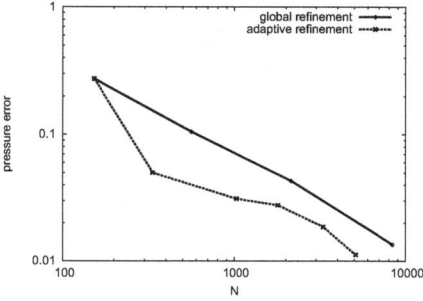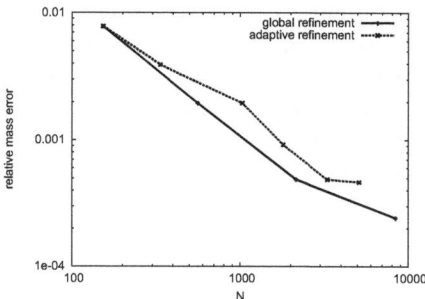

FIGURE 17. Error reduction (left) and error of mass conservation (right).

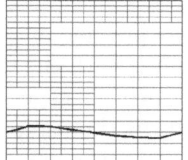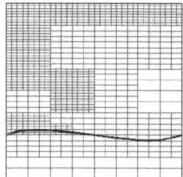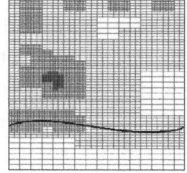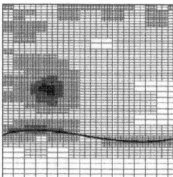

FIGURE 18. Locally adapted meshes with $N = 335$, $N = 1031$, $N = 3309$, and $N = 5123$ cells.

3.6. Non-stationary test case 'Vibrating Thin Plate'

Next, we consider the benchmark configuration shown in Figure 11. Along the upper and lower boundary the usual 'no-slip' condition is used for the velocity. At the (left) inlet a constant parabolic inflow profile is prescribed which drives the flow, and at the (right) outlet the 'do-nothing' boundary condition $\nu\partial_n v - pn = 0$ is realized. This implicitly forces the pressure to have zero mean value at the outlet. The initial condition is zero flow velocity and structure displacement. As material properties, we assume the fluid as incompressible and Newtonian, the cylinder as fixed and rigid, and the structure as (compressible) St. Venant–Kirchhoff (STVK) type.

The first set of computations is done on globally refined meshes for validating the proposed method and its software implementation. Then, for the same configuration adaptive meshes are used where the refinement criteria are either purely heuristic, i.e., based on the cell distance from the interface, or are based on a simplified stationary version of the DWR approach (at every tenth time step) as already used before for the cavity example. In all cases a uniform time-step size of $0.005\,s$ is used. Since in the Eulerian approach the structure deformations are not in a Lagrangian framework, it is not clear how well the mass of the structure is conserved in an Eulerian approach.

Figure 19 illustrates the treatment of corners in the structure by the IP set approach compared to the LS approach. In the Level Set (LS) method the interface is identified by all points for which $\varphi = 0$, while in the IP set method the interface is identified by all points which are on one of the respective isoline segments belonging to the edges of the bar. The differences are visible in the cells that contain the corners.

FIGURE 19. Treatment of corners by the LS method (left) and by the IP set method (right).

Since in the Eulerian approach the structure deformations are not in a Lagrangian framework, it is not immediately clear, due to the coupling with the fluid, how well the mass of the structure is conserved in an Eulerian approach, especially in the course of an instationary simulation comprising hundreds of time steps. In Figure 20, we display the bar's relative mass error as a function of time. Except for certain initial jitters, the relative error is less than 1%.

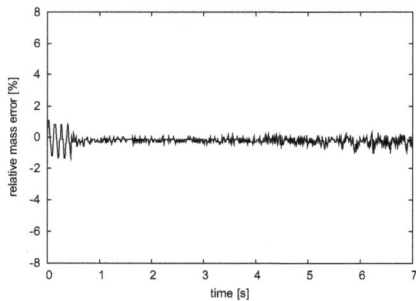

FIGURE 20. Mass conservation error of the IP set method.

Figure 21 illustrates the time dynamics of the structure and the adapted meshes over the time interval $[0, T]$. For both approaches, we obtain a periodic oscillation with maximum amplitudes and frequency, which are quite close to each other: 1.6e-2 versus 1.51e-2 and $6.86s^{-1}$ versus $6.70s^{-1}$.

In the last test case (St. Venant–Kirchhoff material) the fluid is initially in rest and the bar is subjected to a vertical (gravitational) force. This causes the bar to bend downward until it touches the bottom wall. A sequence of snapshots of the transition to steady state obtained by the Eulerian approach for this problem is shown in Figure 22. The position of the trailing tip A is shown in Figure 23. We see that by sensitivity-driven local refinement by the DWR method on only 1900

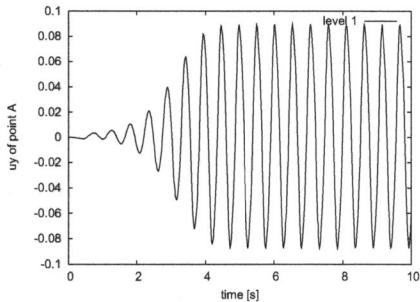 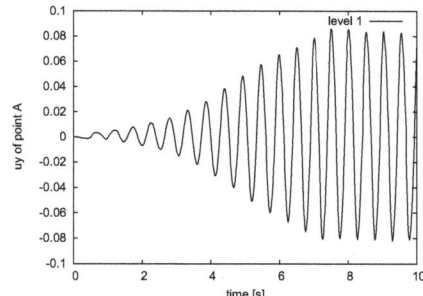

FIGURE 21. Vertical displacement of end point A: ALE approach: max $= 0.0882m$, $\nu = 1.95\,s^{-1}$, Eulerian approach: max $= 0.0822m$, $\nu = 1.92\,s^{-1}$.

cells almost the same accuracy can be achieved as by zonal refinement on 12300 cells. The gain in CPU time needed is almost 85 %.

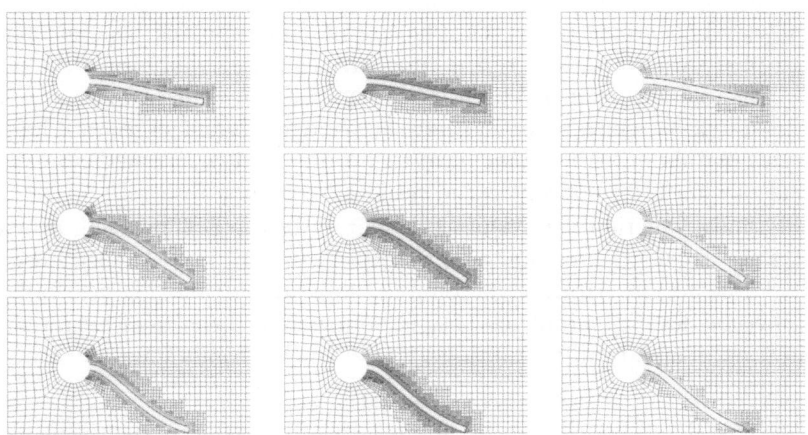

FIGURE 22. Zonal and local refinement: $N \sim 3{,}000$ (left column), $N \sim 12{,}000$ (middle column), $N \sim 1{,}900$ (right column).

3.7. Open problems and further development

In this section, we presented a fully Eulerian variational formulation for 'fluid-structure interaction' (FSI) problems. At several examples the Eulerian approach turns out to yield results which are in good agreement with those obtained by the standard ALE approach. In order to have a 'fair' comparison both methods have been implemented using the same numerical components and software library GASCOIGNE [4]. The method based on the Eulerian approach is inherently more

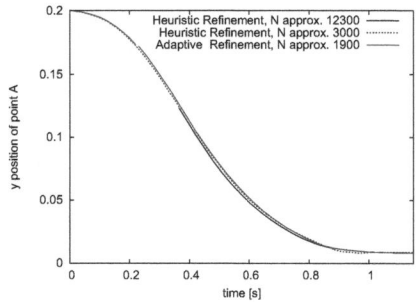 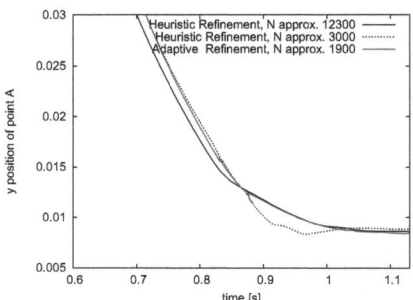

FIGURE 23.   Position $x_A(t)$ of point A: required CPU time: zonal 30 h, local 4 h.

expensive than the ALE method, by about a factor of two, but it can also treat large deformations and topology changes.

One conceptional disadvantage of the Eulerian approach to treating FSI problems is the need for a time-independent outer domain Ω, in which the FSI process takes place. This seems to prevent the use of this approach for simulating flow through blood vessels since here the time-varying vessel wall forms the 'outer domain'. This difficulty can be cured by embedding the whole vessel into an outer softer medium, as seems realistic from the biological point of view. The physical properties of this outer medium have to be provided by biomedical experience. The realization of this concept is the subject of ongoing research.

The fully Eulerian variational formulation of the FSI problem provides the basis for the application of the 'dual weighted residual' (DWR) method for 'goal-oriented' a posteriori error estimation and mesh adaptation. The feasibility of the DWR method for FSI problems has, in a first step, been demonstrated for the computation of steady state solutions. The following topics have to be worked on in the future:

- The DWR method has to be applied also for non-stationary FSI problems, particularly for the simultaneous adaptation of spatial mesh and time step size.
- Another goal is the development of the Eulerian approach for three-dimensional FSI problems and to explore its potential for FSI problems with large deformations and topology changes, such as occurring in heart-valve simulations.
- Application to optimal control problems with FSI

$$\min_{q \in Q} J(u, q)\,! \qquad a(u, q)(\psi) = f(\psi) \; \forall \psi \in V.$$

In the context of the 'all-at-once' approach (KKT System) goal-oriented error estimation is 'relatively cheap'.

References

[1] E. Bänsch and W. Dörfler, *Adaptive finite elements for exterior domain problems.* Numer. Math. **80** (1998), 497–523.

[2] W. Bangerth and R. Rannacher, *Adaptive Finite-Element Methods for Differential Equations.* Lectures in Mathematics, ETH Zürich, Birkhäuser, Basel 2003.

[3] R. Becker and M. Braack, *A finite-element pressure gradient stabilization for the Stokes equations based on local projections.* Calcolo **38** (2001), 173–199.

[4] R. Becker, M. Braack, and D. Meidner: *Gascoigne: A C++ numerics library for scientific computing.* Institute of Applied Mathematics, University of Heidelberg, URL http://www.gascoigne.uni-hd.de/ .

[5] R. Becker and Th. Dunne, *VisuSimple: An interactive visualization utility for scientific computing.* Abschlußband SFB 359, Reactive Flows, Diffusion and Transport (W. Jäger et al., eds.), Springer, Berlin-Heidelberg New York, 2007.

[6] R. Becker, V. Heuveline, and R. Rannacher, *An optimal control approach to adaptivity in computational fluid mechanics.* Int. J. Numer. Meth. Fluids. **40** (2002), 105–120.

[7] R. Becker and R. Rannacher, *An optimal control approach to error estimation and mesh adaptation in finite-element methods.* Acta Numerica 2000 (A. Iserles, ed.), pp. 1-102, Cambridge University Press, 2001.

[8] A. Belmonte, H. Eisenberg, and E. Moses, *From flutter to tumble, Inertial drag and froude similarity in falling paper.* Phys. Rev. Lett. **81** (1998), 345–348.

[9] S. Bönisch, *Adaptive Finite-Element Methods for Rigid Particulate Flow Problems.* Doctoral thesis, Institute of Applied Mathematics, University of Heidelberg, 2006.

[10] S. Bönisch and V. Heuveline, *On the numerical simulation of the instationary free fall of a solid in a fluid. I. The Newtonian case.* Computer & Fluids **36** (2007), 1434–1445.

[11] S. Bönisch and V. Heuveline, *On the numerical simulation of the free fall of a solid in a fluid. II. The viscoelastic case.* SFB Preprint 2004-32, University of Heidelberg, 2004.

[12] S. Bönisch and V. Heuveline, *Advanced flow visualization with HiVision.* Abschlußband SFB 359, Reactive Flows, Diffusion and Transport (W. Jäger et al., eds.), Springer, Berlin-Heidelberg New York, 2007.

[13] S. Bönisch, V. Heuveline, R. Rannacher, *Numerical simulation of the free fall problem.* Proc. Int. Conf. on High Performance Scientific Computing (HPSCHanoi 2003), Hanoi, March 2003, (H.G. Bock, et al., eds.), Springer, Berlin-Heidelberg, 2005.

[14] S. Bönisch, V. Heuveline und P. Wittwer: *Adaptive boundary conditions for exterior flow problems,* J. Math. Fluid Mech. **7** (2005), 85–107.

[15] S. Bönisch, V. Heuveline und P. Wittwer: *Second order adaptive boundary conditions for exterior flow problems: non-symmetric stationary flows in two dimensions,* J. Math. Fluid Mech. **8** (2006), 1–26.

[16] M. Braack and E. Burman: *Local projection stabilization for the Oseen problem and its interpretation as a variational multiscale method,* SIAM J. Numer. Anal. **43** (2006), 2544–2566.

[17] R. Bürger, R. Liu, and W.L. Wendland, *Existence and stability for mathematical models of sedimentation-consolidation processes in several space dimensions*. J. Math. Anal. Appl. **264** (2001), 288–310.

[18] C. Conca, J.S. Martin, and M. Tucsnak, *Existence of the solutions for the equations modelling the motion of rigid body in a viscous fluid*. Commun. Partial Differ. Equations **25** (2000), 1019–1042.

[19] B. Desjardin and M. Esteban, *Existence of weak solutions for the motion of rigid bodies in viscous fluids*. Arch. Rational Mech. Anal. **46** (1999), 59–71.

[20] Th. Dunne, *Adaptive Finite-Element Simulation of Fluid Structure Interaction Based on an Eulerian Formulation*. Doctoral thesis, Institute of Applied Mathematics, University of Heidelberg, 2007.

[21] Th. Dunne and R. Rannacher, *Adaptive finite-element approximation of fluid-structure interaction based on an Eulerian variational formulation*. In 'Fluid-Structure Interaction: Modelling, Simulation, Optimisation' (H.-J. Bungartz and M. Schäfer, eds.), Springer's LNCSE-Series, 2006.

[22] S.B. Field, M. Klaus, M.G. Moore, and F. Nori *Chaotic dynamics of falling disks*. Nature **388** (1997), 252–254.

[23] G.P. Galdi, *On the steady self-propelled motion of a body in a viscous incompressible fluid*. Arch. Rational Mech. Anal. **148** (1999), 53–88.

[24] G.P. Galdi, *On the motion of a rigid body in a viscous liquid: A mathematical analysis with applications*. Handbook of Mathematical Fluid Mechanics (S. Friedlander and D. Serre, eds.), Elsevier, 2001.

[25] G.P. Galdi and A. Vaidya, *Translational steady fall of symmetric bodies in a Navier-Stokes liquid, with application to particle sedimentation*. J. Math. Fluid Mech. **3** (2001), 183–211.

[26] G.P. Galdi, A. Vaidya, M. Pokorny, D.D. Joseph, J. Feng, *Orientation of bodies sedimenting in a second-order liquid at non-zero Reynolds number*. Math. Models Methods Appl. Sci. **12** (2002),1653-1690.

[27] V. Girault and P.-A. Raviart, *Finite-Element Methods for the Navier–Stokes Equations*. Springer: Berlin-Heidelberg-New York, 1986.

[28] R. Glowinski, T.W. Pan, T.I. Hesla, D.D. Joseph, and J. Periaux, *A distributed Lagrange multiplier/fictitious domain method for flow around moving rigid bodies: Application to particulate flow*. Int. J. Numer. Meth. Fluids **30** (1999), 1043–1066.

[29] M.D. Gunzburger, H.C. Lee, and G.A. Seregin, *Global existence of weak solutions for viscous incompressible flows around a moving rigid body in three dimensions*. J. Math. Fluid Mech. **2** (2000), 219–266.

[30] V. Heuveline and R. Rannacher, *Adaptive FEM for eigenvalue problems with application in hydrodynamic stability analysis*. Proc. Int. Conf. 'Advances in Numerical Mathematics', Moscow, Sept. 16–17, 2005 (W. Fitzgibbon et al., eds.), pp. 109–140, Institute of Numerical Mathematics RAS, Moscow, 2006.

[31] V. Heuveline, *HiFlow: A multi-purpose finite-element package*, Rechenzentrum, Universität Karlsruhe, URL `http://hiflow.de/` .

[32] V. Heuveline, *HiVision: A visualization platform*, Rechenzentrum, Universität Karlsruhe, URL `http://hiflow.de/` .

[33] K.-H. Hoffmann and V.N. Starovoitov, *Zur Bewegung einer Kugel in einer zähen Flüssigkeit.* Doc. Math. **5** (2000), 15–21.

[34] J. Hron and S. Turek, *Proposal for numerical benchmarking of fluid-structure interaction between an elastic object and laminar incompressible flow.* In 'Fluid-Structure Interaction: Modelling, Simulation, Optimisation' (H.-J. Bungartz and M. Schäfer, eds.), in Springer's LNCSE-Series, 2006.

[35] J. Hron and S. Turek, *Fluid-structure interaction with applications in biomechanics.* In 'Fluid-Structure Interaction: Modelling, Simulation, Optimisation' (H.-J. Bungartz and M. Schäfer, eds.), in Springer's LNCSE-Series, 2006.

[36] H.H. Hu, *Direct simulation of flows of solid-liquid mixtures.* Int. J. Multiphase Flow **22** (1996), 335–352.

[37] H.H. Hu, D.D. Joseph, and M.J. Crochet, *Direct simulation of fluid particle motions.* Theor. Comp. Fluid Dyn. **3** (1992), 285–306.

[38] T.J.R. Hughes and A.N. Brooks, *Streamline upwind/Petrov–Galerkin formulations for convection dominated flows with particular emphasis on the incompressible Navier–Stokes equation.* Comput. Meth. Appl. Mech. Engrg. **32** (1982), 199–259.

[39] T.J.R. Hughes, L.P. Franc, and M. Balestra, *A new finite-element formulation for computational fluid mechanics: V. Circumventing the Babuska-Brezzi condition: A stable Petrov-Galerkin formulation of the Stokes problem accommodating equal order interpolation.* Comput. Meth. Appl. Mech. Engrg. **59** (1986), 85–99.

[40] C. Liu and N.J. Walkington, *An Eulerian description of fluids containing visco-elastic particles.* Arch. Rat. Mech. Anal. **159** (2001), 229–252.

[41] N.A. Patankar, *A formulation for fast computations of rigid particulate flows.* Center Turbul. Res., Ann. Res. Briefs, 185–196 (2001).

[42] N.A. Patankar, *Physical interpretation and mathematical properties of the stress-DLM formulation for rigid particulate flows.* Int. J. Comp. Meth. Engrg Sci. Mech. **6** (2005), 137–143.

[43] N.A. Patankar and D.D. Joseph, *Modeling and numerical simulation of particulate flows by the Eulerian-Lagrangian approach.* Int. J. Multiphase Flow **27** (2001), 1659–1684.

[44] N.A. Patankar and D.D. Joseph, *Lagrangian numerical simulation of particulate flows.* Int. J. Multiphase Flow **27** (2001), 1685-1706.

[45] N.A. Patankar, P. Singh, D.D. Joseph, R. Glowinski, and T.-W. Pan, *A new formulation of the distributed Lagrange multiplier/fictitious domain method for particulate flows.* Int. J. Multiphase Flow **26** (2000), 1509–1524.

[46] R. Rannacher, *Finite-element methods for the incompressible Navier–Stokes equations.* In 'Fundamental Directions in Mathematical Fluid Mechanics' (G.P. Galdi et al., eds.), pp. 191–293, Birkhäuser, Basel, 2000.

[47] R. Rannacher, *Methods for numerical flow simulation.* In this volume.

[48] R. Rannacher and F.-T. Suttmeier, *Error estimation and adaptive mesh design for FE models in elasto-plasticity.* In Error-Controlled Adaptive FEMs in Solid Mechanics (E. Stein, ed.), pp. 5–52, John Wiley, Chichster, 2002.

[49] N. Sharma and N.A. Patankar, *A fast computation technique for the direct numerical sim,ulation of rigid particulate flows.* J. Comp. Phys. **205** (2005), 439–457.

[50] T.E. Tezduyar, M. Behr, and J. Liou, *A new strategy for finite-element flow computations involving moving boundaries and interfaces-the deforming-spatial-domain/space-time procedures: I. The concept and preliminary tests, II. Computation of free-surface ows, two-liquid ows and ows with drifting cylinders.* Computer Methods in Applied Mechanics and Engineering, 1992.

[51] S.V. Tsynkov, *Numerical solution of problems on unbounded domains. a review.* Appl. Numer. Math. **27** (1998), 465–532.

[52] S.V. Tsynkov, *External boundary conditions for three-dimensional problems of computational aerodynamics.* SIAM J. Sci. Comput. **21** (1999), 166–206.

[53] S. Turek, *Efficient solvers for incompressible flow problems: an algorithmic and computational approach.* Springer, Heidelberg-Berlin-New York, 1999.

[54] S.O. Unverdi and G. Tryggvason, *Computations of multi-fluid flows.* Physica D **60** (1992), 70–83.

[55] VisuSimple, *VisuSimple: An open source interactive visualization utility for scientific computing.* Institute of Applied Mathematics, University of Heidelberg, URL `http://www.visusimple.uni-hd.de/` .

[56] D. Wan and S. Turek, *Direct numerical simulstion of particulate flow via multigrid FEM techniques and the fictitious boundary method.* Int. J. Numer. Meth. Fluids **51** (2006), 531–566.

[57] J. Wang, R. Bai, C. Lewandowski, G.P. Galdi, and D.D. Joseph, *Sedimentation of cylindrical particles in a viscoelastic liquid: shape-tilting.* China Particuology **2** (2004), 13-18.

[58] H.F. Weinberger, *On the steady fall of a body in a Navier–stokes fluid.* Proc. Symp. Pure Mathematics **23** (1973), 421–440.

[59] P. Wittwer, *On the structure of stationary solutions of the Navier-Stokes equations.* Commun. Math. Phys. **226** (2002), 455–474.

Sebastian Bönisch, Thomas Dunne and Rolf Rannacher
Institute of Applied Mathematics
University of Heidelberg
Im Neuenheier Feld 293/294
D-69120 Heidelberg
Germany
e-mail: `sebastian.boenisch@iwr.uni-heidelberg.de`
       `thomas.dunne@iwr.uni-heidelberg.de`
       `rolf.rannachere@iwr.uni-heidelberg.de`

Hemodynamical Flows. Modeling, Analysis and Simulation
Oberwolfach Seminars, Vol. 37, 379–501
© 2008 Birkhäuser Verlag Basel/Switzerland

Numerical Techniques for Multiphase Flow with Liquid-Solid Interaction

Stefan Turek and Jaroslav Hron

The authors acknowledge the contribution by Dmitri Kuzmin, Decheng Wan and other members of the institute and the support by the German Research Foundation (DFG, grant TU 102/11-3 and SFB TRR 30) and partial support by the Nečas Center for Mathematical Modeling (project LC06052)

Introduction

In many fluid applications, particularly in multiphase flow problems with liquid-solid interaction based on the incompressible Navier–Stokes equations, the mathematical description and the numerical schemes have to be designed in such a way that quite complicated constitutive relations and interactions between fluid and solids can be incorporated into existing flow solvers in an accurate and robust manner. In the following sections, first of all we describe finite-element discretization strategies and corresponding solver techniques (approximate Newton methods, multilevel pressure Schur complement techniques, operator-splitting approaches) for the resulting discrete systems of equations. The need for the development of robust and efficient iterative solvers for implicit high-resolution discretization schemes is emphasized and the numerical treatment of extensions of the Navier–Stokes equations (Boussinesq approximation, $k - \varepsilon$ turbulence model) is addressed which is evaluated by simulation results for prototypical applications including multiphase and granular flows. In the second part, a fully monolithic finite-element approach is described for fluid-structure interactions with elastic materials which is applied to several benchmark configurations. In the third part, the concept of FEM fictitious-boundary techniques, together with operator-splitting approaches for particulate flow, is introduced which allows the efficient simulation of systems with many solid particles of different shape and size.

Contents

1. Numerical methods for incompressible flow

1.1. Introduction

For single-phase Newtonian fluids occupying a domain $\Omega \subset \mathbf{R}^d$ $(d = 2, 3)$ during the time interval $(t_0, t_0 + T]$, the incompressible Navier–Stokes equations

$$\frac{\partial \mathbf{u}}{\partial t} - \nu \Delta \mathbf{u} + \mathbf{u} \cdot \nabla \mathbf{u} + \nabla p = \mathbf{f} \quad , \quad \mathrm{Div}\, \mathbf{u} = 0 \qquad (1.1)$$

describe the laminar flow motion which depends on the physical properties of the fluid (viscosity ν) and, possibly, on some external forces $\mathbf{f}$ like buoyancy. The constant density ρ is hidden in the pressure $p(x_1, \ldots, x_d, t)$ which adjusts itself instantaneously so as to render the time-dependent velocity field $\mathbf{u}(x_1, \ldots, x_d, t)$ divergence-free. The problem formulation is completed by specifying the initial and boundary values for each particular application.

Although these equations seem to have a quite simple structure, they constitute a 'grand challenge' problem for mathematicians, physicists, and engineers, and they are (still) object of intensive research activities in the field of *Computational Fluid Dynamics* (CFD), especially since they are the basis for more complex nonlinear, multiphase flow models, particularly arising in bioengineering applications. Incompressible flow problems are interesting from the viewpoint of applied mathematics and scientific computing, since they include the whole range of difficulties which typically arise in the numerical treatment of partial differential equations. Therefore, they provide a perfect starting point for the development of reliable numerical algorithms and efficient software for CFD simulations.

Specifically, the typical problems which scientists and engineers are frequently confronted with, concern the following aspects:

- time-dependent partial differential equations in complex domains,
- strongly nonlinear and stiff systems of strongly coupled equations,
- convection-dominated transport at high Reynolds numbers ($Re \approx \frac{1}{\nu}$),
- saddle-point problems due to the incompressibility constraint,
- local changes of the problem character in space and time.

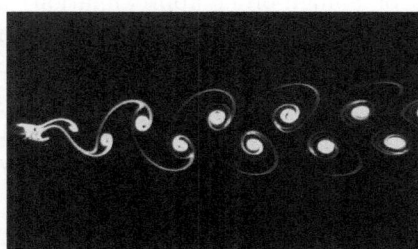 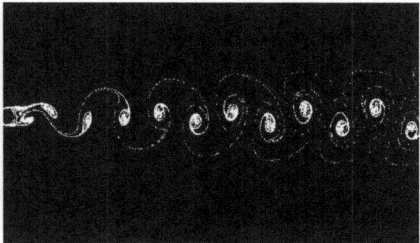

FIGURE 1. Experiment (source: Van Dyke's 'Album of Fluid Motion') vs. numerical simulation (source: 'Virtual Album of Fluid Motion') for flow around a cylinder.

These peculiarities of this 'model problem' impose stringent requirements on virtually all stages of algorithm design: discretization, solver, and software engineering. In particular, the following difficulties must be taken into account:

- nonlinear systems for millions of unknowns (large but sparse matrices),
- conditional stability (explicit schemes) and/or proper time step control,
- anisotropic/unstructured meshes (boundary layers, complex geometries).

Active research aimed at the development of improved numerical methods for the incompressible Navier–Stokes equations has been going on for more than three decades. The number of publications on this topic is enormous (see the book by Gresho et al. [41] for a comprehensive overview). However, in many cases the computational results produced by the available CFD tools are only *qualitatively* correct. A *quantitatively* precise flow prediction for real-life problems requires still that the accuracy of discretization schemes is enhanced and/or the solvers become more efficient. This can be easily demonstrated by benchmark computations [111], especially for non-stationary flows.

Moreover, a current trend in CFD is to combine the 'basic' Navier–Stokes equations (1.1) with more complex models which describe industrial applications involving turbulence, multiphase flow, nonlinear fluids, combustion/detonation, free and moving boundaries, fluid-structure interaction, weakly compressible effects, etc. These extensions, which will be discussed in the present chapter, have one thing in common: they require highly accurate *and* robust discretization techniques as well as efficient solution algorithms for generalized Navier–Stokes-like systems. In order to design and implement such powerful numerical methods for real-life problems, many additional aspects need to be taken into account. The main ingredients of an 'ultimate' CFD code are as follows:

- advanced mathematical methods for PDEs ($\rightarrow$ *discretization*),
- efficient solution techniques for algebraic systems ($\rightarrow$ *solver*),
- reliable and hardware-optimized software ($\rightarrow$ *implementation*).

If all of these components were available, the number of unknowns could be significantly reduced (e.g., via adaptivity or a high-order approximation) and, moreover, discrete problems of the same size could be solved more efficiently. Hence, the marriage of optimal numerical methods and fast iterative solvers would make it possible to exploit the potential of modern computers to the full extent and enhance the performance of incompressible flow solvers (improve the MFLOP/s rates) by *orders of magnitude*. This is why these algorithmic aspects still play an increasingly important role in contemporary CFD research.

In this chapter, we briefly review the Multilevel Pressure Schur Complement (MPSC) approach to the solution of incompressible flow problems and combine it with FEM discretization techniques. We will explain some characteristics of special high-resolution FEM schemes as applied to incompressible flow problems and discuss some computational details regarding the efficient numerical solution of the resulting nonlinear and linear algebraic systems. Furthermore, we will examine different coupling mechanisms between the 'basic' flow model (standard Navier–Stokes equations for velocity and pressure) and additional scalar or vector-valued transport equations, being representative for multiphase flow problems and configurations in biofluid mechanics, in order to capture the relevant physical effects.

1.2. Discretization of the Navier–Stokes equations in time

We consider numerical solution techniques for the incompressible Navier–Stokes equations,

$$\mathbf{u}_t - \nu \Delta \mathbf{u} + \mathbf{u} \cdot \nabla \mathbf{u} + \nabla p = \mathbf{f} \quad , \quad \text{Div}\, \mathbf{u} = 0 \quad , \quad \text{in } \Omega \times (0, T], \qquad (1.2)$$

for given force $\mathbf{f}$ and viscosity ν, with prescribed boundary values on the boundary $\partial\Omega$ and an initial condition at $t = 0$. Solving this problem numerically is still a considerable task in the case of long time calculations and high Reynolds numbers, particularly in 3D and also in 2D if the time dynamics is complex. Examples for 2D and 3D results can be found in www.featflow.de/album.

The common solution approach is a separate discretization in space and time. We first (semi-) discretize in time by one of the usual methods known from the treatment of ordinary differential equations, such as the forward or backward Euler-, the Crank–Nicolson or fractional-step-θ scheme, or others, and obtain a sequence of generalized stationary Navier–Stokes problems.

Basic θ-scheme:

Given $\mathbf{u}^n$ and $k = t_{n+1} - t_n$, then solve for $\mathbf{u} = \mathbf{u}^{n+1}$ and $p = p^{n+1}$

$$\frac{\mathbf{u} - \mathbf{u}^n}{k} + \theta[-\nu\Delta\mathbf{u} + \mathbf{u} \cdot \nabla\mathbf{u}] + \nabla p = \mathbf{g}^{n+1} \quad , \quad \text{Div}\,\mathbf{u} = 0 \quad , \quad \text{in } \Omega \qquad (1.3)$$

with right-hand side $\mathbf{g}^{n+1} := \theta\mathbf{f}^{n+1} + (1 - \theta)\mathbf{f}^n - (1 - \theta)[-\nu\Delta\mathbf{u}^n + \mathbf{u}^n \cdot \nabla\mathbf{u}^n]$.

The parameter θ has to be chosen depending on the time-stepping scheme, e.g., $\theta = 1$ for the backward Euler, or $\theta = 1/2$ for the Crank–Nicolson scheme. The pressure term $\nabla p = \nabla p^{n+1}$ may be replaced by $\theta\nabla p^{n+1} + (1 - \theta)\nabla p^n$, but, with appropriate post-processing, both strategies lead to solutions of the same accuracy. In all cases, we end up with the task of solving, at each time step, a nonlinear saddle point problem of given type which has then to be discretized in space.

In the past, explicit time-stepping schemes have been commonly used in non-stationary flow calculations, but because of the severe stability problems inherent in this approach, the required small time steps prohibit the efficient treatment of long-time flow simulations. Due to the high stiffness, one should prefer implicit schemes in the choice of time-stepping methods for solving this problem. Since implicit methods have become feasible, thanks to more efficient nonlinear and linear solvers, the schemes most frequently used are (still) either the simple first-order backward Euler scheme (BE), with $\theta = 1$, or more preferably the second-order Crank–Nicolson scheme (CN), with $\theta = 1/2$.

These two methods belong to the group of *one-step-θ-schemes*. The CN scheme occasionally suffers from numerical instabilities because of its only weak damping property (not strongly A-stable), while the BE-scheme is of first-order accuracy only (however: it is a good candidate for steady-state simulations). Another method which has proven to have the potential to excel in this competition is the *fractional-step-θ scheme* (FS). It uses three different values for θ and for the time step k at each time level.

For a realistic comparison of all mentioned schemes, we define a macro time step with $K = t_{n+1} - t_n$ as a sequence of 3 time steps of (possibly variable) size k. Then, in the case of the backward Euler or the Crank–Nicolson scheme, we perform three substeps with the same θ ($\theta = 0.5$ or $\theta = 1$) as above and (micro) time step $k = K/3$.

Backward Euler scheme:

$$[I + \tfrac{K}{3}N(\mathbf{u}^{n+\frac{1}{3}})]\mathbf{u}^{n+\frac{1}{3}} + \tfrac{K}{3}\nabla p^{n+\frac{1}{3}} \;=\; \mathbf{u}^n + \tfrac{K}{3}\mathbf{f}^{n+\frac{1}{3}},$$
$$\operatorname{Div}\mathbf{u}^{n+\frac{1}{3}} \;=\; 0,$$

$$[I + \tfrac{K}{3}N(\mathbf{u}^{n+\frac{2}{3}})]\mathbf{u}^{n+\frac{2}{3}} + \tfrac{K}{3}\nabla p^{n+\frac{2}{3}} \;=\; \mathbf{u}^{n+\frac{1}{3}} + \tfrac{K}{3}\mathbf{f}^{n+\frac{2}{3}},$$
$$\operatorname{Div}\mathbf{u}^{n+\frac{2}{3}} \;=\; 0,$$

$$[I + \tfrac{K}{3}N(\mathbf{u}^{n+1})]\mathbf{u}^{n+1} + \tfrac{K}{3}\nabla p^{n+1} \;=\; \mathbf{u}^{n+\frac{2}{3}} + \tfrac{K}{3}\mathbf{f}^{n+1}$$
$$\operatorname{Div}\mathbf{u}^{n+1} \;=\; 0$$

Crank–Nicolson scheme:

$$[I + \tfrac{K}{6}N(\mathbf{u}^{n+\frac{1}{3}})]\mathbf{u}^{n+\frac{1}{3}} + \tfrac{K}{6}\nabla p^{n+\frac{1}{3}} \;=\; [I - \tfrac{K}{6}N(\mathbf{u}^n)]\mathbf{u}^n + \tfrac{K}{6}\mathbf{f}^{n+\frac{1}{3}} + \tfrac{K}{6}\mathbf{f}^n,$$
$$\operatorname{Div}\mathbf{u}^{n+\frac{1}{3}} \;=\; 0,$$

$$[I + \tfrac{K}{6}N(\mathbf{u}^{n+\frac{2}{3}})]\mathbf{u}^{n+\frac{2}{3}} + \tfrac{K}{6}\nabla p^{n+\frac{2}{3}} \;=\; [I - \tfrac{K}{6}N(\mathbf{u}^{n+\frac{1}{3}})]\mathbf{u}^{n+\frac{1}{3}} + \tfrac{K}{6}\mathbf{f}^{n+\frac{2}{3}}$$
$$+ \tfrac{K}{6}\mathbf{f}^{n+\frac{1}{3}},$$
$$\operatorname{Div}\mathbf{u}^{n+\frac{2}{3}} \;=\; 0,$$

$$[I + \tfrac{K}{6}N(\mathbf{u}^{n+1})]\mathbf{u}^{n+1} + \tfrac{K}{6}\nabla p^{n+1} \;=\; [I - \tfrac{K}{6}N(\mathbf{u}^{n+\frac{2}{3}})]\mathbf{u}^{n+\frac{2}{3}} + \tfrac{K}{6}\mathbf{f}^{n+1}$$
$$+ \tfrac{K}{6}\mathbf{f}^{n+\frac{2}{3}},$$
$$\operatorname{Div}\mathbf{u}^{n+1} \;=\; 0.$$

Here and in the following, we use the more compact form for the diffusive and advective part:

$$N(\mathbf{v})\mathbf{u} := -\nu\Delta\mathbf{u} + \mathbf{v}\cdot\nabla\mathbf{u}. \qquad (1.4)$$

For the fractional-step-θ scheme we proceed as follows. Choosing $\theta = 1 - \frac{\sqrt{2}}{2}$, $\theta' = 1 - 2\theta$, and $\alpha = \frac{1-2\theta}{1-\theta}$, $\beta = 1 - \alpha$, the macro time step $t_n \to t_{n+1} = t_n + K$ is split into the three following consecutive substeps (with $\tilde{\theta} := \alpha\theta K = \beta\theta' K$):

Fractional-step-θ scheme:

$$[I + \tilde{\theta}N(\mathbf{u}^{n+\theta})]\mathbf{u}^{n+\theta} + \theta K \nabla p^{n+\theta} = [I - \beta\theta K N(\mathbf{u}^n)]\mathbf{u}^n + \theta K \mathbf{f}^n,$$
$$\text{Div}\,\mathbf{u}^{n+\theta} = 0,$$

$$[I + \tilde{\theta}N(\mathbf{u}^{n+1-\theta})]\mathbf{u}^{n+1-\theta} + \theta' K \nabla p^{n+1-\theta} = [I - \alpha\theta' K N(\mathbf{u}^{n+\theta})]\mathbf{u}^{n+\theta}$$
$$+\theta' K \mathbf{f}^{n+1-\theta},$$
$$\text{Div}\,\mathbf{u}^{n+1-\theta} = 0,$$

$$[I + \tilde{\theta}N(\mathbf{u}^{n+1})]\mathbf{u}^{n+1} + \theta K \nabla p^{n+1} = [I - \beta\theta K N(\mathbf{u}^{n+1-\theta})]\mathbf{u}^{n+1-\theta}$$
$$+\theta K \mathbf{f}^{n+1-\theta},$$
$$\text{Div}\,\mathbf{u}^{n+1} = 0.$$

Being a strongly A-stable scheme, the FS-method possesses the full smoothing property which is important in the case of rough initial or boundary values. Further, it contains only very little numerical dissipation which is crucial in the computation of non-enforced temporal oscillations in the flow. A rigorous theoretical analysis of the FS-scheme applied to the Navier–Stokes problem establishes second-order accuracy for this special choice of θ. Therefore, this scheme combines the advantages of both the classical CN-scheme (second-order accuracy) and the BE-scheme (strongly A-stable), but with the same numerical effort.

These results have been confirmed by extensive numerical simulations which are part of [122]. While that paper examined the accuracy of the above mentioned time-stepping schemes, also the question of fully coupled vs. operator-splitting techniques as well as the treatment of the nonlinearity has been addressed. One of the main statements – see also the results in [124] – was that the fractional-step-θ scheme is our preferred method, together with a fully implicit, fully coupled approach which solves the nonlinear saddle-point problems in each time step via Newton-like techniques together with special multigrid preconditioners ('Multilevel Pressure Schur Complement' [124]). While the question of the total efficiency w.r.t. operator-splitting or fully coupled approaches is hard to answer, this fully implicit approach is definitely the most accurate and most robust – but often most expensive, too – approach which is preferable from a purely mathematical viewpoint since finite-element arguments can be used for this Galerkin approach and, hence, error control and residual based adaptivity become feasible.

A modified fractional-step-θ scheme:
Consider an initial value problem of following form, with $X(t) \in \mathbf{R}^d, d \geq 1$:

$$\begin{cases} \dfrac{dX}{dt} &= f(X,t)\,, \quad \forall t > 0, \\ X(0) &= X_0. \end{cases}$$

Then, the modified θ-scheme (see [135]) with macro time step Δt can be written as three consecutive substeps, where $\theta = 1 - 1/\sqrt{2}$, $X^0 = X_0$, $n \geq 0$ and X^n is known:

$$\frac{X^{n+\theta} - X^n}{\theta \Delta t} = f\left(X^{n+\theta}, t^{n+\theta}\right),$$

$$X^{n+1-\theta} = \frac{1-\theta}{\theta} X^{n+\theta} + \frac{2\theta - 1}{\theta} X^n,$$

$$\frac{X^{n+1} - X^{n+1-\theta}}{\theta \Delta t} = f\left(X^{n+1}, t^{n+1}\right).$$

As shown in [36], the most important properties of this θ-scheme are that

- it is fully implicit;
- it is strongly A-stable;
- it is second-order accurate (in fact, it is "nearly" third-order accurate [36]).

These properties promise some advantageous behavior, particularly in implicit CFD simulations for non-stationary incompressible flow. Applying one step of this scheme to the Navier–Stokes equations, we obtain the following variant of the scheme:

1.
$$\begin{cases} \dfrac{\mathbf{u}^{n+\theta} - \mathbf{u}^n}{\theta \Delta t} + N(\mathbf{u}^{n+\theta})\mathbf{u}^{n+\theta} + \nabla p^{n+\theta} = \mathbf{f}^{n+\theta}, \\[2mm] \mathrm{Div}\,\mathbf{u}^{n+\theta} = 0, \end{cases}$$

2. $\mathbf{u}^{n+1-\theta} = \frac{1-\theta}{\theta}\mathbf{u}^{n+\theta} + \frac{2\theta-1}{\theta}\mathbf{u}^n,$

3.
$$\begin{cases} \dfrac{\mathbf{u}^{n+1} - \mathbf{u}^{n+1-\theta}}{\theta \Delta t} + N(\mathbf{u}^{n+1})\mathbf{u}^{n+1} + \nabla \widetilde{p}^{n+1} = \mathbf{f}^{n+1}, \\[2mm] \mathrm{Div}\,\mathbf{u}^{n+1} = 0, \end{cases}$$

3b. $p^{n+1} = (1-\theta)p^{n+\theta} + \theta \widetilde{p}^{n+1}.$

These three substeps build one macro time step and have to be compared with the previous description of the Backward Euler, Crank–Nicolson and fractional-step-θ scheme which all have been formulated in terms of a macro time step with three substeps, too. Then, the resulting accuracy and numerical cost are better comparable and the rating is fair. The main difference to the previous 'classical' schemes is that substeps 1. and 3. look like an backward Euler step while substep 2. is an extrapolation step only for previously computed data such that no operator evaluations at previous time steps are required.

Substep 3b. can be viewed as postprocessing step for updating the new pressure which however is not a must. In fact, in our numerical tests [135] we omitted this substep 3b. and accepted the pressure from substep 3. as final pressure approximation, that means $p^{n+1} = \widetilde{p}^{n+1}$.

Summarizing, one obtains that the numerical effort of the modified scheme for each substep is cheaper – at least for 'small' time steps (treatment of the nonlinearity) and complex right-hand side evaluations (for instance, in the case of

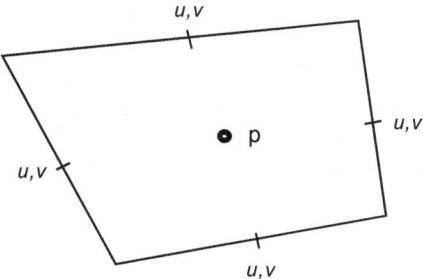

FIGURE 2. Nodal points of the non-conforming finite-element pair $\tilde{Q}_1/Q_0$.

particulate flow which involves collision models for many particles as source terms and CPU-dominant parts) – while the resulting accuracy is similar. Incidentally, the modified θ-scheme is a *Runge–Kutta* one; it has been derived in [36] as a particular case of the fractional-step-θ scheme (see the above reference for details).

1.3. Discretization of the Navier–Stokes equations in space

For the spatial discretization, we choose a finite-element approach. However finite volumes, finite differences or spectral methods are possible, too. A finite-element model of the Navier–Stokes equations is based on a suitable variational formulation. On the finite mesh $\mathcal{T}_h$ (triangles, quadrilaterals or their analogues in 3D) covering the domain Ω with local mesh size h, one defines polynomial trial functions for velocity and pressure. These spaces H_h and L_h should lead to numerically stable approximations as $h \to 0$, i.e., they should satisfy the so-called *inf-sup* (LBB) condition [35]

$$\min_{q_h \in L_h} \max_{\mathbf{v}_h \in H_h} \frac{(q_h, \mathrm{Div}\,\mathbf{v}_h)}{\|q_h\|_0 \, \|\nabla \mathbf{v}_h\|_0} \geq \gamma > 0 \qquad (1.5)$$

with a mesh-independent constant γ. On the other hand, equal order interpolations for velocity and pressure are also admissible provided that an a-priori unstable discretization is stabilized in an appropriate way (see, e.g., [56]).

In what follows, we employ the stable $\tilde{Q}_1/Q_0$ finite-element pair (*rotated bilinear/trilinear* shape functions for the velocities, and a piecewise constant pressure approximation). In the two-dimensional case, the nodal values are the mean values of the velocity vector over the element edges, and the mean values of the pressure over the elements (see Fig. 2).

This non-conforming finite element is a quadrilateral counterpart of the well-known triangular Stokes element of Crouzeix–Raviart [19] and can easily be defined in three space dimensions. A convergence analysis is given in [105] and computational results are reported in [113] and [114]. An important advantage of this finite-element pair is the availability of efficient multigrid solvers which are sufficiently robust in the whole range of Reynolds numbers even on non-uniform and highly anisotropic meshes [112],[125] if appropriate stabilization techniques are applied.

Such stabilization techniques for (non-conforming) FEM discretizations in the context of incompressible flow problems, for instance described by the Navier–Stokes equations or appropriate extensions in multiphase problem settings, are still a challenging task, particularly from a practical point of view regarding the following aspects:

I. Necessity. There are two well-known situations for non-conforming finite-element methods when severe numerical problems may arise: the lack of coercivity for non-conforming low-order approximations for symmetric deformation tensor formulations, mainly visible for the iterative solvers for small Re numbers, and whenever convective operators are dominant, for instance, for medium and high Re numbers or for the treatment of pure transport problems. Then, the standard Galerkin formulation fails and may lead to numerical oscillations and convergence problems of the iterative solvers, too.

II. Robustness. To analyze the quality of FEM stabilization terms for a wide range of Re numbers and for different problem types is one of the central tasks for the use of stabilization techniques as black box tools in future CFD codes. Here, the question of appropriate parameter settings and stabilization terms is always a delicate task, particularly on general meshes.

As a model problem we consider incompressible flow problems described by the generalized Navier–Stokes equations. The Cauchy stress tensor is given by $\sigma = 2\nu(D_{\mathbb{II}}(\mathbf{u}), p)\mathbf{D}(\mathbf{u}) - pI$, where p is the pressure; $\mathbf{D}(\mathbf{u}) = \frac{1}{2}(\nabla\mathbf{u} + \nabla^T\mathbf{u})$ is the rate of the deformation tensor, $\mathbf{u}$ denotes the velocity; $\nu(\cdot)$ is the (nonlinear) viscosity which may depend on the second invariant of the rate deformation tensor $D_{\mathbb{II}}(\mathbf{u}) = \frac{1}{2}\operatorname{tr}(\mathbf{D}^2(\mathbf{u}))$ and the pressure p. Then, depending on the specific viscosity function $\nu(\cdot)$ we consider the following prototypical models (with appropriate parameters ν_0, r). In the case of a Newtonian fluid with constant viscosity, we will replace the deformation tensor $\mathbf{D}(\mathbf{u})$ by the usual formulation with the gradient $\nabla\mathbf{u}$ only.

- Newtonian flow defined for $\nu(z, p) = \nu_0$,
- Non-Newtonian flow due to *Power law*, with $\nu(z, p) = \nu_0 z^{\frac{r}{2}-1}$, resp., *Bingham law* with $\nu(z, p) = \nu_0 z^{-\frac{1}{2}}$,
- Non-Newtonian flow with pressure and shear-dependent viscosity, as described for instance in [50, 79] or in the case of the *Schaeffer model* [110], with $\nu(z, p) = \nu_0 p z^{-\frac{1}{2}}$, for granular powder flow.

In all cases, the velocity $\mathbf{u}$ and the pressure p satisfy the following generalized Navier–Stokes equations:

$$\frac{\partial \mathbf{u}}{\partial t} + \mathbf{u} \cdot \nabla\mathbf{u} - \operatorname{Div}\left(2\nu(D_{\mathbb{II}}(\mathbf{u}), p)\mathbf{D}(\mathbf{u})\right) + \nabla p = \vec{f}, \quad \operatorname{Div}\mathbf{u} = 0, \qquad (1.6)$$

resp., for Newtonian flow with constant viscosity

$$\frac{\partial \mathbf{u}}{\partial t} + \mathbf{u} \cdot \nabla\mathbf{u} - \nu\Delta\mathbf{u} + \nabla p = \vec{f}, \quad \operatorname{Div}\mathbf{u} = 0. \qquad (1.7)$$

In the following, we mainly consider the stationary generalized Navier–Stokes problem (1.6) in a bounded domain $\Omega \subset \mathbf{R}^2$, and we use the rate of the deformation tensor instead of the gradient formulation as in (1.7), unless it is stated explicitly. If we restrict the set V of test functions to be divergence-free and if we take the constitutive laws into account, the above (stationary) equations (1.6) lead to:

$$\int_\Omega 2\nu(D_{\mathbb{II}}(\mathbf{u}), p)\mathbf{D}(\mathbf{u}) : \mathbf{D}(\mathbf{v})\,dx + \int_\Omega (\mathbf{u} \cdot \nabla\mathbf{u})\mathbf{v}\,dx = \int_\Omega \vec{f}\mathbf{v}\,dx, \ \forall \mathbf{v} \in V. \quad (1.8)$$

It is straightforward to penalize the constraint $\text{Div}\,\mathbf{v} = 0$ to derive the equivalent mixed formulations:

Find $(\mathbf{u}, p) \in X \times M$ *such that*

$$\int_\Omega 2\nu(D_{\mathbb{II}}(\mathbf{u}), p)\mathbf{D}(\mathbf{u}) : \mathbf{D}(\mathbf{v})\,dx + \int_\Omega (\mathbf{u} \cdot \nabla\mathbf{u})\mathbf{v}\,dx \ + \int_\Omega p\,\text{Div}\,\mathbf{v}\,dx$$

$$= \int_\Omega \vec{f}\mathbf{v}\,dx, \qquad\qquad \forall \mathbf{v} \in X, \qquad (1.9)$$

$$\int_\Omega q\,\text{Div}\,\mathbf{u}\,dx = 0, \qquad\qquad \forall q \in M,$$

with the spaces $X = [H_0^1(\Omega)]^2$ and $M = L^2(\Omega)$. In some test calculations, we also consider related Stokes problems, which means that we omit the convective term $\int_\Omega (\mathbf{u} \cdot \nabla\mathbf{u})\mathbf{v}\,dx$. For the following analysis, we introduce the bilinear forms:

$$\langle \boldsymbol{A}(\mathbf{w}, q)\mathbf{u}, \mathbf{v}\rangle = \int_\Omega 2\nu(D_{\mathbb{II}}(\mathbf{w}), q)\mathbf{D}(\mathbf{u}) : \mathbf{D}(\mathbf{v})\,dx, \qquad (1.10)$$

$$\langle \boldsymbol{N}(\mathbf{w})\mathbf{u}, \mathbf{v}\rangle = \int_\Omega (\mathbf{w} \cdot \nabla\mathbf{u})\mathbf{v}\,dx \quad , \quad \langle \boldsymbol{B}q, \mathbf{v}\rangle = \int_\Omega q\,\text{Div}\,\mathbf{v}\,dx. \qquad (1.11)$$

Then, we can rewrite our generalized flow problems in the following compact form:

Find $(\mathbf{u}, p) \in X \times M$ *such that*

$$\langle \boldsymbol{A}(\mathbf{u}, p)\mathbf{u}, \mathbf{v}\rangle + \langle \boldsymbol{N}(\mathbf{u})\mathbf{u}, \mathbf{v}\rangle + \langle \boldsymbol{B}p, \mathbf{v}\rangle = \int_\Omega \vec{f}\mathbf{v}dx, \quad \forall \mathbf{v} \in X,$$

$$\langle \boldsymbol{B}q, \mathbf{u}\rangle = 0, \qquad \forall q \in M. \qquad (1.12)$$

For the finite-element discretization, we consider a subdivision $T \in \mathcal{T}_h$ consisting of quadrilaterals in the domain $\Omega_h \in \mathbf{R}^2$, and we employ the non-conforming rotated bilinear *Rannacher–Turek* element. For any quadrilateral T, let (ξ, η) denote a local coordinate system obtained by joining the midpoints of the opposing faces of T. Then, in the *non-parametric* case, we set on each element T

$$\tilde{Q}_1(T) := span\left\{1, \xi, \eta, \xi^2 - \eta^2\right\}. \qquad (1.13)$$

The degrees of freedom are determined by the nodal functionals $\{F_\Gamma^{(a,b)}(\cdot), \Gamma \subset \partial\mathcal{T}_h\}$,

$$F_\Gamma^a := |\Gamma|^{-1}\int_\Gamma v d\gamma \quad \text{or} \quad F_\Gamma^b := v(m_\Gamma) \quad (m_\Gamma \text{ midpoint of edge } \Gamma) \qquad (1.14)$$

such that the finite-element space can be written as

$$W_h^{a,b} := \{v \in L_2(\Omega_h), v \in \tilde{Q}_1(T), \forall T \in \mathcal{T}_h, v \text{ continuous w.r.t. all}$$
$$\text{nodal functionals } F_{\Gamma_{i,j}}^{a,b}(\cdot), \text{ and } F_{\Gamma_{i0}}^{a,b}(v) = 0, \forall \Gamma_{i0}\}. \tag{1.15}$$

Here, $\Gamma_{i,j}$ denotes all inner edges sharing the two elements i and j, while Γ_{i0} denotes the boundary edges of $\partial\Omega_h$. In this paper, we always employ version 'a' with the integral mean values as degrees of freedom. Then, the corresponding discrete functions will be approximated in the spaces

$$V_h := W_h^a \times W_h^a, \ L_h := \left\{q_h \in L^2(\Omega), q_{h|T} = \text{const.}, \forall T \in \mathcal{T}_h\right\}. \tag{1.16}$$

Due to the loss of global continuity of the discrete velocities, the classical (discrete) 'Korn Inequality' $\sum_{T \in \mathcal{T}_h} \|\mathbf{v}\|_{H^1(T)} \leq c (\sum_{T \in \mathcal{T}_h} \|\mathbf{v}\|_{L^2(T)}^2 + \|\mathbf{D}(\mathbf{v})\|_{L^2(T)}^2)^{\frac{1}{2}}$ is not satisfied which is important for problems involving the symmetric part $\mathbf{D}(\mathbf{u})$ of the gradient. Therefore, appropriate edge-oriented stabilization techniques (see [8, 44, 132]) have been recently proposed which directly incorporate the jump across the inter-elementary boundaries. Another source of problems for FEM discretizations, not only for non-conforming finite elements, is the case of dominant convection, that means the case of medium and high Re numbers or if additional tracer equations are included. Consequently, in the following we discuss several stabilizations techniques for both sources of instability and provide a short overview on (typical) stabilization techniques for non-conforming FEM in the case of dominating convection and for satisfying Korn's inequality. Here, we only consider linear stabilization techniques in contrast to high-resolution schemes of FCT-FEM, resp., TVD-FEM type (see [68, 125] for the corresponding details).

FEM upwinding

The main idea is to introduce new edge-centered lumping regions and special lumping operators (for details see [125]). Then, the discrete convective operator $\langle \mathbf{N}(\mathbf{u}_h)\mathbf{v}_h, \mathbf{w}_h\rangle$ is replaced by

$$\langle \tilde{\mathbf{N}}(\mathbf{u}_h)\mathbf{v}_h, \mathbf{w}_h\rangle = \sum_l \sum_{k \in \Lambda_l} \oint_{\Gamma_{lk}} \mathbf{u}_h \cdot \mathbf{n}_{lk} d\gamma [1 - \lambda_{lk}(\mathbf{u}_h)(\mathbf{v}_h(m_k) - \mathbf{v}_h(m_l))]\mathbf{w}_h(m_l).$$
$$\tag{1.17}$$

Based on the local Reynolds number Re_T on each cell T (see [125] for details)

$$Re_T = \frac{\|\mathbf{u}\|_{\infty,T} \cdot h_T}{\nu} \tag{1.18}$$

we can define:

$$\lambda_{lk}(u_h) = \begin{cases} \dfrac{\frac{1}{2} + \delta^* Re_T}{1 + \delta^* Re_T} & \text{if } Re_T \geq 0, \\[2ex] \dfrac{1}{2(1 - \delta^* Re_T)} & \text{otherwise.} \end{cases} \tag{1.19}$$

Here, h_T is a local mesh size parameter on each cell T which can be critical on highly distorted meshes. Moreover, the appropriate choice of the free parameter

δ^* is quite sensitive and can significantly influence the resulting accuracy as the following test calculations will demonstrate.

FEM streamline-diffusion

This method was originally proposed by, for instance, Johnson [61] and by Hughes and Brooks [55], and it has been successfully applied to several classes of problems since streamline-diffusion methods can combine good stability and high accuracy. So, due to its simplicity, streamline-diffusion is a common tool used in many (commercial) CFD codes. It mainly consists of adding the following stabilization term to the original convection operator described by $\langle N(\mathbf{u}_h)\mathbf{v}_h, \mathbf{w}_h \rangle$, that means:

$$\langle \tilde{N}(\mathbf{u}_h)\mathbf{v}_h, \mathbf{w}_h \rangle = \langle N(\mathbf{u}_h)\mathbf{v}_h, \mathbf{w}_h \rangle + \sum_{T \in T_h} \delta_T \int_T (\mathbf{u}_h \cdot \nabla \mathbf{v}_h)(\mathbf{u}_h \cdot \nabla \mathbf{w}_h) dx. \quad (1.20)$$

Here, the critical quantity for an efficient computational treatment is the local damping parameter δ_T. A usual setting is described, for instance, in [125]. The local Reynolds number Re_T can be introduced as before and we can define, for instance,

$$\delta_T = \delta^* \cdot \frac{h_T}{\|\mathbf{u}\|_{\infty,\Omega}} \cdot \frac{2Re_T}{1 + Re_T}. \quad (1.21)$$

Due to the more complex bilinear form, including numerical integration in certain quadrature points, the numerical work for the matrix assembling process is increased and, again, the precise definition of δ^* and h_T, particularly on strongly anisotropic meshes containing large aspect ratios, can be critical. And, finally, the application to non-steady problems and the extension to more complex coupled problems or systems, as for instance Boussinesq approximations or multiphase flow problems is not fully clear.

Edge-oriented FEM stabilization

The main idea is to augment the original finite-element discretization by an interior penalty term involving the jump of the function values or of the gradient of the approximate FEM solution. The jump for a function $\mathbf{u}$ on an edge from ∂T can be defined by

$$[\mathbf{u}] = \begin{cases} \mathbf{u}^+ \cdot \mathbf{n}^+ + \mathbf{u}^- \cdot \mathbf{n}^- & \text{on internal edges,} \\ \mathbf{u} \cdot \mathbf{n} & \text{on Dirichlet boundary edges,} \\ 0 & \text{on Neumann boundary edges,} \end{cases} \quad (1.22)$$

where $\mathbf{n}$ is the outward normal to the edge and $(\cdot)^+$ and $(\cdot)^-$ indicate the value of the generic quantity $(\cdot)$ on the two elements sharing the same edge. In the literature, several jump terms were introduced for different situations:

I. Jump terms including function values:

$$j_1(\mathbf{u}, \mathbf{v}) = \sum_{\text{edge } E} \gamma \nu \frac{1}{|E|} \int_E [\mathbf{u}][\mathbf{v}] d\sigma. \quad (1.23)$$

John et al. [60] introduced the jump term in the non-conforming streamline-diffusion method for convection-dominated problems to achieve the same accuracy as with conforming streamline-diffusion FEM methods. Moreover, to satisfy the discrete Korn inequality, variants have been presented by Hansbo and Larson in [44] and by Turek, Ouazzi and Schmachtel in [132]. And, in a similar way, an analogous approach for a general Korn inequality for piecewise H^1-functions has been introduced by Brenner in [8].

II. Jump terms including the gradient:

$$j_{2,\alpha}(\mathbf{u}, \mathbf{v}) = \sum_{\text{edge } E} \gamma |E|^\alpha \int_E [\nabla \mathbf{u}][\nabla \mathbf{v}] d\sigma,$$

$$j_{3,\alpha}(\mathbf{u}, \mathbf{v}) = \sum_{\text{edge } E} \gamma |E|^\alpha \int_E [\mathbf{n} \cdot \nabla \mathbf{u}][\mathbf{n} \cdot \nabla \mathbf{v}] d\sigma,$$

$$j_{4,\alpha}(\mathbf{u}, \mathbf{v}) = \sum_{\text{edge } E} \gamma |E|^\alpha \int_E [\mathbf{t} \cdot \nabla \mathbf{u}][\mathbf{t} \cdot \nabla \mathbf{v}] d\sigma,$$ (1.24)

$$j_{5,\alpha}(\mathbf{u}, \mathbf{v}) = \sum_{\text{edge } E} \gamma |E|^\alpha \int_E [(\mathbf{t} \cdot \nabla \mathbf{u}) \cdot \mathbf{n}][(\mathbf{t} \cdot \nabla \mathbf{v}) \cdot \mathbf{n}] d\sigma.$$

These terms (with different α) for stabilizing convection dominated problems have been introduced by Burman, Hansbo et al. in a series of papers (see [9, 10]).

III. Jump terms including the divergence:

$$j(\mathbf{u}, \mathbf{v}) = \sum_{\text{edge } E} \gamma |E|^2 \int_E [\text{Div } \mathbf{u}][\text{Div } \mathbf{v}] d\sigma.$$ (1.25)

This approach was also originally proposed by Burman, Hansbo et al. to control the incompressibility condition.

IV. Jump terms including the normal component of function values:

$$j(\mathbf{u}, \mathbf{v}) = \sum_{\text{edge } E} \gamma \nu \frac{1}{|E|} \int_E [\mathbf{n} \cdot \mathbf{u}][\mathbf{n} \cdot \mathbf{v}] d\sigma.$$ (1.26)

To control the non-conformity arising from the pressure term in Darcy's law, this term has been introduced by Burman and Hansbo in [11].

Summarizing this overview, it shows that for different types of problems corresponding jump terms with "free" constant γ and order $|E|^\alpha$ have been introduced. Our own contribution is a new variant which uses only one jump term and one choice for the free parameter γ, and which is nevertheless able to simultaneously treat both types of problems. To be precise, in [131] we proposed the following jump term, resp., discrete stabilization term (with $h_E = |E|$):

$$\langle \mathbf{S}\mathbf{u}_h, \mathbf{v}_h \rangle = \sum_{\text{edge } E} \max(\gamma^* \nu h_E, \gamma h_E^2) \int_E [\nabla \mathbf{u}_h][\nabla \mathbf{v}_h] d\sigma,$$ (1.27)

which will be added to the original bilinear form, resp., discretized stiffness matrices, and which uses only the gradient of the approximate solution. This approach is based on a blending of (1.23), applying an appropriate scaling with h_E^2 due to the gradient of the discrete trial, resp., test functions, together with the formula in (1.24), applying $\alpha = 2$. In the case of nonlinear viscosity ν, this stabilization term may depend in a nonlinear way on the solution, too. The parameters γ, γ^* can be chosen – more or less arbitrarily as numerical tests have shown – in the interval $[0.0001, 0.1]$, with no significant influence on the resulting accuracy, robustness and efficiency. Summarizing, we can differ between the following situations:

1. In the case of treating the convective operators, $\mathbf{u} \cdot \nabla \mathbf{u}$, the upwinding and the streamline-diffusion approach modify the original discrete convective operator $\langle \boldsymbol{N}(\mathbf{u}_h)\mathbf{v}_h, \mathbf{w}_h \rangle$ by an operator $\langle \tilde{\boldsymbol{N}}(\mathbf{u}_h)\mathbf{v}_h, \mathbf{w}_h \rangle$ which may depend in a nonlinear way on the solution $\mathbf{u}_h$ itself via the local Re number; in contrast, the edge-oriented technique always adds a linear operator $\langle \mathbf{S}\mathbf{u}_h, \mathbf{v}_h \rangle$, unless an additional (nonlinear) shock-capturing term is used.

2. In the case of treating the deformation tensor formulation, $\mathbf{D}(\mathbf{u})$, the edge-oriented approach employs an additional stabilization term which mainly depends on the size of the viscosity ν; if ν depends on the solutions $\mathbf{u}_h$ and p_h, then this stabilization can become nonlinear. In the case of upwinding and streamline-diffusion, a corresponding stabilization is only performed if the convective operator is getting dominant. That means that for very low Re numbers, but with deformation tensor, no stabilization is applied since the local Re number vanishes.

In [131], the described edge-oriented stabilization approaches including jump terms of the gradient on the edges of the computational mesh together with low order non-conforming FEM spaces have been numerically analyzed for incompressible flow problems. The computational tests considered cases of various Re numbers (ranging from $Re = 1$, resp., Stokes flow, over medium Re numbers with periodically oscillating flow phenomena up to the case '$Re = \infty$'). Moreover, formulations including the gradient as well as the symmetric deformation tensor ('Korn's inequality') of the velocity have been considered, and we examined non-Newtonian problems with pressure and shear-dependent viscosity, too. The proposed edge-oriented stabilization term involves only the gradient of the discrete test and trial functions (leading to reduced assembling costs), which together with a special parameter setting taking into account the local size of the viscosity, if available, requires only one 'free' parameter γ. The numerical tests for various prototypical problems settings have shown that, in contrast to classical stabilization schemes of upwinding or streamline-diffusion type, the influence of this parameter onto the resulting accuracy, robustness and efficiency is surprisingly small.

Summarizing our computational tests with these special edge-oriented FEM stabilization methods (see [131]), we can state that:

- we can stabilize the lack of coercivity ('Korn's inequality') for problems formulated in terms of the symmetric part of the velocity gradient;

- we can handle problems for most relevant Re numbers, even in the limit of inviscid flow;
- we can successfully realize strategies for simulating non-Newtonian flow with pressure and shear-dependent viscosity;
- we can solve the resulting linear generalized Oseen problems in conjunction with standard multigrid components.

Based on these positive results for a wide range of typical flow simulations, we can state that the proposed stabilisation technique may be an interesting candidate for black-box components in CFD codes. Therefore, one future aim is to study the corresponding numerical behaviour w.r.t. accuracy and robustness in cases of visco-elastic fluids and multiphase flow settings including pure transport problems which are coupled with the Navier–Stokes equations. Moreover, we examine improved iterative solvers which explicitly employ the complete stiffness matrix, based on an extended matrix structures due to the additional coupling of FEM basis functions by the edge-oriented jump terms.

1.4. Pressure Schur complement solvers

Using the notation $\mathbf{u}$ and p also for the coefficient vectors in the representation of the approximate solution, the discretized Navier–Stokes equations may be written as a coupled (nonlinear) algebraic system of the form:

Given $\mathbf{u}^n$ and $\mathbf{g}$, compute $\mathbf{u} = \mathbf{u}^{n+1}$ and $p = p^{n+1}$ by solving

$$A\mathbf{u} + \Delta t B p = \mathbf{g} \quad , \quad B^T \mathbf{u} = 0, \qquad \text{where} \tag{1.28}$$

$$\mathbf{g} = [M - \theta_1 \Delta t N(\mathbf{u}^n)]\mathbf{u}^n + \theta_2 \Delta t \mathbf{f}^{n+1} + \theta_3 \Delta t \mathbf{f}^n . \tag{1.29}$$

Here, M is the (consistent or lumped) mass matrix, B is the discrete gradient operator, and $-B^T$ is the associated divergence operator. Furthermore,

$$A\mathbf{u} = [M - \theta \Delta t N(\mathbf{u})]\mathbf{u}, \qquad N(\mathbf{u}) = K(\mathbf{u}) + \nu L, \tag{1.30}$$

where L is the discrete Laplacian and $K(\mathbf{u})$ is the nonlinear transport operator incorporating a certain amount of artificial diffusion due to some appropriate FEM-stabilization as described before.

The solution of nonlinear algebraic systems like (1.28) is a rather difficult task and many aspects need to be taken into account:

- **treatment of the nonlinearity:** fully nonlinear solution by Newton-like methods or iterative defect correction, explicit or implicit underrelaxation, monitoring of convergence rates, choice of stopping criteria etc.;
- **treatment of the incompressibility:** strongly coupled approach (simultaneous solution for $\mathbf{u}$ and p) vs. segregated algorithms based on operator splitting at the continuous or discrete level (classical projection schemes [14],[140] and pressure correction methods like SIMPLE [28],[92]);
- **complete outer control:** problem-dependent degree of coupling and/or implicitness, optimal choice of linear algebra tools (iterative solvers and underlying smoothers/preconditioners), automatic time step control etc.

This abundance of choices leads to a great variety of incompressible flow solvers which are closely related to one another but exhibit considerable differences in terms of their stability, convergence, and efficiency. The Multilevel Pressure Schur Complement (MPSC) approach outlined below makes it possible to put many existing solution techniques into a common framework and to combine their advantages so as to obtain better run-time characteristics.

The fully discretized Navier–Stokes equations (1.28) as well as the linear subproblems to be solved within the outer iteration loop for a fixed-point defect correction or, with a similar structure, for a Newton-like method admit the representation

$$\begin{bmatrix} A & \Delta tB \\ B^T & 0 \end{bmatrix} \begin{bmatrix} \mathbf{u} \\ p \end{bmatrix} = \begin{bmatrix} \mathbf{g} \\ 0 \end{bmatrix}. \tag{1.31}$$

This is a saddle-point problem in which the pressure acts as the Lagrange multiplier for the incompressibility constraint. In general, we have

$$A = \alpha M + \beta N(\mathbf{u}), \qquad \text{where} \quad \beta = -\theta \Delta t. \tag{1.32}$$

For time-dependent problems, the parameter α is set equal to unity, whereas the steady-state formulation is recovered for $\alpha := 0$ and $\beta := -1$. If the operator A is non-singular, the velocity can be formally expressed as

$$\mathbf{u} = A^{-1}(\mathbf{g} - \Delta tBp) \tag{1.33}$$

and plugged into the discretized continuity equation

$$B^T\mathbf{u} = 0 \tag{1.34}$$

which gives a scalar *Schur complement* equation for the pressure only,

$$B^T A^{-1} Bp = \frac{1}{\Delta t} B^T A^{-1} \mathbf{g}. \tag{1.35}$$

Thus, the coupled system (1.31) can be handled as follows:

1. Solve the Pressure Schur Complement (PSC) equation (1.35) for p.
2. Substitute p into relation (1.33) and compute the velocity $\mathbf{u}$.

It is worth mentioning that the matrix A^{-1} is full and should not be assembled explicitly. Instead, an auxiliary problem is to be solved by a direct method or by inner iterations. For instance, the velocity update (1.33) is equivalent to the solution of the discretized momentum equation $A\mathbf{u} = \mathbf{g} - \Delta tBp$.

Likewise, the matrix $S := B^T A^{-1} B$ is never generated in practice. Doing so would be prohibitively expensive in terms of CPU time and memory requirements. It is instructive to consider a preconditioned Richardson method which yields the following **basic iteration** for the PSC equation:

$$p^{(l+1)} = p^{(l)} - C^{-1} \left[Sp^{(l)} - \frac{1}{\Delta t} B^T A^{-1} \mathbf{g} \right], \qquad l = 0, \ldots, L - 1. \tag{1.36}$$

Here C is a suitable preconditioner which is supposed to be a reasonable approximation to S but be easier to 'invert' in an iterative way.

The number of PSC cycles L can be fixed or chosen adaptively so as to achieve a prescribed tolerance for the residual. The choice $C := S$ and $L = 1$ is equivalent to the coupled solution of the original saddle-point problem (1.31). In principle, this challenging task can be accomplished by a properly configured multigrid method. However, the computational cost per iteration is very high and severe problems are sometimes observed on anisotropic grids that contain cells with high aspect ratios. Moreover, the convergence rates do not improve as the time step Δt is refined. Indeed, note that

$$A = M - \theta \Delta t N(\mathbf{u}) \approx M + \mathcal{O}(\Delta t) \qquad (1.37)$$

for sufficiently small time steps. In this case, A can be interpreted as a non-symmetric (and nonlinear) but well conditioned perturbation of the mass matrix M. On the other hand, the PSC equation (1.35) reveals that

$$\mathrm{cond}(S) = \mathrm{cond}(B^T[M + \mathcal{O}(\Delta t)]^{-1}B) \approx \mathrm{cond}(L) = \mathcal{O}(h^{-2}). \qquad (1.38)$$

It follows that the condition number of the coupled system (1.31) is bounded from below by $\mathcal{O}(h^{-2})$ regardless of the time step. The non-symmetric matrix A acting on the velocity components 'improves' for small Δt but, unfortunately, the overall convergence rates of coupled solvers depend also on the elliptic part $B^T A^{-1} B \approx B^T M^{-1} B$ which is (almost) time step invariant.

In light of the above, the coupled solution strategy is inappropriate for the numerical simulation of non-stationary flows which are dominated by convection and call for the use of small time steps. Hence, the preconditioner C for the Schur complement operator should be designed so as to take relation (1.37) into account. Let us consider 'crude' approximations of the form

$$C := B^T \tilde{A}^{-1} B, \qquad (1.39)$$

where the matrix $\tilde{A}$ should be readily 'invertible' but stay close to A at least in the limit $\Delta t \to 0$. Some typical choices are as follows

$$\tilde{A} := \mathrm{diag}(A), \quad \tilde{A} := M_L, \quad \text{and} \quad \tilde{A} := M - \theta \Delta t \nu L.$$

Incompressible flow solvers based on the Richardson iteration (1.36) with this sort of preconditioning comprise fractional-step projection methods [23],[40], [99],[123], various modifications of the SIMPLE algorithm (for an overview, see Engelman [28] and the literature cited therein) as well as Uzawa-like iterations. They can be classified as *global pressure Schur complement* schemes due to the fact that C is an approximation to the global matrix $S = B^T A^{-1} B$.

On the other hand, coupled solution techniques are to be recommended for the treatment of (quasi-) stationary flows and Navier–Stokes equations combined with RANS turbulence models and/or convection-diffusion equations for other scalar quantities (temperatures, concentrations etc). In this case, it is worthwhile to approximate the pressure Schur complement operator *locally* via direct inversion of small matrix blocks associated with subdomains Ω_i of the domain Ω or, in general, with a subset of the unknowns to be solved for. The resulting *local pressure Schur complement* techniques correspond to

$$C^{-1} := \sum_i (B_{|\Omega_i}^T A_{|\Omega_i}^{-1} B_{|\Omega_i})^{-1}, \tag{1.40}$$

whereby the *patches* Ω_i are usually related to the underlying mesh and can consist of single elements or element clusters. The global relaxation scheme is obtained by embedding these "local solvers" into an outer iteration loop of Jacobi or Gauss–Seidel type. This strategy has a lot in common with *domain decomposition* methods, but is more flexible when it comes to the treatment of boundary conditions. A typical representative of local PSC schemes is the Vanka smoother [139] which is widely used in the multigrid community.

Furthermore, it is possible to combine incompressible flow solvers based on global PSC ("operator splitting") and local PSC ("domain decomposition") methods in a general-purpose CFD code. Indeed, all of these seemingly different solution techniques utilize *additive preconditioners* of the form

$$C^{-1} := \sum_i \alpha_i C_i^{-1}.$$

In the next two sections, we briefly discuss the design of such preconditioners and present the resulting basic iteration schemes which can be used as

- preconditioners for Krylov space methods (CG, BiCGSTAB, GMRES);
- *Multilevel pressure Schur complement* (MPSC) smoothers for multigrid.

The multigrid approach is usually more efficient, as demonstrated by benchmark computations in [111].

1.5. Global MPSC approach

The basic idea behind the family of global MPSC schemes is the construction of globally defined additive preconditioners for the Schur complement operator $S = B^T A^{-1} B$. Recall that the matrix A has the structure

$$A := \alpha M + \beta K(\mathbf{u}) + \gamma L, \tag{1.41}$$

where $\beta = -\theta \Delta t$ and $\gamma = \nu \beta$. Unfortunately, it is hardly possible to construct a matrix $\tilde{A}$ and a preconditioner $C = B^T \tilde{A}^{-1} B$ that would be a sufficiently good approximation to all three components of A and S, respectively. Therefore, one can start with developing individual preconditioners for the reactive (M), convective (K), and diffusive (L) part. In other words, the original problem can be decomposed into simpler tasks by resorting to operator splitting.

Let the inverse of C be composed from those of 'optimal' preconditioners for the limiting cases of a divergence-free L_2-projection (for small time steps), incompressible Euler equations, and a diffusion-dominated Stokes problem

$$C^{-1} = \alpha' C_M^{-1} + \beta' C_K^{-1} + \gamma' C_L^{-1} \approx S^{-1}, \tag{1.42}$$

where the user-defined parameters $(\alpha', \beta', \gamma')$ may toggle between (α, β, γ) and zero depending on the flow regime. Furthermore, it is implied that

C_M is an 'optimal' approximation of the reactive part $B^T M^{-1} B$,

C_K is an 'optimal' approximation of the convective part $B^T K^{-1} B$,

C_L is an 'optimal' approximation of the diffusive part $B^T L^{-1} B$.

The meaning of 'optimality' has to be defined more precisely. Ideally, partial preconditioners should be direct solvers with respect to the underlying subproblem. In fact, this may even be true for the fully 'reactive' case $S = B^T M^{-1} B$. However, if these preconditioners are applied as smoothers in a multigrid context and the convergence rates are largely independent of outer parameter settings as well as of the underlying mesh, then this is already sufficient for optimality of the global MPSC solver. Preconditioners C_M, C_K, and C_L satisfying this criterion are introduced and analyzed in [125].

At high Reynolds numbers, the time steps must remain small due to the physical scales of flow motion. Therefore, the lumped mass matrix M_L proves to be a reasonable approximation to the complete operator A. In this case, our basic iteration (1.36) for the pressure Schur complement equation

$$p^{(l+1)} = p^{(l)} + [B^T M_L^{-1} B]^{-1} \frac{1}{\Delta t} B^T A^{-1} \left[\mathbf{g} - \Delta t B p^{(l)} \right] \qquad (1.43)$$

can be interpreted and implemented as a *discrete projection scheme* such as those proposed in [23],[40],[99]. The main algorithmic steps are as follows [123]:

> 1. Solve the 'viscous Burgers' equation for $\tilde{\mathbf{u}}$
> $$A\tilde{\mathbf{u}} = \mathbf{g} - \Delta t B p^{(l)}.$$
> 2. Solve the discrete 'Pressure-Poisson' problem
> $$B^T M_L^{-1} B q = \frac{1}{\Delta t} B^T \tilde{\mathbf{u}}.$$
> 3. Correct the pressure and the velocity
> $$p^{(l+1)} = p^{(l)} + q, \quad \mathbf{u} = \tilde{\mathbf{u}} - \Delta t M_L^{-1} B q.$$

In essence, the right-hand side of the momentum equation is assembled using the old pressure iterate and the intermediate velocity $\tilde{\mathbf{u}}$ is projected onto the subspace of solenoidal functions so as to satisfy the constraint $B^T \mathbf{u} = 0$.

The matrix $B^T M_L^{-1} B$ corresponds to a mixed discretization of the Laplacian operator [40] so that this method is a discrete analogue of the classical projection schemes derived by Chorin ($p^{(0)} = 0$) and Van Kan ($p^{(0)} = p(t_n)$) via operator splitting for the continuous problem (see [14],[140]). For an in-depth presentation of continuous projection schemes we refer to [98],[99]. Our discrete approach offers a number of important advantages including

- applicability to discontinuous pressure approximations;
- consistent treatment (no splitting) of boundary conditions;
- alleviation of spurious boundary layers for the pressure;
- convergence to the fully coupled solution as l increases;

- remarkable efficiency for non-stationary flow problems.

On the other hand, discrete projection methods lack the inherent stabilization mechanisms that make their continuous counterparts applicable to equal-order interpolations provided that the time step is not too small [98].

In our experience, it is often sufficient to perform exactly $L = 1$ pressure Schur complement iteration with just one multigrid sweep. Due to the fact that the numerical effort for solving the linear subproblems is insignificant, global MPSC methods are much more efficent than coupled solvers in the high Reynolds number regime. However, they perform so well only for relatively small time steps, so that the more robust local MPSC schemes (see the following section) are to be recommended for low Reynolds number flows.

In the case of singular perturbed problems or when the ellipticity is violated (for Newton linearizations, an additional indefinite matrix is added), the development of improved global MPSC schemes is still a big challenge, see the state of the art of multigrid and domain decomposition analysis for convection-dominated problems in [107], [88] and [34]. Moreover, in the last two decades a considerable effort was made to develop and analyse Schur complement preconditioners for some *particular* problems of interest. Sometimes, good approximations can be constructed by thinking in terms of the underlying differential operators, see [78] for a discussion of handling the problem in terms of pseudo-differential operators. Alternatively, one can succeed in building appropriate preconditioners by directly manipulating the matrix blocks, see recent developments in [26]. For example, optimal preconditioners for S are well known for the case of (in-) stationary Stokes equations, while adding convection terms in a system makes the problem of building a robust Schur complement preconditioner much more complicated. Although the problem is far from being completely solved, a notable progress was made in treating it. Below we discuss a few approaches in some detail.

One effective approximation of the Schur complement for the Oseen problem has been recently introduced and analysed in [63, 27]. Several justifications have been given for this approach, the original one being based on the Green's tensor of the Oseen operator on the continuous level. A simpler, heuristic argument is to consider first the case where A and B^T are both squared and commute: $AB^T = B^T A$. If A is non-singular, its inverse must also commute with B^T; hence $A^{-1}B^T = B^T A^{-1}$. If we think of A as representing a second-order differential operator and B is first-order, then, apart from possible problems near the boundary, assuming commutativity is reasonable. It follows from these assumptions that $S^{-1} = (BA^{-1}B^T)^{-1} = A(BB^T)^{-1}$. In practice, however, A and B^T are rectangular and represent operators acting on different function spaces. Note that BB^T is, again, a (scalar) pressure-Poisson-type operator; on the other hand, A is a (vector) convection-diffusion operator acting on the velocities. This suggests introducing a discrete convection-diffusion operator A_p acting on the pressure. Including the necessary mass matrix scalings, the resulting approximation to the inverse of the

Schur complement is thus

$$\hat{S}^{-1} := M_p^{-1} A_p (B M_u^{-1} B^T)^{-1}. \tag{1.44}$$

Numerical experiments show that preconditioner (1.44) may provide mesh-independent convergence rates for Krylov subspace iterative methods. For stationary problems, some dependence on the viscosity is still present: iterations count as $O(\nu^{-\frac{1}{2}})$ even for simple flows. This is however much better than the case of simple preconditioners like a scaled mass matrix and may suffice for applications where Reynolds numbers are low and moderate.

Trying to adopt the preconditioner (1.44) to non-Newtonian flow simulations one finds several problems. One is that derivation of (1.44) substantially uses the fact that the A block consists of n ($n = 2, 3$) independent sub-blocks, corresponding to each velocity component. In our applications velocity components are "mixed" in the left-upper block due to the deformation tensor involved in continuous problem formulation; additional mixing appears if Newton's linearization is applied to obtain the linear problem. Nevertheless, despite the fact that a direct extension of the preconditioner (1.44) to non-Newtonian flows is not clear, this approach can serve as good starting point for further developments.

It was discussed that the building of the preconditioner (1.44) is heavily based on the special structure of the A-block and properties of the underlying PDE. Therefore, for different applications one could be interested in a way of constructing a Schur complement preconditioner entirely from the given matrix. One such approach was suggested in [25]. The originally proposed preconditioner is

$$\hat{S}^{-1} := (BB^T)^{-1}(BAB^T)(BB^T)^{-1}. \tag{1.45}$$

It has been given various justifications. A simple one is to observe that if B was square and invertible, then the inverse of $BA^{-1}B^T$ would be $B^{-T}AB^{-1}$. However, B is rectangular. Therefore it makes sense to replace the inverse of B with the Moore–Penrose pseudoinverse $B^\dagger$. If we assume B to have full rank, the pseudoinverse of B is given by $B^\dagger = B^T(BB^T)^{-1}$, thus justifying (1.45). Originally developed for the Oseen equations the preconditioner (1.45) has a big advantage – it can be directly applied to more complicated problems, in particular for modeling non-Newtonian flows. However, even for the Oseen problem the performance of the preconditioner (1.45) suffers from some mesh and viscosity dependence. For the Oseen problem a relatively simple improvement of (1.45) was suggested recently by Elman and coauthors. Therefore extending this approach to the case of non-Newtonian flows also looks promising. It's worth to mention that being heuristically introduced and justified, none of the approaches (1.44) and (1.45) admits mathematically strict analysis so far. Theoretical development and robust estimates for Schur complement preconditioners for the Oseen system remains to be a challenge for the numerical analysis community.

Another relevant question in the case of non-Newtonian flows is preconditioning a (linear) problem with varying viscosity. Not so much seems to be known here. We mention only recent studies [89] of the Stokes problem with piecewise-constant discontinuous viscosity, which appears in simulations of two-phase flows using continuum surface force models. In these studies a robust preconditioner, with respect to a jump in viscosity and mesh size, for S was constructed and analysed. This research may give a hint on how to treat the case of general varying viscosity and which conditions should be imposed to obtain a robust results.

1.6. Local MPSC approach

The local pressure Schur complement approach is tailored to solving 'small' problems so as to exploit the fast cache of modern processors, in contrast to the readily vectorizable global MPSC schemes. As already mentioned above, the basic idea is to subdivide the complete set of unknowns into patches Ω_i and solve the local subproblems **exactly** within an outer block-Gauss–Seidel/Jacobi iteration loop. Typically, every patch (macroelement) for this 'domain decomposition' method consists of one or several neighboring mesh cells and the corresponding local 'stiffness matrix' C_i is given by

$$C_i := \begin{bmatrix} \tilde{A}_{|\Omega_i} & \Delta t B_{|\Omega_i} \\ B_{|\Omega_i}^T & 0 \end{bmatrix}. \tag{1.46}$$

Its coefficients (and hence the corresponding 'boundary conditions' for the subdomains) are taken from the global matrices, whereby $\tilde{A}$ may represent either the complete velocity matrix A or some approximation of it, for instance, the diagonal part $diag(A)$. The local subproblems at hand are so small that they can be solved directly by Gaussian elimination. This is equivalent to applying the inverse of C_i to a portion of the global defect vector.

The elimination process leads to a fill-in of the matrix, which increases the storage requirements dramatically. Thus, it is advisable to solve the equivalent local pressure Schur complement problem with the compact matrix

$$S_i := B_{|\Omega_i}^T \tilde{A}_{|\Omega_i}^{-1} B_{|\Omega_i} . \tag{1.47}$$

In general, S_i is a full matrix but it is much smaller than C_i, since only the pressure values are solved for. If the patch Ω_i contains just a moderate number of elements, the pressure Schur complement matrix is likely to fit into the processor cache. Having solved the local PSC subproblem, one can recover the corresponding velocity field as described in the previous section. In any case, the basic iteration for a local MPSC method reads

$$\begin{bmatrix} \mathbf{u}^{(l+1)} \\ p^{(l+1)} \end{bmatrix} = \begin{bmatrix} \mathbf{u}^{(l)} \\ p^{(l)} \end{bmatrix} - \omega^{(l+1)} \sum_{i=1}^{N_p} \begin{bmatrix} \tilde{A}_{|\Omega_i} & \Delta t B_{|\Omega_i} \\ B_{|\Omega_i}^T & 0 \end{bmatrix}^{-1} \begin{bmatrix} \delta\mathbf{u}_i^{(l)} \\ \delta p_i^{(l)} \end{bmatrix}, \tag{1.48}$$

where N_p denotes the total number of patches, $\omega^{(l+1)}$ is a relaxation parameter, and the global defect vector restricted to a single patch Ω_i is given by

$$\begin{bmatrix} \delta\mathbf{u}_i^{(l)} \\ \delta p_i^{(l)} \end{bmatrix} = \left(\begin{bmatrix} A & \Delta t B \\ B^T & 0 \end{bmatrix} \begin{bmatrix} \mathbf{u}^{(l)} \\ p^{(l)} \end{bmatrix} - \begin{bmatrix} \mathbf{g} \\ 0 \end{bmatrix} \right)_{|\Omega_i}. \tag{1.49}$$

In practice, we solve the corresponding auxiliary problem

$$
\left[\begin{array}{cc} \tilde{A}_{|\Omega_i} & \Delta t B_{|\Omega_i} \\ B_{|\Omega_i}^T & 0 \end{array} \right] \left[\begin{array}{c} \mathbf{v}_i^{(l+1)} \\ q_i^{(l+1)} \end{array} \right] = \left[\begin{array}{c} \delta \mathbf{u}_i^{(l)} \\ \delta p_i^{(l)} \end{array} \right] \tag{1.50}
$$

and compute the new iterates $\mathbf{u}_{|\Omega_i}^{(l+1)}$ and $p_{|\Omega_i}^{(l+1)}$ as follows:

$$
\left[\begin{array}{c} \mathbf{u}_{|\Omega_i}^{(l+1)} \\ p_{|\Omega_i}^{(l+1)} \end{array} \right] = \left[\begin{array}{c} \mathbf{u}_{|\Omega_i}^{(l)} \\ p_{|\Omega_i}^{(l)} \end{array} \right] - \omega^{(l+1)} \left[\begin{array}{c} \mathbf{v}_i^{(l+1)} \\ q_i^{(l+1)} \end{array} \right]. \tag{1.51}
$$

This two-step relaxation procedure is applied to each patch, so some velocity or pressure components may end up being updated several times. The easiest way to obtain globally-defined solution values at subdomain boundaries is to overwrite the contributions of previously processed patches or to calculate an average over all patch contributions to the computational node.

The resulting local MPSC method corresponds to a simple block-Jacobi iteration for the mixed problem (1.28). Its robustness and efficiency can be easily enhanced by computing the local defect vector (1.49) using the possibly updated solution values rather than the old iterates $\mathbf{u}_{|\Omega_i}^{(l)}$ and $p_{|\Omega_i}^{(l)}$ for the degrees of freedom shared with other patches. This strategy is known as the block-Gauss–Seidel method. Its performance is superior to that of the block-Jacobi scheme, while the numerical effort is approximately the same (for a sequential code).

It is common knowledge that block-iterative methods of Jacobi and Gauss–Seidel type do a very good job as long as there are no strong mesh anisotropies. However, the convergence rates deteriorate dramatically for irregular triangulations which contain elements with high aspect ratios (for example, stretched cells needed to resolve a boundary layer) and/or large differences between the size of two neighboring elements. The use of ILU techniques alleviates this problem but is impractical for strongly coupled systems of equations. A much better remedy is to combine the mesh elements so as to 'hide' the detrimental anisotropies inside of the patches which are supposed to have approximately the same shape and size. Several adaptive blocking strategies for generation of such isotropic subdomains are described in [112],[125].

The global convergence behavior will be satisfactory because only the local subproblems are ill conditioned. Moreover, the size of these local problems is usually very small. Thus, the complete inverse of the matrix fits into RAM and sometimes even into the cache so that the use of fast direct solvers is feasible. Consequently, the convergence rates should be independent of grid distortions and approach those for very regular structured meshes. If hardware-optimized routines such as the BLAS libraries are employed, then the solution of small subproblems can be performed very efficiently. Excellent convergence rates and a high overall performance can be achieved if the code is properly tuned and adapted to the processor architecture in each particular case.

1.7. Multilevel solution strategy

Pressure Schur complement schemes constitute viable solution techniques as such but they are particularly useful as smoothers for a *multilevel* algorithm, e.g., a geometric multigrid method. Let us start with explaining the typical implementation of such a solver for an abstract linear system of the form

$$A_N u_N = f_N. \tag{1.52}$$

It is assumed that there exists a hierarchy of levels $k = 1, \ldots, N$, which may be characterized, for instance, by the mesh size h_k. On each of these levels, one needs to assemble the matrix A_k and the right-hand side f_k for the discrete problem. We remark that only f_N is available a priori, while the sequence of residual vectors $\{f_k\}$ for $k < N$ is generated during the multigrid run. The main ingredients of a (linear) multigrid algorithm are

- matrix-vector multiplication routines for the operators A_k, $k \leq N$;
- an efficient *smoother* (basic iteration scheme) and a *coarse grid solver*;
- *prolongation* I_{k-1}^k and *restriction* I_k^{k-1} operators for grid transfer.

Each k-level iteration $MPSC(k, u_k^0, f_k)$ with initial guess u_k^0 represents a *multigrid cycle* which yields an (approximate) solution of the linear system $A_k u_k = f_k$. On the first level, the number of unknowns is typically so small that the auxiliary problem can be solved directly: $MPSC(1, u_1^0, f_1) = A_1^{-1} f_1$. For all other levels ($k > 1$), the algorithm shown in Table 1 is adopted [125].

After sufficiently many cycles on level N, the desired solution u_N of the generic problem (1.52) is recovered. In the framework of our multilevel pressure Schur complement schemes, there are (at least) two possible scenarios:

Global MPSC approach

Solve the discrete problem (1.52) with

$$A_N := B^T A^{-1} B, \qquad u_N := p, \qquad f_N := \frac{1}{\Delta t} B^T A^{-1} \mathbf{g}.$$

The basic iteration is given by (1.43) and equivalent to a discrete projection cycle, whereby the velocity field $\mathbf{u}$ is updated in a parallel manner (see above). The bulk of CPU time is spent on matrix-vector multiplications with the Schur complement operator $S = B^T A^{-1} B$ which is needed for smoothing, defect calculation, and adaptive coarse grid correction. Unlike standard multigrid methods for scalar problems, global MPSC schemes involve solutions of a viscous Burgers equation and a Poisson-like problem in each matrix-vector multiplication step (to avoid matrix inversion). In the case of highly non-stationary flows, the overhead cost is insignificant but it becomes appreciable as the Reynolds number decreases. Nevertheless, numerical tests indicate that the resulting multigrid solvers are optimal in the sense that the convergence rates are excellent and largely independent of mesh anisotropies.

Local MPSC approach

TABLE 1. Multigrid algorithm

Step 1. Presmoothing: Apply m smoothing steps (PSC iterations) to u_k^0 to obtain u_k^m.

Step 2. Coarse grid correction: Calculate f_{k-1} using the restriction operator I_k^{k-1} via

$$f_{k-1} = I_k^{k-1}(f_k - A_k u_k^m)$$

and let u_{k-1}^i $(1 \leq i \leq p)$ be defined recursively by

$$u_{k-1}^i = MPSC(k-1, u_{k-1}^{i-1}, f_{k-1}), \quad u_{k-1}^0 = 0.$$

Step 3. Relaxation and update: Calculate u_k^{m+1} using the prolongation operator I_{k-1}^k via

$$u_k^{m+1} = u_k^m + \alpha_k I_{k-1}^k u_{k-1}^p, \tag{1.53}$$

where the relaxation parameter α_k may be fixed or chosen adaptively so as to minimize the error $u_k^{m+1} - u_k$ in an appropriate norm, for instance, in the discrete energy norm

$$\alpha_k = \frac{(f_k - A_k u_k^m, I_{k-1}^k u_{k-1}^p)_k}{(A_k I_{k-1}^k u_{k-1}^p, I_{k-1}^k u_{k-1}^p)_k}.$$

Step 4. Postsmoothing: Apply n smoothing steps (PSC iterations) to u_k^{m+1} to obtain u_k^{m+n+1}.

Solve the discrete problem (1.52) with

$$A_N := \begin{bmatrix} A & \Delta t B \\ B^T & 0 \end{bmatrix}, \qquad u_N := \begin{bmatrix} \mathbf{u} \\ p \end{bmatrix}, \qquad f_N := \begin{bmatrix} \mathbf{g} \\ 0 \end{bmatrix}.$$

The basic iteration is given by (1.48) which corresponds to the block-Gauss–Seidel/Jacobi method. The cost-intensive part is the smoothing step, as in the case of standard multigrid techniques for convection-diffusion equations and Poisson-like problems. Local MPSC schemes lead to very robust solvers for coupled problems. This multilevel solution strategy is to be recommended for incompressible flows at low and intermediate Reynolds numbers.

Further algorithmic details (adaptive coarse grid correction, grid transfer operators, nonlinear iteration techniques, time-step control, implementation of boundary conditions) and a description of the high-performance software package FEATFLOW based on MPSC solvers can be found in [125],[126]. Some programming strategies, data structures, and guidelines for the development of a hardware-oriented parallel code are presented in [127],[128],[130].

1.8. Coupling with scalar equations

Both global and local MPSC schemes are readily applicable to the Navier–Stokes equations coupled with various turbulence models and/or scalar conservation laws for temperatures, concentrations, volume fractions, and other scalar variables. In many cases, the quantities of interest must remain strictly non-negative for physical reasons, and the failure to enforce the positivity constraint for the *numerical* solution may have disastrous consequences. Therefore, a positivity-preserving discretization of convective terms is indispensable for such applications. This prerequisite is clearly satisfied by algebraic FEM-FCT and FEM-TVD schemes [68] or by nonlinear versions of the edge-oriented FEM techniques introduced in the previous section.

As a representative example of a two-way coupling between the Navier–Stokes equations and a scalar transport equation, we consider the well-known Boussinesq approximation for natural convection problems. The non-dimensional form of the governing equations for a buoyancy-driven incompressible flow reads [15]

$$\frac{\partial \mathbf{u}}{\partial t} + \mathbf{u} \cdot \nabla \mathbf{u} + \nabla p = \nu \Delta \mathbf{u} + T \mathbf{e}_g, \tag{1.54}$$

$$\frac{\partial T}{\partial t} + \mathbf{u} \cdot \nabla T = d \Delta T, \quad \nabla \cdot \mathbf{u} = 0, \tag{1.55}$$

where $\mathbf{u}$ is the velocity, p is the deviation from hydrostatic pressure and T is the temperature. The unit vector $\mathbf{e}_g$ is directed 'upward' (opposite to the gravitational force) and the non-dimensional diffusion coefficients

$$\nu = \sqrt{\frac{Pr}{Ra}}, \qquad d = \sqrt{\frac{1}{Ra\,Pr}}$$

depend on the Rayleigh number Ra and the Prandtl number Pr. Details of this model and parameter settings for the *MIT benchmark problem* (natural convection in a differentially heated enclosure) can be found in [15].

1.8.1. Finite-element discretization.

After the discretization in space and time, we obtain a system of nonlinear algebraic equations which can be written in matrix form as follows:

$$A_u(\mathbf{u}^{n+1})\mathbf{u}^{n+1} + \Delta t M_T T^{n+1} + \Delta t B p^{n+1} = \mathbf{f}_u, \tag{1.56}$$

$$A_T(\mathbf{u}^{n+1})T^{n+1} = f_T, \qquad B^T \mathbf{u}^{n+1} = 0. \tag{1.57}$$

Here and below the superscript $n+1$ refers to the time level, while subscripts identify the origin of discrete operators (u for the momentum equation and T for the heat conduction equation). Furthermore, the matrices A_u and A_T can be decomposed into a reactive, convective, and diffusive part,

$$A_u(\mathbf{v}) = \alpha_u M_u + \beta_u K_u(\mathbf{v}) + \gamma_u L_u, \tag{1.58}$$

$$A_T(\mathbf{v}) = \alpha_T M_T + \beta_T K_T(\mathbf{v}) + \gamma_T L_T. \tag{1.59}$$

Note that we have the freedom of using different finite-element approximations and discretization schemes for the velocity $\mathbf{u}$ and temperature T. The discrete problem (1.56)–(1.57) admits the representation

$$
\begin{bmatrix} A_u(\mathbf{u}^{n+1}) & \Delta t M_T & \Delta t B \\ 0 & A_T(\mathbf{u}^{n+1}) & 0 \\ B^T & 0 & 0 \end{bmatrix} \begin{bmatrix} \mathbf{u}^{n+1} \\ T^{n+1} \\ p^{n+1} \end{bmatrix} = \begin{bmatrix} \mathbf{f}_u \\ f_T \\ 0 \end{bmatrix} \tag{1.60}
$$

and can be solved in the framework of a global or local MPSC method.

1.8.2. Global MPSC algorithm. Non-stationary flow configurations call for the use of operator splitting tools for the coupled system (1.60). This straightforward approach consists in solving the Navier–Stokes equations for $(\mathbf{u}, p)$ and the energy equation for T in a segregated manner. The decoupled subproblems are embedded into an outer iteration loop and solved sequentially by a global MPSC method (discrete projection). For relatively small time steps, this strategy works very well, and simulation software can be developed in a modular way making use of optimized multigrid solvers. Moreover, it is possible to choose the time step individually for each subproblem.

In the simplest case (just one outer iteration per time step), the sequence of algorithmic steps to be performed is as follows [134]:

> 1. Compute $\tilde{\mathbf{u}}$ from the momentum equation
> $$A_u(\tilde{\mathbf{u}})\tilde{\mathbf{u}} = \mathbf{f}_u - \Delta t M_T T^n - \Delta t B p^n.$$
> 2. Solve the discrete Pressure-Poisson problem
> $$B^T M_L^{-1} B q = \frac{1}{\Delta t} B^T \tilde{\mathbf{u}}.$$
> 3. Correct the pressure and the velocity
> $$p^{n+1} = p^n + q, \quad \mathbf{u}^{n+1} = \tilde{\mathbf{u}} - \Delta t M_L^{-1} B q.$$
> 4. Solve the convection-diffusion equation for T
> $$A_T(\mathbf{u}^{n+1}) T^{n+1} = f_T.$$

Due to the nonlinearity of the discretized convective terms, iterative defect correction or a Newton-like method must be invoked in steps 1 and 4. This algorithm combined with the non-conforming FEM discretization appears to provide an 'optimal' flow solver for unsteady natural convection problems.

1.8.3. Local MPSC algorithm. Alternatively, a fully coupled solution of the problem at hand can be obtained following the local MPSC approach. To this end, a multigrid solver is applied to the suitably linearized coupled system (1.60). Each outer iteration for the nonlinearity corresponds to the following solution update

[112],[134]:

$$
\begin{bmatrix} \mathbf{u}^{(l+1)} \\ T^{(l+1)} \\ p^{(l+1)} \end{bmatrix} = \begin{bmatrix} \mathbf{u}^{(l)} \\ T^{(l)} \\ p^{(l)} \end{bmatrix} - \omega^{(l+1)} [F(\sigma, l)]^{-1} \begin{bmatrix} \delta\mathbf{u}^{(l)} \\ \delta T^{(l)} \\ \delta p^{(l)} \end{bmatrix}, \tag{1.61}
$$

where the global defect vector is given by the relation

$$
\begin{bmatrix} \delta\mathbf{u}^{(l)} \\ \delta T^{(l)} \\ \delta p^{(l)} \end{bmatrix} = \begin{bmatrix} A_u(\mathbf{u}^{(l)}) & \Delta t M_T & \Delta t B \\ 0 & A_T(\mathbf{u}^{(l)}) & 0 \\ B^T & 0 & 0 \end{bmatrix} \begin{bmatrix} \mathbf{u}^{(l)} \\ T^{(l)} \\ p^{(l)} \end{bmatrix} - \begin{bmatrix} \mathbf{f}_u \\ f_T \\ 0 \end{bmatrix} \tag{1.62}
$$

and the matrix to be inverted corresponds to the (approximate) Frechét derivative of the underlying PDE system such that [112],[125]

$$
F(\sigma, l) = \begin{bmatrix} A_u(\mathbf{u}^{(l)}) + \sigma R(\mathbf{u}^{(l)}) & \Delta t M_T & \Delta t B \\ \sigma R(T^{(l)}) & A_T(\mathbf{u}^{(l)}) & 0 \\ B^T & 0 & 0 \end{bmatrix}. \tag{1.63}
$$

The nonlinearity of the governing equations gives rise to the 'reactive' contribution R which represents a solution-dependent mass matrix and may cause severe convergence problems. This is why it is multiplied by the adjustable parameter σ. The Newton method is recovered for $\sigma = 1$, while the value $\sigma = 0$ yields the fixed-point defect correction scheme. In either case, the linearized problem is solved by a fully coupled multigrid solver equipped with a local MPSC smoother of 'Vanka' type [112]. As before, the matrix $F(\sigma, l)$ is decomposed into small blocks C_i associated with individual patches Ω_i. The smoothing of the global residual vector is performed patchwise by solving the corresponding local subproblems.

The size of the matrices to be inverted can be further reduced by resorting to the Schur complement approach. For simplicity, consider the case $\sigma = 0$ (extension to $\sigma > 0$ is straightforward). It follows from (1.56)–(1.57) that

$$
T^{n+1} = A_T^{-1} f_T, \quad \mathbf{u}^{n+1} = A_u^{-1} [\mathbf{f}_u - \Delta t M_T T^{n+1} - \Delta t B p^{n+1}], \tag{1.64}
$$

and the discretized continuity equation can be cast into the form

$$
B^T \mathbf{u}^{n+1} = B^T A_u^{-1} [\mathbf{f}_u - \Delta t M_T A_T^{-1} f_T - \Delta t B p^{n+1}] = 0 \tag{1.65}
$$

which corresponds to the pressure Schur complement equation

$$
B^T A_u^{-1} B p^{n+1} = B^T A_u^{-1} \left[\frac{1}{\Delta t} \mathbf{f}_u - M_T A_T^{-1} f_T \right]. \tag{1.66}
$$

Thus, highly efficient local preconditioners of the form (1.47) can be employed instead of C_i. The converged solution p^{n+1} to the scalar subproblem (1.66) is plugged into (1.64) to obtain the velocity $\mathbf{u}^{n+1}$ and the temperature T^{n+1}.

The advantages of the seemingly complicated local MPSC strategy are as follows. First of all, steady-state solutions can be obtained without resorting to pseudo-time-stepping. Moreover, the fully coupled treatment of dynamic flows makes it possible to use large time steps without any loss of robustness. On the

other hand, the convergence behavior of multigrid solvers for the Newton linearization may turn out to be unsatisfactory and the computational cost per outer
iteration is rather high as compared to the global MPSC algorithm. The performance of both solution techniques as applied to the MIT benchmark problem is
illustrated by the numerical results reported in [134].

1.9. Coupling with $k - \varepsilon$ turbulence model

High-resolution schemes like FCT and TVD (see [68]) or particularly nonlinear
edge-oriented stabilization schemes with shock-capturing play an increasingly important role in simulation of turbulent flows. Flow structures that cannot be resolved on the computational mesh activate the flux limiter which curtails the raw
antidiffusion so as to filter out the small-scale fluctuations. Interestingly enough,
the residual artificial viscosity provides an excellent subgrid scale model for *monotonically integrated* Large Eddy Simulation (MILES), see [6].

In spite of recent advances in the field of LES and DNS (direct numerical
simulation), simpler turbulence models based on Reynolds averaging (RANS) still
prevail in CFD software for simulation of industrial processes. In particular, the
evolution of the turbulent kinetic energy k and of its dissipation rate ε is governed
by two convection-dominated transport equations,

$$\frac{\partial k}{\partial t} + \nabla \cdot \left(k\mathbf{u} - \frac{\nu_T}{\sigma_k} \nabla k \right) = P_k - \varepsilon, \tag{1.67}$$

$$\frac{\partial \varepsilon}{\partial t} + \nabla \cdot \left(\varepsilon\mathbf{u} - \frac{\nu_T}{\sigma_\varepsilon} \nabla \varepsilon \right) = \frac{\varepsilon}{k}(C_1 P_k - C_2 \varepsilon), \tag{1.68}$$

where $\mathbf{u}$ denotes the averaged velocity, $\nu_T = C_\mu k^2/\varepsilon$ is the turbulent eddy viscosity
and $P_k = \frac{\nu_T}{2}|\nabla\mathbf{u} + \nabla\mathbf{u}^T|^2$ is the production term. For the standard $k - \varepsilon$ model,
the default values of the involved parameters are as follows:

$$C_\mu = 0.09, \quad C_1 = 1.44, \quad C_2 = 1.92, \quad \sigma_k = 1.0, \quad \sigma_\varepsilon = 1.3.$$

The velocity field $\mathbf{u}$ is obtained from the incompressible Navier–Stokes equations
with $\nabla \cdot (\nu + \nu_T)[\nabla\mathbf{u} + (\nabla\mathbf{u})^T]$ instead of $\nu\Delta\mathbf{u}$.

We remark that the transport equations for k and ε are strongly coupled and
nonlinear so that their numerical solution is a very challenging task. Moreover,
the discretization scheme must be positivity-preserving because negative values of
the eddy viscosity are unacceptable. Unfortunately, implementation details and
employed 'tricks' are rarely reported in the literature, so that a novice to this
area of CFD research often needs to reinvent the wheel. Therefore, we deem it
appropriate to discuss the implementation of a FEM-TVD algorithm for the $k - \varepsilon$
model in some detail.

1.9.1. Positivity-preserving linearization.
The block-iterative algorithm proposed
in [71],[72] consists of nested loops so that the coupled PDE system is replaced by
a sequence of linear subproblems. The solution-dependent coefficients are 'frozen'
during each outer iteration and updated as soon as new values become available.
The quasi-linear transport equations can be solved by an implicit FEM-FCT or

FEM-TVD scheme but the linearization procedure must be tailored to the need to preserve the positivity of k and ε in a numerical simulation. Due to the presence of sink terms in the right-hand side of both equations, the positivity constraint may be violated even if a high-resolution scheme is employed for the discretization of convective terms. It can be proved that the exact solution to the $k - \varepsilon$ model remains non-negative for positive initial data [86],[87] and it is essential to guarantee that the numerical scheme will also possess this property.

Let us consider the following representation of the equations at hand [75]:

$$\frac{\partial k}{\partial t} + \nabla \cdot (k\mathbf{u} - d_k \nabla k) + \gamma k = P_k, \tag{1.69}$$

$$\frac{\partial \varepsilon}{\partial t} + \nabla \cdot (\varepsilon \mathbf{u} - d_\varepsilon \nabla \varepsilon) + C_2 \gamma \varepsilon = C_1 P_k, \tag{1.70}$$

where the parameter $\gamma = \frac{\varepsilon}{k}$ is proportional to the specific dissipation rate ($\gamma = C_\mu \omega$). The turbulent dispersion coefficients are given by $d_k = \frac{\nu_T}{\sigma_k}$ and $d_\varepsilon = \frac{\nu_T}{\sigma_\varepsilon}$. By definition, the source terms in the right-hand side are non-negative. Furthermore, the parameters ν_T and γ must also be non-negative for the solution of the convection-reaction-diffusion equations to be well-behaved [16]. In our numerical algorithm, their values are taken from the previous iteration and their positivity is secured as explained below. This linearization technique was proposed in [75] and it was noticed that the positivity of the lagged coefficients is even more important than that of the transported quantities and can be readily enforced without violating the discrete conservation principle.

Applying implicit high-resolution schemes to the above equations, we obtain two nonlinear algebraic systems which can be written in the generic form

$$A(u^{(l+1)})u^{(l+1)} = B(u^{(l)})u^{(l)} + q^{(k)}, \qquad l = 0, 1, 2, \ldots \tag{1.71}$$

Here k is the index of the outermost loop in which the velocity $\mathbf{u}$ and the source term P_k are updated. The index l refers to the outer iteration for the $k - \varepsilon$ model, while the index m is reserved for inner flux/defect correction loops. The structure of the matrices A and B is as follows:

$$A(u) = M_L - \theta \Delta t (K^*(u) + T), \tag{1.72}$$

$$B(u) = M_L + (1 - \theta) \Delta t (K^*(u) + T), \tag{1.73}$$

where $K^*(u)$ is the described transport operator incorporating nonlinear antidiffusion and T denotes the standard reaction-diffusion operator which is a symmetric positive-definite matrix with non-negative off-diagonal entries. It is obvious that the discretized production terms $q^{(k)}$ are also non-negative if appropriate positivity-preserving schemes are used. Thus, the positivity of $u^{(l)}$ is inherited by the new iterate $u^{(l+1)} = A^{-1}(Bu^{(l)} + q^{(k)})$ provided that $\theta = 1$ (backward Euler) or the time step is sufficiently small.

1.9.2. Positivity of coefficients. The predicted values $k^{(l+1)}$ and $\varepsilon^{(l+1)}$ are used to recompute the parameter $\gamma^{(l+1)}$ for the next outer iteration (if any). The turbulent eddy viscosity $\nu_T^{(k)}$ is updated in the outermost loop. In the turbulent flow regime

$\nu_T \gg \nu$ and the laminar viscosity ν can be neglected. Hence, we set $\nu_{\text{eff}} = \nu_T$, where the eddy viscosity ν_T is bounded from below by ν and from above by the maximum admissible mixing length $l_{\max}$ (e.g., the width of the computational domain). Specifically, we define the limited mixing length l_* as

$$l_* = \begin{cases} \frac{\alpha}{\varepsilon} & \text{if } \varepsilon > \frac{\alpha}{l_{\max}} \\ l_{\max} & \text{otherwise} \end{cases}, \qquad \text{where } \alpha = C_\mu k^{3/2} \qquad (1.74)$$

and use it to update the turbulent eddy viscosity ν_T in the outermost loop

$$\nu_T = \max\{\nu, l_* \sqrt{k}\} \qquad (1.75)$$

as well as the parameter γ in each outer iteration for the $k - \varepsilon$ model:

$$\gamma = C_\mu \frac{k}{\nu_*}, \qquad \text{where } \nu_* = \max\{\nu, l_* \sqrt{k}\}. \qquad (1.76)$$

In the case of a FEM-TVD method, the positivity proof is only valid for the converged solution to (1.71) while intermediate solution values may be negative. Since it is impractical to perform many defect correction steps in each outer iteration, it is worthwhile to substitute $k_* = \max\{0, k\}$ for k in formulas (1.74)–(1.76) so as to to prevent taking the square root of a negative number. Upon convergence, this safeguard will not make any difference, since k will be non-negative from the outset. The above representation of ν_T and γ makes it possible to preclude division by zero and obtain bounded coefficients without making any ad-hoc assumptions and affecting the actual values of k and ε.

1.9.3. Initial conditions. Another important issue which is rarely addressed in the CFD literature is the initialization of data for the $k - \varepsilon$ model. As a rule, it is rather difficult to devise a reasonable initial guess for a steady-state simulation or proper initial conditions for a dynamic one. The laminar Navier–Stokes equations (1.1) remain valid until the flow gains enough momentum for the turbulent effects to become pronounced. Therefore, the $k - \varepsilon$ model should be activated at a certain time $t_* > 0$ after the startup.

During the 'laminar' initial phase ($t \leq t_*$), a constant effective viscosity ν_0 is prescribed. The values to be assigned to k and ε at $t = t_*$ are uniquely defined by the choice of ν_0 and of the default mixing length $l_0 \in [l_{\min}, l_{\max}]$ where $l_{\min}$ corresponds to the size of the smallest admissible eddies:

$$k_0 = \left(\frac{\nu_0}{l_0}\right)^2, \qquad \varepsilon_0 = C_\mu \frac{k_0^{3/2}}{l_0} \qquad \text{at } t \leq t_*. \qquad (1.77)$$

This strategy was adopted because the effective viscosity ν_0 and the mixing length l_0 are somewhat easier to estimate than k_0 and ε_0. In any case, long-term simulation results are typically not very sensitive to the choice of initial data.

1.9.4. Boundary conditions. At the inlet, Γ_{in} all velocity components and the values of k and ε are given:

$$\mathbf{u} = \mathbf{g}, \qquad k = c_{bc}|\mathbf{u}|^2, \qquad \varepsilon = C_\mu \frac{k^{3/2}}{l_0} \qquad \text{on} \quad \Gamma_{\text{in}}, \qquad (1.78)$$

where $c_{bc} \in [0.001, 0.01]$ is an empirical constant [16] and $|\mathbf{u}| = \sqrt{\mathbf{u} \cdot \mathbf{u}}$ is the Euclidean norm of the velocity. At the outlet Γ_{out}, the normal gradients of all scalar variables are required to vanish, and the 'do-nothing' [125] boundary conditions are prescribed:

$$\mathbf{n} \cdot \mathcal{S}(\mathbf{u}) = \mathbf{0}, \qquad \mathbf{n} \cdot \nabla k = 0, \qquad \mathbf{n} \cdot \nabla \varepsilon = 0 \qquad \text{on} \quad \Gamma_{\text{out}}. \qquad (1.79)$$

Here $\mathcal{S}(\mathbf{u}) = -\left(p + \frac{2}{3}k\right)\mathcal{I} + (\nu + \nu_T)[\nabla\mathbf{u} + (\nabla\mathbf{u})^T]$ denotes the effective stress tensor. The numerical treatment of inflow and outflow boundary conditions does not present any difficulty. In the finite-element framework, relations (1.79) imply that the surface integrals resulting from integration by parts vanish and do not need to be assembled.

At an impervious solid wall Γ_w, the normal component of the velocity must vanish, whereas tangential slip is permitted in turbulent flow simulations. The implementation of the no-penetration (free slip) boundary condition

$$\mathbf{n} \cdot \mathbf{u} = 0 \qquad \text{on} \quad \Gamma_w \qquad (1.80)$$

is nontrivial if the boundary of the computational domain is not aligned with the axes of the Cartesian coordinate system. In this case, condition (1.80) is imposed on a linear combination of several velocity components whereas their boundary values are unknown. Therefore, standard implementation techniques for Dirichlet boundary conditions based on a modification of the corresponding matrix rows [125] cannot be used.

In order to set the normal velocity component equal to zero, we nullify the off-diagonal entries of the preconditioner $A(\mathbf{u}^{(m)}) = \{a_{ij}^{(m)}\}$ in the defect correction loop. This enables us to compute the boundary values of $\mathbf{u}$ explicitly before solving a sequence of linear systems for the velocity components:

$$a_{ij}^{(m)} := 0, \quad \forall j \neq i, \qquad \mathbf{u}_i^* := \mathbf{u}_i^{(m)} + \mathbf{r}_i^{(m)}/a_{ii}^{(m)} \qquad \text{for} \quad \mathbf{x}_i \in \Gamma_w. \qquad (1.81)$$

Next, we project the predicted values $\mathbf{u}_i^*$ onto the tangent vector/plane and constrain the corresponding entry of the defect vector $\mathbf{r}_i^{(m)}$ to be zero:

$$\mathbf{u}_i^{(m)} := \mathbf{u}_i^* - (\mathbf{n}_i \cdot \mathbf{u}_i^*)\mathbf{n}_i, \qquad \mathbf{r}_i^{(m)} := 0 \qquad \text{for} \quad \mathbf{x}_i \in \Gamma_w. \qquad (1.82)$$

After this manipulation, the corrected values $\mathbf{u}_i^{(m)}$ act as Dirichlet boundary conditions for the solution $\mathbf{u}_i^{(m+1)}$ at the end of the defect correction step.

As an alternative to the implementation technique of predictor-corrector type, the projection can be applied to the residual vector rather than to the nodal values of the velocity:

$$a_{ij}^{(m)} := 0, \quad \forall j \neq i, \qquad \mathbf{r}_i^{(m)} := \mathbf{r}_i^{(m)} - (\mathbf{n}_i \cdot \mathbf{r}_i^{(m)})\mathbf{n}_i \qquad \text{for} \quad \mathbf{x}_i \in \Gamma_w. \qquad (1.83)$$

For Cartesian geometries, the algebraic manipulations to be performed affect just the normal velocity component. Note that virtually no extra programming effort is required, which is a significant advantage as compared to another feasible implementation based on local coordinate transformations during the element-by-element matrix assembly [29].

1.9.5. Wall functions. To complete the problem statement, we still need to prescribe the tangential stress as well as the boundary values of k and ε on Γ_w. Note that the equations of the $k - \varepsilon$ model are invalid in the vicinity of the wall where the Reynolds number is rather low and viscous effects are dominant. In order to avoid the need for resolution of strong velocity gradients, *wall functions* can be derived using the boundary layer theory and applied at an internal boundary Γ_δ located at a distance δ from the solid wall Γ_w ([84],[86],[87]).

In essence, a boundary layer of width δ is removed from the actual computational domain Ω and the equations are solved in the reduced domain Ω_δ subject to the following empirical boundary conditions:

$$\mathbf{n} \cdot \mathcal{D}(\mathbf{u}) \cdot \mathbf{t} = -u_\tau^2 \frac{\mathbf{u}}{|\mathbf{u}|}, \qquad k = \frac{u_\tau^2}{\sqrt{C_\mu}}, \qquad \varepsilon = \frac{u_\tau^3}{\kappa \delta} \qquad \text{on} \quad \Gamma_\delta. \qquad (1.84)$$

Here $\mathcal{D}(\mathbf{u}) = (\nu + \nu_T)[\nabla \mathbf{u} + (\nabla \mathbf{u})^T]$ is the viscous part of the stress tensor, the unit vector $\mathbf{t}$ refers to the tangential direction, $\kappa = 0.41$ is the von Kármán constant, and u_τ is the *friction velocity* which is assumed to satisfy

$$g(u_\tau) = |\mathbf{u}| - u_\tau \left(\frac{1}{\kappa} \log y^+ + 5.5 \right) = 0 \qquad (1.85)$$

in the *logarithmic layer*, where the local Reynolds number $y^+ = \frac{u_\tau \delta}{\nu}$ is in the range $20 \leq y^+ \leq 100$, and be a linear function of y^+ in the *viscous sublayer*, where $y^+ < 20$. Note that $\mathbf{u}$ represents the tangential velocity as long as the no-penetration condition (1.80) is imposed on Γ_δ.

Equation (1.85) can be solved iteratively, e.g., by Newton's method [84],

$$u_\tau^{l+1} = u_\tau^l - \frac{g(u_\tau^l)}{g'(u_\tau^l)} = u_\tau^l + \frac{|\mathbf{u}| - u_\tau^l f(u_\tau^l)}{1/\kappa + f(u_\tau^l)}, \qquad l = 0, 1, 2, \ldots \qquad (1.86)$$

where the auxiliary function f is given by

$$f(u_\tau) = \frac{1}{\kappa} \log y_*^+ + 5.5, \qquad y_*^+ = \max \left\{ 20, \frac{u_\tau \delta}{\nu} \right\}.$$

The friction velocity is initialized by the approximation

$$u_\tau^0 = \sqrt{\frac{\nu |\mathbf{u}|}{\delta}}$$

and no iterations are performed if it turns out that $y^+ = \frac{u_\tau^0 \delta}{\nu} < 20$. In other words, $u_\tau = u_\tau^0$ in the viscous sublayer. Moreover, we use $y_*^+ = \max\{20, y^+\}$ in the Newton iteration to guarantee that the approximate solution belongs to the logarithmic layer and remains bounded for $y^+ \to 0$.

The friction velocity u_τ is plugged into (1.84) to compute the tangential stress, which yields a natural boundary condition for the velocity. Integration by parts in the weak form of the Navier–Stokes equations gives rise to a surface integral over the internal boundary Γ_δ which contains the prescribed traction:

$$\int_{\Gamma_\delta} [\mathbf{n} \cdot \mathcal{D}(\mathbf{u}) \cdot \mathbf{t}] \cdot \mathbf{v} \, ds = - \int_{\Gamma_\delta} u_\tau^2 \frac{\mathbf{u}}{|\mathbf{u}|} \cdot \mathbf{v} \, ds. \tag{1.87}$$

The free slip condition (1.80) overrides the normal stress and Dirichlet boundary conditions for k and ε are imposed in the strong sense. For further details regarding the implementation of wall laws we refer to [84],[86],[87].

1.9.6. Underrelaxation for outer iterations.

Due to the intricate coupling of the governing equations, it is sometimes worthwhile to use a suitable underrelaxation technique in order to prevent the growth of numerical instabilities and secure the convergence of outer iterations. This task can be accomplished by limiting the computed solution increments before applying them to the last iterate:

$$u^{(m+1)} := u^{(m)} + \omega^{(m)}(u^{(m+1)} - u^{(m)}), \qquad \text{where} \quad 0 \leq \omega^{(m)} \leq 1. \tag{1.88}$$

The damping factor $\omega^{(m)}$ may be chosen adaptively so as to accelerate convergence and minimize the error in a certain norm [125]. However, fixed values (for example, $\omega = 0.8$) usually suffice for practical purposes. The sort of underrelaxation can be used in all loops (indexed by k, l and m) and applied to selected dependent variables like k, ε or ν_T.

In addition, an *implicit underrelaxation* can be performed in m-loops by increasing the diagonal dominance of the preconditioner [31],[92]:

$$a_{ii}^{(m)} := a_{ii}^{(m)}/\alpha^{(m)}, \qquad \text{where} \quad 0 < \alpha^{(m)} \leq 1. \tag{1.89}$$

Of course, the scaling of the diagonal entries does not affect the converged solution. This strategy proves to be more robust than an *explicit underrelaxation* of the form (1.88). On the other hand, no underrelaxation is needed for moderate time steps which are typically used in dynamic simulations.

1.10. Adaptive time-step control

A remark is in order regarding the time-step selection for implicit schemes. Unlike their explicit counterparts, they are unconditionally stable so that the time step is limited only by accuracy considerations (for non-stationary problems). Thus, it should be chosen adaptively so as to obtain a sufficiently good approximation at the least possible cost. Many adaptive time-stepping techniques have been proposed in the literature. Most of them were originally developed in the ODE context and are based on an estimate of the local truncation error which provides a usable indicator for the step-size control.

The 'optimal' value of Δt should guarantee that the deviation of a user-defined functional J (pointwise solution values or certain integral quantities like lift and drag) from its exact value does not exceed a given tolerance

$$|J(u) - J(u_{\Delta t})| \approx TOL. \tag{1.90}$$

Assuming that the error at the time level t^n is equal to zero, a heuristic error indicator can be derived from asymptotic expansions for the numerical values of J computed using two different time steps. For instance, consider

$$J(u_{\Delta t}) = J(u) + \Delta t^2 e(u) + \mathcal{O}(\Delta t)^4,$$
$$J(u_{m\Delta t}) = J(u) + m^2 \Delta t^2 e(u) + \mathcal{O}(\Delta t)^4,$$

where $m > 1$ is an integer number ($m = 2, 3$). The error term $e(v)$ is supposed to be independent of the time step and can be estimated as

$$e(u) \approx \frac{J(u_{m\Delta t}) - J(u_{\Delta t})}{(m^2 - 1)\Delta t^2}.$$

For the relative error to approach the prescribed tolerance TOL as required by (1.90), the new time step Δt_* should be chosen so that

$$|J(u) - J(u_{\Delta t_*})| \approx \left(\frac{\Delta t_*}{\Delta t}\right)^2 \frac{|J(u_{\Delta t}) - J(u_{m\Delta t})|}{m^2 - 1} = TOL.$$

The required adjustment of the time step is given by the formula

$$\Delta t_*^2 = TOL \frac{(m^2 - 1)\Delta t^2}{|J(u_{\Delta t}) - J(u_{m\Delta t})|}.$$

Furthermore, the solution accuracy can be enhanced by resorting to Richardson's extrapolation (see any textbook on numerical methods for ODEs). The above considerations may lack some mathematical rigor but nevertheless lead to a very good algorithm for automatic time-step control [125]:

---

1. Make one large time step of size $m\Delta t$ to compute $u_{m\Delta t}$.
2. Make m small substeps of size Δt to compute $u_{\Delta t}$.
3. Evaluate the relative changes i.e. $|J(u_{\Delta t}) - J(u_{m\Delta t})|$.
4. Calculate the 'optimal' value Δt_* for the next time step.
5. If $\Delta t_* \ll \Delta t$, reject the solution and go back to step 1.
6. Assign $u := u_{\Delta t}$ or perform Richardson's extrapolation.

Note that the computational cost per time step increases significantly since the solution $u_{m\Delta t}$ may be as expensive to obtain as $u_{\Delta t}$ (due to slow convergence). On the other hand, adaptive time-stepping contributes to the robustness of the code and improves its overall efficiency as well as the credibility of simulation results. Further algorithmic details for this approach can be found in [125].

Another simple strategy for adaptive time-step control was introduced by Valli et al. [137],[138]. Their *PID controller* is based on the relative changes of a suitable indicator variable (temperature distribution, concentration fields, kinetic energy, eddy viscosity etc.) and can be summarized as follows:

1. Compute the relative changes of the chosen indicator variable u

$$e_n = \frac{||u^{n+1} - u^n||}{||u^{n+1}||}.$$

2. If they are too large ($e_n > \delta$), reject u^{n+1} and recompute it using

$$\Delta t_* = \frac{\delta}{e_n} \Delta t_n.$$

3. Adjust the time step **smoothly** so as to approach the prescribed tolerance TOL for the relative changes

$$\Delta t_{n+1} = \left(\frac{e_{n-1}}{e_n}\right)^{k_P} \left(\frac{TOL}{e_n}\right)^{k_I} \left(\frac{e_{n-1}^2}{e_n e_{n-2}}\right)^{k_D} \Delta t_n.$$

4. Limit the growth and reduction of the time step so that

$$\Delta t_{\min} \leq \Delta t_{n+1} \leq \Delta t_{\max}, \qquad m \leq \frac{\Delta t_{n+1}}{\Delta t_n} \leq M.$$

The default values of the PID parameters as proposed by Valli et al. [138] are $k_P = 0.075, k_I = 0.175$ and $k_D = 0.01$. Unlike in the case of adaptive time-stepping techniques based on the local truncation error, there is no need for computing an extra solution with a different time step. Therefore, the cost of the feedback mechanism is negligible. Our own numerical studies [72] confirm that this heuristic control strategy is very robust and efficient.

1.11. Some numerical examples

Flow around a cylinder. The first incompressible flow problem to be dealt with is the well-known benchmark *Flow around a cylinder* developed in 1995 for the priority research program "Flow simulation on high-performance computers" under the auspices of DFG, the German Research Association [111]. This project was intended to facilitate the evaluation of various numerical algorithms for the incompressible Navier–Stokes equations in the laminar flow regime. A quantitative comparison of simulation results is possible on the basis of relevant flow characteristics such as drag and lift coefficients, for which sufficiently accurate reference values are available. Moreover, the efficiency of different solution techniques can be assessed in an objective manner.

Consider the steady incompressible flow around a cylinder with circular cross-section. An in-depth description of the geometrical details and boundary conditions for the 2D/3D case can be found in references [111],[125] which contain all relevant information regarding this benchmark configuration. The flow at $Re = 20$ is actually dominated by diffusion and could be simulated by the standard Galerkin method without any extra stabilization (as far as the discretization is concerned; the iterative solver may require using a stabilized preconditioner). However, it was this 'trivial' steady-state problem that has led us to invent the multi-dimensional flux limiter of TVD type [68] since the need for an improved limiting strategy was apparent.

Furthermore, it is instructive to study the interplay of finite-element discretizations for the convective and diffusive terms. As a matter of fact, discrete upwinding can be performed for the cumulative transport operator or just for the convective part. In the case of the non-conforming $\tilde{Q}_1$-elements, the discrete Laplacian operator originating from the Galerkin approximation of viscous terms is a positive definite matrix but some of its off-diagonal coefficients are negative. Our numerical experiments indicate that it is worthwhile to leave it unchanged and start with a FEM-TVD discretization of the convective term.

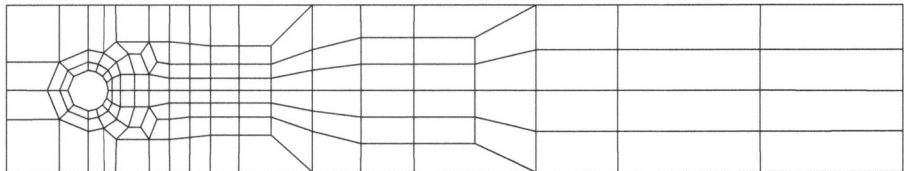

FIGURE 3. Coarse mesh (2D) for the benchmark 'Flow around a cylinder'.

To generate hierarchical data structures for the MPSC algorithms implemented in the software package FEATFLOW [126], we introduce a sequence of successively refined quadrilateral meshes. The elements of the coarse mesh shown in Fig. 3 are subdivided into four subelements at each refinement level, and the 2D mesh is extended into the third dimension for a 3D simulation. The two-dimensional results produced by a global MPSC (discrete projection) method with a FEM-TVD discretization of the convective terms are presented in Table 1. The computational mesh for multigrid level NLEV contains NMT midpoints and NEL elements. For the employed $\tilde{Q}_1/Q_0$ finite-element pair, NMT represents the number of unknowns for each velocity component, while NEL is the number of degrees of freedom for the pressure. It can be seen that the drag and lift coefficients approach the reference values $C_D \approx 5.5795$ and $C_L \approx 0.01061$ as the mesh is refined. The same outcome can be obtained in the local MPSC framework without resorting to pseudo-time-stepping.

In Table 3, we present the drag and lift coefficients for a three-dimensional simulation of the flow around the cylinder. The hexahedral mesh for NLEV=4 consists of 49,152 elements, which corresponds to 151,808 unknowns for each velocity component. In order to evaluate the performance of the global MPSC solver and verify grid convergence, we compare the results to those obtained on a coarser and a finer mesh. All numerical solutions were marched to the steady state by the fully implicit backward Euler method. The discretization of convective terms was performed using (i) finite volume upwinding (UPW), (ii) Samarski's hybrid scheme (SAM), (iii) streamline diffusion stabilization (SD), and (iv) algebraic flux correction (TVD). This numerical study confirms that standard artificial viscosity methods are rather sensitive to the values of the empirical constants, whereas

NLEV	NMT	NEL	C_D	C_L
3	4264	2080	5.5504	$0.8708 \cdot 10^{-2}$
4	16848	8320	5.5346	$0.9939 \cdot 10^{-2}$
5	66976	33280	5.5484	$0.1043 \cdot 10^{-1}$
6	267072	133120	5.5616	$0.1056 \cdot 10^{-1}$
7	1066624	532480	5.5707	$0.1054 \cdot 10^{-1}$
8	4263168	2129920	5.5793	$0.1063 \cdot 10^{-1}$

TABLE 2. Global MPSC method / TVD-FEM discretization.

FEM-TVD performs remarkably well. The reference values $C_D \approx 6.1853$ and $C_L \approx 0.95 \cdot 10^{-2}$ for this 3D configuration were calculated in [59] by an isoparametric high-order FEM.

NLEV	UPW-1st	SAM-1.0	SD-0.25	SD-0.5	TVD
3	6.08/ 1.01	5.72/ 0.28	5.78/-0.44	5.98/-0.52	5.80/ 0.36
4	6.32/ 1.20	6.07/ 0.62	6.13/ 0.26	6.26/ 0.18	6.14/ 0.46
5	6.30/ 1.20	6.14/ 0.83	6.17/ 0.70	6.23/ 0.64	6.18/ 0.80

TABLE 3. Global MPSC method: 3D simulation, $C_D/(C_L \cdot 100)$.

Backward facing step. Let us proceed with a three-dimensional test problem which deals with a turbulent flow over a backward facing step at $Re = 44,000$, see [84] for details. Our objective is to validate the implementation of the $k - \varepsilon$ model as described above. As before, the incompressible Navier–Stokes equations are discretized in space using the non-conforming $\tilde{Q}_1/Q_0$ finite-element pair, while conforming Q_1 (trilinear) elements are employed for the turbulent kinetic energy and its dissipation rate. All convective terms are handled by the fully implicit FEM-TVD method. The velocity-pressure coupling is enforced in the framework of a global MPSC formulation.

Standard wall laws are applied on the boundary except for the inlet and outlet. The stationary distribution of k and ε in the middle cross-section ($z = 0.5$) is displayed in Fig. 4. The variation of the friction coefficient

$$c_f = \frac{2\tau_w}{\rho_\infty u_\infty^2} = \frac{2u_\tau^2}{u_\infty^2} = \frac{2k}{u_\infty^2}\sqrt{C_\mu}$$

along the bottom wall is presented in Fig. 5 (left). The main recirculation length $L \approx 6.8$ is in a good agreement with the numerical results reported in the literature [84]. Moreover, the horizontal velocity component (see Fig. 5, right) assumes positive values at the bottom of the step, which means that the weak secondary vortex is captured as well. The parameter settings for this three-dimensional simulation were as follows:

$$\delta = 0.05, \quad c_{bc} = 0.0025, \quad \nu_0 = 10^{-3}, \quad l_0 = 0.02, \quad l_{max} = 1.0.$$

The computational mesh shown in Fig. 6 contains 57,344 hexahedral cells (pressure unknowns), which corresponds to 178,560 faces (degrees of freedom for each velocity component) and 64,073 vertices (nodes for k and ε).

1.12. Application to more complex flow models

In this section, we discuss several incompressible flow models which require highly accurate FEM techniques and which contain generalized Navier–Stokes problems of the form

$$\rho \left(\frac{\partial \mathbf{u}}{\partial t} + \mathbf{u} \cdot \nabla \mathbf{u} \right) = \mathbf{f} + \mu \Delta \mathbf{u} - \nabla p \quad, \quad \text{Div } \mathbf{u} = 0 \qquad (1.91)$$

complemented by additional PDEs which describe physical processes like

1. Heat transfer in complex geometries
   $\rightarrow$ *Ceramic plate heat exchanger*
2. Multiphase flow with chemical reaction
   $\rightarrow$ *Gas-liquid reactors*
3. Nonlinear fluids/granular flow
   $\rightarrow$ *Sand motion in silos*
4. Free and moving boundaries
   $\rightarrow$ *Level set FEM methods.*

Although these typical flow configurations (to be presented below) differ in their complexity and cover a wide range of Reynolds numbers, all of them require an accurate treatment of diffusive and convective phenomena. Numerical artefacts such as small-scale oscillations/ripples may cause an abnormal termination of the simulation run due to division by zero, floating-point overflow etc. However, in the worst case they can be misinterpreted as physical phenomena and eventually result in making wrong decisions for the design of industrial equipment. Therefore, non-physical solution behavior should be avoided at any cost, and the use of high-resolution FEM discretization schemes is recommendable. Moreover, simulation results must be validated by comparison with experimental data and/or numerical solutions computed on a finer mesh.

1.12.1. Heat transfer in 'Plate Heat Exchangers'. The first example deals with the development of optimization tools for a constellation described by 'coupled stacks' with different layers (see Fig. 7). This example is quite typical for a complex flow model in the laminar regime – small Re numbers due to slow fluid velocities and very small diameters – which however requires unstructured FEM approaches due to complicated small-scale geometrical details and which is coupled with additional tracer equations which often are described by a convection-diffusion equation with a small diffusion coefficient or by a pure transport problem. Hence, the appropriate treatment on unstructured meshes and the implicit handling with large time steps – the underlying flow field is (almost) quasi-stationary – are the critical aspects

Distribution of k in the cutplane $z = 0.5$

Distribution of ε in the cutplane $z = 0.5$

FIGURE 4. Backward facing step: stationary solution, $Re = 44,000$.

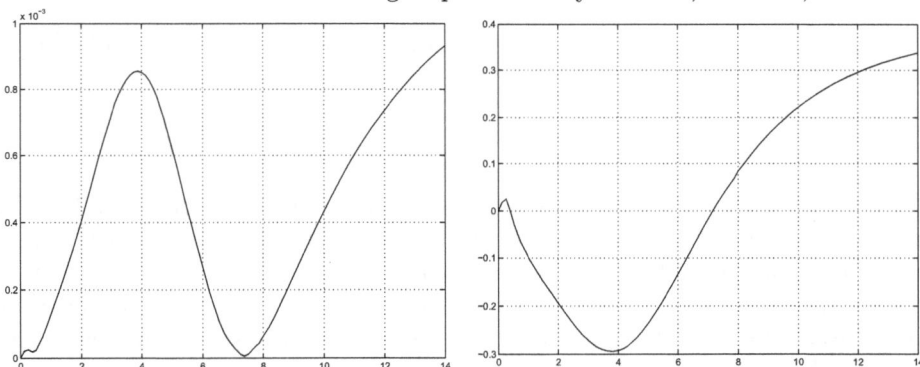

FIGURE 5. Distribution of c_f (left) and u_x (right) along the bottom wall.

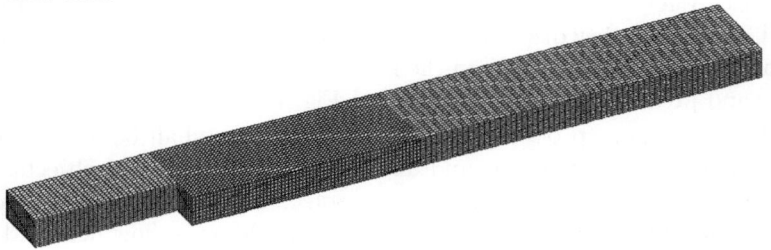

FIGURE 6. Hexahedral computational mesh for the 3D simulation.

for an accurate and efficient methodology which shall be exemplarily illustrated in the following.

a)   b)   c)

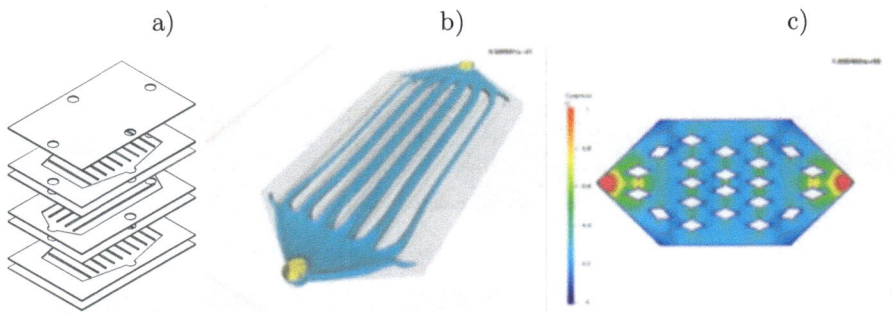

FIGURE 7. Plate heat exchanger: a) geometric configuration; b) typical flow pattern; c) velocity field for an 'optimal' distribution of internal objects.

The aim of the underlying numerical study is the understanding and improvement of the

- internal flow characteristics
- heat transfer characteristics

in the shown configuration which can be described by a Boussinesq model. Restricting to one stack only, the internal geometry between 'inflow' and 'outflow' holes has to be analyzed and channel-like structures or many internal 'objects' have to be placed to achieve a homogeneous flow field and correspondingly a homogeneous distribution of tracer substances in the interior. In order to determine the optimal shape and distribution of internal obstacles, the simulation software to be developed must be capable of resolving all small-scale details. Therefore, the underlying numerical algorithm must be highly accurate and, moreover, sufficiently flexible and robust. Algebraic FCT/TVD schemes and nonlinear edge-oriented FEM techniques belong to the few discretization techniques that do meet these requirements. Some preliminary results based on the incompressible Navier–Stokes equations for velocity and pressure only are presented in Fig. 7.

The following step beyond 'manual optimization' shall be a fully automatic optimization of shape, number, and distribution of the internal objects. Furthermore, the temperature equations are to be solved for each stack as well as for the whole system taking into account heat transfer both in the flow field and between the walls. In addition, chemical reaction models should be included, which give rise to another set of coupled convection-reaction-diffusion equations. Last but not least, accurate FEM techniques have to be employed in the postprocessing step, whereby the 'residence time distribution' inside of each stack is measured by solving a pure transport equation for passive tracers: Since the flow field is almost stationary and allows large time steps due to the large viscosity parameters, the nonlinear transport equation has to be treated in an implicit way which is a quite

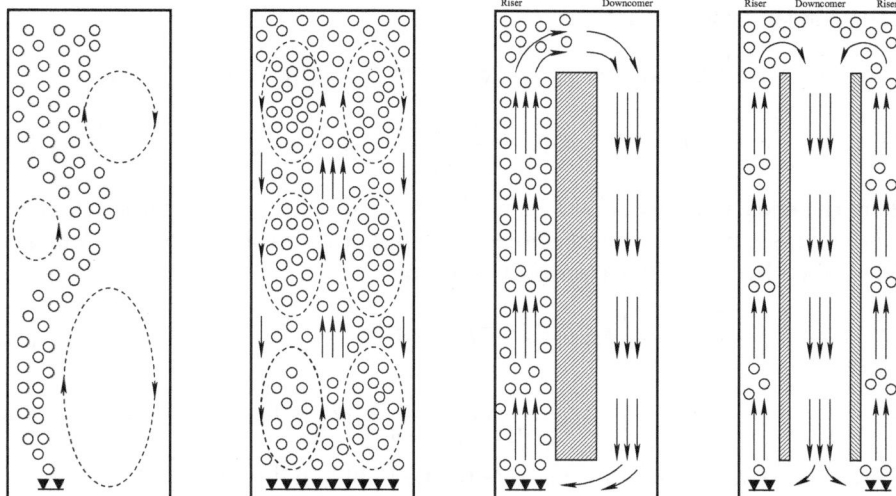

FIGURE 8. Bubble columns (left) and airlift loop reactors (right).

typical requirement for the accurate and efficient treatment for such type of flow problems.

1.12.2. Bubbly flow in gas-liquid reactors. Bubble columns and airlift loop reactors are widely used in industry as contacting devices which enable gaseous and liquid species to engage in chemical reactions. The liquid is supplied continuously or in a batch mode and agitated by bubbles fed at the bottom of the reactor. As the bubbles rise, the gaseous component is gradually absorbed into the bulk liquid where it may react with other species. The geometric simplicity of bubble columns makes them rather easy to build, operate and maintain. At the same time, the prevailing flow patterns are very complex and unpredictable, which represents a major bottleneck for the design of industrial units. By insertion of internal parts, bubble columns can be transformed into airlift loop reactors which exhibit a stable circulation pattern with pronounced *riser* and *downcomer* zones (see Fig. 8). Hence, shape optimization appears to be a promising way to improve the reactor performance by adjusting the geometry of the internals.

   In the present chapter, we adopt a simplified two-fluid model which is based on an analog of the Boussinesq approximation (1.55) for natural convection problems. At moderate gas holdups, the gas-liquid mixture behaves as a weakly compressible fluid which is driven by the bubble-induced buoyancy. Following Sokolichin et al. [116],[117] we assume the velocity $\mathbf{u}_L$ of the liquid phase to be divergence-free. The dependence of the effective density $\tilde{\rho}_L$ on the local gas holdup ϵ is taken into account only in the gravity force, which is a common practice for single-phase flows induced by temperature gradients. This leads to the following generalization of the Navier–Stokes equations:

$$\frac{\partial \mathbf{u}_L}{\partial t} + \mathbf{u}_L \cdot \nabla \mathbf{u}_L = -\nabla p_* + \nabla \cdot \left(\nu_T [\nabla \mathbf{u}_L + \nabla \mathbf{u}_L^T] \right) - \epsilon \mathbf{g},$$

$$\text{Div } \mathbf{u}_L = 0, \qquad p_* = \frac{p - p_{\text{atm}}}{\rho_L} + \mathbf{g} \cdot \mathbf{e}_g - gh, \tag{1.92}$$

where the eddy viscosity $\nu_T = C_\mu k^2/\varepsilon$ is a function of the turbulent kinetic energy k and its dissipation rate ε (see above). Recall that the evolution of these quantities is described by two scalar transport equations,

$$\frac{\partial k}{\partial t} + \nabla \cdot \left(k\mathbf{u}_L - \frac{\nu_T}{\sigma_k} \nabla k \right) = P_k + S_k - \varepsilon, \tag{1.93}$$

$$\frac{\partial \varepsilon}{\partial t} + \nabla \cdot \left(\varepsilon \mathbf{u}_L - \frac{\nu_T}{\sigma_\varepsilon} \nabla \varepsilon \right) = \frac{\varepsilon}{k}(C_1 P_k + C_\varepsilon S_k - C_2 \varepsilon), \tag{1.94}$$

where the extra source terms are due to the bubble-induced turbulence

$$P_k = \frac{\nu_T}{2} |\nabla \mathbf{u} + \nabla \mathbf{u}^T|^2, \qquad S_k = -C_k \epsilon \nabla p \cdot \mathbf{u}_{\text{slip}}.$$

The involved slip velocity $\mathbf{u}_{\text{slip}}$ is proportional to the pressure gradient

$$\mathbf{u}_{\text{slip}} = -\frac{\nabla p}{C_W}$$

and the 'drag' coefficient $C_W \approx 5 \cdot 10^4 \, \frac{kg}{m^3 s}$ is determined from empirical correlations for the rise velocity of a single bubble in a stagnant liquid [116].

The gas density ρ_G is related to the common pressure p by the ideal gas law $p = \rho_G \frac{R}{\eta} T$, which enables us to express the local gas holdup ϵ and the interfacial area a_S per unit volume as follows [70],[72]:

$$\epsilon = \frac{\tilde{\rho}_G RT}{p\eta}, \qquad a_S = (4\pi n)^{1/3}(3\epsilon)^{2/3}.$$

The *effective* density $\tilde{\rho}_G = \epsilon \rho_G$ and the number density n (number of bubbles per unit volume) satisfy the following continuity equations:

$$\frac{\partial \tilde{\rho}_G}{\partial t} + \nabla \cdot (\tilde{\rho}_G \mathbf{u}_G) = -m_{\text{int}}, \tag{1.95}$$

$$\frac{\partial n}{\partial t} + \nabla \cdot (n\mathbf{u}_G) = 0. \tag{1.96}$$

The interphase momentum transfer is typically dominated by the drag force, and the density of gas is much smaller than that of liquid, so that the inertia and gravity terms in the momentum equation for the gas phase can be neglected [116],[117]. Under these (quite realistic) simplifying assumptions, the gas phase velocity $\mathbf{u}_G$ can be computed from the algebraic slip relation

$$\mathbf{u}_G = \mathbf{u}_L + \mathbf{u}_{\text{slip}} + \mathbf{u}_{\text{drift}}, \qquad \mathbf{u}_{\text{drift}} = -d_G \frac{\nabla n}{n},$$

where the drift velocity $\mathbf{u}_{\text{drift}}$ is introduced to model the bubble path dispersion by turbulent eddies. It is usually assumed that $d_G = \nu_T/\sigma_G$, where the Schmidt number σ_G equals unity. Substitution into (1.95)–(1.96) yields

$$\frac{\partial \tilde{\rho}_G}{\partial t} + \nabla \cdot (\tilde{\rho}_G(\mathbf{u}_L + \mathbf{u}_{\text{slip}}) - \nu_T \nabla \tilde{\rho}_G) = -m_{\text{int}}, \qquad (1.97)$$

$$\frac{\partial n}{\partial t} + \nabla \cdot (n(\mathbf{u}_L + \mathbf{u}_{\text{slip}}) - \nu_T \nabla n) = 0. \qquad (1.98)$$

Note that the contribution of $\mathbf{u}_{\text{drift}}$ gives rise to diffusive terms in both equations and it is implied that $\tilde{\rho}_G \nabla n/n = \nabla \tilde{\rho}_G$. Strictly speaking, this relation is valid only for an (almost) constant bubble mass $m = \tilde{\rho}_G/n$ but can also be used in the framework of 'operator-splitting', whereby convection-diffusion and reaction-absorption processes are decoupled from one another.

The sink term m_{int} in equations (1.95) and (1.97) is due to the reaction-enhanced mass transfer. It is proportional to the interfacial area a_S and can be modeled in accordance with the standard two-film theory. The effective concentrations of all species in the liquid phase are described by extra convection-reaction-diffusion equations [67],[69],[70],[72]. If the coalescence and breakup of bubbles cannot be neglected, equation (1.96) should be replaced by a detailed *population balance model* for the bubble size distribution [32]. In any case, we end up with a very large system of convection-dominated PDEs which are strongly coupled and extremely sensitive to non-physical phenomena that may result from an improper discretization of the convective terms.

In order to implement the above *drift-flux model* in a finite-element code, we need to collect all the numerical tools presented so far:

- high-resolution discretization schemes for the convective terms;
- MPSC solvers for the incompressible Navier–Stokes equations;
- block-iterative coupling mechanisms (Boussinesq approximation);
- implementation techniques for the $k - \varepsilon$ turbulence model;
- adaptive time-stepping (PID control of the local gas holdup).

The segregated algorithm proposed in [72] consists of nested loops for the intimately coupled subproblems which are solved sequentially using solution values from the previous outer iteration to evaluate the coefficients and source/sink terms. In each time step, the outermost loop is responsible for the coupling of all relevant equation blocks and contains another outer iteration loop for the equations of the $k - \varepsilon$ turbulence model which are closely related to one another and must be solved in a coupled fashion. The buoyancy force in the Navier–Stokes equations is evaluated using the gas holdup from the previous outer iteration and fixed-point defect correction is employed for all nonlinear convective terms, which gives rise to another sequence of outer iterations. The iterative process is repeated until the residual of the momentum equation and/or the relative changes of all variables become small enough.

Operator-splitting tools are employed to separate convection-diffusion and absorption-reaction processes at each time step. First, all scalar quantities are transported without taking the sources/sinks into account. The homogeneous equations are decoupled and can be processed in parallel. An implicit time discretization of Crank–Nicolson or backward Euler type is performed for all equations. The value of the implicitness parameter θ and of the local time step can be selected individually for each subproblem so as to maximize accuracy and/or stability. The communication between the subproblem blocks takes place at the end of the common macro time step Δt_n which is chosen adaptively so as to control the changes of the gas holdup distribution. The flow chart of algorithmic steps to be performed is as follows [72]:

1. Recover the pressure gradient ∇p via L_2-projection.
2. Compute the associated slip velocity $\mathbf{u}_{\text{slip}} = -\frac{\nabla p}{c_W}$.
3. Solve the homogeneous continuity equation for $\tilde{\rho}_G$.
4. Update the number density n according to (1.98).
5. Convert $\tilde{\rho}_G$ and n into ϵ and a_S; evaluate m_{int}.
6. Solve the transport equations for concentrations.
7. Solve the ODE systems for absorption-reaction.
8. Enter the inner loop for the $k - \varepsilon$ model (1.93)–(1.94).
9. Compute the turbulent eddy viscosity $\nu_T = C_\mu \frac{k^2}{\varepsilon}$.
10. Insert ν_T and ϵ into (1.92) and evaluate the residual.
11. If converged, then proceed to the next time step.
12. Solve the Navier–Stokes equations and go to 1.

The first example deals with the locally aerated bubble column that was investigated in detail by Becker et al. [3]. The snapshots of the meandering bubble swarm displayed in Fig. 9 are in a good agreement with experimental data. The evolution of the gas holdup in the middle cross section of a prototypical airlift loop reactor is shown in Fig. 10. Aeration takes place at the bottom of the riser section where both phases flow upward. At the upper surface, the bubbles escape, while the liquid is diverted into the gas-free downcomer so as to form a closed loop. The two-phase flow reaches a steady state within a few seconds after the startup (see the right diagram). Computational results for the reaction-enhanced absorption of CO_2 in a locally aerated bubble column filled with an aqueous solution of NaOH are presented in [69],[72].

1.12.3. Nonlinear (granular) flow. Another interesting example for the need of special FEM techniques is the numerical simulation of nonlinear incompressible fluids,

$$\frac{\partial \mathbf{u}}{\partial t} + (\mathbf{u} \cdot \nabla)\mathbf{u} - \text{Div}\,\mathbf{T} + \nabla p = \mathbf{f} \quad , \quad \text{Div}\,\mathbf{u} = 0 \tag{1.99}$$

where $\mathbf{T}$ is the stress tensor. Denoting the deformation rate by $\mathbf{D}$ and the rotation rate by $\mathbf{W}$,

$$\mathbf{D} = \frac{1}{2}(\nabla \mathbf{u} + \nabla \mathbf{u}^T) \quad , \quad \mathbf{W} = \frac{1}{2}(\nabla \mathbf{u} - \nabla \mathbf{u}^T), \tag{1.100}$$

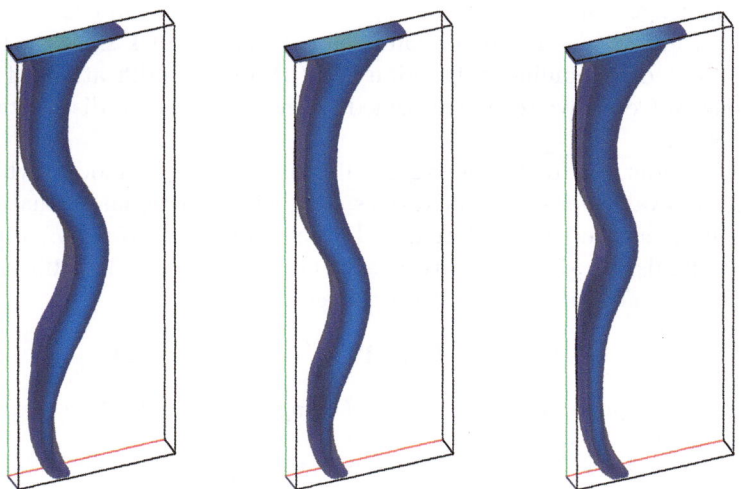

FIGURE 9. Gas holdup distribution in a flat bubble column.

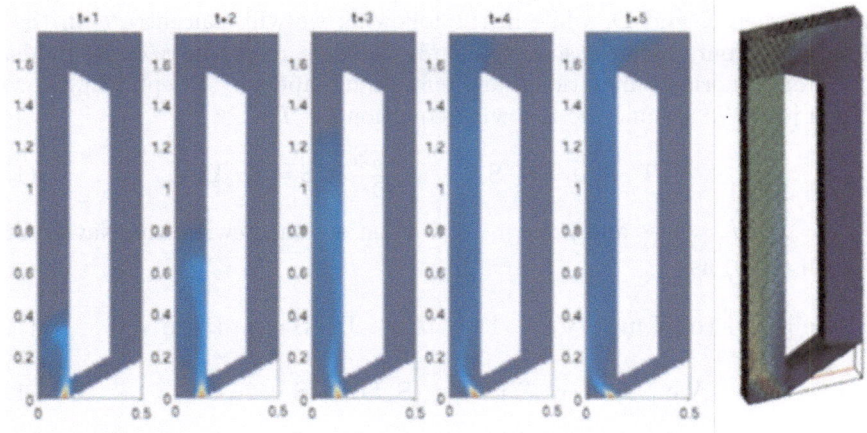

FIGURE 10. Gas holdup distribution in an airlift loop reactor.

we obtain as special case of (1.99) the case of Newtonian fluids via:

$$\mathbf{T} = 2\nu\mathbf{D} = \nu(\nabla\mathbf{u} + \nabla\mathbf{u}^T). \tag{1.101}$$

The most popular variant of nonlinear fluids are *Power-Law* models, with a viscosity ν in (1.101) which depends in a nonlinear way on $\mathbf{D} = \mathbf{D}(\mathbf{u})$,

$$\mathbf{T} = 2\nu(\mathbf{D})\,\mathbf{D} \quad, \quad \nu(\mathbf{D}) = \nu_0(\epsilon_1 + \epsilon_2|\mathbf{D}|)^{\alpha}, \tag{1.102}$$

with specific parameters ν_0, ϵ_1, ϵ_2 and α. Fluids with $\alpha > 0$ are called *shear thickening*, in contrast to the case $\alpha < 0$ (*shear thinning*), both of which lead to numerical challenges. Such *differential type* models – see also *Bingham* or *Reiner–Rivlin* fluids [79],[103] – do not require implicit calculations of $\mathbf{T}$ since the tensor $\mathbf{T}$ can be represented as (nonlinear) function of $\mathbf{D}(\mathbf{u})$. Only modifications of existing Navier–Stokes solvers have to be performed, without additional discretizations of equations for $\mathbf{T}$.

This is in contrast to the more general type of *rate-type* models which couple (nonlinear) evaluations and derivatives of $\mathbf{T}$ with functional evaluations and derivatives in space and time of $\mathbf{u}$ und $\mathbf{D}$. Examples are *Rivlin–Ericksen* and *second-grade* fluids [62],[79] and particularly *Oldroyd* models [38],[103]. Defining the *objective time derivative* of the tensor $\mathbf{T}$ as

$$\frac{\mathcal{D}_a \mathbf{T}}{\mathcal{D}t} = \frac{\partial \mathbf{T}}{\partial t} + (\mathbf{u} \cdot \nabla)\mathbf{T} + \mathbf{TW} + (\mathbf{TW})^T - a[\mathbf{TD} + (\mathbf{TD})^T], \qquad (1.103)$$

with $-1 \leq a \leq 1$, a general description for *Oldroyd* models [49],[102] reads

$$\lambda_1 \frac{\mathcal{D}_a \mathbf{T}}{\mathcal{D}t} + \mathbf{T} + \gamma(\mathbf{T}, \mathbf{D}) = 2\mu(\lambda_2 \frac{\mathcal{D}_a \mathbf{D}}{\mathcal{D}t} + \mathbf{D}), \qquad (1.104)$$

with *relaxation time* λ_1, *retardation time* λ_2 and viscosity parameter μ which may additionally depend on $\mathbf{D}$. In variants with $\gamma(\mathbf{T}, \mathbf{D}) \neq 0$ (*Oldroyd 8-constants* model, *Larson* model, *Phan–Thien–Tanner* model), there is another nonlinear relation between $\mathbf{T}$ and $\mathbf{D}$, while in the following we will concentrate on the case $\gamma(\mathbf{T}, \mathbf{D}) = 0$, containing the so-called *Jeffrey* models, resp., the *Maxwell* fluids. Examples for numerical simulations can be found in papers by Joseph [62], Glowinski [38] and [49]. We assume the following equation for $\mathbf{T}$,

$$\mathbf{T} = 2\mu_v \mathbf{D} + \mathbf{S} \quad , \quad \lambda \frac{\mathcal{D}_a \mathbf{S}}{\mathcal{D}t} + \mathbf{S} = 2\mu_e \mathbf{D} \qquad (1.105)$$

with $\lambda = \lambda_1$, $\mu_v = \mu \frac{\lambda_2}{\lambda_1}$ and $\mu_e = \mu - \mu_v$. Then, we can rewrite the Navier–Stokes model in (1.99) as

$$\text{Re}\,[\frac{\partial \mathbf{u}}{\partial t} + (\mathbf{u} \cdot \nabla)\mathbf{u}] + \nabla p - (1 - \alpha)\Delta \mathbf{u} - \text{Div}\,\mathbf{S} = \mathbf{f}, \text{ Div }\mathbf{u} = 0, \qquad (1.106)$$

$$\text{We}\,[\frac{\partial \mathbf{S}}{\partial t} + (\mathbf{u} \cdot \nabla)\mathbf{S} + \beta_a(\mathbf{S}, \mathbf{D})] + \mathbf{S} = 2\alpha \mathbf{D}, \qquad (1.107)$$

with the following definitions ($-1 \leq a \leq 1$, characteristic length L, resp., velocity $\mathbf{V}$):

$$\beta_a(\mathbf{S}, \mathbf{D}) = \mathbf{SW} + (\mathbf{SW})^T - a[\mathbf{SD} + (\mathbf{SD})^T], \qquad (1.108)$$

$$\alpha = \frac{\mu_e}{\mu_e + \mu_v} \quad , \quad \text{Re} = \frac{LV\rho}{\mu_e + \mu_v} \quad , \quad \text{We} = \frac{V\lambda}{L}. \qquad (1.109)$$

As a result, we obtain a coupled system consisting of a Navier–Stokes-like system (1.106) in $\mathbf{u}$, p with additional term Div $\mathbf{S}$. Furthermore, there is a nonlinear non-stationary tensor-valued reaction-transport equation in (1.107) which couples both tensors $\mathbf{S}$ and $\mathbf{D}(\mathbf{u})$. To solve this highly complex system, there is need for appropriate discretization techniques in space and time (implicit Euler, Crank–Nicolson or fractional-step method together with LBB-stable FEM spaces). In particular, high-resolution FEM techniques are strongly required for the tensor-valued transport problems since a monotone, oscillation-free and highly accurate discretization of the tensor $\mathbf{T}$ is very important. Moreover, the mentioned MPSC techniques for solving the saddle point problems and for the coupling between the different equations are important research aspects for this kind of problem.

Finally, we mention the hypoplastic model for granular material of Kolymbas [64] for the numerical simulation of dry cohesionless material, for instance, for the flow of sand in silos. This approach contains components from *rate-type* as well as from *differential-type* models such that the relation to the described *Oldroyd* model gets immediately visible:

$$\mathrm{Re}\,[\frac{\partial \mathbf{u}}{\partial t} + (\mathbf{u} \cdot \nabla)\mathbf{u}] \;=\; -\nabla p + \mathrm{Div}\,\mathbf{T} + \mathbf{f}\,,\; \mathrm{Div}\,\mathbf{u} = 0, \qquad (1.110)$$

$$\frac{\partial \mathbf{T}}{\partial t} + (\mathbf{u} \cdot \nabla)\mathbf{T} \;=\; -[\mathbf{TW} - \mathbf{WT}] \qquad (1.111)$$

$$+ C_1\,\frac{1}{2}(\mathbf{TD} - \mathbf{DT}) + C_2 tr(\mathbf{TD}) \cdot \mathbf{I}$$

$$+ C_3\sqrt{tr\mathbf{D}^2}\,\mathbf{T} + C_4\frac{\sqrt{tr\mathbf{D}^2}}{tr\mathbf{T}}\mathbf{T}^2$$

$$+ \nu(\mathbf{D})\,[\frac{\partial \mathbf{D}}{\partial t} + (\mathbf{u} \cdot \nabla)\mathbf{D} + \mathbf{DW} - \mathbf{WD}].$$

Here, $\mathbf{I}$ denotes the unit matrix, $\nu(\mathbf{D})$ a nonlinear (tensor) function of *Power-Law* type, and C_i are specific material constants. Again, the robust and accurate treatment of the convective terms in absence of any second-order elliptic diffusive term is a very important aspect for the numerical simulations.

1.12.4. Free surface/interface flows. The final example shortly illustrates the need for implicit TVD or edge-oriented FEM discretizations for incompressible (laminar) flow problems which contain free surfaces, resp., free interfaces:

$$\rho_i\left(\frac{\partial \mathbf{v}}{\partial t} + \mathbf{v} \cdot \nabla\mathbf{v}\right) - \nabla \cdot (2\mu_i\mathbf{S}) + \nabla p = \rho_i\mathbf{g}\,, \qquad (1.112)$$

$$\nabla \cdot \mathbf{v} = 0 \quad \text{in } \Omega_i\,, \quad i = 1, 2, \ldots \qquad (1.113)$$

with deformation tensor $\mathbf{S} = 1/2\left(\nabla\mathbf{v} + (\nabla\mathbf{v})^T\right)$, densities ρ_i and (dynamic) viscosities μ_i of the i-th fluid, gravity $\mathbf{g}$ and $\Gamma = \bar{\Omega}_1 \cap \bar{\Omega}_2$ describing the interface, $\Omega = \Omega_1 \cup \Omega_2$ the complete domain with boundary Σ. This system is completed by the boundary conditions

$$\mathbf{v} = \mathbf{b} \quad \text{on } \Sigma\,, \qquad (1.114)$$

and the initial values

$$\mathbf{v}|_{t=0} = \mathbf{v}^0 \quad \text{in } \Omega\,; \qquad \Gamma|_{t=0} = \Gamma^0\,. \qquad (1.115)$$

Moreover, the conditions at the (free) interface read:

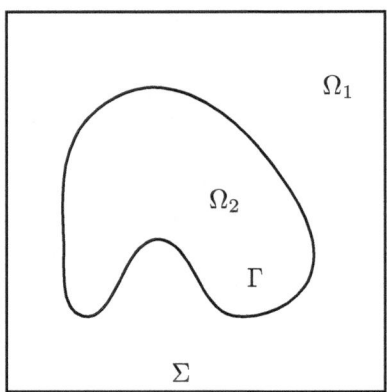

FIGURE 11. Problem description.

1) *Mass balance.* The interface moves with the fluid,

$$V = \mathbf{v}_1 \cdot \mathbf{n} = \mathbf{v}_2 \cdot \mathbf{n}, \tag{1.116}$$

with interface velocity V, normal vector $\mathbf{n}$ on the interface at Γ and $\mathbf{v}_i$ as velocity of the i-th fluid.

2) *Momentum balance.* The following relation holds for the stress tensor $\mathbf{T}_i = -p\mathbf{I} + 2\mu_i \mathbf{S}$,

$$(\mathbf{T}_1 - \mathbf{T}_2) \cdot \mathbf{n} = \kappa\sigma\mathbf{n}, \tag{1.117}$$

with pressure p, surface tension coefficient σ and curvature κ of the interface. These conditions read:

$$[\mathbf{v}]\,|_\Gamma \cdot \mathbf{n} = 0\,, \qquad -\left[-p\mathbf{I} + 2\mu(\mathbf{x})\mathbf{S}\right]|_\Gamma \cdot \mathbf{n} = \kappa\sigma\mathbf{n}\,, \tag{1.118}$$

with $[\cdot]|_\Gamma = \lim_{\mathbf{x}\in\Omega_2\to\Gamma}(\cdot) - \lim_{\mathbf{x}\in\Omega_1\to\Gamma}(\cdot)$, $\rho(\mathbf{x}) = \rho_1$ in Ω_1 and ρ_2 in Ω_2, and $\mu(\mathbf{x}) = \mu_1$ in Ω_1 and μ_2 in Ω_2.

Beside classical front tracking methods, the (implicit) reconstruction of free interfaces via an 'indicator function' φ, which satisfies $\varphi = 0$ directly at the interface, is an alternative procedure. In contrast to the well-known VOF approach [48], which uses a discontinuous indicator function φ, the Level-Set method is our method of choice, particularly in the FEM context.

In both approaches, VOF as well as Level-Set, the indicator function φ and hence the position of the free interface due to the isoline $\varphi = 0$ is advected via the following pure transport equation:

$$\frac{\partial\varphi}{\partial t} + \mathbf{v} \cdot \nabla\varphi = 0, \tag{1.119}$$

while, in addition, the Level-Set function φ has the 'signed distance' property:

$$\frac{\partial\varphi}{\partial t} + \mathbf{v} \cdot \nabla\varphi = 0 \quad , \quad |\nabla\varphi| = 1. \tag{1.120}$$

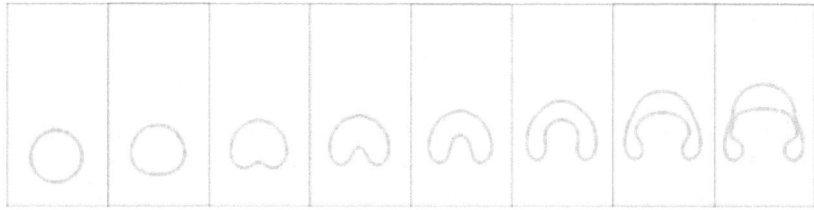

FIGURE 12. Rising bubble via the Level-Set method.

That means that φ is a function with zero-level at the interface, and the evaluation of φ provides the (signed) distance to the interface Γ. The advantage is that this special approach, $|\nabla\varphi| = 1$, leads to a smooth function φ so that higher-order discretization schemes become feasible. However, this additional problem is not trivial to solve, and a typical approach for this so-called 'reinitialisation' step is a non-stationary formulation of $|\nabla\varphi| = 1$ as pure transport equation. Denoting the solution of the transport problem (1.119) in each time step by φ_{old}, we are looking for a distance function φ, with the same zero-level as φ_{old}, as steady-state solution of:

$$\frac{\partial\varphi}{\partial t} = \text{sign}(\varphi_{old})(1 - |\nabla\varphi|). \qquad (1.121)$$

This is equivalent to solving

$$\frac{\partial\varphi}{\partial t} + \text{sign}(\varphi_{old})|\nabla\varphi| = \text{sign}(\varphi_{old}) \qquad (1.122)$$

or finally:

$$\frac{\partial\varphi}{\partial t} + \text{sign}(\varphi_{old})\frac{\nabla\varphi \cdot \nabla\varphi}{|\nabla\varphi|} = \frac{\partial\varphi}{\partial t} + \mathbf{w} \cdot \nabla\varphi = \text{sign}(\varphi_{old}). \qquad (1.123)$$

Here, we define the 'transport direction' $\mathbf{w} = \text{sign}(\varphi_{old})\frac{\nabla\varphi}{|\nabla\varphi|}$ and solve this (nonlinear) problem in every time step until steady state which makes it obvious that robust and highly accurate FEM discretizations methods together with efficient solvers are needed. Numerical studies show that a monotone, oscillation-free and accurate discretization of the indicator function φ and hence of the interface position is a very crucial point which is mainly responsible for the quality of free interface simulations.

1.13. Conclusions

Even for laminar, single-phase flow models, there is still a big need for better mathematical approaches regarding discretization and solution aspects if much 'better' CFD simulations tools w.r.t. accuracy, flexibility, robustness and particularly total efficiency have to be derived. Then, the next laborious step will be to develop professional software for grand-challenge industrial problems. However, the benchmarks and numerical studies show that we *must* spend such efforts in improving the 'basic tools', before stepping to more complex simulations! Otherwise, there might be no chance to tackle successfully much more complex simulation problems, providing not only qualitatively, but also quantitatively accurate results. So, the numerical main ingredients of our research studies for the near future read:

- **Higher-order FEM spaces:** Preliminary tests, by other CFD groups as well as internally, show the much higher accuracy of higher-order polynomials. Even for a biquadratic Q_2 element with piecewise P_1 pressure elements (see next chapter), the gain already can be huge. However, at the moment it is not clear how these approaches behave for non-smooth solutions with strong gradients, and particularly the appropriate stabilization of the convective terms for higher Re numbers is not obvious. Here, we plan to extend the described TVD and particularly the edge-oriented FEM methodology to the case of biquadratic FEM which is principally possible. This step together with the development of corresponding hierarchical solvers of multigrid type, which are absolutely necessary to exploit the potentially higher accuracy of higher-order elements, will be examined in the very near future.

- **Nonlinear solvers:** Since TVD and nonlinear edge-oriented stabilization techniques are *per se* nonlinear discretizations, appropriate Newton-like methods have to be applied if implicit treatments – as for instance in the 'reinitialization step' or for 'tracer transport' in slow flow – are necessary. However, due to the 'discrete and discontinuous' structure of such techniques (TVD), the problem of deriving derivatives being part of the (approximate) Jacobian matrices is still unsolved and requires intensive studies.

- **Multigrid solvers:** Analogously to the nonlinear considerations, the numerical behaviour of standard (geometric) multigrid of the resulting linear subproblems is not yet clear and justifies further efforts.

- **Tensor-valued transport operators:** As described for nonlinear fluids, additional hyperbolic problems with transport operators may appear which however are defined for tensor-values functions, in contrast to the typical scalar quantities, such that the extension of the described FEM techniques for these cases has to be examined in the near future.

- **A-posteriori error control:** Since the described stabilization techniques are all FEM methods, appropriate a-posteriori error control mechanisms for Galerkin-type discretizations should be applicable. Particularly, residual-based error control via dual problems for user-specific quantities in the spirit of Rannacher and collaborators shall be examined and generalized to these new classes of FEM methods to achieve 'optimal' control on refining and coarsening the computational meshes.

However, beside all these mathematical efforts in this work, results of computational studies (see [127] for a critical discussion) show that on recent processors holds: *Data moving is costly, and not data processing*! This is the major reason why traditional numerical approaches for PDE's and their realization as PDE software have massive problems in achieving a significant percentage of the possible peak performance. On the other hand, special tools for the Numerical Linear Algebra and for the complete solution part of the arising huge linear system of equations can be significantly improved: Via cache-based techniques through *exploiting locally structured features*. We do not talk about improvements of 20% – 200%; in fact, we aim to get overall speed-ups of size 10 – 1000, already on modern single processors, while further significant improvements should become possible due to corresponding optimal parallel strategies!

To realize these aims, very special 'local structures' have to be incorporated into complex CFD simulation tools such that modern numerical components can be applied, as, for instance, *adaptive meshing* and *a-posteriori error control* or generalized *multigrid/domain decomposition* solvers. Furthermore, the same philosophy has to be applied to fully implicit Navier–Stokes solvers of MPSC-type which again are optimized with respect to performing more 'arithmetic operation intensive' work (Numerical Linear Algebra, multigrid, Krylov-space solvers) instead of 'memory access expensive' tasks (assembling of stiffness matrices, defects and residuals, modifying the mesh). It is obvious that the necessary work for designing and particularly for realizing such software concepts is not an easy job; the related FEAST project (see [128]) is based on such strategies, at least regarding the discussed data structures and applied solvers. Anyway, the technological trends strongly indicate that such numerical and computational labour is necessary: although the development of such software tools may take longer, the gain in future will be huge, especially for 'real-life' applications in 3D which at the moment often tend to be impossible due to huge performance problems. However, all these software engineering aspects have always to be accompanied by corresponding efforts in modern Numerics which also must explicitly take into account the recent hardware developments. Otherwise, it may happen that many 'nice' mathematical developments get impracticable if much less than one percent of the possible computing power is achieved.

2. FSI for fluid – elastic solid configurations

2.1. Overview

Both problems of viscous fluid flow and of elastic body deformation have been studied separately for many years in great detail. But there are many problems encountered in real life where an interaction between those two medias is of great importance. Typical example of such a problem is the area of aero-elasticity. Another important area where such interaction is of great interest is biomechanics. Such interaction is encountered especially when dealing with the blood circulatory system. The problem of a pulsative flow in an elastic tube, the flow through the heart flaps, the flow in the heart chambers are some of the examples. In all these cases we have to deal with large deformations of a deformable solid interacting with an unsteady, often periodic, fluid flow. The ability to model and predict the mechanical behavior of biological tissues is very important in several areas of bioengineering and medicine. For example, a good mathematical model for biological tissue could be used in such areas as early recognition or prediction of heart muscle failure, advanced design of new treatments and operative procedures, and the understanding of atherosclerosis and associated problems. Other possible applications include development of virtual-reality programs for training new surgeons or designing new operative procedures (see [85, 93]), and last but not least the design of medical instruments or artificial replacements with optimal mechanical and other properties as close as possible to the original parts (see [147]). These are some of the areas where a good mathematical model of soft tissue with reliable and fast numerical solution is essential for success.

2.1.1. Fluid-structure models. There have been several different approaches to the problem of fluid-structure interaction. Most notably the work of [97, 95, 96, 94] where an immersed boundary method was developed and applied to a three-dimensional model of the heart. In this model they consider a set of one-dimensional elastic fibers immersed in three-dimensional fluid region and using parallel supercomputer they were able to model the pulse of the heart ventricle. Their method can capture the anisotropy caused by the muscle fibers.

A fluid-structure model with the wall modeled as a thin shell was used to model the left heart ventricle in [17, 18] and [101, 100]. In [46, 47] a similar approach was used to model a flow in a collapsible tubes. In these models the wall is modeled by a two-dimensional thin shell which can be modified to capture the anisotropy of the muscle. In reality the thickness of the wall can be significant and very important. For example in arteries the wall thickness can be up to 30% of the diameter and its local thickening can be the cause of an aneurysm creation. In the case of heart ventricle the thickness of the wall is also significant and also the direction of the muscle fibers changes through the wall.

2.1.2. Mixture models for perfusion. Another class of models which fall into the fluid structure interaction problems are the fluid-solid mixture models used for simulation of soft tissue perfusion like muscles or cartelage. Mixture theory was first applied to swelling and diffusion in rubber materials [21, 104], mechanics of skin [90], compression of cartilage [119, 66, 106] and blood perfusion through biological tissues in [141, 142]. (see, for example, [33] and [82]) The basic idea of mixture theory is the assumption of co-occupancy, i.e., at each spatial point there is certain fraction of each constituent (with associated fields) and there are prescribed balance equations for each constituent of the mixture as is usual for a single continuum, with additional terms representing the interaction between constituents within the mixture.

There have been several numerical studies of mixture models. One-dimensional diffusion of fluid through an isotropic material is solved in [115], for transversely isotropic materials in [21] and in [106] a one-dimensional diffusion through isotropic stretched slab is solved using a velocity boundary condition. Finite-element solutions of mixture models for the small deformation, linear elastic case are presented in [66, 142] and for nonlinear large deformation description of various soft tissues in [119, 118, 120, 24, 143, 1, 74].

2.1.3. Theoretical results. The theoretical investigation of fluid structure interaction problems is complicated by the need of mixed description. While for the solid part the natural view is the material (Lagrangian) description, for the fluid it is the spatial (Eulerian) description. In the case of their combination some kind of mixed description (usually referred to as the arbitrary Lagrangian–Eulerian description or ALE) has to be used which brings additional nonlinearity into the resulting equations.

In [73] a time dependent, linearized model of interaction between a viscous fluid and an elastic shell in small displacement approximation and its discretization is analyzed. The problem is further simplified by neglecting all changes in the geometry configuration. Under these simplifications, by using energy estimates they are able to show that the proposed formulation is well posed and a global weak solution exists. Further they show that an independent discretization by standard mixed finite elements for the fluid and by non-conforming discrete Kirchhoff triangle finite elements for the shell together with backward or central difference approximation of the time derivatives converges to the solution of the continuous problem.

In [108] a steady problem of equilibrium of an elastic fixed obstacle surrounded by a viscous fluid is studied. Existence of an equilibrium state is shown with the displacement and velocity in $C^{2,\alpha}$ and pressure in $C^{1,\alpha}$ under assumption of small data in $C^{2,\alpha}$ and the domain boundaries of class C^3.

For a basic introduction and complete reference of continuum theory see [43, 121, 80, 45]. Its application in biomechanics are presented, e.g., in [33] and [81]. We will mention in the following sections the basic notation and setup used in this work.

A numerical solution of the resulting equations of the fluid structure interaction problem poses a great challenge since it includes the features of nonlinear elasticity, fluid mechanics and their coupling. The easiest solution strategy, mostly used in the available software packages, is to decouple the problem into the fluid part and solid part, for each of those parts to use some well established method of solution; then the interaction is introduced as external boundary conditions in each of the subproblems. This has the advantage that there are many well-tested finite-element based numerical methods for separate problems of fluid flow and elastic deformation; on the other hand, the treatment of the interface and the interaction is problematic. The approach presented here treats the problem as a single continuum with the coupling automatically taken care of as internal interface, which in our formulation does not require any special treatment.

2.2. Continuum description

Let $\Omega \subset \mathbb{R}^3$ be a reference configuration of a given body. Let $\Omega_t \subset \mathbb{R}^3$ be a configuration of this body at time t. Then a one-to-one, sufficiently smooth mapping $\vec{\chi}_\Omega$ of the reference configuration Ω to the current configuration

$$\vec{\chi}_\Omega : \Omega \times [0, T] \mapsto \mathbb{R}^3, \tag{2.1}$$

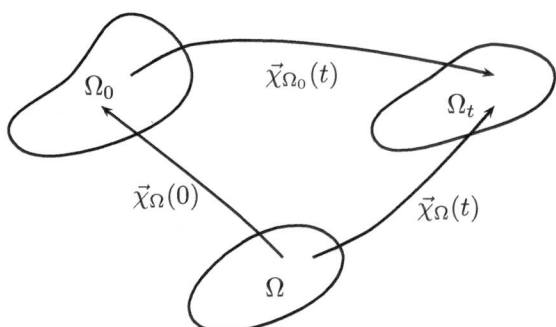

FIGURE 13. The referential domain Ω, initial Ω_0 and current state Ω_t and relations between them. The identification $\Omega \equiv \Omega_0$ is adopted in this text.

describes the motion of the body, see Figure 13. The mapping $\vec{\chi}_\Omega$ depends on the choice of the reference configuration Ω which can be fixed in various ways. Here we think of Ω to be the initial (stress-free) configuration Ω_0. Thus, if not emphasized, we mean by $\vec{\chi}$ exactly $\vec{\chi}_\Omega = \vec{\chi}_{\Omega_0}$.

If we denote by $\vec{X}$ a material point in the reference configuration Ω, then the position of this point at time t is given by

$$\vec{x} = \vec{\chi}(\vec{X}, t). \tag{2.2}$$

Next, the mechanical fields describing the deformation are defined in a standard manner. The displacement field, the velocity field, deformation gradient and its determinant are

$$\vec{u}(\vec{X}, t) = \vec{\chi}(\vec{X}, t) - \vec{X}, \qquad \vec{v} = \frac{\partial \vec{\chi}}{\partial t}, \qquad \boldsymbol{F} = \frac{\partial \vec{\chi}}{\partial \vec{X}}, \qquad J = \det \boldsymbol{F}. \tag{2.3}$$

Let us adopt the following useful notations for some derivatives. Any field quantity φ with values in some vector space Y (i.e., scalar-, vector- or tensor-valued) can be expressed in the Eulerian description as a function of the spatial position $\vec{x} \in \mathbb{R}^3$,

$$\varphi = \tilde{\varphi}(\vec{x}, t) : \Omega_t \times [0, T] \mapsto Y.$$

Then we define the following notation for the derivatives of the field φ:

$$\frac{\partial \varphi}{\partial t} := \frac{\partial \tilde{\varphi}}{\partial t}, \qquad \nabla \varphi = \frac{\partial \varphi}{\partial \vec{x}} := \frac{\partial \tilde{\varphi}}{\partial \vec{x}}, \qquad \operatorname{div} \varphi := \operatorname{tr} \nabla \varphi. \tag{2.4}$$

In the case of Lagrangian description we consider the quantity φ to be defined on the reference configuration Ω; then for any $\vec{X} \in \Omega$ we can express the quantity φ as

$$\varphi = \bar{\varphi}(\vec{X}, t) : \Omega \times [0, T] \mapsto Y,$$

and we define the derivatives of the field φ as

$$\frac{d\varphi}{dt} := \frac{\partial \bar{\varphi}}{\partial t}, \qquad \operatorname{Grad} \varphi = \frac{\partial \varphi}{\partial \vec{X}} := \frac{\partial \bar{\varphi}}{\partial \vec{X}}, \qquad \operatorname{Div} \varphi := \operatorname{tr} \nabla \varphi. \tag{2.5}$$

These two descriptions can be related to each other through the relations

$$\bar{\varphi}(\vec{X}, t) = \tilde{\varphi}(\vec{\chi}(\vec{X}, t), t), \tag{2.6}$$

$$\frac{d\varphi}{dt} = \frac{\partial \varphi}{\partial t} + (\nabla \varphi)\vec{v}, \qquad \operatorname{Grad} \varphi = (\nabla \varphi)\boldsymbol{F}, \qquad \int_{\Omega_t} \varphi \, dv = \int_{\Omega} \varphi J \, dV, \tag{2.7}$$

$$\frac{d\boldsymbol{F}}{dt} = \nabla \vec{v}, \qquad \frac{\partial J}{\partial \boldsymbol{F}} = J \boldsymbol{F}^{-T}, \qquad \frac{dJ}{dt} = J \operatorname{div} \vec{v}. \tag{2.8}$$

For the formulation of the balance laws we will need to express time derivatives of some integrals. The following series of equalities obtained by using the previously stated relations will be useful:

$$\frac{d}{dt} \int_{\Omega_t} \varphi \, dv = \frac{d}{dt} \int_{\Omega} \varphi J \, dV = \int_{\Omega} \frac{d}{dt}(\varphi J) \, dV = \int_{\Omega_t} \left(\frac{d\varphi}{dt} + \varphi \operatorname{div} \vec{v} \right) dv$$

$$= \int_{\Omega_t} \left(\frac{\partial \varphi}{\partial t} + \operatorname{div}(\varphi \vec{v}) \right) dv = \int_{\Omega_t} \frac{\partial \varphi}{\partial t} dv + \int_{\partial \Omega_t} \varphi \vec{v} \cdot \vec{n} \, da \tag{2.9}$$

$$= \frac{\partial}{\partial t} \int_{\Omega_t} \varphi \, dv + \int_{\partial \Omega_t} \varphi \vec{v} \cdot \vec{n} \, da.$$

The Piola identity is used, $\operatorname{Div}(J\boldsymbol{F}^{-T}) = \vec{0}$, which can be checked by differentiating the left-hand side and using (2.8) together with an identity obtained by differentiating the relation $\boldsymbol{F}\boldsymbol{F}^{-1} = \boldsymbol{I}$.

2.2.1. Balance laws in the ALE formulation. The Eulerian (or spatial) description is well suited for a problem of fluid flowing through some spatially fixed region. In such a case the material particles can enter and leave the region of interest. The fundamental quantity describing the motion is the velocity vector.

On the other hand, the Lagrangian (or referential) description is well suited for a problem of deforming a given body consisting of a fixed set of material particles. In this case the actual boundary of the body can change its shape. The fundamental quantity describing the motion in this case is the vector of displacement from the referential state.

In the case of fluid-structure interaction problems we can still use the Lagrangian description for the deformation of the solid part. The fluid flow now takes place in a domain with boundary given by the deformation of the structure which can change in time and is influenced back by the fluid flow. The mixed ALE description of the fluid has to be used in this case. The fundamental quantity describing the motion of the fluid is still the velocity vector but the description is accompanied by a certain displacement field which describes the change of the fluid domain. This displacement field has no connection to the fluid velocity field and the purpose of its introduction is to provide a transformation of the current fluid domain and corresponding governing equations to some fixed reference domain. This method is sometimes called a pseudo-solid mapping method [109].

Let $\mathcal{P} \subset \mathbb{R}^3$ be a fixed region in space (a control volume) with the boundary $\partial \mathcal{P}$ and unit outward normal vector $\vec{n}_\mathcal{P}$, such that $\mathcal{P} \subset \Omega_t$ for all $t \in [0, T]$. Let ϱ

denote the mass density of the material. Then the balance of mass in the region $\mathcal{P}$ can be written as

$$\frac{\partial}{\partial t} \int_{\mathcal{P}} \varrho dv + \int_{\partial \mathcal{P}} \varrho \vec{v} \cdot \vec{n}_{\mathcal{P}} da = 0. \tag{2.10}$$

If all the fields are sufficiently smooth, this equation can be written in local form with respect to the current configuration as

$$\frac{\partial \varrho}{\partial t} + \mathrm{div}(\varrho \vec{v}) = 0. \tag{2.11}$$

It will be useful to derive the mass balance equation from the Lagrangian point of view. Let $\mathcal{Q} \subset \Omega$ be a fixed set of particles. Then $\vec{\chi}(\mathcal{Q}, t) \subset \Omega_t$ is a region occupied by these particles at the time t, and the balance of mass can be expressed as

$$\frac{d}{dt} \int_{\vec{\chi}(\mathcal{Q}, t)} \varrho dv = 0, \tag{2.12}$$

which in local form with respect to the reference configuration can be written as

$$\frac{d}{dt}(\varrho J) = 0. \tag{2.13}$$

In the case of an arbitrary Lagrangian–Eulerian description we take a region $\mathcal{Z} \subset \mathbb{R}^3$ which is itself moving independently of the motion of the body. Let the motion of the control region $\mathcal{Z}$ be described by a given mapping

$$\vec{\zeta}_{\mathcal{Z}} : \mathcal{Z} \times [0, T] \mapsto \mathbb{R}^3, \qquad \mathcal{Z}_t \subset \Omega_t \quad \forall t \in [0, T],$$

with the corresponding velocity $\vec{v}_{\mathcal{Z}} = \frac{\partial \vec{\zeta}_{\mathcal{Z}}}{\partial t}$, deformation gradient $\boldsymbol{F}_{\mathcal{Z}} = \frac{\partial \vec{\zeta}_{\mathcal{Z}}}{\partial \vec{X}}$ and its determinant $J_{\mathcal{Z}} = \det \boldsymbol{F}_{\mathcal{Z}}$. The mass balance equation can be written as

$$\frac{\partial}{\partial t} \int_{\mathcal{Z}_t} \varrho dv + \int_{\partial \mathcal{Z}_t} \varrho(\vec{v} - \vec{v}_{\mathcal{Z}}) \cdot \vec{n}_{\mathcal{Z}_t} da = 0, \tag{2.14}$$

this can be viewed as an Eulerian description with a moving spatial coordinate system or as a grid deformation in the context of the finite-element method. In order to obtain a local form of the balance relation we need to transform the integration to the fixed spatial region $\mathcal{Z}$,

$$\frac{\partial}{\partial t} \int_{\mathcal{Z}} \varrho J_{\mathcal{Z}} dv + \int_{\partial \mathcal{Z}} \varrho(\vec{v} - \vec{v}_{\mathcal{Z}}) \cdot \boldsymbol{F}_{\mathcal{Z}}^{-T} \vec{n}_{\mathcal{Z}} J_{\mathcal{Z}} da = 0, \tag{2.15}$$

then the local form is

$$\frac{\partial}{\partial t}(\varrho J_{\mathcal{Z}}) + \mathrm{div}\left(\varrho J_{\mathcal{Z}}(\vec{v} - \vec{v}_{\mathcal{Z}}) \cdot \boldsymbol{F}_{\mathcal{Z}}^{-T}\right) = 0. \tag{2.16}$$

The two previous special formulations can be now recovered. If the region $\mathcal{Z}$ is not moving in space, i.e., $\mathcal{Z} = \mathcal{Z}_t, \forall t \in [0, T]$, then $\vec{\zeta}_{\mathcal{Z}}$ is the identity mapping, $\boldsymbol{F}_{\mathcal{Z}} = \boldsymbol{I}, J_{\mathcal{Z}} = 1, \vec{v}_{\mathcal{Z}} = \vec{0}$, and (2.16) reduces to (2.11). While, if the region $\mathcal{Z}$ moves exactly with the material, i.e., $\vec{\zeta}_{\mathcal{Z}} = \vec{\chi}|_{\mathcal{Z}}$, then $\boldsymbol{F}_{\mathcal{Z}} = \boldsymbol{F}, J_{\mathcal{Z}} = J, \vec{v}_{\mathcal{Z}} = \vec{v}$, and (2.16) reduces to (2.13).

The balance of linear momentum is postulated in a similar way. Let $\boldsymbol{\sigma}$ denote the Cauchy stress tensor field, representing the surface forces per unit area, $\vec{f}$ be the body forces acting on the material per its unit mass. Then the balance of linear momentum in the Eulerian description is stated as

$$\frac{\partial}{\partial t} \int_{\mathcal{P}} \varrho \vec{v} dv + \int_{\partial \mathcal{P}} \varrho \vec{v} \otimes \vec{v} \vec{n}_{\mathcal{P}} da = \int_{\partial \mathcal{P}} \boldsymbol{\sigma}^T \vec{n}_{\mathcal{P}} da + \int_{\mathcal{P}} \varrho \vec{f} dv. \qquad (2.17)$$

The local form of the linear momentum balance is

$$\frac{\partial \varrho \vec{v}}{\partial t} + \operatorname{div}(\varrho \vec{v} \otimes \vec{v}) = \operatorname{div} \boldsymbol{\sigma}^T + \varrho \vec{f}, \qquad (2.18)$$

or with the use of (2.11) we can write

$$\varrho \frac{\partial \vec{v}}{\partial t} + \varrho (\nabla \vec{v}) \vec{v} = \operatorname{div} \boldsymbol{\sigma}^T + \varrho \vec{f}. \qquad (2.19)$$

From the Lagrangian point of view the momentum-balance relation is

$$\frac{d}{dt} \int_{\vec{\chi}(Q,t)} \varrho \vec{v} dv = \int_{\partial \vec{\chi}(Q,t)} \boldsymbol{\sigma}^T \vec{n}_{\vec{\chi}(Q,t)} da + \int_{\vec{\chi}(Q,t)} \varrho \vec{f} dv. \qquad (2.20)$$

Let us denote by $\boldsymbol{P} = J \boldsymbol{\sigma}^T \boldsymbol{F}^{-T}$ the first Piola–Kirchhoff stress tensor [43], then the local form of the momentum balance is

$$\frac{d}{dt} (\varrho J \vec{v}) = \operatorname{Div} \boldsymbol{P} + \varrho J \vec{f}, \qquad (2.21)$$

or using (2.13) we can write

$$\varrho J \frac{d\vec{v}}{dt} = \operatorname{Div} \boldsymbol{P} + \varrho J \vec{f}. \qquad (2.22)$$

In the arbitrary Lagrangian–Eulerian formulation we obtain

$$\frac{\partial}{\partial t} \int_{\mathcal{Z}_t} \varrho \vec{v} dv + \int_{\partial \mathcal{Z}_t} \varrho \vec{v} \otimes (\vec{v} - \vec{v}_{\mathcal{Z}}) \vec{n}_{\mathcal{Z}_t} da = \int_{\partial \mathcal{Z}_t} \boldsymbol{\sigma}^T \vec{n}_{\mathcal{Z}_t} da + \int_{\mathcal{Z}_t} \varrho \vec{f} dv, \qquad (2.23)$$

which in the local form gives

$$\frac{\partial \varrho J_{\mathcal{Z}} \vec{v}}{\partial t} + \operatorname{div} \left(\varrho J_{\mathcal{Z}} \vec{v} \otimes (\vec{v} - \vec{v}_{\mathcal{Z}}) \boldsymbol{F}_{\mathcal{Z}}^{-T} \right) = \operatorname{div} \left(J_{\mathcal{Z}} \boldsymbol{\sigma}^T \boldsymbol{F}_{\mathcal{Z}}^{-T} \right) + \varrho J_{\mathcal{Z}} \vec{f}, \qquad (2.24)$$

or with the use of (2.16) we can write

$$\varrho J_{\mathcal{Z}} \frac{\partial \vec{v}}{\partial t} + \varrho J_{\mathcal{Z}} (\nabla \vec{v}) \boldsymbol{F}_{\mathcal{Z}}^{-T} (\vec{v} - \vec{v}_{\mathcal{Z}}) = \operatorname{div} \left(J_{\mathcal{Z}} \boldsymbol{\sigma}^T \boldsymbol{F}_{\mathcal{Z}}^{-T} \right) + \varrho J_{\mathcal{Z}} \vec{f}. \qquad (2.25)$$

In the case of angular momentum balance we assume that there are no external or internal sources of angular momentum, then it follows that the Cauchy stress tensor has to be symmetric, i.e., $\boldsymbol{\sigma} = \boldsymbol{\sigma}^T$. Assuming an isothermal condition the energy balance is satisfied and the choice of the constitutive relations for the materials has to be compatible with the balance of entropy [121].

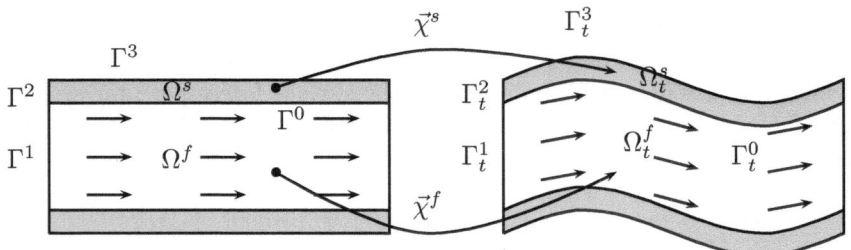

FIGURE 14. Undeformed (original) and deformed (current) configurations.

2.3. Fluid structure interaction problem formulation

At this point we make a few assumptions that allow us to deal with the task of setting up a tractable problem. Let us consider a flow between thick elastic walls as shown in Figure 14. We will use the superscripts ^{s} and ^{f} to denote the quantities connected with the solid and fluid. Let us assume that both materials are incompressible and all the processes are isothermal, which is a well-accepted approximation in biomechanics, and let us denote the constant densities of each material by ϱ^f, ϱ^s.

2.3.1. Monolithic description.

We denote by Ω_t^f the domain occupied by the fluid and Ω_t^s by the solid at time $t \in [0, T]$. Let $\Gamma_t^0 = \bar{\Omega}_t^f \cap \bar{\Omega}_t^s$ be the part of the boundary where the solid interacts with the fluid and $\Gamma_t^i, i = 1, 2, 3$, be the remaining external boundaries of the solid and the fluid as depicted in Figure 14.

Let the deformation of the solid part be described by the mapping

$$\vec{\chi}^s : \Omega^s \times [0, T] \mapsto \mathbb{R}^3, \tag{2.26}$$

with the corresponding displacement $\vec{u}^s$ and the velocity $\vec{v}^s$ given by

$$\vec{u}^s(\vec{X}, t) = \vec{\chi}^s(\vec{X}, t) - \vec{X}, \qquad \vec{v}^s(\vec{X}, t) = \frac{\partial \vec{\chi}^s}{\partial t}(\vec{X}, t). \tag{2.27}$$

The fluid flow is described by the velocity field $\vec{v}^f$ defined on the fluid domain Ω_t^f,

$$\vec{v}^f(\vec{x}, t) : \Omega_t^f \times [0, T] \mapsto \mathbb{R}^3. \tag{2.28}$$

Further we define the auxiliary mapping, denoted by $\vec{\zeta}^f$, to describe the change of the fluid domain and corresponding displacement $\vec{u}^f$ by

$$\vec{\zeta}^f : \Omega^f \times [0, T] \mapsto \mathbb{R}^3, \qquad \vec{u}^f(\vec{X}, t) = \vec{\zeta}^f(\vec{X}, t) - \vec{X}. \tag{2.29}$$

We require that the mapping $\vec{\zeta}^f$ is sufficiently smooth, one to one and has to satisfy

$$\vec{\zeta}^f(\vec{X}, t) = \vec{\chi}^s(\vec{X}, t), \quad \forall (\vec{X}, t) \in \Gamma^0 \times [0, T]. \tag{2.30}$$

In the context of the finite-element method this will describe the artificial mesh deformation inside the fluid region and it will be constructed as a solution to a suitable boundary value problem with (2.30) as the boundary condition.

The momentum and mass balance of the fluid in the time-dependent fluid domain analogous to (2.16) are

$$\varrho^f \frac{\partial \vec{v}^f}{\partial t} + \varrho^f (\nabla \vec{v}^f)(\vec{v}^f - \frac{\partial \vec{u}^f}{\partial t}) = \operatorname{div} \boldsymbol{\sigma}^f, \qquad \operatorname{div} \vec{v}^f = 0 \qquad \text{in } \Omega_t^f, \qquad (2.31)$$

together with the momentum and mass balance of the solid in the solid domain

$$\varrho^s \frac{\partial \vec{v}^s}{\partial t} + \varrho^s (\nabla \vec{v}^s)\vec{v}^s = \operatorname{div} \boldsymbol{\sigma}^s, \qquad \operatorname{div} \vec{v}^s = 0 \qquad \text{in } \Omega_t^s. \qquad (2.32)$$

The interaction is due to the exchange of momentum through the common part of the boundary Γ_t^0. On this part we require that the forces are in balance and, simultaneously, the no-slip boundary condition holds for the fluid, i.e.,

$$\boldsymbol{\sigma}^f \vec{n} = \boldsymbol{\sigma}^s \vec{n} \quad \text{on } \Gamma_t^0, \qquad\qquad \vec{v}^f = \vec{v}^s \quad \text{on } \Gamma_t^0. \qquad (2.33)$$

The remaining external boundary conditions can be of the following kind: a natural boundary condition on the fluid inflow and outflow part Γ_t^1 with p_B given value. Alternatively we can prescribe a Dirichlet-type boundary condition on the inflow or outflow part Γ_t^1,

$$\boldsymbol{\sigma}^f \vec{n} = p_B \vec{n} \quad \text{on } \Gamma_t^1, \qquad\qquad \vec{v}^f = \vec{v}_B \quad \text{on } \Gamma_t^1, \qquad (2.34)$$

where $\vec{v}_B$ is given. The Dirichlet boundary condition is prescribed for the solid displacement at the part Γ_t^2 and the stress free boundary condition for the solid is applied at the part Γ_t^3,

$$\vec{u}^s = \vec{0} \quad \text{on } \Gamma_t^2, \qquad\qquad \boldsymbol{\sigma}^s \vec{n} = \vec{0} \quad \text{on } \Gamma_t^3. \qquad (2.35)$$

We introduce the domain $\Omega = \Omega^f \cup \Omega^s$, where Ω^f, Ω^s are the domains occupied by the fluid and solid in the initial undeformed state, and two fields defined on this domain as

$$\vec{u} : \Omega \times [0, T] \to \mathbb{R}^3, \qquad\qquad \vec{v} : \Omega \times [0, T] \to \mathbb{R}^3,$$

such that the field $\vec{v}$ represents the velocity at the given point and $\vec{u}$ the displacement on the solid part and the artificial displacement in the fluid part, taking care of the fact that the fluid domain is changing with time,

$$\vec{v} = \begin{cases} \vec{v}^s & \text{on } \Omega^s, \\ \vec{v}^f & \text{on } \Omega^f, \end{cases} \qquad\qquad \vec{u} = \begin{cases} \vec{u}^s & \text{on } \Omega^s, \\ \vec{u}^f & \text{on } \Omega^f. \end{cases} \qquad (2.36)$$

Due to the conditions (2.30) and (2.33) both fields are continuous across the interface Γ_t^0 and we can define global quantities on Ω as the deformation gradient and its determinant,

$$\boldsymbol{F} = \boldsymbol{I} + \nabla \vec{u}, \qquad\qquad J = \det \boldsymbol{F}. \qquad (2.37)$$

Using this notation the solid balance laws (2.32) can be expressed in the Lagrangian formulation with the initial configuration Ω^s as reference,

$$J \varrho^s \frac{d\vec{v}}{dt} = \operatorname{Div} \boldsymbol{P}^s, \qquad J = 1 \qquad \text{in } \Omega^s. \qquad (2.38)$$

The fluid equations (2.31) are already expressed in the arbitrary Lagrangian–Eulerian formulation with respect to the time dependent region Ω_t^f; now we transform the equations to the fixed initial region Ω^f by the mapping ζ^f defined by (2.29):

$$\varrho^f \frac{\partial \vec{v}}{\partial t} + \varrho^f (\nabla \vec{v}) \boldsymbol{F}^{-1}(\vec{v} - \frac{\partial \vec{u}}{\partial t}) = J^{-1} \operatorname{Div}(J \boldsymbol{\sigma}^f \boldsymbol{F}^{-T}), \quad \operatorname{Div}(J \vec{v} \boldsymbol{F}^{-T}) = 0 \quad \text{in } \Omega^f.$$
(2.39)

It remains to prescribe some relation for the mapping ζ^f. In terms of the corresponding displacement $\vec{u}^f$ we formulate some simple relation together with the Dirichlet boundary conditions required by (2.30), for example

$$\frac{\partial \vec{u}}{\partial t} = \Delta \vec{u} \quad \text{in } \Omega^f, \qquad \vec{u} = \vec{u}^s \quad \text{on } \Gamma^0, \qquad \vec{u} = \vec{0} \quad \text{on } \Gamma^1.$$
(2.40)

Other choices are possible. For example, the mapping $\vec{u}^f$ can be realized as a solution of the elasticity problem with the same Dirichlet boundary conditions [109]. Then, the complete set of the equations can be written as

$$\frac{\partial \vec{u}}{\partial t} = \begin{cases} \vec{v} & \text{in } \Omega^s, \\ \Delta \vec{u} & \text{in } \Omega^f, \end{cases}$$
(2.41)

$$\frac{\partial \vec{v}}{\partial t} = \begin{cases} \frac{1}{J \varrho^s} \operatorname{Div} \boldsymbol{P}^s & \text{in } \Omega^s, \\ -(\nabla \vec{v}) \boldsymbol{F}^{-1}(\vec{v} - \frac{\partial \vec{u}}{\partial t}) + \frac{1}{J \varrho^f} \operatorname{Div}(J \boldsymbol{\sigma}^f \boldsymbol{F}^{-T}) & \text{in } \Omega^f, \end{cases}$$
(2.42)

$$0 = \begin{cases} J - 1 & \text{in } \Omega^s, \\ \operatorname{Div}(J \vec{v} \boldsymbol{F}^{-T}) & \text{in } \Omega^f, \end{cases}$$
(2.43)

with the initial conditions

$$\vec{u}(0) = \vec{0} \quad \text{in } \Omega, \qquad\qquad \vec{v}(0) = \vec{v}_0 \quad \text{in } \Omega, \qquad (2.44)$$

and boundary conditions

$$\vec{u} = \vec{0}, \quad \vec{v} = \vec{v}_B \quad \text{on } \Gamma^1, \qquad \vec{u} = \vec{0} \quad \text{on } \Gamma^2, \qquad \boldsymbol{\sigma}^s \vec{n} = \vec{0} \quad \text{on } \Gamma^3. \qquad (2.45)$$

2.3.2. Constitutive equations. In order to solve the balance equations we need to specify the constitutive relations for the stress tensors. For the fluid we use the incompressible Newtonian relation

$$\boldsymbol{\sigma}^f = -p^f \boldsymbol{I} + \mu(\nabla \vec{v}^f + (\nabla \vec{v}^f)^T), \qquad (2.46)$$

where μ represents the viscosity of the fluid and p^f is the Lagrange multiplier corresponding to the incompressibility constraint.

For the solid part we assume that it can be described by an incompressible hyper-elastic material. We specify the Helmholtz potential Ψ, and the solid Cauchy stress tensor and the first Piola–Kirchhoff stress tensor are given by

$$\boldsymbol{\sigma}^s = -p^s \boldsymbol{I} + \varrho^s \frac{\partial \Psi}{\partial \boldsymbol{F}} \boldsymbol{F}^T, \qquad\qquad \boldsymbol{P}^s = -J p^s \boldsymbol{F}^{-T} + J \varrho^s \frac{\partial \Psi}{\partial \boldsymbol{F}}, \qquad (2.47)$$

where p^s is the Lagrange multiplier corresponding to the incompressibility constraint.

The Helmholtz potential can be expressed as a function of different quantities

$$\Psi = \hat{\Psi}(\boldsymbol{F}) = \hat{\Psi}(\boldsymbol{I} + \nabla\vec{u}),$$

but due to the principle of material frame indifference the Helmholtz potential Ψ depends on the deformation only through the right Cauchy–Green deformation tensor $\boldsymbol{C} = \boldsymbol{F}^T \boldsymbol{F}$ [43]

$$\Psi = \tilde{\Psi}(\boldsymbol{C}). \tag{2.48}$$

A certain coerciveness condition is usually imposed on the form of the Helmholtz potential

$$\bar{\Psi}(\nabla\vec{u}(\vec{X},t)) \geq a \left\|\nabla\vec{u}(\vec{X},t)\right\|^2 - b(\vec{X}), \tag{2.49}$$

where a is a positive constant and $b \in L^1(\Omega^s)$. With this assumption and using the integral identity (2.58) we can derive an energy estimate of the form

$$\frac{c}{2} \left\|\vec{v}(T)\right\|^2_{L^2(\Omega_T)} + \int_0^T \mu \left\|\nabla\vec{v}\right\|^2_{L^2(\Omega_t^f)} dt + a \left\|\nabla\vec{u}(T)\right\|^2_{L^2(\Omega^s)}$$

$$\leq \|b\|_{L^1(\Omega^s)} + \frac{1}{2} \left\|\vec{v}_0\right\|^2_{L^2(\Omega^f)} + \frac{\beta}{2} \left\|\vec{v}_0\right\|^2_{L^2(\Omega^s)}. \tag{2.50}$$

where $c = \min(1, \beta)$.

Typical examples for the Helmholtz potential used for isotropic materials like rubber is the Mooney–Rivlin material

$$\tilde{\Psi} = c_1(I_C - 3) + c_2(II_C - 3), \tag{2.51}$$

where $I_C = \operatorname{tr}\boldsymbol{C}$, $II_C = \frac{1}{2}(\operatorname{tr}^2\boldsymbol{C} - \operatorname{tr}\boldsymbol{C}^2)$, $III_C = \det\boldsymbol{C}$ are the invariants of the right Cauchy–Green deformation tensor $\boldsymbol{C}$ and c_i are some material constants. A special case of neo-Hookean material is obtained for $c_2 = 0$. With a suitable choice of the material parameters the entropy inequality and the balance of energy are automatically satisfied.

2.3.3. Weak formulation. We non-dimensionalize all the quantities by a given characteristic length L and speed V as follows:

$$\hat{t} = t\frac{V}{L}, \qquad \hat{\vec{x}} = \frac{\vec{x}}{L}, \qquad \hat{\vec{u}} = \frac{\vec{u}}{L}, \qquad \hat{\vec{v}} = \frac{\vec{v}}{V},$$

$$\hat{\boldsymbol{\sigma}}^s = \boldsymbol{\sigma}^s\frac{L}{\varrho^f V^2}, \qquad \hat{\boldsymbol{\sigma}}^f = \boldsymbol{\sigma}^f\frac{L}{\varrho^f V^2}, \qquad \hat{\mu} = \frac{\mu}{\varrho^f V L}, \qquad \hat{\Psi} = \Psi\frac{L}{\varrho^f V^2},$$

further using the same symbols, without the hat, for the non-dimensional quantities and denoting by $\beta = \frac{\varrho^s}{\varrho^f}$ the densities ratio. The non-dimensionalized system with the choice of material relations, (2.46) for viscous fluid and (2.47) for the hyper-elastic solid is

$$\frac{\partial \vec{u}}{\partial t} = \begin{cases} \vec{v} & \text{in } \Omega^s, \\ \Delta \vec{u} & \text{in } \Omega^f, \end{cases} \tag{2.52}$$

$$\frac{\partial \vec{v}}{\partial t} = \begin{cases} \frac{1}{\beta} \operatorname{Div}\left(-Jp^s \mathbf{F}^{-T} + \frac{\partial \Psi}{\partial \mathbf{F}}\right) & \text{in } \Omega^s, \\ -(\nabla \vec{v})\mathbf{F}^{-1}(\vec{v} - \frac{\partial \vec{u}}{\partial t}) & \text{in } \Omega^f, \\ + \operatorname{Div}\left(-Jp^f \mathbf{F}^{-T} + J\mu \nabla \vec{v} \mathbf{F}^{-1} \mathbf{F}^{-T}\right) & \end{cases} \tag{2.53}$$

$$0 = \begin{cases} J - 1 & \text{in } \Omega^s, \\ \operatorname{Div}(J\vec{v}\mathbf{F}^{-T}) & \text{in } \Omega^f, \end{cases} \tag{2.54}$$

and the boundary conditions

$$\boldsymbol{\sigma}^f \vec{n} = \boldsymbol{\sigma}^s \vec{n} \quad \text{on } \Gamma_t^0, \qquad\qquad \vec{v} = \vec{v}_B \quad \text{on } \Gamma_t^1, \tag{2.55}$$

$$\vec{u} = \vec{0} \quad \text{on } \Gamma_t^2, \qquad\qquad \boldsymbol{\sigma}^f \vec{n} = \vec{0} \quad \text{on } \Gamma_t^3. \tag{2.56}$$

Let $I = [0, T]$ denote the time interval of interest. We multiply the equations (2.52)–(2.54) by the test functions $\vec{\zeta}, \vec{\xi}, \gamma$ such that $\vec{\zeta} = \vec{0}$ on Γ^2, $\vec{\xi} = \vec{0}$ on Γ^1, and integrate over the space domain Ω and the time interval I. Using integration by parts on some of the terms and the boundary conditions we obtain

$$\int_0^T \int_\Omega \frac{\partial \vec{u}}{\partial t} \cdot \vec{\zeta} \, dV \, dt = \int_0^T \int_{\Omega^s} \vec{v} \cdot \vec{\zeta} \, dV \, dt - \int_0^T \int_{\Omega^f} \nabla \vec{u} \cdot \nabla \vec{\zeta} \, dV \, dt, \tag{2.57}$$

$$\int_0^T \int_{\Omega^f} J \frac{\partial \vec{v}}{\partial t} \cdot \vec{\xi} \, dV \, dt + \int_0^T \int_{\Omega^s} \beta J \frac{\partial \vec{v}}{\partial t} \cdot \vec{\xi} \, dV \, dt$$
$$+ \int_0^T \int_{\Omega^f} J \nabla \vec{v} \mathbf{F}^{-1}(\vec{v} - \frac{\partial \vec{u}}{\partial t}) \cdot \vec{\xi} \, dV \, dt - \int_0^T \int_\Omega Jp\mathbf{F}^{-T} \cdot \nabla \vec{\xi} \, dV \, dt \tag{2.58}$$
$$+ \int_0^T \int_{\Omega^s} \frac{\partial \Psi}{\partial \mathbf{F}} \cdot \nabla \vec{\xi} \, dV \, dt + \int_0^T \int_{\Omega^f} J\mu \nabla \vec{v} \mathbf{F}^{-1} \mathbf{F}^{-T} \cdot \nabla \vec{\xi} \, dV \, dt = 0,$$

$$\int_0^T \int_{\Omega^s} (J - 1)\gamma \, dV \, dt + \int_0^T \int_{\Omega^f} \operatorname{Div}(J\vec{v}\mathbf{F}^{-T})\gamma \, dV \, dt = 0. \tag{2.59}$$

Let us define the spaces

$$U = \{\vec{u} \in L^\infty(I, [W^{1,2}(\Omega)]^3), \vec{u} = \vec{0} \text{ on } \Gamma^2\},$$
$$V = \{\vec{v} \in L^2(I, [W^{1,2}(\Omega_t)]^3) \cap L^\infty(I, [L^2(\Omega_t)]^3), \vec{v} = \vec{0} \text{ on } \Gamma^1\},$$
$$P = \{p \in L^2(I, L^2(\Omega))\},$$

then the variational formulation of the fluid-structure interaction problem is to find $(\vec{u}, \vec{v} - \vec{v}_B, p) \in U \times V \times P$ such that equations (2.57), (2.58) and (2.59) are satisfied for all $(\vec{\zeta}, \vec{\xi}, \gamma) \in U \times V \times P$ including appropriate initial conditions.

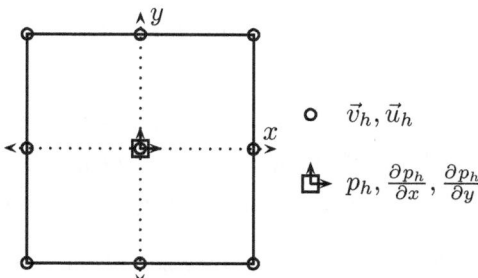

FIGURE 15. Location of the degrees of freedom for the Q_2, P_1^{dis} element.

2.3.4. Discretization. In the following, we restrict ourselves to two dimensions which allows systematic tests of the proposed methods in a very efficient way, particularly in view of grid-independent solutions. The time discretization is done by the Crank-Nicholson scheme which is only conditionally stable but which has better conservation properties than for example the implicit Euler scheme [30, 65]. The Crank–Nicholson scheme can be obtained by dividing the time interval I into the series of time steps $[t^n, t^{n+1}]$ with step length $k_n = t^{n+1} - t^n$. Assuming that the test functions are piecewise constant on each time step $[t^n, t^{n+1}]$, writing the weak formulation (2.57)-(2.58) for the time interval $[t^n, t^{n+1}]$, approximating the time derivatives by the central differences $\frac{\partial f}{\partial t} \approx \frac{f(t^{n+1}) - f(t^n)}{k_n}$, and approximating the time integration for the remaining terms by the trapezoidal quadrature rule as

$$\int_{t^n}^{t^{n+1}} f(t)dt \approx \frac{k_n}{2}(f(t^n) + f(t^{n+1})),$$

we obtain the time-discretized system. The last equation corresponding to the incompressibility constraint is taken implicitly for the time t^{n+1} and the corresponding term with the Lagrange multiplier p_h^{n+1} in the equation (2.58) is also taken implicitly.

The discretization in space is done by the finite-element method. We approximate the domain Ω by a domain Ω_h with polygonal boundary and by $\mathcal{T}_h$ we denote a set of quadrilaterals covering the domain Ω_h. We assume that $\mathcal{T}_h$ is regular in the usual sense that any two quadrilaterals are disjoint or have a common vertex or a common edge. By $\bar{T} = [-1, 1]^2$ we denote the reference quadrilateral. Our treatment of the problem as one system suggests that we use the same finite elements on both, the solid part and the fluid region. Since both materials are incompressible, we have to choose a pair of finite-element spaces known to be stable for the problems with incompressibility constraint. One possible choice is the conforming biquadratic, discontinuous linear Q_2, P_1^{dis} pair, see Figure 15 for the location of the degrees of freedom. This choice results in 39 degrees of freedom per element in the case of our displacement, velocity, pressure formulation in two dimensions and 112 degrees of freedom per element in three dimensions.

The spaces U, V, P on an interval $[t^n, t^{n+1}]$ would be approximated in the case of the Q_2, P_1^{dis} pair as

$$U_h = \{\vec{u}_h \in [C(\Omega_h)]^2, \vec{u}_h|_T \in [Q_2(T)]^2 \quad \forall T \in \mathcal{T}_h, \vec{u}_h = \vec{0} \text{ on } \Gamma_2\},$$

$$V_h = \{\vec{v}_h \in [C(\Omega_h)]^2, \vec{v}_h|_T \in [Q_2(T)]^2 \quad \forall T \in \mathcal{T}_h, \vec{v}_h = 0 \text{ on } \Gamma_1\},$$

$$P_h = \{p_h \in L^2(\Omega_h), p_h|_T \in P_1(T) \quad \forall T \in \mathcal{T}_h\}.$$

Let us denote by $\vec{u}_h^n$ the approximation of $\vec{u}(t^n)$, $\vec{v}_h^n$ the approximation of $\vec{v}(t^n)$ and p_h^n the approximation of $p(t^n)$. Further we will use following shorthand notation:

$$\boldsymbol{F}^n = \boldsymbol{I} + \nabla \vec{u}_h^n, \quad J^n = \det \boldsymbol{F}^n \quad J^{n+\frac{1}{2}} = \frac{1}{2}(J^n + J^{n+1}),$$

$$(f, g) = \int_\Omega f \cdot g \, dV, \quad (f, g)_s = \int_{\Omega^s} f \cdot g \, dV, \quad (f, g)_f = \int_{\Omega^f} f \cdot g \, dV,$$

f, g being scalars, vectors or tensors.

Writing down the discrete equivalent of the equations (2.57)–(2.59) yields

$$\left(\vec{u}_h^{n+1} - \vec{u}_h^n, \vec{\eta}\right) - \frac{k_n}{2}\left\{\left(\vec{v}_h^{n+1} + \vec{v}_h^n, \vec{\eta}\right)_s + \left(\nabla \vec{u}_h^{n+1} + \nabla \vec{u}_h^n, \nabla \vec{\eta}\right)_f\right\} = 0, \qquad (2.60)$$

$$\left(J^{n+\frac{1}{2}}(\vec{v}_h^{n+1} - \vec{v}_h^n), \vec{\xi}\right)_f + \beta \left(\vec{v}_h^{n+1} - \vec{v}_h^n, \vec{\xi}\right)_s - k_n \left(J^{n+1} p_h^{n+1}(\boldsymbol{F}^{n+1})^{-T}, \nabla \vec{\xi}\right)_s$$

$$+ \frac{k_n}{2}\left\{\left(\frac{\partial \Psi}{\partial \boldsymbol{F}}(\nabla \vec{u}_h^{n+1}), \nabla \vec{\xi}\right)_s + \left(J^{n+1} \nabla \vec{v}_h^{n+1}(\boldsymbol{F}^{n+1})^{-1} \vec{v}_h^{n+1}, \vec{\xi}\right)_f\right.$$

$$\left. + \mu \left(J^{n+1} \nabla \vec{v}_h^{n+1}(\boldsymbol{F}^{n+1})^{-1}, \nabla \vec{\xi}(\boldsymbol{F}^{n+1})^{-1}\right)_f\right\}$$

$$+ \frac{1}{2}\left((J^{n+1} \nabla \vec{v}_h^{n+1}(\boldsymbol{F}^{n+1})^{-1} + J^n \nabla \vec{v}_h^n(\boldsymbol{F}^n)^{-1})(\vec{u}_h^{n+1} - \vec{u}_h^n), \vec{\xi}\right)_f$$

$$+ \frac{k_n}{2}\left\{\left(\frac{\partial \Psi}{\partial \boldsymbol{F}}(\nabla \vec{u}_h^n), \nabla \vec{\xi}\right)_s + \left(J^n \nabla \vec{v}_h^n(\boldsymbol{F}^n)^{-1} \vec{v}_h^n, \vec{\xi}\right)_f\right.$$

$$\left. + \mu \left(J^n \nabla \vec{v}_h^n(\boldsymbol{F}^n)^{-1}, \nabla \vec{\xi}(\boldsymbol{F}^n)^{-1}\right)_f\right\} = 0,$$

$$\qquad (2.61)$$

$$\left(J^{n+1} - 1, \gamma\right)_s + \left(J^{n+1} \nabla \vec{v}_h^{n+1}(\boldsymbol{F}^{n+1})^{-1}, \gamma\right)_f = 0. \qquad (2.62)$$

Using the basis of the spaces U_h, V_h, P_h as the test functions $\vec{\zeta}, \vec{\xi}, \gamma$ we obtain a nonlinear algebraic set of equations. In each time step we have to find $\vec{X} = (\vec{u}_h^{n+1}, \vec{v}_h^{n+1}, p_h^{n+1}) \in U_h \times V_h \times P_h$ such that

$$\vec{\mathcal{F}}(\vec{X}) = \vec{0}, \qquad (2.63)$$

where $\vec{\mathcal{F}}$ represents the discrete version of the system (2.60–2.62).

2.3.5. Solution algorithm. The system (2.63) of nonlinear algebraic equations is solved using the Newton method as the basic iteration. One step of the Newton iteration can be written as

$$\vec{X}^{n+1} = \vec{X}^n - \left[\frac{\partial \vec{\mathcal{F}}}{\partial \vec{X}}(\vec{X}^n)\right]^{-1} \vec{\mathcal{F}}(\vec{X}^n). \tag{2.64}$$

This basic iteration can exhibit quadratic convergence provided that the initial guess is sufficiently close to the solution. To ensure the convergence globally, some improvements of this basic iteration are used. The damped Newton method with line search improves the chance of convergence by adaptively changing the length of the correction vector. The solution update step in the Newton method (2.64) is replaced by $\vec{X}^{n+1} = \vec{X}^n + \omega \delta \vec{X}$, where the parameter ω is determined such that a certain error measure decreases. One of the possible choices for the quantity to decrease is

$$f(\omega) = \vec{\mathcal{F}}(\vec{X}^n + \omega \delta \vec{X}) \cdot \delta \vec{X}. \tag{2.65}$$

Since we know $f(0) = \vec{\mathcal{F}}(\vec{X}^n) \cdot \delta \vec{X}$, and $f'(0) = \left[\frac{\partial \vec{\mathcal{F}}}{\partial \vec{X}}(\vec{X}^n)\right] \delta \vec{X} \cdot \delta \vec{X} = \vec{\mathcal{F}}(\vec{X}^n) \cdot \delta \vec{X}$, computing $f(\omega_0)$ for $\omega_0 = -1$ or ω_0 determined adaptively from previous iterations, we can approximate $f(\omega)$ by a quadratic function

$$f(\omega) = \frac{f(\omega_0) - f(0)(\omega_0 + 1)}{\omega_0^2}\omega^2 + f(0)(\omega + 1).$$

Then setting $\tilde{\omega} = \frac{f(0)\omega_0^2}{f(\omega_0) - f(0)(\omega_0 + 1)}$, the new optimal step length $\omega \in [-1, 0]$ is

$$\omega = \begin{cases} -\dfrac{\tilde{\omega}}{2} & \text{if } \dfrac{f(0)}{f(\omega_0)} > 0, \\[2mm] -\dfrac{\tilde{\omega}}{2} - \sqrt{\dfrac{\tilde{\omega}^2}{4} - \tilde{\omega}} & \text{if } \dfrac{f(0)}{f(\omega_0)} \leq 0. \end{cases} \tag{2.66}$$

This line search can be repeated with ω_0 taken as the last ω until, for example, $f(\omega) \leq \frac{1}{2}f(0)$. By this we can try to enforce a monotone convergence of the approximation $\vec{X}^n$.

An adaptive time-step selection was found to help in the nonlinear convergence. A heuristic algorithm was used to correct the time-step length according to the convergence of the nonlinear iterations in the previous time step. If the convergence was close to quadratic, i.e., only up to three Newton steps were needed to obtain the required precision, the time step could be slightly increased, otherwise the time-step length was reduced.

The structure of the Jacobian matrix $\frac{\partial \vec{\mathcal{F}}}{\partial \vec{X}}$ is

$$\frac{\partial \vec{\mathcal{F}}}{\partial \vec{X}}(\vec{X}) = \begin{pmatrix} S_{uu} & S_{uv} & 0 \\ S_{vu} & S_{vv} & B_u + B_v \\ B_u^T & B_v^T & 0 \end{pmatrix}, \tag{2.67}$$

1. Let $\vec{X}^n$ be some starting guess.
2. Set the residuum vector $\vec{R}^n = \mathcal{F}(\vec{X}^n)$ and the tangent matrix $\boldsymbol{A} = \frac{\partial \vec{\mathcal{F}}}{\partial \vec{X}}(\vec{X}^n)$.
3. Solve for the correction $\delta\vec{X}$

$$\boldsymbol{A}\delta\vec{X} = \vec{R}^n.$$

4. Find an optimal step length ω.
5. Update the solution $\vec{X}^{n+1} = \vec{X}^n - \omega\delta\vec{X}$.

FIGURE 16. One step of the Newton method with the line search.

and it can be computed by finite differences from the residual vector $\vec{\mathcal{F}}(\vec{X})$,

$$\left[\frac{\partial\vec{\mathcal{F}}}{\partial\vec{X}}\right]_{ij}(\vec{X}^n) \approx \frac{[\vec{\mathcal{F}}]_i(\vec{X}^n + \alpha_j\vec{e}_j) - [\vec{\mathcal{F}}]_i(\vec{X}^n - \alpha_j\vec{e}_j)}{2\alpha_j}, \tag{2.68}$$

where $\vec{e}_j$ are the unit basis vectors in $\mathbb{R}^n$ and the coefficients α_j are adaptively taken according to the change in the solution in the previous time step. Since we know the sparsity pattern of the Jacobian matrix in advance, it is given by the used finite-element method, this computation can be done in an efficient way so that the linear solver remains the dominant part in terms of the cpu time. However, the resulting nonlinear and linear solution behavior is quite sensitive w.r.t. the parameters [51].

2.3.6. Multigrid solver. The solution of the linear problems is the most time-consuming part of the solution process. A good candidate seems to be a direct solver for sparse systems like UMFPACK [22]; while this choice provides very robust linear solvers, its memory and CPU time requirements are too high for larger systems (i.e., more than 20000 unknowns). Large linear problems can be solved by Krylov space methods (BiCGStab, GMRes[2]) with suitable preconditioners. One possibility is the ILU preconditioner with special treatment of the saddle point character of our system, where we allow certain fill-in for the zero diagonal blocks [7]. The alternative option for larger systems is the multigrid method presented in this section.

We utilize the standard geometric multigrid approach based on a hierarchy of grids obtained by successive regular refinement of a given coarse mesh. The complete multigrid iteration is performed in the standard defect-correction setup with the V or F-type cycle. While a direct sparse solver [22] is used for the coarse grid solution, on finer levels a fixed number (2 or 4) of iterations by local MPSC schemes (Vanka-like smoother [139, 125]) is performed. Such iterations can be

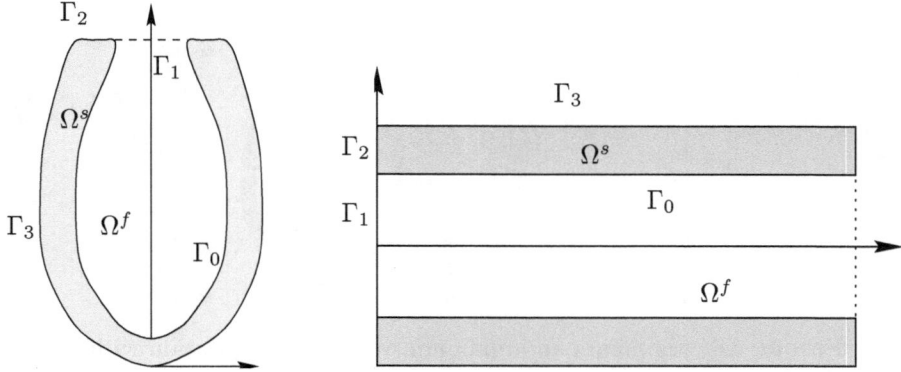

FIGURE 17. Schematic view of the ventricle and elastic tube geometries.

written as

$$
\begin{bmatrix} \mathbf{u}^{l+1} \\ \mathbf{v}^{l+1} \\ p^{l+1} \end{bmatrix} = \begin{bmatrix} \mathbf{u}^{l} \\ \mathbf{v}^{l} \\ p^{l} \end{bmatrix} - \omega \sum_{\text{Patch } \Omega_i} \begin{bmatrix} S_{\mathbf{uu}|\Omega_i} & S_{\mathbf{uv}|\Omega_i} & 0 \\ S_{\mathbf{vu}|\Omega_i} & S_{\mathbf{vv}|\Omega_i} & kB_{|\Omega_i} \\ c_{\mathbf{u}}B^T_{s|\Omega_i} & c_{\mathbf{v}}B^T_{f|\Omega_i} & 0 \end{bmatrix}^{-1} \begin{bmatrix} \mathbf{def}^l_{\mathbf{u}} \\ \mathbf{def}^l_{\mathbf{v}} \\ def^l_p \end{bmatrix}.
$$

The inverse of the local systems (39×39) can be done by hardware optimized direct solvers. The full nodal interpolation is used as the prolongation operator $\boldsymbol{P}$ with its transposed operator used as the restriction $\boldsymbol{R} = \boldsymbol{P}^T$.

2.4. Applications

In this section we present a few example applications to demonstrate the presented methods. As a motivation we consider the numerical simulation of the cardiovascular hemodynamics which has become a useful tool for deeper understanding of the onset of diseases of the human circulatory system, as for example blood cell and intima damages in stenosis, aneurysm rupture, evaluation of the new surgery techniques of heart, arteries and veins.

In order to test the proposed numerical method simplified two-dimensional examples which include some of the important characteristics of the biomechanical applications are computed. The first example is a flow in an ellipsoidal cavity and the second is a flow through a channel with elastic walls. In both cases the flow is driven by changing fluid pressure at the inflow part of the boundary while the elastic part of the boundary is either fixed or stress-free.

The constitutive relations used for the materials are the incompressible Newtonian model (2.46) for the fluid and the hyper-elastic neo-Hookean material (2.51) with $c_2 = 0$ for the solid. This choice includes all the main difficulties the numerical method has to deal with, namely the incompressibility and large deformations.

2.4.1. Flow in an ellipsoidal cavity. The motivation for our first test is the left heart ventricle which is approximately ellipsoidal void surrounded by the heart

S. Turek and J. Hron

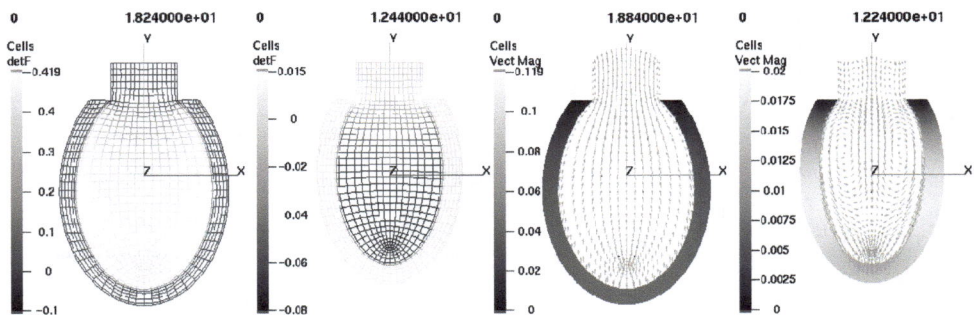

FIGURE 18. Maximum and minimum volume configuration with the fluid flow

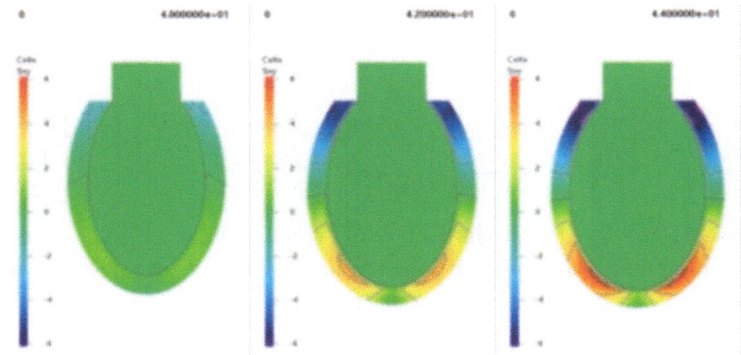

FIGURE 19. Shear stress distribution in the wall during the period.

muscle. In our two-dimensional computations we use an ellipsoidal cavity, see Figure 17, with prescribed time-dependent natural boundary condition at the fluid boundary part Γ^1:

$$p(t) = \sin t \quad \text{on } \Gamma^1. \tag{2.69}$$

The material of the solid wall is modeled by the simple neo-Hookean constitutive relation (2.51) with $c_2 = 0$.

Figures 18 and 19 show the computational grid for the maximal and minimal volume configuration of the cavity and the velocity field of the fluid for the same configurations.

One of the important characteristics is the shear stress exerted by the fluid flow on the wall material. Figure 19 shows the distribution of the shear stress in the domain for three different times.

In Figures 20 and 21 the volume change of the cavity as a function of the time and the average pressure inside the cavity versus the volume of the cavity is

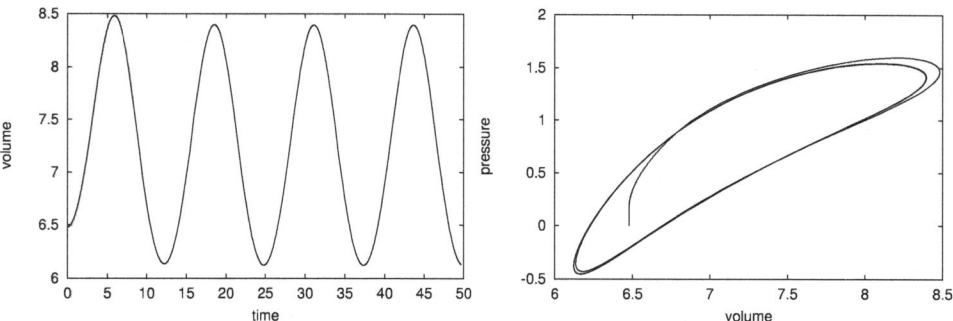

FIGURE 20. Volume of the fluid inside and the pressure-volume diagram for the ellipsoidal cavity test.

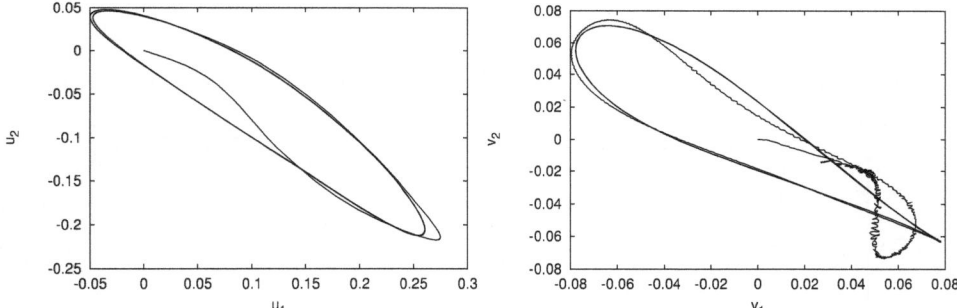

FIGURE 21. The displacement trajectory and velocity of a point at the fluid solid interface (inner side of the wall) for the ellipsoidal cavity test.

shown together with the trajectory and velocity of a material point on the solid-fluid interface. We can see that after the initial cycle which was started from the undeformed configuration the system comes to a time periodic solution.

2.4.2. Flow in an elastic channel. The second application is the simulation of a flow in an elastic tube or, in our 2-dimensional case, a flow between elastic plates. The flow is driven by a time-dependent pressure difference between the ends of the channel of the form (2.69). Such a flow is also interesting to investigate in the presence of some constriction as a stenosis, which is shown in Figure 25.

For the flow in the channel without any constriction the time dependence of the fluid volume inside the channel is shown together with the pressure volume diagram in the figure and the trajectory and velocity of a material point on the solid fluid interface in Figures 23 and 24. The velocity field is shown in Figure 22 at different stages of the pulse.

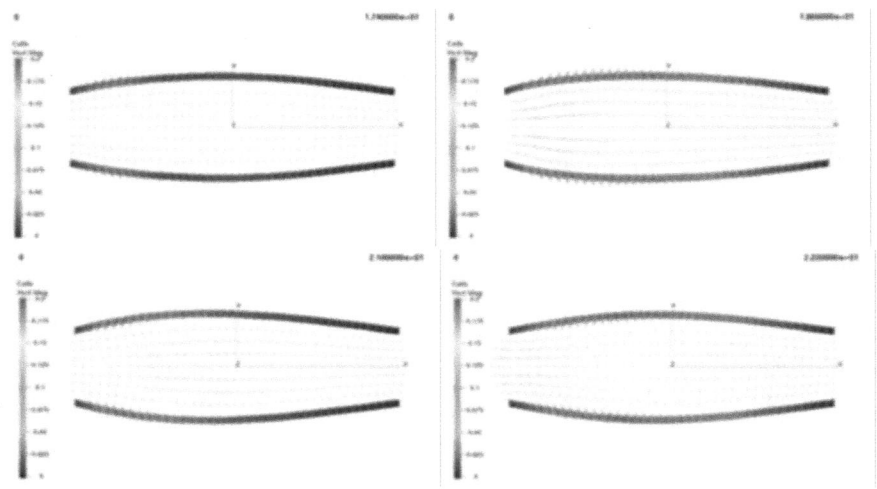

FIGURE 22. Velocity field during one pulse in channel without an obstacle

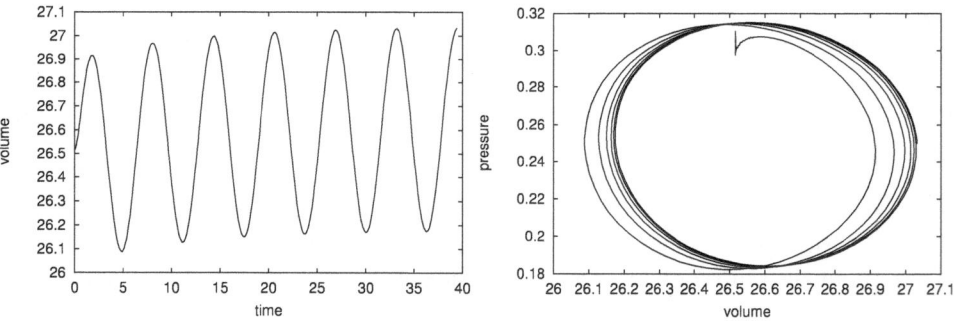

FIGURE 23. Volume of the fluid in the channel and the pressure-volume diagram.

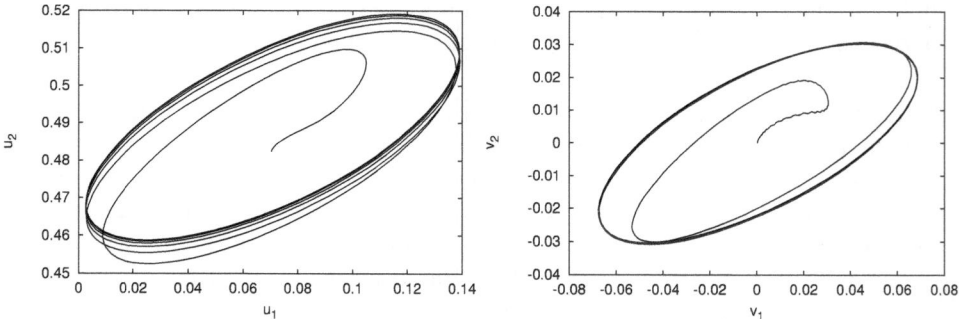

FIGURE 24. Displacement trajectory and velocity of a point at the fluid-solid interface (inner side of the wall).

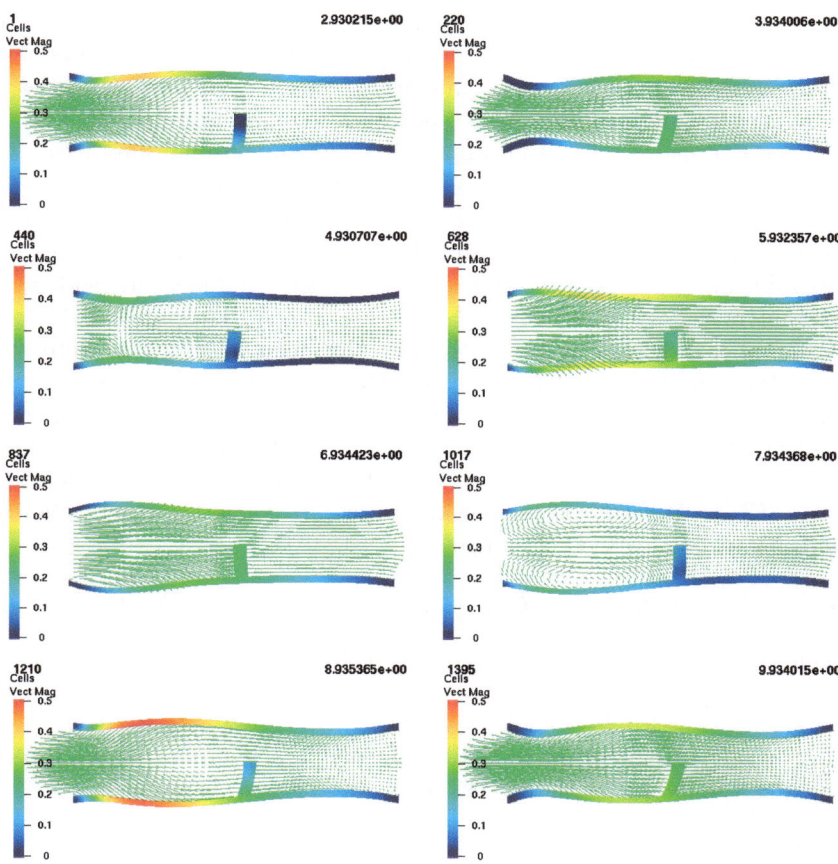

FIGURE 25. Fluid flow and pressure distribution in the wall during one pulse for the example flow in a channel with constriction.

Finally in Figure 25 the velocity field in the fluid and the pressure distribution throughout the wall is shown for the computation of the flow in a channel with elastic obstruction. In this example the elastic obstruction is modeled by the same material as the walls of the channel and is fixed to the elastic walls. Both ends of the walls are fixed at the inflow and outflow and the flow is again driven by a periodic change of the pressure at the left end.

2.5. FSI benchmark

We finally consider the problem of viscous fluid flow interacting with an elastic body which is being deformed by the fluid action. As mentioned in the previous part such a problem is encountered in many real-life applications of great importance. In order to analyze different ways to solve such problems a simple benchmark configuration with known solution can be a useful tool. One suitable simple numerical benchmark was proposed in [129]. The configurations consist of laminar incompressible channel flow around an elastic object which results in self-induced oscillations of the structure. Moreover, characteristic flow quantities and corresponding plots are provided for a quantitative comparison.

We consider the flow of an **incompressible Newtonian fluid** interacting with an **elastic solid**. We denote by Ω_t^f the domain occupied by the fluid and Ω_t^s by the solid at the time $t \in [0, T]$. Let $\Gamma_t^0 = \bar{\Omega}_t^f \cap \bar{\Omega}_t^s$ be the part of the boundary where the elastic solid interacts with the fluid. The fluid is considered to be **Newtonian**, **incompressible** and its state is described by the velocity and pressure fields $\vec{v}^f, p^f$. The constant density of the fluid is ϱ^f and the viscosity is denoted by ν^f. The Reynolds number is defined by $\text{Re} = \frac{2r\bar{V}}{\nu^f}$, with the mean velocity $\bar{V} = \frac{2}{3}v(0, \frac{H}{2}, t)$, r radius of the cylinder and H height of the channel (see Fig. 26). The structure is assumed to be **elastic** and **compressible**. Its configuration is described by the displacement $\vec{u}^s$, with velocity field $\vec{v}^s = \frac{\partial \vec{u}^s}{\partial t}$. The material is specified by giving the Cauchy stress tensor $\boldsymbol{\sigma}^s$ (the second Piola–Kirchhoff stress tensor is then given by $\boldsymbol{S}^s = J\boldsymbol{F}^{-1}\boldsymbol{\sigma}^s\boldsymbol{F}^{-T}$) by the following constitutive law for the **St. Venant–Kirchhoff** material ($\boldsymbol{E} = \frac{1}{2}(\boldsymbol{F}^T\boldsymbol{F} - \boldsymbol{I})$):

$$\boldsymbol{\sigma}^s = \frac{1}{J}\boldsymbol{F}\left(\lambda^s(\text{tr }\boldsymbol{E})\boldsymbol{I} + 2\mu^s\boldsymbol{E}\right)\boldsymbol{F}^T, \qquad \boldsymbol{S}^s = \lambda^s(\text{tr }\boldsymbol{E})\boldsymbol{I} + 2\mu^s\boldsymbol{E}. \qquad (2.70)$$

The boundary conditions on the fluid-solid interface are assumed to be

$$\boldsymbol{\sigma}^f\vec{n} = \boldsymbol{\sigma}^s\vec{n}, \quad \vec{v}^f = \vec{v}^s \qquad\qquad \text{on } \Gamma_t^0, \qquad (2.71)$$

where $\vec{n}$ is a unit normal vector to the interface Γ_t^0. This implies the no-slip condition for the flow, and that the forces on the interface are in balance.

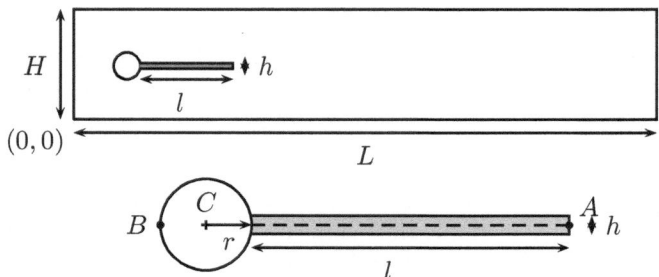

FIGURE 26. Computational domain with the details of the structure part.

The domain is based on the 2D version of the well-known CFD benchmark in [133] and shown here in Figure 26. By omitting the elastic bar behind the cylinder one can exactly recover the setup of the *flow around cylinder* configuration which allows for validation of the flow part by comparing the results with the older flow benchmark. The setting is intentionally non-symmetric [133] to prevent

		value [m]
channel length	L	2.5
channel width	H	0.41
cylinder center position	C	$(0.2, 0.2)$
cylinder radius	r	0.05

		value [m]
elastic structure length	l	0.35
elastic structure thickness	h	0.02
reference point (at $t = 0$)	A	$(0.6, 0.2)$
reference point	B	$(0.2, 0.2)$

TABLE 4. Overview of the geometry parameters.

parameter	FSI1	FSI2	FSI3
ϱ^s $[10^3 \frac{kg}{m^3}]$	1	10	1
ν^s	0.4	0.4	0.4
μ^s $[10^6 \frac{kg}{ms^2}]$	0.5	0.5	2.0
ϱ^f $[10^3 \frac{kg}{m^3}]$	1	1	1
ν^f $[10^{-3} \frac{m^2}{s}]$	1	1	1
U $[\frac{m}{s}]$	0.2	1	2

parameter	FSI1	FSI2	FSI3
$\beta = \frac{\varrho^s}{\varrho^f}$	1	10	1
ν^s	0.4	0.4	0.4
$Ae = \frac{E^s}{\varrho^f U^2}$	3.5×10^4	1.4×10^3	1.4×10^3
$Re = \frac{Ud}{\nu^f}$	20	100	200
$\bar{U}$	0.2	1	2

TABLE 5. Parameter settings for the full FSI benchmarks.

the dependence of the onset of any possible oscillation on the precision of the computation.

2.5.1. Boundary and initial conditions. A parabolic velocity profile is prescribed at the left channel inflow

$$v^f(0, y) = 1.5\bar{U}\frac{y(H - y)}{\left(\frac{H}{2}\right)^2} = 1.5\bar{U}\frac{4.0}{0.1681}y(0.41 - y), \qquad (2.72)$$

such that the mean inflow velocity is $\bar{U}$ and the maximum of the inflow velocity profile is $1.5\bar{U}$. The *no-slip* condition is prescribed for the fluid on the other boundary parts. i.e., top and bottom wall, circle and fluid-structure interface Γ_t^0.

The outflow condition can be chosen by the user, for example *stress-free* or *do-nothing* conditions. The outflow condition effectively prescribes some reference value for the pressure variable p. While this value could be arbitrarily set in the incompressible case, in the case of a compressible structure this will have influence on the stress and consequently the deformation of the solid. In this description, we set the reference pressure at the outflow to have *zero mean value*.

Suggested starting procedure for the non-steady tests is to use a smooth increase of the velocity profile in time as

$$v^f(t, 0, y) = \begin{cases} v^f(0, y)\frac{1 - \cos(\frac{\pi}{2}t)}{2} & \text{if } t < 2.0, \\ v^f(0, y) & \text{otherwise,} \end{cases} \qquad (2.73)$$

where $v^f(0, y)$ is the velocity profile given in (2.72).

level	#refine	#el	#dof
0+0	0	62	1338
1+0	1	248	5032
2+0	2	992	19488
3+0	3	3968	76672
4+0	4	15872	304128

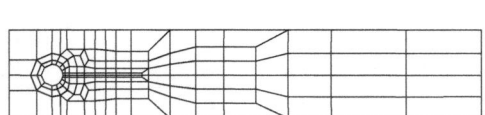

FIGURE 27. Example of a coarse mesh and the number of degrees of freedom for refined levels.

level	nel	ndof	ux of A $[\times 10^{-3}]$	uy of A $[\times 10^{-3}]$	drag	lift
2 + 0	992	19488	0.022871	0.81930	14.27360	0.76178
3 + 0	3968	76672	0.022775	0.82043	14.29177	0.76305
4 + 0	15872	304128	0.022732	0.82071	14.29484	0.76356
5 + 0	63488	1211392	0.022716	0.82081	14.29486	0.76370
6 + 0	253952	4835328	0.022708	0.82086	14.29451	0.76374
ref.			0.0227	0.8209	14.295	0.7638

TABLE 6. Results for **FSI1**.

2.5.2. Computational results. The mesh used for the computations is shown in Fig. 27. The following FSI tests are performed for two different inflow speeds. FSI1 is resulting in a steady state solution, while FSI2, FSI3 result in periodic solutions. The computed values are summarized in Table 6 for the test FSI1 and in Figures 28, 29 for the tests FSI2 and FSI3.

2.6. Conclusion

In this section we presented a general formulation of the dynamic fluid-structure interaction problem suitable for applications with finite deformations and laminar flows. While the presented example calculations are simplified to allow initial testing of the numerical methods the formulation is general to allow immediate extension to more realistic material models. For example, in the case of material anisotropy one can consider

$$\tilde{\Psi} = c_1(I_C - 3) + c_2(II_C - 3) + c_3(|\boldsymbol{F}\vec{a}| - 1)^2,$$

with $\vec{a}$ being the preferred material direction. The term $|\boldsymbol{F}\vec{a}|$ represents the extension in the direction $\vec{a}$. In [57, 58] a similar material relation of the form

$$\tilde{\Psi} = c_1 \left(\exp\left(b_1(I_C - 3)\right) - 1\right) + c_2 \left(\exp\left(b_2(|\boldsymbol{F}\vec{a}| - 1)\right) - 1\right)$$

has been proposed to describe a passive behavior of the muscle tissue. Adding to any form of Ψ a term like $f(t, \vec{x})(|\boldsymbol{F}\vec{a}| - 1)$ one can model the active behavior of a material and then the system can be coupled with additional models of chemical and electric activation of the active response of the tissue, see [82]. In the same manner the constitutive relation for the fluid can be directly extended to the

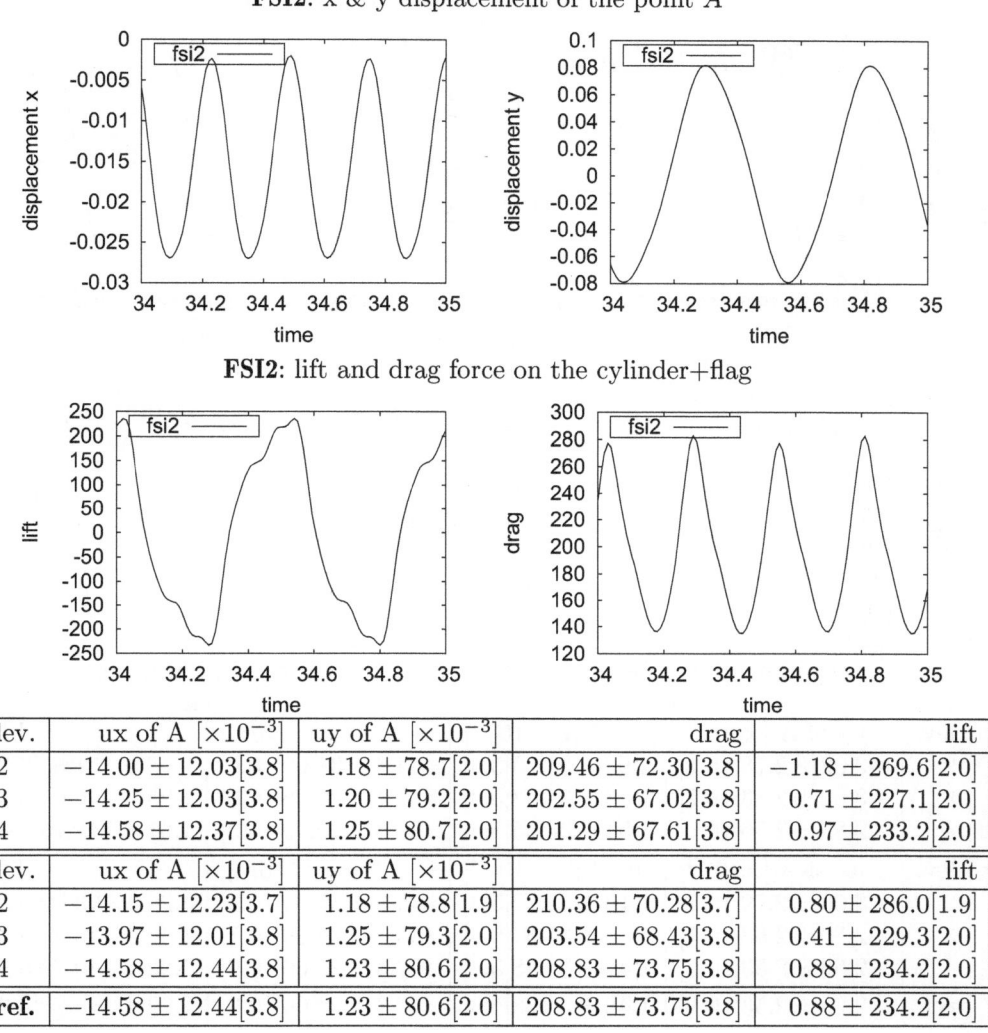

FSI2: x & y displacement of the point A

FSI2: lift and drag force on the cylinder+flag

lev.	ux of A $[\times 10^{-3}]$	uy of A $[\times 10^{-3}]$	drag	lift
2	$-14.00 \pm 12.03[3.8]$	$1.18 \pm 78.7[2.0]$	$209.46 \pm 72.30[3.8]$	$-1.18 \pm 269.6[2.0]$
3	$-14.25 \pm 12.03[3.8]$	$1.20 \pm 79.2[2.0]$	$202.55 \pm 67.02[3.8]$	$0.71 \pm 227.1[2.0]$
4	$-14.58 \pm 12.37[3.8]$	$1.25 \pm 80.7[2.0]$	$201.29 \pm 67.61[3.8]$	$0.97 \pm 233.2[2.0]$
lev.	ux of A $[\times 10^{-3}]$	uy of A $[\times 10^{-3}]$	drag	lift
2	$-14.15 \pm 12.23[3.7]$	$1.18 \pm 78.8[1.9]$	$210.36 \pm 70.28[3.7]$	$0.80 \pm 286.0[1.9]$
3	$-13.97 \pm 12.01[3.8]$	$1.25 \pm 79.3[2.0]$	$203.54 \pm 68.43[3.8]$	$0.41 \pm 229.3[2.0]$
4	$-14.58 \pm 12.44[3.8]$	$1.23 \pm 80.6[2.0]$	$208.83 \pm 73.75[3.8]$	$0.88 \pm 234.2[2.0]$
ref.	$-14.58 \pm 12.44[3.8]$	$1.23 \pm 80.6[2.0]$	$208.83 \pm 73.75[3.8]$	$0.88 \pm 234.2[2.0]$

FIGURE 28. Results for **FSI2** with time step $\Delta t = 0.002, \Delta t = 0.001$.

power-law models used to describe the shear-thinning property of the blood. Further extension to visco-elastic models and coupling with the mixture-based model for soft tissues together with models for chemical and electric processes involved in biomechanical problems would allow to perform realistic simulations for real applications.

To obtain the solution approximation the discrete systems resulting from the finite-element discretization of the governing equations need to be solved which requires sophisticated solvers of nonlinear systems and fast solvers for very large

FSI3: x & y displacement of the point A

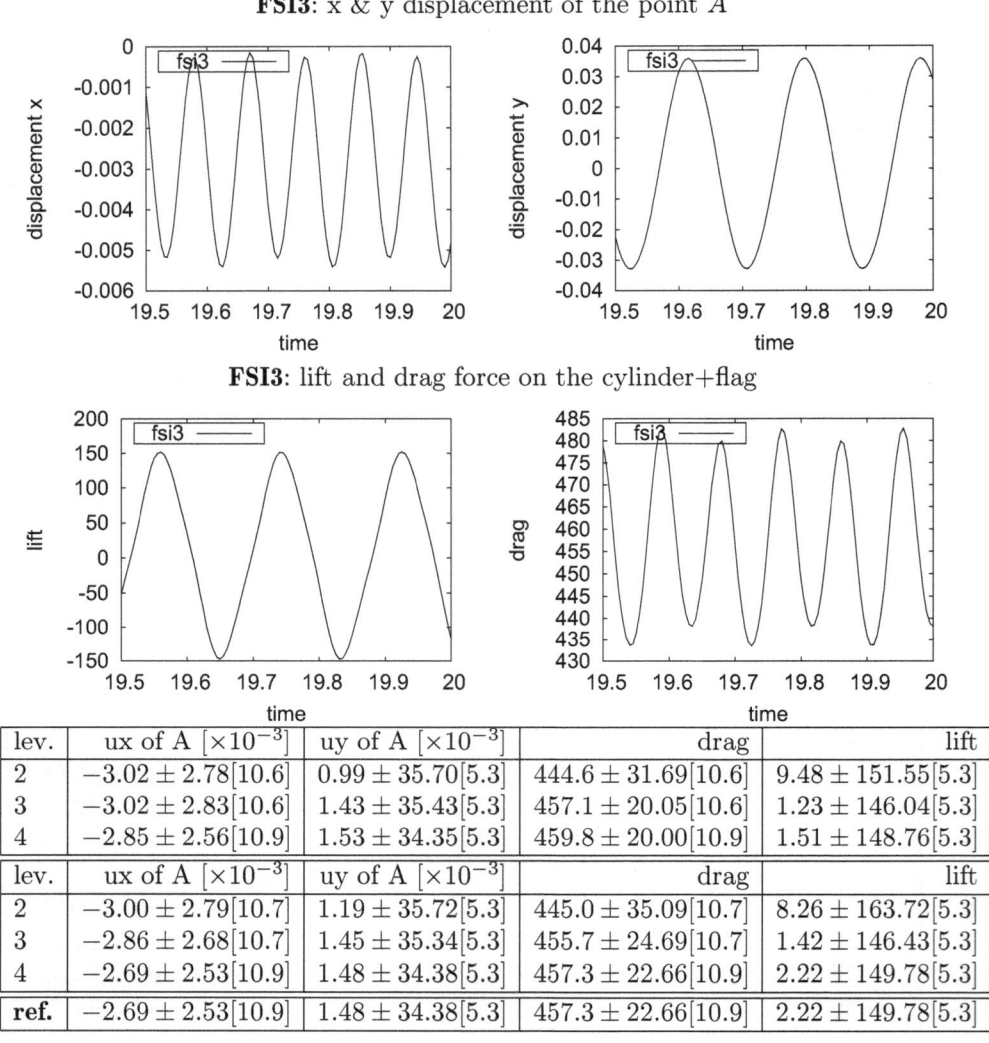

FSI3: lift and drag force on the cylinder+flag

lev.	ux of A $[\times 10^{-3}]$	uy of A $[\times 10^{-3}]$	drag	lift
2	$-3.02 \pm 2.78[10.6]$	$0.99 \pm 35.70[5.3]$	$444.6 \pm 31.69[10.6]$	$9.48 \pm 151.55[5.3]$
3	$-3.02 \pm 2.83[10.6]$	$1.43 \pm 35.43[5.3]$	$457.1 \pm 20.05[10.6]$	$1.23 \pm 146.04[5.3]$
4	$-2.85 \pm 2.56[10.9]$	$1.53 \pm 34.35[5.3]$	$459.8 \pm 20.00[10.9]$	$1.51 \pm 148.76[5.3]$
lev.	ux of A $[\times 10^{-3}]$	uy of A $[\times 10^{-3}]$	drag	lift
2	$-3.00 \pm 2.79[10.7]$	$1.19 \pm 35.72[5.3]$	$445.0 \pm 35.09[10.7]$	$8.26 \pm 163.72[5.3]$
3	$-2.86 \pm 2.68[10.7]$	$1.45 \pm 35.34[5.3]$	$455.7 \pm 24.69[10.7]$	$1.42 \pm 146.43[5.3]$
4	$-2.69 \pm 2.53[10.9]$	$1.48 \pm 34.38[5.3]$	$457.3 \pm 22.66[10.9]$	$2.22 \pm 149.78[5.3]$
ref.	$-2.69 \pm 2.53[10.9]$	$1.48 \pm 34.38[5.3]$	$457.3 \pm 22.66[10.9]$	$2.22 \pm 149.78[5.3]$

FIGURE 29. Results for **FSI3** with time step $\Delta t = 0.001, \Delta t = 0.0005$.

linear systems. The computational complexity increases tremendously for full 3D problems and with more complicated models like visco-elastic materials for the fluid or solid components. The main advantage of the presented numerical method is its accuracy and robustness with respect to the constitutive models. The possible directions of improving the efficiency of the solvers include development of fast linear solvers based on multigrid ideas, spatial and temporal adaptivity and effective use of parallel computations.

3. Numerical techniques for fluid-rigid solid configurations

In this section, we investigate the numerical simulation of particulate flow using a new moving-mesh method combined with the multigrid fictitious-boundary method (FBM) [136, 146, 144]. With this approach, the mesh is dynamically relocated through a (linear) partial differential equation to capture the surface of the moving particles with a relatively small number of grid points. The complete system is realized by solving the mesh movement and the partial differential equations of the flow problem alternately via an operator-splitting approach. The flow is computed by a special ALE formulation with a multigrid finite-element solver, and the solid particles are allowed to move freely through the computational mesh which is adaptively aligned by the moving-mesh method in every time step. One important aspect is that the data structure of the undeformed initial mesh, in many cases a tensor-product mesh or a semi-structured grid consisting of many tensor-product meshes, is preserved, while only the spacing between the grid points is adapted in each time step so that the high efficiency of structured meshes can be exploited. Numerical results demonstrate that the interaction between the fluid and the particles can be accurately and efficiently handled by the presented method. It is also shown that the presented method significantly improves the accuracy of the previous multigrid FBM to simulate particulate flow with many moving rigid particles.

3.1. Introduction

The numerical simulation of particulate flow or the motion of small rigid particles in a viscous liquid is one of the main focuses of engineering research and still a challenging task in many applications. Depending on the area of application, these types of problems arise frequently in numerous settings, such as sedimenting and fluidized suspensions, lubricated transport, hydraulic fracturing of reservoirs, slurries, understanding solid-liquid interaction, etc.

Several numerical simulation techniques for particulate flows have been developed over the past decade. In these methods, the fluid flow is governed by the continuity and momentum equations, while the particles are governed by the equation of motion for a rigid body. The flow field around each individual particle is resolved, the hydrodynamic force between the particle and the fluid is obtained from the solutions. Hu, Joseph and coworkers [52, 53] as well as Maury [83] developed a finite-element method based on unstructured grids to simulate the motion of a large number of rigid objects in Newtonian and visco-elastic fluids. This approach is based on an arbitrary Lagrangian–Eulerian (ALE) technique. Both the fluid and solid equations of motion are incorporated into a single coupled variational equation. The hydrodynamic forces and torques acting on the particles are eliminated in the formulation. The nodes on the particle surface move with the particle, while the nodes in the interior of the fluid are computed using Laplace's equation to guarantee a smoothly varying distribution of the nodes. At each time step, a new mesh is generated when the old one becomes too distorted, and the flow field is projected onto the new mesh. In this scheme, the positions of the particles and grid nodes are updated explicitly, while the velocities of the fluid and the solid particles are determined implicitly. In the case of 2D, the remeshing of the body-fitted meshes can be done by available grid generation software, but in the more interesting case of a full 3D simulation, the problem of efficient body-fitted grid generation is not yet solved in a satisfying manner yet.

In a series of papers by Glowinski and coauthors [37, 91, 38, 36], they proposed a distributed Lagrange multiplier (DLM)/fictitious-domain method for the direct numerical simulation of a large number of rigid particles in fluids. In the DLM method, the entire fluid-particle domain is assumed to be a fluid and the particle domain is constrained to move with the rigid motion. The fluid-particle motion is treated implicitly using a combined weak formulation in which the mutual forces cancel. This formulation permits the use of a fixed structured grid thus eliminating the need for remeshing the domain. In [136, 146, 144, 145], we presented a similar but different multigrid fictitious-boundary method (FBM) for the detailed simulation of particulate flow. The method is based on a fixed (unstructured) FEM background grid. The motion of the solid particles is modeled by the Newton–Euler equations. Based on the boundary conditions applied at the interface between the particles and the fluid which can be seen as an additional constraint to the governing Navier–Stokes equations, the fluid domain can be extended into the whole domain which covers both fluid and particle parts. The FBM starts with a coarse mesh which may contain already many of the geometrical fine-scale details, and employs a (rough) boundary parametrization which sufficiently describes all large-scale structures with regard to the (geometric) boundary conditions. Then, all fine-scale features are treated as interior objects such that the corresponding components in all matrices and vectors are unknown degrees of freedom which are implicitly incorporated into all iterative solution steps.

An advantage of these fictitious-domain methods over the generalized standard Galerkin finite-element method is that they allow a fixed grid to be used, eliminating the need for remeshing, and they can be handled independently from the flow features. Much progress has been made for adopting the fictitious-domain, resp., boundary methods to simulate particulate flow, yet the quest for more accurate and efficient methods remains active, particularly for many particles of different shape and size: an underlying problem when adopting the fictitious-domain methods is that the boundary approximation is of low accuracy only. Particularly in 3D, the ability of the fictitious-domain methods to deal accurately with the interaction between fluid and rigid particles is greatly limited unless very fine meshes are used. One remedy could be to preserve the mesh topology, for instance, as generalized tensor product or block-structured meshes, while a local alignment of the positions of the grid points with the physical boundary of the solid is achieved by a moving-mesh process such that the boundary approximation error can be significantly decreased.

Many researchers have come to recognize mesh adaption as an effective tool for simulating sharp fronts or moving interfaces and reducing numerical dispersion and oscillation. It has been demonstrated that significant improvements in accuracy and efficiency can be gained by adapting the mesh nodes so that they remain concentrated in regions of sharp fronts or interfaces. There are many existing mesh-adaptive methods to achieve this purpose. Generally speaking, mesh adaptivity is usually applied in the form of local mesh refinement or through a continuous mesh mapping. In locally adaptive mesh-refinement methods [4], an adaptive mesh is obtained by adding or removing points to achieve a desired level of accuracy. This allows a systematic error analysis. However, local refinement methods require more complicated data structures, compared to simple tensor-product meshes, and fairly technical methods to communicate information among different levels of hierarchical refinements. In the mapping approach [54, 13], the mesh points are moved continuously in the whole domain to concentrate in regions where the solution has the largest variations or moving interfaces locate. These solution-adaptive or geometry-adaptive meshes can be used to compute accurately the sharp variation or the moving interface problems. They also have the additional advantage of allowing the use of standard solvers since all computations are performed in the same logical domain using a uniform mesh.

Over the past decade, several mesh-adaptive techniques have been developed, namely the so-called h-, p- and r- methods. The first two do static remeshing; here, the h-method does automatic refinement or coarsening of the spatial mesh based on a-posteriori error estimates or error indicators, and the p-method takes higher- or lower-order approximations locally as needed. In contrast, the r-method (also known as moving-mesh method) relocates grid points in a mesh having a fixed number of nodes in such a way that the nodes remain concentrated in regions of rapid variation of the solution or corresponding moving interfaces. The r-method is often a dynamic method which means that it uses time-stepping or pseudo-time-stepping approaches to construct the desired transformation. The r-method or moving-mesh method differs from the h-, and p-methods in that the former keeps the same number of mesh points, resp., unknowns throughout the entire solution process, while the latter have to treat hanging node problems. Thus, the size of computation and the data structures are fixed, which enables the r-method much easier to be incorporated into most CFD codes without the need for changing the system matrix structures and adding special interpolation procedures. The r-method has received more attention due to some new developments which clearly demonstrate its potential for problems such as those having moving interfaces [5, 76, 77, 12, 39]. Nevertheless, it is clear that the preferred method of choice in future might be of r-h-p-type, that means combining all these adaptive techniques.

The primary objective of this contribution is to combine the multigrid fictitious-boundary method (FBM) [136, 146, 144] with a prototypical version of a new moving-mesh method described in [39] for the simulation of particulate flow and to analyse the accuracy of the proposed combining method, comparing its results with the pure multigrid fictitious-boundary method without mesh adaptation. As we have shown in [144], the use of the multigrid FBM does not require to change the mesh during the simulations, although the rigid particles vary their positions. The advantage is that no expensive remeshing has to be performed while a fixed mesh can be used such that in combination with appropriate data structures and fast CFD solvers very high efficiency rates can be reached. However, the accuracy for capturing the surfaces of solid particles is only of first order which might lead to accuracy problems. For a better approximation of the particle surfaces, we adopt a deformed grid, created from an arbitrary block-structured mesh, in which the topology is preserved: only the grid spacing is changed such that the grid points are concentrated near the surface of the rigid particles. Here, the solution of additional linear Poisson problems in every time step is required for generating the deformed grid, which means that the additional work is significantly less than the main fluid-particle part.

This section is organized as follows. In Subsection 3.2, the physical models together with collision models for particulate flow are described. The moving-mesh method and examples are presented in Subsection 3.3. The detailed numerical schemes and algorithms including the multigrid FBM, ALE formulation, time and space discretizations, as well as the complete numerical algorithm are given in Subsection 3.4. Prototypical numerical experiments are implemented and computational results will be presented in Subsection 3.5. Improvements of accuracy over the previous pure multigrid FBM will be emphasized. The concluding remarks will be given in Subsection 3.6.

3.2. Description of the physical models

3.2.1. Governing equations. In our numerical studies of particle motion in a fluid, we will assume that the fluids are immiscible and Newtonian. The particles are assumed to be rigid. Let us consider the unsteady flow of N particles with mass M_i $(i = 1, \ldots, N)$ in a fluid with density ρ_f and viscosity ν. Denote $\Omega_f(t)$ as the domain occupied by the fluid at time t, and $\Omega_i(t)$ as the domain occupied by the ith particle. So, the motion of an incompressible fluid is governed by the following Navier–Stokes equations in $\Omega_f(t)$,

$$\rho_f \left(\frac{\partial \mathbf{u}}{\partial t} + \mathbf{u} \cdot \nabla \mathbf{u} \right) - \nabla \cdot \sigma = 0, \qquad \text{Div } \mathbf{u} = 0 \qquad \forall t \in (0, T), \qquad (3.1)$$

where σ is the total stress tensor in the fluid phase defined as

$$\sigma = -p\,\mathbf{I} + \mu_f \left[\nabla \mathbf{u} + (\nabla \mathbf{u})^T \right]. \qquad (3.2)$$

Here, $\mathbf{I}$ is the identity tensor, $\mu_f = \rho_f \cdot \nu$, p is the pressure and $\mathbf{u}$ is the fluid velocity. Let $\Omega_T = \Omega_f(t) \cup \{\Omega_i(t)\}_{i=1}^N$ be the entire computational domain which shall be independent of t. Dirichlet- and Neumann-type boundary conditions can be imposed on the outer boundary $\Gamma = \partial\Omega_f(t)$. Since $\Omega_f = \Omega_f(t)$ and $\Omega_i = \Omega_i(t)$ are always depending on t, we drop t in all following notation.

The equations that govern the motion of each particle are the following Newton–Euler equations, i.e., the translational velocities $\mathbf{U}_i$ and angular velocities ω_i of the ith particle satisfy

$$M_i \frac{d\,\mathbf{U}_i}{d\,t} = (\Delta M_i)\,\mathbf{g} + \mathbf{F}_i + \mathbf{F}'_i, \qquad \mathbf{I}_i \frac{d\,\omega_i}{d\,t} + \omega_i \times (\mathbf{I}_i\,\omega_i) = T_i, \qquad (3.3)$$

where M_i is the mass of the ith particle; $\mathbf{I}_i$ is the moment of the inertia tensor; ΔM_i is the mass difference between the mass M_i and the mass of the fluid occupying the same volume; $\mathbf{g}$ is the gravity vector; $\mathbf{F}'_i$ are collision forces acting on the ith particle due to other particles which come close to each other. We assume that the particles are smooth without tangential forces of collisions acting on them; the details of the collision model will be discussed in the following subsection. $\mathbf{F}_i$ and T_i are the resultants of the hydrodynamic forces and the torque about the center of mass acting on the ith particle which are calculated by

$$\mathbf{F}_i = (-1) \int_{\partial\Omega_i} \sigma \cdot \mathbf{n}\, d\Gamma_i, \qquad T_i = (-1) \int_{\partial\Omega_i} (\mathbf{X} - \mathbf{X}_i) \times (\sigma \cdot \mathbf{n})\, d\Gamma_i, \qquad (3.4)$$

where σ is the total stress tensor in the fluid phase defined by Eq. (3.2), $\mathbf{X}_i$ is the position of the mass center of the ith particle, $\partial\Omega_i$ is the boundary of the ith particle, $\mathbf{n}$ is the unit normal vector on the boundary $\partial\Omega_i$ pointing outward to the flow region. The position $\mathbf{X}_i$ of the ith particle and its angle θ_i are obtained by integration of the kinematic equations

$$\frac{d\,\mathbf{X}_i}{d\,t} = \mathbf{U}_i, \qquad \frac{d\,\theta_i}{d\,t} = \omega_i. \qquad (3.5)$$

No-slip boundary conditions are applied at the interface $\partial\Omega_i$ between the ith particle and the fluid, i.e., for any $\mathbf{X} \in \bar{\Omega}_i$, the velocity $\mathbf{u}(\mathbf{X})$ is defined by

$$\mathbf{u}(\mathbf{X}) = \mathbf{U}_i + \omega_i \times (\mathbf{X} - \mathbf{X}_i). \qquad (3.6)$$

3.2.2. Collision models. For handling more than one particle, a collision model is needed to prevent the particles from interpenetrating each other. Glowinski, Joseph and coauthors [38, 36] proposed repulsive force models in which an artificial short-range repulsive force between particles is introduced keeping the particle surfaces more than one element (the range of the repulsive force) apart from each other. In these models, overlapping of the regions occupied by the rigid bodies is not allowed since conflicting rigid-body motion constraints from two different particles are not imposed at the same velocity nodes. However, in numerical calculations, the overlapping of particles might happen: For solving this problem, Joseph et al. [150] suggested a modified repulsive force model in which the particles are allowed to come arbitrarily close and even to overlap slightly each other. When conflicting rigid body motion constraints from two different particles are applied onto a velocity node, then the constraint from the particle that is closer to that node is used. A repulsive force is only applied when the particles overlap each other.

Following such ideas, we examine a modified collision model with a new definition of short-range repulsive forces which cannot only prevent the particles from getting too close, it can also deal with the case of particles overlapping each other when numerical simulations bring the particles very close due to unavoidable numerical truncation errors. For the particle-particle collisions, the repulsive force is determined as

$$
\mathbf{F}_{i,j}^P = \begin{cases} 0, & \text{for } d_{i,j} > R_i + R_j + \rho, \\[2mm] \dfrac{1}{\epsilon_P'}(\mathbf{X}_i - \mathbf{X}_j)(R_i + R_j - d_{i,j}), & \text{for } d_{i,j} < R_i + R_j, \\[2mm] \dfrac{1}{\epsilon_P}(\mathbf{X}_i - \mathbf{X}_j)(R_i + R_j + \rho - d_{i,j})^2, & \text{for } R_i + R_j \leq d_{i,j} \leq R_i + R_j + \rho, \end{cases}
$$

where R_i and R_j are the radii of the ith and jth particle, $\mathbf{X}_i$ and $\mathbf{X}_j$ are the coordinates of their mass centers, $d_{i,j} = |\mathbf{X}_i - \mathbf{X}_j|$ is the distance between their mass centers, ρ is the range of the repulsive force (usually $\rho = 0.5 \sim 2.5\Delta h$, Δh is the local mesh size), ϵ_P and ϵ_P' are small positive stiffness parameters for particle-particle collisions. If the fluid is sufficiently viscous, and $\rho \simeq \Delta h$ as well as ρ_i/ρ_f are of order 1 (ρ_i is the density of the ith particle, ρ_f is the fluid density), then we can take $\epsilon_P \simeq (\Delta h)^2$ and $\epsilon_P' \simeq \Delta h$ in the calculations. For the particle-wall collisions, the corresponding repulsive force reads

$$
\mathbf{F}_i^W = \begin{cases} 0, & \text{for } d_i' > 2R_i + \rho, \\[2mm] \dfrac{1}{\epsilon_W'}(\mathbf{X}_i - \mathbf{X}_i')(2R_i - d_i'), & \text{for } d_i' < 2R_i, \\[2mm] \dfrac{1}{\epsilon_W}(\mathbf{X}_i - \mathbf{X}_i')(2R_i + \rho - d_i')^2, & \text{for } 2R_i \leq d_i' \leq 2R_i + \rho, \end{cases}
$$

where $\mathbf{X}_i'$ is the coordinate vector of the center of the nearest imaginary particle P_i' located on the boundary wall Γ w.r.t. the ith particle, $d_i' = |\mathbf{X}_i - \mathbf{X}_i'|$ is the distance between the mass centers of the ith particle and the center of the imaginary particle P_i', ϵ_W is a small positive stiffness parameter for particle-wall collisions, usually $\epsilon_W = \epsilon_P/2$ and $\epsilon_W' = \epsilon_P'/2$ in our calculations. Then, the total repulsive forces (i.e., collision forces) exerted on the ith particle by the other particles and the walls can be expressed as follows,

$$
\mathbf{F}_i' = \sum_{j=1, j\neq i}^N \mathbf{F}_{i,j}^P + \mathbf{F}_i^W . \tag{3.7}
$$

3.3. Moving-mesh method

In this section, we briefly describe the moving-mesh method which will be adopted and coupled with the multigrid fictitious-boundary method (FBM) [136, 146, 144] to solve numerically the particulate flow equations. The details of the moving-mesh method can be found in [39].

The moving-mesh problem can be treated to constructing a transformation φ, $x = \varphi(\xi)$ from the computational space (with coordinate ξ) to the physical space (with coordinate x). There are several types of moving-mesh methods, generally computing x by minimizing a variational form or computing the mesh velocity $v = x_t$ using a Lagrangian-like formulation. The moving-mesh method we will employ belongs to the velocity-type class, which is based on Liao's work [5, 76, 77, 12] and Moser's work [20]. This method has several advantages: only linear Poisson problems on fixed meshes are to be solved, monitor functions can be obtained directly from error distributions or distance functions, mesh tangling can be prevented, and the data structure from the starting mesh is preserved.

Suppose $g(x)$ (area function) to be the area distribution on the undeformed mesh, while $f(x)$ (monitor function) in contrast describes the absolute mesh size distribution of the target grid, which is independent of the starting grid and chosen according to the need of the physical problem. Then, the transformation φ can be computed via the following four steps:

1. Compute the scaling factors c_f and c_g for the given monitor function $f(x) > 0$ and the area function g using

$$c_f \int_\Omega \frac{1}{f(x)}dx = c_g \int_\Omega \frac{1}{g(x)}dx = |\Omega|, \qquad (3.8)$$

where $\Omega \subset \mathbb{R}^2$ is a computational domain, and $f(x) \approx$ local mesh area. Let $\tilde{f}$ and $\tilde{g}$ denote the reciprocals of the scaled functions f and g, i.e.,

$$\tilde{f} = \frac{c_f}{f}, \qquad\qquad \tilde{g} = \frac{c_g}{g}. \qquad (3.9)$$

2. Compute a grid-velocity vector field $v : \Omega \to \mathbb{R}^n$ by solving the Poisson problem

$$-\mathrm{div}(v(x)) = \tilde{f}(x) - \tilde{g}(x), \quad x \in \Omega, \quad \text{and} \quad v(x) \cdot \mathfrak{n} = 0, \quad x \in \partial\Omega, \qquad (3.10)$$

where $\mathfrak{n}$ is the outer normal vector of the domain boundary $\partial\Omega$ which may consist of several boundary components.

3. For each grid point x, solve the following ODE system

$$\frac{\partial\varphi(x,t)}{\partial t} = \eta(\varphi(x,t),t), \quad 0 \le t \le 1, \quad \varphi(x,0) = x, \qquad (3.11)$$

with

$$\eta(y,s) := \frac{v(y)}{s\tilde{f}(y) + (1-s)\tilde{g}(y)}, \quad y \in \Omega, \quad s \in [0,1]. \qquad (3.12)$$

4. Get the new grid points via

$$\varphi(x) := \varphi(x,1). \qquad (3.13)$$

Here, we give two examples for the generation of deformed grids using the described moving-mesh method. Fig. 30 shows the case of two objects with one rectangle and one ellipse inside a square domain. The starting mesh presented in Fig. 30(a) is equidistant, and we want to generate a deformed mesh which can align the grid lines near both the surface of the ellipse and the rectangle. We choose the

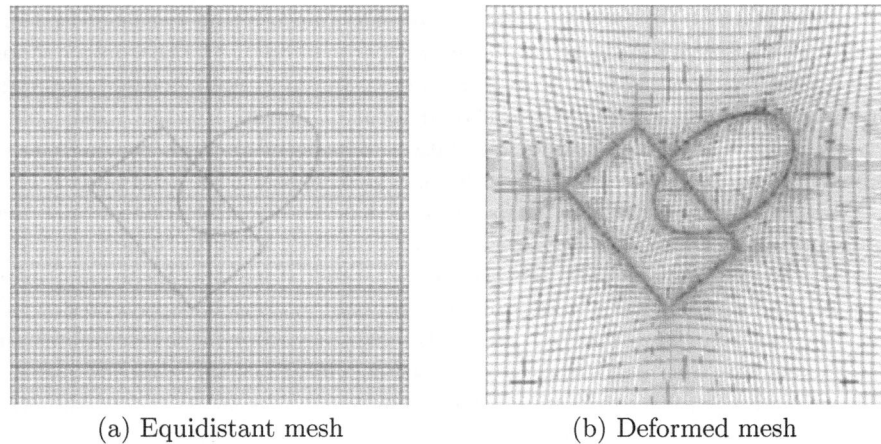

(a) Equidistant mesh   (b) Deformed mesh

FIGURE 30. Example of a deformed mesh generated from an equidistant mesh

monitor function $f(x)$ as a function of Δd, here Δd is the minimum distance of grid points to the both surfaces of the two solid bodies (an ellipse, resp., a rectangle). Fig. 30(b) shows the generated deformed mesh. We can see that the grid lines are concentrated around the surfaces of the two solid bodies.

Fig. 31 presents another case with two ellipses in a long channel. The starting mesh and positions of the two ellipses are given in Fig. 31(a). If we choose the monitor function $f(x)$ as in the above case, i.e., to be a function of Δd, with Δd being the minimum distance of grid points to both surfaces of the two solid ellipses, then the generated deformed mesh is shown in Fig. 31(b): We can see that there are still too many grid lines remaining in the lower part of the channel; however, there is no solid body. Moreover, the grid points in the gap between the two ellipses are not distributed very well. Next, we try to use another monitor function $f(x)$, a digit filter function $\Delta d_1 \times \Delta d_2$, here Δd_1 and Δd_2 are the distances of the grid points to the surface of both ellipses; then, a much better deformed mesh can be generated. We can see in Fig. 31(c) that the grid lines are more concentrated around the surfaces of both ellipses and in the region of the gap between the two ellipses, and there are less grid lines staying in the down part of the channel, too. It is clearly shown that the quality of the grid distribution is depending on the choice of the monitor function which is subject of further research.

3.4. Numerical method

3.4.1. Multigrid FEM fictitious-boundary method. The details of the multigrid FEM fictitious-boundary method have been presented in [136, 146, 144]. For illustration, a brief description is given below.

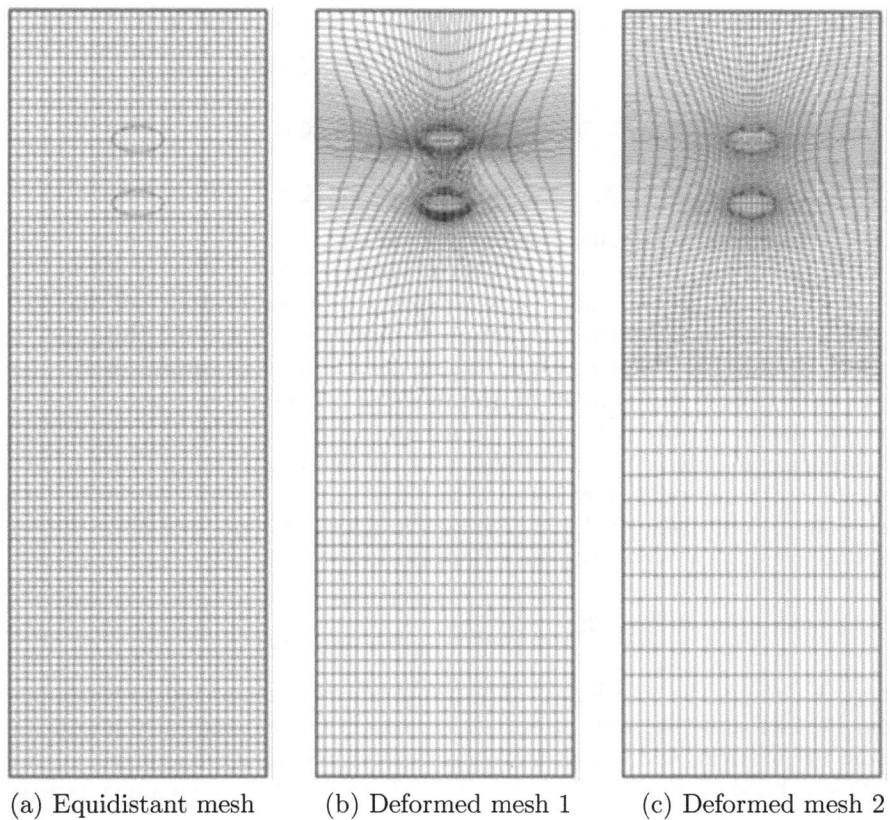

(a) Equidistant mesh   (b) Deformed mesh 1   (c) Deformed mesh 2

FIGURE 31. Second example of deformed meshes generated from
a equidistant mesh

The fictitious-boundary method (FBM) is based on a multigrid FEM background grid which covers the whole computational domain Ω_T and can be chosen
independently from the particles of arbitrary shape, size and number. It starts
with a coarse mesh which may already contain many of the geometrical details of
Ω_i $(i = 1, \ldots, N)$, and it employs a fictitious-boundary indicator (see [136]) which
sufficiently describes all fine-scale structures of the particles with regard to the
fluid-particle matching conditions of Eq. (3.6). Then, all fine-scale features of the
particles are treated as interior objects such that the corresponding components
in all matrices and vectors are unknown degrees of freedom which are implicitly
incorporated into all iterative solution steps (see [146]). Hence, by making use of
Eq. (3.6), we can perform calculations for the fluid in the whole domain Ω_T. The
considerable advantage of the multigrid FBM is that the total mixture domain Ω_T
does not have to change in time, and can be meshed only once. The domain of
definition of the fluid velocity $\mathbf{u}$ is extended according to Eq. (3.6), which can be

seen as an additional constraint to the Navier–Stokes equations (3.1), i.e.,

$$
\begin{cases}
\text{Div } \mathbf{u} = 0 & (a) \quad \text{for } \mathbf{X} \in \Omega_T, \\[2mm]
\rho_f \left(\dfrac{\partial \mathbf{u}}{\partial t} + \mathbf{u} \cdot \nabla \mathbf{u} \right) - \nabla \cdot \sigma = 0 & (b) \quad \text{for } \mathbf{X} \in \Omega_f, \\[2mm]
\mathbf{u}(\mathbf{X}) = \mathbf{U}_i + \omega_i \times (\mathbf{X} - \mathbf{X}_i) & (c) \quad \text{for } \mathbf{X} \in \bar{\Omega}_i, \; i = 1, \dots, N.
\end{cases}
\tag{3.14}
$$

For the study of interactions between the fluid and the particles, the calculation of the hydrodynamic forces acting on the moving particles is very important. From Eq. (3.4), we can see that the surface integrals on the wall surfaces of the particles should be conducted for the calculation of the forces $\mathbf{F}_i$ and T_i. However, in the presented multigrid FBM method, the shapes of the wall surface of the moving particles are implicitly imposed in the fluid field. If we reconstruct the shapes of the wall surface of the particles, it is not only a time-consuming work, but also the accuracy is only of first order due to a piecewise constant interpolation from our indicator function. For overcoming this problem, we perform the hydrodynamic force calculations using a volume-based integral formulation. To replace the surface integral in Eq. (3.4) we introduce a function α_i,

$$
\alpha_i(\mathbf{X}) = \begin{cases} 1 & \text{for } \mathbf{X} \in \Omega_i, \\[2mm] 0 & \text{for } \mathbf{X} \in \Omega_T \setminus \Omega_i, \end{cases}
\tag{3.15}
$$

where $\mathbf{X}$ denotes the coordinates. The importance of such a definition can be seen from the fact that the gradient of α_i is zero everywhere except at the wall surface of the ith particle, and equals to the normal vector $\mathbf{n}_i$ of wall surface of the ith particle defined on the grid, i.e., $\mathbf{n}_i = \nabla \alpha_i$. Then, the hydrodynamic forces acting on the ith particle can be computed by

$$
\mathbf{F}_i = - \int_{\Omega_T} \sigma \cdot \nabla \alpha_i \, d\Omega, \qquad T_i = - \int_{\Omega_T} (\mathbf{X} - \mathbf{X}_i) \times (\sigma \cdot \nabla \alpha_i) \, d\Omega.
\tag{3.16}
$$

The integral over each element covering the whole domain Ω_T can be calculated with a standard Gaussian quadrature of sufficiently high order. Since the gradient $\nabla \alpha_i$ is non-zero only near the wall surface of the ith particle, thus the volume integrals need to be computed only in one layer of mesh cells around the ith particle, which leads to a very efficient treatment.

The algorithm of the (classical) multigrid FEM fictitious-boundary method for solving the coupled system of fluid and particles can be summarized as follows:

1. Given the positions and velocities of the particles, solve the fluid equations Eqs. (3.14) (a) and (b) in the corresponding fluid domain involving the position of the particles for the fictitious-boundary conditions.
2. Calculate the corresponding hydrodynamic forces and the torque acting on the particles by using Eq. (3.16), and compute the collision forces by Eq. (3.7).
3. Solve Eq. (3.3) to get the translational and angular velocities of the particles, and then obtain the new positions and velocities of the particles by Eq. (3.5).

4. Use Eq. (3.14) (c) to set the new fluid domain and fictitious-boundary conditions, and solve for the new velocity and pressure of the fluid phase as described in step 1.

3.4.2. ALE formulation of the FBM. For a better approximation of the solid surfaces, we adopt the above described moving-mesh method such that we can preserve the mesh topology as generalized tensor product or block-structured meshes, while a local alignment with the rigid-body surface is reached. The moving-mesh method is sometimes referred to as quasi-Lagrangian method. When the moving-mesh method is applied to the multigrid FBM, a mesh velocity $\mathbf{W}_m$ in the convective term in Eq. (3.14b) should be introduced, i.e.,

$$\rho_f \left[\frac{\partial \mathbf{u}}{\partial t} + (\mathbf{u} - \mathbf{W}_m) \cdot \nabla \mathbf{u} \right] - \nabla \cdot \sigma = 0 \text{ for } \mathbf{X} \in \Omega_f, \qquad (3.17)$$

so that no interpolation routines from old to new meshes is required since the grid topology is preserved.

In the literature this is also referred to as an Arbitrary Lagrangian–Eulerian (ALE) formulation. Note that the mesh velocities $\mathbf{W}_m$ do not appear in the continuity equation, as a pressure-Poisson problem is solved to satisfy the continuity equation in an outer loop. Care has to be taken to satisfy the geometric conservation law (GCL), where the mesh velocity $\mathbf{W}_m$ must be equal to the movement of the mesh velocity $\Delta \mathbf{x}$ during the time step. Therefore, the mesh velocities $\mathbf{W}_m$ should be calculated according to the nodal movement from the previous time step by

$$\mathbf{W}_m = \frac{1}{\Delta t}(\mathbf{x}^{n+1} - \mathbf{x}^n) \qquad (3.18)$$

where Δt is the time-step size and n denotes the time-step number.

In each time step, a new deformed mesh is generated based on a starting mesh, then the system matrices are updated and the mesh velocity according to the new position of the deformed mesh nodes will be calculated. Since the moving-mesh method keeps the same number of mesh points throughout the entire solution process, the size of computation and the data structure are fixed, which enables this method to be much easier to incorporate into most CFD codes without the need for changing the system matrix structures and for special interpolation procedures.

3.4.3. Time discretization by fractional-step-θ scheme. We first semi-discretize the Eqs. (3.14) (a) and (3.17) in time by the fractional-step-θ scheme (see the previous contribution in this book). Given $\mathbf{u}^n$ and the time step $K = t_{n+1} - t_n$, then solve for $\mathbf{u} = \mathbf{u}^{n+1}$ and $p = p^{n+1}$. In the fractional-step-θ-scheme, one macro time step

$t_n \to t_{n+1} = t_n + K$ is split into three consecutive substeps with $\tilde{\theta} := \alpha\theta K = \beta\theta' K$,

$$\mathbf{u}^{n+\theta} + \theta K \nabla p^{n+\theta} = [I - \beta\theta K N(\mathbf{u}^n)]\mathbf{u}^n$$
$$\text{Div } \mathbf{u}^{n+\theta} = 0,$$

$$[I + \tilde{\theta}N(\mathbf{u}^{n+1-\theta})]\mathbf{u}^{n+1-\theta} + \theta' K \nabla p^{n+1-\theta} = [I - \alpha\theta' K N(\mathbf{u}^{n+\theta})]\mathbf{u}^{n+\theta}$$
$$\text{Div } \mathbf{u}^{n+1-\theta} = 0,$$
(3.19)

$$[I + \tilde{\theta}N(\mathbf{u}^{n+1})]\mathbf{u}^{n+1} + \theta K \nabla p^{n+1} = [I - \beta\theta K N(\mathbf{u}^{n+1-\theta})]\mathbf{u}^{n+1-\theta}$$
$$\text{Div } \mathbf{u}^{n+1} = 0,$$

where $\theta = 1 - \frac{\sqrt{2}}{2}$, $\theta' = 1 - 2\theta$, and $\alpha = \frac{1-2\theta}{1-\theta}$, $\beta = 1 - \alpha$, $N(\mathbf{v})\mathbf{u}$ is a compact form for the diffusive and convective part,

$$N(\mathbf{v})\mathbf{u} := -\nu \, \nabla \cdot \left[\nabla\mathbf{u} + (\nabla\mathbf{u})^T\right] + (\mathbf{v} - \mathbf{W}_m) \cdot \nabla\mathbf{u}. \tag{3.20}$$

Therefore, from Eq. (3.19) in each time step, we have to solve nonlinear problems of the following type,

$$[I + \theta_1 K N(\mathbf{u})]\mathbf{u} + \theta_2 K \nabla p = \mathbf{f}, \quad \mathbf{f} := [I - \theta_3 K N(\mathbf{u}^n)]\mathbf{u}^n, \quad \text{Div } \mathbf{u} = 0. \tag{3.21}$$

For Eq. (3.14) (c), we simply take an explicit expression like

$$\mathbf{u}^{n+1} = \mathbf{U}_i^n + \omega_i^n \times (\mathbf{X}^n - \mathbf{X}_i^n). \tag{3.22}$$

3.4.4. Space discretization by finite-element method. If we define a pair $\{\mathbf{u}, p\} \in H := \mathbf{H}_0^1(\Omega) \times L := \mathrm{L}_0^2(\Omega)$, and bilinear forms $a(\mathbf{u}, \mathbf{v}) := (\nabla\mathbf{u}, \nabla\mathbf{v})$ and $b(p, \mathbf{v}) := -(p, \text{Div } \mathbf{v})$, a weak formulation of the Eq. (3.21) reads as follows,

$$\begin{cases} (\mathbf{u}, \mathbf{v}) + \theta_1 K \left[a(\mathbf{u}, \mathbf{v}) + n(\mathbf{u}, \mathbf{u}, \mathbf{v})\right] + \theta_2 K \, b\,(p, \mathbf{v}) = (\mathbf{f}, \mathbf{v}), & \forall \mathbf{v} \in H \\ b\,(q, \mathbf{u}) = 0, & \forall q \in L. \end{cases}$$
(3.23)

Here, $\mathrm{L}_0^2(\Omega)$ and $\mathbf{H}_0^1(\Omega)$ are the usual Lebesgue and Sobolev spaces, $n(\mathbf{u}, \mathbf{u}, \mathbf{v})$ is a trilinear form defined by

$$n(\mathbf{u}, \mathbf{v}, \mathbf{w}) := \int_\Omega [u_i - (w_m)_i] \left(\frac{\partial v_j}{\partial x_i} + \frac{\partial v_i}{\partial x_j}\right) w_j \, dx. \tag{3.24}$$

To discretize Eq. (3.23) in space, we introduce a quadrilateral mesh T_h for the whole computational domain Ω_T, where h is a parameter characterizing the maximum width of the elements of T_h. To obtain the fine mesh T_h from a coarse mesh T_{2h}, we simply connect opposing midpoints. In the fine grid T_h, the old midpoints of the coarse mesh T_{2h} become vertices. We choose the $\tilde{Q}1/Q0$ element pair which uses rotated bilinear shape functions for the velocity spanned by $\langle x^2 - y^2, x, y, 1\rangle$ in 2D and piecewise constants for the pressure in cells. The nodal values are the mean values of the velocity vector over the element edges and the mean values of the pressure over the elements rendering this approach non-conforming. This $\tilde{Q}1/Q0$ element pair has several important features which have been described in a previous chapter of this contribution. If we choose finite-dimensional spaces H_h and L_h and define a pair $\{\mathbf{u}_h, p_h\} \in H_h \times L_h$, the discrete version of (3.23) reads,

$$
\begin{cases}
(\mathbf{u}_h, \mathbf{v}_h) + \theta_1 K \left[a_h(\mathbf{u}_h, \mathbf{v}_h) + \tilde{n}_h(\mathbf{u}_h, \mathbf{u}_h, \mathbf{v}_h) \right] & \\
\qquad + \theta_2 K \, b_h(p_h, \mathbf{v}_h) = (\mathbf{f}, \mathbf{v}_h), & \forall \, \mathbf{v}_h \in H_h \\
b_h(q_h, \mathbf{u}_h) = 0, & \forall \, q_h \in L_h,
\end{cases}
\tag{3.25}
$$

where $a_h(\mathbf{u}_h, \mathbf{v}_h) := \sum_{T \in T_h} a(\mathbf{u}_h, \mathbf{v}_h)_{|T}$ and $b_h(p_h, \mathbf{v}_h) := \sum_{T \in T_h} b(p_h, \mathbf{v}_h)_{|T}$. Note that $\tilde{n}_h(\mathbf{u}_h, \mathbf{u}_h, \mathbf{v}_h)$ is a new convective term which includes streamline-diffusion stabilizations defined by

$$
\tilde{n}_h(\mathbf{u}_h, \mathbf{v}_h, \mathbf{w}_h) := \sum_{T \in T_h} n(\mathbf{u}_h, \mathbf{v}_h, \mathbf{w}_h)_{|T} + \sum_{T \in T_h} \delta_T (\mathbf{u}_h \cdot \nabla \mathbf{v}_h, \mathbf{u}_h \cdot \nabla \mathbf{w}_h)_{|T}, \tag{3.26}
$$

here δ_T is a local artificial viscosity which is a function of a local Reynolds number Re_T,

$$
\delta_T := \delta^* \cdot \frac{h_T}{\|\mathbf{u}\|_\Omega} \cdot \frac{2 Re_T}{1 + Re_T}, \qquad Re_T = \frac{\|\mathbf{u}\|_T \cdot h_T}{\nu}, \tag{3.27}
$$

where $\|\mathbf{u}\|_\Omega$ means the maximum norm of velocity in Ω_T, $\|\mathbf{u}\|_T$ is an averaged norm of velocity over T, h_T denotes the local mesh size of T, and δ^* is an additional free parameter which can be chosen arbitrarily ($\delta^* = 0.1$ is used in our calculations, see [124]). Obviously, for small local Reynolds numbers, with $Re_T \to 0$, δ_T is decreasing such that we reach in the limit case the second-order central discretization. Vice versa, for convection dominated flows with $Re_T \gg 1$, we add an diffusion term of size $O(h)$ which is aligned to the streamline direction $\mathbf{u}_h$.

3.4.5. Discrete projection scheme. For solving the discrete nonlinear problems after time and space discretizations, we have to take the following points into account, i.e., treatment of the nonlinearity, treatment of the incompressibility, and complete outer control like convergence criteria for the overall outer iteration, number of splitting steps, convergence control, embedding into multigrid, etc. In general, there are (at least) two possible approaches for solving the discrete problems [122]. One is the so-called full Galerkin schemes: first, we treat the nonlinearity by an outer nonlinear iteration of fixed-point- or quasi-Newton-type or by linearization via extrapolation in time, and then we obtain linear subproblems (Oseen equations) which can be solved by a direct coupled or a splitting approach separately for velocity and pressure. Typical schemes are preconditioned GMRES-like or multigrid solvers based on smoothers/preconditioners like Vanka, SIMPLE or local pressure Schur complement (see [124]). The disadvantage of these approaches is the high numerical cost for small time steps which are typical for particulate flow. Another possibility are the projection-type schemes: First, we split the coupled problem and obtain definite problems in $\mathbf{u}$ (Burgers equations) as well as in p (Pressure-Poisson problems). Then, we treat the nonlinear problems in $\mathbf{u}$ by an appropriate nonlinear iteration or linearization technique while optimal multigrid solvers are used for the Poisson-like problems. Classical schemes belonging to this class are the Chorin and van Kan projection schemes and the discrete projection method, all of them are well suited for dynamic configurations which require small time steps (see [123]).

In this chapter, based on the latter approach combined with multigrid methods, we adopt the discrete projection method (DPM) as special variant of the more general multigrid pressure Schur complement (MPSC) schemes to solve the discrete nonlinear problems after time and space discretizations. A detailed description of DPM and MPSC schemes has been presented in [124]: We first perform as outer iteration a fixed-point iteration, applied to the fully nonlinear momentum equations. Then, in the inner loop, we solve the corresponding velocity equations involving linear transport-diffusion problems. Finally, the pressure is updated via a pressure-Poisson-like problem, and the corresponding velocity field is adjusted. Since every time step requires the solution of linearized Burgers equations and Poisson-like problems, an optimized multigrid approach is used. The most important components are matrix-vector multiplication, smoothing operator and grid transfer routines (prolongation and restriction) for the underlying FEM spaces which have been realized in FEATFLOW (see [124] for the details).

3.4.6. Numerical algorithm. The whole algorithm of the multigrid FEM-FBM and moving-mesh methods for the numerical simulation of rigid particulate flows can be summarized as follows:

1. Given (initial) particle positions $\mathbf{X}_i$ and velocity $\mathbf{U}_i$ in the overall domain Ω_T. Next, we assume that we have finished the calculations at time t_n.
2. Generate a new deformed mesh via the four steps of Eq. (3.8)–(3.13).
3. Compute the mesh velocity $\mathbf{W}_m$ by using Eq. (3.18) based on the new and the previous deformed mesh.
4. Set the fictitious-boundary conditions by using Eq. (3.14) (c) with the 'old' particle positions $\mathbf{X}_i^n$ and the velocity $\mathbf{U}_i^n$ at time t_n.
5. By using the FBM and implementing the discrete projection scheme, build the system matrix and solve the fluid ALE equations of Eqs. (3.25) to get the fluid velocity $\mathbf{u}^{n+1}$ and the pressure p^{n+1} at time t_{n+1} on the full computational domain Ω_T.
6. Calculate the hydrodynamic forces $\mathbf{F}_i^{n+1}$ and T_i^{n+1} exerted on every particle ($i = 1, \ldots, N$) by using the volume integration formulation of Eq. (3.16).
7. When two particles come too close, the time step has to be reduced. Then, we adopt several substeps with $\Delta t_k = K/\Lambda$ ($k = 1, \ldots, \Lambda$, Λ is the number of substeps calculations, $K = t_{n+1} - t_n$) for calculating the collisions and updating the particle positions and velocities during the collisions. Set $\mathbf{U}_i^{n,0} := \mathbf{U}_i^n$ and $\mathbf{X}_i^{n,0} := \mathbf{X}_i^n$.
8. Determine the number of substep calculations Λ by

$$\Lambda = \begin{cases} 1, & \text{if } (d_{i,j})_{\min} \geq (R_i + R_j)_{\max}, \\ \mathrm{MIN}\left\{10, 1 + \mathrm{MAX}\left(\dfrac{|d_{i,j} - R_i - R_j|}{\varrho}\right)\right\}, & \text{if } (d_{i,j})_{\min} < (R_i + R_j - \varrho)_{\max}, \end{cases}$$

$$\tag{3.28}$$

where ϱ is the maximum penetration distance to be allowed (maximal overlapping range).

9. By using the Newton–Euler equations of Eq. (3.5) to calculate the motion of the solid particles, we obtain the new interim particle position $\mathbf{X}_i^{n+1/2,k}$ and velocity $\mathbf{U}_i^{n+1/2,k}$ as well as the new angular velocity ω_i^{n+1} and angle θ_i^{n+1} by

$$\mathbf{U}_i^{n+1/2,k} = \mathbf{U}_i^{n,k} + \left(\frac{\Delta M_i \, \mathbf{g}}{M_i} + \frac{\mathbf{F}_i^n + \mathbf{F}_i^{n+1}}{2\,M_i} \right) \Delta t_k \,, \tag{3.29}$$

$$\mathbf{X}_i^{n+1/2,k} = \mathbf{X}_i^{n,k} + \left(\frac{\Delta M_i \, \mathbf{g}}{M_i} + \frac{\mathbf{F}_i^n + \mathbf{F}_i^{n+1}}{2\,M_i} \right) (\Delta t_k)^2 \,, \tag{3.30}$$

$$\omega_i^{n+1} = \omega_i^n + \left(\frac{T_i^n + T_i^{n+1}}{2\,\mathbf{I}_i} \right) K \,, \qquad \theta_i^{n+1} = \theta_i^n + \left(\frac{\omega_i^n + \omega_i^{n+1}}{2} \right) K \,. \tag{3.31}$$

10. Use the collision model of Eqs. (3.2.2) and (3.2.2) to calculate the repulsive forces $(\mathbf{F}_i')^{n+1,k}$ with the interim particle position $\mathbf{X}_i^{n+1/2,k}$.

11. Update the particle positions and the velocity by the repulsive forces to obtain the new particle position $\mathbf{X}_i^{n+1,k}$ and the velocity $\mathbf{U}_i^{n+1,k}$ at time t_{n+1} by

$$\mathbf{U}_i^{n+1,k} = \mathbf{U}_i^{n+1/2,k} + \frac{(\mathbf{F}_i')^{n+1}}{M_i} \Delta t_k \,, \quad \mathbf{X}_i^{n+1,k} = \mathbf{X}_i^{n+1/2,k} + \frac{(\mathbf{F}_i')^{n+1}}{M_i} (\Delta t_k)^2 . \tag{3.32}$$

12. Set $\mathbf{U}_i^{n,k+1} := \mathbf{U}_i^{n+1,k}$ and $\mathbf{X}_i^{n,k+1} := \mathbf{X}_i^{n+1,k}$ if $k < \Lambda$, and repeat the steps $9-12$.

13. Set $\mathbf{U}_i^{n+1} := \mathbf{U}_i^{n+1,\Lambda}$ and $\mathbf{X}_i^{n+1} := \mathbf{X}_i^{n+1,\Lambda}$.

14. Advance to the next new time step, set $t_n := t_{n+1}$ and repeat the steps $2-14$.

3.5. Numerical results

First of all, a benchmark configuration of 2D flow around a circular body in a channel is given to assess the suitability and accuracy of the hydrodynamic force calculations based on the combination of the multigrid FBM and the new moving-mesh technique. Then, three cases of a single moving particle in the fluid are presented to further validate the improvement of accuracy and efficiency using the presented method. Finally, the drafting, kissing and tumbling of two disks in a channel and the sedimentation of 120 circular particles in a cavity are provided to show that the presented method can be easily implemented for the simulation of particulate flow with large number of particles. Since we have used a prototypical implementation of the new mesh deformation method only, efficiency considerations are of only preliminary type while we concentrated more on the validation and accuracy aspects of the new FEM-ALE fictitious-boundary method for particulate flow.

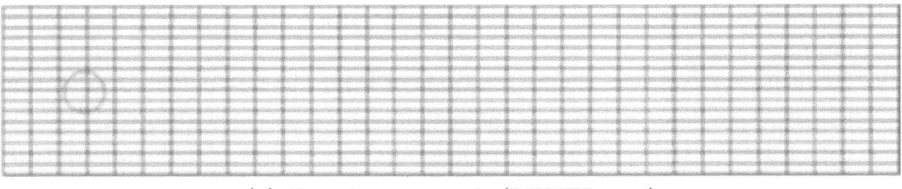

(a) Equidistant mesh (LEVEL = 4)

(b) Deformed mesh (LEVEL = 4)

FIGURE 32. Different meshes adopted for flow around a fixed circular cylinder

3.5.1. Benchmark experiment. We first consider the benchmark case of flow around a fixed circular cylinder in a channel as described in [148]. The channel height is $H = 0.41$, length $L = 2.2$, the cylinder diameter $D = 0.1$. The center point of the cylinder is located at $(0.2, 0.2)$. The Reynolds number is defined by $Re = \bar{U}D/\nu$ with the mean velocity $\bar{U} = 2U(0, H/2, t)/3$. The kinematic viscosity of the fluid is given as $\nu = \mu_f/\rho_f = 10^{-3}$ and its density $\rho_f = 1$. The inflow profiles are parabolic $U(0, Y, t) = 6.0\bar{U}Y(H - Y)/H^2$ with $\bar{U} = 0.2$ such that the resulting Reynolds number is $Re = 20$ which leads to steady-state solutions.

TABLE 7. Drag and lift coefficients for flow around a fixed circular cylinder with Re = 20

by using equidistant meshes

LEVEL	NVT	NEL	Drag C_d	Lift C_l	VEF(%)	CPU
4	561	512	4.44688	-0.0649815	91.643	0.8
5	2145	2048	5.31808	-0.3508040	97.697	3.0
6	8385	8192	5.50358	-0.0093675	99.261	12.0
7	33153	32768	5.50585	0.0312388	99.842	54.6
8	131841	131072	5.53049	0.0239737	99.958	375.0
reference values			$C_d = 5.5795$		$C_l = 0.010618$	

Fig. 32 shows the equidistant and deformed (static) meshes employed in our calculations, in which the circle shows the position of the cylinder. The shown meshes are successively refined by connecting opposite midpoints. The deformed

TABLE 8. Drag and lift coefficients for flow around a fixed circular
cylinder with Re = 20

by using deformed meshes

LEVEL	NVT	NEL	Drag C_d	Lift C_l	VEF(%)	CPU
4	561	512	6.25486	0.0682610	99.332	1.3
5	2145	2048	5.72950	0.0297392	99.788	4.0
6	8385	8192	5.61971	0.0291702	99.946	14.0
7	33153	32768	5.58139	0.0118296	99.985	76.0
8	131841	131072	5.57706	0.0104031	99.997	402.0
reference values			$C_d = 5.5795$		$C_l = 0.010618$	

mesh is generated from the equidistant mesh, but it has more grid nodes and
elements concentrated and aligned around the surface of the cylinder. In Tables 7
and 8, the characteristics of these meshes after several global refinements are given.
The meaning of "LEVEL" is the number of refinements, "NVT" the number of
vertices, and "NEL" the number of elements, "VEF" the ratio of the effective
cylinder area covered by the fixed mesh through the fictitious-boundary method
w.r.t. the real cylinder area, "CPU" the typical CPU time needed (DELL Precision
Workstation 670, 2.66 GHz) for reaching steady-state.

Tables 7 and 8 present the computed results by using the equidistant mesh
and deformed mesh, respectively. The corresponding reference results of drag and
lift coefficients for this benchmark problem are also listed in these tables for comparison. From the tables, we can see that all results of drag and lift coefficients
are converging w.r.t. mesh refinement except those for the lift coefficients when
using equidistant mesh. The results for the deformed meshes are much better and
more accurate than those for the equidistant meshes. For the deformed mesh cases,
on LEVEL = 6 with only NEL = 8192, quite satisfying results have already been
obtained. Obviously, the accuracy is improved by using the grid-deformation techniques. Moreover, the effective area ratio of the cylinder captured by the deformed
mesh lines has reached more than 99% in LEVEL = 4, while it requires LEVEL = 6
for equidistant meshes. For CPU time, about 10% time is increased for the cases
of deformed meshes over the equidistant mesh on the same level due to additional
time needed for the deformed-mesh generation. However, the main CPU time is
still needed for the fluid-solid part, not for the part of deformed-mesh generation.
Moreover, in the deformed-mesh cases, we can adopt a lower-level mesh which can
obtain high accuracy results but only small CPU time increases compared to the
case of equidistant meshes. For example, for the deformed mesh in LEVEL = 7,
very accurate results of force calculations are possible and only 76 sec CPU is
needed, while for the equidistant mesh in LEVEL = 8, the CPU time of 375 sec
is required but the results of force calculations are still not so accurate. Hence, it
shows that the deformed mesh can significantly improve the force calculations.

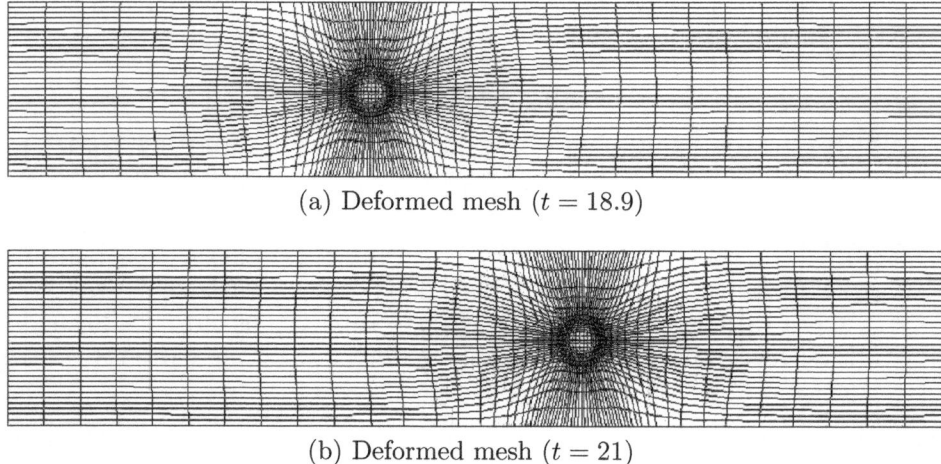

(a) Deformed mesh ($t = 18.9$)

(b) Deformed mesh ($t = 21$)

FIGURE 33. Deformed meshes for one oscillating circular cylinder
in a channel.

3.5.2. One oscillating circular cylinder in a channel. Next, an oscillating circular
cylinder in a channel with a prescribed velocity is considered. The channel height
is $H = 0.41$, length $L = 2.2$, the cylinder diameter $D = 0.1$. The center point
of the cylinder is located initially at $(1.1, 0.2)$. The prescribed velocity for the
cylinder is given by $u = 2\pi f A \cos(2\pi f t)$, $A = 0.25$, $f = 0.25$, $v = 0$, and no-slip
velocity conditions are imposed at the two walls, inlet and outlet of the channel.
The kinematic viscosity of the fluid is given by $\nu = \mu_f/\rho_f = 10^{-3}$ and its density
$\rho_f = 1$. The fluid in the channel is initially at rest.

The deformed meshes are generated by using the moving-mesh method in
every time step in order to align the mesh around the surface of the moving
cylinder. Fig. 33 gives two snapshot results at $t = 18.9$ and $t = 21$ of the deformed
meshes, and Fig. 34 presents the corresponding local vector field, norm of velocity
and local vorticity distribution, respectively. These pictures show that the flow in
the channel is disturbed by the oscillating cylinder, and the vortex is generated
periodically in the wake of the cylinder. Fig. 35 illustrates the comparison of the
computed drag coefficient C_d by the presented method with the reference result
based on the body-fitted mesh (see [146]). The results calculated from LEVEL = 5
to LEVEL = 8 and the parameters of the number of elements "NEL" for each
refinement level are all shown together. From the comparisons, we can see that
the presented results are identical with increasing the mesh refinement, and they
agree very well with the reference results. On the deformed mesh on LEVEL = 5
with 2048 elements, very good results have already been reached.

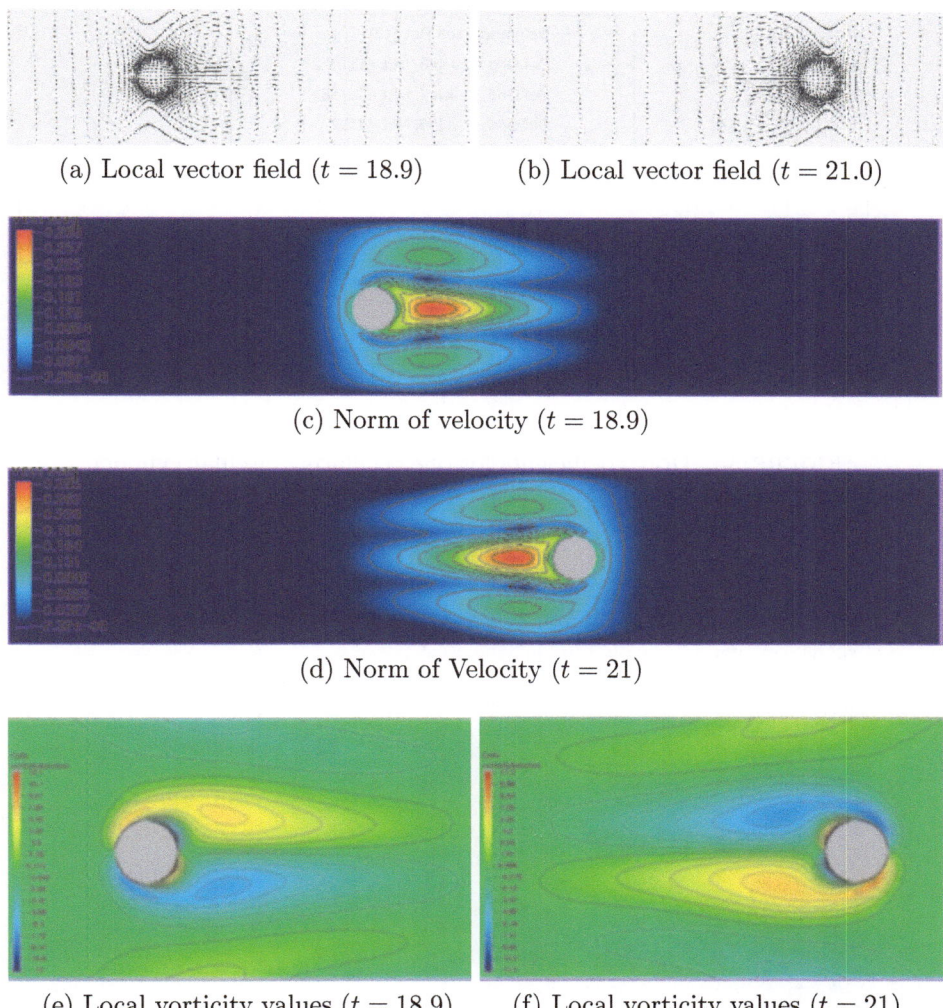

(a) Local vector field ($t = 18.9$)   (b) Local vector field ($t = 21.0$)

(c) Norm of velocity ($t = 18.9$)

(d) Norm of Velocity ($t = 21$)

(e) Local vorticity values ($t = 18.9$)   (f) Local vorticity values ($t = 21$)

FIGURE 34. Oscillating circular cylinder in a channel.

3.5.3. One 2D circular ball falling down in a channel. The computational domain is a channel of width 2 and height 6. A rigid circular ball with diameter $d = 0.25$ and density $\rho_p = 1.5$ is located at $(1, 4)$ at time $t = 0$, and it is falling down under gravity in an incompressible fluid with density $\rho_f = 1$ and viscosity $\nu = 0.1$, with gravitational acceleration constant $g = -980$. We suppose that the ball and the fluid are initially at rest. The simulation is carried out on fixed equidistant meshes and moving deformed meshes, respectively, each of them having two different level,

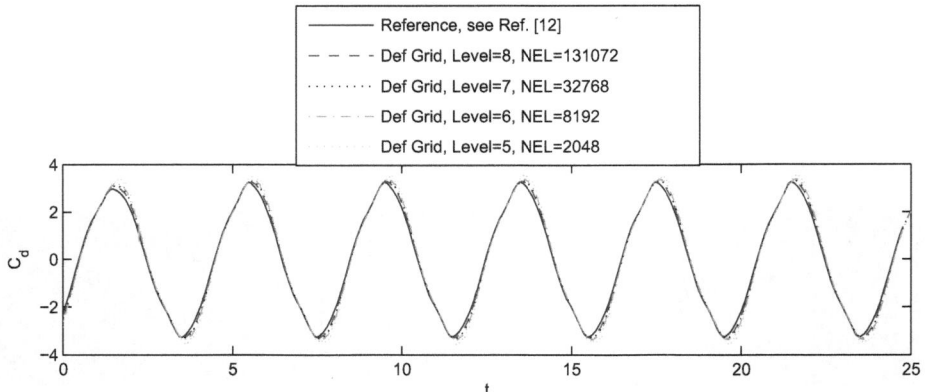

FIGURE 35. Drag coefficients for one oscillating circular cylinder in a channel.

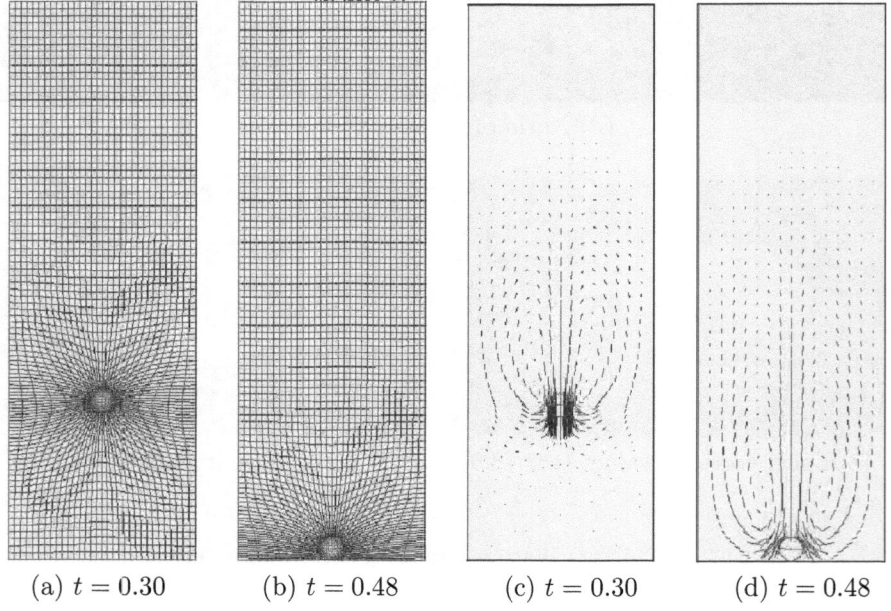

(a) $t = 0.30$ (b) $t = 0.48$ (c) $t = 0.30$ (d) $t = 0.48$

FIGURE 36. One 2D circular ball falling down in a channel.

i.e., LEVEL $= 6$ with 12545 nodes and 12288 elements, as well as LEVEL $= 7$ with 49665 nodes and 49152 elements which provide sufficiently accurate results.

Fig. 36 gives two snapshots at $t = 0.30$ and $t = 0.48$ on the deformed mesh and the corresponding vector field, respectively. Fig. 37 presents the comparison

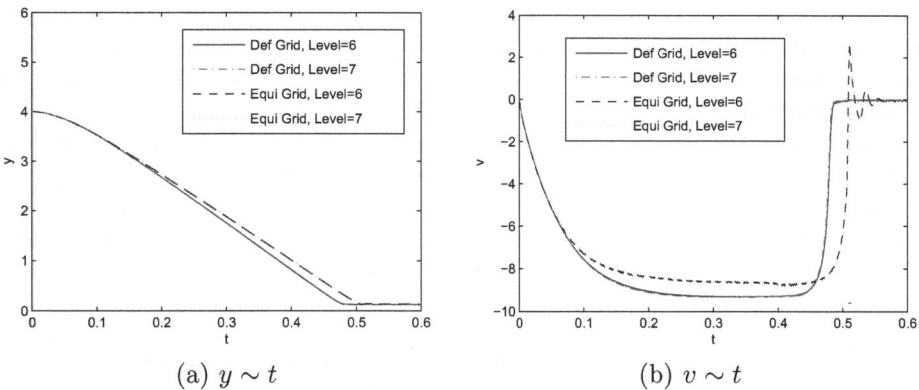

(a) $y \sim t$ (b) $v \sim t$

FIGURE 37. Time history of y-position and v-velocity component of the center of the falling ball.

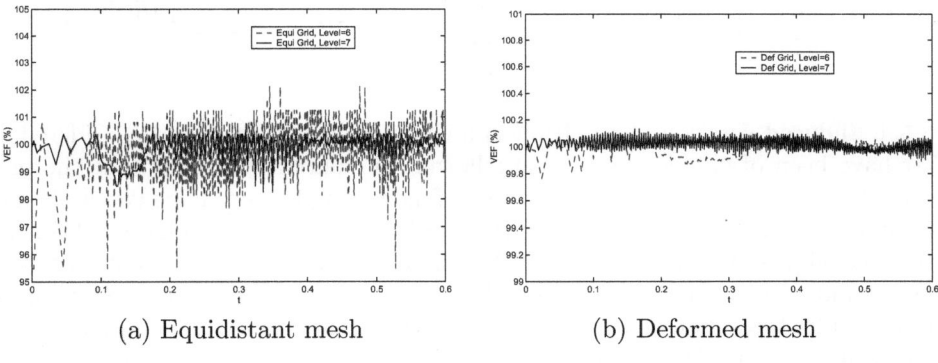

(a) Equidistant mesh   (b) Deformed mesh

FIGURE 38. Time history of the effective area ratio of the falling ball.

of the time history of the y-coordinate and v-velocity component of the center of the ball by using equidistant meshes and deformed meshes, each of them are calculated by the mesh levels LEVEL = 6 and LEVEL = 7, respectively, both leading to convergent results with refining the mesh. If we compare these results with those obtained by Glowinski in [36], we can see that the results of the deformed meshes are much closer to his results than those of the equidistant meshes, which proves again that the accuracy is improved when the moving deformed meshes are employed. Fig. 38 shows the ratio (%) of the effective area of the falling ball covered by the underlying fixed equidistant mesh and the adaptively deformed mesh compared with the real area of the disk, respectively. Again we can see that the deformed mesh can much better capture the falling ball than the equidistant

mesh, which also explains why the deformed mesh can improve the accuracy of
the simulation. Table 9 shows the typical CPU time for one time step and memory (in MByte) needed (DELL Precision Workstation 670, 2.66 GHz) by using
equidistant and deformed meshes, respectively. We can also see the linear relation
between CPU and storage cost w.r.t. the mesh size due to the optimized multigrid
components. The CPU time of the deformed mesh cases is increased by appr. 10%
to 20% compared to the equidistant mesh cases, since the deformed mesh generation, mesh velocities and system matrix update are needed to implement in every
time step. However, the computer memory storage required for both cases is not
significantly changed.

TABLE 9. CPU time for one time step and storage for one falling ball.

LEVEL	Equidistant mesh		Deformed mesh	
	CPU time	Storage (MB)	CPU time	Storage (MB)
4	0.11	0.15	0.13	0.19
5	0.53	0.71	0.61	0.76
6	1.89	2.97	2.36	3.10
7	7.36	11.82	8.95	12.83

3.5.4. Induced rotation of an airfoil in a channel. The rigid bodies considered so
far have been of circular form. The following simulation will show that the presented method can deal with rigid bodies of more complicated shape very well, and
provides better results compared to that without using the moving-grid deformation techniques. We consider a NACA0012 airfoil that has a fixed center of mass
and is induced to rotate freely around its center of mass due to hydrodynamical
forces under the action of an incompressible viscous flow in a channel. The channel
has width 20 and height 4. The density of the fluid is $\rho_f = 1$ and the density of the
airfoil is $\rho_p = 1.1$. The viscosity of the fluid is $\nu_f = 10^{-2}$. The initial condition for
the fluid velocity is $\mathbf{u} = 0$ and the boundary conditions are given as $\mathbf{u} = 0$ when
$y = 0$ or 4 and $\mathbf{u} = 1$ when $x = 0$ or 20 for $t \geq 0$. The initial angular velocity and
angle of incidence of the airfoil are zero. The airfoil length is 1.0089304 and the
fixed center of mass of the airfoil is at $(0.420516, 2)$. Hence the Reynolds number
is about 101 with respect to the length of the airfoil and the inflow speed. The
detailed shape of the NACA0012 is described as follows (for $0 \leq X \leq 1.0089304$):

$$Y = 0.6 \cdot \{\, 0.2969 \cdot \sqrt{X} + X \cdot [\, -0.126 + X \cdot [\, -0.3516 + X \cdot (\, 0.2843 - 0.1015 \cdot X \,)\,]\,]\,\}.$$

The simulation is realized on both fixed equidistant meshes and moving deformed meshes, each of them having two different levels, i.e., LEVEL = 7 with
41409 nodes and 40960 elements, as well as LEVEL = 8 with 164737 nodes and
163840 elements. The deformed mesh is generated in each time step in order to
always keep the grid alignment around the surface of the induced rotating airfoil.

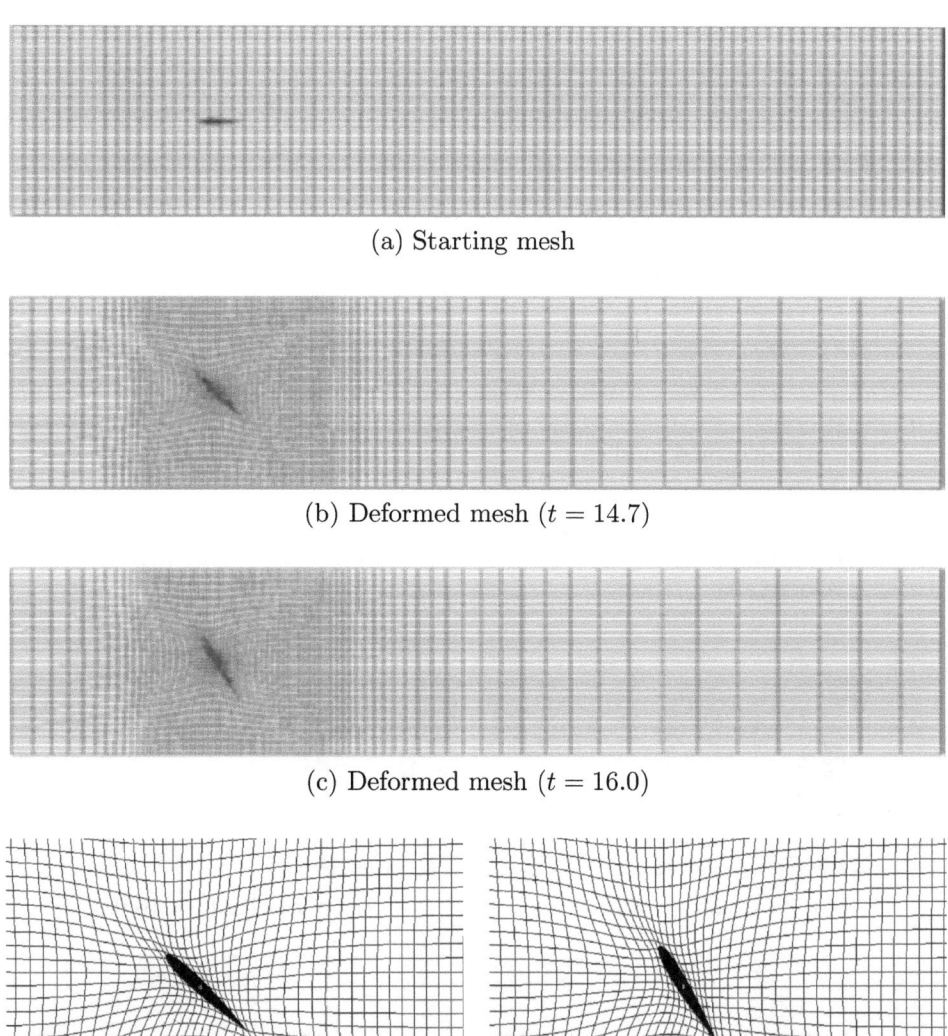

(a) Starting mesh

(b) Deformed mesh ($t = 14.7$)

(c) Deformed mesh ($t = 16.0$)

(d) Zoomed mesh ($t = 14.7$)   (e) Zoomed mesh ($t = 16.0$)

FIGURE 39. Initial mesh and deformed meshes during the simulation of the induced rotation of a NACA0012 airfoil in a channel.

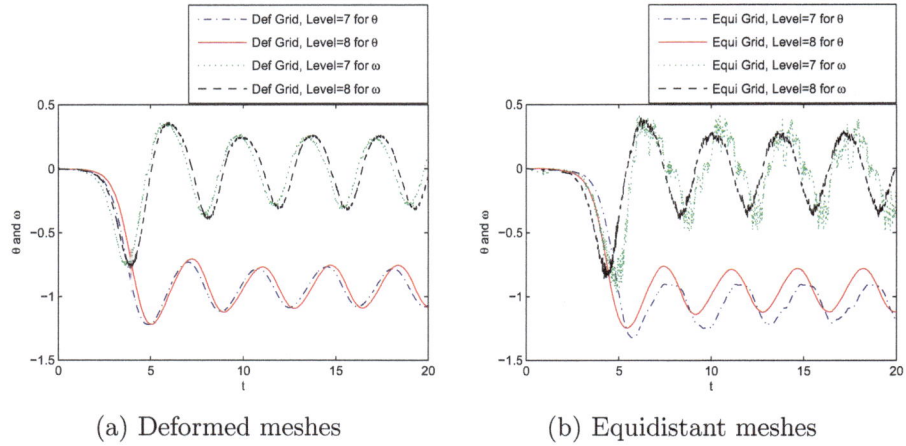

(a) Deformed meshes   (b) Equidistant meshes

FIGURE 40. Time history of the angle of incidence θ and angular velocity ω.

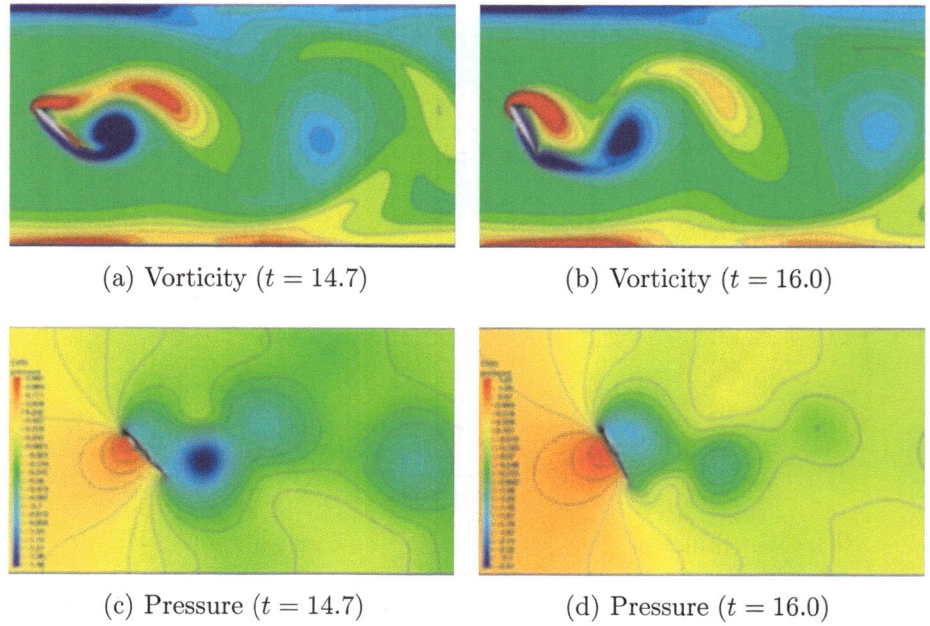

(a) Vorticity ($t = 14.7$)   (b) Vorticity ($t = 16.0$)

(c) Pressure ($t = 14.7$)   (d) Pressure ($t = 16.0$)

FIGURE 41. Local vorticity and pressure values for the rotating NACA0012 airfoil.

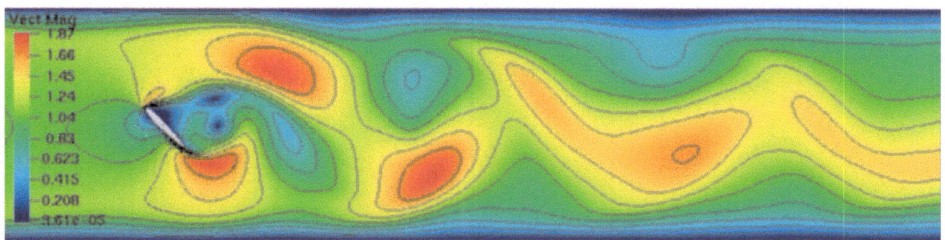

(a) Norm of velocity ($t = 14.7$)

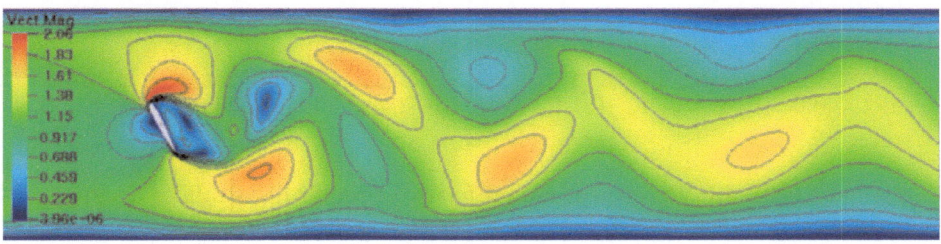

(b) Norm of velocity ($t = 16$)

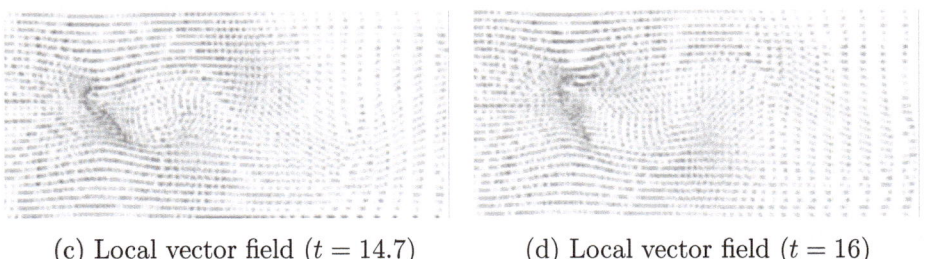

(c) Local vector field ($t = 14.7$)   (d) Local vector field ($t = 16$)

FIGURE 42. Induced rotation of a NACA0012 airfoil in a channel.

Fig. 39 (a) shows the starting equidistant mesh to generate the deformed meshes. In Fig. 39 (b) and (c), two deformed meshes at $t = 14.7$ and $t = 16.0$ are presented, their local zooms are illustrated in Fig. 39 (d) and (e). Fig. 40 plots the time history of the angle of incidence θ and the angular velocity ω of the airfoil calculated by using the equidistant and the deformed meshes, each of them is performed on the two levels LEVEL $= 7$ and LEVEL $= 8$, respectively. The local vorticity distribution, local pressure contour, norm of velocity and local vector field for $t = 14.7$ and $t = 16.0$ are given in Fig. 41 and Fig. 42. From these figures and pictures, we can see that the airfoil quickly reaches a periodic motion and intends to keep its broadside perpendicular to the inflow direction which is a stable

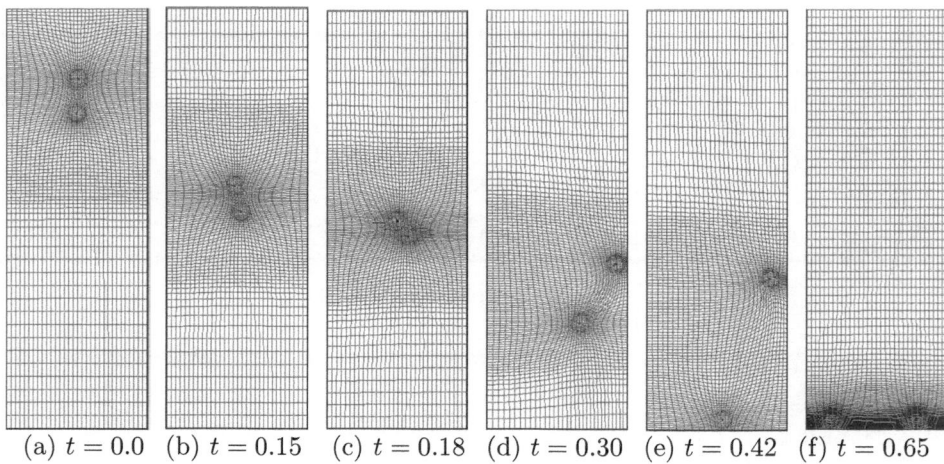

(a) $t = 0.0$ (b) $t = 0.15$ (c) $t = 0.18$ (d) $t = 0.30$ (e) $t = 0.42$ (f) $t = 0.65$

FIGURE 43. Deformed meshes for two circular disks falling down in a channel.

position for a non-circular rigid body setting in a channel at moderate Reynolds numbers. We observe that the results of the deformed meshes converge better to a mesh independent solution than those on the equidistant meshes, and are in excellent agreement with those obtained by Glowinski, Joseph and coauthors [38, 36]. The results of the equidistant meshes exhibit much more numerical oscillations and loose stability since they cannot catch very well the velocity field close to the leading edge of the airfoil, which causes the numerical solution blew up near the leading edge of the airfoil. Obviously, good results and significant accuracy improvements are achieved by using the moving-grid deformation techniques. It illustrates that the presented method can easily handle more complex shapes of rigid bodies and can obtain more accurate results than those without employing the moving grid deformation techniques.

3.5.5. Drafting, kissing and tumbling of two disks in a channel. We will carry on the cases of multiple particles in a fluid to show that the presented method can be easily applied to more realistic particulate flow with several particles.

When two particles are dropped close to each other, they interact by undergoing "drafting, kissing and tumbling" [149], which is often chosen to examine the complete computational model of particulate flow, including the prevention of collisions. Therefore, we also study the sedimentation of two circular particles in a two-dimensional channel, comparing the results w.r.t. two different levels of mesh refinement and regarding the results in [36]. The computational domain is a channel of width 2 and height 6. Two rigid circular disks with diameter $d = 0.25$ and density $\rho_p = 1.5$ are located at $(1, 5)$ (No.1 disk) and $(1, 4.5)$ (No.2 disk) at time $t = 0$, and they are falling down under gravity in an incompressible fluid

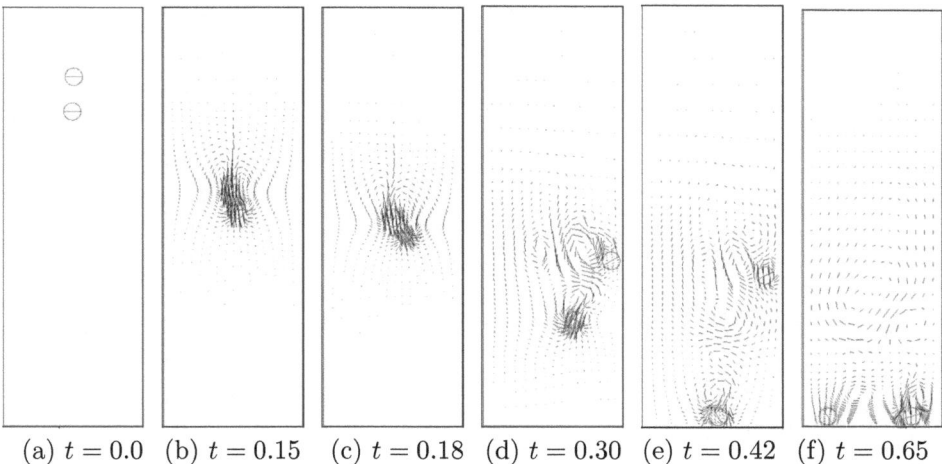

(a) $t = 0.0$ (b) $t = 0.15$ (c) $t = 0.18$ (d) $t = 0.30$ (e) $t = 0.42$ (f) $t = 0.65$

FIGURE 44. Vector fields for two circular disks falling down in a channel.

with density $\rho_f = 1$ and viscosity $\nu = 0.01$, with gravitational acceleration constant $g = -980$. We suppose that the disks and the fluid are initially at rest. The simulation is carried out on moving deformed meshes, having two different levels, i.e., LEVEL = 7 with 49665 nodes and 49152 elements, as well as LEVEL = 8 with 197633 nodes and 196608 elements which gives sufficiently accurate solutions.

Fig. 43 shows the moving deformed meshes employed in simulation of the two falling disks. The grid lines are always concentrated around the surfaces of the two disks and in the region of the gap between the two disks, and move with the two disks during the computations. Fig. 44 presents the corresponding computed vector fields. From these figures, we can see that the disk in the wake (No.1 particle) falls more rapidly than the disk No.2 in front since the fluid forces acting on it are smaller. The gap between them decreases, and they almost touch ("kiss") each other at time $t = 0.15$. After touching, the two disks fall together until they tumble ($t = 0.18$) and subsequently they separate from each other ($t = 0.30$). The tumbling of the disks takes place because the configuration, when both are parallel to the flow direction, is unstable. The No.1 disk is touching first the bottom wall at $t = 0.42$, while the No.2 disk reaches the bottom wall at $t = 0.65$. In Fig. 45, several quantities are plotted. These are the time histories of the x and y-coordinate of the two disk centers, u and v-component of the disk translational velocities, obtained for the two deformed meshes, LEVEL = 7 and LEVEL = 8, respectively. We can see that the results computed on the two different deformed meshes are essentially indistinguishable. All results compare qualitatively well with those presented in [91, 36, 150, 151].

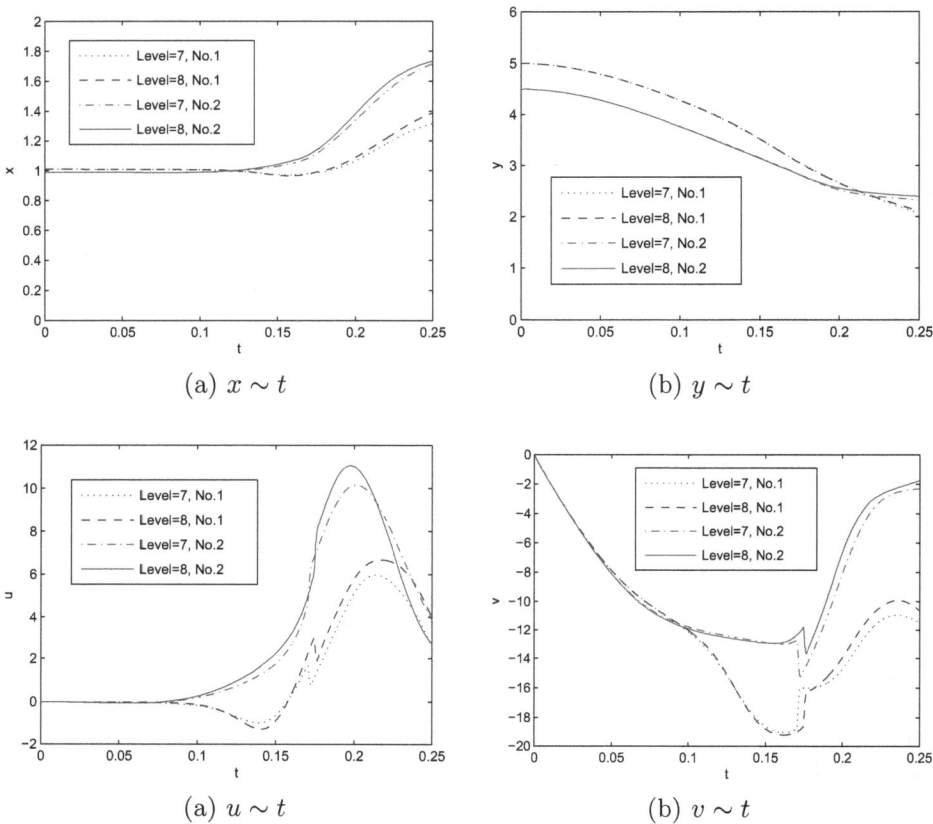

(a) $x \sim t$ (b) $y \sim t$

(a) $u \sim t$ (b) $v \sim t$

FIGURE 45. Time history of two circular disks falling down in a channel: (a) x-coordinate, (b) y-coordinate of the center of the two disks, and (c) u-component, (d) v-component of the translational velocity of the center of the two disks. Dotted line for No.1 disk and dot-dashed line for No.2 disk by deformed mesh LEVEL = 7, as well as dashed line for No.1 disk, and solid line for No.2 disk by deformed mesh LEVEL = 8.

3.5.6. Sedimentation of 120 circular particles. Finally, we consider the sedimentation of 120 circular particles with identical size falling down in a closed rectangular cavity. The width and height of the cavity are 4 and 6. The 120 particles are placed at the top of the cavity with 8 rows, while in each row the number of particles is 15. The diameter of the particles is 0.24. The range of the repulsive force is chosen as $\rho = 0.0167$. The position of the particles at time $t = 0$ is shown in Fig. 50(a). The particles and the fluid are at rest at $t = 0$. The density of the fluid is $\rho_f = 1$ and the density of the particles is $\rho_i = 1.1$ ($i = 1, \ldots, 120$). The viscosity of the fluid is $\nu = 10^{-2}$ (all quantities in non-dimensional form). The parameter ϵ_P in the collision model has been taken as 10^{-6}, and $\epsilon_W = \epsilon_P/2$, $\epsilon'_P = \epsilon_P$, $\epsilon'_W = \epsilon_W$. The moving meshes are shown in Figs. 46–47. We can see that grid lines are moved and concentrated around the surfaces of each particle and in the regions of the gap between particles, and can always keep alignment with the surfaces of each particle even when the particles move. The corresponding snapshots for the evolution of the vector fields of the 120 sedimenting circular particles are illustrated in Figs. 50 and 51. These figures clearly show the development of the Rayleigh–Taylor instability. This instability develops into a fingering and textbook phenomenon, and many symmetry-breaking and other bifurcation phenomena including drafting, kissing and tumbling take place at various scales in space and time. Many complex vortices have been formed which pull the particles downward and mix each other. Finally, the particles settle at the bottom of the cavity and the fluid returns to rest.

3.6. Conclusions

We demonstrated the combination of the multigrid fictitious-boundary method and a new moving-mesh method for the simulation of particulate flow with moving rigid particles of different shape and size. The new approach directly improves the accuracy upon the previous pure multigrid FBM based on static non-aligned meshes. Moreover, it is computationally cheap and simple to implement. Since the size of the problem and the data structure of the moving deformed meshes are fixed, this enables the proposed method to be incorporated into most CFD codes without the need for changing the matrix structures and for adding special interpolation procedures. Numerical tests demonstrate that it is suitable to accurately and efficiently perform the direct numerical simulation of particulate flow with large number of moving particles. One classical CFD benchmark experiment, three numerical examples of single moving particles in a fluid as well as the drafting, kissing and tumbling of two disks in a channel and the sedimentation of 120 particles in a cavity have been presented to show that the new method can significantly improve the accuracy for dealing with the interaction between the fluid and the particles, and can be easily applied to more realistic flow configurations with many moving particles.

In the future, we will combine these mesh-deformation techniques in the context of the FEM-ALE fictitious-boundary approach with special hardware-oriented implementation features for generalized tensor-product meshes: As long as the (local) data structure of such highly-structured meshes with different local spacing is preserved, special cache and pipelining-oriented techniques can be developed which can significantly exploit the available supercomputing power of modern processors and which allow special hardware components like GPU-computing. Preliminary results are already available and will be presented in future.

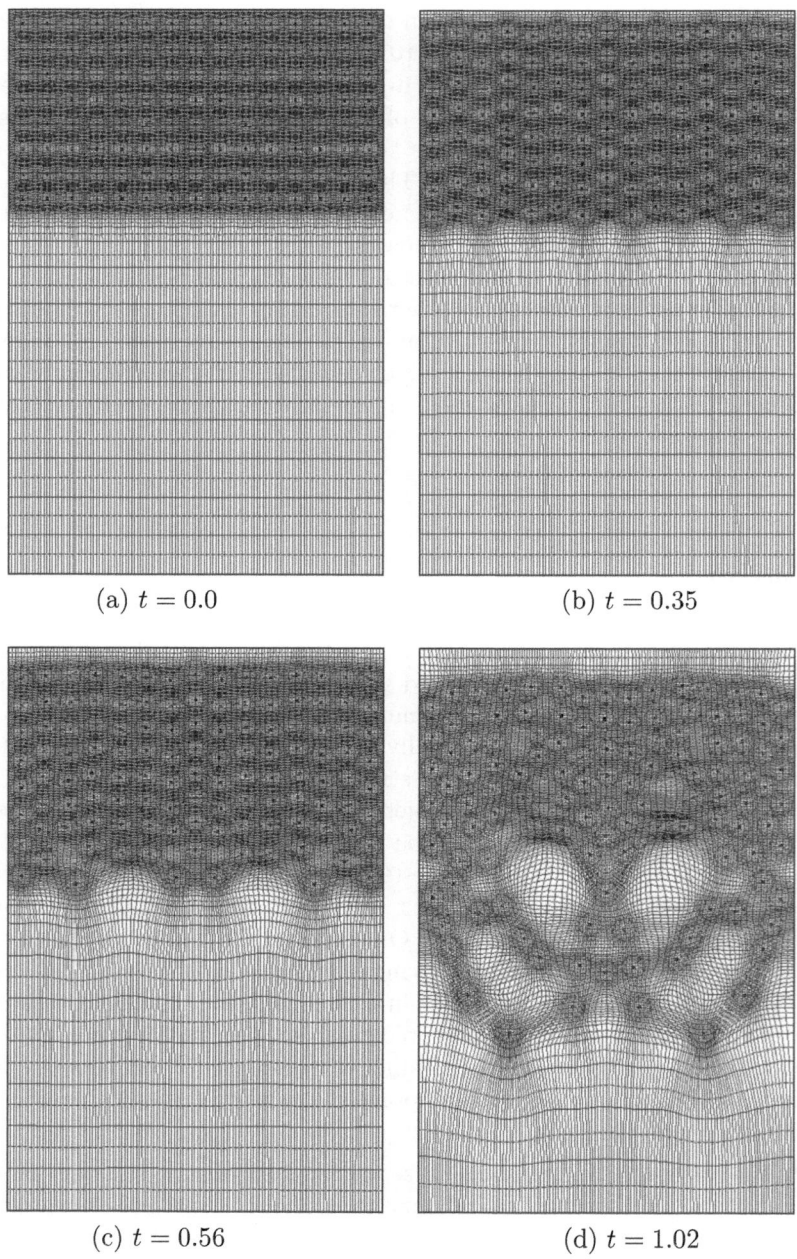

(a) $t = 0.0$ (b) $t = 0.35$

(c) $t = 0.56$ (d) $t = 1.02$

FIGURE 46. Deformation meshes for 120 particles falling down in
a cavity.

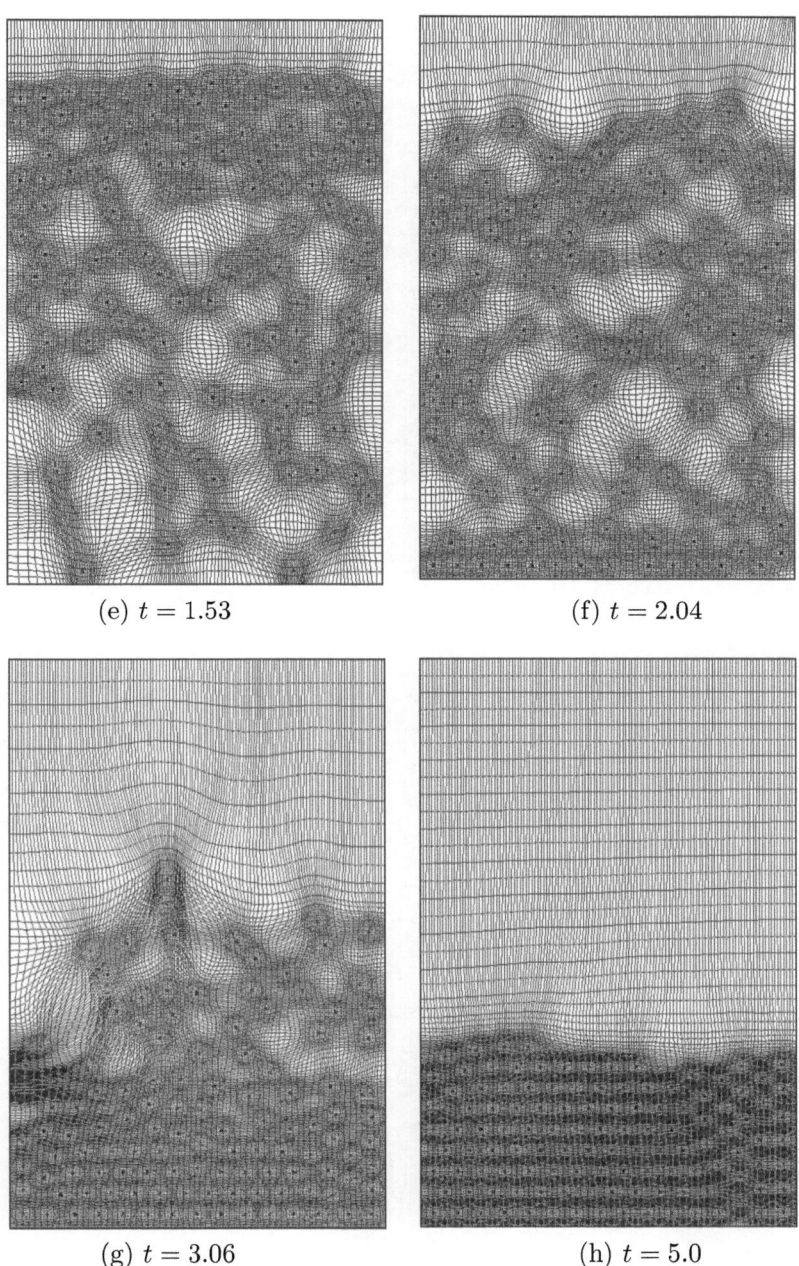

(e) $t = 1.53$ (f) $t = 2.04$

(g) $t = 3.06$ (h) $t = 5.0$

FIGURE 47. Deformation meshes for 120 particles falling down in a cavity (cont.).

S. Turek and J. Hron

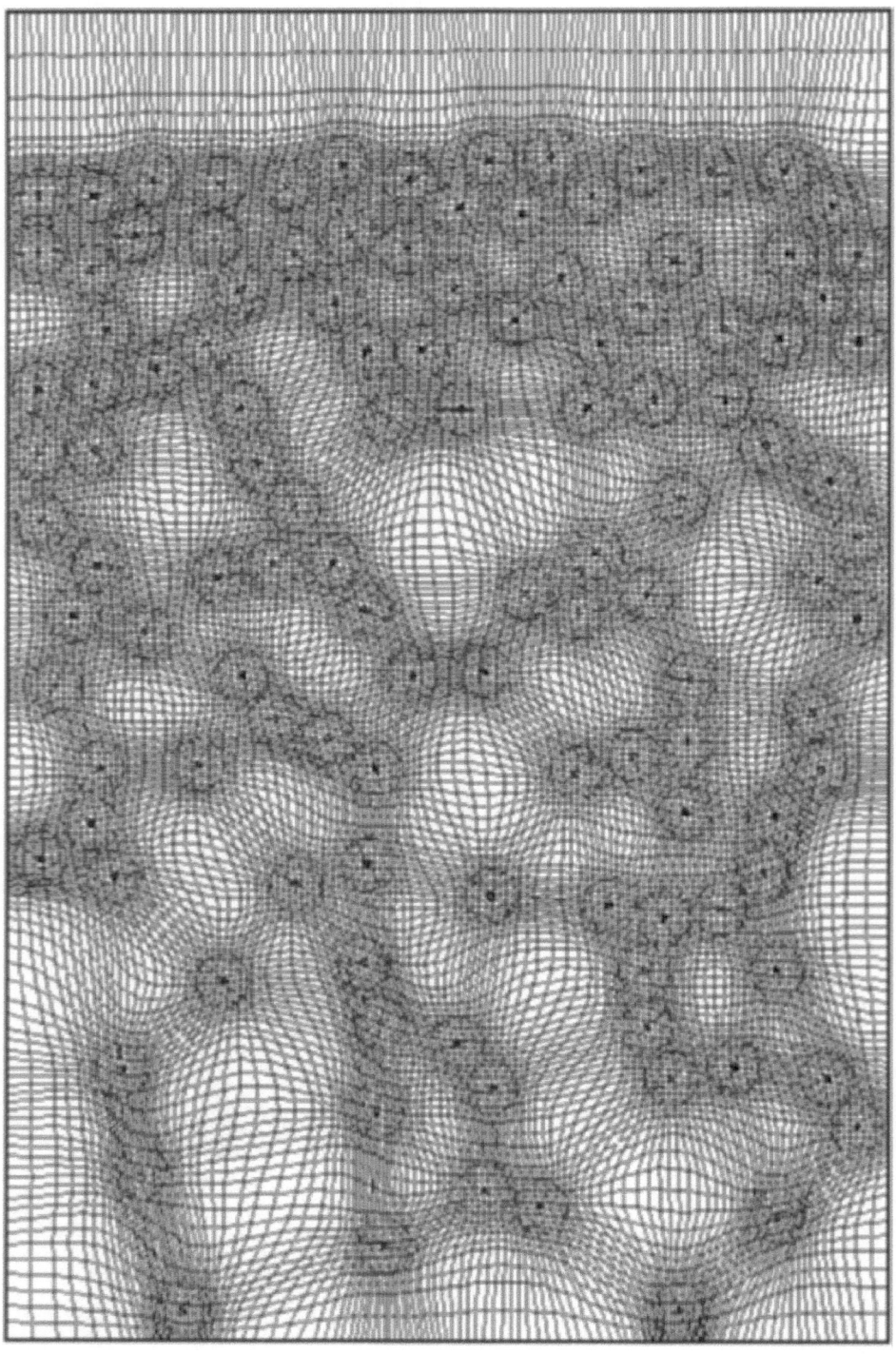

FIGURE 48. Zoomed mesh at $t = 1.53$.

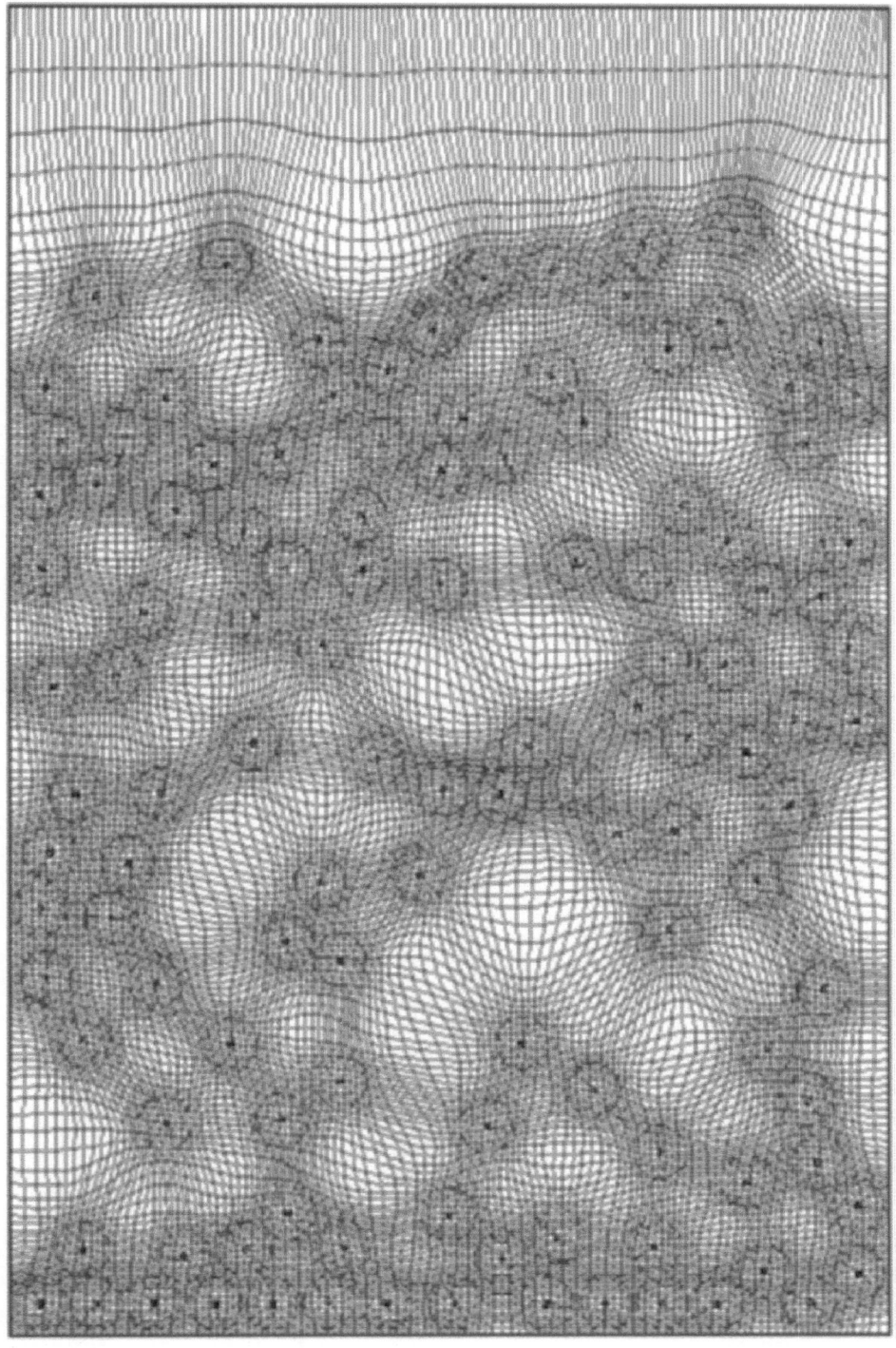

FIGURE 49. Zoomed mesh at $t = 2.04$.

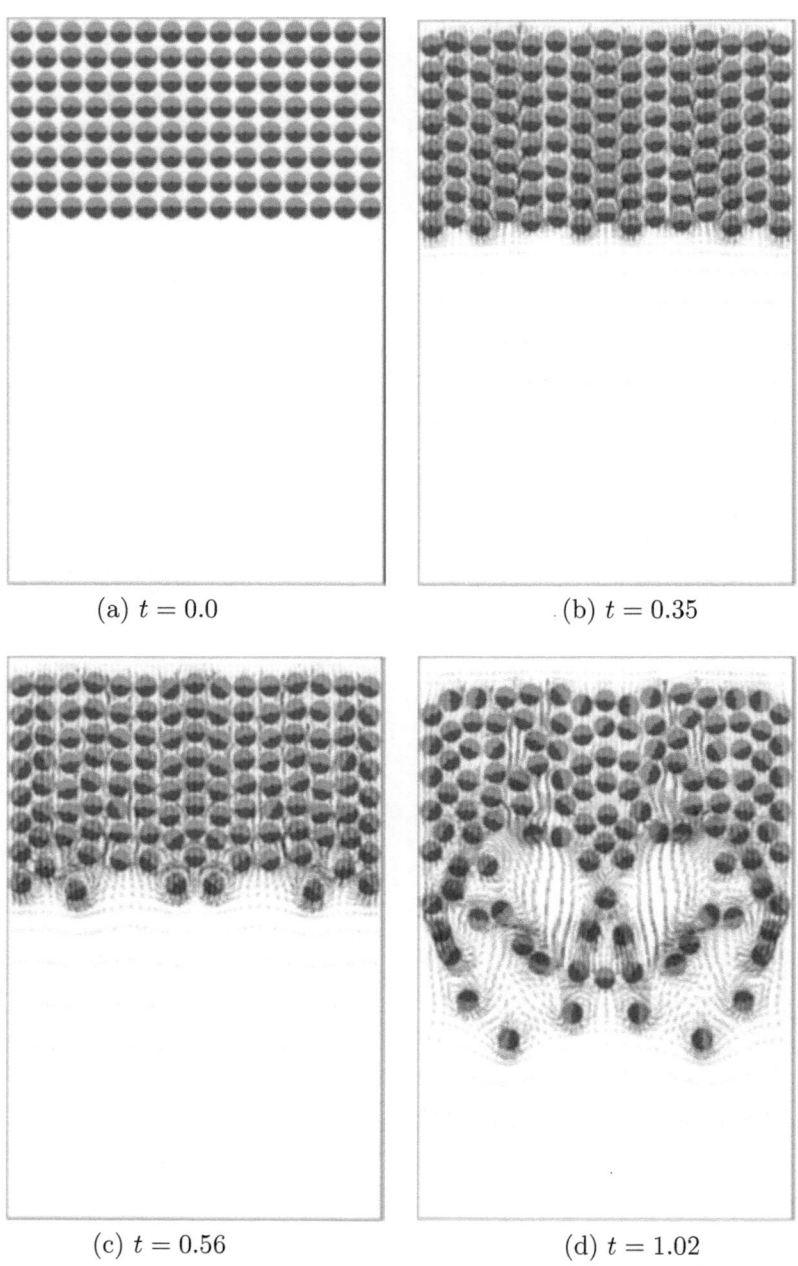

(a) $t = 0.0$ (b) $t = 0.35$

(c) $t = 0.56$ (d) $t = 1.02$

FIGURE 50. Velocity fields for 120 particles falling down in a cavity.

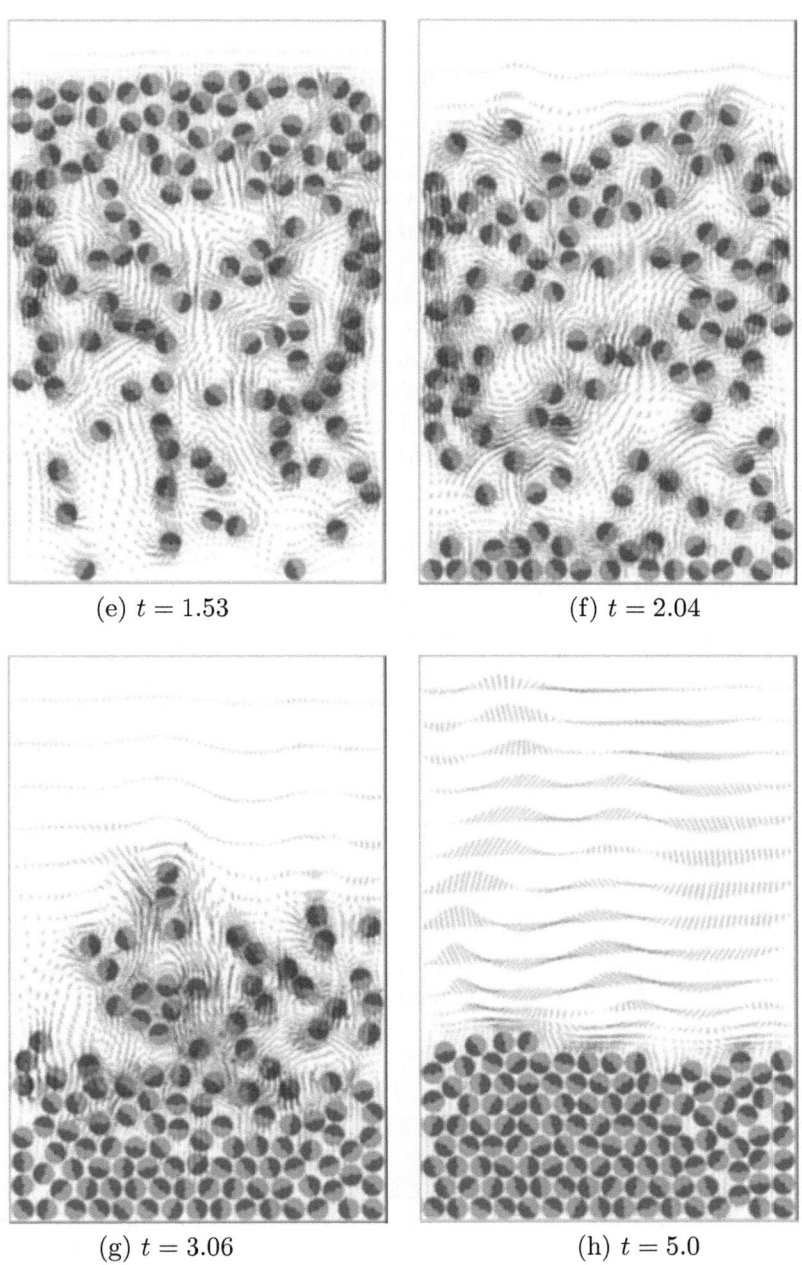

(e) $t = 1.53$ (f) $t = 2.04$

(g) $t = 3.06$ (h) $t = 5.0$

FIGURE 51. Velocity fields for 120 particles falling down in a cavity (cont.).

References

[1] E.S. Almeida and R.L. Spilker, Finite-element formulations for hyperelastic transversely isotropic biphasic soft tissues, *Comp. Meth. Appl. Mech. Engng.* **151** (1998), 513–538.

[2] R. Barrett, M. Berry, T.F. Chan, J. Demmel, J. Donato, J.Dongarra, V. Eijkhout, R. Pozo, C. Romine and H. Van der Vorst, *Templates for the solution of linear systems: Building blocks for iterative methods*, SIAM, Philadelphia, PA, second edition, 1994.

[3] S. Becker, A. Sokolichin and G. Eigenberger, Gas-liquid flow in bubble columns and loop reactors: Part II. *Chem. Eng. Sci.* **49** (1994), 5747–5762.

[4] M.J. Berger and P. Collela, Local Adaptive Mesh Refinement for Shock Hydrodynamics, *J. Comput. Phy.* **82** (1989), 64–84.

[5] P.B. Bochev, G. Liao and G.C. de la Pena, Analysis and Computation of Adaptive Moving Grids by Deformation, *Numerical Methods for Partial Differential Equations* **12** (1996), 489.

[6] J.P. Boris, F.F. Grinstein, E.S. Oran and R.J. Kolbe, New insights into Large Eddy Simulation, *Fluid Dynamics Research* **10** no. 4–6 (1992), 199–227.

[7] R. Bramley and X. Wang, *SPLIB: A library of iterative methods for sparse linear systems*, Department of Computer Science, Indiana University, Bloomington, IN, 1997. http://www.cs.indiana.edu/ftp/bramley/splib.tar.gz.

[8] S. Brenner, Korn's inequalities for piecewise H^1 vector fields, *Math. Comp.* **73** (2004), 1067–1087.

[9] E. Burman and P. Hansbo, Edge stabilization for Galerkin approximations of convection-diffusion-reaction problems, *Comp. Meth. Mech. Eng.* **193** (2004), 1437–145.

[10] E. Burman and P. Hansbo, A stabilized non-conforming finite-element method for incompressible flow, *Comp. Meth. Mech. Eng.* **195** no. 23–24 (2006), 2881–2899.

[11] E. Burman and P. Hansbo, Stabilized Crouzeix-Raviart element for the Darcy-Stokes problem, *Num. Meth. for Part. Diff. Equ.* **21** no. 5 (2005), 986–997.

[12] X.X. Cai, D. Fleitas, B. Jiang and G. Liao, Adaptive Grid Generation Based on Least-Squares Finite-Element Method, *Computers and Mathematics with Applications* **48** no. 7–8 (2004), 1077–1086.

[13] H.D. Ceniceros and T.Y. Hou, An Efficient Dynamically Adaptive Mesh for Potentially Singular Solutions, *J. Comput. Phy.* **172** (2001), 609–639.

[14] A.J. Chorin, Numerical solution of the Navier–Stokes equations, *Math. Comp.* **22** (1968), 745–762.

[15] M.A. Christon, P.M. Gresho and S.B. Sutton, Computational predictability of natural convection flows in enclosures. In: K.J. Bathe (ed), Proc. First MIT Conference on *Computational Fluid and Solid Mechanics*, Elsevier, 2001, 1465–1468.

[16] R. Codina and O. Soto, Finite-element implementation of two-equation and algebraic stress turbulence models for steady incompressible flows. *Int. J. Numer. Methods Fluids* **30** bo. 3 (1999), 309–333.

[17] K.D. Costa, P.J. Hunter, J.M. Rogers, J.M. Guccione, L.K. Waldman and A.D. McCulloch, A three-dimensional finite-element method for large elastic deformations

of ventricular myocardum: I – Cylindrical and spherical polar coordinates, *Trans. ASME J. Biomech. Eng.* **118** no. 4 (1996), 452–463.

[18] K.D. Costa, P.J. Hunter, J.S. Wayne, L.K. Waldman, J.M. Guccione and A.D. Mc-Culloch, A three-dimensional finite-element method for large elastic deformations of ventricular myocardum: II – Prolate spheroidal coordinates, *Trans. ASME J. Biomech. Eng.* **118** no. 4 (1996), 464–472.

[19] M. Crouzeix and P.A. Raviart, Conforming and non-conforming finite-element methods for solving the stationary Stokes equations, *R.A.I.R.O.* **R-3** (1973), 77–104.

[20] B. Dacorogna and J. Moser, On a Partial Differential Equation Involving the Jacobian Determinant, *Annales de le Institut Henri Poincaré* **7** (1990), 1–26.

[21] F. Dai and K.R. Rajagopal, Diffusion of fluids through transversely isotropic solids, *Acta Mechanica* **82** (1990), 61–98.

[22] T.A. Davis and I.S. Duff, A combined unifrontal/multifrontal method for unsymmetric sparse matrices, *ACM Trans. Math. Software* **25** no. 1 (1999), 1–19.

[23] J. Donea, S. Giuliani, H. Laval and L. Quartapelle, Finite-element solution of the unsteady Navier–Stokes equations by a fractional step method. *Comput. Meth. Appl. Mech. Engrg.* **30** (1982), 53–73.

[24] P.S. Donzelli, R.L. Spilker, P.L. Baehmann, Q. Niu and M.S. Shephard, Automated adaptive analysis of the biphasic equations for soft tissue mechanics using a posteriori error indicators, *Int. J. Numer. Meth. Engng.* **34** no. 3 (1992), 1015–1033.

[25] H.C. Elman, Preconditioning for the steady-state Navier–Stokes equations with low viscosity, *SIAM J. Sci. Comput.* **20** (1999), 1299–1316.

[26] H.C. Elman, V.E. Howle, J. Shadid, R. Shuttleworth, and R. Tuminaro, Block preconditioners based on approximate commutators, *SIAM J. Sci. Comput.* **27** (2005), 1651–1668.

[27] H.C. Elman, D.J. Silvester and A.J. Wathen, Performance and analysis of saddle point preconditioners for the discrete steady-state Navier–Stokes equations, *Numer. Math.* **90** (2002), 665–688.

[28] M.S. Engelman, V. Haroutunian and I. Hasbani, Segregated finite-element algorithms for the numerical solution of large–scale incompressible flow problems. *Int. J. Numer. Meth. Fluids* **17** (1993), 323–348.

[29] M.S. Engelman, R.L. Sani and P.M. Gresho, The implementation of normal and/or tangential boundary conditions in finite-element codes for incompressible fluid flow. *Int. J. Numer. Meth. Fluids* **2** (1982), 225–238.

[30] C. Farhat, M. Lesoinne and N. Maman, Mixed explicit/implicit time integration of coupled aeroelastic problems: three-field formulation, geometric conservation and distributed solution, *Int. J. Numer. Methods Fluids* **21** no. 10 (1995), 807–835. Finite-element methods in large-scale computational fluid dynamics, Tokyo, 1994.

[31] J.H. Ferziger and M. Peric, *Computational Methods for Fluid Dynamics.* Springer, 1996.

[32] C. Fleischer, *Detaillierte Modellierung von Gas-Flüssigkeits-Reaktoren.* Fortschr.-Ber. VDI-Reihe 3, Nr. 691, Düsseldorf: VDI Verlag, 2001.

[33] Y.C. Fung, *Biomechanics: Mechanical properties of living tissues*, Springer-Verlag, New York, NY, 2nd edition, 1993.

[34] M. Garbey, Yu. Kuznetsov and Yu. Vassilevski, Parallel Schwarz method for a convection-diffusion problem, *SIAM J. Sci. Comp.* **22** (2000), 891–916.

[35] V. Girault and P.A. Raviart, *Finite-Element Methods for Navier–Stokes Equations*, Springer Verlag, Berlin–Heidelberg, 1986.

[36] R. Glowinski, Finite-Element Methods for Incompressible Viscous Flow, In: *Handbook of Numerical Analysis*, Vol. IX, P.G. Ciarlet and J.L. Lions (eds), North-Holland, Amsterdam, 701–769, 2003.

[37] R. Glowinski, T.W. Pan, T.I. Hesla and D.D. Joseph, A Distributed Lagrange Multiplier/Fictitious-Domain Method for Particulate Flows, *Int. J. Multiphase Flow* **25** (1999), 755–794.

[38] R. Glowinski, T.W. Pan, T.I. Hesla, D.D. Joseph and J. Periaux, A Fictitious-Domain Approach to the Direct Numerical Simulation of Incompressible Viscous Flow Past Moving Rigid Bodies: Application to Particulate Flow, *J. Comput. Phys.* **169** (2001), 363–426.

[39] M. Grajewski, M. Köster, S. Kilian and S. Turek, Numerical Analysis and Practical Aspects of a Robust and Efficient Grid Deformation Method in the Finite-Element Context, Submitted to *SISC* (2006).

[40] P.M. Gresho, On the theory of semi-implicit projection methods for viscous incompressible flow and its implementation via a finite-element method that also introduces a nearly consistent mass matrix, Part 1: Theory, Part 2: Implementation. *Int. J. Numer. Meth. Fluids* **11** (1990), 587–659.

[41] P.M. Gresho, M.S. Engelman and R.L. Sani, *Incompressible Flow and the Finite-Element Method: Advection-Diffusion and Isothermal Laminar Flow*. Wiley, 1998.

[42] F.F. Grinstein and C. Fureby, On Monotonically Integrated Large Eddy Simulation of turbulent flows based on FCT algorithms. In this volume.

[43] M.E. Gurtin, *Topics in Finite Elasticity*, SIAM, Philadelphia, PA, 1981.

[44] P. Hansbo and M.G. Larson, A simple non-conforming bilinear element for the elasticity problem, Chalmers Finite-Element Center, 2001, Chalmers University of Technology, Göteborg Sweden, h1750-P, Preprint, 2001-01, ISSN 1404-4382.

[45] P. Haupt, *Continuum Mechanics and Theory of Materials*, Springer, Berlin, 2000.

[46] M. Heil, Stokes flow in collapsible tubes: Computation and experiment. *J. Fluid Mech.*, 353:285–312, 1997.

[47] M. Heil, Stokes flow in an elastic tube – a large-displacement fluid-structure interaction problem, *Int. J. Num. Meth. Fluids* **28** no. 2 (1998), 243–265.

[48] C.W. Hirt et al., An arbitrary Lagrangian-Eulerian computing method for all flow speeds. *J. Comput. Phys.* **135** (1997), 203.

[49] J. Hron, Numerical simulation of visco-elastic fluids. Contribution to: WDS'97, Freiburg, 1997.

[50] J. Hron, J. Málek, J. Nečas and K.R. Rajagopal, Numerical simulations and global existence of solutions of two-dimensional flows of fluids with pressure- and shear-dependent viscosities, *Mathematics and Computers in Simulation* **61** (2003), 297–315.

[51] J. Hron and S. Turek, A monolithic FEM/multigrid solver for ALE formulation of fluid structure interaction with application in biomechanics, In: H.-J. Bungartz and

M. Schäfer (eds), *Fluid-Structure Interaction: Modelling, Simulation, Optimisation*, LNCSE, Springer, 2006.

[52] H.H. Hu, D.D. Joseph and M.J. Crochet, Direct Simulation of Fluid Particle Motions, *Theor. Comp. Fluid Dyn.* **3** (1992), 285–306.

[53] H.H. Hu, N.A. Patankar and M.Y. Zhu, Direct Numerical Simulations of Fluid-Solid Systems Using the Arbitrary Lagrangian-Eulerian Techniques, *J. Comput. Phys.* **169** (2001), 427–462.

[54] W. Huang, Practical Aspects of Formulation and Solution of Moving-Mesh Partial Differential Equations, *J. Comput. Phy.* **171** (2001), 753–775.

[55] T.J.R. Hughes and A.N. Brooks, *A multidimensional upwind scheme with no crosswind diffusion. Finite-element methods for convection-dominated flows*, 1979, New York, ASME, 1979.

[56] T.J.R. Hughes, L.P. Franca and M. Balestra, A new finite-element formulation for computational fluid mechanics: V. Circumventing the Babuska–Brezzi condition: A stable Petrov–Galerkin formulation of the Stokes problem accommodating equal order interpolation. *Comp. Meth. Appl. Mech. Engrg.* **59** (1986), 85–99.

[57] J.D. Humphrey, R.K. Strumpf and F.C.P. Yin, Determination of a constitutive relation for passive myocardium: I. A new functional form, *J. Biomech. Engng.* **112** no. 3 (1990), 333–339.

[58] J.D. Humphrey, R.K. Strumpf and F.C.P. Yin. Determination of a constitutive relation for passive myocardium: II. Parameter estimation. *J. of Biomech. Engng.* **112** no. 3 (1990), 340–346.

[59] V. John, Higher-order finite-element methods and multigrid solvers in a benchmark problem for the 3D Navier–Stokes equations. *Int. J. Numer. Meth. Fluids* **40** (2002), 775–798.

[60] V. John, J.M. Maubach and L. Tobiska, Non-conforming streamline-diffusion-finite-element-method for convection-diffusion problems, *Numer. Math.* **78** (1997), 165–188.

[61] C. Johnson, *Numerical solution of partial differential equations by the finite-element method*, Cambridge, Cambridge University Press, 1987.

[62] D.D. Joseph, R. Glowinski, J. Feng and T.W. Pan, A three-dimensional computation of the force and torque on an ellipsoid settling slowly through a viscoelastic fluid. *J. Fluid Mech.* **283** (1995), 1–16.

[63] D. Kay, D. Loghin, and A.J. Wathen, A preconditioner for the steady-state Navier–Stokes equations, *SIAM J. Sci. Comput.* **24** (2002), 237–256.

[64] D. Kolymbas, An outline of hypoplasticity, *Archive of Applied Mechanics* **61** (1991), 143–151.

[65] B. Koobus and C. Farhat, Second-order time-accurate and geometrically conservative implicit schemes for flow computations on unstructured dynamic meshes. *Comput. Methods Appl. Mech. Engrg.* **170** no. 1–2 (1999), 103–129.

[66] M.K. Kwan, M.W. Lai and V.C. Mow. A finite-deformation theory for cartilage and other soft hydrated connective tissues – I. Equilibrium results. *J. Biomech.* **23** no. 2 (1990), 145–155.

[67] D. Kuzmin, Numerical simulation of mass transfer and chemical reactions in gas-liquid flows. In: *Proceedings of the 3rd European Conference on Numerical Mathematics and Advanced Applications*, World Scientific, 2000, 237–244.

[68] D. Kuzmin and S. Turek, High-resolution FEM-TVD schemes based on a fully multidimensional flux limiter. *J. Comput. Phys.* **198** (2003), 131–158.

[69] D. Kuzmin and S. Turek, Efficient numerical techniques for flow simulation in bubble column reactors. In: *Preprints of the 5th German-Japanese Symposium on Bubble Columns*, VDI/GVC, 2000, 99–104.

[70] D. Kuzmin and S. Turek, Finite-element discretization tools for gas-liquid flows. In: M. Sommerfeld (ed.), *Bubbly Flows: Analysis, Modelling and Calculation*, Springer, 2004, 191–201.

[71] D. Kuzmin and S. Turek, Multidimensional FEM-TVD paradigm for convection-dominated flows. *Proceedings of the IV. European Congress on Computational Methods in Applied Sciences and Engineering (ECCOMAS 2004)*, Book of Abstracts, Volume II, 2004.

[72] D. Kuzmin and S. Turek, Numerical simulation of turbulent bubbly flows. Technical report **254**, University of Dortmund, 2004. To appear in: *Proceedings of the 3rd International Symposium on Two-Phase Flow Modelling and Experimentation*, Pisa, September 22–24, 2004.

[73] P. Le Tallec and S. Mani, Numerical analysis of a linearised fluid-structure interaction problem. *Num. Math.* **87** no. 2 (2000), 317–354.

[74] M.E. Levenston, E.H. Frank and A.J. Grodzinsky, Variationally derived 3-field finite-element formulations for quasistatic poroelastic analysis of hydrated biological tissues. *Comp. Meth. Appl. Mech. Engng.* **156** (1998), 231–246.

[75] A.J. Lew, G.C. Buscaglia and P.M. Carrica, A note on the numerical treatment of the k-epsilon turbulence model. *Int. J. Comput. Fluid Dyn.* **14** no. 3 (2001), 201–209.

[76] G. Liao and B. Semper, A Moving Grid Finite-Element Method Using Grid Deformation, *Numerical Methods for Partial Differential Equations* **11** (1995), 603–615.

[77] F. Liu, S. Ji and G. Liao, An Adaptive Grid Method and its Application to Steady Euler Flow Calculations, *SIAM Journal on Scientific Computing* **20** no. 3 (1998), 811–825.

[78] D. Loghin and A.J. Wathen, Schur complement preconditioning for elliptic systems of partial differential equations, *Numer. Linear Algebra Appl.* **10** (2003), 423–443.

[79] J. Málek, K.R. Rajagopal and M. Ruzicka, Existence and regularity of solutions and the stability of the rest state for fluids with shear dependent viscosity. *Mathematical Models and Methods in Applied Sciences* **5** (1995), 789–812.

[80] F. Maršík, *Termodynamika kontinua*, Academia, Praha, 1. edition, 1999.

[81] F. Maršík and I. Dvořák, *Biothermodynamika*, Academia, Praha, 2. edition, 1998.

[82] W. Maurel, Y. Wu, N. Magnenat Thalmann and D. Thalmann, *Biomechanical models for soft tissue simulation*, ESPRIT basic research series, Springer-Verlag, Berlin, 1998.

[83] B. Maury, Direct Simulations of 2D Fluid-Particle Flows in Biperiodic Domains, *J. Comput. Phy.* **156** (1999), 325–351.

[84] G. Medić and B. Mohammadi, NSIKE - an incompressible Navier–Stokes solver for unstructured meshes. *INRIA Research Report* **3644** (1999).

[85] M.I. Miga, K.D. Paulsen, F.E. Kennedy, P.J. Hoopes, D.W. Hartov and A. Roberts, Modeling surgical loads to account for subsurface tissue deformation during stereotactic neurosurgery, In *IEEE SPIE Proceedings of Laser-Tissue Interaction IX, Part B: Soft-tissue Modeling* **3254** (1998), 501–511.

[86] B. Mohammadi, A stable algorithm for the k-epsilon model for compressible flows. *INRIA Research Report* **1355** (1990).

[87] B. Mohammadi and O. Pironneau, *Analysis of the k-epsilon turbulence model.* Wiley, 1994.

[88] M.A. Olshanskii and A. Reusken, Convergence analysis of a multigrid method for a convection-dominated model problem, *SIAM J. Num. Anal.* **43** (2004), 1261–1291.

[89] M.A. Olshanskii and A. Reusken, Analysis of a Stokes interface problem, *Numer. Math.* **103** (2006), 129–149.

[90] C.W.J. Oomens and D.H. van Campen, A mixture approach to the mechanics of skin, *J. Biomech.* **20** no. 9 (1987), 877–885.

[91] N.A. Patankar, P. Singh, D.D. Joseph, R. Glowinski and T.W. Pan, A New Formulation of the Distributed Lagrange Multiplier/Fictitious-Domain Method for Particulate Flows, *Int. J. Multiphase Flow* **26** (2000), 1509–1524.

[92] S.V. Patankar, *Numerical Heat Transfer and Fluid Flow.* McGraw-Hill, 1980.

[93] K.D. Paulsen, M.I. Miga, F.E. Kennedy, P.J. Hoopes, A.Hartov and D.W. Roberts, A computational model for tracking subsurface tissue deformation during stereotactic neurosurgery, *IEEE Transactions on Biomedical Engineering* **46** no. 2 (1999), 213–225.

[94] C.S. Peskin, Numerical analysis of blood flow in the heart, *J. Computational Phys.* **25** no. 3 (1977), 220–252.

[95] C.S. Peskin, The fluid dynamics of heart valves: experimental, theoretical, and computational methods, In: *Annual review of fluid mechanics*, Vol. 14, 235–259, Annual Reviews, Palo Alto, Calif., 1982.

[96] C.S. Peskin and D.M. McQueen, Modeling prosthetic heart valves for numerical analysis of blood flow in the heart, *J. Comput. Phys.* **37** no. 1 (1980), 113–132.

[97] C.S. Peskin and D.M. McQueen, A three-dimensional computational method for blood flow in the heart. I. Immersed elastic fibers in a viscous incompressible fluid, *J. Comput. Phys.* **81** no. 2 (1989), 372–405.

[98] A. Prohl, *Projection and Quasi-Compressibility Methods for Solving the Incompressible Navier–Stokes Equations.* Advances in Numerical Mathematics. B.G. Teubner, Stuttgart, 1997.

[99] L. Quartapelle, *Numerical Solution of the Incompressible Navier–Stokes Equations.* Birkhäuser Verlag, Basel, 1993.

[100] A. Quarteroni, Modeling the cardiovascular system: a mathematical challenge, In: B. Engquist and W. Schmid (eds), *Mathematics Unlimited – 2001 and Beyond*, 961–972, Springer-Verlag, 2001.

[101] A. Quarteroni, M. Tuveri and A. Veneziani, Computational vascular fluid dynamics: Problems, models and methods, *Computing and Visualization in Science* **2** no. 4 (2000), 163–197.

[102] A. Quarteroni, M. Tuveri and A. Veneziani, Computational vascular fluid dynamics. *Comp. Vis. Science* **2** (2000), 163–197.

[103] K.R. Rajagopal, Mechanics of non-Newtonian fluids, Recent Developments in Theoretical Fluid mechanics. Pitman Research Notes in Mathematics 291, eds. G.P. Galdi and J. Necas, 129–162, Longman, 1993.

[104] K.R. Rajagopal and L. Tao, *Mechanics of mixtures*, World Scientific Publishing Co. Inc., River Edge, NJ, 1995.

[105] R. Rannacher and S. Turek, A simple non-conforming quadrilateral Stokes element. *Numer. Meth. PDEs* **8** no. 2 (1992), 97–111.

[106] R.A. Reynolds and J.D. Humphrey, Steady diffusion within a finitely extended mixture slab, *Math. Mech. Solids* **3** no. 2 (1998), 127–147.

[107] A. Reusken, Convergence analysis of a multigrid method for convection-diffusion equations, *Numer. Math.* **91** (2002), 323–349.

[108] M. Rumpf, On equilibria in the interaction of fluids and elastic solids, In: *Theory of the Navier–Stokes equations*, 136–158, World Sci. Publishing, River Edge, NJ, 1998.

[109] P.A. Sackinger, P.R. Schunk and R.R. Rao, A Newton-Raphson pseudo-solid domain mapping technique for free and moving boundary problems: a finite-element implementation, *J. Comput. Phys.* **125** no. 1 (1996), 83–103.

[110] D.G. Schaeffer, Instability in the evolution equation describing incompressible granular flow, *J. of Differential Equations* **66** (1998), 19–50.

[111] M. Schäfer and S. Turek (with support of F. Durst, E. Krause, R. Rannacher), Benchmark computations of laminar flow around cylinder, In E.H. Hirschel (ed.), *Flow Simulation with High-Performance Computers II*, Vol. 52 von *Notes on Numerical Fluid Mechanics*, Vieweg, 1996, 547–566.

[112] R. Schmachtel, *Robuste lineare und nichtlineare Lösungsverfahren für die inkompressiblen Navier–Stokes-Gleichungen*. PhD thesis, University of Dortmund, 2003.

[113] P. Schreiber, *A new finite-element solver for the non-stationary incompressible Navier–Stokes equations in three dimensions*, PhD Thesis, University of Heidelberg, 1996.

[114] P. Schreiber and S. Turek, An efficient finite-element solver for the non-stationary incompressible Navier–Stokes equations in two and three dimensions, Proc. Workshop *Numerical Methods for the Navier–Stokes Equations*, Heidelberg, Oct. 25–28, 1993, Vieweg.

[115] J.J. Shi, *Application of theory of a Newtonian fluid and an isotropic nonlinear elastic solid to diffusion problems*, PhD thesis, University of Michigan, Ann Arbor, 1973.

[116] A. Sokolichin, *Mathematische Modellbildung und numerische Simulation von Gas-Flüssigkeits-Blasenströmungen*. Habilitation thesis, University of Stuttgart, 2004.

[117] A. Sokolichin, G. Eigenberger and A. Lapin, Simulation of buoyancy driven bubbly flow: established simplifications and open questions. *AIChE Journal* **50** no. 1 (2004), 24–45.

[118] R.L. Spilker and J.K. Suh, Formulation and evaluation of a finite-element model for the biphasic model of hydrated soft tissues, *Comp. Struct., Front. Comp. Mech.* **35** no. 4 (1990), 425–439.

[119] R.L. Spilker, J.K. Suh and V.C. Mow, Finite-element formulation of the nonlinear biphasic model for articular cartilage and hydrated soft tissues including strain-dependent permeability, *Comput. Meth. Bioeng.* **9** (1988), 81–92.

[120] J.K. Suh, R.L. Spilker and M.H. Holmes, Penalty finite-element analysis for non-linear mechanics of biphasic hydrated soft tissue under large deformation, *Int. J. Numer. Meth. Engng.* **32** no. 7 (1991), 1411–1439.

[121] C. Truesdell, *A first course in rational continuum mechanics*, Vol. 1, Academic Press Inc., Boston, MA, second edition, 1991.

[122] S. Turek, A comparative study of time stepping techniques for the incompressible Navier–Stokes equations: From fully implicit nonlinear schemes to semi-implicit projection methods, *Int. J. Numer. Meth. Fluids* **22** (1996), 987–1011.

[123] S. Turek, On discrete projection methods for the incompressible Navier–Stokes equations: An algorithmical approach. *Comput. Methods Appl. Mech. Engrg.* **143** (1997), 271–288.

[124] S. Turek, *Efficient Solvers for Incompressible Flow Problems*, Springer Verlag, Berlin-Heidelberg-New York, 1999.

[125] S. Turek, *Efficient Solvers for Incompressible Flow Problems: An Algorithmic and Computational Approach*, LNCSE 6, Springer, 1999.

[126] S. Turek et al., *FEATFLOW: finite-element software for the incompressible Navier–Stokes equations*. User manual, University of Dortmund, 2000. Available at *http://www.featflow.de.*

[127] S. Turek, C. Becker and S. Kilian, Hardware-oriented numerics and concepts for PDE software. Special Journal Issue for PDE Software, Elsevier, International Conference on Computational Science ICCS2002, Amsterdam, 2002, FUTURE 1095 (2003), 1–23.

[128] S. Turek, C. Becker and S. Kilian, Some concepts of the software package FEAST. In: J.M. Palma, J. Dongarra and V. Hernandes (eds), *VECPAR'98 – Third International Conference for Vector and Parallel Processing*, Lecture Notes in Computer Science, Springer, Berlin, 1999.

[129] S. Turek and J. Hron, Proposal for numerical benchmarking of fluid-structure interaction between an elastic object and laminar incompressible flow, In: H.-J. Bungartz and M. Schäfer (eds), *Fluid-Structure Interaction: Modelling, Simulation, Optimisation*, LNCSE, Springer, 2006.

[130] S. Turek and S. Kilian, An example for parallel ScaRC and its application to the incompressible Navier–Stokes equations, Proc. *ENUMATH'97*, World Science Publ., 1998.

[131] S. Turek and A. Ouazzi, Unified edge-oriented stabilization of non-conforming finite-element methods for incompressible flow problems, to appear in *J. Numer. Math.* (2007).

[132] S. Turek, A. Ouazzi and R. Schmachtel, Multigrid methods for stabilized non-conforming finite elements for incompressible flow involving the deformation tensor formulation, *J. Numer. Math.* **10** (2002), 235–248.

[133] S. Turek and M. Schäfer, Benchmark computations of laminar flow around cylinder. In: E.H. Hirschel (ed), *Flow Simulation with High-Performance Computers II*, Vol. 52 of *Notes on Numerical Fluid Mechanics*, Vieweg, 1996. Co. F. Durst, E. Krause, R. Rannacher.

[134] S. Turek and R. Schmachtel, Fully coupled and operator-splitting approaches for natural convection. *Int. Numer. Meth. Fluids* **40** (2002), 1109–1119.

[135] S. Turek, L. Rivkind, J. Hron and R. Glowinski, Numerical Study of a Modified Time-Steeping theta-scheme for incompressible flow simulations, *Journal of Scientific Computing* **28** (2006), 533–547.

[136] S. Turek, D.C. Wan and L.S. Rivkind, The Fictitious-Boundary Method for the Implicit Treatment of Dirichlet Boundary Conditions with Applications to Incompressible Flow Simulations. Challenges in Scientific Computing, *Lecture Notes in Computational Science and Engineering* **35**, Springer, 37–68, 2003.

[137] A.M.P. Valli, G.F. Carey and A.L.G.A. Coutinho, Finite-element simulation and control of nonlinear flow and reactive transport. In: *Proceedings of the 10th International Conference on Numerical Methods in Fluids*. Tucson, Arizona, 1998, 450–45.

[138] A.M.P. Valli, G.F. Carey and A.L.G.A. Coutinho, Control strategies for timestep selection in simulation of coupled viscous flow and heat transfer. *Commun. Numer. Methods Eng.* **18** no. 2 (2002), 131–139.

[139] S.P. Vanka, Implicit multigrid solutions of Navier–Stokes equations in primitive variables. *J. Comp. Phys.* **65** (1985), 138–158.

[140] J. Van Kan, A second-order accurate pressure-correction scheme for viscous incompressible flow. *SIAM J. Sci. Stat. Comp.* **7** (1986), 870–891.

[141] W.J. Vankan, J.M. Huyghe, J.D. Janssen and A. Huson, Poroelasticity of saturated solids with an application to blood perfusion, *Int. J. Engng. Sci.* **34** no. 9 (1996), 1019–1031.

[142] W.J. Vankan, J.M. Huyghe, J.D. Janssen and A. Huson, Finite-element analysis of blood flow through biological tissue, *Int. J. Engng. Sci.* **35** no. 4 (1997), 375–385.

[143] M.E. Vermilyea and R.L. Spilker, Hybrid and mixed-penalty finite elements for 3-d analysis of soft hydrated tissue, *Int. J. Numer. Meth. Engng.* **36** no. 24 (1993), 4223–4243.

[144] D.C. Wan and S. Turek, Direct Numerical Simulation of Particulate Flow via Multigrid FEM Techniques and the Fictitious-Boundary Method, *Int. J. Numer. Method in Fluids* **51** (2006), 531–566.

[145] D.C. Wan and S. Turek, An Efficient Multigrid-FEM Method for the Simulation of Liquid-Solid Two Phase Flows, *J. for Comput. and Appl. Math.*, in press (2007).

[146] D.C. Wan, S. Turek and L.S. Rivkind, An Efficient Multigrid FEM Solution Technique for Incompressible Flow with Moving Rigid Bodies, In: *Numerical Mathematics and Advanced Applications, ENUMATH* 2003, Springer, 844–853, 2004.

[147] C. Zoppou, S.I. Barry and G.N. Mercer, Dynamics of human milk extraction: a comparitive study of breast-feeding and breast pumping, *Bull. Math. Biology* **59** no. 5 (1997), 953–973.

[148] M. Schäfer and S. Turek, Benchmark computations of laminar flow around cylinder, In: E.H. Hirschel (ed), *Flow Simulation with High-Performance Computers* II. Vol. 52 of Notes on Numerical Fluid Mechanics, Vieweg, 547–566, 1996.

[149] A. Fortes, D.D. Joseph and T. Lundgren, Nonlinear mechanics of fluidization of beds of spherical particles, *J. Fluid Mech.* **177** (1987), 497–483.

[150] P. Singh, T.I. Hesla and D.D. Joseph, Distributed Lagrange multiplier method for particulate flows with collisions, *Int. J. Multiphase Flow* **29** (2003), 495–509.

[151] C. Diaz-Goano, P. Minev and K. Nandakumar, A Lagrange multiplier/fictitious-domain approach to particulate flows, In: W. Margenov and Yalamov (eds), *Lecture Notes in Computer Science* **2179**, Springer, 409–422, 2001.

Stefan Turek
Institute of Applied Mathematics
University of Dortmund
Vogelpothsweg 87
D-44227 Dortmund
Germany
e-mail: `ture@featflow.de`

Jaroslav Hron
Mathematical Institute
Charles University
Sokolovska 83
186 75 Praha
Czech Republic
e-mail: `hron@karlin.mff.cuni.cz`

Oberwolfach Seminars (OWS)

The workshops organized by the *Mathematisches Forschungsinstitut Oberwolfach* are intended to introduce students and young mathematicians to current fields of research. By means of these well-organized seminars, also scientists from other fields will be introduced to new mathematical ideas. The publication of these workshops in the series *Oberwolfach Seminars* (formerly *DMV seminar*) makes the material available to an even larger audience.

OWS 38: Bobenko, A.I. / Schröder, P. / Sullivan, J.M. / Ziegler, G.M. (Eds.), Discrete Differential Geometry (2008). ISBN 978-3-7643-8620-7

Discrete differential geometry is an active mathematical terrain where differential geometry and discrete geometry meet and interact. It provides discrete equivalents of the geometric notions and methods of differential geometry, such as notions of curvature and integrability for polyhedral surfaces. Current progress in this field is to a large extent stimulated by its relevance for computer graphics and mathematical physics. This collection of essays, which documents the main lectures of the 2004 Oberwolfach Seminar on the topic, as well as a number of additional contributions by key participants, gives a lively, multi-facetted introduction to this emerging field.

OWS 37: Galdi, G.P. / Rannacher, R. / Robertson, A.M. / Turek, S., Hemodynamical Flows (2008). ISBN 978-3-7643-7805-9

OWS 36: Cuntz, J. / Meyer, R. / Rosenberg, J.M., Topological and Bivariant K-theory (2007). ISBN 978-3-7643-8398-5

Topological K-theory is one of the most important invariants for noncommutative algebras. Bott periodicity, homotopy invariance, and various long exact sequences distinguish it from algebraic K-theory. We describe a bivariant K-theory for bornological algebras, which provides a vast generalization of topological K-theory. In addition, we discuss other approaches to bivariant K-theories for operator algebras. As applications, we study K-theory of crossed products, the Baum-Connes assembly map, twisted K-theory with some of its applications, and some variants of the Atiyah-Singer Index Theorem.

OWS 35: Itenberg, I. / Mikhalkin, G. / Shustin, E., Tropical Algebraic Geometry (2007). ISBN 978-3-7643-8309-1

Tropical geometry is algebraic geometry over the semifield of tropical numbers, i.e., the real numbers and negative infinity enhanced with the (max,+)-arithmetics. Geometrically, tropical varieties are much simpler than their classical counterparts. Yet they carry information about complex and real varieties.
These notes present an introduction to tropical geometry and contain some applications of this rapidly developing and attractive subject. It consists of three chapters which complete each other and give a possibility for non-specialists to make the first steps in the subject which is not yet well represented in the literature. The intended audience is graduate, post-graduate, and Ph.D. students as well as established researchers in mathematics.

OWS 34: Lieb, E.H. / Seiringer, R. / Solovej, J.P. / Yngvason, J., The Mathematics of the Bose Gas and its Condensation (2005). ISBN 978-3-7643-7336-8

OWS 33: Kreck, M. / Lück, W., The Novikov Conjecture: Geometry and Algebra (2004). ISBN 978-3-7643-7141-8

DMV 32: Bolthausen, E. / Sznitman, A.-S., Ten Lectures on Random Media (2002). ISBN 978-3-7643-6703-9

DMV 31: Huckleberry, A. / Wurzbacher, T. (Eds.), Infinite Dimensional Kähler Manifolds (2001). ISBN 978-3-7643-6602-5

DMV 30: Scholz, E. (Ed.), Hermann Weyl's *Raum—Zeit—Materie* and a General Introduction to His Scientific Work (2001). ISBN 978-3-7643-6476-2

DMV 29: Kalai, G. / Ziegler, G.M. (Eds.), Polytopes — Combinatorics and Computation (2000). ISBN 978-3-7643-6351-2

DMV 28: Cercignani, C. / Sattinger, D., Scaling Limits and Models in Physical Processes (1998). ISBN 978-3-7643-5985-0